International Encyclopedia of
WOMEN SCIENTISTS

International Encyclopedia of
WOMEN SCIENTISTS

ELIZABETH H. OAKES

Facts On File, Inc.

International Encyclopedia of Women Scientists

Facts On File, Inc.
132 West 31st Street
New York NY 10001

Library of Congress Cataloging-in-Publication Data

Oakes, Elizabeth H., 1951–
    International encyclopedia of women scientists / Elizabeth H. Oakes.
        p. cm.
    Includes bibliographical references and index.
    ISBN 0-8160-4381-7
    1. Women scientists—Biography—Encyclopedias. I. Title.
    Q141.O27 2001
    500'.82'0922—dc21
    [B]   2001023100

Facts On File books are available at special discounts when purchased in bulk quantities for businesses, associations, institutions or sales promotions. Please call our Special Sales Department in New York at 212/967-8800 or 800/322-8755.

You can find Facts On File on the World Wide Web at http://www.factsonfile.com

Text design by Erika K. Arroyo
Cover design by Cathy Rincon
Chronology by Dale Williams

Printed in the United States of America.

VB FOF 10 9 8 7 6 5 4 3 2 1

This book is printed on acid-free paper.

CONTENTS

ACKNOWLEDGMENTS

I am particularly grateful to the writers and researchers who assisted me on this project—Bill Baue, Rebecca Stanfel, Mariko Fujinaka, and Jack Rummel: Their commitment to excellence in every aspect of their work is always appreciated. For assistance with photographs, I would like to thank those scientists who graciously responded to my requests and the many libraries and archives who helped. I would also like to express thanks to the University of Montana Mansfield Library, where much of the research for this book was done, and to the authors of the many science reference books I consulted, especially Lisa Yount.

Finally, I would like to thank my editor, Frank K. Darmstadt, for his work and support throughout this project.

INTRODUCTION

Here are the stories of 500 women scientists from all scientific disciplines and all periods of history, as far back as 400 B.C., who have contributed significantly to their fields, sometimes against quite impossible odds. From the example of the Greek physician Agnodice—who in the fourth century B.C. cut her hair short and dressed as a man in order to avoid arrest for breaking the law against women practicing medicine—to contemporary examples of women turned down for jobs and discouraged from pursuing their education, we see that women scientists have often faced formidable obstacles. While it is important to remember these obstacles, and you will find them here in the stories of women's lives, the focus of this book is on the achievements of women scientists in their chosen fields . . . and there are many.

Among the firsts in this book, you'll find the first female physics professor, the first female research scientist at General Electric Corporation, the first woman to win a Nobel Prize, and the first woman to head a branch of the U.S. military. You'll also read about the discovery of the polio vaccine, the creation of the Apgar Score System for evaluating the health of newborns, and the development of the first computer languages. All these scientific achievements and many more are presented in basic, everyday language that makes even the most complex concepts accessible.

THE SCIENTISTS

International Encyclopedia of Women Scientists includes the well-known scientific "greats" of history as well as contemporary scientists whose work is just verging on greatness. Among these are many minority scientists who have often been excluded from books such as this. A majority of the 500 scientists in this book represent the traditional scientific disciplines of physics, chemistry, biology, astronomy, and the earth sciences. A smaller number represent mathematics, computer science, philosophy of science, medicine, engineering, anthropology, and psychology. In addition to the biographical entries, the book contains 100 illustrations.

To compile the entrant list, I relied largely on the judgment of other scientists, consulting established reference works, such as the *Dictionary of Scientific Biography*, science periodicals, awards lists, and publications from science organizations and associations. Despite this process, I cannot claim to present the "most important" historical and contemporary figures. Time constraints and space limitations prevented the inclusion of many deserving scientists.

THE ENTRIES

Entries are arranged alphabetically by surname, with each entry given under the name by which the scientist is most commonly known. The typical entry provides the following information:

Entry Head: Name, birth/death dates, nationality, and field(s) of specialization.
Essay: Essays range in length from 500 to 1,000 words, with most totaling around 750 words. Each contains basic biographical information—date and place of birth, family information, educational background, positions held, prizes awarded, etc.—but the greatest attention is given to the scientist's work. Names in small caps within the essays provide easy reference to other scientists represented in the book.

In addition to the alphabetical list of scientists, readers searching for names of individuals from specific countries or scientific disciplines can consult one of the following indexes found at the end of the book:

Field of Specialization Index: Groups entrants according to the scientific field(s) in which they worked.
Nationality Index: Organizes entrants by country of birth and/or citizenship.
Subject Index: Lists page references for scientists and scientific terms used in the book.

LIST OF ENTRIES

ENTRIES A-Z

Adamson, Joy
(1910–1980)
Austrian
Naturalist

Best known for her book *Born Free: A Lioness of Two Worlds,* which detailed her experiences raising a lion cub in Africa, Joy Adamson was an artist and naturalist who did much to further the cause of wildlife preservation. Adamson raised a number of wild animals on game reserves in Kenya, where she spent the better part of her life. A film version of the highly popular *Born Free* was produced in 1964 and eventually led to a television series.

Born Friederike Viktoria Gessner on January 20, 1910, Adamson grew up in Troppau, Silesia, an area of Austria that later became part of Slovakia. Adamson's father, Viktor Gessner, was an architect and urban planner, and her mother, the former Traute Greipel, came from a wealthy family of paper manufacturers. Adamson demonstrated an interest in animals and creative pursuits from a young age; after shooting and killing a deer on the family's estate—hunting was a popular pastime on the estate—a teenaged Adamson swore she would never again kill for sport.

Adamson had varied interests and studied such subjects as psychoanalysis, painting, metalwork, music, dressmaking, and archaeology at schools in Vienna. Though Adamson planned to pursue a career in medicine, she did not take her final exam. Instead, in 1935, Adamson married Victor von Klarwill, an Austrian businessman. Because Adamson's new husband was Jewish, the couple decided to move to Kenya to escape the growing Nazi movement. Adamson went ahead of von Klarwill, and during her journey she met botanist Peter Bally. After divorcing von Klarwill, Adamson married Bally in 1938. Bally traveled through Kenya to study plant specimens, and Adamson accompanied him. She began to paint the plants Bally collected, eventually completing about 700 paintings.

Adamson's second marriage ended in divorce in 1942, and a year later she married George Adamson, a game warden. During the following years, Adamson continued her paintings of flowers and plants and also began to paint portraits of tribal members. Then, in 1956, George Adamson killed a lioness that attacked him. After discovering that the lioness was protecting three cubs, George Adamson brought home the cubs. Two were sent to the Rotterdam Zoo, but Adamson kept the third cub and named her Elsa. Adamson and her husband raised Elsa and trained her to live in the wild. Adamson chronicled these experiences in the book, *Born Free,* which was published in 1960. Elsa eventually had three cubs of her own and began to visit the Adamsons. When Elsa died at the age of five, the Adamsons trained her three cubs and set them free in Serengeti National Park. Adamson wrote about the cubs in *Living Free* and *Forever Free,* sequels to *Born Free.*

During the 1960s, Adamson worked to increase awareness of wildlife endangerment and the need for preservation, capitalizing on the popularity of her books. In 1961, Adamson established the Elsa Wild Animal Appeal Fund in the United Kingdom. Chapters in the United States and Canada followed. Adamson was also a founder of the World Wildlife Fund and among the first to boycott apparel made from animal fur. In 1962, she traveled around the world to speak about wildlife preservation. The proceeds from her activities funded the establishment of wildlife reserves and conservation efforts.

Though little was known about the behavior of cheetahs, Adamson raised and trained a cheetah, named Pippa, in the late 1960s. She detailed her experiences with Pippa in two books, *The Spotted Sphinx,* published in 1969, and *Pippa's Challenge,* published in 1972. Adamson moved to an estate outside of Nairobi in 1971, and in 1976 she focused on raising a leopard cub named Penny. This experience, too, led to a book, *Queen of Sheba: The Story of an African Leopard,* which was published in 1980.

Adamson was the recipient of numerous honors and awards for her efforts to advance the wildlife preservation movement. She was presented the Award of Merit from Czechoslovakia in 1970, the Joseph Wood Krutch Medal of the U.S. Humane Society in 1971, and the Austrian Cross of Honor for science and art in 1976. Adamson also received the 1947 Gold Grenfell Medal from the Royal Horticultural Society for her illustrations of East African plant life. Adamson was murdered by a former servant on January 3, 1980 in the Shaba Game Reserve in northern Kenya.

Agassiz, Elizabeth Cabot Cary
(1822–1907)
American
Naturalist and Educator

Although Elizabeth Cabot Cary Agassiz received no formal education, she collaborated with her husband—the famed naturalist Louis Agassiz—to publish important works on natural history. In addition to participating together in several of his expeditions, the couple cofounded the Anderson School of Natural History. After her husband's death, Agassiz established the Harvard Annex, later named Radcliffe College, which she served as its first president. As one of the nation's elite colleges, Radcliffe served as a testament to Agassiz's commitment to women's higher education.

Born in Boston, Massachusetts, on December 5, 1822, Elizabeth Cabot Cary was the second of the seven children of Thomas and Mary Cushing Perkins Cary. Although she never attended school because of her frail health, Elizabeth Cary was tutored by a governess at home. She showed no early interest in science, but she was exposed to languages, music, and art. In 1846, she met Louis Agassiz, then a professor of natural history at the University of Neuchâtel, Switzerland. Agassiz emigrated to the United States shortly thereafter, accepting a position at Harvard University as the chair of natural history at the Lawrence Scientific School. Cary married Agassiz in 1850. While the couple had no children together, Elizabeth Cary Agassiz became mother to her husband's three children from a prior marriage.

To help support her new family, Elizabeth Agassiz launched a girls' school in her Cambridge home in 1856. Although she did not teach any classes herself, she sat in on the natural history lectures her husband delivered to the school's pupils. Her interest in the subject was sparked by this experience. After closing the school in 1863, she devoted herself to collaborating with Louis Agassiz on a number of scientific endeavors.

In 1859, Elizabeth Agassiz published her first book, *A First Lesson in Natural History,* which incorporated a number of her husband's theories. *Seaside Studies,* cowritten with her stepson Alexander Agassiz, appeared in 1865. A well-regarded textbook and field guide on marine zoology, *Seaside Studies* discussed a range of topics, such as the distribution of sea life and the embryology of various marine species. Together with Louis, Elizabeth Agassiz embarked on the Thayer expedition in 1865 to study the fauna of Brazil. Her copious notes about the voyage provided the basis for *A Journey in Brazil,* a book authored jointly by Louis and Elizabeth in 1868.

The couple's collaboration continued during the Hassler expedition (1871–72), a deep-sea dredging effort along the Atlantic and Pacific coasts of the United States. In 1873, the duo founded the Anderson School of Natural History on Penikese Island in Buzzard's Bay, Massachusetts. A summer school and a marine laboratory, Anderson accepted both male and female students (which was rather uncommon for the time). In 1873, Louis Agassiz died of a cerebral hemorrhage. Although his untimely death ended the couple's fruitful joint ventures, Elizabeth Agassiz turned to new projects of her own. In 1885, she published *Louis Agassiz: His Life and Correspondence,* a two-volume biography of her husband, which provided essential information about his theories.

Long interested in women's education, Elizabeth Agassiz devoted the remainder of her life to championing higher education for women. Although she did not believe in the coeducation of men and women, she was an ardent proponent of women's rights to equal educational opportunities. After traveling to Oxford and Cambridge to gather information, Agassiz founded the all-women's Harvard Annex in 1879, which shared the resources and faculty of Harvard. Agassiz served as its first president. The institution was rechristened Radcliffe College in 1893 and was formally linked to Harvard at that time. Agassiz remained president until 1899 when she retired. A scholarship and student hall were named in her honor. After suffering an initial cerebral hemorrhage in 1904, Elizabeth Agassiz died of a second one in 1907.

Agnesi, Maria Gaetana
(1718–1799)
Italian
Mathematician

Maria Gaetana Agnesi is best remembered for her pioneering two-volume work, *Analytical Institutions,* in which she synthesized and clarified existing information about algebra as well as integral and differential calculus. *Analytical Institutions* was translated into several languages, and it became the standard calculus textbook in Europe for over a century after Agnesi's death. In her most famous work, she discussed the formulation of a cubic curve, now known as the witch of Agnesi.

Born in Milan, Italy, on May 16, 1718, Maria Gaetana Agnesi was the eldest child of Pietro and Anna Fortunato

Brivio Agnesi. A professor of mathematics at the University of Bologna, Agnesi's father would remarry twice after her mother's death in 1732. The family would eventually grow to include 21 children.

Because he came from a wealthy merchant family, Pietro Agnesi could hire the finest tutors for his children. Although most women of this era received at best a strict convent education, Maria Agnesi was schooled in Greek, Hebrew, Spanish, and other languages. At the age of nine, she recited from memory Horace's defense of higher education for women. Her father held regular intellectual gatherings at his home, and Maria Agnesi was called upon to debate topics of philosophy and science with some of the leading thinkers of the day. Despite her unconventional education, Maria Agnesi's first ambition was to become a nun. However, her father forbade her from joining a convent, so Agnesi devoted herself to the study of mathematics and to supervising the household. With her tutor, Ramiro Rampinelli, Agnesi delved into calculus texts. In 1738, she published *Propositiones Philosophicae,* which included almost 200 theses on science and philosophy that she had defended at her father's soirees.

That same year, at the age of 20, Agnesi began writing *Analytical Institutions* as a calculus textbook for her younger brothers. It would take her nearly 10 years to complete and would eventually consist of two massive volumes—the first on algebra and geometry, the second on differential and integral calculus. She oversaw the work's publication in her home, soliciting feedback from fellow mathematicians. Upon its publication in 1748, the book revolutionized the study of calculus. In addition to integrating much of the contemporary information about calculus that had previously existed only in scattered fragments, *Analytical Institutions* introduced new ideas and methods. Of most significance was the cubic curve she discussed which now bears her name—the witch of Agnesi. This moniker derives from the fact that Agnesi actually misnamed her insight—referring to it as the *versiera* (which translates from colloquial Italian as "witch") rather than the proper *versoria.*

Agnesi's work brought her immediate recognition. After being elected a member of the Accademia delle Scienze, an honorary society of intellectuals, she was nominated to become a professor at the University of Bologna by Pope Benedict XIV in 1752. However, she never served on the faculty. That same year, Agnesi's father died, and she abandoned mathematics to pursue her religious calling. For the rest of her life, Agnesi cared for the sick and the poor. In 1771, she was named the director of a home for the elderly by Cardinal Pozzobonelli. Although several mathematicians would occasionally send her work to review, she always responded that she was no longer concerned with such topics. After suffering for years from dropsy, also known as edema, Maria Gaetana Agnesi died on January 9, 1799. In accordance with her wishes, she was buried with the poor in a common grave. *Analytical Institutions* remained the standard calculus textbook in Europe for nearly 100 years after her death.

Agnodice
(c. 400 B.C.)
Greek
Physician

Agnodice is recognized as the first female gynecologist. There is, however, little concrete information about her life. Hyginus, a Latin author of the first century A.D., provides the only existing account of Agnodice's life and work. Subsequent scholarship has suggested, though, that she may be a mythical figure rather than a flesh-and-blood physician. In any event, the story of her courageous decision to practice medicine in ancient Athens has earned her a respected place in the history of medicine.

No details about Agnodice's family are extant, though general facts about women's role in ancient medicine are known. While Greek women had few privileges, they were allowed to practice midwifery and healing. The famed medical pioneer Hippocrates (c. 460 B.C.) did not accept women at his primary medical school located on the island of Cos. However, he did allow women to attend another of his schools in Asia Minor, where they could study gynecology and obstetrics. After Hippocrates' death, women were barred from the practice of medicine, possibly because Athenian rulers discovered that women gynecologists were performing abortions. To deter those who might violate this ban, the death penalty was instituted as punishment. As many Athenian women were reluctant to visit male doctors, given the social codes of the period, women's access to medical care decreased after the imposition of the ban, and female mortality rates rose.

Agnodice came of age in this political and social climate. Determined to become a doctor, Agnodice flouted the laws barring her entrance into the profession. She disguised herself as a man, cutting her hair and dressing in male garments. After attending classes taught by the renowned Herophilos at the University of Alexandria, Agnodice began to practice as a gynecologist. According to Hyginus's account, Agnodice went to assist a woman in labor. Thinking Agnodice was a male doctor, the woman refused her help. But Agnodice "lifted up her clothes and revealed herself to be a woman and was thus able to treat her patient," wrote Hyginus.

When news of her skills and gender spread, Greek women flocked to Agnodice, at last able to confide in a female doctor. Unaware of Agnodice's gender, male doctors grew jealous of her popularity and accused her of seducing women (and of the women feigning illness to visit

Agnodice). Agnodice was dragged before an Athenian court. She was forced to reveal her gender to avoid the death penalty for corrupting women. Despite her confession, the male physicians became more outraged and accused her more forcefully for breaking the law forbidding women from studying medicine. Agnodice was rescued from her plight when the wives of leading men arrived in the court. According to Hyginus, they declared that "you men are not spouses but enemies, since you are condemning her who discovered health for us."

The truth of Hyginus's tale of Agnodice cannot be confirmed. Nonetheless, later women seeking entrance into the medical profession employed Agnodice as a symbol for the precedent of women in medicine. Agnodice also underscores the powerful impulse of women to be treated by female physicians—an argument Victorian women raised to justify the need for women gynecologists more than 2,000 years after Agnodice's time.

Ajakaiye, Deborah Enilo
(1940–   )
Nigerian
Geologist

In her pursuit to study geology, Deborah Enilo Ajakaiye had to break through a number of barriers. She became the first female professor of physics in Africa, the first woman dean of science in West Africa, the first female fellow and president of the Nigerian Mining and GeoSciences Society, and the first woman fellow of the Nigerian Academy of Sciences. Ajakaiye has authored five books and more than 43 major papers. A pioneer in the technique of geovisualization, Ajakaiye has played an important role in detecting resource deposits in Nigeria.

Born in 1940, Ajakaiye was the fifth of six children. Raised in the tin-mining city of Jos in the Oyo province of Nigeria, Ajakaiye received an education from an early age. Unlike many other Nigerian families, Ajakaiye was fortunate to have parents who believed in creating opportunities for all their children—both male and female. Equality was emphasized in the family home. Chores were apportioned equally among the children, with both the boys and girls having to clean the house and prepare meals.

In 1957, Ajakaiye enrolled at the University College in Ibadan. Although she initially planned to major in mathematics, she ultimately decided to "study something I could touch." She graduated with a degree in physics in 1962. After obtaining a master's degree at the University of Birmingham in England, she returned to Nigeria for her doctorate. In 1970, Ajakaiye received her Ph.D. in geophysics from Adhadu Bello University. Her dissertation examined the mining area of Jos.

Ajakaiye remained in academia, becoming the first female full professor of physics in Africa in 1980. She has taught at both the Adhadu Bello University and the University of Jos, and she has served as the dean of natural sciences at the latter institution. Along with her impressive background in theoretical areas, she has sought to remain involved in practical problems facing Nigeria as well. "I chose . . . geophysics because I felt this field could make possible significant contributions to the development of my country," she wrote in a 1993 paper for the American Association for the Advancement of Science. Not only could her area of expertise help pinpoint mineral deposits under Nigeria's soil, but geophysics could also locate sources of groundwater. In addition to her work with geovisualization (whereby computers produce three-dimensional images of potential deposits beneath the earth), Ajakaiye has also conducted a survey for a geophysical map. This project, in which Ajakaiye led several of her female students, mapped the topography of northern Nigeria. More recently, Ajakaiye collaborated with scholars at Rice University and the University of Houston to determine structural styles in the Niger Delta.

Ajakaiye's efforts to benefit both her scientific field and her nation have been recognized. She was honored by the Nigerian Mining and GeoSciences Society for her "distinguished contributions to mining and geosciences in Africa[,]" becoming the first woman to win the award. Ajakaiye was also named a fellow of the Geological Society of London, the first black African to attain this distinction. Despite her many successes, Ajakaiye has remained modest and committed to equal educational opportunities for women.

Alexander, Hattie Elizabeth
(1901–1968)
American
Physician

Hattie Elizabeth Alexander's greatest contribution to medical science was her discovery of a treatment for influenzal meningitis, a common and often fatal disease that afflicted infants and children. Her work in this area led her to investigate bacterial resistance to antibiotics and to conclude that such resistance was caused by genetic mutation. She was also a pioneer in research on DNA. The author of more than 60 professional articles, Alexander enjoyed the distinction of being named the first woman president of the American Pediatric Society.

The second of Elsie May Townsend and William Bain Alexander's eight children, Hattie Elizabeth Alexander was born on April 5, 1901, in Baltimore, Maryland. A mediocre student at Goucher College, Alexander nonetheless hoped to study medicine. From 1923 to 1926, she worked as a bacteriologist for the U.S. Public Health Service Laboratory in Washington, D.C. After being admitted to Johns Hopkins University's medical school, Alexander excelled and earned her M.D. in 1930. Upon completing her formal education, Alexander accepted two internships in succession—the first at the Harriet Lane Home in Baltimore (1930–31), and the second at the Babies Hospital of the Columbia-Presbyterian Medical Center in New York City (1931–33). As an intern, Alexander gained her first exposure to the fatal course of influenzal meningitis.

In 1933, Alexander embarked on a series of overlapping positions at the Babies Hospital, the Vanderbilt Clinic of the Columbia-Presbyterian Medical Center, and Columbia University's College of Physicians and Surgeons. Appointed at Babies Hospital as an adjunct assistant pediatrician in 1933 and an attending pediatrician in 1938, Alexander devoted her early career to combating influenzal meningitis. The disease was caused by *Hemophilus influenzae,* a deadly bacteria that inflamed the meninges (the tissues surrounding the brain and spinal cord). Alexander sought to create a serum (the component of the blood containing antibodies) that would combat the illness. She injected bacteria from the spinal fluid of children suffering from the disease into healthy rabbits. The rabbits produced an antibody to influenzal meningitis that proved effective in treating the disease in humans. Alexander announced the first complete cure of infants suffering from this disease in 1939.

In the early 1940s, Alexander turned her attention to treating influenzal meningitis with antibiotics in conjunction with the rabbit serum. As a result of this combined treatment, the rate of infant deaths caused by the disease dropped 80 percent by 1942. After serving as an instructor in children's diseases at Columbia, Alexander was named associate professor in 1948. Alexander's investigation of supplementary drug treatment for influenzal meningitis led her to confront antibacterial resistance, the phenomenon in which bacteria become immune to the effects of antibiotics. Theorizing that such resistance was caused by the bacteria's genetic mutations, Alexander spearheaded early research into microbiological genetics. In 1950, she successfully used DNA to alter the hereditary makeup of *Hemophilus.* Alexander attained the rank of full professor in 1958 and retired as a professor emerita in 1966.

Alexander received a number of awards in recognition of her accomplishments. After receiving the E. Mead Johnson Award for Research in Pediatrics in 1942, Alexander became the first woman to earn the Oscar B. Hunter Memorial Award of the American Therapeutic Society (1962). In 1964, she also became the first woman elected president of the American Pediatric Society. In addition to the numerous lives she saved with her treatment of influenzal meningitis, Alexander is credited with playing an essential role

in the acceptance of DNA research in medical fields. She died of cancer on June 24, 1967.

Alvariño, Angeles
(1916–   )
Spanish/American
Marine Biologist

Angeles Alvariño De Leira, who is known by the professional name Angeles Alvariño, is a fishery research biologist and oceanographer who has specialized in the study of marine zooplankton, animal plankton such as jellyfish and sea anemones. In particular, Alvariño investigated the geographic distribution and ecology of zooplankton and significantly advanced scientists' understanding of the little-studied marine organisms. Alvariño's research led to the detection of 22 undiscovered ocean species.

Born on October 3, 1916, in El Ferrol, Spain, Alvariño possessed a bright and inquisitive mind. Alvariño's mother was Maria del Carmen Gonzalez Diaz-Saavedra de Alvariño. Her father, Antonio Alvariño Grimaldos, was a physician who allowed Alvariño to roam freely in his personal library. It was in the library that Alvariño discovered her interest in natural history. Though Alvariño hoped to follow in her father's footsteps, her father encouraged her to choose a different path—he did not wish his daughter to suffer the pain and disappointment caused by the inability to help critical patients.

Alvariño enrolled at the Lycée in 1930 and studied a wide range of subjects, including humanities, social science, and physical and natural sciences. After completing two dissertations in both letters and science, Alvariño graduated summa cum laude from the University of Santiago de Compostela, located in Spain, in 1933. Though Alvariño was still interested in pursuing a career in medicine, her father continued to object, and so Alvariño decided to study natural sciences. After entering the University of Madrid in 1934, Alvariño was forced to postpone her studies when the Spanish Civil War caused the university to close its doors from 1936 to 1939, but after resuming her schooling, she earned her master's degree in natural sciences in 1941. A year earlier, Alvariño had married Sir Eugenio Leira Manso, captain of the Spanish Royal Navy and knight of the Royal and Military Order of Saint Hermenegild.

After receiving her master's degree, Alvariño was an instructor of natural sciences at assorted colleges in El Ferrol. Interested in pursuing research, she joined the Spanish Department of Sea Fisheries as a research biologist in 1948. Though women were not allowed to participate in the Spanish Institute of Oceanography in Madrid, Alvariño nevertheless became involved with oceanography studies and research and, thanks to her outstanding work, was officially admitted as a student researcher in 1950. Meanwhile, Alvariño had taken courses at the University of Madrid (later known as the University Complutense), and in 1951, after the completion of three theses, in chemistry, plant ecology, and experimental psychology, she was awarded a doctoral certificate. A year later Alvariño became a marine biologist and oceanographer with the Spanish Institute of Oceanography.

In 1953, Alvariño was awarded a British Council Fellowship and traveled to the Marine Biological Laboratory, located in Plymouth, England, to study zooplankton. There she met Frederick Stratten Russell, a British marine biologist and expert on jellyfish. Russell encouraged Alvariño to focus her investigations on chaetognaths (arrow worms), hydromedusae (jellyfish), and siphonophores (transparent, floating water organisms). To better study these organisms and gather samples, Alvariño created plankton nets that were then taken out to sea by Spanish fishing boats and research ships.

Alvariño traveled to the United States in 1956 when she was awarded a Fulbright Fellowship. She conducted research at the Woods Hole Oceanographic Institute in Massachusetts. In 1958, Alvariño was offered a position as a biologist at the Scripps Institute of Oceanography in La Jolla, California. She continued her studies of zooplankton, and her work resulted in a doctoral degree from the University of Madrid in 1967. Alvariño left Scripps in 1969 and became a fisheries biologist with National Marine Fisheries Service's Southwest Fisheries Science Center (SWFSC) in La Jolla. Alvariño remained with the SWFSC until her retirement in 1987, studying chaetognaths, hydromedusae, and siphonophores and their connection to the survival of larval fish.

Alvariño has also been associated with the University of San Diego and San Diego State University. Though retired since 1987, she has continued her investigations of zooplankton. A pioneering marine biologist, Alvariño has contributed to more than 100 scientific articles. In 1993, King Juan Carlos I and Queen Sophia of Spain awarded Alvariño the Great Silver Medal of Galicia. Alvariño and her husband reside in La Jolla, California. The couple has one child.

Anastasi, Anne
(1908–   )
American
Psychologist

Anne Anastasi was a key figure in the development of psychology as a quantitative behavioral science. Most notably, she was a pioneer in the field of psychometrics—the study of how psychological traits are influenced, formed, and measured. Rated by her peers in 1987 as the most prominent living woman in psychology in the English-speaking world, Anastasi authored the definitive work on psychological testing.

Born on December 19, 1908, in New York City, Anastasi was the only child of Anthony and Theresea Gaudiosi. Her father died when she was only one year old, and she was educated at home by her grandmother until she began the sixth grade. She struggled in high school for two months before she dropped out and decided to enter college early.

Anastasi was admitted to Barnard College in 1924 when she was only 15. Although she planned to study mathematics, she became enthralled by psychology during a class taught by the department head, Harry Hollingworth. After reading an article by Charles Spearman in which he applied complex mathematical concepts to questions of psychology, she became convinced that she could meld the study of mathematics with psychology. Anastasi entered Barnard's Honors Program in psychology during her sophomore year and received her bachelor's degree in 1928 at the age of 19. A member of Phi Beta Kappa, she also won the Caroline Duror Graduate Fellowship. She then enrolled at Columbia University, proceeding directly to the Ph.D. track because she had already taken a number of graduate classes in psychology (obviating the need for her to obtain her master's degree first). After specializing in the field of differential psychology—the area of study that explores individual and group differences in behavior—she received her Ph.D. in 1930.

The same year she earned her doctorate, Anastasi returned to Barnard as a professor. In 1933, she married John Porter Foley Jr., a fellow Columbia graduate student. Four years later, she wrote what is now considered a classic textbook, *Differential Psychology*. In 1939, she left Barnard to become an assistant professor and the sole member of the fledgling psychology department at Queens College of the City of New York. She remained there for eight years before accepting a position as an associate professor of psychology in the Graduate School of Arts and Sciences at Fordham University in 1947. She was promoted to full professor at Fordham in 1951 and remained at the school for the duration of her career.

Throughout her professional life, Anastasi's research consistently concentrated on the nature and measurement of psychological traits. Her most influential work, *Psychological Testing* (1982), investigated the role of education and heredity on the development of human personality traits. She demonstrated how difficult measuring these traits can be because of variables such as training, culture, and language differences. Anastasi transcended the "nature-nurture" debate in her work, arguing that psychologists are misguided in their efforts to explain human behavior solely in terms of heredity or environment. She was emphatic that neither "nature" nor "nurture" exists independently and that psychologists should seek to understand the interplay of the two forces.

Anastasi retired from Fordham in 1979, though she remains active in her field as both an author and a speaker. In addition to being lauded as the leading woman psychologist in the English-speaking world, she has received a number of other honors. In 1972, she became the first woman in 50 years to be elected president of the American Psychological Association. Her lifelong achievements were recognized in 1987 when President Ronald Reagan awarded her the National Science Medal.

Ancker-Johnson, Betsy
(1929–)
American
Physicist

Betsy Ancker-Johnson overcame a number of obstacles to become an internationally renowned solid-state physicist. Despite being told by her professors in graduate school that women cannot think analytically, Ancker-Johnson made important contributions to the study of plasmas in solids as both a faculty member and an industrial researcher. An adept manager, Ancker-Johnson attained high positions at both the Boeing Aerospace Company and General Motors. She has also served as the assistant secretary for science and technology at the U.S. Department of Commerce and has been the associate laboratory director of physics research at Argonne National Laboratory.

Betsy Ancker was born to Clinton and Fern Ancker in St. Louis, Missouri, on April 29, 1929. After completing high school, Ancker enrolled at Wellesley College, where she received her B.A. in 1949. While at college, Ancker-Johnson was encouraged to pursue her interest in physics, and after graduating, she departed for Tübingen University in Germany for graduate studies in physics. Although she hoped to explore a new culture at the same time she earned her doctorate, Ancker's experience at Tübingen was painful. Faculty members and fellow students ridiculed her goals, claiming that she must be husband hunting. Undeterred, she pressed on with her studies, earning her Ph.D. in physics in 1953.

Upon returning to the United States, Ancker confronted similar prejudice in her search for employment. Unable to secure a job in industrial research, she accepted a post as a lecturer at the University of California, Berkeley, in 1953. While volunteering at the Inter-Varsity Christian Fellowship in 1954, Ancker met her future husband, Harold Johnson, a mathematics professor. After leaving the University of California in 1954, she worked as a senior research physicist at Sylvana Corporation's Microwave Physics Laboratory (now a branch of GTE) from 1956 to 1958. She married Johnson in 1958, changing her name to Ancker-Johnson. The couple had four children—Ruth, David, Paul, and Martha.

Following a stint with RCA from 1958 to 1961, Ancker-Johnson transferred to the Boeing Corporation's research laboratories in 1961, where she remained until 1970, investigating the instabilities that can occur in plasmas in solids. Her most significant discovery occurred in 1967 when she detected microwave emissions from an electron-hole plasma. This breakthrough led others to realize that solid-state plasmas may serve as microwave sources of radiation. In 1970, Ancker-Johnson was named supervisor of solid-state and plasma electronics at Boeing, and she was promoted to manager of the company's advanced energy systems department in 1973. While at Boeing, she received a number of patents for devices that included a solid density probe, a solid signal generator, and a solid-state amplifier and phase detector.

Ancker-Johnson's career took another turn in 1973 when she became the first female scientist to be appointed assistant secretary for science and technology at the U.S. Department of Commerce. After four years in that capacity—during which she oversaw a budget of more than $230 million—she became the associate laboratory director for physics research at Argonne National Laboratory in 1977.

Former assistant secretary for science and technology at the U.S. Department of Commerce, Betsy Ancker-Johnson. *(U.S. Department of Commerce)*

Two years later, Ancker-Johnson reentered the business sector when she joined General Motors (GM). As GM's vice president in charge of environmental activities, she was the first woman vice president in the entire auto industry. At GM, she was responsible not only for dealing with waste emissions from the company's plants worldwide but for automobile safety, fuel economy, and noise emissions as well. She remained with General Motors until her retirement in 1992.

Ancker-Johnson has remained active in the numerous professional organizations to which she belongs. In addition to being elected to the National Academy of Engineering in 1975, she was named a fellow of the American Physical Society and of the Institute of Electrical and Electronic Engineers. The recipient of several honorary degrees, Ancker-Johnson has actively promoted the role of women scientists. As part of this effort, she published *Nobel Prize Women in Science: Their Lives, Struggles, and Momentous Discoveries* in 1993.

Andersen, Dorothy Hansine
(1901–1963)
American
Physician and Pathologist

Although Dorothy Hansine Andersen's gender was used to deny her an appointment as a surgeon, she overcame the obstacles placed in her path to become a pioneering medical researcher. Andersen is best remembered for discovering and naming the disorder known as cystic fibrosis as well as for developing a simple procedure for detecting the disease. Andersen was also an expert on cardiac defects, and she served as a consultant to the Armed Forces Institute of Pathology during World War II.

Dorothy Hansine Andersen was born on May 15, 1901, in Asheville, North Carolina. The only child of Hans Peter and Mary Louise Mason Andersen, Dorothy Andersen learned to be self-sufficient at an early age. Her father died when she was only 13, leaving her to care single-handedly for her invalid mother. Dorothy Andersen moved with her mother to Saint Johnsbury, Vermont, where she nursed her mother while attending school.

Andersen graduated from high school in 1918 and entered Mount Holyoke College. After her mother died in 1920, she completed her undergraduate degree without the financial or emotional support of any relatives. In 1922, Andersen began medical school at Johns Hopkins University, where she worked in the laboratory of Dr. FLORENCE SABIN, the first woman to enroll, graduate, and teach at Johns Hopkins. Andersen completed her M.D. in 1926 and subsequently interned at the University of Rochester. Despite her stellar academic record, however, Andersen was excluded from residency positions in surgery and pathol-

ogy because such disciplines typically barred women. Determined to pursue a career in pathology nevertheless, Andersen took a post as a research assistant in pathology at Columbia University's College of Physicians and Surgeons. She began a doctoral program in endocrinology at Columbia and earned the degree of doctor of medical science in 1935.

Upon receiving her doctorate, Andersen accepted an appointment as a pathologist at the Babies Hospital of the Columbia-Presbyterian Medical Center in New York City, where she would remain for the duration of her career. She immediately began studying heart defects in infants. In 1935, while performing a postmortem examination on a child who had supposedly died of celiac disease (a nutritional disorder), Andersen noted a lesion on the child's pancreas. Following this lead, she searched for similar cases in autopsy files and painstakingly discovered an unnamed disease. The congenital disorder she uncovered affected the mucous glands and pancreatic enzymes and caused abnormal digestion and difficulty in breathing. In 1938, she presented a paper on the disease—which she called cystic fibrosis—to the American Pediatric Society.

As a medical researcher, Andersen was expected simply to describe cystic fibrosis. Never one to follow the status quo, however, Andersen searched for both a diagnosis and a cure for the disease. She ultimately discovered that those afflicted by cystic fibrosis also had an increase in sweat salinity, and thus she pioneered the simple "sweat test" used to detect the disease today.

In 1945, Andersen was named assistant pediatrician at Babies Hospital. During World War II, she was called upon to apply her extensive knowledge of cardiac anatomy as a consultant to the Armed Forces Institute of Pathology. In 1952, Andersen was appointed chief of pathology at Babies Hospital, and she was promoted to full professor in 1958. She continued her quest to find a cure for cystic fibrosis, and in 1959, she published her final paper—on the occurrence of the disease in young adults.

Andersen was frequently derided for her unconventional lifestyle. In addition to flouting the conservative dress code of the day, she took up "unfeminine" hobbies such as hiking, canoeing, and carpentry. Despite these personal attacks, Andersen's important contributions to medicine were recognized by her peers. In 1948, she received the Borden Award for research in nutrition. She was honored with the Elizabeth Blackwell Award in 1952, the Distinguished Service Medal of the Columbia-Presbyterian Medical Center in 1963, was named an honorary fellow of the American Academy of Pediatrics, and served as the honorary chair of the Cystic Fibrosis Research Foundation. Diagnosed with lung cancer in 1960, Andersen died on March 3, 1962.

Anderson, Elizabeth Garrett
(1836–1917)
British
Physician

Elizabeth Garrett Anderson achieved the distinction of being the first British woman to be certified as a physician. However, her determination to break new ground did not end with this accomplishment. She also became the first woman to earn a medical degree from the Sorbonne, the first woman to be elected to a British school board, and the first female mayor in Britain. In addition to serving as both dean and president of the London School of Medicine for Women, Anderson founded the New Hospital for Women—which was renamed the Elizabeth Garrett Anderson Hospital after her death.

The second child of Newson and Louisa Dunnell Garrett, Elizabeth Garrett was born on June 9, 1836, in London. When she was still a child, the family moved to Aldeburgh, Suffolk, England. At the age of 13, Elizabeth Garrett was sent to attend boarding school with her sister. Although she learned French and improved her writing, science and mathematics were not part of the curriculum, and it was expected that young Elizabeth would return home and await marriage. However, she felt strongly about constructing a meaningful life for herself apart from marriage.

Elizabeth Garrett's life would change in 1859 when she made the acquaintance of ELIZABETH BLACKWELL, America's first female doctor. Emboldened by Blackwell's example, Garrett sought to pursue a career in medicine. Her family was initially vehemently opposed to this plan. But, struck by her commitment, Garrett's father soon became her staunchest supporter. However, no British university would admit female medical students. Undeterred, Garrett began medical training as a nurse at Middlesex Hospital in 1860. The resident physicians at Middlesex were impressed by her aptitude and allowed her to attend operations and follow on rounds. Threatened by her success, the male students successfully petitioned to have her dismissed.

Garrett remained undaunted. Although the Medical Act of 1858 clearly stated that a person must be licensed by a qualifying board of a British university to be listed on the medical register, Garrett discovered a loophole. The Licentiate of the Society of Apothecaries (L.S.A.) entitled its holders to practice medicine, and the charter of the Society of Apothecaries had not specifically barred women from taking its licensing exam. To qualify, a candidate had to serve as a physician's apprentice for five years and attend a set number of courses. In 1861, Garrett began the slow process of fulfilling these requirements, and in 1865 she passed the examination. Although the L.S.A. degree was not as esteemed as a medical degree, Garrett was now entitled to practice medicine in England. In 1866, she entered

Elizabeth Garrett Anderson, Britain's first certified female physician as well as its first female mayor. *(National Library of Medicine, National Institutes of Health)*

the Medical Register. (The Society of Apothecaries immediately enacted restrictions against women after Garrett passed the exam.)

Garrett founded St. Mary's Dispensary for Women in 1866 (which would be renamed the New Hospital for Women and Children in 1872). Despite her achievements, Garrett still desired the prestige conferred by a medical degree. After learning that the University of Paris had begun to admit women, Garrett studied French, reviewed her coursework, and took several examinations. In 1870, Garrett became the first woman to earn a medical degree from the Sorbonne.

That same year, Garrett applied for a position at the Shadwell Hospital for Children, but one of the hospital's board members—James Skelton Anderson—vowed to block her efforts. After meeting Garrett personally, however, he dropped his objections. In fact, he was so impressed by the young physician that he eventually proposed to her. The couple was married in 1871 and had three children, two of whom survived to adulthood—Louisa and Alan. With SOPHIA JEX-BLAKE, Elizabeth Garrett Anderson cofounded the London School of Medicine for Women in 1877. (By 1878, a woman's right to a medical education was secured). She was named dean in 1883, a position she held until 1903, when she became president.

In 1902, the Andersons retired to Aldeburgh, where James was elected mayor. Upon his death, Elizabeth Garrett Anderson completed his term and was resoundingly re-elected. Her legacy was profound. In addition to opening the practice of medicine to women in Britain, she served as a shining example that women could be successful in any sphere. In 1914, her own daughter, Dr. Louisa Garrett Anderson, left for France to head a military hospital during World War I. Elizabeth Garrett Anderson died on December 17, 1917, at the age of 81.

Anderson, Gloria Long
(1938–)
American
Chemist

Gloria Long Anderson is a respected chemist and educator who has spent more than 30 years at Morris Brown College, a historically black college located in Atlanta, Georgia. At one point, Anderson served as the college's dean of academic affairs. During her long academic career, Anderson continued her chemistry research, particularly in regard to fluorine-19 chemistry. An active member of the chemistry community, Anderson has been involved with numerous scientific organizations. In addition, Anderson has worked to advance opportunities for minorities and was involved with the Corporation for Public Broadcasting to raise awareness regarding minorities and women in the workplace.

Born Gloria Long on November 5, 1938, in Altheimer, Arkansas, Anderson was the daughter of Charley Long and Elsie Lee Foggie. Anderson attended the University of Arkansas at Pine Bluff, which was then called the Arkansas Agricultural, Mechanical and Normal College. A bright and dedicated student, Anderson received a Rockefeller Scholarship from 1956 to 1958. After earning her bachelor's degree and graduating summa cum laude in 1958, Anderson began graduate studies at Atlanta University. In 1960, she married Leonard Sinclair Anderson, and a year later she was awarded her master's degree. In 1962, Anderson taught chemistry at South Carolina State College. From 1962 to 1964, she was an instructor at Morehouse College in Atlanta. Anderson then attended the University of Chicago to work on her Ph.D. in organic chemistry. She served as a research and teaching assistant and gained her doctorate in 1968.

With her Ph.D. in hand, Anderson became an associate professor and chair of the chemistry department at Morris Brown College in 1968. She steered her energies toward teaching but also continued her research, specializing in fluorine-19 chemistry. Fluorine-19 was a fluorine isotope that had magnetic properties, and fluorine itself was a highly reactive element that was used in a number of industrial applications. The study of fluorine-19 chemistry rose in significance in the early 1940s when fluorine compounds were found to have a number of commercial applications. Anderson studied fluorine-19 chemistry using nuclear magnetic resonance spectroscopy, which enabled her to investigate the structure and nucleic responses of molecules placed in a strong magnetic field.

In addition to chairing the chemistry department, Anderson served as the Fuller E. Callaway professor of chemistry at Morris Brown from 1973 to 1984. From 1984 to 1989, Anderson was the dean of academic affairs. She returned to her position as Callaway professor in 1990. In the mid-1990s, Anderson served as interim president at Morris Brown as well as dean of science and technology. Though most of Anderson's career has been at Morris Brown, she was affiliated with a number of other organizations. Anderson was a research fellow and research consultant with the National Science Foundation in the early 1980s, and in 1984 she was a research fellow in the Rocket Propulsion Lab of the Air Force Office of Scientific Research. The following year, Anderson was a United Negro Fund Distinguished Scholar. In 1990, she became a research consultant with BioSPECS, located in the Netherlands. Anderson has also worked with the Atlanta University Center Research Committee and the Office of Naval Research. In addition to her fluorine-19 chemistry investigations, Anderson has studied antiviral drugs and the synthesis of solid rocket propellants.

Anderson has been recognized for her outstanding work as a chemist and an educator. She was the Outstanding Teacher at Morris Brown in 1976, received the Teacher of the Year Award in 1983, and was inducted into the Faculty/Staff Hall of Fame at Morris Brown, also in 1983. In 1987, Anderson was given the Alumni All-Star Excellence in Education Award by the University of Arkansas at Pine Bluff. She served on the board of the Corporation for Public Broadcasting from 1972 to 1979. Anderson is a member of numerous scientific organizations, including the American Institute of Chemists, the American Chemical Society, the Georgia Academy of Science, and the National Institute of Science.

Anning, Mary
(1799–1847)
British
Paleontologist

Mary Anning transcended several sizable barriers to become one of the world's greatest fossil finders. Not only did she lack any formal training in her field, but she faced considerable prejudice on account of her gender and low social status. But Anning persevered, and later she discovered a nearly complete plesiosaur skeleton and, in her crowning achievement, the first pterodactyl fossil. She also uncovered one of the first specimens of ichthyosaurus. Her work made considerable contributions to the fledgling discipline of paleontology.

Mary Anning was born in 1799 in Lyme Regis, England, to Richard and Mary Anning. Although her parents had 10 children, only two survived to adulthood—Mary and her brother Joseph. Employed as a cabinetmaker, Richard Anning collected fossils as a hobby. The seaside cliffs of Lyme Regis were rife with fossils, especially those from the Jurassic period (the geological epoch of about 180 million years ago, which was characterized by the dominance of the dinosaurs), and Richard passed along his fossil-hunting passion to his wife and children. Unfortunately, he gave them little that was tangible. His death in 1810 left the family poor and debt-ridden.

After Richard's death, the Anning family supported itself by collecting fossils, which they sold to museums and private collectors. Young Mary made her first significant discovery sometime between 1809 and 1811, when she uncovered a nearly intact specimen of ichthyosaurus, an extinct marine reptile with a fishlike body and dolphinlike head. The Annings sold this skeleton as well, but they remained indigent until Thomas Birch (a professional fossil collector who had made the family's acquaintance) sought to alleviate their hardship. In the early 1820s, he auctioned off a number of his most prized fossils, donating the proceeds to the Annings.

Anning made another important find in 1924 when she unearthed an exceptional plesiosaurus skeleton. This discovery of the well-preserved plesiosaurus—a marine reptile characterized by a small head and long neck—was one of the era's most significant developments in paleontology. Her 1828 find of the first pterodactyl specimen (a flying reptile) solidified her reputation as a genuine fossilist in the eyes of the scientific community. Despite her many accomplishments, however, Anning's discoveries were usually sold to museums and private collections without due credit being accorded to her as the finder of the fossils.

Anning had taken charge of the family's fossil enterprise in the late 1820s, after her brother Joseph abandoned the collecting business for a career as an upholsterer, and her fossil shop in Lyme became a tourist attraction. More importantly, though, Anning won the respect of fellow paleontologists who visited her there. She received an annuity from the British Association for the Advancement of Science in 1838 as well as a stipend from the Geological Society of London, and she was named the first honorary member of the Dorset County Museum and of the Geological Society. She died of breast cancer in 1847, whereupon her obituary was published in the *Quarterly Journal of the Geological Society*. Although her achievements were largely forgotten in the decades after her death, Anning's important contributions to the field of paleontology have recently been excavated. Mary Anning is now recognized not only as an expert fossil locator but also as an astute anatomist.

Apgar, Virginia
(1909–1974)
American
Physician

Physician Virginia Apgar's best-known contribution to the medical sciences was the development of a test designed to evaluate the health of a newborn. This breakthrough system helped save the lives of many infants and became a standard procedure in hospitals worldwide. Apgar made numerous other advances in obstetric anesthesia and raised the awareness Americans had regarding birth defects.

Born on June 7, 1909, in Westfield, New Jersey, Apgar was raised in a family environment that encouraged the love of music as well as science. Apgar's mother was Helen Apgar and her father, Charles Apgar, was a businessman who had many hobbies, including astronomy and wireless telegraphy. Charles Apgar influenced his daughter to take an interest in science.

In the 1920s, Apgar entered Mount Holyoke College, where she was a very active student. Apgar not only waitressed and worked in the school library and other facilities to fund her tuition, but she also participated in numerous extracurricular activities, such as the college newspaper, orchestra, theater, and tennis. After graduating in 1929, Apgar entered medical school at Columbia University's College of Physicians and Surgeons. She received her medical degree in 1933 and hoped to pursue a career in surgery. A professor encouraged Apgar to instead investigate anesthesiology, then a new field. Apgar became only the 50th physician to achieve certification in the new specialty.

After completing her education, Apgar took a teaching position at Columbia University's medical school in 1936. Two years later, she became the director of the anesthesia division at Presbyterian Hospital. By 1949, Apgar had successfully built an anesthesiology department at the school, thereby lending the new field credibility, and was named full professor. In her research, Apgar concentrated on the effects of anesthesia during birth. She observed that newborn infants were generally taken directly to the hospital nursery with no examination—primary attention was given instead to the mothers. Because of the cursory treatment given to the newborns, health problems generally were not noticed

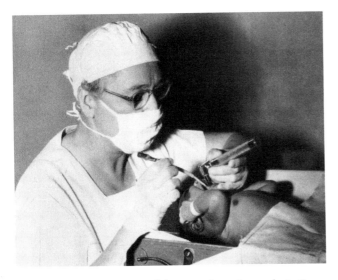

Virginia Apgar, who developed the Apgar Score System for testing the health of newborns immediately following birth. *(National Library of Medicine, National Institutes of Health)*

until they became critical or fatal. Apgar thus developed a simple test, which came to be known as the Apgar Score System, that nurses could administer immediately after birth. Using a scale ranging from zero to two, the test evaluated the newborn's skin color, muscle tone, breathing, heart rate, and reflexes. The resultant total score, known as the Apgar Score, indicated the physical condition of the infant. A low score was indicative of problems requiring immediate attention, while a high score signified a healthy baby. Apgar first introduced the system in 1952 and published it in 1953.

After many years researching obstetric anesthesia and working as a physician, Apgar returned to school, earning a master's degree in public health from Johns Hopkins University in 1959. Also that year Apgar became a senior executive with the charitable organization National Foundation—March of Dimes. Apgar's job was to raise funds to support research on birth defects and to increase public support and awareness. In 1967, Apgar became the director of the organization's research department.

In addition to her pioneering work related to the Apgar Score System, Apgar also conducted research on neonatal acid-base status and introduced the anterior approach to the stellate ganglion. Among the many honors Apgar received were the Elizabeth Blackwell Citation from the New York Infirmary in 1960, the Distinguished Service Award of the American Society of Anesthesiologists in 1961, and the Gold Medal for Distinguished Achievement in Medicine, awarded in 1973 by Columbia University's College of Physicians and Surgeons. Apgar was also named Woman of the Year in science by Ladies' Home Journal in 1973 and was honored with a U.S.

stamp in her name in 1994. Apgar died from liver disease at the age of 65 in 1974.

Arber, Agnes Robertson
(1879–1960)
British
Botanist

A devoted researcher, botanist Agnes Robertson Arber spent her career studying the anatomy and morphology of plants. Over the course of nearly 60 years, Arber published more than 80 scientific works that greatly enhanced scientists' knowledge and understanding of plants. Arber's first published book, *Herbals; Their Origin and Evolution: A Chapter in the History of Botany: 1470 to 1670,* became a widely read standard.

Agnes Robertson was born on February 23, 1879, in London, England. Her parents, Henry Robert Robertson and Agnes Lucy Turner, were artistic and raised their children in a creative and intellectual environment. Arber's father was an artist, and her sister Janet went on to become an artist as well. Her brother Donald pursued academia, later becoming a professor of Greek at Trinity College, Cambridge. Arber was greatly influenced by the stimulating family atmosphere and also gained an interest in art, which she later incorporated into her scientific research.

In 1897, Arber entered University College in London. Two years later, she was awarded a B.Sc. degree with first-class honors. Arber then attended Newnham College, Cambridge, on a scholarship. Newnham was one of two women's colleges at Cambridge. In 1901, she passed the first part of the natural science Tripos, the honors examination, and a year later she passed the second part, which covered botany and geology. In 1903, Arber received a Quain Studentship in biology and returned to University College to conduct research and teach. Arber was granted a D.Sc. degree in 1905 from the University of London. Arber received a lectureship in botany at University College from 1908 to 1909 and continued to develop her skills as a plant morphologist and anatomist. In 1909, Agnes married the much older E. A. Newell Arber, a paleobotanist at Trinity College. Their daughter, Muriel, was born in 1913. Newell Arber died when Muriel was only five years old.

Following her association with University College, Arber became affiliated with Newnham College. She was able to use Newnham's Balfour Laboratory facilities to conduct her research until 1927, when Newnham shut down the laboratory. Arber then constructed a laboratory in a spare room in her home, which allowed her to work in solitude and whenever she pleased. At Balfour, Arber had for a period shared the facilities with botanist Edith

Saunders, and Arber discovered that she preferred to work independently. So dedicated to her research was Arber that she chose to live modestly and pursue her work rather than supplement her income by teaching. Arber's book on herbals provided information on the history of botany in addition to detailed information and renderings of numerous plants. It was published in 1912, and a second edition came out in 1938. Arber published three additional books: one on water plants, which was published in 1920; one on monocotyledons, published in 1925; and a book on cereal, bamboo, and grass, which came out in 1934. Arber also produced more than 80 journal articles.

Arber's last work to include original scientific research was published in the early 1940s. She later turned toward philosophical, mystical, and historical subjects. Her final books included *The Natural Philosophy of Plant Form,* published in 1950, and *The Mind and the Eye,* published in 1954. Though Arber's later written works deviated from a purely scientific approach, she was widely recognized and respected as an outstanding botanist who had contributed greatly to the understanding of plant form. Arber was elected to the Royal Society in 1948, becoming only the third female to receive the honor, and she was also the first woman to achieve the Linnean Medal, presented by the Linnean Society of London.

Auerbach, Charlotte
(1899–1994)
German
Geneticist

Known to many scientists as the "mother of chemical mutagenesis," Charlotte Auerbach conducted extensive research on mutagens and other genes with the desire to unlock the mysteries of genetic mutation. Auerbach significantly advanced scientists' understanding of mutagenesis through her experiments on the fruit fly, or *Drosophila,* and bread mold, or *Neurospora.* One of Auerbach's goals was to make science palatable and understandable to the general public, and to this end she wrote and published two books on genetics for this audience.

Auerbach was born on May 14, 1899, in Krefeld, Germany, to a German-Jewish family. She was surrounded by scientific influences—her father was a chemist, an uncle a physicist, and her grandfather an anatomist who identified a part of the human intestine later named *Auerbach's plexus* in his honor.

After earning degrees from a number of universities in Germany, including those at Würzburg, Freiburg, and Berlin, Auerbach was forced to flee the country in 1933 because of the rise of Nazism. Auerbach journeyed to Edinburgh, Scotland, where she studied and worked at the In-

stitute of Animal Genetics. She received her Ph.D. in 1935 and spent the rest of her career at the Institute, taking leave only temporarily for sabbaticals in the United States and Japan.

A few years after earning her doctorate, Auerbach became interested in mutagenesis when she learned of the research of American geneticist Hermann Joseph Muller. Muller, who had demonstrated in 1927 that X rays acted as mutagenic agents, spent a year at the Institute of Animal Genetics. During that time, Muller succeeded in persuading Auerbach to research mutation to learn how genes behave. Rather than study X rays, Auerbach researched the effects of mustard gas on the fruit fly. Mustard gas, an agent used in chemical warfare during World War I, was believed to induce effects similar to those of X rays. Through rigorous and thorough research, Auerbach found that mustard gas was a powerful chemical mutagen and was able to produce observable mutations in the fruit fly and in mice—the first documented success with chemical mutagens. She also observed that chemical mutagens acted more slowly than X rays, which produced an immediate effect. Because of the sensitive nature of her studies, secrecy shrouded her research, providing Auerbach with ample time to conduct her experiments and formulate her theories. Auerbach published her results in 1947, the same year she was named a lecturer at the institute.

Continuing with her studies of genetic mutation, Auerbach became involved with the study of DNA, which had been found to be the carrier of genetic code in the early 1950s. Auerbach's knowledge of chemical mutagenesis allowed her to determine that the initial step in the chemical mutagenesis process was a chemical change in DNA. In 1958, the same year she became a reader at the institute, Auerbach traveled to the United States, where she was a visiting professor at the Oakridge National Laboratory in Tennessee. There she experimented with the bread mold *Neurospora* to study spontaneous mutations.

Auerbach studied many other chemical mutagens during her career, including sulfur mustard, formaldehyde, nitrogen mustard, and diepoxybutane. She published two genetics books for the general public—*Genetics in the Atomic Age,* published in 1956, and *The Science of Genetics,* published in 1961. Auerbach was also a dedicated educator and authored two textbooks on mutagenesis. She became a full professor in 1967 and retired two years later. Among the many awards she received were the Keith Medal from the Royal Society of Edinburgh in 1947, the Darwin Medal from the Royal Society of London in 1977, and honorary degrees from universities including those at Dublin, Cambridge, and Leiden. Auerbach remained involved in the scientific community during the years before her death, working to increase awareness of environmental mutagens and to increase the understanding of genetics among laypeople.

Avery, Mary Ellen
(1920–   )
American
Pediatrician

Pediatrician Mary Ellen Avery made breakthrough discoveries regarding respiratory conditions of newborns. Avery is attributed with determining the primary cause of respiratory distress syndrome (RDS), a severe condition that affected some premature infants. Avery found that the onset of RDS was tied to the lack of pulmonary surfactant—a substance naturally produced by the lungs—in undeveloped lungs. After discovering the cause of RDS, Avery led the hunt for the prevention and successful treatment of the often fatal condition.

Born on May 6, 1927, in Camden, New Jersey, Avery was the second daughter born to William Clarence Avery, proprietor of a manufacturing firm in Philadelphia, and Mary Catherine Miller Avery, a high school vice principal. As a child, Avery was inspired to pursue medicine and pediatrics by Emily Bacon, a female neighbor who was a pediatrician. Bacon frequently discussed medicine with Avery and occasionally took her to the hospital.

Avery attended Wheaton College, located in Norton, Massachusetts, and studied chemistry. She graduated summa cum laude in 1948 with a bachelor's degree. Avery then entered medical school at Johns Hopkins University, earning her medical degree in 1952. A bout of tuberculosis following her medical schooling helped spark Avery's interest in studying the lungs. She completed her pediatrics residency at Johns Hopkins in 1957.

After finishing her residency, Avery went to Harvard Medical School in 1957 as a research fellow. It was at Harvard that Avery made a breakthrough discovery concerning RDS. During her research, Avery compared sets of lungs from healthy laboratory animals to sets from infants who had died as a result of RDS. She observed that the healthy lungs were coated with foamy matter while the lungs from deceased newborns were not. Avery theorized that this foamy substance must play a crucial role in RDS—immature lungs, Avery believed, lacked essential pulmonary surfactant, which caused the foam. The results of her research were published in the *American Journal of Diseases of Children* in 1959.

During the 1960s, Avery was on the pediatrics faculty at Johns Hopkins University. In 1969, she traveled to Montreal, Canada, to work as a professor and chairman of the pediatrics department at McGill University. Since 1974, Avery has been associated with Harvard Medical School. Continuing with her search for effective therapies and treatments for RDS over the years, Avery researched the effects of glucocorticoid hormones on lung development. These studies laid the groundwork for the use of glucocorticoids in pregnant women with a high risk for giving birth

prematurely. Avery's studies also helped lead to the development of replacement surfactant in 1991.

In addition to her pioneering research studies, Avery established the Division of Newborn Medicine, originally known as the Joint Program in Neonatology, at Harvard Medical School and Children's Hospital, Boston, in 1974. Avery also penned a classic textbook, *The Lung and Its Disorders in the Newborn Infant,* as well as a number of other books on diseases of the newborn. Among the many honors and awards bestowed upon Avery were the American Lung Association's Edward Livingston Trudeau Medal in 1984; the VIRGINIA APGAR Award, given by the American Academy of Pediatrics in 1991; the National Medal of Science in 1991; the Karolinska Institute in Stockholm's Marta Philipson Award for Pediatric Research in 1998; and the Massachusetts Medical Society's Lifetime Achievement Award in 1999. Avery has also received numerous honorary degrees and is a member of the National Academy of Sciences. Avery showed no signs of slowing down and continued to work at Harvard Medical School in the late 1990s.

Founder of the Joint Program in Neonatology at Harvard Medical School and Children's Hospital, Boston, Mary Ellen Avery is famous for her groundbreaking discovery of the primary causes of respiratory distress syndrome in infants. *(The Alan Mason Chesney Medical Archives of the Johns Hopkins Medical Institutions)*

Ayrton, Hertha (Phoebe Sarah Marks)
(1854–1923)
British
Electrical Engineer

Best known for her studies of the electric arc, Hertha Ayrton was one of the first female electrical engineers. Though Ayrton often assisted her physicist husband with research, it was Ayrton who made pioneering discoveries regarding the electric arc, which was commonly used in lighting in the late 1800s. Ayrton also carried out studies on ripples in sand and developed a fan that was used to free areas such as mines and bunkers from poisonous gases.

Hertha Ayrton was born Phoebe Sarah Marks on April 28, 1854, in Portsea, England. Of Jewish descent, Ayrton's father, Levi Marks, was a jeweler and clock maker who had fled from Poland. Levi Marks died in 1861, leaving his wife, Alice Marks, and eight children. To support the large family, Ayrton's mother worked as a seamstress, while Ayrton, the eldest daughter, cared for her siblings. As a result, Ayrton did not begin her formal education until the age of nine, when she enrolled at a school in London operated by her aunt.

In 1876, Ayrton began attending Girton College, a women's school affiliated with Cambridge University. She studied mathematics and changed her name to Hertha. Though she passed the mathematical Tripos examination in 1880, Ayrton did not receive a degree, as the policy at Cambridge was to award only certificates and not degrees to female students. Ayrton thus passed an examination through the University of London, which awarded her a bachelor's degree.

Soon after graduation, Ayrton created her first major invention—a drafting tool that divided a line into equal parts. The tool could also be used to enlarge and reduce figures and proved useful for engineers and architects, among others. Ayrton secured a patent for her device in 1884. It was also during this period that Ayrton began to attend Finbury Technical College. There she attended the lectures of physicist William Edward Ayrton. The couple was married in 1885 and later had a daughter.

Ayrton helped her husband with his projects in physics and electricity, which led to her work on the electric arc beginning in 1893. One problem with the electric arc was that it was sometimes inconsistent—it would begin to flicker and make hissing noises. The arc was an electron stream that flowed between two carbon electrodes separated by a crater. Through a series of experiments, Ayrton concluded that the hissing and flickering was caused by the contact of oxygen with the carbon in the electrodes in the crater. Ayrton wrote a series of articles on the electric arc for *The Electrician* beginning in 1895, and in 1899 she presented a paper on "The Hissing Arc" to the Institution of Electrical Engineers. In 1902, Ayrton developed her articles into a book.

Because of William Ayrton's failing health, the Ayrtons moved to the coast in the early 1900s, hoping the new climate and ocean air might have recuperative powers. Hertha Ayrton was forced to abandon her electrical studies because of the absence of a laboratory, but she soon became interested in a new subject—the ripples and ridges formed in the sand on the beaches. To uncover the cause of these ripples, Ayrton constructed a number of glass tanks in which she placed sand and water. She then put the tanks on rollers to simulate ocean waves. Ayrton observed that the constant motion of waves over the same location created ripples and that the ripples pushed the sand into two formations between the tops of the waves. These experiments laid the foundation for the development of a fan used to dispel toxic gases and bring in fresh air. This fan was used in military bunkers as well as in mines and factories.

Hertha Ayrton received a number of awards and honors for her work. She was the first woman to be elected a member of the Institution of Electrical Engineers (1899) and was the first woman to present her own paper before the Royal Society of London (1904). The Royal Society awarded Ayrton the Hughes Medal in 1906 for her work on the electric arc and sand ripples. Despite these honors, many in the scientific community continued to doubt Ayrton's abilities and research because of her gender. Ayrton thus became an advocate for women's equality and lobbied for women's right to vote, which came to fruition in 1918, when British women achieved the right to vote. Ayrton died five years later, on August 23, 1923.

B

Bailey, Florence Merriam (Florence Augusta Merriam)
(1863–1948)
American
Naturalist

Florence Merriam Bailey was a pioneer in the study of birds in their natural habitats. Bailey fought for the protection of bird life and was among the first to realize the environmental and biological importance of birds. Devoting her life and career to observing birds, Bailey wrote more than 100 articles on her bird research. Bailey was also active with the Audubon Society.

She was born Florence Augusta Merriam on August 8, 1863, in Locust Grove, in upstate New York. Her father, Clinton Levi Merriam, was a congressman, and her mother, Caroline Hart Merriam, was a college graduate who was an amateur astronomer. Bailey's interest in wildlife and nature began at an early age, and she spent much time exploring the surrounding woods and studying animals and birds. Her interest in nature was encouraged by her father, who was interested in natural history and was a friend of naturalist John Muir as well as her older brother, Clinton Hart Merriam. Hart, as he was called, planned to enter medical school and instructed his sister in anatomy and biology.

After completing studies at a private school in Utica, New York, Bailey enrolled at Smith College, a women's school, in 1882. By this time, Bailey had come to view birds as valuable treasures and was one of the first to promote the study of birds in the field rather than killing birds and studying primarily their bones and feathers. It was during her college years that Bailey became more actively involved in her work with birds and bird protection. Dismayed by the fashion trend of using bird feathers and stuffed birds to adorn ladies' hats, Bailey began to write articles to educate people on the negative aspects of the practice, which led to the deaths of 5 million birds a year. Bailey established a

Smith College chapter of the Audubon Society and organized events, including bird walks and a letter-writing campaign, to discourage the killing of birds for fashion's sake. Though Bailey completed school in 1886, she did not receive a degree as she did not follow a degree course. Smith College later awarded Bailey a bachelor's degree in 1921.

A bout of tuberculosis soon after her graduation led Bailey to California to recuperate, and she spent most of the 1890s traveling about the western part of the United States, rigorously studying the nesting, feeding, and mating habits of birds in their habitats. Bailey's first published book, *Birds Through an Opera Glass,* was released in 1889 and was a collection of articles she had completed for the *Audubon Magazine.* Bailey used her own name rather than follow the common practice of adopting a male pseudonym. She wrote several books during her travels as well, including *A-Birding on a Bronco* (1896) and *Birds of Village and Field* (1898).

After her journeys, Bailey settled in Washington, D.C., to live with her brother Hart, who had become the first chief of the U.S. Biological Survey in 1885 as well as a founder of the National Geographic Society in 1888. In 1899, at the age of 37, Bailey married one of her brother's field naturalists, Vernon Bailey. Together the couple traveled and studied in the West, Vernon concentrating on mammals and Florence on birds.

In addition to her more than 100 articles, Bailey completed 10 books, including the well-known *Handbook of Birds of the Western United States* (1902) and *Birds of New Mexico* (1928). Devoted to raising awareness among the general public about the value of birds, Bailey spent many years teaching ornithology classes to adults, primarily through the Audubon Society of the District of Columbia, a chapter she helped to found. Bailey was the first woman to become an associate member of the American Ornithologists' Union (1885) and was also the first female fellow (1920) and the first woman to receive the society's Brewster

Medal (1931). Bailey continued to work and teach into her later years, and her last major book was published in 1939. She died on September 22, 1948.

Baker, Sara Josephine
(1873–1945)
American
Physician

Influenced by her Quaker upbringing and the death of her father and brother in a typhoid epidemic when she was 16, Sara Josephine Baker was determined to pursue a career in medicine and became the first woman to receive a doctorate in public health. She revolutionized infant care in New York City when she was appointed chief of the city's newly created division of child hygiene in 1908. In this position, she reduced the city's infant mortality rate to the lowest of all major cities worldwide by taking such progressive measures as establishing milk distribution stations, setting up programs to train and license midwives, and creating a foster care system for infants. In addition to her work as a doctor, she was active in the women's movement.

The daughter of Orlando Daniel Mosser Baker, a lawyer, and his wife, who was among the first women to attend Vassar College, Baker was born on November 15, 1873, in Poughkeepsie, New York. She was raised among affluent people, but the influence of her Quaker father and Aunt Abby instilled in her the strength to pursue her own path in life, even when it meant overcoming great obstacles.

Baker abandoned her plans to follow in her mother's footsteps and attend Vassar College when she lost her father and brother to typhoid. Instead, she enrolled in New York Women's Medical College, graduating second in her class in 1898. After serving an internship at New England Hospital for Women and Children in Boston, where she worked with residents of the city's most impoverished slums, she moved to New York City and opened a clinic near Central Park West. Establishing a practice as a female doctor was difficult, and Baker was unable to make ends meet. She eventually closed the clinic and went to work for the city department of health as a medical inspector. This job, which involved examining sick children in schools and working to prevent the spread of contagious disease, led to a job in 1902 in which Baker was given the difficult task of searching for sick infants in the Hell's Kitchen area of New York and trying to prevent some of the 1,500 deaths occurring each week from dysentery.

When the New York City Department of Health established its first division of child hygiene in 1908, there was no one more qualified to run it than Sara Josephine Baker. As the first woman in the country to hold an executive position in a health department, Baker took it upon herself to see that preventive health care and health education became the responsibility of government. At milk stations set up throughout the city, nurses examined infants, distributed inexpensive, high-quality milk, and encouraged mothers to bring their children for regular checkups. Her first 15 milk stations were credited in 1911 with saving the lives of more than one thousand babies.

Baker also established a mandatory licensing program for midwives that resulted in lower rates of infection for home deliveries than for hospital deliveries. Her Little Mothers' League trained young girls in infant care since older siblings were often responsible for caring for infants while their mothers worked. In addition, Baker was responsible for introducing the concept of prenatal care and for creating a foster care system that placed babies in homes where they would receive significantly better care than was available in institutions.

As a pioneer in the women's movement, Baker single-handedly lobbied New York University Medical School,

The first woman to receive a doctorate in public health, Sara Josephine Baker was chief of New York City Department of Health's first division of child hygiene in 1908. *(National Library of Medicine, National Institutes of Health)*

where she eventually earned her doctorate in public health, to admit women. She helped to found the College Equal Suffrage League and marched in the first annual Fifth Avenue Suffrage Parade. Between 1922 and 1924, Baker served as the U.S. representative on the health committee of the League of Nations. Active in children's and women's health issues throughout her life, she served on numerous committees and was president for one term of the American Medical Women's Association. Sara Josephine Baker died of cancer in New York City on February 22, 1945.

Bari, Nina Karlovna
(1901–1961)
Russian
Mathematician

Nina Karlovna Bari's work on trigonometric series is acknowledged to be the foundation of function and trigonometric series theory. The first woman to attend Moscow State University, Bari refined the constructive method of proof to obtain results in function theory. Bari was a prolific author who wrote 55 publications, including her seminal book *A Treatise on Trigonometric Series*.

Born in Moscow on November 19, 1901, Bari was the daughter of Olga and Karl Adolfovich Bari. Her father was a physician. Russian education was segregated by gender at this time, and the most rigorous classes were reserved for boys. But Bari flouted convention, took the boy's high school exam, and passed. She enrolled at the Faculty of Physics and Mathematics of Moscow State University in the wake of the 1917 Russian Revolution, becoming the first woman to attend this prestigious institution. She graduated in 1921, just three years after entering the university.

Bari pursued a teaching career after leaving Moscow State. She was a lecturer at the Moscow Forestry Institute, the Moscow Polytechnic Institute, and the Sverdlov Communist Institute. She also won the distinction of receiving the only paid research fellowships at the Research Institute of Mathematics and Mechanics, where she studied under the brilliant mathematician Nikolai Nikolaevich Luzin. With other elite students, Bari was part of a group nicknamed Luzitania in honor of its mentor. Luzin inspired Bari to explore function theory. (In fact, he rejected any area of mathematical study except function theory.) Bari's thesis examined trigonometric series and functions. In 1922, she presented the centerpiece of her dissertation to the Moscow Mathematical Society, becoming the first woman to address this organization. She successfully defended her thesis in 1926 and was awarded the Glavnauk Prize for her research.

Bari studied in Paris in 1927, attending the Sorbonne and the Collège de France. After a period at the Polish Mathematical Congress in Lvov, Poland, she won a Rockefeller Grant that enabled her to return to Paris. Her interest in working abroad was motivated by the disintegration of the Luzitania movement. Her mentor's radical ideas and often abrasive personality had alienated his colleagues. (He left Moscow State University in 1930 for a position at the Academy of Science's Steklov Institute, was accused of ideological sabotage 1936, and was ultimately forced to withdraw from academia and research.) In 1929, Bari returned to Russia to join the faculty of Moscow State University. She was promoted to full professor in 1932. Three years later, she was awarded the degree of Doctor of the Physical-Mathematical Sciences, which was a more prestigious research degree than the standard Ph.D.

Bari remained at Moscow State, leading function theory work at the university with her colleague D. E. Men'shov. In 1952, Bari published one of her most important works—an article entitled "On Primitive Functions and Trigonometric Series Converging Almost Everywhere." She presented her ideas at the 1956 Third All-Union Congress in Moscow and at the 1958 International Congress of Mathematicians in Edinburgh, Scotland. Later in life, Bari married Viktor Vladmirovich Nemytski, a fellow Soviet mathematician. The details of their marriage have been lost.

In 1960, Bari completed her final work, a 900-page book on cutting-edge trigonometric series theory. This monograph, *A Treatise on Trigonometric Series*, was widely admired and is now considered a standard reference textbook for those studying function and trigonometric series theory. Bari died on Moscow on July 15, 1961, when she fell in front of a moving train. It is suspected that her death was a suicide. Her colleagues at Moscow State believed that Luzin had been her lover and that she was seriously depressed after he died in 1960.

Barney, Ida
(1886–1982)
American
Astronomer

Educated as a mathematician, Ida Barney used her ability to make complex mathematical computations and her tolerance for tedious, detailed work to contribute a vast array of new information to the field of astronomy. Over a period of 23 years, she cataloged the positions, magnitudes, and proper motions of more than 150,000 stars, which involved taking more than half a million measurements. The 22 volumes she published during her lifetime contain this information and remain an important contribution to our understanding of celestial bodies.

Ida Barney was born in 1886 in New Haven, Connecticut, to Eben Barney and Ida Bushnell Barney. An excellent student who was inducted into the honor societies of Phi Beta Kappa and Sigma Xi, she received a B.A. degree from Smith College in 1908 and a Ph.D. in mathematics from Yale University in 1911. For 10 years following the completion of her education, she taught mathematics at Rollins College, Smith College, Lake Erie College, and again at Smith College.

During the 1920s, many universities received increased funding to conduct massive projects in astronomy, and the Yale Observatory was among them. In 1922, Barney joined the staff there and began work under Frank Schlesinger. Since he did not believe women could make theoretical contributions in science, she was put in charge of the tedious and time-consuming task of measuring stars on photographic plates and using mathematical computations to translate their positions to celestial coordinates. Barney succeeded in this massive undertaking, while Schlesinger developed several new devices that made the process of measuring the stars and computing their locations easier.

The first innovation was a projection device that was designed to reduce both strain on the eyes and accidental errors that resulted from the observer bumping the microscope. This new mechanism allowed images to be viewed and bisected in a small screen so that the scientist was not always required to be peering through a microscope. This development increased both the speed and accuracy of measurements taken by Barney and other scientists. The second innovation involved making use of the emerging technology in computer science. The catalogs published by Barney and Schlesinger between 1939 and 1943 were the first ones for which IBM punch card machines were used for many of the computations.

When Schlesinger retired from the Yale Observatory in 1941, Barney became the sole supervisor of the project and the sole author of subsequent volumes on the stars she was measuring. From the time she took over direction of the project, all measurements of the photographic plates were carried out at the IBM Watson Scientific Laboratory with a new electronic device. The device automatically centered images, resulting in higher accuracy in determining their positions and less eye fatigue.

Ida Barney was awarded the ANNIE J. CANNON Prize of the American Astronomical Society in 1952, an award that was given every three years to a woman of any nationality who had made outstanding contributions to the field of astronomy. After her retirement in 1955, several more catalogs of stars were published under her name. She died on March 7, 1982, at the age of 95. The "Yale positions," as detailed in her numerous volumes, have been hailed as highly accurate and, therefore, extremely important to the future determinations of the motions of these stars.

Barton, Clara
(1821–1912)
American
Nurse and Humanitarian

Best known for establishing the American Red Cross, the U.S. arm of the International Red Cross, Clara Barton was a humanitarian who tirelessly worked to help victims of natural disasters. She was affiliated with the American Red Cross for more than two decades. Barton also worked as a battlefield nurse during the Civil War, and her fearless dedication resulted in the moniker "Angel of the Battlefield."

Born Clarissa Harlowe Barton on December 25, 1821, in the small town of North Oxford, Massachusetts, Barton was the youngest of five children born to a middle-class farming family. Barton's father, Stephen Barton, was not only a farmer but also a businessman and state legislator. Barton's compassion for her fellow humans was largely influenced by her father. Barton's mother, Sarah Stone, was an intense and eccentric woman with a fierce temper. With many daily family chores, Barton developed a strong work ethic at an early age. She was also active in the community,

Clara Barton, known as "Angel of the Battlefield" for service during the Civil War, is best known for establishing the American Red Cross. *(National Library of Medicine, National Institutes of Health)*

tutoring underprivileged children and helping to nurse the sick.

Because teaching was one of the few respectable professions for women in the early 1800s, Barton became a teacher in 1839. After several years of teaching, she journeyed to New York to attend the Clinton Liberal Institute. When she finished her schooling, Barton returned to teaching, working at the first public school in New Jersey. In 1854, however, Barton, fighting depression and a nervous breakdown, left teaching for good. She moved to Washington, D.C., where she secured a job as a copy clerk in the U.S. Patent Office, making her the first female clerk in the federal government.

Though her job as a copyist was not necessarily fulfilling, Barton's life was changed forever with the outbreak of the Civil War in 1861. Hoping to help the troops, she requested permission to become a battlefield nurse. Women were not allowed near the battlefield, but Barton was extremely persistent, and in 1862 she was given a pass to the front lines. For three years, Barton traveled to battlefields with wagons of medical and food supplies. She fed wounded soldiers while also tending to their injuries. In 1865, the Civil War ended, and Barton spent the year attempting to identify dead soldiers and locate missing soldiers. Through her tireless efforts, she managed to identify 22,000 men.

In the mid- to late 1860s, Barton traveled across the United States to deliver lectures about her experiences during the Civil War. Though Barton's lectures were well received, the physical and emotional pressures pushed her dangerously close to a breakdown in 1868. On the advice of her doctor, she traveled to Europe to recuperate. In 1869, Barton was in Geneva, Switzerland, where she met a group affiliated with the International Red Cross. One of the group, Dr. Louis Appia, asked for Barton's assistance in persuading the U.S. government to endorse the Treaty of Geneva. Among the provisions included in the treaty was the establishment of the Red Cross, an international wartime relief organization. Barton agreed, but before she could return to the United States, the Franco-Prussian War commenced, and she joined the Red Cross workers on the front lines.

Barton returned to the United States in 1873, but she soon suffered a nervous breakdown and was unable to work for two years. By 1877, Barton's health was stabilized, and she began her efforts to form a U.S. branch of the Red Cross. Dr. Appia appointed Barton the U.S. representative, and she worked to gain support from the government and generate publicity and national support. Barton also expanded the scope of the Red Cross to include peacetime efforts, such as helping victims of natural disasters, diseases, and major accidents. In 1881, Barton was elected president of the American Red Cross and succeeded in forming the first chapters. A year later, the U.S. government ratified the Treaty of Geneva, and from 1883, Barton dedicated her career to the Red Cross. Barton was both an administrator and a hands-on relief work participant. She was involved in helping victims of a Michigan forest fire, Ohio and Mississippi River floods, a drought in Texas, an Illinois tornado, and much more.

Though Barton was intensely committed to the Red Cross, infighting within the organization led her to resign in 1904. A year later, the U.S. government assumed control of the national organization. Barton moved to Glen Echo, Maryland, and wrote *A Story of the Red Cross*, published in 1905, and *Story of My Childhood,* published in 1907. Never married, Barton died of pneumonia on April 12, 1912.

Bascom, Florence
(1862–1945)
American
Geologist

Regarded as one of the first female geologists in the United States, Florence Bascom opened many doors for future women scientists. A lifelong college educator, Bascom instructed female students to become professional geologists. Among her students were paleontologist JULIA ANNA GARDNER, petrologist Anna Jonas Stose, crystallographer Mary Porter, and Scripps College's Isabel Fothergill Smith. Bascom also carried out research on the formation of the Piedmont Mountains in the eastern United States.

Born in Williamstown, Massachusetts, on July 14, 1862, Bascom, the youngest of six children, grew up in an educated and enlightened environment. Her mother, Emma Curtiss

The first woman and the first geologist to be awarded a Ph.D. by Johns Hopkins University, Florence Bascom could not officially attend classes, so her seat was placed behind a screen. *(Courtesy of Sophia Smith College)*

Bascom, was a teacher and suffragist, and her father, John Bascom, was a professor at Williams College who also supported women's right to vote. John Bascom was hired as the president of the University of Wisconsin in 1874, and the family moved to Madison, Wisconsin.

The University of Wisconsin opened its doors to women students in 1875, and in 1877 Bascom began her studies there. She received two bachelor's degrees, one in arts and one in literature, in 1882, and two years later she earned her bachelor's degree in science. Influenced by her father's friend, a geology professor at Ohio State University, Bascom became increasingly interested in geology, particularly petrography, the study of rocks, with an emphasis on the configuration of rock layers that formed mountains. In 1887, Bascom was awarded a master's degree in geology. Two years later, she became one of the first women allowed to take classes at Johns Hopkins University. Bascom received her Ph.D. in 1893. Her dissertation argued that rocks believed to be sediments were, in fact, metamorphosed lava flows.

During her years of schooling, Bascom also spent time teaching. She taught at the Hampton Institute for Negroes and American Indians (now Hampton University) from 1884 to 1885, at Rockford College from 1887 to 1889, and at Ohio State University immediately following graduation, from 1893 to 1895. In 1895, Bascom began teaching at Bryn Mawr College, where she remained for the rest of her career. Geology was not a department at Bryn Mawr but was considered part of the natural sciences. As a result, Bascom initially worked out of a storage area in the newly established sciences building. Within two years, she had gathered a collection of minerals, rocks, and fossils and founded the school's department of geology. Bascom was strongly committed to educating future female geologists, and when Bryn Mawr attempted to lessen geology from a major to an elective, Bascom protested by threatening to quit. Rather than lose her, Bryn Mawr conceded to her wishes. She became a full professor in 1906.

In addition to training young geologists, Bascom worked as an assistant geologist for the U.S. Geological Survey beginning in 1896. She became a full geologist in 1909, and from 1909 to 1938, Bascom spent summers traveling through the Piedmont Mountains, which ran from Alabama to New York and were part of the Appalachian range. There she collected rock samples and studied the rock layers, afterward completing detailed reports about her findings. These reports were part of the U.S. Geological Survey's efforts to create an extensive map of the nation.

Not only was Bascom the first woman to be hired by the U.S. Geological Survey, but she was also the first woman to present a paper to the Geological Society of Washington, in 1901. She was also the first female fellow of the Geological Society of America, which she joined in 1894. Bascom served as associate editor of *American Geol-*

ogist from 1896 to 1905. After retiring from Bryn Mawr in 1928, Bascom served as vice president of the Geological Society of America in 1930 and continued to work until her death in 1945.

Bassi, Laura Maria Catarina
(1711–1788)
Italian
Physicist

A devoted educator, Laura Maria Catarina Bassi was the world's first female physics professor. Knowledgeable not only in physics but in anatomy, mathematics, and a number

Laura Bassi, the world's first female physics professor, taught in Bologna, Italy, in the mid-1700s. *(The Schlesinger Library, Radcliffe Institute, Harvard University)*

of languages, including Latin, Greek, and French, Bassi was well known throughout Europe for her intelligence and accomplishments, and students and scholars traveled to Italy to attend her lectures. Bassi was active with the Academy of Science of Bologna, Italy.

Born on October 31, 1711, in Bologna, Italy, Bassi grew up in a wealthy family. Bassi's father was a lawyer, and Bassi was educated at home, taught by the family physician, Dr. Gaetano Tacconi, a university professor and member of the Academy of Science. Bassi showed high intelligence from an early age, and she excelled in her studies, which included such subjects as philosophy, mathematics, anatomy, natural history, French, and Latin.

In 1733, when Bassi was 21 years of age, she took part in a public debate with five distinguished philosophers. Bassi was influenced to join the debate by her family and friends. Her participation in the debate piqued the curiosity of so many spectators that the debate had to be moved from the university hall to the great hall of the Palace of the Senators. One member of the intrigued crowds was the archbishop of Bologna, Cardinal Lambertini, who later became Pope Benedict XIV. Lambertini was impressed with Bassi's knowledge and visited her the day after the debate to offer congratulations and persuade Bassi to continue her education. Less than a month following the debate, Bassi received a doctorate in philosophy from the University of Bologna. The ceremony took place at the Hall of Hercules in the Communal Palace. The Bologna Senate also provided funds that enabled Bassi to continue her studies.

Bassi became a professor of physics at the University of Bologna after passing a rigorous public examination. Though the University of Bologna held a liberal view toward female scholars, Bassi was the first female professor hired in the physics department. She continued her own studies while teaching physics. She studied mathematics and Greek and continued to expand her knowledge in mechanics and hydraulics. Bassi was particularly interested in Newtonian physics. She also taught anatomy and became the chair of the anatomy department. In 1738, Bassi married Giuseppe Veratti, a physician and fellow professor. Together the couple had 12 children.

After her marriage, Bassi remained active in academia and continued to teach. Bassi successfully lobbied for increased responsibilities at the university as well as a higher salary in order to finance laboratory equipment costs. In the 1740s, Bassi was involved with the Academy of Science of Bologna. She presented a number of papers at the Academy, including "On the Compression of Air," presented in 1746, "On the Bubbles Observed in Freely Flowing Fluid," given in 1747, and "On the Bubbles of Air that Escape from Fluids," presented in 1748.

Though Bassi was not particularly concerned with publishing, she wrote a number of scientific papers, including several on physics, chemistry, mathematics, mechanics, and technology. Bassi knew and corresponded with scholars across Europe, including the philosopher Voltaire. In addition to her academic work, Bassi wrote poetry and worked in various ways to help the poor. She was appointed the chair of experimental physics at Bologna in 1776, two years before her death.

Bechtereva, Natalia Petrovna
(1924–)
Russian
Neurologist

Understanding the relationship of mind and brain, that is, of perception, thinking, and emotion to the electrochemical functioning of the brain, has been the lifework of Natalia Bechtereva. This effort resulted in Bechtereva's explanation in 1968 of "the error detector," the way the brain, through neural network processing in the cortical and subcortical regions, realizes that an error in thinking has been made. Bechtereva was able to map this interplay of neural networks in tests given to patients while electrodes were implanted in their brains. Through this and other cognitive experiments, she postulated a theory about how thinking works in the brain.

Natalia Bechtereva was born in Leningrad (now St. Petersburg), Russia, which was then part of the relatively new Union of Soviet Socialist Republics (U.S.S.R.), founded in the ruins of czarist Russia after World War I and the overthrow of Czar Nicholas II by the Bolsheviks in 1917. (The exact date of her birth is unknown.) She is a widow with one son who was born in 1949.

Her studies in biological sciences were interrupted by World War II, which required mobilization of all available manpower in the Soviet Union to head off invading Nazi armies. She received her B.S. degree in biological sciences in 1951 when she was 27 years old and earned an M.D. degree in 1959. In 1954, while she was still a medical student, she was appointed to the position of senior scientific worker at Leningrad's Polenov Neurological Institute. By 1962, she was head of Polenov's Physiology Lab and deputy director of the institute. In 1962, she was appointed head of Polenov's Department of Human Neurophysiology. It was during the 1960s that Bechtereva began to make the first major breakthroughs in understanding the physiological workings of the human brain.

Blending mathematics, physics, and physiology, Bechtereva developed a complex approach to brain functioning research. She sought to link the structural-functional organization of the brain with the cerebral mechanisms of thinking, memory, and emotion. As a result of her studies, Bechtereva suggested that thinking processes were supported by a linked neural network in the cortex and subcortex. This system operated on two levels, one a fixed, or rigid, system that is always present during thinking and other "flexible" elements that kick in only to complete specific tasks such as recognizing the semantic meaning of speech. This network is globally integrated in the brain so that if one part of the brain is injured or impaired, other parts of the brain/neural network can often compensate to allow thinking, memory, and emotion to continue, albeit sometimes not in exactly the same way or with the same results as before.

Bechtereva's approach yielded several surprising findings. Observing the physiology of emotions, she discovered a protective cerebral mechanism that can be set in motion as a counterweight to negative emotions. This mechanism is a physiological process that was observed under clinical conditions. Bechtereva also has developed therapies to fight chronic brain diseases. By stimulating different parts of the brain with electrical impulses given by brain probes, spinal cord, visual, and acoustic neural networks have been reintegrated in ways that help patients who suffer from certain kinds of motor problems and with problems in seeing and hearing.

Since 1990, Bechtereva has been director of the Institute of the Human Brain and head of the institute's Neurophysiology of Thinking and Consciousness Lab, an organization in St. Petersburg that is part of the Russian Academy of Sciences. She is supervising the lab's use of noninvasive positron emission tomography technology to develop an even more complete understanding of the global neural networks that give rise to consciousness. For her efforts in understanding brain function, she has been awarded Germany's Hans Berger Medal (1970), the U.S. Cybernetic Society's McCulloch Medal (1972), and elected to the Soviet (now Russian) Academy of Sciences (1981) and the U.S. Academy of Medicine and Psychiatry (1997). From 1983 to 1994, she was editor-in-chief of the *International Journal of Psychophysiology.*

Bell Burnell, Susan Jocelyn
(1943–   )
British
Astronomer

In 1967, Jocelyn Bell (she later added Burnell to her surname when she married), a 24-year-old doctoral student, jolted the world of astronomy when she discovered a new type of star whose existence was previously thought to be impossible. These stars periodically emitted very precise pulses of radio waves and were quickly named *pulsars.* For this discovery, Bell's supervisor, Dr. Anthony Hewish, won the Nobel Prize in physics in 1974.

Jocelyn Bell was born in Belfast, Northern Ireland, on July 15, 1943, into a Quaker family. Susan's father, Philip, was an architect and her mother, Allison, stayed home to raise Jocelyn, the oldest, and her sisters and brother. When Bell was 11, she flunked an exam that divided all British youth into two categories: those who would continue their studies in a university and those who would go into trades. In response to this setback, the Bells decided to send Jocelyn to a private Quaker school in England. There she began to get interested in science, especially astronomy.

In 1961, Bell enrolled as a physics student at the University of Glasgow, where she earned a B.S. with honors in 1965. Later that year, she was accepted into Cambridge University, where she was put to work on the tedious chore of sorting through charts of data from Cambridge's new radio telescope. Her faculty advisor, Anthony Hewish, was studying the radio frequency signals of quasars, distant stars and galaxies that are characterized by the high rate of speed at which they are moving away from our galaxy.

The Cambridge radio telescope scanned the whole sky every four days and every day churned out data contained on 100 feet of paper. Bell had the job of looking for, and determining the location of, anomalies in the data lines. About two months into her work in 1967, she noticed what she called "a bit of scruff" on the paper only half an inch long. She realized that this scruff was appearing at regular intervals, which meant that it was always encountered at the same spot in space. On more thorough examination, Bell discovered that the scruff was a pulse signal that was later measured to occur precisely every 1.3373011 seconds, accurate to one-millionth of a second. She later discovered other radio transmissions from different objects in space that had similar pulsing radio signals but never of the same frequency.

For a while, Anthony Hewish, Jocelyn Bell, and the rest of the team did not know the source of these pulsations. Some team members half-jokingly speculated that they were signals made by intelligent beings on planets in other solar systems—LGMs they called them, for "little green men." But this theory was quickly struck down after measurements determined that the pulsating signals had to come from a star. Later, most astronomers came to agree with the theory that pulsars are signals emitted from small, superdense neutron stars, collapsed suns that are typically 15 to 20 kilometers in diameter.

A great controversy arose in 1974 when Hewish, but not Bell, was awarded the Nobel Prize for physics for the discovery of pulsars. British astronomer Fred Hoyle asserted that "there has been a tendency to misunderstand the magnitude of Miss Bell's achievement. . . . [which] came from a willingness to contemplate as a serious possibility a phenomenon that all past experience suggested was impossible." Hoyle further added that Hewish had "pinched" the Nobel Prize by failing to give Bell credit for her work.

Soon after her discovery, Bell married Martin Burnell, a British civil servant, and over the next 10 years moved with him as he was given jobs around the United Kingdom.

With each move, Bell Burnell would take a new job at a university or astronomical lab. She also gave birth to a son and for several years, worked part time. In 1991, Bell Burnell, now divorced, became professor of physics at the Open University, an institution that offers university-level classes via correspondence and computer to working-class and part-time students in Europe. Mindful of her own near-miss of a college education, Bell Burnell finds the Open University job satisfying. "It's very rewarding," she says, "[The students] are so keen, very committed. . . . I've had students on oil rigs, lighthouse keepers, . . . mums out of work." For her work on pulsars and other astronomical issues, Bell Burnell has been awarded the Franklin Institute's Michelson Medal (1973), the Royal Astronomical Society's Hershel Medal (1989), and the National Radio Astronomy's Jansky Award (1995).

Bellow, Alexandra
(1935–   )
Romanian/American
Mathematician

Alexandra Bellow is a pioneer in the field of ergodic theory. This complex area of mathematical theory is concerned with the long-term averages of the successive values of a function on a set when the set is mapped into itself. It also looks at whether these averages equal a reasonable function on that set. Ergodic theory has far-reaching applications for the study of statistical mechanics, number theory, and probability, and it can even be used to explore the concept of entropy in physics. A professor at Northwestern University, Bellow has done much significant work in the field.

Born on August 30, 1935, in Bucharest, Romania, Alexandra Bellow was the daughter of two physicians—Dumitru and Florica Bagdasar. The communist takeover of Romania cast a long shadow over Bellow's childhood. Although her father initially supported the communists, and was even appointed minister of health, he was soon accused of "defection" and imprisoned. Bellow's mother replaced her husband in his post, but she too was removed from it in 1948.

Bellow studied mathematics at the University of Bucharest. In 1956, she married Cassius Ionescu Tulcea, a professor of mathematics at the university. She earned the equivalent of a master's degree in 1957 and then moved to the United States with her new husband to continue her education at Yale University. She was awarded her Ph.D. in mathematics from Yale in 1959 for her thesis "Ergodic Theory of a Random Sequence."

After graduating, Bellow served as a research associate at Yale from 1959 to 1961 before taking a similar position at the University of Pennsylvania from 1961 to 1962. She then accepted an assistant professorship at the University of Illinois in 1962 and was promoted to associate professor in 1964. She left the University of Illinois in 1967 to become a professor of mathematics at Northwestern University. Together with Tulcea, Bellow published *Topics in the Theory of Lifting* in 1969. The couple was divorced that same year.

In 1971, Bellow began to concentrate on ergodic theory, writing a number of articles on the topic. She married the Nobel Prize–winning author Saul Bellow in 1974. They spent 1975 in Jerusalem, where she taught mathematics at the University of Jerusalem. Back at Northwestern in 1976, Bellow continued her research. With D. Kolzow, she edited the proceedings of a conference on measure theory in 1975. She also edited the *Transactions of the American Mathematical Society* from 1974 to 1977, and she served as associate editor of the *Annals of Probability* from 1979 to 1981 and of *Advances in Mathematics* in 1979. Moreover, she cowrote an influential paper with Harry Furstenberg in 1979, which applied number theory to ergodic theory. In it, the pair proposed the Bellow-Furstenberg theorem. She was a Fairchild Scholar at the California Institute of Technology in 1980.

In the 1980s and 1990s, Bellow pursued her work on ergodic theory and continued to bring together mathematicians to explore the field. With De Paul University professor Roger Jones, she organized a conference—"Almost Everywhere: Convergence in Probability and Ergodic Theory"—in 1989. That same year, she married mathematician and civil engineer Alberto Calderón (she had divorced Saul Bellow in 1986). She gave the EMMY NOETHER Lecture for the Association of Women Mathematicians in 1991. Subsequently, she collaborated with Jones and other mathematicians on eight papers that explore partial sequences of observations. In this work, Bellow sought to identify when averages based on partial observations are likely valid for entire populations, when they are not valid, and by how much they may be off.

Bellow retired from Northwestern in 1997, though she continued to research and publish on ergodic theory. She has won several awards for her contributions, including a 1987 fellowship from the Alexander von Humboldt Foundation. She is considered to be a leading figure in the development of ergodic theory.

Benedict, Ruth Fulton
(1887–1948)
American
Anthropologist

A pioneer woman anthropologist, Ruth Benedict studied the Indians of the American Southwest and later in life wrote movingly against racism. As she explored the beliefs and mythologies of other cultures, she also urged that her readers not forget the common humanity of all peoples. She was a leading advocate of a research approach that blended anthropology, psychology, and sociology, which she believed led to a fuller and more nuanced understanding of cultures different from our own.

Ruth Fulton was born in New York City on June 5, 1887, the daughter of Frederick Fulton, a surgeon, and Beatrice Fulton, a homemaker and teacher. Ruth was still a toddler when her father died, a loss that not only affected her but sent her mother into a long period of depression and mourning. She spent her childhood and youth drifting with her younger sister, Margery, between her grandparents' farm and various cities where her mother took teaching jobs. She was a solitary and lonely child, whose isolation was made more acute by partial deafness that was the result of a bout of measles. In 1914, when she was 27, she married Stanley Benedict, a biochemist. The marriage was not a success, and after 1930, they were permanently separated.

Benedict graduated from Vassar College in 1909. To make a living, she taught English in secondary schools, but she seems to have wanted to make a name for herself as a poet and published a number of her poems in magazines under the pseudonym Anne Singleton. After 10 years of this life, she decided to refocus and, in 1919, began taking classes at the leftist New School for Social Research in New York City. She discovered the discipline of anthropology and Columbia University's charismatic anthropology professor, Franz Boas. Benedict soon transferred to Columbia and earned a Ph.D. from that institution in 1923.

Upon completing her dissertation, Benedict was hired as a teacher at Columbia. Her work during the 1920s and early 1930s centered on studies of two pueblo Indian groups in New Mexico, the Cochiti and the Zuni. Two books came out of this study, *Tales of the Cochiti Indians* (1931) and *Zuni Mythology* (1935). From her studies of the Zuni and Cochiti people and from her readings in anthropology, Benedict published a more general exploration of how cultures affect individual human behavior. This book, *Patterns of Culture* (1934), asserted that each culture has a "personality," traits that are distinctive to it and that heavily influence how individual members of that culture act on a daily basis. This theory has generally fallen into disrepute now, but it was quite popular for several decades after Benedict first expounded it.

In 1936, Benedict became head of Columbia's anthropology department. She had already gathered a group of students around her. One of these, MARGARET MEAD, would continue in Benedict's footsteps during the 1950s and 1960s. In 1940, in response to the ideology of racial superiority espoused by the Nazis and to some extent the Japanese military regime, Benedict published a book, *Race: Science and Politics,* that demolished these beliefs, which she termed "racist," a word coined by her. During World War II, Benedict advised the U.S. Army about how to deal with the people it encountered in the war zones, and after

the war, she authored *The Chrysanthemum and the Sword,* an objective study of Japanese culture that tried to humanize the Japanese for American readers and make Japan understandable to a nation that had just emerged from a life-and-death struggle with it.

Benedict was editor of *The Journal of American Folklore* from 1925 to 1940. She was also president of the American Anthropological Association in 1947–48. She died suddenly of a heart attack on September 17, 1948.

Benerito, Ruth Mary Roan
(1916–   )
American
Chemist

During her 43-year career as a chemist, Ruth Benerito has studied the caloric and fat content needed to keep intravenously fed medical patients alive. She has also worked with the cotton textile industry to devise new ways to treat cotton fabric so that it will not easily soil or wrinkle. These projects have led to contributions in the analysis and production of fat emulsions, epoxides, metallic salts, and diepoxy compounds.

Ruth Mary Roan was born in New Orleans, Louisiana, on January 22, 1916, to John Edward Roan, a civil engineer and railroad executive, and Bernadette Elizardi, an artist. She was the third of six siblings and, from an early age, displayed an interest and talent for science and mathematics. She graduated from high school when she was 14. In 1950, Roan married Frank H. Benerito.

In 1935, Benerito enrolled in Sophie Newcomb College, the women's college of Tulane University in New Orleans. She completed her B.S in chemistry in three years and then spent a year in graduate chemistry studies at Bwyn Mawr in Pennsylvania. The difficulty of finding work during the depression forced her to return to New Orleans in 1937. She managed to get a job as a science teacher at a New Orleans area high school and finished her master's degree in chemistry at Tulane in 1938. Benerito then secured a teaching job at Randolph-Macon Women's College in Virginia in 1940. She taught chemistry there until 1943 when she returned to New Orleans to teach at Sophie Newcomb and Tulane. She was employed as a full-time teacher until 1953. During this time, she studied for her Ph.D. from the University of Chicago, which she was awarded in 1948.

In 1953, Benerito left full-time teaching to take a job with the U.S. Department of Agriculture (USDA). She was hired as a research chemist at the USDA's Southern Regional Research Center (SRRC) in New Orleans, one of four such centers whose goal was to study and promote regional agricultural products. At the SRRC's Oilseed Laboratory, Benerito was project leader of the Intravenous Fat Program and devised new ways to analyze proteins and fats in seeds such as peanuts, pecans, and cottonseed. This work led to suggestions about ways in which seeds could be treated so that they could be used to help in the intravenous feeding of medical patients.

In 1958, Benerito began a long involvement in research about ways to treat cotton fabrics when she took the job of research leader of the SRRC's Physical Chemistry Research Group, Cotton Chemical Reactions Laboratory. Here Benerito worked on refining epoxides that would make cotton fabrics crease resistant when they dried. She later devised an epoxide that would keep cottons crease resistant when they were dry or wet. These discoveries enabled the cotton fabric industry to remain competitive with manufacturers of synthetic fabrics.

During this and other work, Benerito was granted more than 50 patents, and she also continued to work as an adjunct professor of chemistry at Tulane University. In 1964, she was awarded the Distinguished Service Award from the USDA. In 1968, she was given the Federal Women's Award and the Southern Chemist Award, and in 1970, she won the American Chemical Society's Garvan Award. She retired from government service in 1986 after a career as "an outstanding and inspiring teacher of chemistry, a brilliant research scientist, and an inspiring and untiring leader of research" (from the Southern Chemist Award).

Bennett, Isobel
(1909–   )
Australian
Marine Biologist

Although she does not have an undergraduate, much less a graduate, degree from any university and never formally studied her discipline, Isobel Bennett over a career of 39 years has built an international reputation as a marine biologist. She was an informal leader on the University of Sydney's research ship, *Thistle,* and she developed a specialty in the study of intertidal areas, those zones of beach and marsh that periodically flood with the rise and fall of the ocean's tide.

Born in Brisbane, Australia, in 1909, Bennett was the oldest of four children. Not much information is known about Bennett's family other than the facts that her father had a hard time earning a living and her mother died when she was nine years old. By the time Bennett was 16, she had to go to work to help support her brothers and sisters. Between 1925 and 1932, she worked at a variety of secretarial jobs. She found herself out of work with the advent of the worldwide economic depression in 1932. Having some savings and suddenly free time, Bennett and her sister took a cruise to Norfolk Island, an Australian possession several hundred miles east of Brisbane. On that trip, Bennett befriended William J. Dakin, a professor of zoology at the

University of Sydney. At the end of the cruise, Dakin presented Bennett with a timely job offer: to assist him on a history of whaling that he was writing. Bennett accepted; she would remain at the University of Sydney for the rest of her career.

Marine biology was Dakin's specialty within the field of zoology, and it also was a field that appealed to Bennett. As she helped Dakin in his lab, he in turn trained her in his discipline. In fact, she was getting an on-the-job postgraduate seminar on marine biology. Having a good memory, she quickly familiarized herself with the sea creatures found in the waters off Australia. She regularly went out on expeditions on the *Thistle* and helped students and faculty members alike collect and classify marine specimens. Back at the university, she reorganized and cataloged the department's library.

Within 10 years, Bennett had become an expert in intertidal marine biology, especially of Australia and Antarctica. She visited Antarctica with four other Australian women scientists in 1957 and wrote numerous scientific papers and nine books about the intertidal zone. Her books, many of them illustrated with her own photography, include *Fringe of the Sea* (1967), *On the Seashore* (1969), *Great Barrier Reef* (1971), and *Shores of Macquarie Island* (1971). Some of these works have become standard textbooks on their subjects.

Over the years, Bennett won much recognition for her work. The University of Sydney awarded her an honorary master's degree in 1962. In 1982, she received the Mueller Medal, which is given by the Australia and New Zealand Association for the Advancement of Science. She was only the second woman to have received this award. She has also won awards from the Royal Zoological Society of New South Wales for two of her books. Semiretired, Bennett still takes a keen interest in new developments in marine science.

Berkowitz, Joan B.
(1931–   )
American
Physical Chemist

Joan B. Berkowitz has made contributions to a number of research fields during her career as a physical chemist. In addition to her focus on environmental management and hazardous waste, Berkowitz has investigated the thermodynamics of high-temperature vaporization, the electrochemistry of flames, and the oxidation of alloys. In 1989, Berkowitz formed Farkas, Berkowitz, and Company, a consulting firm that specializes in environmental projects, including hazardous waste management.

Born on March 13, 1931, in Brooklyn, New York, Berkowitz displayed an independent and bright nature from an early age. Her father, Morris Berkowitz, was a salesman for the Englander Mattress Company. Her mother, Rose Gerber Berkowitz, was a housewife who became interested in women's rights and suffrage. Influenced by her mother's beliefs, Berkowitz determined by the age of 12 that she would pursue a career for herself.

Berkowitz received a scholarship to attend Swarthmore College, and in 1952, she graduated with a bachelor's degree in chemistry. Though she hoped to pursue graduate studies in physical chemistry at Princeton University, where her boyfriend since high school was studying mathematics, Princeton's chemistry department did not admit women at that time. Berkowitz thus attended the University of Illinois at Urbana. In 1955, she received her Ph.D. in physical chemistry.

After earning her doctorate, Berkowitz journeyed to Yale University to complete postdoctoral work from 1955 to 1957. Furthering her doctoral work on electrolytes, Berkowitz investigated polyelectrolytes and ionic solutions at Yale. She began working for Arthur D. Little, Inc., based in Cambridge, Massachusetts, in 1957. Two years later, Berkowitz married her high school boyfriend, Arthur P. Mattuck. Mattuck had completed his studies and was a professor at the Massachusetts Institute of Technology. The couple had one daughter. Berkowitz and Mattuck divorced in 1977.

At Arthur D. Little, Berkowitz worked in basic and applied research programs involving high-temperature chemistry and environmental science. She studied the kinetics of oxidation of transition metals, notably molybdenum, tungsten, and zirconium, and she also researched the properties, such as strength and hardness, and reactions of chromium-based alloys to high temperatures. As a result of these studies, Berkowitz developed reusable molds for castings from molybdenum and tungsten. She obtained a patent for these molds, which were later used in space programs to make vehicles.

Though Berkowitz made contributions to the space program, she also investigated environmental issues. One of her projects involved the study of the processes used for cleaning coal in power plant boilers. Berkowitz investigated the effects of these processes on particulate emissions. She also looked into problems involving limestone injection wet scrubbing.

During her long career at Arthur D. Little, Berkowitz was involved with numerous outside organizations. From 1963 to 1968, she was an adjunct professor of chemistry at Boston University. Beginning in 1972, Berkowitz focused more heavily on environmental issues and hazardous waste. She played a role in the production of the U.S. Environmental Protection Agency's first report on hazardous waste and in 1975 served on a team evaluating physical, chemical, and biological methods used for treating hazardous wastes. In 1980, Berkowitz was promoted to vice president at Arthur D. Little.

Berkowitz left Arthur D. Little and moved to Washington, D.C., in 1986, to head Risk Science International. In 1989, Berkowitz cofounded Farkas, Berkowitz, and Company, with Allen Farkas. Berkowitz has been very active in the chemical and environmental communities. She is a member of the American Chemical Society, the Electrochemical Society, the Air Pollution Control Association, and the American Institute of Chemists. Berkowitz also served as president of the Electrochemical Society, making her the first female president of the organization, from 1979 to 1980. For her pioneering work, Berkowitz received the Achievement Award of the Society of Women Engineers in 1985.

Bernstein, Dorothy Lewis
(1914–1988)
American
Mathematician

Mathematician Dorothy Lewis Bernstein's research involved a mathematical function known as the Laplace transform. Named after French mathematician and astronomer Pierre-Simon de Laplace, the function had applications in the solving of partial differential equations and in operational calculus. Bernstein was also an educator and did much to advance the relatively new field of computer science, particularly as it applied to mathematics. She was a firm believer in the practical application of mathematics.

Bernstein was born on April 11, 1914, in Chicago, Illinois. Her parents, Jacob and Tillie Bernstein, were Russian immigrants. Bernstein spent the bulk of her childhood in Milwaukee, Wisconsin. She began attending the University of Wisconsin at Madison in 1930. A bright and dedicated student, Bernstein earned both her bachelor's and master's degrees in 1934 after passing an oral examination and completing her thesis, which dealt with the complex roots of polynomials. She graduated summa cum laude.

After earning her degrees, Bernstein stayed at the University of Wisconsin for an additional year, during which she worked as a teaching fellow. She was then awarded a scholarship to pursue her doctorate in mathematics at Brown University, located in Rhode Island. Bernstein faced a rather unsupportive crowd at Brown. Not only were her teaching responsibilities limited to only three female students, but Bernstein was counseled by a dean to avoid searching for teaching positions in certain areas of the United States because of her religion (Jewish) and her gender. In addition, Bernstein's doctoral examination was particularly demanding. Her adviser later told her that the degree of difficulty was prompted by Bernstein's gender and her education—universities in the East did not feel Midwestern schools lived up to their standards.

Despite the hardships, Bernstein managed to acquire a teaching position at Mount Holyoke College in Massachusetts, where she taught from 1937 to 1940. In 1939, Bernstein earned her Ph.D. from Brown with a dissertation on the Laplace transform. After teaching at Mount Holyoke, she traveled across the country, gaining valuable training on the education of mathematics students. Bernstein taught at the University of Wisconsin at Madison in 1941 then worked as a research associate at the University of California at Berkeley during the summer of 1942. In 1943, she secured a teaching position at the University of Rochester in New York and was promoted to assistant professor in 1946. It was at Rochester that Bernstein became more involved with computers. She undertook a research project investigating digital computers and their abilities to process massive amounts of data at high speeds to perform difficult mathematical problems. The project, which was affiliated with the Office of Naval Research, resulted in Bernstein's 1950 publication of *Existence Theorems in Partial Differential Equations.*

Bernstein remained at the University of Rochester until 1957. By that point, she had become a full professor. She journeyed to the University of California in Los Angeles as a visiting professor from 1957 to 1958, and in 1959, she was offered a professorship at Goucher College in Baltimore, Maryland. During her 20 years at Goucher, Bernstein built a strong mathematics department and was instrumental in adding applied mathematics and computer science to the undergraduate curriculum. She was also responsible for establishing an internship program for students majoring in math. Bernstein was the chair of the department from 1960 to 1970 and the director of the computer center from 1961 to 1967.

A strong advocate for computer science, Bernstein lobbied for the instruction of computer programming and computer use in mathematics classes at high schools. This work was conducted through her involvement with the National Science Foundation. She also helped found the Maryland Association for Educational Use of Computers in 1972. Bernstein became the first female president of the Mathematical Association of America in 1979 and was also involved with the American Mathematical Society and the Society of Industrial and Applied Mathematics. Bernstein retired from Goucher in 1979 and died in 1988.

Bertozzi, Andrea
(1965–   )
American
Mathematician

Andrea Bertozzi is a mathematician who is known for her interdisciplinary work with computer scientists, physicists, and engineers. Much of her work has, in one way or

another, examined the behavior of thin liquid films on hard surfaces. In tandem with physicists and engineers, she has worked at the Argonne National Laboratory and at Duke University constructing mathematical models that explain this and other physical phenomena.

Born in 1965, in Boston, Massachusetts, to William and Norma Bertozzi, Andrea was encouraged by both of her parents to study and attend university. Her father, a professor of physics at the Massachusetts Institute of Technology, encouraged her to pursue her interest in the sciences. In 1991, she married Bradley Koetje, a management consultant.

Bertozzi knew from an early age that she was interested in mathematics. Even in the first grade, she was captivated by the rudimentary math that was being taught and pushed to learn more. By high school, she had begun to learn advanced math and was concentrating on theory and abstract concepts, which she found to be the most interesting part of mathematics.

After graduating from high school in Lexington, Massachusetts, in 1983, Bertozzi enrolled in Princeton University to study mathematics. She also studied a considerable amount of physics, although she took no degree in that subject. She earned her B.A. in math in 1987 and remained

Andrea Bertozzi, a mathematician known for her interdisciplinary work with computer scientists, physicists, and engineers, shown here with Tupper, a male Samoyed. *(Courtesy of Andrea Bertozzi)*

at Princeton to complete an M.S. in 1988 and a Ph.D. in 1991.

After completing her Ph.D., Bertozzi took a position as L. E. Dickson Instructor of Mathematics at the University of Chicago. At Chicago, Bertozzi first became interested in the mathematics of thin films. She began working with a group of physicists who were studying mathematical models that described the behavior of phenomena that were similar to thin films. Gradually, the problem centered specifically on a mathematical description of liquids flowing on a solid surface. This was an area of mathematics that had not received much attention but had been researched by physicists since the 1960s.

Bertozzi remained at the University of Chicago until 1995 when she was offered the position of associate professor at Duke University in Durham, North Carolina. Then during 1995–96, she worked at the Argonne National Laboratory, located outside of Chicago in Argonne, Illinois. Here, as a MARIA GOEPPERT-MAYER Distinguished Scholar, she continued her work in the field of scientific computing, which she had begun at the University of Chicago.

The purpose of scientific computing is to create computer models that simulate physical processes on the computer. In this way, virtual experiments that can mimic actual physical conditions are created. At Argonne, Bertozzi continued her study of the mathematical-physical properties of thin liquids on dry surfaces. This problem, which seems relatively simple, is actually complicated. A liquid applied to a dry surface will not spread evenly but will pool and spread onto the surface in fingerlike rivulets. Bertozzi worked on a set of partial differential equations, also called evolution equations because this kind of math describes an event occurring over time, that fit a model for film-coating behavior into mathematical terms.

This work, although basic research, may someday be helpful for industries such as the microchip-manufacturing sector, which needs to understand this coating process in making their complicated and delicate product.

After her year at the Argonne Lab, Bertozzi returned to her job as associate professor of mathematics at Duke University in 1996. In 1998, she became associate professor of mathematics and physics, and in 1999, she became a full professor in both disciplines. Currently, she is director of Duke's Center for Nonlinear and Complex Systems, an interdisciplinary research center that includes scientists from the disciplines of math, biology, engineering, medical sciences, and environmental studies. In addition to her studies of thin films on hard surfaces, Bertozzi works in more general problems of fluid dynamics.

Bertozzi was recognized for her work by the Sloan Foundation, which awarded her a research fellowship in 1995. In 1996, she was presented the Presidential Early Career Award for Scientists and Engineers by the U.S. Office of Naval Research. Cambridge University Press published her book, coauthored with Andrew Majda, *Vorticity and Incompressible Flow*, in 2000.

Bilger, Leonora Neuffer
(1893–1975)
American
Chemist

In a career that spanned 48 years, Leonora Neuffer Bilger built a reputation as a gifted teacher of chemistry and an able administrator of the University of Hawaii's chemistry department. She also contributed to the understanding of nitrogen compounds by her decades-long work studying these substances.

Born on February 3, 1893, in Boston, Massachusetts, Leonora Neuffer Bilger was the daughter of George and Elizabeth Neuffer. Around the turn of the century, the family moved to the booming industrial river town of Cincinnati, Ohio. Neuffer attended primary and secondary schools in Cincinnati, and in 1909, she enrolled as an undergraduate in chemistry at the University of Cincinnati. She earned her B.A. in 1913, a master's degree in chemistry from the University of Cincinnati in 1914, and a Ph.D. from the same institution in 1916 for her study of hydroxylamines and hydroxamic acids.

After attaining her doctorate, she took the position of professor and chair of the chemistry department of Sweet Briar College in Virginia in 1916. Neuffer stayed at Sweet Briar for two years before returning to the University of Cincinnati to join that institution's faculty of chemistry. At Cincinnati, she directed chemical research at the university's basic chemistry laboratory. In 1924, Neuffer won the Sarah Berliner Fellowship to further her study of asymmetric nitrogen compounds at Cambridge University in England. She remained in England for a year. On her return, she took a two-year leave of absence from the University of Cincinnati to serve as a visiting professor at the University of Hawaii. When she returned to Cincinnati, she married Earl M. Bilger, another member of Cincinnati's chemistry department faculty. In 1929, she and her husband accepted permanent positions at Hawaii's department of chemistry.

The study of nitrogen compounds continued to occupy Leonora Neuffer Bilger's time in Hawaii, but she also became increasingly involved as a teacher and administrator. In 1943, she was made head of the chemistry department at Hawaii; she would remain as department head until 1954. In the late 1940s, the University of Hawaii decided that it needed to build a new chemistry laboratory to replace its small and aging structure. Because of her competence and her experience as head of research at the University of Cincinnati's chemical laboratory, Bilger was chosen by the faculty as the lead consultant in this project.

To see what sorts of facilities other universities had and to gather information about cutting-edge laboratory design from other chemists, Bilger visited 25 chemical laboratories around the United States. She used this knowledge to work with the project architects to design a 70,000-square-foot, $1.5-million facility that was opened for staff and students in 1951. The building, Bilger stated, would "provide an environment that arouses the enthusiasm of large numbers of students and research workers" (*Journal of Chemical Education*). This structure was eventually named after her.

For her lifetime of teaching, administration, and research, Leonora Neuffer Bilger was awarded the American Chemical Society's Garvan Award in 1953, the same year she was also appointed senior professor at the University of Hawaii. Bilger remained senior professor until 1958 and was professor emerita from 1960 to 1964. She died on February 19, 1975, at the age of 82.

Birman, Joan S.
(1927–   )
American
Mathematician

Joan Birman fell in love with patterns early in life, later focusing her mathematical research on knot and braid theory. Birman appreciated the beauty of these low-dimensional topologies, likening them to art, but she also developed the practical applications of knot theory in collaboration with biologists, who utilized her work to determine the knotted structures of DNA.

Birman was born on May 30, 1927. As first-generation Americans, her parents instilled in Birman and her three sisters a strong work ethic, especially in education, which they believed to be the key to personal betterment. Birman's love of patterns developed at an early age: Instead of playing games with marbles, she marveled over the swirl of patterns inside the globes as they rolled; and when her mother sewed her plaid skirts, Birman matched the patterns together seamlessly.

In elementary school, the addition and multiplication of odd numbers peaked her mathematical inquisitiveness as she tried to anticipate if the answer would be odd or even. She loved high school geometry, competing with her fellow students at the all-girls school she attended to solve theorems, but college calculus confused her, as it did not require the same kind of spatial visualization as geometry. She attended Barnard College of Columbia University, earning her B.A. in 1948. She remained at the university for graduate study, earning her M.S. in physics in 1950.

Birman utilized her expertise as a systems analyst in the aircraft industry and at Stevens Institute of Technology for several years before devoting her life to parenting, as she raised three children. In 1961, she recommenced her

education with part-time doctoral work. She studied at the Courant Institute of Mathematical Sciences of New York University under Wilhelm Magnus. This was long before nontraditional students gained credence, but the fact that Birman was an older woman did not faze Magnus: He took his student seriously, encouraging her to pave the way for women to become research mathematicians. Birman earned her Ph.D. in 1968.

Birman returned to teach at her alma mater, and she chaired the department of mathematics at Barnard College from 1973 through 1991, with a two-year break from 1987 through 1989. She was honored with a Sloan Foundation Fellowship from 1974 through 1976 as well as a Guggenheim Fellowship in 1994 through 1995. She also spent time as a Visiting Member at the prestigious Institute for Advanced Study in Princeton, and then she returned to her professorship at Barnard.

Birman's publications have made a strong impact on her field. Her book, *Braids, Links, and Mapping Class Groups*, is an influential text, and her article "New Points in View of Knot Theory," which appeared in the April 1993 edition of the *Bulletin of the American Mathematical Society*, earned her the 1995 Chauvenet Prize in expository writing from the Mathematical Association of America. She has also delivered lectures in 13 countries, notably at the 1990 meeting of the International Congress of Mathematicians in Kyoto, Japan, when she discussed the work of Vaughan Jones.

Also in 1990, Birman used her personal funds to establish the Ruth Lyttle Satter Prize in Mathematics in memory of her sister, who was a research botanist at the University of Connecticut. The American Mathematical Society awards the $4,000 prize every two years to a woman who made an outstanding contribution to the mathematical research in the previous five years. MARGARET DUSA MCDUFF received the first prize in 1991, and 1993 recipient LAI-SANG YOUNG served on the selection committee that chose 1995 recipient SUN-YOUNG ALICE CHANG. In honor of Birman's 70th birthday, Barnard College and Columbia University hosted a Conference in Low-Dimensional Topology in 1998, featuring Vaughan Jones as a speaker, among others. Outside of mathematics, Birman enjoys cooking, which she finds not only relaxing but also inspirational to her mathematical creativity.

Bishop, Katharine Scott
(1889–1975)
American
Physician

Working in the male-dominated field of medical research at a time when few women had professional pursuits,

Katharine Scott Bishop enjoyed many accomplishments during her career. Best known for codiscovering vitamin E and identifying its critical role in biological reproduction, Bishop also practiced medicine and worked as an anesthesiologist and educator.

Born Katharine Scott on June 23, 1889, in New York City, Bishop was the daughter of Walter and Katherine Emma Scott. After completing her studies at the Latin School in Somerville, Massachusetts, she attended Wellesley College. Bishop earned her undergraduate degree from Wellesley in 1910 and proceeded to enroll in premedical classes at Radcliffe College. She then attended Johns Hopkins Medical School and was awarded her medical degree in 1915.

After earning her medical degree, Bishop journeyed to California to teach histology, the study of the microscopic structure of animal and plant tissues, at the University of California Medical School from 1915 to 1923. In addition to her teaching, Bishop conducted research with Herbert McLean Evans, an anatomist. Together the researchers discovered and investigated vitamin E in 1922, originally referring to it as *substance X*. The substance was found to play a vital role in the ability for rats to reproduce—a discovery the scientists made when they successfully deprived rats of substance X. Vitamin E is mostly found in foods of plant origin and is a fat-soluble vitamin. The body is unable to produce vitamin E, and thus an outside dietary source is needed, which Bishop and Evans recognized during their extensive research. The discovery of vitamin E by Bishop and Evans led to further studies of the important class of vitamins that later came to be known as antioxidants. This group of substances also includes vitamin C and beta carotene, or vitamin A.

Bishop and Evans published their findings in the *Journal of the American Medical Association* in 1923 in a joint article entitled "Existence of a Hitherto-Unknown Dietary Factor Essential for Reproduction." Evans continued to work with the substance, and later, in 1935, Evans and other researchers succeeded in isolating vitamin E and the actual factor, called tocopherol, responsible for reproduction. Bishop, meanwhile, married attorney Tyndall Bishop and had a position as a histopathologist at the George William Hooper Institute of Medical Research in San Francisco from 1924 to 1929. During this period, Bishop published a number of articles on physiology and histology.

Bishop chose to focus on raising her two daughters during the 1930s. She also spent two years at the University of California Medical School studying public health. In the mid-1930s, her husband grew ill, and Bishop practiced general medicine and anesthesiology to support the family. She also worked as an anesthesiologist at St. Luke's Hospital in San Francisco. After the death of Tyndall Bishop in 1938, Bishop accepted a position at Alta Bates Hospital in Berkeley, California, in 1940.

An active member of the medical community, Bishop belonged to several professional organizations, including the American Association of Anatomists, the Association for the Advancement of Science, and the Society for Experimental Biology and Medicine. She practiced medicine at Alta Bates Hospital until her retirement in 1953. Bishop spent her remaining years at her home in Berkeley. She died on September 20, 1975.

Blackburn, Elizabeth Helen
(1948–   )
Australian/American
Molecular Biologist

A renowned molecular biologist and biochemist, Elizabeth H. Blackburn is best known for her discovery of telomerase, an enzyme needed for the reproductive process of chromosomes. Blackburn's studies significantly advanced the scientific community's understanding of deoxyribonucleic acid (DNA), and her work with telomerase has opened the doors to new research, particularly in regard to cancer research and gerontology studies.

Born on November 26, 1948, in Hobart, Australia, Blackburn developed an interest in medicine and biology at an early age. She was no doubt inspired by her physician parents, Harold and Marcia (Jack) Blackburn. Blackburn attended the University of Melbourne and earned her B.S. degree in 1970. She received her master's degree a year later. Blackburn then left her native Australia to pursue a doctorate at Cambridge University in England. She earned her Ph.D. in molecular biology in 1975 with a dissertation on the sequencing of nucleic acids.

With her doctorate in hand, Blackburn traveled to the United States. She began a fellowship in biology at Yale University in 1975, the same year she married American biologist John Sedat. Blackburn had met Sedat at Cambridge, where Sedat had been conducting postdoctoral research in biology. At Yale, Blackburn investigated the structure and replication of chromosomes and began to work with telomeres, which are found at the ends of chromosomes and help stabilize gene cells.

Blackburn left Yale in 1977 and moved to California to work as a research fellow at the University of California, San Francisco. A year later, she was offered a position as assistant professor at the University of California, Berkeley. Continuing her investigations of telomeres, Blackburn observed a connection between the size of a telomere and the capacity of a chromosome to divide and replicate. Through continued efforts, she discovered that cells carried out a process to replace missing telomeres, without which gene cells could not survive. In particular, in 1985, Blackburn and her graduate assistant, Carol W. Greider, succeeded in isolating telomerase, the enzyme responsible for synthesizing new telomeres, thus helping chromosomes to replicate. Telomerase also regulates the length of telomeres. Blackburn's findings made the creation of artificial telomeres and chromosomes possible, greatly advancing genetic research.

In 1986, Blackburn became a full professor at the University of California, Berkeley. Four years later, in 1990, she became a professor in the Department of Microbiology and Immunology, as well as in Biochemistry and Biophysics, at the University of California, San Francisco. In 1993, she was named chair of the Department of Microbiology and Immunology, becoming the first female to earn the distinction. During these years, Blackburn continued her research on telomeres and telomerase. In 1990, Blackburn and several of her students published a paper on the effects of defective telomerase. Faulty telomerase causes telomeres to shrink, which in turn affects genetic reproduction. These findings had applications in cancer research, as cancer cells possess markedly long telomeres. Blackburn and Greider published *Telomeres (Monograph 29)*, a book of essays on telomeres, in 1995.

Blackburn has received much recognition for her pioneering work on telomeres. In 1988, she was awarded the Eli Lilly Award for Microbiology, and in 1990, she received the National Academy of Science's Molecular Biology Award. Three years later, Blackburn was elected a foreign associate of the National Academy of Science. She was given an honorary doctorate from Yale University in 1991, and in 1992, she was elected a fellow of the Royal Society of London. Though committed to her career, Blackburn's priorities are balanced between work and family life, which includes Sedat and their son. Blackburn discussed the importance of motherhood and family in the on-line article, "Balancing Family and Career: One Way That Worked."

Blackwell, Elizabeth
(1821–1910)
British/American
Physician

With her graduation from medical school in 1849, Elizabeth Blackwell began a nearly 40-year career in medicine, a sojourn that was remarkable not only for the energy and dedication of its practitioner but also because Blackwell was the first licensed woman M.D. in the Western world. Blocked at nearly every point in her early career by male doctors who could not accept a woman in their ranks, Blackwell carved out a niche for herself and other women physicians in what had been an all-male profession.

Blackwell was born in Bristol, England, in 1821 into a family of religious activists and social reformers. Her father,

Samuel Blackwell, was a wealthy sugar refiner and Dissenter, that is, a member of one of the several Protestant sects that had arisen in England in the 18th and 19th centuries whose views opposed those of the established Church of England. Blackwell's mother, Hannah Lane Blackwell, was also a supporter of social reform and Dissenter religious belief.

Because Samuel Blackwell strongly believed that women should receive the same education as men, he engaged private tutors to teach all 12 of his children. Elizabeth received an excellent classical education in England until a fire at Samuel Blackwell's refinery compelled the family to move to the United States in 1832. Blackwell, who set up another sugar refinery in New York, was able to continue his children's first-class education in that city. Eventually, not only Elizabeth but her sisters Anna, Ellen, and Emily earned university degrees. Emily would eventually follow in Elizabeth's footsteps by becoming a medical doctor.

When Elizabeth was 17, a financial panic bankrupted her father's New York business, and again the family moved, this time to Cincinnati, Ohio. Samuel Blackwell died soon after the move to Cincinnati, and Elizabeth was forced to work as a teacher in a boarding

Elizabeth Blackwell became the first licensed female medical doctor in the Western world when she graduated from medical school in 1849. *(National Library of Medicine, National Institutes of Health)*

school founded by her sisters. After several years of teaching in Cincinnati and western Kentucky, Blackwell decided to apply for medical school. According to comments she made in her autobiography, *Pioneering Work in Opening the Medical Profession to Women,* she "hated everything connected with the body, and could not bear the sight of a medical book." She was rejected by Harvard, Yale, and numerous other colleges before finally being accepted at Geneva College in upstate New York, mainly because the acceptance committee thought her application was a prank. When she arrived in 1847, the teachers were startled but kept to their word, and she was admitted.

Blackwell performed well at Geneva College. In 1848, she served a year's residency at Philadelphia Hospital, an institution serving mainly the poor in Philadelphia. Here, according to Blackwell, "the young resident physicians, unlike their chief, were not friendly. When I walked in, they walked out." Blackwell persevered and was awarded a degree in 1849. From 1849 to 1851, she went to Europe to further her medical education.

On her return to the United States in 1851, Blackwell began lecturing about medicine and hygiene on the abolitionist and suffragist circuit that had been established by some of the reformers who had been friends of her father's when she was a child living in New York. A group of Quaker women who were active in this circle offered financial aid, and in 1853, Blackwell was able to open a clinic for poor women and children. Blackwell, along with Polish immigrant physician MARIE ZAKRZEWSKA and Blackwell's younger sister Emily, staffed the clinic. The cases were those common to the urban poor—typhus, cholera, food poisoning, as well as other common ailments such as broken bones, cancer, and infections. By 1868, Blackwell had managed to add a medical college, the Women's Medical College of the New York Infirmary, to the clinic. It would be a major training hospital for future generations of women physicians.

In 1869, one year after having established Women's Medical College, Blackwell returned to England to live and work. She established a successful practice there but, as she got older, gradually moved into retirement in the Scottish Highlands.

Blackwell not only personally broke the gender barrier in Western medicine but was influential in training successive generations of women physicians to continue the revolution she had charted. By founding the New York Infirmary, she also challenged other doctors to become engaged in bringing medical services to the poor, a struggle she continued almost until her death at the age of 89 in 1910. Hers was a legacy of social activism and personal commitment that has flowered into organizations such as Doctors Without Borders and other engaged medical groups.

Blagg, Mary Adela
(1858–1944)
British
Astronomer

Mary Blagg played an instrumental role in the standardization of lunar nomenclature in the early 20th century. A self-taught astronomer, Blagg pursued knowledge in astronomy and the sciences until she became an accepted and skilled astronomer and an expert on the nomenclature of lunar formations. Blagg was a member of the subcommittee of the International Astronomical Union and worked on gathering and standardizing lunar terminology. She made significant advances in astronomy during a time when few women were accepted as professionals in scientific communities.

Born in 1858 in Cheadle, North Staffordshire, in England, Blagg grew up in privileged circumstances. Blagg's father worked as a lawyer, and Blagg attended a private boarding school in London. Education and higher learning were not customarily pursued by women during those times, and women of Blagg's circumstance and background generally pursued worthy causes. Blagg followed convention, but she also yearned to educate herself further, and she studied her brother's textbooks to teach herself mathematics. Her curious and inquisitive nature drew her to the sciences, and further individual studies provided Blagg with a competent grasp of the basics of astronomy.

Blagg decided to continue studying astronomy after attending a lecture given by J. A. Hardcastle, a British astronomer. Blagg became involved with the Council of the International Association of Academies and the International Astronomical Union beginning in the early 1900s. A committee delegated with the task of standardizing lunar nomenclature was formed in 1907 by the Council of the International Association of Academies. The standardization of this nomenclature was necessary for scientists to locate, study, and discuss unique and specific features of planets or satellites. The committee was not successful in publishing a report explaining lunar nomenclature because of a series of deaths of committee members. Fortunately, however, Mary Blagg, a committee member, had managed to make considerable progress.

In 1919, the International Astronomical Union was formed at a meeting in Brussels. The organization was established to regulate planetary and satellite nomenclature. During the 1919 meeting, a new committee was established and assigned the task of standardizing lunar and Martian nomenclatures. Mary Blagg and a number of other astronomers were appointed to this committee and began work on the development of a standard for lunar nomenclature. The committee was headed by astronomer H. H. Turner. During this period, Blagg worked as an assistant to Turner, who was investigating variable stars, stars that had variable brightness because of inner changes or the occasional concealment of mutually revolving stars. Together Blagg and Turner published a series of 10 papers detailing their work, which greatly furthered astronomers' understanding of variable stars. In 1935, the nomenclature committee published its report, *Named Lunar Formations*, which was coauthored by Blagg and K. Muller. The report was the first orderly listing of lunar nomenclature.

Blagg made significant contributions to astronomy while working with Turner. In addition to assisting him with his study of variable stars and working on lunar nomenclature, Blagg also discovered new elements for a number of stars, including Lyrae, RT Cygni, V Cassiopeiae, and U Persei. She also investigated light waves, greatly furthering the work of other astronomers on the subject.

Mary Blagg was respected by her peers for her knowledge and dedication, and she was awarded for her outstanding work by the Royal Astronomical Society, which elected her a member. Her skills and avid interest in astronomy allowed Blagg to work with the best astronomers of her day. After Blagg's death in 1944, a lunar crater was named in her honor.

Blodgett, Katharine Burr
(1898–1979)
American
Physicist

The first woman to win a Ph.D. from Cambridge University and the first woman research scientist at General Electric Corporation (GE) were one in the same person: Katharine Burr Blodgett, one of American's trailblazing women physicists. At General Electric, Blodgett developed nonreflecting glass for cameras and perfected methods for applying and measuring extremely thin surface films.

Born on January 10, 1898, in Schenectady, New York, Blodgett was the daughter of George Bedington Blodgett, a lawyer who was the head of GE's patent department, and Katharine Buchanen Blodgett. George Blodgett died before Katharine's birth, and Mrs. Blodgett moved Katharine and her older brother to New York City, then France, where Blodgett learned to speak, read, and write in French.

For her secondary education, Blodgett returned to New York City to attend the private Rayson School run by three English sisters. She did so well at Rayson that she won a scholarship to attend Bryn Mawr, a women's college in Pennsylvania. There, under the tutelage of two inspiring professors—Charlotte Scott in math and James Barnes in physics—Blodgett graduated second in her class in 1917 with a degree in physics. During a visit to GE's plant and headquarters while she was still an undergraduate, Blodgett met and befriended Irving Langmuir, the future Nobel laureate and one of GE's principal research scientists. Lang-

Katharine Burr Blodgett, the first female research scientist at General Electric Corporation and the first woman to win a Ph.D. from Cambridge University, which she did in 1926. *(AIP Emilio Segré Visual Archives, Physics Today Collection)*

muir advised Blodgett, who was hoping for employment at GE's labs after graduation, to seek postgraduate education. This Blodgett did by earning an M.S. from the University of Chicago in 1918 for her study of the gas absorption potential of coconut charcoals, research with direct application to the development of gas masks to counter poison gas attacks during World War I.

Langmuir was sufficiently impressed with Blodgett's skill and dedication to hire her immediately as his assistant at GE. For the first several years they worked on improvements to GE's electric lightbulbs. Blodgett was lucky to find work in a place where her father had, in a manner, paved her way. "It was virtually impossible," she said later, "for women scientists to find professional-level jobs at corporations at that time." But because Blodgett was personally known to many GE executives, she was allowed to work there. No one at GE would ever regret this decision.

In 1924, Blodgett won a place as a physics doctoral student at Sir Ernest Rutherford's Cavendish Laboratory, one of the most prestigious centers of scientific learning in the world. Her doctoral dissertation was about the behavior of electrons in ionized mercury vapor. In 1926, with Ph.D. in hand, Blodgett returned to GE to begin work on the study of surface chemistry, a subject that would occupy her attention for much of the rest of her career. She succeeded in finding a technique to apply superthin layers, or films, of fatty acids, from 4 to 44 molecules thick, to the surface of metals. She noticed that these films gave off different colors at different thicknesses. To vastly simplify the measurement of layer thickness, she devised a gauge that would read the film's color, thus decoding its thickness to one-millionth of an inch (approximately 4.4 molecules thick). Later she applied this technology to developing nonreflective glass, which was achieved by laying an adhesive soapy film 44 molecules thick (four-millionths of an inch) onto a glass lens surface. The thickness was exactly the length of one-quarter of a wave of light and prevented light refraction, which was the cause of reflection. The invention was used for improved camera and motion projection lenses.

Blodgett won many awards for her work during her career. The American Association of University Women gave her its Achievement Award in 1944. She also won the American Chemical Society's Garvan Award in 1951 and the Photographic Society of America's Progress Medal in 1972. At her death in 1979, one of her coworkers, Vincent J. Shaefer, recalled that "the methods she developed have become classical tools of the science and technology of surfaces and films. She will be long—and rightly—hailed for the simplicity, elegance, and the definitive way in which she presented them to the world."

Blum, Lenore Epstein
(1942–   )
American
Mathematician

A mathematical researcher, Lenore Blum has also been active in promoting women mathematicians and encouraging women to enter mathematics and computer sciences. She has also been an inspiring teacher of mathematics and a committed administrator who has led several university mathematics departments. In her field, Blum is mainly known for her work with computational mathematics, a mix of math and logic whose goal is to better understand and describe mathematically theoretical problems posed by the structure of computing devices.

Born in 1942 in New York City, Blum lived in her home city for only nine years before she moved with her family to Caracas, Venezuela, where her father had taken a job. One of two siblings, she was an intellectually precocious child who was encouraged to study by both parents, but especially by her mother who was a schoolteacher.

In Caracas, Blum briefly attended a Spanish-language Venezuelan school but then transferred to the American School, which is a private institution. Because the tuition was expensive at the American School, Blum's mother arranged to take a job there as a teacher to pay for her daughters' education. Blum graduated from high school early and, at age 16, returned to the United States to enroll in the Carnegie Institute of Technology in Pittsburgh.

In high school, Blum had been interested in art and mathematics. At Carnegie Institute, she intended to join these two pursuits in the study of architecture. Blum majored in architecture for two years but, missing the purity of pure mathematical exploration, decided to switch to mathematics as a major.

In 1960, when she was 18 and a college junior, Blum married Manuel Blum, then a graduate student at the Massachusetts Institute of Technology (MIT). To be with her husband, she transferred to Simmons College, a women's college in Boston. Here she continued her study of mathematics, but finding that Simmons did not offer advanced math, she began taking classes in advanced algebra at MIT.

After graduation from Simmons in 1962, Blum enrolled in MIT's graduate school and became interested in new approaches to some of the preexisting problems of algebra. Picking up on work done by several other mathematicians, Blum began using new techniques of logic in mathematical solutions. She followed this approach to complete her Ph.D. in mathematics, which she earned from MIT in 1970.

After completing her doctorate, Blum moved to the University of California at Berkeley to study with the logician Julia Robinson. While working at Berkeley on a postdoctoral fellowship, she became involved in efforts to push for more tenured positions for women mathematicians at American universities. She was an early activist in the newly formed Association of Women in Mathematics. Blum became the third president of this organization in the mid-1970s.

In 1973, Blum was appointed lecturer of mathematics at Mills College, a San Francisco area women's college. She helped redesign Mills College's mathematics program, eventually becoming head of the department and a full professor of mathematics. She remained at Mills College until 1986.

In 1988, Blum was hired as a research scientist at the International Computer Science Institute, a San Francisco area think tank. From 1992 to 1997, she was deputy director of the University of California at Berkeley's Mathematics Sciences Research Institute.

Blum has been honored for her work in mathematics with a Letts-Villard Research Professorship at Mills College. She also presented a paper at the 1990 International Congress of Mathematicians, and she has served as vice president of the American Mathematics Society. Currently, Blum is professor of mathematics at Carnegie-Mellon University in Pittsburgh, Pennsylvania.

Boden, Margaret
(1936–   )
British
Psychologist

Since the mid-1960s, Margaret Boden has devoted her career to the idea that study of the computer programming subdiscipline called artificial intelligence can indicate a lot about the workings of the human mind. Boden has drawn analogies between how the mind organizes itself to complete tasks and the way computer scientists attempt to program their machines to do simple tasks such as responding to questions or assembling parts of an object.

Boden was born on November 26, 1936. There is not much information available about her family other than it was working-class, and because of the rigid English class system, there was little expectation within Boden's family that she would go to university. "I'd never expected to go to college," Boden has said, "[because] neither of my parents did," but access to a college education was changing in England after World War II.

Merit scholarships opened college doors for many people like Boden. She did well in secondary school and in 1955 won a scholarship to Cambridge University, one of Britain's elite universities. Later, in her early thirties, Boden married. She has two children and was divorced in 1981.

As a undergraduate at Cambridge, Boden studied philosophy and medicine. From 1959 to 1962, she worked with mental patients in Birmingham and taught philosophy at the University of Birmingham. Neither of these jobs fully engaged Boden's imagination, and she decided to expand her education by embarking on a doctoral program in social and cognitive psychology at Harvard University. Boden began her doctoral studies in 1962 still unsure of exactly what direction she would take within her discipline. A chance reading of a book by George A. Miller, entitled *Plans and the Structure of Behavior* (1960), provided her with inspiration. The book discussed the idea of applying "the notion of a computer program to the whole of psychology," a concept, Boden says, that hit her "like a flash of lightning."

Boden applied this concept to the theories of psychology then current in an attempt to see how these preexisting theories of human perception and thinking could be translated into a logical program that would drive a computer to understand or operate in the world on the terms given by each psychological theory. This study grew into her dissertation, which in turn became her first book, *Purposive Explanation in Psychology*, published in 1972.

By 1965, although still engaged on writing her dissertation for Harvard, Boden had returned to Britain to teach at Sussex University. She completed her second book, *Artificial Intelligence and Natural Man* (1977), which was an elaboration of her basic idea that insights into the human thought process could be derived from continuing work into computer artificial intelligence studies. Her third book, *Piaget*, published in 1979, was an examination of the theories of childhood development of Swiss psychologist Jean Piaget. Boden dissected the philosophical and biological assumptions that underlie Piaget's theories and compared his ideas to work in artificial intelligence. In her fourth book, *The Creative Mind* (1970), Boden explores the insights into human creativity that can be gained through an examination of programmers' attempts to install creativity in their machines.

Boden's awards include membership in the British Academy of Science and the American Association for Artificial Intelligence. In 1987, she became the first dean of the University of Sussex's School of Cognitive and Computing Sciences.

Bodley, Rachel Littler
(1831–1888)
American
Chemist and Botanist

Educated in chemistry and botany, Rachel Bodley devoted a large part of her career to teaching chemistry to female medical students. Later, she moved into administration and used her position to champion the importance of training women to be doctors. In 1881, she compiled one of the first studies that tracked the careers of female graduates of an American medical school, a work that was later published as a pamphlet entitled *The College Story*.

The third of five children, Rachel Bodley was born in 1831 to Anthony Bodley and Rebecca Talbot Bodley of Cincinnati, Ohio. Her father was a carpenter, and her mother was a teacher who ran a private school in Cincinnati. The fact that both of Bodley's parents worked was an unusual arrangement for the time and may have inspired Bodley to later break social taboos by pursuing a career in science.

Bodley began her education at her mother's school and continued it at Wesleyan Female College in Cincinnati, the first chartered college for women in the United States, from which she received a diploma in 1849. Upon graduation from Wesleyan, Bodley taught science there for 11 years. In 1860, she studied advanced chemistry and physics for two years at Polytechnic College in Philadelphia while simultaneously studying anatomy and physiology at the Female Medical College in the same city.

Bodley returned to Cincinnati to continue her teaching career, serving as an instructor of natural sciences at the Cincinnati Female Seminary. In 1865, she moved back to Philadelphia where she was named as the first professor of chemistry at the Female Medical College, later renamed Woman's Medical College. After teaching for nine years, Bodley was appointed dean of the school in 1874, a position she held until her sudden death by heart attack in 1888.

Despite her heavy load as a teacher and administrator, Bodley found time to work on botany studies, a discipline that had been one of her loves since childhood. In Cincinnati, and later in Philadelphia, she added to the botanical knowledge of North American plant species by collecting and classifying local flora.

Bodley also devoted considerable effort to community work outside the domain of the Female Medical College. She served on secondary school boards in Philadelphia twice (1882–85 and 1887–88) and was appointed an outside inspector of Philadelphia's charitable institutions by the Pennsylvania Board of State Charities in 1883.

Although Bodley was not a leader in the advancement of theoretical or experimental knowledge in chemistry or botany, numerous institutions recognized her contributions to science and the education of women in scientific and medical disciplines. In 1871, she was inducted as a member into the Academy of Natural Sciences of Philadelphia, and in 1876, she was made a corresponding member of the New York Academy of Sciences and a charter member of the American Chemical Society. She was also awarded an honorary M.D. degree by the Female Medical College in 1879, and in 1880, she became a member of the Franklin Institute located in Philadelphia. During her lifetime, several of her lectures were published and distributed nationwide. Through her teaching, lecturing, and engagement in scientific societies, Bodley demonstrated that women had the potential to play as valuable a role as men in the scientific endeavor.

Boivin, Marie Anne Victoire Gallain
(1773–1841)
French
Physician

A midwife by training, Marie Boivin came to be regarded as the most knowledgeable European obstetrician and gynecologist of the early 1800s in spite of the fact that she was denied entrance into medical school in France because of her gender. Undeterred, Boivin, under the tutelage of her mentor MARIE-LOUISE LACHAPELLE, taught herself the techniques of birth delivery and became an expert on the diseases of the female reproductive organs. She wrote numerous books and pamphlets within her medical discipline, several of which were considered the leading texts in her field even through the mid-19th century.

Marie Anne Victoire Gallain was born on April 9, 1773, in the small town of Montreuil, which was near the royal palace of Versailles in northern France. There is little information about her father, mother, or siblings other than the fact that Boivin lived with a sister, who ran a hospital in Estampes, a small town near Paris, during the French Revolution. With the end of the revolution, she returned to Montreuil and in 1797, at the age of 24, married Louis Boivin, a government bureaucrat. She gave birth to a daughter in 1798 and was widowed that same year. To earn a living, Boivin apprenticed herself to the well-known midwife Marie-Louise Lachapelle at the Hospice de la Mater-

nité in Paris. She received a degree in midwifery in 1800 and helped Lachapelle set up a formal school of midwifery at the hospice.

Boivin then practiced her trade for a year in Versailles until the death of her daughter prompted her to return to Paris and the Hospice de la Maternité. Boivin remained at the Maternité as supervisor-in-chief until 1811 when she and Lachapelle had a falling out over the publication of Boivin's first book, *Mémorial de l'art des accouchements* ("About the Art of Childbirth"), a case textbook for midwives.

L'art des accouchements was lavishly illustrated with more than a hundred precise drawings that showed the various possible positions of the fetus in the womb. Accompanying text detailed the symptoms to be aware of and course of action to take for each. Boivin originally had not considered her illustrations suitable for publication (she considered them too shockingly blunt for the general public), but François Chaussier, the directing physician at the Maternité, insisted that they be included. Because no such work had been written in Europe since 1688, there was great need for Boivin's book. It was hugely popular and was translated into German and Italian. Probably envious about the success of Boivin's book, Lachapelle fired her.

From 1811 until her death, Boivin directed several hospitals and maternity wards in Paris, and she also continued to write prolifically about obstetrics and gynecology. She continued her work and study as a gynecologist and became the leading expert in France about pathologies of women's reproductive organs. She translated into French two important English works about hemorrhaging of the uterus, Edward Rigby's *Treatise on Hemorrhages of the Uterus* (translated in 1818) and Duncan Stewart's *Treatise of Uterine Hemorrhage* (translated in 1820). In 1818, she wrote her own book on uterine hemorrhaging, which was a history of the thought and treatment on the subject from antiquity until the early 19th century. Finally, in 1827, she published an important work on the hydatiform mole, a condition of abnormal pregnancy in which the fetus degenerates into a mass of cysts. The work was called *Nouvelle recherches dur l'origine, la nature et le traitement de la mole vesiculaire ou grossesse hydatique.*

During her lifetime, Boivin won several awards in recognition of her work. The king of Prussia presented her with the Order of Civil Merit in 1814. For her work about uterine hemorrhaging, Boivin was awarded a prize in 1819 from the Medical Society of Paris, which had assumed that she was a man (she had given her initials, not her full name). However, because she was a woman, she was never admitted to the French Royal Academy of Medicine, to which she famously replied, "The midwives of the academy didn't need me." Her works about diseases of the uterus were considered the best available in Europe until the mid-1800s. She died in semipoverty on May 16, 1841.

Bondar, Roberta Lynn
(1945–)
Canadian
Neurologist

Trained as a neurologist and physician, Roberta Bondar has made a name for herself as an astronaut in the U.S. National Aeronautics and Space Administration (NASA) and as an expert in space medicine. She has investigated brain physiology and conducted other medical and material science experiments as a payload specialist on NASA's space lab.

Born on December 4, 1945, in Sault Saint Marie, Ontario, Canada, Roberta Bondar attended public primary and secondary schools in her hometown. In 1964, she enrolled in the University of Guelph in Guelph, Ontario. As an undergraduate, she studied zoology and agriculture and earned a B.S. in agriculture in 1968. Following graduation, Bondar enrolled in the University of Western Ontario where she studied experimental pathology. She won her M.S. in that subject in 1968 and then continued her education at the University of Toronto, where she earned a Ph.D. in neurobiology in 1974. Bondar then enrolled in medical school at McMaster University and received a medical degree from that institution in 1977. She did her internship at the Toronto General Hospital and finished her residency in neurology at the University of Western Ontario.

Between 1980 and 1983, Bondar traveled widely to study and work. She was a fellow in neurology at the Royal College of Physicians and Surgeons of Canada in 1981 and worked in neuro-ophthalmology at Tufts New England Medical Center in Boston and at the Playfair Neuroscience Unit in Toronto's Western Hospital. In 1982, she was hired as an assistant professor of neurology at McMaster University; she also served as director of McMaster's Multiple-Sclerosis Clinic.

In 1983, Bondar applied for, and was chosen as a member of, the Canadian Astronaut Program. She trained on this program from 1984 to 1992 while also holding down other jobs. She taught a course in the biomedicine of space for Canada's Department of National Defense and worked as a clinical neurologist at the University of Ottawa's medical school. She helped the Canadian Parliament as chair of the Canadian Life Sciences Subcommittee for the Space Station from 1985 to 1989. She also served on the Council on Science and Technology for the premier of Ontario in 1988–89.

In the later 1980s and early 1990s, Bondar's duties increasingly brought her to the Johnson Space Center in Houston, Texas. In Houston, Bondar began intensive training with other astronauts in preparation for a space flight onboard the space shuttle *Discovery*. After years of training in Canada, she was finally selected as a crew member for a flight that was slated for January 1992. This flight, the first

International Microgravity Laboratory Mission, had as one of its main goals the study of the effects of low gravity on the physiological processes of the human body. To aid this study, Bondar prepared a series of experiments to measure blood flow in the brain during microgravity (as low-gravity conditions are called). As payload specialist, she was also put in charge of material's experiments on the mid-deck of the space station.

For her work on the space station, Bondar won the NASA Space Medal in 1992. That same year, she was also given a Presidential Citation by the American Academy of Neurology and was awarded the Order of Canada by the Canadian government. Bondar has been given honorary doctorates by more than 20 universities, including McGill University in 1992 and the University of Montreal in 1994. Currently, she is a professor at the Centre for Advanced Technology and Education in Canada and is a visiting research scientist at the Universities Space Station Association at the Johnson Space Center in Houston.

Boole, Mary Everest
(1832–1916)
British
Mathematician

Largely self-taught in mathematics, Mary Everest Boole was a prolific writer and thinker who developed theories of mathematical education as well as the psychology of learning. Boole assisted her well-known mathematician and logician husband, George Boole, who was credited with developing a calculus of symbolic logic. Mary Everest Boole spent many years writing and lecturing about psychology, philosophy, and the educational and mental processes of children.

Born Mary Everest in 1832 in Warwickshire, Gloucestershire, in England, Boole was the older of two children born to Reverend Thomas Roupell Everest and Mary Ryall. Boole's family included many accomplished academics, and she grew up in an intellectually active home. Boole's uncle, Sir George Everest, had worked as a surveyor in India, and Mount Everest had been named in his honor. Another uncle was vice president of Queen's College in Cork, Ireland. Because of an illness afflicting Boole's father, the family moved to Poissy, France, in 1837 so that he could be treated by Samuel Hahnemann, the founder of homeopathy, a medical system in which Reverend Everest strongly believed. Boole was tutored by Monsieur Deplace, who introduced her to mathematics and to a teaching method that profoundly affected her; Deplace presented Boole with a problem and then asked a succession of questions, prompting Boole to logically and systematically discover the answer.

Boole and her family returned to England when she was 11 years of age. In order to assist her father, Boole left school and took on such duties as helping her father prepare sermons and teaching Sunday school classes. During these years, Boole did not abandon her studies; she used the resources in her father's library to teach herself the basics of calculus, but she yearned to learn and understand more. In 1850, at the age of 18, Boole was introduced to George Boole, who was then 35 years old and a well-respected mathematician, during a visit to her aunt and uncle in Cork, Ireland. After returning home, Boole began corresponding with George Boole about mathematics and science. Their relationship grew, and after the death of her father in 1855, the two married. Boole moved to Cork, and the couple had five daughters in nine years. During that time, Boole collaborated with and assisted her husband in his work, which included the publication of *Laws of Thought,* a book that presented his ideas that algebraic formulas could be applied to both qualitative and quantitative problems.

In 1864, George Boole died of pneumonia, and Boole was left to raise five children, including one who was only six months old, alone. Boole moved her family to London and found a job as a librarian at Queen's College, the first women's college in England. Though Boole would have preferred to teach, women were not allowed to teach at the college at that time. Boole also managed a boardinghouse for students and began leading weekly gatherings to discuss mathematics, philosophy, psychology, and other topics of interest. Boole completed a book, *The Message of Psychic Science for Mothers and Nurses,* which outlined her ideas about developing mental intelligence. The controversy surrounding the subject matter of the book caused Boole to lose her job as a librarian and the lease at the boardinghouse in 1873.

Boole began working as a secretary to James Hinton, an ear surgeon who had been a friend of her father, in 1873. Hinton also had an interest in psychology and had written numerous books, and following his death in 1875, Boole continued to advance his ideas. Boole also became interested in evolution and believed that basic ideas about the universe could be expressed with numbers and symbols.

From the age of 50, Boole began writing a series of books and articles. Though her ideas regarding psychology and the unconscious were largely dismissed, Boole's theories concerning the mental process of children eventually gained wide acceptance. Among her published works were *Lectures on the Logic of Arithmetic* (1903) and *The Preparation of the Child for Science* (1904). Her controversial book, *The Message of Psychic Science for Mothers and Nurses,* was not published until 1883. Boole was also responsible for inventing curve stitching, now called string geometry, which helped children learn about geometrical units such as angles. Boole remained active until her death in 1916. Her complete works were published in 1931 as

Collected Works and included more than 1,500 pages and four volumes.

Bozeman, Sylvia
(1947–   )
American
Mathematician

An educator and researcher, Sylvia Bozeman has spent much of her life engaged in making mathematics and the sciences accessible to women and minorities in the United States. She has made the recruitment of minorities into the sciences part of her career's work. As a mathematics researcher, she has spent considerable time exploring functional analysis, integral equations, and the mathematics of data compression and decompression for files of computer data. She has also developed and coordinated numerous programs that promote mathematics as a career to elementary, high school, and college students.

Born in 1947 in rural central Alabama, Sylvia Trimble grew up in the hamlet of Camp Hill, a town near Alabama's border with Georgia. In 1968, she married Robert Bozeman, who is also a mathematician. They have two children.

She knew even in elementary school that she had an affinity for math. In high school, she wanted to take as many math courses as she could, but few were available at her school. She was lucky that her math teacher recognized her desire to learn. He tutored her and several other students in trigonometry at night. On her own, Bozeman also studied advanced geometry from books provided by her teacher.

Following her graduation from high school in 1964, Bozeman enrolled in Alabama A&M University. There she expanded her knowledge of mathematics, making up for the deficiencies of her high school math curriculum. Bozeman's enthusiasm for mathematics came to the attention of the chairman of the physics department who helped her get into a National Aeronautics and Space Administration math and computer project. In the summer of 1967, she attended a mathematics seminar at Harvard University that helped her sharpen her calculus and computer-programming skills.

After graduation from Alabama A&M with a B.A. in mathematics in 1968, Bozeman enrolled in the graduate school of Vanderbilt University in Nashville, Tennessee. She completed a master's degree in mathematics at Vanderbilt in 1970, writing her master's thesis on a branch of algebra called group theory and concentrating on prime-order groups.

After completing her master's degree, Bozeman took a few years off from academics to begin raising a family. Around 1972, she moved with her husband to Atlanta and began teaching mathematics at Spelman College, a historically black institution. At Spelman, she was influenced by ETTA FALCONER, the chair of the math department. In 1976, Bozeman decided to return to graduate work in mathematics. That year, she enrolled as a doctoral student at Emory University in Atlanta. Concentrating on functional analysis, a mix of algebra and a kind of geometry called topology, she earned her Ph.D. in 1980.

Bozeman quickly gained a tenured professorship after completing her doctorate, and by 1982, she was chairperson of Spelman's math department. Through the 1980s and 1990s, she developed programs such as the Spelman Summer Science and Mathematics Institute, which trains high school teachers in new techniques in math and especially targets teachers in minority schools. In 1993, she was named director of Spelman's Center for the Scientific Application of Mathematics, a multidisciplinary research organization whose goal is to increase the number of African Americans in research science positions.

In recognition of her efforts, Bozeman has received the White House Initiative Faculty Award for Excellence in Science and Technology (1988) and the Distinguished College and University Teaching Award from the Southeastern Section of the Mathematical Association of America (1995). As of 2000, Bozeman still teaches at Spelman College.

Brandegee, Mary Katharine Layne
(1844–1920)
American
Botanist

Trained as a physician, Katharine Brandegee is best known for her studies in botany in California and other western states. She was a pioneering botanist in this region and did extensive work cataloging and describing the taxonomy and beneficial effects of numerous medicinal and other plants found there.

Born in western Tennessee in 1844, Mary Katharine Layne was the daughter of Marshall Layne, a farmer, and Mary Morris Layne, a homemaker. Marshall Layne was a restless man and moved his family numerous times in a general westward trek across the continent. Like thousands of other American migrants, Layne was drawn to California by the fabulous gold strike of 1849. Yet he did not come as a miner; instead he decided to stick to farming. By the early 1850s, Layne had settled with his wife and 10 children on a farm near the town of Folsom, California, in the foothills of the Sierra Nevada east of Sacramento. Here Katharine Layne finished her secondary education. In 1866, when she was 22, Katharine Layne married an Irish immigrant named Hugh Curran. The marriage lasted eight years, until Curran's death in 1874.

Upon the death of her husband, Katharine Curran decided to seek an education in medicine. To achieve this, she moved to San Francisco in 1875 and began to study medicine at the newly established University of California's medical department. She earned her M.D. degree from this institution in 1878.

While still in medical school, Curran expanded her innate curiosity for plants. She became especially interested in medicinal plants and began an informal study of this category of plant life in California. Over the next five years, as she became increasingly acquainted with California's native medicinal and nonmedicinal plants, her interests shifted from the practice of medicine to the study of botany. She mentored herself to the preeminent authority on California botany, Dr. Hans Herman Behr. Curran gradually worked herself into a position at the botany section of the California Academy of Sciences. By 1883, she had become the academy's curator of botany, a position she held until 1893. When she resigned the curatorship, she made sure that the position went to her protégée, ALICE EASTWOOD.

In 1889, Katharine Curran married for the second time, on this occasion to Townshend Brandegee, a civil engineer and amateur plant collector. The couple were drawn to each other by their shared interest in science and plant life in general. Katharine would often accompany her husband on engineering jobs into the California countryside to collect plants. In the early 1890s, they won approval from the California Academy of Sciences to begin the publication of the academy's bulletins, which announced news about the academy and scientific discoveries that occurred in that state. Over the next 10 years, Katharine Brandegee served as the editor of the bulletins. She and Townshend Brandegee also founded another publication on their own, *Zoe*, a journal that reported about botanical discoveries in the West.

Katharine Brandegee was instrumental in figuring out the ecological zone boundaries of different species of plants. Her early work systematized what had been a rather hodgepodge understanding of California and western plant species. She died in Berkeley, California, at the age of 76, in 1920.

Braun, Emma Lucy
(1889–1971)
American
Botanist and Ecologist

Trained as a geologist and botanist, Lucy Braun made a name for herself as a plant ecologist. She used her knowledge of geology and botany to classify habitat locations and catalog what types of plants lived in certain geologic formations. In her mid-career, she made a careful study of the taxonomy of vascular plants of the Ohio and Kentucky region. Her later studies of forest makeup resulted in an authoritative book about deciduous forests in the eastern part of the United States.

Born in Cincinnati on April 19, 1889, Lucy Braun was the daughter of George Frederick Braun, a school principal, and Emma Moriah Wright Braun, a teacher in the same school that her husband headed. Braun and her older sister, Annette, both attended their parents' school, and both developed an early interest in nature while accompanying their parents on trips to the woods to collect and identify trees and plants. Lucy Braun began a personal collection of plants that she dried and cataloged while still in high school. At the end of her life, this collection numbered more than 11 thousand plants, and at her death it was presented to the herbarium of the Smithsonian Institution where still it remains. Braun never married.

Braun earned a liberal arts degree in 1910 from the University of Cincinnati. She remained at that institution for her M.A. in geology, which she completed in 1912. During the summer of 1912, she studied plant ecology at the University of Chicago with Henry C. Cowles, one of the foremost experts in that field in the United States. She received her Ph.D. in botany from the University of Cincinnati in 1914.

Braun spent her entire 61-year teaching and research career at the University of Cincinnati. She began in 1910 as an assistant in geology and advanced to instructor in botany (1917), assistant professor of botany (1923), associate professor (1927), professor (1946), then professor emeritus (1948–73). Her period as professor emeritus was almost exclusively devoted to research.

Braun's first studies during the late teens and 1920s, which included her dissertation, were about physiographic ecology, especially the kinds of vegetation that grew on conglomerate rock in the Cincinnati area and an unglaciated dolomite area in nearby Adams County, Ohio. She also studied the migration of plant colonies during the North American glaciers of the most recent Ice Age. She matched isolated contemporary plant ecologies with geologic records of glacial retreat to chart this plant migration. And she also was the first person to conduct a comparative study in which she matched the flora of the Cincinnati region to catalog studies made a hundred years earlier. This pioneering effort opened up the field of comparative ecology and demonstrated how studies over time could examine the changes in plant ecology that resulted from human factors such as population growth and industrial development.

During the 1930s, she began studying the composition of forests in the Illinoian Till Plain, which included southwestern Ohio. She later branched out to study forests in the Cumberland Plateau and the Appalachians of eastern Kentucky. She eventually expanded this forest study to include all of the forests of the eastern United States. These studies resulted in her massive book, *Deciduous Forests of Eastern North America*, published in 1950.

For her groundbreaking work in plant ecology, Braun was given the Mary Soper Pope Award (1952) by the Cranbrook Institute of Science. She won a Certificate of Merit (1956) from the Botanical Society of America. In 1950, she was the first woman elected president of the Ecological Society of America. She died of heart failure in 1971 at the age of 81.

Breckenridge, Mary
(1881–1965)
American
Nurse Midwife

Mary Breckenridge established the practice of nurse-midwifery in the United States with her founding of the Frontier Nursing Service (FNS) in eastern Kentucky in the late 1920s. She used the British combination of nursing and midwifery as her model but tailored her approach to rural Appalachia, where nurse-midwives could mobilize on horseback to reach even the most remote patients. In launching the Frontier School of Midwifery and Family Nursing, Breckenridge invoked the symbolism of the banyan tree as an analogy for her educational model: Just as the ancient tree would send down roots from its branches to extend itself, so too would the teaching of nurse-midwifery extend the area that could be covered with competent health and obstetrical care.

Breckenridge was born in 1881 into a prominent American family that had produced a U.S. vice president as well as a congressman and diplomat. However, such prominence could not shield her from tragedy, as her husband and two children died prematurely. Breckenridge transformed her grief into action as she dedicated herself to nursing as a means of preventing childhood sickness and death.

During World War I, Breckenridge served as a public health nurse, learning the value of mobility in caring for the wounded in wartime situations. Traveling throughout England and France, she observed European methods of maternity care and trained as a nurse-midwife. After the war, she imported the concept back to the United States. The same mobility that benefited battlefield care could be applied to rural situations, where much of the population lived great distances from the nearest hospital, so Breckenridge brought the medical care to them. She commenced her work in 1925 in southeastern Kentucky. By 1928, she had named her concept the Frontier Nursing Service.

In the beginning, Breckenridge funded the FNS personally, inviting British nurse-midwives to the United States to provide the initial expertise. With a central hospital as its hub, nursing stations fanned out across the countryside. Nurses on horseback extended the range of the FNS. Within the first five years, the FNS covered 700

Mary Breckenridge, who founded the Frontier Nursing Service in Kentucky in the late 1920s, establishing the practice of nurse-midwifery in the United States. *(National Library of Medicine, National Institutes of Health)*

square miles to serve more than 1,000 families. Dr. Louis Dublin conducted a study of the FNS's first 1,000 births; the lack of any maternal deaths due to pregnancy or labor attested to the competence of the nurse-midwives.

Steady demand and clear success promoted continuing growth for the FNS, which formed the core of what later became the American Association of Nurse-Midwives. Breckenridge raised the funds necessary to found the Frontier School of Midwifery and Family Nursing. In 1939, with the threat of World War II, the British nurse-midwives returned to England with the FNS concept in tow, establishing the Frontier Graduate School of Midwifery with just two students to begin with. Over the next half-century, the number of students graduated from FNS programs grew to more than 500.

Breckenridge died in 1965, able to pronounce her FNS initiative a success: "The glorious thing about it is that it has worked." The FNS hospital in Hyden, Kentucky, was renamed the Mary Breckenridge Hospital in honor of its founder. In further tribute, the United States Postal Service created a 77-cent postage stamp bearing the likeness of Mary Breckenridge. And in 1998, children's

author Rosemary Wells wrote a book about Breckenridge and the FNS entitled *Mary on Horseback: Three Mountain Stories,* with an afterword briefly recounting Breckenridge's biography and the birth of the FNS.

Brill, Yvonne Claeys

(1924–)
American
Aerospace Engineer

Propellants and propulsion systems were the focus of Yvonne Claeys Brill's accomplished career. She holds the patent for a hydrazine resistojet, a single propellant rocket system she developed in the 1970s that is still in use today. Her work has included stints with private industry and with government, including a position from 1981 to 1983 with the National Aeronautics and Space Administration's (NASA) space shuttle program.

Born on December 30, 1924, in Winnipeg, Canada, to parents who discouraged her interest in science, Brill succeeded despite the lack of family support in obtaining her B.S. degree in mathematics at the University of Manitoba in 1945. In order to find a job in her field, she moved to Santa Monica, California, where she went to work as a mathematician for Douglas Aircraft Company. There, she studied aircraft propeller noise but found the work less than challenging. With hopes of finding more interesting work, she enrolled in the graduate program in chemistry at the University of Southern California, attending graduate classes at night while holding down her job during the day. Following World War II, she transferred to the aerodynamics department at Douglas but soon accepted a position as a research analyst with the RAND think tank in Santa Monica. At RAND, one of Brill's mentors helped her to make her big breakthrough into the propellant department, where she studied rocket and missile designs and propellant formulas while continuing her graduate studies at night.

Brill earned her M.S. degree in chemistry in 1951, the same year she met and married her husband, a research chemist. In 1952, the couple moved to Connecticut, where Brill accepted a position as a staff engineer at the United Aircraft Research Laboratory in East Hartford. Her work there focused on rocket and ramjet engines. She changed jobs in 1955 to work at the Wright Aeronautical Division of Curtiss-Wright Corporation, where she developed high-energy fuels and studied state-of the-art turbojet and turbofan engines that were adapted for advanced aircraft. Between 1957 and 1966, Brill had three children and served as a part-time consultant on rocket propellants to FMC Corporation in Princeton.

When she resumed full-time work in 1966, Brill pursued the most rewarding research of her career at RCA Astro-Electronics (now GE Astro). It was here that she developed the hydrazine resistojet thruster, which represented a monumental advance in the field of single-propellant rockets. She was appointed manager of NOVA Propulsion in 1978 and then went on to perform early work on the Mars Observer Spacecraft, which was launched in 1992.

Brill left RCA to serve for two years as director of the Solid Rocket Motor Fuel program in NASA's shuttle program. After one more brief stint back at RCA, she worked in London from 1986 to 1991 as space segment engineer with INMARSAT before retiring. Since then, she has served as a consultant for Telespace, Ltd., in Skillman, New Jersey.

The author of 40 publications, Brill's illustrious career has brought her many honors, including the 1993 Resnik Challenger Medal, the 1986 Society of Women Engineers (SWE) Achievement Award, and the Diamond Super Woman Award from *Harper's Bazaar* and DeBeers Corporation, among others. She is a fellow of the American Institute of Aeronautics and Astronautics and of the SWE and a member of the National Academy of Engineering and the British Interplanetary Society.

Britton, Elizabeth Gertrude Knight
(1858–1934)
American
Botanist

Considered in her day one of the leading experts in her field of bryology (the study of mosses) in the United States, Elizabeth Britton wrote more than 300 papers about ferns, mosses, flowering plants, and wildflower preservation. She became a specialist in bryology, taught botany in an unofficial capacity at Columbia College, and capped her career by playing a central role in the founding of the New York Botanical Garden and the Wild Flower Preservation Society of America.

Born on January 9, 1858, in New York City, Elizabeth Knight was the daughter of James Knight, a wealthy manufacturer and plantation owner, and Sophie Ann Compton Knight, a homemaker. Because her father's family had considerable business interests in Cuba, Knight lived on that Caribbean island until she was 11. Her first botanical interests were developed in Cuba where she studied the island's animals and tropical plants. Knight was sent back to New York for her secondary education in 1869. She attended a private school in Manhattan. In 1885, at age 27, she married Nathaniel Lord Britton, a professor of geology at Columbia College. She had no children.

After her graduation from secondary school, Knight attended New York's Normal College, an institution whose students were primarily women and that trained elementary and secondary school teachers. When she graduated from Normal College in 1875 at age 17, she was immediately given a position there as a teacher; she was the Normal School's botany teacher from 1883 to 1885.

Even as she worked as a teacher at Normal College, the study of botany consumed all of Knight's spare time in the late 1870s. In a summer field trip to Newfoundland in 1879, Knight found a rare grass fern, *Schizaea pusilla pursh.*, growing on a lake shore. This discovery was important within the world of North American botany because it confirmed a finding of the same grass made 60 years previously. News of Knight's work was published by none other than America's leading botanist, Asa Gray. As a result of this and other fieldwork, Knight was elected to the Torrey Botanical Club of New York City in 1879, a group of professional and amateur botanists who maintained their own herbarium and met regularly to exchange information. It was through the Torrey Botanical Club that Knight met Nathaniel Britton, whom she married in 1885.

From the 1880s through the 1920s, Elizabeth Britton devoted a considerable part of her professional life to the collection and study of mosses. She worked hard to revise moss genera so that classification of these fit into new findings that came to her through others and through her own research. She also sorted through the collection of ferns and mosses gathered by Henry Hurd Rusby, a botanist who had been employed by the Parke Davis drug company to gather herbal plants in South America. To complete this work, she and her husband traveled to London in 1888 to consult the holdings of the Linnaean Society. There, because she was a woman, she was banned from working on the main floor; instead she completed her studies upstairs.

In 1885, Britton became unofficial curator of the Columbia College herbarium. By the 1890s, she unofficially supervised doctoral students at Columbia who were working in her field of mosses and ferns. She never received pay or an official appointment for these efforts. In 1891, after a trip to the British Botanic Gardens at Kew, Britton spearheaded a committee that succeeded in establishing the New York Botanical Gardens (NYBG). She worked as director of the NYBG's gardens for 33 years. In 1902, Britton organized the founding of the Wild Flower Preservation Society of America. She became secretary-treasurer of that organization.

In recognition of her intensive and varied work in the field of botany, Britton became one of 25 founding members of the Botanical Society of America in 1893. She was given the high honor of being appointed to an international committee to determine moss nomenclature by the Botanical Congress in 1905. In 1906, her name was included with a star in the first edition of *American Men of Science*. The "starred" scientists were those whom the editors considered the top 1,000 scientists in the United States. Elizabeth Britton died of a stroke at age 76 on February 25, 1934, in New York City.

Brooks, Harriet
(1876–1933)
Canadian
Physicist

An early researcher in the field of atomic physics, Harriet Brooks had to fight for her right to practice her profession at a time when women were generally not welcomed on university faculties. Despite such opposition, Brooks distinguished herself in a short career by playing a pivotal role in the discovery of a new element, the gas radon.

The second child of George Brooks, a traveling salesman for a flower company, and Elizabeth Brooks, a homemaker, Harriet Brooks was born in Exeter, Ontario, on July 2, 1876. Brook's elementary and secondary education was punctuated by the frequent moves the family made as a result of her father's sales job. Nonetheless, Brooks stood out as a student. She won a scholarship to attend McGill University, one of the leading institutions of higher learning in Canada. She studied mathematics, physics, and languages at McGill and graduated with honors from there in 1898. In 1907, Brooks married Frank Pitcher, a physicist. She had two children.

Immediately after her graduation, Brooks began teaching physics part-time at the Royal Victoria College, a recently founded women's college affiliated with McGill University. She also started work at the Macdonald Laboratory, McGill's atomic physics workshop. Fortunately for Brooks, the lab had just acquired a new director in Ernest Rutherford the year she graduated. Rutherford was a New Zealand-born British physicist who had no problems working with women scientists like Brooks. At the Macdonald Lab, Brooks joined with Rutherford and other scientists in their famous experiments to examine and classify the types of rays being emitted by radium, which had just been discovered by Pierre and MARIE CURIE in Paris. The Rutherford team was able to determine that radium emitted several different kinds of radiation, which were named alpha and beta rays. While working in the Macdonald Lab, Brooks also earned her M.A. in physics from McGill in 1901, becoming the first woman to win a master's degree at that institution.

Brooks's specialty during her time at the Macdonald Lab was the study of radon, one of the unusual "emanations" that resulted from the radioactive breakdown of radium. Through her experiments, Brooks was able to determine that radon was not an isotope, or slightly different atomic variant, of radium but was instead a separate element. Her master's thesis, and a paper published in 1901, were significant because they showed for the first time that one element can change into another through the process of radioactive decay. In her investigations of the decay of radium into radon, Brooks stumbled on an atomic phenomenon that she labeled "recoil," which occurs in the emission of particles from radioactive atoms. Through these recoil studies, Brooks explored the process of radioactive decay of radium and the element actinium, a silvery, metallic radioactive element discovered by André-Louis Debierne at Madame Curie's laboratory in Paris in 1899.

After completing her master's degree in Canada, Brooks spent a year's study at the famous Cavendish Lab at Cambridge University in England. At Cavendish, Brooks fell in love with a physicist from Columbia University and followed him back to New York City in 1905. For a while, she took a teaching job at Barnard, Columbia's women's college. However, when she announced her engagement, she was fired from Barnard's faculty because of the policy then in place at most American universities forbidding a husband and wife from working at the same institution. Brooks vehemently protested, saying, "It is a duty I owe to my profession and my sex to show that a woman has a right to practice her profession and cannot be condemned to abandon it merely because she marries." Probably as a result of this controversy, Brooks broke off her engagement. She traveled to Europe and studied for a time at Marie Curie's Radium Institute. In 1907, after marrying a fellow physicist, she returned to Canada and quit her career to raise her children.

Because of her truncated career, which she abandoned at age 31, Brooks did not garner the awards and recognition she deserved. However, she was remembered by her mentor, Nobel Prize winning physicist Ernest Rutherford, as "next the Mme. Curie,. . . the most prominent woman physicist in the [study] of radioactivity." She died at the relatively young age of 56 on April 17, 1933.

Brown, Rachel Fuller
(1898–1980)
American
Biochemist

Rachel Fuller Brown, working in Albany, New York, collaborated long-distance with ELIZABETH HAZEN, who worked in New York City, to discover the antifungal antibiotic that they christened *nystatin* after the New York State Department of Health, which employed them both. Instead of pocketing the proceeds from this discovery, Brown and Hazen established a nonprofit corporation to distribute the profits from nystatin sales to scientists conducting research.

Brown was born on November 23, 1898, in Springfield, Massachusetts. Her family moved to Webster Groves, Missouri, where her father, real estate and insurance agent George Hamilton Brown, left her mother, Annie Fuller. Fuller returned the family to Springfield, where she raised the family alone from 1912 on. The generosity of a wealthy family friend financed Brown through Mount Holyoke College in western Massachusetts, near Springfield.

There, Brown double majored in history and chemistry for her 1920 A.B. She then did graduate study at the

University of Chicago for her master's degree in organic chemistry. She worked, teaching chemistry and physics at the Francis Shimer School, to earn money for doctoral work on organic chemistry and bacteriology. She submitted her thesis in 1926 but did not receive her Ph.D. until seven years later when she finally got the chance to take her oral examination, which she readily passed.

In the meanwhile, she had taken on a job at the New York State Department of Health, where she remained for 42 years. Her first major accomplishment there was to help develop a pneumonia vaccine that continues to be used. In 1948, she and Hazen commenced their work searching for a fungal antibiotic by testing soil samples. A sample taken at a friend of Hazen's dairy farm in Virginia yielded a specific actinomycete microorganism, later named *Streptomyces norsei,* that produced two antifungal elements—one too toxic for humans, but the other quite safe, as it turned out.

Brown prepared an antibiotic of small white crystals from the second antifungal substance that proved effective against fatal fungal attacks on the lungs and central nervous system as well as against more common attacks, such as vaginal yeast infections and athlete's foot. Brown and Hazen presented their results at the 1950 meeting of the National Academy of Sciences, and they established a non-profit organization, called the Research Corporation, through which they applied for a patent. After Food and Drug Administration approval, while the patent was pending, E. R. Squibb and Sons secured the license to develop and market a commercial version of the antibiotic that it called *Mycostatin,* first available in 1954. Patent number 2,797,183 officially registered Brown and Hazen's antifungal antibiotic on June 25, 1957.

The Research Corporation earned more than $13 million in royalties until the patent expired; Brown and Hazen donated all of those funds to support scientific research. Interestingly, nystatin proved effective against fungal attacks not only on humans but also on trees, treating Dutch elm disease, and on artwork, restoring paintings damaged by water and mold. The pair also collaborated to discover two other antibiotics, phalamycin and capacidin. Brown died on January 14, 1980. Fourteen years later, the National Inventors Hall of Fame inducted both Brown and Hazen into its legions in recognition of their significant contribution to medical science with the development of nystatin.

Browne, Barbara Moulton
(1915–1997)
American
Bacteriologist

Barbara Moulton Browne is best remembered for her 1960 testimony to a Senate subcommittee regarding her concerns about the manner in which the federal Food and Drug Administration (FDA) approved or rejected new drugs. Shortly before the hearing, Browne had resigned her post at the FDA in order to voice her concerns about the cozy relationships between drug manufacturers and many FDA employees responsible for overseeing the drug evaluation process. Due in large part to her comments, Congress enacted a law placing tighter controls on the FDA's procedures for approving new drugs.

Born on August 26, 1915, in Chicago, Illinois, Barbara Moulton Browne was the younger of her parents' two children. Harold Moulton, her father, was a professor of economics at the University of Chicago. The family later moved to Washington, D.C., when Harold was named president of the Brookings Institution.

Browne had a rather peripatetic undergraduate career, taking classes at both Smith College and the University of Vienna before eventually obtaining her bachelor's degree from the University of Chicago in 1937. After graduating, Browne spent two more years at the University of Chicago, exploring the subject of bacteriology in general and infectious diseases in particular. Recapitulating her father's geographical pattern, she then enrolled at George Washington University in Washington, D.C., from which she earned a master's degree in 1940 and a medical degree in 1944. She completed her surgical residency at Chicago's St. Luke's Hospital and at Suburban Hospital in Bethesda, Maryland, between 1945 and 1947.

At the end of her residency, Browne again returned to George Washington University when she took a position as an anatomy instructor. After a year, she went into general practice, simultaneously serving at the student health service of Washington State College. Browne left Washington, D.C., again in 1950 when she went back to Illinois to become Illinois State Normal University's assistant director of student health. She subsequently transferred to Chicago's Municipal Contagious Diseases Hospital, where she was named assistant medical director. She also acted as a medical instructor at the University of Illinois in 1953.

Browne joined the FDA in 1955 as a medical officer. At the time, the FDA was a subdivision of the Department of Health, Education, and Welfare. Browne's duties included examining the veracity of claims made by drug manufacturers on behalf of their new products. But she quickly became frustrated by the tremendous influence those drug companies routinely wielded over the process as well as by the FDA's own internal procedures that tipped the scales heavily in favor of approval. (A medical officer was allowed to approve a drug without obtaining the permission of her superiors, but rejecting a drug required the blessing of at least three and sometimes as many as five higher-ups, including the FDA commissioner.) Unwilling to accede to this one-sided scheme that favored drug company profits over consumer safety, Browne resigned in 1960. Later that year, she appeared as a key witness before the Kefauver

Senate Subcommittee on Monopoly and Antitrust, which was probing the conduct of the drug industry and of the FDA's antibiotics division chief (who was suspected to have taken over a million dollars in drug company money over a seven-year period). Thanks in no small part to Browne's testimony, Congress revamped the drug approval process, reducing the power drug companies could exercise over it.

In 1961, Browne found employment with the Bureau of Deceptive Practices at the Federal Trade Commission (FTC). Working as a medical officer with the Division of Scientific Opinions, Browne continued to call attention to practices that she believed posed the risk of fraud or deception to consumers. In 1962, she married E. Wayles Browne Jr., who was an economist on the Kefauver Subcommittee when Browne testified before it.

Browne remained at the FTC for the rest of her career. She died on May 12, 1997, of Alzheimer's disease. Her efforts were widely recognized during her lifetime. A member of many scientific and professional organizations, including the American Public Health Association, the American Society of Microbiology, the American Society of Hematology, and the American Medical Women's Association, Browne was also named a fellow of the American Association for the Advancement of Science in 1963. In addition, she received the Federal Woman's Award in 1967. Her impact can still be felt in the more stringent rules controlling government approval of new medicines and in the myriad consumer protection groups that now exist.

Browne, Marjorie Lee
(1914–1979)
American
Mathematician

Marjorie Lee Browne and EVELYN BOYD GRANVILLE were the first two African-American women to receive Ph.D. degrees in mathematics in the United States. Browne then devoted her life to teaching: Not only did she become a college professor, but she also ran programs to teach secondary school teachers as a means of strengthening the whole scope of mathematical education. She also supported the entrance of women and minorities into the field of mathematics.

Browne was born on September 9, 1914, in Memphis, Tennessee. Her mother died when she was two years old, and Brown was raised by her stepmother, Lottie Taylor Lee. Her father, Lawrence Johnson Lee, was a transportation mail clerk who had attended some college—an anomaly for an African American in the early 20th century. Browne attended LeMoyne High School, a private school for African Americans established by the Methodists after the Civil War, and she graduated in 1931. She then matriculated at Howard University, supporting herself through scholarships, loans, and jobs. She graduated cum laude with a B.S.

degree in mathematics in 1935, after which she taught mathematics and physics for a year at the Gilbert Academy in New Orleans.

Browne proceeded to the University of Michigan, where she earned her M.S. degree in 1939. From 1942 through 1945, she worked as an instructor at Wiley College in Marshall, Texas. She then returned to the University of Michigan as a teaching fellow, earning her Ph.D. in mathematics in 1949. That year, Granville also earned her doctorate from Yale University, making these two the first African-American doctors of philosophy in mathematics in the United States. Browne wrote her dissertation on one-parameter subgroups in certain topological and matrix groups, and she edited her dissertation into a paper published in *the American Mathematical Monthly* in 1955 entitled "A Note on the Classical Groups."

Although Browne applied to many colleges and universities, most rejected her politely with racist undertones. North Carolina Central University (NCCU) accepted her, and she remained there throughout her career, rising to the status of professor and heading the mathematics department from 1951 until 1970. Browne considered herself a "pre-Sputnik mathematician," referring to the time when mathematicians and scientists conducted "pure" research, before the space race transformed the laboratory and the library into marketplaces. Browne continued to focus her research on topology, regardless of its commercial applicability.

Throughout her career, Browne received numerous fellowships and awards. From 1952 through 1953, she received a fellowship from the Ford Foundation to attend Cambridge University in England. From 1957 on, she served as the principle investigator and lecturer for the Summer Institute for Secondary School Science and Mathematics Teachers at NCCU, funded by the National Science Foundation (NSF). The NSF supported her throughout her career: In 1964 through 1965, she directed the first Undergraduate Research Participation Program at NCCU; she studied differential topology as an NSF fellow at Columbia University from 1965 through 1966; and she studied computing and numerical analysis at the University of California at Los Angeles as an NSF Faculty Fellow.

In 1960, Browne received a $60,000 grant from IBM to set up a digital computer center, one of the first at a minority university. In 1974, Brown was the first to receive the W. W. Rankin Memorial Award from the North Carolina Council of Teachers of Mathematics. Browne died of an apparent heart attack at home in Durham, North Carolina, on October 19, 1979. At the time, she was writing a monograph applying a postulational approach to the development of the real number system.

The year of her death, four of her former students established the Marjorie Lee Browne Trust Fund at NCCU to support the Marjorie Lee Browne Memorial Scholarship and the Marjorie Lee Browne Distinguished Alumni Lecture Series. In 1996, the National Association of Mathematics, a group devoted to the advancement of African Americans in the field of mathematics, renamed its lecture series as the Granville-Browne Session of Presentations by Recent Doctoral Recipients in the Mathematical Sciences to honor the first two African-American women doctors in mathematics.

Burbidge, Eleanor Margaret Peachey
(1919–   )
British
Astronomer

In a career that has taken her to universities and research institutes around the world, Margaret Burbidge has contributed important insights about the nature of stars and galaxies. Her earliest work examined how stars were formed. Later, she studied the formation and physics of galaxies and put together a detailed examination of quasars. She accomplished this while also teaching and acting as an administrator of centers of astronomical studies.

Margaret Peachey was born on August 12, 1919, in Manchester, England. Her father was a chemistry teacher at a university in Manchester, and her mother also held a degree in chemistry. Her family moved to London in the early 1920s when her father established a laboratory there. While a doctoral student during World War II, she married Geoffrey Burbidge, a fellow astrophysicist. She has one daughter.

Burbidge entered the University of London as a science student and graduated with honors from there in 1939. She immediately began to work on a doctorate in astrophysics from the University of London and attained her Ph.D. in 1943.

From 1946 to 1951, Burbidge served as acting director of the University of London Observatory. During the early 1950s she and her husband traveled to the United States to work, first at the University of Chicago's Yerkes Observa-

E. Margaret Burbidge, whose work as an astronomer has contributed numerous insights into the nature of stars and galaxies.   *(AIP Emilio Segrè Visual Archives)*

tory, then at the observatory at Harvard University. By 1953, Burbidge was back in England where she worked with her husband, the astronomer Fred Hoyle, and physicist William A. Fowler on the problem of stellar formation, the evolution of stars from nebulous gases to fully formed then collapsed objects. Burbidge and her team looked at a number of different stars and used a spectroscope to break down these stars' light into separate, discrete bands. This spectroscopic information gave the team an idea of the chemical composition of the stars. When Burbidge and the others examined the stars they were studying, they saw that some were richer in light elements such as hydrogen and helium, while others contained a higher percentage of heavier elements such as iron. Because they knew that suns burn energy through nuclear fusion of elements, mainly hydrogen and helium, they theorized that the stars with the lightest overall atomic composition were younger stars, while the heaviest were older. In Margaret Burbidge's words, this information about nucleosynthesis, as the solar process of going from lighter to heavier elements was called, was "very relevant [in deciphering] the formation and early history of the solar system and to the question of whether planetary systems about stars are common and rare."

In the late 1950s, the Burbidges returned to the United States. While her husband worked at the Mount Wilson Observatory outside Los Angeles, Margaret Burbidge studied galaxies at the California Institute of Technology. She was especially interested in how, like solar systems, galaxies rotate around an axis. The rotation of a galaxy counterbalances its gravity, which without the rotation would collapse in on itself. Burbidge also calculated the weight and mass of the galaxies she studied.

From 1962 to 1990, Burbidge worked at the University of California at San Diego (UCSD), first, because of nepotism rules that prevented her and her husband from holding positions in the same department, as a professor of chemistry, then, once this rule was abolished, as a full professor of physics. She and her husband coauthored a book, *Quasi-Stellar Objects* (1967), a study of quasars, that is, starlike objects that "vary in light and radio flux" over time. From 1979 to 1988, she directed UCSD's Center for Astrophysics and Space Sciences. She also briefly served as director of England's Royal Greenwich Astronomy (1972–73) before returning to San Diego for good.

For her impressive scientific achievements, Burbidge was awarded the American Astronomical Society's Helen B. Warner Prize in 1959, and she was elected a fellow of the Royal Society in 1964, the American Academy of Sciences in 1968, and the U.S. National Academy of Sciences in 1978. In 1985, she was presented the U.S. National Science Medal, and she was given the Albert Einstein World Science Medal in 1988. As a sign of the increased acceptance of women in the sciences, Burbidge turned down the 1971 ANNIE JUMP CANNON Prize, which is given annually by the American Astronomical Society to women only. "It is high time," Burbidge said, "that discrimination in favor of, as well as against, women in professional life be removed."

Burton, Leone
(1936–)
Australian
Mathematician and Educator

Trained as a mathematician and educator, Leone Burton has become an authority on the teaching of mathematics and the study of how particular societies at different times choose to emphasize certain approaches in mathematics over others. Burton has also written about why women have been underrepresented in mathematics, and she has spoken on math and math education at numerous international education conferences.

Born in Sydney, Australia, in 1936, Burton is the daughter of Scottish parents who had immigrated to Australia in the mid-1930s. At least one of her parents was Jewish, and the family's move was prompted by the rise of Nazism in Germany and official anti-Semitism throughout much of Europe in the 1930s.

In Sydney, Burton attended girls' schools for her primary and secondary education. She did not take immediately to math and, in fact, was not tremendously motivated by school. After graduation from high school in 1954, she attended an art school in Australia for a few years but dropped out when she fell ill. After recovering, Burton worked for a year to save money to travel.

Burton traveled throughout the United States, making her way from California to New York. In New York, she left for England where she settled in with an uncle and aunt in London. Her first real exposure to mathematics came from the uncle, Hyman Levy, her mother's brother, in whose house she was staying. Levy was a retired mathematician and left-leaning political activist. During her stay with the Levys, Burton became much more acquainted with mathematical thought and leftist politics.

In 1959, she decided to return to university and enrolled as an undergraduate in the University of London. Here she started out as a history major but gradually shifted her interest to math and philosophy. Knowing that these two subjects had been historically closely related, Burton immersed herself in readings about the two disciplines, including the famous work by Bertram Russell and Alfred Whitehead, *Principia Mathematica*. She also studied the social ground out of which Western mathematics had sprung. In 1963, she was awarded a B.A. in mathematics from the University of London.

After completing her undergraduate degree, Burton worked for several years as a teacher—first in a high school, then at the primary level. She was disheartened to find that the way math was taught at the secondary level emphasized rote learning rather than an exploration of ideas. This disillusionment prompted her to begin studying the educational system itself.

In 1968, she earned a B.A. in education from the University of London and began teaching education at Battersea College. Here she taught people who wanted to become mathematics teachers and took part in reforming the educational establishment from within. She also began teaching mathematics education at a number of English universities.

By 1980, Burton had earned her Ph.D. in education at the University of London. In books, magazine pieces, and journal articles, she has explored such topics as ethno-mathematics, the ways mathematics are expressed in particular cultures. To do this, she has examined different number systems invented by various cultures as well as the concepts of zero and infinity, and more advanced mathematical systems.

In recognition of her work, Burton was invited to serve on the steering committee of the International Organization of Women and Mathematics Education from 1984 to 1988. She has also spoken at many other international education conferences. Currently, she is a professor of education in mathematics and science at the University of Birmingham in England.

-C-

Caldicott, Helen
(1938–   )
Australian
Physician

A pediatrician who has worked extensively with children who have cystic fibrosis, Helen Caldicott is best known as an antinuclear activist. She has used her knowledge of the medical effects of radiation and her impassioned rhetoric as an activist to inspire people around the world to lobby against nuclear armament and nuclear energy. In the early 1970s, she organized opposition to the nuclear tests France was conducting in the South Pacific, which were in violation of the International Atmospheric Test Ban Treaty of 1962. Her speeches publicizing the violation and educating the public about the effects of radiation, particularly on children, resulted in the French government ending these tests. She also was responsible for reviving the U.S.-based organization Physicians for Social Responsibility and leading it during a period of rapid expansion in the early 1980s.

The daughter of Philip Broinowski, a factory manager, and Mary Mona Enyd Coffey Broinowski, an interior designer, Caldicott was born in Melbourne, Australia, on August 7, 1938. She attended public schools with the exception of four years spent at Fintona Girls School, a private secondary school in Adelaide. As an adolescent, she read Nevil Shute's novel, *On The Beach,* and was significantly affected by its depiction of nuclear holocaust. She entered the University of Adelaide medical school when she was just 17, graduating in 1961 with a B.S. in surgery and an M.B. in medicine (the equivalent of an American M.D.). The following year, she married William Caldicott, a pediatric radiologist, and the couple had three children—Philip, Penny, and William Jr.

After completing a three-year fellowship in nutrition at Harvard Medical School in Boston, Caldicott returned to Adelaide and accepted a position in the renal unit of Queen Elizabeth Hospital. There, she completed her residency and a two-year internship in pediatrics. She also established a clinic for the treatment of cystic fibrosis.

It was after working with children in the cystic fibrosis clinic and having children of her own that Caldicott began to organize others and speak out against nuclear proliferation. Her first challenge concerned France's illegal testing over Mururoa, a French colony in the South Pacific. Caldicott learned that in 1972, following five years of testing by France, there were higher than normal radiation levels in drinking water and rain in Australia. She set about to educate the public on the effects of radiation. Her speeches in opposition to the testing inspired a mass popular movement that resulted in the Australian government taking legal action against France through the International Court of Justice and in France putting an end to its atmospheric testing.

Caldicott also led a struggle against the commercial uranium industry. In 1975, the Australian Council of Trade Unions passed a resolution banning the mining, transport, and sale of uranium, and the government implemented an export ban. But these measures prevailed only until 1982, when international pressure forced the ban to be lifted.

In the years following 1975, Caldicott and her family spent an increasing amount of time in the United States. She held appointments at the Children's Hospital Medical Center in Boston and became an instructor in pediatrics at Harvard Medical School. In 1978, she became involved with Physicians for Social Responsibility, a group whose membership grew rapidly following the near meltdown of Pennsylvania's Three Mile Island nuclear reactor on March 28, 1979. In 1980, Caldicott stopped practicing medicine in order to devote all her time to leading the organization. Her work involved lots of travel and public speaking to raise awareness among the general population. The organization made a documentary film, *Eight Minutes to Midnight,* which was often part of Caldicott's presentation and

Pediatrician Helen Caldicott, who is best known as an antinuclear activist and leader of Physicians for Social Responsibility. *(Photo by David Young, Carolyn Johns/All One Voice, Courtesy of W. W. Norton & Company)*

was nominated for an Academy Award in 1982. She also wrote *Nuclear Madness: What You Can Do!* with coauthors Nancy Herrington and Nahum Stiskin.

Eventually Physicians for Social Responsibility lobbied for a more mainstream platform than what Caldicott espoused, and she resigned as president in 1983. She went on to help found the Medical Campaign Against Nuclear War, the Women's Party for Survival, and the Women's Action for Nuclear Disarmament, among other organizations. She has written two other books: *Missile Envy: The Arms Race and Nuclear War,* which came out in 1984, and *If You Love This Planet: A Plan to Heal the Earth,* which was published in 1992. She also ran for Parliament in Australia in 1990, losing by a very small margin.

Her many awards and honors include the Humanist of the Year Award from the American Association of Humanistic Psychology in 1982, the International Year of Peace Award from the Australian government in 1986, and a nomination for the Nobel Peace Prize in 1985. In the late 1990s, she was living in Canberra, Australia, with her husband.

Caldwell, Mary Letitia
(1890–1972)
American
Chemist

Affiliated with Columbia University for more than 40 years, Mary Caldwell devoted her career to research and teaching. Her research centered on the study of enzymes, especially a family of enzymes called the amylases, but her teaching was equally important to her. She was a mentor to numerous students, male and female, but because many of her students were women, she is known for her role in encouraging women to enter the field of chemistry.

Caldwell was born on December 18, 1890, in Bogotá, Colombia, while her parents, Milton Caldwell and Sarah Adams Caldwell, were living in that South American country. Milton Caldwell, an Episcopal minister, had been sent there as a missionary preacher. Mary Caldwell and her four siblings grew up in Colombia, probably schooled by their mother, until they reached high-school age. At that point, the family returned to the United States so that the children could receive their secondary schooling in their home country.

In 1909, Caldwell enrolled in Western College for Women in Oxford, Ohio. At this time, because of formal and informal bans on women as students, it was still difficult, and sometimes impossible, for women to attend many colleges and universities in the United States. Western College for Women offered a general liberal arts education, which included science courses such as chemistry and physics. Caldwell won an A.B. degree from Western College in 1913. For the next five years, she taught at her alma mater, first as an instructor, then an assistant professor in chemistry. In 1918, she decided to continue her education by enrolling in the master's program in chemistry at Columbia University in New York City. She received her M.S. degree from that institution in 1918. Columbia offered her a fellowship to help her get her doctorate, which she was awarded in 1921.

At Columbia, Caldwell developed a close professional relationship with Henry Sherman, who taught what was then called nutritional and biological chemistry but which today we know as biochemistry. During her master's and Ph.D. studies, she began to examine the way the amylase enzymes work in plants, animals, and the human body. In animals, the family of amylase enzymes are found mainly in saliva and the pancreas. Their function is to help the body digest food by breaking down carbohydrates, which are converted into the sugars glucose and maltose. Caldwell worked mainly with the pig pancreas, which produces relatively more amylase than other animals. However, she found it difficult to follow the chain of chemical reactions of amylase in her study because the other chemicals that were being used to track this process were not pure enough. Therefore, her first order of business was to refine the process of producing these support chemicals so that a true picture of the role of amylase in the pancreas and saliva could be seen. This she succeeded in doing during the 1920s and 1930s. Her techniques soon became standard practice for any biochemist engaged in the study of amylases.

From 1921 to 1959, Caldwell taught chemistry at Columbia, beginning as an instructor and working her way up to full professor, a position she achieved in 1948 (she was the only woman full or associate professor of chemistry at Columbia at that time). One of her main accomplishments was to mentor 18 women through their Ph.D. degrees at Columbia. For her efforts as a research scientist and educator, she was awarded the American Chemical Society's Garvan Medal in 1960. In 1961, Columbia University granted her an honorary doctorate of sciences in recognition of her dedication to that institution. She died on July 3, 1972, in Fishkill, New York.

Cambra, Jessie G.
(1919–   )
American
Engineer

Jessie G. Cambra, a pioneer in the development of California's transportation systems, forged a trail for women to follow in both her educational and professional pursuits. The first woman to graduate from the University of California at Berkeley School of Engineering, Cambra went on to be the first female engineer licensed by examination in California. She made major contributions to many public works projects during her 30-year career, but she is probably best known for designing and supervising the first successful highway reconstruction project in California and the first computerized traffic signal system at a major arterial intersection.

Born to Blanch Preneville, a bookkeeper, and Andrew Giambroni, a businessman and banker, on September 15, 1919, in Oakland, California, Cambra was one of five children. She has commented that having brothers instilled in her the belief that she could do anything they could do, including pursuing the male-dominated field of engineering. In 1942, she graduated from the University of California at Berkeley School of Engineering with a B.S. degree in civil engineering. One year later, on November 6, 1943, she married Manuel S. Cambra. The couple had two sons, the first born in 1947 and the second in 1957.

When Jessie Cambra entered the job market, engineers were in sharp demand since so many men were enlisted to fight in World War II. She was hired immediately as a field engineer by a San Francisco firm called Bechtel, McCone & Parons. Five months after starting, she was promoted to assistant civil engineer as a result of her outstanding work on a major Standard Oil Company refinery. In 1944, Cambra left this job to go to work for Alameda County, California. One year later, she was promoted to civil engineer and became the first licensed woman engineer in the state.

As an engineer with the county, Cambra worked on road design and drainage systems. She was chosen to join the Public Works Association in 1947 and later became its first female director. Cambra was promoted to senior civil engineer in 1951, which meant that she took charge of several road and bridge construction projects. In less than two years, she was promoted again to supervising civil engineer, and in 1953 she was appointed principal civil engineer in charge of the engineering division of the Alameda County Road Department. For the next 20 years, Cambra managed her staff through the transition from manual to computerized operation. She initiated numerous technological advances, including the development of a computer program that improved efficiency and increased the number of design options available to engineers in the department.

In 1960, Cambra was the first female to join the County Engineers Association of California. She served as a representative to the California transportation commission of the California State Legislature. When the deputy director of the Road Department where she worked became terminally ill, Cambra was appointed assistant chief to act in his place. In November 1974, she was appointed deputy director of public works and became head of the Alameda County Road Department, which had a budget of $12 million, 200 employees, and 547 miles of county road to manage. In addition to overseeing the planning and construction of boulevards, concrete-reinforced bridges, rehabilitation projects, and signalized intersections in Alameda County, Cambra started and administered the Federal Aid to Urban Highways Program.

In 1977, Cambra received the Samuel A. Greeley Award for her outstanding public service in the field of public works, and in 1978 she received recognition from the Hayward Boy's Club for her fund-raising efforts. She was also ranked by the American Public Works Association as a top-ten engineer and named as an Outstanding Alumna by Tau Beta Pi, her engineering fraternity at the University of California at Berkeley.

Cambra retired from the Alameda County job in 1980 and opened her own business as a consulting engineer. She worked on estimates and prepared qualifying plans for public works subcontractors and tract developments. In a career that included many firsts for women, Cambra certainly proved that women could be as successful as men in the field of engineering.

Canady, Alexa I.
(1950–   )
American
Neurosurgeon

On the first day of her neurosurgery residency at the University of Minnesota, a high-level administrator breezed past her and quipped, "Oh, you must be our new equal-opportunity package," Alexa Irene Canady has recalled. Despite obstacles like this, Canady succeeded in becoming the first African-American female neurosurgeon in the United States and has been honored along the way by many professional and academic organizations. Her career has included several teaching positions and an appointment as director of neurosurgery at Children's Hospital of Michigan in Detroit.

Born November 7, 1950, in Lansing, Michigan, to Elizabeth Hortense Golden Canady, an educational administrator, and Clinton Canady Jr., a dentist, Canady has three brothers: Clinton III, Alan, and Mark. She attended a secondary school where she was the only black girl and described the racism she encountered there in an interview with Brian Lanker, which was published in *I Dream A World: Portraits of Black Women Who Changed the World*. Canady stated, "During the second grade, I did so well on the California reading test that the teacher thought it was inappropriate for me to have done that well. She lied about what scores were mine, and ultimately, she was fired." Despite such experiences, Canady's academic prowess was unscathed, and she was recognized as a National Achievement Scholar by the time she reached high school in the 1960s.

At the University of Michigan, she started out studying mathematics but switched her emphasis after attending a minority health careers program. She graduated with a B.S. in 1971 and an M.D. in 1975. While in school, she received the American Medical Women's Association citation and was elected to Alpha Omega Alpha, an honorary medical society. She completed an internship at New Haven Hospital in Connecticut in 1975–76 before moving to the University of Minnesota for her neurosurgery residency, which she completed in 1981. Immediately following her residency, Canady was awarded a fellowship in pediatric neurosurgery at Children's Hospital in Philadelphia, where she also taught neurosurgery at the University of Pennsylvania College of Medicine.

In 1982, Canady returned to Michigan and took a job in neurosurgery at Henry Ford Hospital in Detroit before transferring the following year to pediatric neurosurgery at Children's Hospital of Michigan. In 1986, she was appointed assistant director of neurosurgery, and in 1987, she became the director. In addition to treating patients and serving in these administrative positions, Canady has also taught at Children's Hospital of Michigan and was named Teacher of the Year there in 1984. In 1985, she took a

position as a clinical instructor at Wayne State University School of Medicine and then in 1987, she accepted a clinical associate professorship there. In June 1988, she married George Davis.

In 1986, Canady was named Woman of the Year by the Detroit chapter of the National Association of Negro Business and Professional Women's Club. That same year, she was the recipient of the Candace Award, given by the National Coalition of 100 Black Women. Canady has continued to teach neurosurgery and to treat patients while also mentoring minority students pursuing careers in medicine.

Cannon, Annie Jump
(1863–1941)
American
Astronomer

For 44 years, Annie Jump Cannon devoted herself to the study and classification of stars. Her almost single-handed compilation of star positions and type into a massive star catalog (she personally accounted for the study of 350,000 stars) has not been, and probably never will be, equaled by any individual. To any outsider this could seem to be tedious work, but for Cannon it was a joy. "Each new [star] spectrum," she said, "is the gateway to a wonderful world."

The oldest of three children, Cannon was born in Delaware on December 11, 1863. Her family was large (seven, including her four half-brothers and -sisters) and, because her father, Wilson, owned a shipyard, financially secure. Wilson Cannon was a politically progressive man. He had cast the deciding vote in the state senate that kept Delaware in the Union at the beginning of the Civil War, and he believed that his daughters deserved a good education. Mary Cannon, Annie's mother, was an amateur astronomer and taught her daughter to identify stars and constellations from an observatory she set up in the attic of their house.

In Delaware, Cannon studied at the Wilmington Conference Academy. In 1880, when she was 16, she entered Wellesley, a private university for women that had been founded only five years before. At Wellesley, under the influence of SARAH WHITING, a professor of physics, she continued her studies in physics and astronomy in a much more rigorous way. It was during this time that Cannon learned about spectroscopy, a technique for determining a star's physical and directional properties by studying its light after it had been refracted through a glass prism.

Cannon received a B.S from Wellesley in 1884 and returned to Wilmington to be with her family. An outgoing and vivacious personality, Cannon remained in Wilmington for nine years. She engaged in the activities typical of a woman in her twenties at that time, yet, perhaps because of partial deafness that had resulted from an infection during her days

Annie Jump Cannon, whose work formed the core of *The Henry Draper Catalogue,* which was published between 1918 and 1949 and is still a primary reference for astronomers. *(Harvard College Observatory)*

magnifying glasses at lines of refracted starlight. In fact, Edward C. Pickering, the director of the Harvard Observatory, specifically chose women for this job because he believed that they had more patience than men for this task. After dismissing a male assistant from the job of star classification, Pickering stated that his maid could do a better job than the young man. To prove his point, Pickering hired the maid, WILLIAMINA FLEMING, who later would be a colleague of Annie Cannon's and whom Cannon would succeed in 1911 as the observatory's curator of photographs.

Cannon's work resulted in the multivolume star directory, *The Henry Draper Catalogue,* published between 1918 and 1949, one of the most comprehensive inventories of star positions and spectra ever assembled. She also published under her own name numerous short papers about her work and was especially interested in variable stars, suns whose light intensity varies with time.

For her efforts, Cannon was made William Cranch Bond Astronomer at Harvard in 1938, which was one of the first times that a woman had been appointed to an honorary academic of scientific position at that university. She also won honorary degrees from the University of Groningen in Germany in 1921 and Oxford University in 1925. She became an honorary member of the Royal Astronomical Society in 1914 and won numerous awards, including the Nova Medal of the American Association of Variable Star Observers (1922) and the Draper Medal of the U.S. National Academy of Sciences (1931). Annie Jump Cannon died of heart failure at 77 in 1941.

Carothers, E. (Estella) Eleanor
(1882–1957)
American
Zoologist and Geneticist

In a peripatetic career that took her from her home state of Kansas to Pennsylvania to Iowa and back to Kansas, Eleanor Carothers focused her studies on the cellular development and genetic makeup of grasshoppers. She also taught zoology at the University of Pennsylvania before devoting more than 36 years as a laboratory researcher.

E. Eleanor Carothers was born in Newton, Kansas, a small prairie town near Wichita, on December 4, 1882. There is no information about her family other than the names of her parents, Z. W. Carothers and Mary Bates Carothers. She apparently never married nor had children.

Carothers began her studies at Nickerson Normal College but transferred to the University of Kansas where she completed her undergraduate degree, a liberal arts course of studies in which she would have had the standard science courses such as chemistry and biology. She graduated from the University of Kansas in 1911 and the following year earned a master's degree in zoology from the

at Wellesley, she never felt as though she fit in among the young people in her hometown. Unlike many of her friends, she did not marry, and after the death of her mother in 1893, she returned to Wellesley to begin work on a master's degree in astronomy.

In 1896, while she was working on her master's degree, Cannon was hired by the Harvard Observatory in Boston. She was put to work using her expertise in photographic development and spectroscopy to examine and classify stars. She would examine each star's light spectrum to determine its surface temperature, and she refined a classification system that separated the stars into categories from most to least hot. From the spectrum she also could determine what chemical elements the star was made of, the star's size, and the speed at which it moved through space.

Analyzing star spectra was demanding work and meant that Cannon had to spend hours peering through

University of Kansas. Carothers won the University of Pennsylvania's Pepper Fellowship in 1913, which helped her begin her doctoral studies at that institution. She studied and did lab work for three years at the University of Pennsylvania, completing her Ph.D. in zoology in 1916. As part of her studies, she did fieldwork in the American Southwest in 1915, probably collecting grasshoppers, which were becoming the focus of her research.

From 1914 to 1936, Eleanor Carothers served as an assistant professor of zoology at the University of Pennsylvania. Teaching took part of her time and lab work all the time that remained. A lot of Carothers's lab work was done on an independent contract basis for the Woods Hole Marine Biological Laboratory in Massachusetts. Carothers's zoological specialty was entomology, the study of insects, which she narrowed down even further to the study of Orthoptera, an insect order characterized by biting mouths, two pairs of wings, and incomplete metamorphosis. These include grasshoppers, crickets, and mantises. Carothers chose to study grasshoppers. Her work on grasshoppers centered on the genetics and cytology of these creatures. She looked at the differences in heteromorphic homologous chromosomes among different species of grasshoppers and gathered data and offered suggestions about the influence of cytology on heredity. Carothers published papers that summarized the findings of her research in the *Journal of Morphology, Quarterly Review of Biology, Proceedings of the Entomological Society,* and *Biological Bulletin.* Descriptions of Carothers's work for Woods Hole were published annually by the Marine Biological Laboratory.

In 1936, Carothers left the University of Pennsylvania to return to the Midwest where she became a research associate at the University of Iowa's zoology department. At the University of Iowa, Carothers did research on the physiology and cytology of normal cells. Funding for this research came from a grant that she received from the Rockefeller Foundation. Carothers worked at the University of Iowa until 1941 when she moved to Kingman, Kansas, a small town on the Great Plains not far from where she was born. From Kingman, she continued to work as a researcher for the Marine Biological Lab at Woods Hole until her death in 1957.

For her work in zoology and genetics, Carothers was cited with a star, an indication of especially high status, in the 1927 *American Men of Science.*

Carr, Emma Perry
(1880–1972)
American
Chemist

A desire to understand the molecular structure of organic compounds led Emma Carr to study the new technique of spectroscopy. During her career she became known as the leading American specialist in ultraviolet spectroscopy. Her research in this field yielded valuable new information about the chemical structure of complicated organic substances such as hydrocarbons. In addition, Carr was the guiding force in building a well-regarded chemistry department at Mount Holyoke College.

Emma Carr was born on July 23, 1880, in Holmesville, Ohio, the third child of Edmund Cone Carr, a physician, and Mary Jack Carr. Her family encouraged their children to pursue college education. Her brother James followed in his father's footsteps by becoming a doctor. Another brother, Edmund, became a businessman. Emma was the only one of three sisters who earned an advanced degree and went into a profession. She never married.

Carr's university education began with a year's study at Ohio State University in 1898–99. She then transferred to Mount Holyoke College in Massachusetts where she studied chemistry and also worked as an assistant in the chemistry department to help pay for her tuition. She attended the University of Chicago for one year (1904–05) and attained a B.S. in chemistry there. Carr returned to Mount Holyoke to teach for two years before going back to the University of Chicago to get a Ph.D. in chemistry. She was helped in her doctoral work by Mary E. Wolley Fellowship and Lowenthal Fellowship grants. She was awarded a Ph.D. in 1910.

After getting her doctorate, Carr returned to Mount Holyoke as an associate professor of chemistry. By 1913, at the age of 33, she had been appointed to full professor and head of the chemistry department. Carr proved a popular

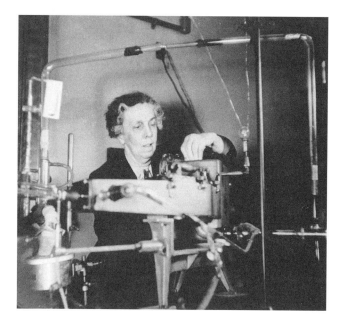

Emma Perry Carr, who was the leading American specialist in ultraviolet spectroscopy during her career. *(The Mount Holyoke College Archives and Special Collections)*

administrator and teacher. She set up exacting but exciting standards for undergraduates and made herself accessible to the students, living as a head of one residence hall for a while and frequently attending dinners with the students in their dormitories. She also plunged into research. Early on, Carr decided to investigate the use of a newly invented instrument, the spectrograph. Before 1920, this instrument, which had originated in Europe, was generally unavailable in the United States. Carr recalled that "I knew nothing about the technique except what I had read in the foreign journals but we went ahead and ordered our first Hilger spectrograph and began work in 1913." She and her students used an ultraviolet spectrograph to get absorption information about cyclopropane, pentenes, and other organic compounds. The spectrograph allowed them to gather more exact information about how these complex molecules bonded.

Carr had wanted to travel to Europe to study how European scientists worked with spectrographs, but she had to wait until the end of World War I to make her first professional visit there. In 1919, she studied at Alfred Walter Stewart's laboratory at Queen's University in Belfast, Northern Ireland, to learn more about the latest theories and techniques in ultraviolet spectroscopy. By 1924, Carr was so proficient in this technique that she was asked to be one of three specialists who would put together a book of absorption spectra data for the International Critical Tables group. She spent a year doing this work at the labs of Victor Henri at the University of Zurich. She also represented the United States at meetings of the International Union of Pure and Applied Chemists in 1926, 1927, and 1937. By the 1940s, Carr's work with ultraviolet spectroscopy on carbon-carbon bonds caught the attention of the petroleum industry. She gave talks about her work to several petroleum industry groups. Carr taught at Mount Holyoke until 1946. During her time as chairwoman of the department, 43 of her undergraduate students went on to get Ph.D.s.

Emma Carr was repeatedly honored for her work. In 1937, she was the first woman to be presented the American Chemical Society's (ACS) Garvin Award. She also received honorary degrees from Allegheny College (1939), Russell Sage College (1941), and Hood College (1957). In 1957, she and her friend and colleague, MARY SHERRILL, were awarded the James Flack Norris Award for excellence in teaching by the northeastern section of the ACS. Carr died on January 7, 1972, at the age of 92, in Evanston, Illinois.

Carson, Rachel Louise
(1907–1964)
American
Marine Biologist and Ecologist

A gifted writer as well as a diligent scientist, Rachel Carson wrote radio scripts, magazine articles, and books that made science accessible to ordinary men and women around the world. Most of Carson's writings focus on the sea and sea life, but later in her career she wrote one of the most important books ever published about humankind's relationship to nature. This work, *Silent Spring,* published in 1962, caused a furor, coined a new word, *ecology,* and jump-started a nascent movement against industrial pollution of the earth's environment.

Carson was born on May 27, 1907, in Springdale, Pennsylvania, a coal-mining town in the western part of that state. Her father, Robert Warden Carson, tended a small farm and sold insurance and real estate. Rachel's mother, Maria Frazier Carson, stayed home to raise Rachel and her two siblings, Robert and Marian. Considerably younger than her brother and sister, Rachel developed into a shy child who liked to stay on the farm and help with the animals. It was here that her love of the natural world began.

In 1925, Carson won a scholarship to attend the Pennsylvania College for Women. She began studying English with the intention of becoming a writer but switched her major to zoology after discovering an affinity for the life sciences. Especially curious about the sea and sea life, she was inspired by the lines of a favorite romantic epic, *Locksley Hall,* "For the wind arises, roaring seaward, and I go."

Rachel Carson, whose book, *Silent Spring,* is often cited as the inspiration for the creation of the Environmental Protection Agency. *(Beinecke Rare Book & Manuscript Library, Yale University)*

In 1929, after she graduated with honors from college, Carson saw the sea for the first time as a summer apprentice at Woods Hole Marine Biological Laboratory. She began graduate work in aquatic biology at Johns Hopkins University in Baltimore, where she earned an M.A. in zoology in 1932.

After teaching at universities for several years, Carson was confronted with a series of personal crises. In 1935, her father died, which prompted her mother to move in with her. The next year Carson's sister Marian died; Carson and her mother decided to adopt and raise Marian's two young daughters. Suddenly in need of secure employment, Carson took a job as junior aquatic biologist at the U.S. Bureau of Fisheries. She was only the second woman employed by that agency in a nonclerical position.

Carson's boss quickly realized that his new assistant had a talent for explaining scientific work to the general public. For the next 16 years, she worked for the Bureau of Fisheries (later renamed the Fish and Wildlife Service) writing and editing pamphlets and books published by that agency. But it was in books she wrote for publication outside the bureau that Carson's talent was best displayed. In 1941, her first book, entitled *Under the Sea Wind,* about oceanic aquatic life, was published by Simon & Schuster. In the foreword, Carson described her passion for the sea: "To stand at the edge of the sea, to sense the ebb and flow of the tides, to feel the breath of a mist moving over a great salt marsh . . . is to have knowledge of things that are nearly as eternal as any earthly life can be."

Carson wrote two other, more scientifically detailed books about the sea in the late 1940s and 1950s. *The Sea Around Us,* published in 1951, became a best-seller and won the National Book Award. It also allowed Carson to quit her government job to pursue writing full time. *The Edge of the Sea* (1955), about life on the ocean's shores, was also a best-seller.

After receiving a letter from a friend, Olga Owens Huckins, Carson embarked on her most famous book in 1957. Huckins wrote Carson to tell her about how the pesticide DDT, sprayed from airplanes, had killed a number of birds at her bird sanctuary. Prompted by this warning, Carson began researching the harmful effects on humans and wildlife of human-made pollutants such as insecticides and how these compounds upset the fragile interconnected balance of plant, insect, bird, and animal life. "There would be no peace for me if I kept silent," Carson wrote. "As cruel a weapon as the caveman's club, the chemical barrage has been hurled against the fabric of life." The work resulted in *Silent Spring,* her fourth book, published in 1962.

As a result of her criticism of the chemical industry, Carson found herself subjected to a barrage of attacks, which slandered her personally and questioned her scientific credentials. Even though she was ill with cancer, she defended herself and her work skillfully. In response to her book, President John F. Kennedy appointed a panel whose findings supported her allegations. The federal Environmental Protection Agency was created in 1970 as a direct result of her work. For *Silent Spring* and her other books, Carson was awarded the National Audubon Society's Audubon Medal, the Schweitzer Medal from the Animal Welfare Institute, and was elected to the American Academy of Arts and Sciences. She died of breast cancer on April 14, 1964.

Caserio, Marjorie Constance Beckett
(1929–   )
British/American
Chemist

As a researcher and a teacher, Marjorie Caserio has made significant contributions to the discipline of chemistry. A specialist in physical organic chemistry, Caserio has studied the reactions of carbocyclic ring compounds, bonding and reaction mechanisms of organic sulfur compounds, and has coauthored a standard chemistry textbook, *Basic Principles of Organic Chemistry* (1964). She has also blazed trails for women who want to teach chemistry at the university level.

Born in Cricklewood, England, on February 26, 1929, to Herbert Cardoza Beckett and Doris May House Beckett, Marjorie Beckett was the second of two children. Her father owned a business that manufactured hotel and restaurant equipment, and the family was prosperous. Beckett was encouraged to pursue a basic education, but university studies did not seem to be included in her family's plans for her. She married Fred Caserio, also a chemist, in 1957 and has two children, Alan and Brian.

After graduating from a well-regarded private girls' school in 1944, Marjorie Caserio entered Chelsea College, a technical school that offered vocational studies, courses that prepared its students for university, and an undergraduate college program. Caserio finished her preuniversity courses in 1946 and applied for, but was rejected by, the University of London. She then studied chemistry at Chelsea College. She won her B.S. from that institution in 1950. Because she received so little support and encouragement from colleges and businesses in the United Kingdom, Caserio applied for a grant that would fund study for a master's degree in chemistry at a university in the United States. She got this grant and studied at Bryn Mawr College in Pennsylvania under physical chemist Ernst Berliner. She was awarded an M.A. from Bryn Mawr in 1951. Caserio returned to the United Kingdom in 1951, but again, because she was a woman, she had a hard time finding a job. She worked for a year as the only woman chemist at a research institute but felt stifled and isolated. In 1953, she returned to Bryn Mawr to study for a Ph.D., which she won in 1956.

Caserio's first job after attaining her Ph.D. was as a postdoctoral researcher at the California Institute of Technology, a position she held from 1956 to 1965. Working with John D. Roberts, a physical organic chemist, Caserio studied the way three- and four-membered carbocyclic ring compounds reacted with other chemicals as well how diazomethane reacted with alcohols. In 1965, Caserio was finally able to land a tenure-track teaching position. She was given an assistant professorship at the University of California (U.C.) at Irvine, near her home of Laguna Beach. She became a full professor there in 1972. At U.C. Irvine, Caserio studied the reaction of allenes to other chemicals. She also was one of the first researchers to use nuclear magnetic resonance and ion cyclotron resonance techniques in her studies. Caserio began to take on work in the administration of studies in California. She was chairperson of the U.C. Irvine chemistry department from 1986 to 1990. In 1990, she became vice chancellor of academic affairs at the University of California at San Diego.

For recognition of her efforts to advance knowledge in chemistry, Caserio won a Distinguished Teaching Award from U.C. Irvine in 1974. She also was awarded a John S. Guggenheim Fellowship in 1975, the same year she won the Garvan Award from the American Chemical Institute. Now retired from academia and research, she believes that "nothing we ever do is lost. It comes back to profit us in unexpected ways. It has always been a most gratifying experience to meet the occasional student years later, who volunteers how much s/he enjoyed [a] course or the book [I wrote]. . . . This is what makes a career as an educator so worthwhile."

Chang, Sun-Young Alice
(1948–)
Chinese/American
Mathematician

Sun-Young Alice Chang received the 1995 Ruth Lyttle Satter Prize in Mathematics from the American Mathematical Society (AMS) in recognition of her "deep contributions" to the understanding of mathematics. This prize specifically honored women for their scientific research, and Chang encouraged women to follow in her footsteps by pursuing advanced study and research in the sciences.

Chang was born on March 24, 1948, in Ci-an, China. She attended the National University of Taiwan, where she earned her B.S. in 1970. She then traveled to the United States to pursue her doctorate at the University of California at Berkeley. She received her Ph.D. in 1974, the year after she entered a marriage that has produced two children. Chang then filled a series of assistant professorships, first at the State University of New York at Buffalo

for one year after receiving her doctorate. In 1975, the University of California at Los Angeles (UCLA) appointed her as its Hedrick Assistant Professor of Mathematics, a position she retained until 1977. That year, she moved to the University of Maryland at College Park, where she remained until 1980 when she returned to UCLA as an associate professor, eventually rising to the rank of full professor. She held a concurrent professorship at the University of California at Berkeley from 1988 through 1989. For the next two years, she served as the vice president of the AMS.

During her first year at UCLA, she filled a Sloan Fellowship for the National Academy of Sciences. A decade later, she served on the academy's Board of Mathematical Sciences from 1990 through 1992. During that same period, she was a member of the Advisory Panel for the Mathematical Sciences of the National Science Foundation. At that time, Chang also served on the selection committee for the Noether Lectures of the Association for Women in Mathematics, from 1991 through 1994. She herself was selected as the featured speaker at the meeting of the International Congress of Mathematicians held in Berkeley in 1986.

Chang focused her research on geometry and topology, specifically studying nonlinear partial differential equations and isospectral geometry. In January 1995, Chang received the Ruth Lyttle Satter Prize in Mathematics from the American Mathematical Society at its 101st annual meeting held in San Francisco. Mathematician JOAN BIRMAN of Columbia University established the prize in 1990, named in the memory of her sister, who was a research botanist at the University of Connecticut. The AMS awarded the $4,000 prize every two years to a woman who made an outstanding contribution to the mathematical research in the previous five years. DUSA MCDUFF received the first prize in 1991, and 1993 recipient LAI-SANG YOUNG served on the selection committee that chose Chang.

Chang produced her prizewinning work in collaboration with Paul Yang, Tom Branson, and Matt Gursky, who she thanked in her acceptance speech. This team studied partial differential equations on Riemannian manifolds, specifically focusing on extremal problems in spectral geometry and the compactness of isospectral metrics within a fixed conformal class on a compact 3-manifold. Chang focused subsequent research on extremal functions of Sobolev inequalities. After receiving the Satter Prize, Chang served a three-year term on the Editorial Boards Committee of the AMS that ended in 1998. In 1996 through 1997, Chang participated in the University of Texas's Distinguished Lecturer Series, which addressed graduate students. Chang has encouraged women in particular to pursue mathematics at the graduate level.

Chase, Mary Agnes Meara
(1869–1963)
American
Botanist

Although lacking a college degree and university training, Agnes Chase made a name for herself as a botanical illustrator and practicing botanist. Her first work in the field of botany was as an illustrator for a private botanist and the Field Museum of Natural History in Chicago. Chase eventually used this skill to get a job with the U.S. Department of Agriculture (USDA). There she learned about the science of botany on the job and eventually became senior USDA botanist.

Mary Agnes Meara was born in Iroquois County, Illinois, on April 29, 1869, the daughter of an Irish immigrant and blacksmith, Martin John Chase, and Mary Brannick Meara. She was the second youngest of six children. After the death of her father when she was two, Agnes Meara moved with her family to Chicago. Meara studied for a time at a Chicago elementary school but was forced into the job market at a relatively early age to help support her family. She never attended college.

While a proofreader and typesetter at a newspaper called the *School Herald,* Meara met and later married (in 1888) the paper's editor, William Ingraham Chase. Unfortunately, William Chase, who was 15 years older than Agnes Meara, already suffered from an advanced case of tuberculosis. He died within a year of their marriage. While working as a proofreader at another paper, the *Inter-Ocean,* Chase met the Reverend Ellsworth Hill, a retired minister and part-time bryologist (student of mosses). Chase began accompanying Hill on his field expeditions, and Hill began teaching Chase the basics of botany, especially the identification of mosses. When he discovered that she had a talent for drawing, he put her to work as an unpaid illustrator of the species he collected. This job lead to another unpaid illustrating position, this time with Charles Frederick Millspaugh, the curator of botany at the Field Museum of Natural History. In 1901, the Reverend Mr. Hill got Chase a better-paying job with the USDA as a meat inspector at the Chicago stockyards. Because of her experience as a botanical illustrator, Chase then applied for and won a job in 1903 as a botanical artist at the USDA's Bureau of Plant Industry in Washington, D.C. She would spend the rest of her working career at the Washington headquarters of the USDA.

In 1905, Chase was promoted as illustrator for Albert Spear Hitchcock, a senior USDA botanist whose specialty was agrostology, the study of grasses. Again Chase was fortunate in having a mentor who was unconcerned with her lack of formal education. Hitchcock trained Chase in agrostology, and she gradually left illustration work behind for the practice of botany. In 1907, she was promoted to the position of scientific assistant; by 1923, she had become assistant botanist, then in 1925 associate botanist. Following Hitchcock's death in 1935, Chase was made senior botanist for agrostology. In 1937, she was appointed custodian of grasses for the U.S. National Museum, a job that required her to oversee the U.S. grass herbarium.

Chase was immensely proud of her career, and she also took politics very seriously. At a time when women were discouraged from participating in politics, she became actively engaged as a campaigner for socialist causes, women's suffrage, and prohibitionism. She once promised to burn any of Woodrow Wilson's speeches that had the words *freedom* and *liberty* in them so long as women were denied the right to vote. She was an early member of the National Association for the Advancement of Colored People and the International League of Peace and Freedom.

For her scientific work, Chase was given a Certificate of Merit in 1956 by the Botanical Society of America. She was awarded a medal by the government of Brazil in 1958 for her work in botany in that country, and in 1961, she was made a fellow of the Linnaean Society in London. She also finally received her university degree, an honorary D.Sc. from the University of Illinois, in 1958. A woman of huge energy, Chase continued to take frequent botanical field trips after her retirement from government service in 1939. She died of heart failure on September 14, 1963, at age 94.

Chasman, Renate Wiener
(1932–1977)
German/Israeli
Physicist

As a young refugee from Nazi Germany, Renate Wiener Chasman eventually moved to Israel where she studied physics at the university level. Most of her contributions to the field were made in the United States, the country in which she lived and worked for most of her adult life. Chasman is best known for her research on atomic particle physics. She studied beta decay in atomic nuclei and later worked on several generations of particle accelerators at the Brookhaven National Laboratory in New York State.

Born on January 10, 1932, in Berlin, Germany, to Hans Wiener and Else Scheyer Wiener, Chasman grew up in a Jewish family that suffered the indignities and threats to livelihood and life imposed by the Nazi regime that took power in Germany in 1933. Her father, Hans, was a lawyer and founding member of Germany's Social Democratic party. Even though he wanted to stay in Germany and conduct a political struggle against the Nazis, by 1938 he knew that he and his family were in danger of being arrested and sent to a concentration camp. The family fled to Sweden in December 1938.

After her arrival in Sweden, Renate Chasman and her sister lived in a girls' home in the north of Sweden for several years. By the time she was of high school age, she had moved back to the capital of Stockholm, where she attended a public school. By the time she entered high school, Chasman knew she wanted to study mathematics and physics. She was encouraged in her intellectual pursuit by her high school math teacher, Gunnar Almquist, who taught her advanced math and physics.

After graduation from high school in 1950, Chasman decided to move to Israel and attend Hebrew University in Jerusalem. She earned her M.S. in physics from Hebrew University in 1955, and in 1959, she was awarded a Ph.D. in experimental physics from that same institution.

During her doctoral studies, Chasman concentrated on problems associated with beta decay in the atomic nucleus. She developed what are termed Wiener coefficients to express phenomena of parity nonconservation in beta decay mathematically.

Chasman's work on beta decay gained the attention of CHIEN-SHIUNG WU, a professor of physics at Columbia University in New York City and a renowned woman physicist. Wu invited Chasman to work with her at Columbia, so in 1959, she moved to the United States. Chasman worked as a research associate in Wu's lab until 1962 when she moved to Yale University in New Haven, Connecticut, with her new husband, Chellis Chasman, whom she met while working at Columbia and married in 1962. The Chasmans stayed at Yale for only a year before moving again, this time to the Brookhaven National Laboratory on Long Island. Chasman was to remain at Brookhaven for the rest of her career.

By 1965, Chasman was working at Brookhaven's Accelerator Department. She began to conduct theoretical research on the building of a new and improved particle accelerator, a device used to study atomic properties by smashing atomic nuclei and observing the reaction of the nuclei's constituent particles. Chasman was a key figure in the redesign of the accelerator's injector, which boosted the power of the machine. Chasman's later work at Brookhaven involved theoretical work on the design of lattice configurations that would work with the design of superconducting proton storage rings used in proton-proton colliders. She also worked to convert electron synchrotrons to dual uses by tapping ultraviolet and X-ray emissions to study phenomena in solid-state physics, chemistry, and biology.

In recognition of her work, Chasman was selected to sit on a review committee at the Fermi National Accelerator Lab in Illinois to work as an adviser and visiting scientist at the European CERN accelerator in Switzerland. She died of complications of melanoma on October 17, 1977.

Châtelet, Gabrielle-Emilie du
(1706–1749)
French
Physicist and Mathematician

A daughter of the French aristocracy during the height of the Bourbon monarchy in the 18th century, Emilie du Châtelet's intellectual audacity stood in marked contrast to the roles women were expected to play in that era. At the age of 27, she became a close friend of the philosopher François Voltaire, who introduced her to the works of Sir Isaac Newton. Châtelet passed Newton's ideas on to the French public by translating some of his works. She also published several books in which she speculated about the nature of fire and offered an overview of the advances made in physics during her lifetime.

Born on December 17, 1706, in Paris, Gabrielle-Emilie Le Tonnelier de Breteuil was the daughter of Louis-Nicolas Tonnelier de Breteuil, chief of protocol at Louis XIV's royal court, and Gabrielle Anne de Froulay. Both of her parents were from the nobility, and the family owned extensive tracts of land in several provinces of France. In 1725, she married Florent-Claude du Châtelet, like Emilie an aristocrat. They had three children.

Emilie du Châtelet was unusual for her time in that she was given a thorough and extensive education as a child by private tutors. This was the same education that was provided to boys from wealthy or aristocratic families and included studies in Latin and other European languages, mathematics, and, unusual even for boys, physics.

Châtelet's marriage to her husband was arranged for money and family prestige. There was apparently little love between the two of them. This was not unusual for aristocrats of that time, and it left Emilie du Châtelet with lots of time to read and pursue her studies. The defining moment in her intellectual and emotional life came in November 1733 when she became reacquainted with the famous writer and philosopher, François Voltaire (she had briefly met him as a child). Voltaire and Châtelet immediately recognized that they shared an intellectual and physical attraction. They became lovers, and Voltaire moved into the Châtelet estate at Cirey. This arrangement was acceptable to Florent-Claude du Châtelet because he was often away from the estate with his military career and he also took lovers.

During the first year he stayed at Cirey, Voltaire introduced Châtelet to the writing of Isaac Newton. Voltaire had just published his controversial book, *Lettres Anglais ou philosophiques* (1733), which advocated social and political liberalism and got him into hot water with the conservative monarchy in Paris. For Voltaire, Newton's scientific approach was just as revolutionary as political liberalism. It asserted that scientific truth had to be derived from observation and measurement of the physical world and that theories about the material world should be based on these measurable observations. This approach, backed up by impressive results, confronted a long-standing approach that had been espoused by the French mathematician René Descartes that the world was made from a predetermined, mathematical order that could be understood, without experimentation, through logic and mathematical formula. Emilie du Châtelet immediately rejected her old scientific training and embraced Newton's ideas.

Châtelet's earliest work was a book about experiments she conducted with fire. The work, *Dissertation sur la nature et la propagation du feu* (1737), was an attempt to understand the process by which mass is converted into energy. The next year, to help spread word of Newton's insights, Châtelet and Voltaire collaborated on a book about the English scientist entitled *Eléments de la philosophie de Newton* (1738), a work that was officially attributed to Voltaire but that was probably more the product of Châtelet. She followed this in 1740 with a physics history and text called *Institutions de physique,* which examined Newton's work as well as that of the German mathematician and philosopher Gottfried Leibniz. Châtelet spent the last part of her life translating Newton's *Philosophie naturalis principia mathematica* into French. This book was published posthumously in 1759.

Emilie du Châtelet was very much a player in the debate about science and, through her books, an educator to the general public about work that was being done in the sciences during her lifetime. For her efforts, she was voted membership in the Bologna Institute in Italy. She assessed herself accurately when she said, "Judge me for my own merits, or lack of them, but do not look upon me as a mere appendage to this great general or that renowned scholar. . . . I am in my right a whole person, responsible for myself alone."

Chinn, May Edward
(1896–1980)
American
Physician

May Edward Chinn confronted dual discrimination as an African-American woman physician, though she persevered to break down barriers preventing her from practicing medicine and conducting research. She was the first African-American woman to graduate from Bellevue Hospital Medical College and the first African-American woman to intern at Harlem Hospital. In her private practice, she provided care for patients who would not otherwise receive treatment due to racism or classism. Later in her career, she performed pioneering research on cancer, helping to develop the Pap smear test for cervical cancer.

Chinn was born on April 15, 1896, in Great Barrington, Massachusetts. Her father, William Lafayette, was the son of a plantation slave and her owner; at the age of 11, he escaped from this Virginia plantation. Her mother, Lulu Ann, was the daughter of a slave and a Chickahominy Native American. She worked as the live-in cook at the Long Island mansion of the Tiffany family of jewelers, who treated Chinn as a family member. Growing up, she attended musical concerts in New York City and learned to play piano, accompanying the singer Paul Robeson in the early 1920s. The Tiffany family also taught her the German and French languages.

Chinn's mother, who valued education, saved enough money from cooking to send Chinn to the Bordentown Manual and Training Industrial School, a New Jersey boarding school, until Chinn contracted osteomyelitis of the jaw. Chinn remained in New York City after her surgery there, but she was too poor to finish high school. Despite her lack of a diploma, she took the entrance examination to Columbia Teachers College and passed it, matriculating in 1917.

Chinn studied her first love, music, until a professor mocked her race as unfit for playing classical music. At the same time, she received high praise for a scientific paper she wrote on sewage disposal, so she changed her major to science. In her senior year, she secured a full-time position as a lab technician in clinical pathology, so she completed her course work at night to graduate with a bachelor's degree in science in 1921. She proceeded to study medicine at Bellevue Medical College, becoming its first African-American woman graduate in 1926.

Rockefeller Institute was prepared to offer Chinn a research fellowship until it learned of her race. Harlem Hospital was the only medical institution in the city that offered Chinn an internship. Although Chinn broke a barrier as the first African-American woman to intern there and to accompany paramedics on ambulance calls, she confronted another obstacle when the hospital refused her practicing privileges there. Chinn established a private practice instead, seeing patients in her office and performing procedures in their homes. This experience prompted her to earn a master's degree in public health from Columbia University in 1933.

In 1940, Harlem Hospital finally granted Chinn admitting privileges, in part due to Mayor Fiorello La Guardia's push for integration in the wake of the 1935 Harlem riots. Then, in 1944, the Strang Clinic hired Chinn to conduct research on cancer, and she remained there for the next 29 years. The Society of Surgical Oncology invited her to become a member, and in 1975, she established a society to promote African-American women to attend medical school. She maintained her private practice until the age of 81. While attending a reception at Columbia University in honor of a friend, Chinn collapsed and died on December 1, 1980.

Clapp, Cornelia M.
(1849–1934)
American
Zoologist and Marine Biologist

Clapp was an early leader in the investigation of the biological sciences in the United States. Beginning as a gym and math teacher at Mount Holyoke College, she followed her boundless curiosity to become a professor of zoology. She worked in the field of embryology until the opening of Woods Hole Marine Biology Laboratory in Massachusetts, at which time she threw herself into marine biology. She was also a gifted teacher who mentored several generations of women in the biological sciences.

Born on March 17, 1849, in Montague, Massachusetts, to Richard C. Clapp and Eunice Amelia Slate Clapp, Cornelia Maria Clapp was the oldest child in a family of seven. Because both of her parents were teachers, education was given a high priority in the family. Along with her three brothers and three sisters, she attended primary and secondary schools in Montague. She never married.

In 1868, at 19, she enrolled in Mount Holyoke Seminary, an early women's college. At Mount Holyoke, she earned a general liberal arts degree in 1871. In 1874, after she became interested in biological science, Clapp enrolled in the late (d. 1873) famed Harvard biologist Louis Agassiz's school on Penikese Island. She studied there, absorbing Agassiz's methods, for a year. In 1889, Clapp earned a Ph.D. from Syracuse University, and in 1896, she earned another Ph.D. from the University of Chicago.

Clapp began her teaching career immediately after graduating from Mount Holyoke in 1871. For a year, she taught Latin at Potter Hall, a boys' school in Pennsylvania. She then returned to Mount Holyoke to teach but found "to my consternation, I didn't know what I was to teach." Her first year she taught math, but by the second year she had switched to biology. She also taught gymnastics until 1891. During her first year as a teacher at Mount Holyoke, Clapp collaborated with her former teacher and now colleague, LYDIA SHATTUCK, on the study of amoebas. They studied these and other microorganisms that they had collected at a pond under the school's microscope.

Clapp's year on Penikese Island was a turning point in her life and career. Here she received confirmation that she was on the right track: it was best to study nature not out of texts but directly by observation. "I had an opening of doors at Penikese," she later remembered. ". . . Everybody was talking. Discussions in every corner. I felt my mind going in every direction." After her stay in Penikese, Clapp relied much less on standard natural history texts to teach her courses at Mount Holyoke. Instead, she began to immerse her students in direct obser-

vation. She had them study the development of a chicken embryo by obtaining a hen and removing an egg every day of the 21-day gestation cycle. The embryo was removed from the egg so that the students could study its daily development. At Penikese, Clapp also got hooked on marine life as a research subject. By the late 1880s, she had made this the primary subject of her studies. She spent her first summer of study at Woods Hole in 1888 and came back each summer until her death.

Because Clapp did not publish many papers about her research, it is not possible to assess her accomplishments as a research scientist. However, it is clear that she was a dedicated and inspiring teacher, and a persistent, if unpublished, student of marine biology. Her peers recognized her effort and talent. She was a member of the American Association for the Advancement of Science, Society of American Zoologists, and Association of American Anatomists. In 1906, she was included in *American Men of Science*, one of only six women selected that year. And in 1923, Mount Holyoke named its new science laboratory building, Clapp Laboratories, after her. She died on December 31, 1934, at the age of 85.

Clark, Eugenie
(1922–   )
American
Marine Biologist

Fascinated by fish and underwater life from an early age, Eugenie Clark has built a career around the study of fish and marine mammal behavior. She was the first to discover the mechanism by which fish such as blowfish puff themselves up for self-protection, and she has discovered and classified many kinds of rare and exotic fish in the Red Sea and Micronesia. She has devoted considerable time to studying shark behavior, and she has become well known to lovers of marine life through her popular writings for a general audience about life in the sea.

The only child of Charles Clark and Yumico Mitomi Clark, Eugenie Clark was born on May 4, 1922, in New York City. Her father was a barber who died when Eugenie was two. Yumico Clark then raised Eugenie by herself while working as a swimming teacher and a sales clerk behind the counter of a newspaper and cigar stand in a New York athletic club. One of Clark's first memories of being entranced with fish came when her mother would leave her at the New York Aquarium for several hours on Saturday mornings. While her mother was at work, Eugenie stared transfixed at "the glass tanks with moving creatures in them. . . . I brought my face as close as possible to the glass and pretended I was walking on the bottom of the sea." In 1951, Clark married Ilias Papakonstantinou, a Greek-born doctor. She had four children with Papakonstantinou, her daughters Hera and Aya, and her sons Tak and Niki.

Clark entered Hunter College in New York City in 1938 and received a B.S. in zoology there in 1942. She had planned to get a job after graduation as an ichthyologist, but the entry of the United States into World War II prevented this from happening. Instead, while she worked as a chemist in a plastics company in New Jersey, she continued her education at night at New York University. She received a master's degree from there in 1949 and a Ph.D. in marine biology in 1950.

After the war, Clark did her first diving at the Scripps Institute at La Jolla, California. In 1949, she was sent by the U.S. Navy to the newly acquired Pacific islands of Guam, Saipan, and the Palaus to inventory the kinds of fish life found in the waters around these islands. The navy wanted to know especially which fish were poisonous. From 1950 to 1952, Clark was involved in a study of the marine life of the Red Sea from the small Egyptian port town of Ghardaqa. She collected 300 species of fish, including 40 poisonous ones. In 1952, she won grants that allowed her to take time to write the first of her books. Entitled *Lady with a Spear,* it was about her experiences cataloging fish in the Red Sea. From 1955 to 1975, Clark received 15 grants from the National Science Foundation and other scientific organizations to study fish life around the world.

Clark's book was a commercial success, and in 1954, two of its readers, William H. and Anne Vanderbilt, offered Clark a marine biology lab of her own on Florida's west coast. Clark accepted and in 1955 opened the Cape Haze Marine Laboratory in Placida, Florida. Here Clark first began an intensive study of shark behavior. She devised a series of experiments, the first of their kind, designed to show that sharks had intelligence equal to other animal forms such as rats or pigeons. She devised a contraption in the shark tanks that the sharks had to bump with their noses if they wanted food. Within only a few days the sharks had learned how to ask for food. Clark also studied the sexual behavior of groupers and learned that this type of fish can actually change its gender in as fast as 10 seconds, a behavior that helps ensure its survival as a species.

From 1968 to 1992, Clark taught marine biology as a professor at the University of Maryland. She has won a gold medal from the Society of Women Geographers (1975), the John Stoneman Marine Environmental Award (1982), and the Explorer's Club Lowell Thomas Award (1986). She also wrote another popular science book in 1968 entitled *The Lady and the Shark,* telling the general public about her research and trying to allay fear about the shark as a dangerous species. "I think I have tried to give a better reputation to sharks," Clark has said, "and I feel this is my most worthwhile contribution."

Clarke, Edith
(1883–1959)
American
Electrical Engineer

Edith Clarke helped to initiate the technological revolution as a human computer, prefiguring modern computers. In order to facilitate the computation process, Clarke devised charts graphing the functions of technical equations to eliminate the need to solve the problem every time anew. Like a personal computer, she had a strong memory, aiding in her ability to perform long, involved solutions. She entered the field of engineering as the one discipline that proved as interesting intellectually as duplicate whist, an intricate card game.

Clarke was born on a farm near Ellicott City, Maryland, on February 10, 1883, one of nine children born to Susan Dorsey Owings. When her mother and father, John Ridgely Clarke, a lawyer, died in 1897, Clarke attended boarding school and then in 1904 spent her inheritance educating herself at Vassar College. She graduated Phi Beta Kappa and with honors with A.B. degrees in both mathematics and astronomy in 1908. She taught mathematics for the next three years, first at a private girls' school in San Francisco and then at Marshall College in Huntington, West Virginia.

In 1911, Clarke returned to her education with the School of Civil Engineering at the University of Wisconsin. However, she forsook her education for work: Her job as a computing assistant under research engineer George A. Campbell for the American Telephone and Telegraph company (AT&T) proved more interesting than school. Clarke spent World War I leading a group of women calculators for the Transmission Department of AT&T while simultaneously taking night courses on radio at Hunter College and electrical engineering at Columbia University.

Clarke continued her graduate education at the Massachusetts Institute of Technology starting in 1918; the next year, she became the university's first woman to receive a master's degree in electrical engineering. Though she could not secure a position as an engineer, she did commence her 26-year relationship with General Electric (GE) by leading a group of human "computers" in the turbine engineering department. In 1921, she devised a solution to electrical power transmission line problems with a "graphical calculator," which she patented. That year, she took a leave of absence from GE to take up a professorship of physics at Constantinople Women's College (now Istanbul American College); upon her return, GE promoted her to the position of engineer.

Clarke focused her research on the transmission of electrical energy, writing extensively on the topic. Her first publication, "Transmission Line Calculator," (which appeared in the June 1923 *General Electric Review*), exemplified her methodology of expediting the work in her industry by charting key calculations. In 1932, she won the distinction of writing the best paper in the Northeastern District of the American Institute of Electrical Engineers (AIEE). She became the first woman to read her paper, entitled "Three-Phase Multiple-Conductor Circuits," before the congregated AIEE (which elected her a member in 1948.) However, her paper had not only intellectual interest but also practical application, addressing the issue of circuit overloading by considering the benefits and detriments of various combinations of multiple conductors.

In 1943, Clarke published the first volume of her landmark textbook, *Circuit Analysis of AC Power Systems*, adding a second volume in 1950. Five years earlier, she had retired from GE to a farm in Maryland, but she soon found herself back in the scientific community as the first woman appointed by the University of Texas as an electrical engineering professor. She retired in 1955, but the previous year she received the Achievement Award from the Society of Women Engineers. Clarke died on October 29, 1959, in Olney, Maryland.

Clay-Jolles, Tettje Clasina
(1881–1972)
Dutch
Physicist

Tettje Clasina Clay-Jolles led a short but significant career as a physicist before she devoted herself to familial duties. She collaborated with her husband to discover that atmospheric radiation varies according to geographic latitude, an assertion that was hotly contested but ultimately proved correct. Clay-Jolles also was one of the first woman scientists in the Netherlands.

Clay-Jolles was born in 1881 in Assen, in the Netherlands. Her mother was Eva Dina Halbertsma, her father was Maurits Aernout Diederick Jolles, and her two older sisters were Hester and Leida. Clay-Jolles was the first and only girl to attend the local gymnasium, or secondary school. At the end of her six years there, she took both the alpha and beta series examinations, testing her knowledge of the humanities and the sciences, respectively; she passed both tests, an unusual occurrence.

Clay-Jolles continued her education at the prestigious University of Groningen, commuting from Assen daily by train. In 1903, she transferred to the University of Leiden, where she became one of a very few women studying physics. She studied under Heike Kamerlingh Onnes, who directed her doctoral research on low-temperature physics. Clay-Jolles met and fell in love with another one of Kamerlingh Onnes's students, Jacob Clay, and the couple married in June 1908. She continued writing her thesis until December of that year, when she devoted her life to child-rearing. The couple had three children—a daughter and two sons—that Clay-Jolles raised over the next dozen years.

The Institute of Technology in Bandung, Java, appointed Jacob Clay as a professor of physics in 1920, so he moved his family there. Clay-Jolles worked as an assistant in a well-appointed laboratory conducting research on vacuum pumps, a technology that she had studied in her days as a graduate student. She assisted her husband by editing (and typing) all of his publications. In 1921, the Nobel laureate Hendrik Antoon Lorentz hired her to edit a volume of his lectures for publication, an assignment that acknowledged her expertise as an exacting scientist and scholar.

Throughout the 1920s, Clay-Jolles collaborated with her husband studying the nature of cosmic rays, radiation in the ultraviolet solar spectrum, and the intensity of atmospheric radiation. This last topic proved controversial, as Clay-Jolles and her husband contended that geographic latitude determined variations in radiation readings, attributing this phenomenon to the differing degree of ultraviolet penetrability in the upper atmosphere and the ozone layer. Jan Boerema and Maarten Pieter Vrij contended that ultraviolet radiation penetrated the atmosphere at the tropics, in contrast to the husband-and-wife team's findings.

Attempting to solve this dispute, Clay-Jolles and her husband measured ultraviolet light at their location in the tropics with a cadmium electrocell, then compared their readings to those recorded by their former colleague from Leiden, Cornelius Braak, who had been the director of the Batavia Observatory. In 1933, Clay-Jolles published these findings in the East Indian scientific journal, *Natuurkundig Tijdschrift voor Nederlandsch-Indië,* opposite Maarten Vrij's publication of his findings. That same year, Clay-Jolles and her husband jointly published an article entitled "Measurements of Ultraviolet Sunlight in the Tropics" in the *Proceedings of the Amsterdam Academy of Sciences.* Four years earlier, in 1929, Jacob Clay had accepted a position as a professor of experimental physics at the University of Amsterdam. Clay-Jolles abandoned her scientific career when the family returned to their homeland. She died in Amsterdam in 1972.

Claypool, Edith Jane
(1870–1915)
British/American
Pathologist and Zoologist

A selfless and dedicated scientist, Edith Jane Claypool was one of the first women to enter the field of medical pathology. She did valuable research on the cell structure of human blood and tissue that helped doctors diagnose certain infections that appeared similar to tuberculosis. She also worked on a project to develop an improved vaccine for typhoid fever. During this experimentation, she became infected and died of the disease.

One of a pair of twins, Edith Claypool was born in Bristol, England, on January 1, 1870. In 1879, at age nine, she moved with her twin sister, Agnes (later Agnes Mary Claypool Moody, also a well-known scientist), and her mother and father to the United States. Her father, Edward Waller Claypool, was a professor of natural sciences who had been hired to teach at Buchtel College in Akron, Ohio. Her mother, Jane Trotter Claypool, was a homemaker. Edith Claypool never married.

Edith and her sister were given a primary and secondary education at home by their parents, and in 1888, Edith enrolled in Buchtel College where she earned an undergraduate degree in biology in 1892. The sisters, who were close, then both enrolled in master's programs at Cornell University in Ithaca, New York. They also both took summer seminars at Woods Hole Marine Biology Lab in Massachusetts and eventually ended up teaching together at Wellesley College. At Cornell, Edith began studying blood cells. She won her M.A. from that institution in 1893. By 1899, Edith Claypool had decided to become a doctor. She entered Cornell University's medical school and spent two years there before moving to Los Angeles, California, to take care of her dying mother. She finished her medical degree at the University of California, Southern Branch (later UCLA), in 1904.

Claypool's career began the year after she finished her M.A. at Cornell. Following an established professional path for women scientists and professionals, she took a teaching job at Wellesley College in 1894. Few, if any, jobs were open to women at the older, traditional men's universities, and fewer jobs still were available to women in business and the professions. Claypool taught zoology, physiology, and histology (animal cellular research) at Wellesley College, and for several years, she also served as head of the department of zoology. By 1899, she realized that her true vocation lay in medicine, and she resigned from Wellesley to enroll in the medical school at Cornell.

After she moved to California in 1901, Claypool again took up teaching for a time at the Throop Polytechnic Institute in Pasadena while she decided what move to make next. In 1902, at the same time she enrolled at the University of California medical school, she also began working part-time as a pathologist at a hospital in Los Angeles. She stayed at this hospital after she received her M.D. As a pathologist working for other doctors and surgeons, Claypool gained valuable experience with work being done on vaccines and diagnosing bacterial diseases. By 1912, she felt it was time to push herself into new challenges again and left her pathology job in Los Angeles for laboratory research work at the University of California at Berkeley.

For three years at the department of pathology in Berkeley, Claypool worked on techniques to diagnose and cure various lung infections. Her work on a new typhoid vaccine began after the outbreak of World War I in 1914. This disease was killing thousands of Allied troops in France, and Claypool and her colleagues were trying to produce a vaccine that was more effective than the one then available. Her death at the too-young age of 45 shocked and saddened her friends and associates, who set up a memorial research fund in her name. The Edith J. Claypool Memorial Fund gives annual grants for research into infectious diseases.

Cleopatra the Alchemist
(c. fourth century)
Egyptian
Alchemist, Chemist, and Philosopher

Historically, little is known about Cleopatra. She probably lived in the thriving multicultural city of Alexandria sometime during the fourth century. At least one thing about her is known for sure: She was not the much more famous woman who shares her name, the Cleopatra who was a princess of the Ptolemaic dynasty and who loved and betrayed the Roman general Marc Anthony. Cleopatra the Alchemist is known for one surviving work, *A Dialogue of Cleopatra and the Philosophers,* a record of a conversation she had with a priest and a philosopher about the nature of the alchemical endeavor. In all likelihood, this book is not a verbatim record of an actual conversation but a work of art in which such a conversation is simulated so that the author can make known her beliefs.

Because so little is known about the actual Cleopatra, it is impossible to know what education she received. However, assuming she lived in Alexandria, some speculations can be made about the milieu she lived in and her likely education. Although the origins of alchemy undoubtedly run far back in several cultures, including Western culture, western alchemy seems to have come together as a coherent idea around the third century B.C. Originally, the main focus of alchemy was to find chemicals that would simulate gold when applied to various metals. Thus alchemy was, from the beginning, an early form of experimentation in the science of chemistry. It also seems to have been a con game in that in its earliest manifestations, its practitioners seem to have been engaged in trying to pass off alloys of other metals as gold. Later, these mundane concerns disappeared, and alchemy became both a philosophical quest and an ongoing chemical experiment. Also mixed into this practice was worship of the Greek god Hermes, mythical inventor of arts and sciences. The practice was strictly controlled by priests, and their findings were kept secret. Thus Cleopatra probably came from a priestly family. She may even have been a priestess in one of the pagan Greek sects whose most important deity was Hermes.

From about the third century B.C., the center of the western alchemical world seems to have become Alexandria. The city itself was founded in 332 B.C. by Alexander the Great. It quickly attracted not just Greeks but peoples from the whole of the Mediterranean basin—Jews, Arabs, Romans, Phoenicians, and others. Other well-known alchemists of this period were MARIA THE JEWESS and Sophia the Egyptian; both may have been colleagues of Cleopatra's.

By Cleopatra's time, alchemists were no longer swindlers, but philosophers. Their studies were based on the Greek idea that all substances possessed an eternal and unchanging form beneath the surface of things. This essence, called the *prima materia,* appeared to change with the addition of other, baser elements. It was the alchemist's task to strip away these baser materials and find this essence of existence, which was often identified with mercury (although this was not the actual element mercury, but a more mystical substance). If an alchemist could obtain this *prima materia,* then he or she could make any other metal, including gold, merely by adding the proper "coloring" agents to it.

Cleopatra's project, as displayed in *A Dialogue of Cleopatra and the Philosophers,* was more spiritual. She viewed alchemy as an organic process and mixed the ideas of purification of materials with images of the purification of the soul. She frequently used sexual symbolism—conception and birth—to speak of this process. Another of Cleopatra's important legacies are the alchemical symbols and drawings she left on a single sheet of papyrus called the "Chrysopoeia." This rare document shows the Ouroboros, a snake eating its tail, which is the symbol of eternity. It also includes sketches of an ancient still used to distill compounds and another ancient chemical device called the *kerotakis.*

Cleopatra's work, and the work of succeeding alchemists, laid much of the groundwork for the eventual evolution of the science that we today know as chemistry.

Cobb, Jewel Plummer
(1924–)
American
Cellular Biologist

A biologist by training, Jewel Cobb has been a researcher, teacher, and administrator during her career. As a researcher, she has done valuable work studying the growth of cancer cells and evaluating treatments to combat cancer. For many years, she was also a teacher of biology, and during the final phase of her career, she served as a university administrator. She ended her career running California State University at Fullerton, an important part of the University of California system.

Born on January 17, 1924, in Chicago, Illinois, Jewel Plummer was the daughter of Frank V. Plummer and Carriebel Cole Plummer. Plummer's family was upper middle class and comfortable. Her father was a physician, and her mother was a dancer who taught dance and physical education in Chicago's schools. Plummer's father especially encouraged her interest in science when she was young. In 1954, Plummer married Roy Cobb, an insurance salesman. She has a son, Jonathan, born in 1957.

In 1941, Cobb graduated from a public high school in Chicago and enrolled at the University of Michigan. Her time at Michigan was marked by racial tension. Black students there were put into a segregated dormitory and could not eat at popular student restaurants near the campus. After a year at Michigan, Cobb transferred to the historically black Talladega College in Alabama, from which she graduated with a B.A. in biology in 1944. Cobb then went north to New York University (NYU) to work on her master's and doctoral degrees. She was awarded an M.A. in 1947 and, in 1950, won her Ph.D. in cell physiology.

Jewel Plummer Cobb, who is renowned as both a cancer researcher and college administrator. *(Courtesy of Jewel Plummer Cobb)*

At NYU, Cobb studied skin pigment formation in vitro. She continued in vitro experimentation in her first job after completing a Ph.D., as a researcher for the National Cancer Institute at New York's Harlem Hospital. For two years (1950–52), she studied the growth of cancer cells and the effects of anticancer chemotherapy agents against these cells. Cobb briefly left New York to take a teaching job at the University of Illinois medical school in Chicago in 1952. However, by 1954, she had returned to New York to work at NYU's Tissue Culture Research Laboratory. In what she terms, "a most exciting phase of cell research," Cobb continued her work on cancer cell growth and the new chemotherapy drugs that were being invented to beat cancer. Her early work on drugs such as 6-mercatopurine and actinomycin D established a baseline of knowledge on which other researchers would build in the following years.

Cobb began the most intensive teaching phase of her career in 1960 when she left the NYU research effort to take a job as professor of biology at Sarah Lawrence College in Bronxville, New York, just outside of New York City. She remained at Sarah Lawrence until 1969, when she accepted a job as dean of Connecticut College in New London. Cobb found that she like being an administrator. In 1976, she took the job of dean at Douglass College, the women's college of New Jersey's Rutgers University. In 1981, she accepted the final appointment of her career, the post of president of California State University at Fullerton. At Cal State Fullerton, she ran a university with 24,000 students and 1,000 faculty members, and she lobbied the California legislature for funding for her institution. Cobb retired from that job and from her full-time career in education and science in 1990.

For her work as a scientist and educator, Cobb has been elected a fellow of the New York Academy of Sciences and the American Association for the Advancement of Science. She has also had student dormitories named after her at Douglass College and Cal State Fullerton. She is the recipient of more than 20 honorary doctorate degrees.

Cohn, Mildred
(1913–   )
American
Biochemist

A student of and partner with several future Nobel Prize–winning chemists, Mildred Cohn contributed to humankind's understanding of biochemical processes through her study of chemical reactions within animal cells. She pioneered the use of nuclear magnetic resonance (NMR) devices to study biochemical mechanisms within enzyme reactions. She also studied energy transduction and cellular reactions to the organic substance adenosine triphosphate (ATP).

Born in New York City on July 12, 1913, Cohn was the second of two children of Isidore Cohn and Bertha Klein Cohn, both of whom immigrated from Russia to the United States just after the turn of the century. The education of their children was important to the Cohns. With her parents' encouragement and because she was a bright child, Mildred moved rapidly through the New York public school system. She graduated from high school in the Bronx in 1927 when she was 14 years old. In 1938, Cohn married Henry Primakoff. They have three children.

Because Cohn came from a poor family, she decided to go to Hunter College in Manhattan. Part of the City College system, it charged its students no tuition. Cohn had realized in high school that she wanted to become a scientist, but she was not sure which scientific discipline to pursue. She liked physics and chemistry equally well. Because Hunter offered a major in chemistry but not physics, she decided on the former. Graduating from Hunter in 1931 at the beginning of the Great Depression, Cohn tried but failed to get a scholarship for graduate chemistry studies. She then took the savings she had accumulated from several part-time jobs and entered Columbia University in New York City. There she studied under future Nobel laureate Harold Urey, who inspired her with his passion for chemistry. Because of lack of money, Cohn had to drop out of graduate school after one year. She then took a job with the National Advisory Committee for Aeronautics, a scientific research arm of the federal government. By 1934, she had accumulated enough money to return to graduate studies at Columbia. She won her Ph.D. in organic chemistry in 1937.

After completing her doctorate, Cohn landed a job with biochemist Vincent du Vigneaud at Washington University in St. Louis. Cohn was valuable to du Vigneaud because she had experience at Columbia working with isotopic tracers to examine biochemical reactions in human and animal cells. At du Vigneaud's lab, she worked intensively on an assignment given her by her boss, charting the progress of isotopic tracers through the metabolism of sulfur-amino acids. This study enabled Cohn to determine what kinds of chemical reactions were occurring during this cellular process. Cohn worked with du Vigneaud for nine years and followed him when he was offered a job in the New York area.

In 1946, Cohn returned to Washington University to take a job with a pair of future Nobel laureates, Carl and GERTY CORI, biochemists who were working on the study of enzymes. This time, Cohn extracted a promise that she could do research on projects of her own choosing. She began using an NMR to track phosphorus as it reacted with ATP, a study that revealed considerable information about the biochemistry of ATP. By 1960, Cohn had moved again, this time to the Department of Biophysics at the medical school of the University of Pennsylvania. She became a full professor there in 1961 and taught and did research full time until 1982, when she retired.

For a career of outstanding work, Cohn was awarded the Garvan Medal from the American Chemical Society in 1963, the Franklin Institute's Cresson Medal in 1975, and the National Medal of Science in 1983. She holds numerous honorary doctorates from American universities.

Colborn, Theodora
(1927–   )
American
Zoologist and Ecologist

Trained in her youth as a pharmacist, Theodora Colborn returned to university studies in her fifties to earn a degree in zoology. She was a leader in a series of studies that examined pollutants that seemed to be causing reproduction problems among animals that ate fish from the Great Lakes. As a researcher at the World Wildlife Fund, Colborn has urged that the federal government and state legislatures do more to determine the cause of these pollutants and take actions to stop their use.

Theodora Colborn was born on March 28, 1927, in New Jersey. She graduated from Rutgers University with a degree in pharmacy, and for 15 years, she worked as a pharmacist in her home state. In 1964, she decided to change her life radically by moving to Colorado to become a sheep farmer. In Colorado, she began to connect deeply with nature, and by 1981, she had become interested enough in the ecological challenges facing humans and other species that she enrolled in a doctoral zoology program at the University of Wisconsin at Madison. She won her Ph.D. in zoology in 1984.

After earning her Ph.D., Colborn moved to Washington, D.C., and began working with environmental groups. At the Conservation Foundation (which later merged with the World Wildlife Fund), she collaborated on a book that assessed the environmental condition of the Great Lakes. Colborn noted that the amount of industrial pollution that ended up in the lakes had decreased considerably since the 1960s. Nonetheless, after a review of scores of scientific papers that had been written about pollution and wildlife along the lake, Colborn noticed that a number of animals that ate fish and lived near the lakes were experiencing problems reproducing healthy offspring. A handful of zoological and ecological studies had examined this problem, and a surprising number of them had determined that these species were producing a high number of sick or deformed babies.

The cause of this high mortality among the young of these species, Colborn speculated, might be found in chemicals that affected the hormone production in these animals. Too much or too little hormone production during pregnancy can cause birth defects or the birth of animals with insufficient immune protection. In the latter case, offspring are much more likely to die of diseases that would cause only mild illnesses in healthy animals. Colborn then began looking at the kinds of pollutants that still could be found in the waters of the Great Lakes.

In 1991, she met and corresponded with a group of scientists to examine which pollutants might affect hormone production in animals. After sifting through stacks of scientific data, Colborn's group narrowed down to a list of 51 chemicals that were likely to have an adverse effect on the hormones of the affected animals. These were chemicals that are used in plastics, pesticides, and common household objects. What these chemicals do when ingested by an animal is mimic the action of a hormone, thus upsetting the delicate hormonal balance in the creature that is needed to produce healthy offspring. The parents of the offspring appear normal; the poisonous effects of the chemicals can be seen only in the offspring.

As happened with RACHEL CARSON in the 1960s, industry groups and other scientists were quick to dispute Colborn's findings. The argument made against her was that the data was insufficient to clearly establish a link between these chemicals and the birth defects. In the meantime, Colborn continues her studies and believes that this conclusive link will be established.

Colden, Jane
(1724–1766)
American
Botanist

Given a general education that far surpassed that of most women of her time and encouraged in the study of botany by her father, Jane Colden became one of the most celebrated early North American botanists. She was well versed in the botanical classification system established by Carl von Linné (also known as Linnaeus), and she discovered and classified a number of plants that grew in New York, Connecticut, New Jersey, and Pennsylvania.

Born on March 27, 1724, in Coldengham, the estate of her family in New York's Hudson River valley, Jane Colden grew up in an illustrious family. Her father, Cadwallader Colden, was a Scot who had been trained in London as a physician. He emigrated to Philadelphia in 1710 to establish his medical practice. In 1715, he returned to England for a year, during which time he married Alice Christie. Cadwallader and Alice Colden both returned to the British North American colonies in 1716; by the early 1720s, they were living in the province of New York where Cadwallader had built a successful political career. He became surveyor general of New York and eventually would become lieutenant governor of the province. Jane Colden was the second of 10 children. She did not marry until she was 35, which was late for a woman of her time. Her husband was the physician William Farquhar.

The Colden household was typical of the small group of colonial aristocratic intellectuals to which they belonged. There was a fairly substantial library that contained works of philosophy and science as well as some literature. Jane Colden was educated at home in Coldengham by her mother, Alice. There were no public schools then in that relatively remote region of New York. Thus only children of the wealthy received an education, and usually of these privileged children, only the boys were educated. Cadwallader Colden, however, believed that all his children deserved an education. After Jane had received her basic education from her mother, her father gave her an advanced education that centered on the sciences and, in particular, botany.

Colden began her training in botany by reading the works of Linnaeus, which her father had translated from Latin to English. From her late teens, that is, from around the early 1740s, until 1759, Colden sought out plants from the region in which she lived. Because it was then deemed unsuitable for a woman to embark alone on collection expeditions, she probably seldom ventured out to collect plants by herself. But she undoubtedly accompanied others when they went into the fields and woods, and she certainly solicited plant samples from her friends and neighbors. Because of her frequent correspondence with American botanists John and William Bartram, and English botanists Alexander Garden, John Ellis, and others, Colden was aware of the discoveries being made during her life. She compiled a catalog of more than 300 local plants and is credited with the discovery of the northern golden thread and the gardenia, which she named after Alexander Garden. Both of these were at about the same time discovered and named by other, male botanists, so Colden received no credit for these finds.

Colden seems to have ended her botanical research after her marriage to William Farquhar. She died on March 10, 1766, after giving birth to a baby that also died. Her work was praised by Alexander Garden as being "extremely accurate." She was the only woman botanist whose work was included in Linnaeus's botanical masterwork *Species Plantarum*.

Cole, Rebecca J.
(1846–1922)
American
Physician

Rebecca J. Cole was the first African-American woman to graduate from the Women's Medical College and the second African-American woman in the United States to earn a medical degree, a major accomplishment in the wake of the Civil War. Cole devoted her career, which lasted for 50 years, to improving the health care of women and children, especially those in destitute circumstances.

Cole was born on March 16, 1846, in Philadelphia, Pennsylvania. She was the second of five children. At a time when education was discouraged for African Americans, Cole studied at the Institute for Colored Youth (now Cheney University). She graduated in 1863, and then she taught for a year before pursuing medical studies at the Women's Medical College (now part of the Allegheny University of the Health Sciences) in Pennsylvania. When she received her M.D. in 1867, she was only the second African-American woman to do so (Rebecca Lee Crumpler had graduated from the New England Female Medical College in Boston a mere three years earlier, in 1864).

The year she earned her medical degree, she moved to New York City, where she went to work as a resident physician for the New York Infirmary for Women and Children. This hospital had been founded in the 1850s by women physicians, who both owned and operated it. Hospital co-founder ELIZABETH BLACKWELL, the first white woman to earn a medical degree in the United States, assigned Cole to the position of sanitary visitor, whereby she educated the poor in their homes throughout the city, demonstrating proper hygiene, prenatal and infant care, and health maintenance. With the mushrooming urban population, Cole's services were in constant demand, and her caregiving and information dissemination clearly improved the conditions of the poverty-stricken families she visited. Although this work was both physically and emotionally arduous, Cole persevered in this position because she saw that her efforts produced tangible results, raising the standard of living for the families she visited.

After her time in New York City, Cole moved to Columbia, South Carolina, where she practiced medicine briefly. She then moved to Washington, D.C., to serve as a superintendent of the Government House for Children and Old Women. After this stint, she returned to her hometown, Philadelphia, where she set up a private practice in the city's impoverished South section. In 1873, she collaborated with Charlotte Abby, another female physician, to establish the Women's Directory Center to serve the underprivileged. The mission of this initiative was to provide medical and legal services to women and children throughout the city of Philadelphia who could not otherwise afford these services.

In 1896, Cole founded the National Association of Colored Women. That year, Cole published an article entitled "First Meeting of the Women's Missionary Society of Philadelphia" in the October and November edition of *The Women's Era*. Cole practiced medicine for 50 years, continuing to provide care for the destitute and overlooked until her death. Cole died in Philadelphia on August 14, 1922.

Colmenares, Margarita Hortensia
(1957–   )
American
Environmental Engineer

Margarita Colmenares achieved many firsts in her career as an environmental engineer: She was the first Hispanic engineer to serve a White House Fellowship, and she was the first woman president of the Society of Hispanic Professional Engineers (SHPE). In addition, she was often the first woman or the first Latina that many of her colleagues ever worked with; her expertise, efficiency, and excellent communication skills convinced them of the complete competence of women and Hispanics as workers in one of the most challenging professions, engineering.

Colmenares was born on July 20, 1957, in Sacramento, California, the eldest of five children. Her parents, Luis S. Colmenares and Hortensia O. Colmenares, were immigrants from Oaxaca, Mexico, and believed in education as the key to self-betterment. Despite the financial strain, they sent their daughter to a private all-girls Catholic school, where Colmenares founded a Mexican-American student coalition.

She commenced her undergraduate career studying business at California State University in Sacramento but found she was more interested in engineering, so she transferred to Sacramento City College to study chemistry, physics, and calculus, prerequisites for degrees in engineering. She simultaneously worked part-time at the California Department of Water Resources inspecting dams and water-purifying plants. The Engineering School at Stanford University accepted her transfer application, and she financed her study there with five different scholarships. Besides her studies, she also found time to teach, direct, and perform Mexican folk dance with the Stanford Ballet Folklorico. In 1981, when she earned her B.S. in civil engineering, her mother also received her bachelor's degree in education after 12 years of part-time study.

Between her junior and senior years at Stanford, Colmenares worked for the Chevron Corporation in Texas and California through its Cooperative Education Program. After graduation, she remained with Chevron, where she worked her way up through the ranks from recruiting coordinator at the San Francisco office to field construction engineer in Salt Lake City to foreign trade representative to compliance specialist in Houston. In 1986, she became the lead engineer directing an $18-million environmental cleanup project at the Chevron refinery in El Segundo, California, where she then remained after 1989 as an air-quality specialist.

In 1982, Colmenares founded the San Francisco chapter of the SHPE, and seven years later, in 1989, the SHPE elected her as its first woman president. She filled this position for one year while still working, but she convinced Chevron to name her executive-on-loan for the next year, granting her a paid leave of absence. Chevron granted her another paid leave from 1991 through 1992, when she became one of 16 to win a White House Fellowship, the first Hispanic engineer to receive this honor. The White House assigned her to the Department of Education, where she served as a special assistant to deputy secretary David T. Kearns.

After returning to Chevron in 1992 to work on international operations in Latin America, the Department of Education hired Colmenares back in 1994 as the director of corporate liaison. In this position, she sought to enhance educational opportunities in collaboration with the industry sector.

Although Colmenares's career is still in its early stages, she has already received numerous honors: In 1989 alone, *Hispanic Engineer* magazine granted her its Community Service Award, *Hispanic* magazine named her its Outstanding Hispanic Woman of the Year, and the SHPE named her its Hispanic Role Model of the Year. *Hispanic Business* magazine listed her twice (in 1990 and 1992) as one of its 100 most influential Hispanics in the United States. In another first, she was the youngest recipient of the California Community College League's Outstanding Alumni Award.

Colwell, Rita Rossi
(1934–   )
American
Microbiologist

Beginning her career as a geneticist, Rita Colwell eventually focused her scientific curiosity on the fields of marine microbiology and biotechnology. She has studied the microbiology of the oceans and estuaries, looked at how pathogens such as *Escherichia coli* live in water, and explored ways for humans to harvest marine life such as seaweed. She is also deeply involved in efforts to understand and preserve healthy marine ecology.

Born on November 23, 1934, in Beverly, Massachusetts, Rita Rossi is the daughter of Louis Rossi and Louise DiPalma Rossi. She married another scientist, Jack Colwell, in 1956. She has two daughters.

Colwell excelled in chemistry in high school, and even though she was advised by her high school counselors that chemistry was not a field for women, she enrolled in Purdue University in 1952 intent on becoming a chemist. After taking large and poorly taught chemistry classes at Purdue, she decided to switch her major to bacteriology; she was awarded a B.S. in this field in 1956. After finishing her undergraduate degree, Colwell was accepted into the doctoral program in genetics at the University of Washington. She received her Ph.D. in that field in 1961.

From 1961 to 1964, Colwell taught at the University of Washington as an associate professor. She then moved to Georgetown University in Washington, D.C., where she served as an assistant, then associate professor from 1964 to 1972. She completed her career at the University of Maryland where she began her career as a professor of microbiology.

Beginning with her doctoral studies at the University of Washington and continuing through her work at Georgetown University and the University of Maryland, Colwell focused on the study of marine bacteria. She studies all types, those that are harmful to fish life and humans, and those which are beneficial to these lifeforms. The dangerous organisms she has studied are the bacteria *E. coli,* sometimes found in estuary shellfish, and harmful, even deadly, if eaten by humans. She has also looked at the ecology of *Vibrio chlorae* and the process by which it transmits the disease cholera. She has devised computer programs that help marine microbiologists identify these bacteria. Many of her studies looking at bacteria and the state of the marine ecological systems in which these bacteria are found have centered on the Chesapeake Bay, the most prominent estuary that borders her home state, Maryland. Her studies on the Chesapeake Bay are thorough; they are concerned with not only the bacteria in the bay but the health of its fish and shellfish and the effects of chemical and animal-produced waste that have ended up in the bay.

As director of the University of Maryland Sea Grants Program, Colwell has studied and promoted studies that look at the possibility of harvesting medicinal bacterial and plant life from the sea. She believes that the seas are greatly underused as sources of compounds that could yield new drugs, and she has also studied new ways to harvest fish from sea farms and use marine biotechnology to recycle waste. Colwell became director of the National Science Foundation in 1998.

For her work in marine microbiology, Colwell has been awarded the Fisher Award from the American Association of Microbiologists (1985), the Gold Medal of the International Biotechnology Institute (1990), and the Purkinje Gold Medal Achievement Award from the Czechoslovakian Academy of Sciences (1991). Her alma mater, Purdue University, awarded her an honorary doctorate in 1993.

Colwin, Laura North Hunter
(1911–)
American
Biologist

Fusing her early interest in protozoology with the study of embryology, Laura Colwin became known for her work on the fertilization and reproduction of microscopic organisms such as Protozoa. These are unicellular creatures, many of which occupy an ecological niche as parasites of larger creatures. She also worked on embryology problems for worms and sea creatures such as sea cucumbers. During these studies, she discovered that some of the existing ideas about how these creatures are fertilized were incorrect; her work gave a more accurate description of their reproduction mechanisms.

Born on July 5, 1911, in Philadelphia, Pennsylvania, Laura Hunter was the daughter of Robert John Hunter, a physician, and Helen Virginia North Hunter, a housewife. Both of Hunter's parents were well educated (Helen Hunter had majored in German and Latin at Bryn Mawr), and Laura was encouraged to pursue her studies. From an early age, she learned about the natural world by attending lectures and exhibits with her parents at the museum of the University of Pennsylvania. In 1940, she married Arthur Lentz Colwin, an embryologist.

Colwin attended a public primary school in Philadelphia, but at 13 she enrolled in the Phoebe Anna Thorne School, a private school for girls. The Thorne School was informally affiliated with Bryn Mawr, a women's college near Philadelphia that Colwin's mother had attended. Like her mother, Laura Colwin decided to attend Bryn Mawr, enrolling in an undergraduate program there in 1928. After taking an introductory biology course, Colwin realized that she wanted to study that discipline. She graduated with a major in biology from Bryn Mawr in 1932. She has described this undergraduate experience as "lively, austere, merry . . . irritating [and] invigorating," something that "marked [her] for life." Hunter then enrolled in a doctoral program at the University of Pennsylvania. She decided to specialize in protozoology and began studying cilia, protozoans with hairlike strings that float from their bodies and help propel them through their liquid environment. Her mentor in these studies was the protozoologist D. H. Wenrich. She won an M.A. degree from the University of Pennsylvania in 1934 and her Ph.D. in protozoology from the same institution in 1938.

Colwin's interest in marine life began when she attended the Woods Hole Marine Biological Laboratory (MBL) in Massachusetts in 1933. She returned to Woods Hole in 1938. A summer institute for the study of ocean life, Woods Hole introduced Colwin to the study and collection of marine organisms. From 1936 to 1940, Colwin taught biology at Pennsylvania College for Women (now called Chatham College). In 1940, after she married Arthur Colwin, she took a job as an instructor of zoology at Vassar College in New York, where she remained until 1943. At the end of World War II, Colwin got a job as an untenured instructor at Queen's College, where her husband also worked. She remained at Queens College until she retired in 1973.

At Queens College, and during the summers at the MBL in Woods Hole, Colwin concentrated on her research, which she did in collaboration with her husband. Her first project was to study the development of the acorn worm (*Saccoglossus kowalevskii*) from an embryo into the early part of its life cycle. In doing this work, she realized that information about how the acorn worm's eggs are fertilized was inaccurate. She pioneered the study of this process in the acorn worm and other organisms through the use of an electron microscope, taking thousands of photos of the fertilization process through this device. In this way, she noted that the sperm of these creatures does not penetrate the egg, but that sperm and egg fuse.

For her years of work at their university, the administrators of Queen's College finally made Laura Colwin a full tenured professor of biology in 1966. She is also a member of the American Society for Developmental Biology and the New York Academy of Sciences. From 1971 to 1975, she was a trustee of the MBL at Woods Hole.

Comstock, Anna Botsford
(1854–1930)
American
Entomologist

The first woman to become a professor at Cornell University, Anna Botsford Comstock was a prominent entomologist and science educator. She was instrumental in advancing the study of nature in schools, and she authored a best-selling book on natural cycles and processes—*Handbook of Nature Study*. In addition, she edited a scholarly journal and cofounded Comstock Publishing Company with her husband, John Henry Comstock.

The only child of Marvin and Phoebe Botsford, Anna Comstock was born in 1854 in New York State. When she enrolled at Cornell University in 1874, she intended to study English. However, she took a class in invertebrate zoology taught by John Henry Comstock and became fascinated by the subject of entomology (the study of insects). She married her professor in 1878, and the couple moved to Washington, D.C.

While her husband served as chief entomologist at the U.S. Department of Agriculture, Anna Comstock helped him with clerical, editorial, and laboratory work. She also illustrated his books on entomology with detailed drawings that won her praise. The couple returned to Cornell in 1880, and Comstock completed her bachelor's degree in natural history five years later. She then immersed herself in matters of science education. In 1895, she became a spokesperson for the nature study movement, which sought to instruct children about the relationship between farming and the study of nature. She also lectured about science education at several teachers' colleges in New York State.

Anna Comstock made history in 1899 when she was appointed assistant professor at Cornell, becoming the first woman to join the university's faculty. Unfortunately, several of Cornell's conservative trustees objected to her rank, and she was demoted to the position of lecturer in 1900. In 1911, she wrote her most famous work, *Handbook of Nature Study*, in which she discussed natural events and cycles. She was reinstated as an assistant professor at Cornell in 1913 and was promoted to full professor in 1920. Comstock exerted considerable influence on her field through her writing. Along with her seminal book, she authored many articles and contributed to a diverse array of publications. In addition, she edited the journal *Nature Study Review* from 1917 to 1923.

Comstock retired from full-time teaching in 1922, although she remained active as a part-time professor at Cornell until 1930. Her legacy was profound. Through her tireless efforts in the nature study movement, she influenced science education policy. Moreover, her writings popularized natural history for a lay audience. The *Handbook of Nature Study* became a best-seller, printed in 25 editions and eight languages. The book was especially popular among gardeners, farmers, teachers, and others interested in the outdoors. Moreover, the Comstocks left behind a successful publishing enterprise, Comstock Publishing Company, which printed books on entomology. The business was later incorporated into Cornell University Press. Anna Comstock also furthered the cause of women in science. Through her struggle to remain a professor at Cornell University, she broke down barriers and paved the way for future women to enter academia. She died in August of 1930 at the age of 75.

Conway, Lynn Ann
(1938–)
American
Electrical Engineer

Lynn Ann Conway pioneered significant advancements in the architecture of computers that revolutionized their construction; in essence, her two major innovations sped up the arrival of the digital age. She first worked with a team that integrated computer circuitry design, streamlining the process that previously required multiple engineers, each with specialized knowledge. Second, she simplified the process for fabricating computer chip prototypes, thereby greatly reducing the time it took to test newly invented software and hardware. She collaborated with Carver Mead to publish this protocol, called very large scale integrated (VLSI) circuit design, in the 1980 textbook *Introduction to VLSI Systems*, which instantly became a classic.

Conway was born on January 2, 1938, in Mount Vernon, New York. Her excellent performance in high-school mathematics and physics courses prompted her to major in physics at the Massachusetts Institute of Technology (MIT), which she left in her third year to travel the United States. In the early 1960s, she finished her undergraduate degree at Columbia University, where she earned her B.S. in 1962. She remained there for graduate work, receiving her M.S. in electrical engineering in 1963.

Conway impressed a visiting professor with a software system design, leading him to convince IBM to offer her a research position at its Yorktown Heights, New York, research facility in 1964. The next year, she worked on the IBM-ACS (Advanced Computer Systems) team that developed the first "superscalar" computers; although IBM abandoned ACS as incompatible with its 360 line, her invention of "dynamic instruction scheduling" survived as the platform technology incorporated by Intel, Hewlett-Packard, Sun, and Compaq in their processors. In 1968, IBM fired Conway when it learned of a medical condition she had for which she was participating in controversial experimental treatments that ultimately solved her problem.

In 1969, Conway secured a position as a senior staff engineer with Memorex Corporation, where she remained until 1973, when the Xerox Corporation hired her as research engineer at its prestigious Palo Alto Research Center. There, she collaborated with Mead and a team of colleagues to simplify the relationship between digital system architecture and microelectronics, resulting in the development of the VLSI chip design methodology.

Lynn Conway, a pioneer in the advancement of computer architecture. *(Courtesy of Lynn Conway)*

As a visiting associate professor of electrical engineering and computer science at MIT from 1978 though 1979, she designed a VLSI course that became the prototype for similar courses nationwide, using her book as the text. Her MIT students designed their own chips by her methodology, then sent the blueprints to Xerox over the ARPAnet (a predecessor to the Internet); Xerox contracted with Hewlett-Packard Research to fabricate the chips, which it returned to the students six weeks later, an amazingly quick turnaround time.

In 1983, the United States Department of Defense hired Conway as the chief scientist and assistant director of strategic computing for the Defense Advisory Research Projects Agency. She helped orchestrate the department's Strategic Computing Initiative during her two years there, then she moved on to the University of Michigan in 1985 as a professor of electrical engineering and computer science and as the associate dean of the College of Engineering.

Conway's many honors include the 1984 Harold Pender Award from the Moore School at the University of Pennsylvania, the 1985 John Price Wetherill Medal from the Franklin Institute, the 1985 Meritorious Civilian Service Award, presented by the Secretary of Defense, and the 1990 National Achievement Award from the Society of Women Engineers. In 1998, she retired to emerita status at the University of Michigan, though she remained active conducting research in her field.

Conwell, Esther Marly
(1922–   )
American
Physicist

Esther Conwell was honored early in her career with the Society of Women Engineers Achievement Award in 1960. Almost four decades later, in 1997, she became the first woman to receive the Edison Medal from the Institute of Electrical Engineers in recognition of her career of meritorious achievement in the electrical arts and sciences. Conwell focused her research on solid-state physics, conducting pioneering work on the transmission of electronic signals through semiconductors, research that ushered in the technological and digital age.

Conwell was born on May 23, 1922, in New York City. She remained in the city for her undergraduate studies at Brooklyn College, where she received her bachelor's degree in 1942. She then moved upstate to do graduate work at the University of Rochester, where she conducted her research under Victor Weisskopf. Together, they devised the Conwell-Weisskopf theory, describing how "impurity ions," which carry electricity by emitting electrons, actually impede the flow of electrons. Her master's thesis was considered sensitive material, and thus it was locked away until after World War II, when it was finally published in 1950. She received her master's degree in 1945.

Conwell performed doctoral work at the University of Chicago, where she earned her Ph.D. in physics in 1948 while simultaneously teaching physics at Brooklyn College from 1946 until 1951. She then spent a short year on the technical staff at Bell Telephone Laboratories before transferring to GTE Laboratories, where she remained for the next two decades. She started as an engineering specialist, before GTE promoted her to manage the physics department in 1963 when she returned from a year as a visiting professor at the École Normale Supérieure in Paris. Over the next seven years, she directed the solid-state physics group and subsequently managed the electro-optics program.

In 1972, the Massachusetts Institute of Technology appointed Conwell to the Abby Rockefeller Mauze Chair for one semester, and thereafter the Xerox Corporation hired her as a principal scientist. In 1981, Xerox promoted her to the rank of research fellow. While at Xerox, she focused much of her research on the physics of xerography, specifically on how photoconductors transport electrical charges. In 1990, while still working for Xerox, she taught at the University of Rochester as an adjunct professor and acted as the associate director of the university's Center for Photoinduced Charge Transfer, which was funded by the National Science Foundation. When she retired from Xerox 1998, she joined the university full-time in its chemistry department.

Conwell was the only professor at the University of Rochester to be a member of both the National Academy of Sciences and the National Academy of Engineering. She was also a member of the American Academy of Arts and Sciences and the American Physical Society. Throughout her career, she received four patents for her inventions; her publications included more than 200 papers, one monograph on high field transport in semiconductors in the solid-state physics series, and she edited two other books.

Cori, Gerty Theresa Radnitz
(1896–1957)
American
Biochemist

With her husband, Carl Cori, Gerty Cori established how the body stores and uses food, a process called *metabolism* that was not well understood until the Coris began studying it in the late 1920s. For this work, and for her later study of enzymes, proteins that cause physical changes in the body, Gerty Cori, with Carl Cori and an Argentine

doctor named Bernardo Houssay, was awarded the 1947 Nobel Prize for medicine.

Gerty Cori was born Gerty Theresa Radnitz in Prague on August 15, 1896. She was the oldest of three daughters of Otto and Martha Radnitz, German-speaking Jews who lived in what was then the Austro-Hungarian Empire, which was ruled by the last of the Hapsburg dynasty from the imperial capital of Vienna. Otto Radnitz was a prosperous chemist who managed a sugar-beet refining business. Otto arranged for Gerty to be tutored at home. When she was 10, she was sent to a girls school for the rest of her secondary education. (Later, after she had become an American citizen, Cori had one child, Carl Thomas, born in 1936.)

Although her father had not intended for her to continue her studies in a university, Gerty Radnitz insisted on studying at a medical school. After studying a year for a difficult entrance examination, Cori passed the exam and

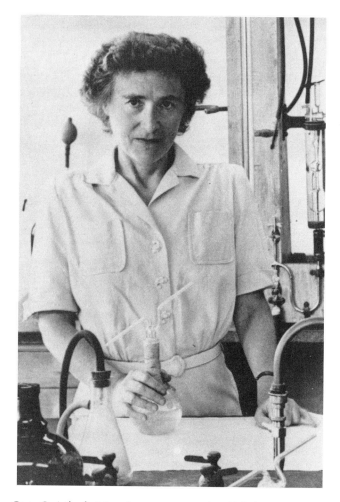

Gerty Cori, the first American woman to win a Nobel Prize in science. *(Bernard Becker Medical Library, Washington University School of Medicine)*

entered Carl Ferdinand University in Prague in 1914. Even though World War I began in the summer of her first year, the city was physically untouched by the war and Gerty Radnitz studied here for six years, finally earning her medical degree in 1920. She also met Karl Cori, a medical student from the Adriatic port city of Trieste. The two realized that they shared remarkably similar imaginations and were married in the summer after their graduation.

Gerty Cori's first job after medical school was as a researcher at Vienna's Karolinen Children's Hospital. Karl Cori had taken a job as a teacher at the University of Graz. Both Coris were by now disillusioned about their chances of professional success and personal happiness in war-ravaged Europe. Anti-Semitism was still an obstacle to their careers, one that could have affected Karl because Gerty was from a Jewish family. In 1922, Karl took a job as a pathologist at the New York State Institute for Malignant Diseases in Buffalo. Six months later, he persuaded his bosses to hire Gerty as the institute's assistant pathologist.

Even though they were supposed to be working to find parasites that the institute's director believed caused cancer, Gerty and Carl (he changed the spelling of his name after he moved to the United States) instead began devoting most of their time trying to figure out how the body converted carbohydrates into energy for the body. In 1922, Canadian physician Frederick Banting had discovered the hormone insulin, which he began using to treat people with diabetes. Banting and other doctors knew that insulin worked for diabetics, but they did not know why it worked. Gerty Cori, with Carl, decided this was a promising area for research.

By studying the effects of glucose, the sugar converted by the body from carbohydrates, and glycogen, a variant of glucose stored in the body's fat and muscles, in laboratory animals, they described how the body cycles these substances to store and use energy. This discovery, known as the Cori cycle, was announced in 1929. In 1931, Gerty and Carl moved to Washington University in St. Louis. As before, Carl was hired as a professor, while Gerty was given a more lowly job even though she was a full partner with her husband in their research. In St. Louis, they refined the study of glucose and glycogen and identified several previously unknown enzymes that aid the body in burning carbohydrates as fuel.

For her work in understanding the biochemistry of the human body, Cori was elected to the National Academy of Science and the National Science Foundation board. In 1947, after she had received the Nobel Prize, she was finally made a full professor at Washington University. That same year, she learned that she had myelofibrosis, a cancer of the bone marrow. She struggled valiantly, continuing her work for the next 10 years before succumbing to the disease in 1957. A good summation of her contribution to science can be found in the words of the Nobel Foundation: She unraveled "the intricate patterns of chemical reactions in the living cells, where everything appears to depend on everything else."

Cowings, Patricia Suzanne
(1948–   )
American
Research Psychophysiologist

Astronauts nicknamed Patricia Cowings the "baroness of barf," because she induced nausea in them in order to study their physiological and psychological reactions to space motion sickness. She then trained them in biofeedback techniques to control their biological responses to "zero-gravity sickness syndrome." Cowings was the first woman that the National Aeronautics and Space Administration (NASA) trained to be an astronaut, though she never served as one in space. Cowings overcame obstacles due not only to her gender and race, as an African American, but also due to her youth, as she earned her doctorate at the age of 23 and was not taken seriously early in her career.

Cowings was born on December 15, 1948, in New York City. Her father, Albert S. Cowings, owned and ran a grocery store, and her mother, Sadie B. Cowings, was an assistant teacher after earning her A.A. degree at age 65. With three brothers (now a two-star army general, a jazz vocalist, and a musician), Cowings learned early on that she could do anything that males could. By the age of nine, she became interested in science and particularly in becoming an astronaut, in part due to her love of *Star Trek*. She hated mathematics, though, and had to discipline herself to master statistics, which she called "sadistics."

Cowings studied psychology at the State University of New York at Stony Brook, where she earned a research assistantship in 1968. She graduated cum laude with psychology honors in 1970 and proceeded to graduate study at the University of California at Davis, where she was a graduate research assistant. She received the Distinguished Scholarship Award in 1971. That summer and the next, she participated in the NASA-Summer Student Program, commencing her lifelong relationship with the governmental agency. She wrote a paper proposing self-regulation to overcome biomedical problems in space for her professor, Dr. Hans Mark, who directed NASA's Ames Research Center (ARC) at the time. Upon her graduation in 1973 with both an M.A. and a Ph.D. in psychology, she received a National Research Council Post-Doctoral Associateship, which funded her research at ARC for two years, until 1975.

After spending two years as a research psychologist at San Jose State University Foundation, Cowings returned to ARC in 1977 as a research psychologist and a principal investigator in the Psychophysiological Research Laboratory. She was a scientist astronaut candidate early in her career and received training as a payload specialist, though she never traveled on a mission.

Cowings's research focused on training astronauts to control their biological and psychological reactions to space travel. She conducted an experiment conducted in 12 half-hour sessions training 50 volunteers to control their bodies, raising their own body temperature and relaxing certain muscles when they experienced motion sickness. These techniques aided 65 percent of this group to suppress motion-sickness symptoms altogether, and another 20 percent experienced improvements. None in the control group, which received no training, had relief from their symptoms. Cowings's techniques were finally employed in the September 1992 Spacelab-J mission with the space shuttle *Endeavor.*

In January 1997, Cowings spent a month in Star City, Russia, training Russian cosmonauts autogenic-feedback training exercises with the assistance of her husband, fellow ARC scientist William B. Toscano. The couple has one son, Christopher Michael Cowings Toscano. The next month, she received the Government Award for Outstanding Technical Contribution at the Black Engineer of the Year Conference. This award recognized her as one of the year's top 30 African Americans working in the fields of science and technology. Earlier in her career, she had won the 1993 NASA Individual Achievement Award and the 1991 Black United Fund of Texas Award.

Cox, Geraldine Anne Vang
(1944–   )
American
Environmental Scientist

Geraldine V. Cox has specialized in developing policies regarding environmental issues, especially hazardous waste and water pollution. In response to the explosion of a plant in Bhopal, India, Cox drafted the guidelines in 1985 for the Community Awareness and Emergency Response, a protocol that was adopted first by the chemical industry in the United States and later served as the blueprint for federal law and international standards enforced by the United Nations.

Cox was born on January 10, 1944, in Philadelphia, Pennsylvania. She spent her entire academic career at Drexel University, where she studied environmental science. She earned both her bachelor's and master's degrees there before receiving her Ph.D. in 1970. She worked in the chemical industry for six years before filling a White House Fellowship as special assistant to the secretary of the U.S. Department of Labor for one year in 1976.

After graduating from Drexel, Cox accepted a position as technical coordinator of environmental programs at the Raytheon Company, commencing her career of interacting between the private and public sectors on environmental issues, from water pollution to environmental health to ecological damage assessment. Cox commenced her contributions to the public sector in 1974 by joining the Program Committee of the Water Pollution Control Federation, a membership she maintained until 1979. In 1975, she founded the Marine Water Quality Committee, chairing it until 1980.

After her stint at the White House, Cox became a member of the National Academy of Sciences Environmental Measurement Panel of the National Bureau of Standards from 1977 until 1980. Also in 1977, she worked for the American Petroleum Institute as an environmental scientist. In 1979, the Chemical Manufacturing Association appointed her as its vice president and technical director, continuing her role as liaison between the industry and the government.

In 1991, Cox accepted a vice presidency at Fluor Daniel, a subsidiary of the international Fluor Corporation. In the wake of the opening up of Eastern Europe and the former Soviet Union, Eurotech Ltd. was established in 1995 to market state-of-the-art technologies from these regions through environmentally sound strategies; Cox became the publicly held company's chief operating officer. She simultaneously helped start up AMPOTECH Corporation, which sought to decrease greenhouse gas emissions by producing power with hydroponics and other advanced technologies. Cox served as the company's chairman and chief executive officer.

Cox has received much recognition for her initiatives that have improved the quality of our environment and reversed the destruction of it. In 1984, the Society of Women Engineers granted her its highest honor, its Achievement Award, in recognition of her work on controlling water pollution. The United States Coast Guard also recognized Cox's efforts to stem the tide of water pollution with its 1991 Meritorious Public Service Award, the highest honor bestowed on civilians by the Coast Guard. She had worked with the Coast Guard throughout the 1980s, chairing its Marine Occupational Safety and Health Committee and working as a member of its Transportation Advisory Committee. In addition, Cox served numerous other organizations, including the American Society for Testing and Materials, the Water Pollution Control Federation, and the American National Standards Institute.

Cox, Gertrude Mary
(1900–1978)
American
Statistician

Gertrude Mary Cox exerted incalculable influence on the field of statistics. She coauthored the classic textbook in experimental statistics and acted as a trailblazer for women in the field, becoming a full professor and heading an academic department at a time when women with doctorates (which Cox lacked) struggled to secure promotions. Cox also established several institutes for the study of statistics.

Cox was born on January 13, 1900, in Dayton, Iowa. Her parents were John William Allen and Emmaline (Maddy) Cox. She intended to become a deaconess in the Methodist Episcopal Church after graduating from Perry High School in 1918. She worked at an orphanage in Montana in preparation for her career of social service, but in 1925, she changed the course of her future by attending Iowa State College in Ames to study mathematics. After earning her B.S. in 1929, she remained at the college to conduct graduate study in statistics under George Snedecor. In 1931, she earned the first M.S. in statistics awarded by Iowa State.

The University of California at Berkeley employed Cox as a graduate assistant while she studied psychological statistics there. Two years after leaving Iowa State, Cox returned at Snedecor's invitation to work in the new statistical laboratory. Cox's work designing statistical experiments diverted her attention from doctoral studies, but the college appointed her as an assistant professor in 1939 regardless. She eventually obtained her status as a doctor at Iowa State's centennial celebration on March 22, 1958, when it granted her an honorary doctorate of science.

In 1940, the North Carolina State College School of Agriculture in Raleigh solicited nominations for candidates to head its new department of experimental statistics from Snedecor, who left Cox's name off the list despite her obvious qualifications, assuming the hiring committee would not consider a woman. At her request, he appended her name, and that year she became North Carolina State's first female full professor and the first woman department head simultaneously.

Cox proved herself in short order, establishing the North Carolina State Institute of Statistics in 1944, acting as its first director. Within two years, the University of North Carolina joined the institute to teach statistical theory, leaving North Carolina State responsible for teaching methodology only. The main mission of the institute was to spread throughout the South the "gospel according to St. Gertrude," or her application of statistical analysis to diverse fields, such as plant genetics, quality control, and agricultural economics. She accomplished this objective through work conferences and summer conferences that brought statisticians to North Carolina from around the world.

In 1947, she helped found the Biometrics Society, an extension of the work she performed as the editor of both *Biometrics Bulletin* and *Biometrics* for the decade between 1945 and 1955. In the middle of this tenure, she collabo-

rated with William G. Cochran to revise their class notes from Iowa State into a textbook, entitled *Experimental Designs,* which became a classic in the field immediately after its 1950 publication.

Before retiring from North Carolina State in 1960, she coordinated the formation of the Research Triangle Institute, a consortium of the University of North Carolina, North Carolina State, and Duke University. Upon her retirement, she took up the directorship of the institute's statistics section. She retired from this duty in 1964, only to take up another post at the University of Cairo in Egypt, where she spent a year helping to found its Institute of Statistics. She spent the rest of her retirement traveling, continuing to promote statistics internationally, notably in Thailand.

In 1959, the year before her first retirement, she received the Oliver Max Gardner Award, and the year of her retirement, Gamma Sigma Delta granted her its distinguished service award. Long after her final retirement, she continued to contribute to her field, acting as the president of the Biometrics Society from 1969 through 1970. Cox died of leukemia in Durham, North Carolina, on October 17, 1978.

Cox's influence on the field of statistics continued to be acknowledged long after her death. In 1989, the American Statistical Association's Committee on Women in Statistics established the Gertrude Cox Scholarship, a $1,000 prize to promote women to enter graduate programs in statistically oriented fields. A decade later, in 1999, the Research Triangle Institute honored its founder by naming a new building after Gertrude Cox.

Cremer, Erika
(1900–1996)
German/Austrian
Chemist

Trained as a physical chemist, Erika Cremer worked with the Nobel Prize-winning chemist Otto Hahn on early isotope separation experiments in Germany. Because she was a woman, she was told that she would never get a faculty position in Germany. However, during World War II, Germany and Austria suffered from an extreme manpower shortage, and Cremer landed a job as a chemistry professor and researcher at the University of Innsbruck. There, at the end of the war, she developed the first methods of gas chromatography.

Born in Munich, Germany, on May 20, 1900, Erika Cremer was the only daughter of Max Cremer, a scientist and the inventor of the glass electrode, and Elisabeth Rothmund Cremer, a housewife. During her secondary schooling, Cremer's father moved to Berlin to teach. The switch to the more rigid Prussian schools was difficult for Cremer, but she managed to adapt and graduated from high school in 1921. She never married.

Cremer entered the University of Berlin in 1921. At that time, German universities, and especially the science departments of these universities, admitted few women students. There was almost no chance for a woman to develop a career as a professor at the university level in the sciences. German universities simply did not hire women as university professors. Ignoring these obstacles, Cremer plunged into her work. The teachers were some of the most talented scientists of their day, men like Fritz Haber (Nobel Prize, 1918), Max Planck (Nobel Prize, 1918), and Albert Einstein (Nobel Prize, 1921). Cremer studied under physical chemist Max Boderstein and concentrated on the kinetics of chemical reactions. She won her Ph.D. with honors in 1927.

Because of her well-received Ph.D. dissertation, Cremer was invited to work at the University of Leningrad by Nikolai Semenov, who would receive the Nobel Prize in 1957 for his work on kinetics in chemical reactions. Perhaps because of the political baggage attached to this offer, Cremer turned Semenov down. Instead, in the late 1920s and early 1930s, she did research work at the Kaiser Wilhelm Institute for Physical Chemistry and the Institute for Physical Chemistry at the University of Freiburg. During this time, she applied the new quantum theories of physics to the study of photochemistry. She then returned to Berlin to work with Fritz Haber, but Haber's Institute for Chemistry was closed by the Nazi government in 1933. After four years without work, Cremer found a job in 1937 with Otto Hahn at the Kaiser Wilhelm Institute for Physical Chemistry. Here she worked with Hahn on isotope separations and learned from him in 1938 about the first theoretical speculations concerning the possibility of nuclear fission.

The beginning of World War II in 1940 opened up a job for Cremer at the University of Innsbruck in Austria. There she could at last pursue her own projects, one of which was to see if it would be possible to separate and analyze gases through chromatography. From 1940 to 1944, she worked on the theoretical as well as practical sides of this project. The practical element consisted of building a device with long tubes that would allow gases to flow through a stationary, nonvolatile solid or liquid. As the gases passed through the tubes, they would separate and their composition would be read by the temperature and thermal conductivity. Her first experiment involved the separation of nitrogen and carbon dioxide.

Cremer remained at the University of Innsbruck until she retired in 1971. She received the Wilhelm Exner Medal in 1958, the Erwin Schrödinger Prize from the Austrian Academy of Sciences (1970), and the M. S. Tswett Chromatography Award (1972) from the U.S.S.R. Academy of Sciences. She died in Innsbruck in 1996.

Crosby, Elizabeth Caroline
(1888–1983)
American
Neuroanatomist

In a career that spanned more than 60 years, Elizabeth Crosby taught thousands of medical students, trained 39 doctoral candidates, and supervised more than 30 postdoctoral scientists in her laboratory at the University of Michigan Medical School. She is known for her work studying the human brain and nervous system.

Born in Petersburg, Michigan, on October 25, 1888, Elizabeth Caroline Crosby was the daughter of Lewis Frederick Crosby and Frances Kreps Crosby. She never married, but during the 1940s, she adopted two young girls whom she raised as her children. At the time of her death, she was survived by five grandchildren and one great-grandchild.

She studied at the public schools in Petersburg and, in 1907, began her undergraduate studies at Adrian College in Adrian, Michigan. After receiving her B.S. in mathematics in only three years at Adrian, she continued her studies at the University of Chicago. She completed her master's degree in anatomy at Chicago in 1912 and her Ph.D. in neuroanatomy in 1915. Her mentor at Chicago was C. Judson Herrick. Crosby's dissertation was a study of the forebrain of the alligator.

Between 1915 and 1918, Crosby suspended her research career to care for her ailing mother. At the same time, she held down the jobs of teacher, principal, and superintendent at Petersburg's high school. On the death of her mother in 1918, she was appointed instructor of anatomy at the medical school of the University of Michigan in Ann Arbor. Crosby worked on a steady stream of research about mammalian brains during her career at Michigan. At the same time, she maintained a demanding course load. She usually taught anatomy courses in the mornings from Monday to Saturday, and in the afternoons she held back at least two hours for discussions with graduate students. The rest of the afternoons and a good portion of the evenings were reserved for work in her laboratory. By the late 1930s, Crosby's work had become so well known that she was frequently tempted with offers to work at other universities. During 1939–40, she went to Aberdeen, Scotland, to help Marischal College establish a neuroanatomy teaching and research program.

At the University of Michigan Medical School, she frequently did rounds with medical students, guiding them in clinical work that involved patients with neurological disorders. By 1936, she had become a full professor of anatomy at the University of Michigan and the first woman professor at Michigan's medical school. Told by a mentor at the University of Michigan that is was not proper for women to speak at professional meetings, Crosby shunned speaking engagements until she was 58. In 1946, she gave her first public talk about anatomy as the first woman recipient of the Henry Russell Lectureship. In the early 1960s, she collaborated with another anatomist to write a standard text on neuroanatomy, and in 1982, at the age of 94, she saw her book *Comparative Correlative Neuroanatomy of the Vertebrate Telencephalon* published. Toward the end of her career, one of her colleagues asserted that "no one in the world rivals Doctor Crosby's knowledge of the entire nervous system of animals throughout the vertebrate phylum."

Crosby won numerous awards during her long career. These include the Henry Gray Award in Neuroanatomy given by the American Association of Neuroanatomists (1972) and the prestigious National Medal of Science (1979) given to Crosby by President Jimmy Carter. She died in 1983.

Curie, Marie Sklodowska
(1867–1934)
Polish/French
Physicist

Of Polish origin but for most of her life a resident of France, Curie was the first woman to be awarded a Nobel Prize. The award in 1903 for her work in physics was shared with her husband, Pierre Curie, and Henri Becquerel for their discovery that the atoms of certain elements emit particles from their nuclei, a property that Marie Curie named *radioactivity*. Curie won an unprecedented second Nobel Prize in 1911, this time in chemistry, for isolating a quantity of the radioactive element radium.

Marie Sklodowska was born in Warsaw, Poland, on November 7, 1867, the youngest of five children. She came from a family that prized education: Her father, Wladislaw, and mother, Bronislawa, both were teachers. At that time, Poland was ruled by Russia, and Wladislaw Sklodowska's participation in anti-Russian activities got him fired from his job in 1873. For almost 15 years, the family was desperately poor, but still Marie Sklodowska graduated from secondary school at the top of her class in 1883.

Sklodowska did not have the money to begin her university studies immediately. Instead, she took a job as a governess for a wealthy Polish family and used her income to help support her sister, Bronia, who had traveled to Paris to study medicine. By 1891, Marie joined her sister in the French capital to begin her own studies. Marie Sklodowska, one of only a very few female students at the Sorbonne, graduated first in her class in physics in 1893. Now aided by a scholarship from the Polish government, she immediately began work on a master's in mathematics at the Sorbonne, which she completed in 1894. She would later add a Ph.D in physics, also from the Sorbonne, in 1903.

In 1894, while she was completing her master's degree, Marie Sklodowska met Pierre Curie, a French physicist who had been working on magnetism and piezoelectricity, the relationship between electrical currents and the structure of crystals such as quartz. Curie was an idealistic young scientist. "It is necessary to make a dream of life," he wrote Sklodowska, "and make a dream of reality." Marie Sklodowska fell in love with him, and in 1895 they married and began working together as well as raising a family (they had two children, Irène, born in 1897, and Eve, born in 1904). In 1896, stimulated by Wilhelm Roentgen's discovery of X rays and the discovery by Henri Becquerel of "rays" being emitted from uranium, the Curies turned their

Marie Curie, the first woman to win a Nobel Prize and the first scientist of either gender to win the prestigious prize twice. Shown here with her husband, Pierre Curie, with whom she shared the 1903 Nobel Prize in physics. *(AIP Emilio Segrè Visual Archives)*

attention to finding out more about the properties of this phenomenon and these substances.

Marie Sklodowska, now Marie Curie, speculated that the ability of a substance to emit radioactivity, a term she coined in 1898, was a property of the atoms of the elements of that substance. Thus she and Pierre attempted to separate the various elements of pitchblende, which was known to contain uranium but was so radioactive that the Curies suspected that it also contained other as yet undiscovered radioactive elements. Even as she labored to extract these still mysterious substances from pitchblende, Marie also worked with her husband to establish their existence theoretically. This they did in 1898 when they announced the theoretical discovery of two new elements—polonium, named after Marie Curie's native land, and radium. Working long hours in a derelict shack provided them by the stingy French government, Marie Curie eventually managed to extract a small amount of radium in 1902.

From 1900 to 1906, in addition to the long hours spent in the laboratory, Marie Curie was forced to work as a teacher at a secondary school for girls in order to help support her family. Not even the Ph.D. and Nobel Prize awarded her in 1903 prompted the French intellectual establishment to give her a university teaching job or well-paying job as a scientist. This changed only with Pierre's sudden death in a street accident in 1906, after which Marie Curie was appointed to Pierre's job of professor of physics at the Sorbonne. Belatedly realizing its negligence of Curie's work, the French government funded the Institute of Radium, with Marie Curie as its head, in 1912. Curie devoted the last part of her life to making the institute a center of atomic research and a training ground for new scientists, many of them women, to follow in her footsteps. At the end of her life, she wrote, "I am one of those who think that science has great beauty. A scholar in the laboratory is not just a technician; he is also a child face to face with natural phenomena that impress him like a fairy tale." Marie Curie died in 1934 from leukemia, which was the result of her long exposure to radiation.

D

Daly, Marie Maynard
(1921–)
American
Biochemist

Marie M. Daly was the first African-American woman to earn a doctorate in chemistry. She spent her career studying the heart and arteries, specifically investigating harmful effects on them. Late in her career, she established a scholarship at her alma mater, Queens College, to help poor students afford a college education.

Daly was born on April 16, 1921, in Corona, a neighborhood of Queens, in New York City. Her mother was Helen Page, and her father, Ivan C. Daly, moved from the British West Indies to the United States, where he attended Cornell University to study chemistry on a scholarship until his funds ran out after the first semester. He became a postal clerk in New York City. Daly attended Hunter College High School, an all-girls school with an all-female faculty, where she determined to fulfill her father's destiny by becoming a chemist herself. She studied chemistry at the newly formed Queens College, graduating with honors in 1942.

The college offered Daly a fellowship and a part-time job as a laboratory assistant while she pursued a master's degree at New York University, which she earned in one year. She continued to work at Queens College, tutoring chemistry, until she saved enough money to enroll at Columbia University, where she studied biochemistry under MARY L. CALDWELL. In 1948, she became the first African-American woman to earn her Ph.D. in chemistry.

Daly accepted a temporary position at Howard University, where she taught introductory physical science under Herman Branson, while she awaited a grant from the American Cancer Society to fund her research at the Rockefeller Institute of Medicine in New York City (now Rockefeller University.) She obtained this fellowship, which lasted

from 1948 through 1951, when she became an assistant in general physiology under A. E. Mirsky.

In 1955, Daly became an associate in biochemistry at the College of Physicians and Surgeons at Columbia University, where she worked with Quentin B. Deming. This team discovered a correlation between the incidence of heart attacks and cholesterol. Daly later investigated the correlation between cigarette smoking and lung diseases. She also studied the role of the kidneys in human metabolism as well as hypertension and atherosclerosis.

In 1958, Daly accepted a concurrent position as an established investigator for the American Heart Association through 1963. In 1960, she became an assistant professor of biochemistry at the Albert Einstein College of Medicine at Yeshiva University in New York. In 1961, Daly married Vincent Clark. In 1962, she was appointed to another concurrent position as a career scientist with the Health Research Council of New York, a post she retained for the next decade. In 1971, Yeshiva University promoted her to associate professor of biochemistry and medicine, a position she retained until her 1986 retirement.

In 1988, Daly established a scholarship fund at Queens College, which she earmarked for minority students who intended to pursue physics or chemistry. She dedicated this scholarship to the memory of her father, as it would keep students the likes of her father from falling through the cracks.

Darden, Christine
(1942–)
American
Aeronautical Engineer

Christine Darden conducts research for the National Aeronautics and Space Administration (NASA) on the effects of the sonic booms created by supersonic aircraft, such as their contribution to noise pollution and their potential depletion of the ozone layer. She also investigates means of reducing, if not altogether eliminating, these negative effects. In order to facilitate her research, she wrote a computer software program that simulates a sonic boom under experimental conditions in a wind tunnel.

Darden was born Christine Voncile Mann on September 10, 1942, in Monroe, North Carolina, the youngest of five children. Her parents, both youngest children of large families, attended college with the financial support of their older siblings. Her mother, Desma Cheney, became an elementary-school teacher, and her father, Noah Horace Mann Sr., became an insurance agent with North Carolina Mutual Life. They encouraged their daughter's interest in education and supported her through Allen High School, a Methodist boarding school in Asheville, North Carolina.

Mann studied mathematics at Hampton Institute, earning her B.S. in 1962. Upon graduation, she taught high-school mathematics. In 1963, she married Walter L. Darden Jr., who was attending Virginia State College. In 1965, she obtained a position at the college as a research assistant in aerosol physics, analyzing air quality in search of specific types of pollutants. The next year, she became an instructor in mathematics at the college. She also conducted graduate study, earning her M.S. from Virginia State in 1967.

That year, the Darden family moved to Hampton, Virginia, where Walter taught middle-school science. In addition to raising two children, Christine Darden applied for three jobs, accepting the best-paying offer as a data analyst with NASA at Langley Research Center. Her duties consisted of doing very routine calculations for engineers, but as the work integrated more and more computer technology, Darden began to write computer programs for the engineers. NASA promoted her to the rank of aerospace engineer in 1973.

Darden then took advantage of NASA's incentives to continue her education by conducting doctoral study in mathematics and engineering science with George Washington University in Washington, D.C. She earned her Ph.D. in 1983 and further continued her education by attending management classes conducted by NASA to promote employees into research administration positions. Darden achieved this goal with her 1989 promotion to lead the Sonic Boom Team.

Darden led her team in designing and testing new wing designs and nose-cone shapes in attempts to improve aerodynamics and decrease the effects of sonic booms. She published her findings in more than 40 journal articles. Two of her better-known publications were "The Importance of Sonic Boom Research in the Development of Future High Speed Aircraft," in the winter 1992 issue of the *Journal of the National Technical Association,* and "Study of the Limitations of Linear Theory Methods As Applied to Sonic Boom Pressure Signatures," in the November–December 1993 issue of the *Journal of Aircraft.* Darden also generated several mathematical algorithms specifically oriented toward sonic boom research.

Darden has been recognized for the excellence of her research with the 1985 Dr. A. T. Weathers Technical Achievement Award from the National Technical Association. In addition, Langley Research Center awarded her with its Certificate of Outstanding Performance three times, in 1989, 1991, and 1992.

Daubechies, Ingrid
(1954–   )
Belgian
Mathematician

In a discipline noted for its often abstruse theoretical speculations, Ingrid Daubechies has applied high-level mathematical thought to problems that have very practical applications. Her solutions to the problems of wavelets have been used by software engineers and other applied scientists to aid in the easier recognition of certain kinds of signals and in the storage of data.

Ingrid Daubechies was born on August 17, 1954, in Houthalen, Belgium, a mining town in eastern Belgium near the German border. Daubechies's father was a mining engineer who worked in the area's main industry, and her mother, although university trained, stayed at home to raise Ingrid and her brother. Both of Daubechies's parents encouraged her in her studies and helped tutor her. Her father was especially helpful in her science and math studies. Daubechies is married to Robert Calderbank, a scientist. She has two children.

Ingrid Daubechies, whose mathematical analysis of wavelets has been used by software engineers to improve signal recognition and data storage. *(Courtesy of Ingrid Daubechies)*

In the town of Turnhout, near the Dutch border, Daubechies attended an all-girls high school. In 1972, she enrolled at Vrije University in Brussels where she majored in physics. As an undergraduate, Daubechies was interested in quantum mechanics, the theories formulated in the 1920s by Werner Heisenberg, Niels Bohrs, and others, and subsequently developed by other physicists, that explained the relationship between electrons and nuclei at the atomic level. Because of her natural aptitude and the confidence instilled in her by her parents and at her secondary school, Daubechies had no trouble competing with her mostly male fellow students at the university. She completed her education at the University of Marseilles in France where she switched to mathematics. Her mentor in Marseilles was mathematician Alex Grossmann. While at Marseilles, Daubechies began to build an international reputation through the publication of her work. This solid research won her a postdoctoral fellowship at Princeton University in New Jersey.

At Princeton in the early 1980s, Daubechies began working on the problem that would garner her great acclaim. This was an examination of the phenomenon known as wavelets. The mathematical analysis of wavelets is similar to the procedures used in Fourier analysis. Daubechies had earlier begun work on wavelets in Marseilles, but in Princeton, she had time to begin trying to figure out solutions. One month before her fellowship was over, she finished a paper, later published in a scientific journal, that would mark a major step forward in the understanding of this mathematical problem.

After her two years at Princeton, Daubechies returned to Belgium to become a professor at her alma mater, Vrije University in Brussels. Besides teaching, she continued mathematical research and focused on a problem known as Weyl quantization. In 1986, Daubechies returned to the United States to teach and do research at New York University. Here she continued work on wavelets and developed tables of coefficients that applied to wavelets and that would allow engineers to simplify their calculations in areas such as the storage of data on computer disks and the interpretation of wave signals such as sound. Daubechies's solution, known as "Daubechies wavelets," was tremendously helpful in devising compressed but easily accessible computer files. From 1991 to 1993, Daubechies worked on wavelet problems at the Bell Labs in New Jersey. In 1992, she expanded on this work in a book entitled *Ten Lectures on Wavelets*. In the mid-1990s, she returned to Princeton University where she is a professor of mathematics.

Daubechies has won numerous awards for her work, including the Louis Enpain Prize (1984) from the Belgium National Science Foundation and the MacArthur Award (1993) from the MacArthur Foundation in the United States.

Davis, Margaret B.
(1931–)
American
Paleoecologist

Early in Margaret B. Davis's career as a paleoecologist, a discipline that studies past ecologies of the earth by geological evidence in fossils, she challenged the existing paradigm governing the correction factors employed when studying pollen production by various tree species 10,000 years ago. Paleoecologists had been utilizing correction factors between 4:1 and 35:1. Davis threw a monkey wrench in these calculations by suggesting correction factors of as much as 24,000:1.

Margaret Bryan Davis was born on October 23, 1931, in Boston, Massachusetts. She remained in Boston growing up and attended Radcliffe College, where she studied floral physiology and ecology as well as stratigraphic pollen deposits from the late Quaternary period under the renowned paleobotanist Elso Borghoorn. She earned her A.B. degree in biology in 1953.

Davis won a Fulbright Fellowship from 1953 through 1954 to study palynology, or the research of pollen from ancient plants, at the University of Copenhagen under Johannes Iversen of the Danish Geological Survey. She conducted research on glacial plant pollen deposits, publishing her findings in her first paper, "Interglacial Pollen Spectra from Greenland," in a 1954 issue of a Danish geological journal. When she returned to Boston, she married Rowland Davis in 1956 (the couple divorced in 1970), and she based her doctoral dissertation on her European fieldwork to earn her Ph.D. in biology from Harvard University in 1957.

Harvard retained Davis as a National Science Foundation postdoctoral fellow for two years, after which she transferred this fellowship to the California Institute of Technology to focus on geoscience for the next two years. Davis then spent a year at Yale University as a research fellow, studying the correlation between vegetation composition and pollen sedimentation in lakes. In 1961, she joined the Department of Botany at the University of Michigan as a research associate, commencing a dozen-year relationship with the university.

In a 1963 issue of the *American Journal of Science*, Davis published her article, "On the Theory of Pollen Analysis," which turned her field on its head by bringing much more precision to the analysis of pollen records as a means of interpreting the history of plant migration. She also mapped the migration of certain tree species across eastern North America over the past 14,000 years. In 1964, she obtained a concurrent position as an associate research biologist with the Great Lakes Research Division. The University of Michigan promoted her to an associate professorship in the Department of Zoology in 1966, and in

1970, both institutions erased the "associate" before her title, promoting her to the status of full professor and research biologist.

Davis returned to Yale in 1973 as a professor of biology, remaining there for three years. In 1976, the University of Minnesota appointed her as a professor and head of the Department of Ecology and Behavioral Biology. In 1983, the university named her the Regents Professor of Ecology, a title she retained thereafter. In 1989, Davis published her prediction that sugar maple trees will disappear from their southernmost range and will migrate eastward in Minnesota, according to data compiled from the National Aeronautics and Space Administration and the National Oceanographic and Atmospheric Administration.

Throughout Davis's career, she published more than 65 papers, and she served as president of the American Quaternary Association from 1978 through 1980 and of the Ecological Society of America from 1987 through 1988. She has received numerous honors, including the Ecological Society of America's Eminent Ecologist Award as well as the 1993 Nevada Medal, which awarded her $5,000 for her contributions to the understanding of the history, present, and future of environmental change.

Davis, Marguerite
(1887–1967)
American
Chemist

Marguerite Davis discovered vitamins A and B, working in collaboration with Elmer Verner McCollum, a biochemist and colleague at the University of Wisconsin at Madison. Scientists had come to the realization that certain elements contained in specific foods proved necessary for sustenance and proper nutrition, but the identity of these elements remained a mystery until McCollum and Davis's discovery.

Davis was born on September 16, 1887, in Racine, Wisconsin. She hailed from a family of academics and activists: Her father, Jefferson J. Davis, was a physician and a botanist who worked as a professor at the University of Wisconsin. Her grandmother, Amy Davis Winship, was a women's rights campaigner and a social worker. In 1906, Davis matriculated at the school where her father taught, but she transferred from the University of Wisconsin to the University of California at Berkeley two years later, and two years after that, in 1910, she earned her bachelor of science degree there.

Davis returned to the University of Wisconsin to commence graduate studies, but she never did complete the required coursework to receive her master's degree. Instead, she moved to New Brunswick, New Jersey, to work for the Squibb Pharmaceutical Company. Before long, however, she returned to her home state and to the University of Wisconsin, where she installed herself as a chemical researcher in the same laboratory as McCollum. The biochemist had been trying to identify the chemical keys to nutrition.

Several researchers were hot on this trail already: Christiaan Eijkman, a Dutch physician, and Sir Frederick Gowland Hopkins, a British biochemist, had come to the realization that certain elements in food, which they could not yet put their fingers on, must be required by the body for survival. Casimir Funk, a Polish-American biochemist, took their conjectures a step further by hypothesizing that these elements were in fact amines, and thus he tacked the Latin word for life, *vita*, in front, dubbing them *vitamines*. McCollum followed these men's trail by trying to isolate the most basic food components necessary to keep animals alive, though his efforts did not yield a clear answer.

In 1913, McCollum and Davis happened upon an element in butterfat that fit the life-sustaining criteria they were searching for. Their discovery differed from a similar discovery by Eijkman, so they distinguished the discoveries by calling theirs *fat-soluble A* and Eijkman's *water-soluble B*. These substances became known as vitamins A and B, dropping the "e" from the end of the word when it was discovered that not all of the substances were amines. Lafayette Benedict Mendel, a physiological chemist from Yale University, discovered vitamins A and B simultaneously, though McCollum and Davis are credited with the discovery.

Davis went on to establish the University of Wisconsin's nutrition laboratory, and then she moved to Rutgers University, which commissioned her to establish a nutrition lab there. In 1940, she retired to her birthplace, where she remained active as a chemistry consultant. She died in Racine on September 19, 1967, three days after reaching her 80th birthday.

DeWitt, Lydia Maria Adams
(1859–1928)
American
Anatomist and Pathologist

An interest in medicine led Lydia DeWitt into the fields of pathology and anatomy around the turn of the 20th century. Her initial work focused on microscopic anatomy, especially nerve endings and the relationship of nerve endings to muscle fibers. DeWitt's later work centered on public health issues. She worked on methods to diagnose and treat diphtheria, typhoid, and tuberculosis, endemic and life-threatening diseases during the early 20th century.

Born on February 1, 1859, in Flint, Michigan, Lydia Adams was the daughter of Oscar Adams, a lawyer, and Elizabeth Walton Adams. She had two siblings. Adams's mother died in 1864, and she was raised by her stepmother, who was her mother's sister. In 1878, at age 19, she married Alton DeWitt, a school superintendent. She had two children, Stella and Clyde.

DeWitt graduated from a public secondary school in Flint and, in 1886, also graduated with a teaching degree from Ypsilanti Normal College. Between 1886 and 1895, she taught at several public schools in Michigan. In 1895, DeWitt returned to college and enrolled in the University of Michigan's combined medical and science program. She received an M.D. degree in 1898 and a B.S. degree in 1899. In 1906, she went to the University of Berlin to study anatomy for a year.

The University of Michigan was DeWitt's first professional home. She began as a demonstrator in anatomy in 1896, then from 1897 to 1902 she served as an assistant in histology. She was an instructor of histology from 1902 to 1910. Much of DeWitt's research work at the University of Michigan between 1896 and 1910 was done in association with Carl Huber. This included comparative studies of nerve endings of different species. Later, on her own, she investigated the transmission of nerve impulses to heart muscles and the functioning of the pancreas. Her studies of the pancreas were important steps in the understanding of the functions of this organ. By 1900, medical science knew there was a link between the pancreas and diabetes. However, because the pancreas secretes a variety of enzymes, scientists had not been able to isolate these into discrete groups. By 1906, DeWitt had managed to isolate a group of pancreatic cells known as the *islets of Langerhans*. She concluded that this group of enzymes contained the critical ingredients necessary for controlling the metabolism of carbohydrates in the body—in short, the raw material that could offer a cure to diabetes. Because of the inadequate equipment at her lab, she could not carry her experiments to the next steps of isolating insulin, a procedure that was completed in 1921–22 by Canadian researchers J. J. R. McLeod and Frederick Banting.

In 1910, DeWitt left Michigan to become an instructor of pathology at Washington University in St. Louis. There she worked with George Dock, one of her mentors at Michigan, on studies of diphtheria and ways to diagnose typhoid more accurately. As a result of this work, in 1912 DeWitt was given a job as assistant professor of pathology at the University of Chicago's Sprague Memorial Institute to work on chemical treatments for tuberculosis. In 1918, she became an associate professor. She remained at the University of Chicago until her retirement in 1926.

For her work in medicine, DeWitt was given an honorary doctorate from the University of Michigan. She was also elected president of the Chicago Pathological Society in 1924. She died on March 10, 1928, at her daughter's home in Winter, Texas.

DeWitt-Morette, Cécile Andrée Paule
(1922–   )
French
Physicist

A key researcher in mathematical and theoretical physics, Cécile Andrée Paule DeWitt-Morette has had a rich and varied international career. As a graduate and postdoctoral student, she was fortunate to have worked with many Nobel laureates in physics in both France and the United States. In the 1950s, she founded the first European summer institute for theoretical physics, and from 1945, she has worked at a number of American and European universities as a theoretical physicist.

Born on December 21, 1922, in Paris, Morette is the daughter of André Morette and Marie-Louise Ravaudet. Her father was an engineer and industrial manager who restored a large steel complex in Normandy during the 1920s. Her mother was an educated woman who had prepared for university before marrying André Morette. In 1951, Morette married Bryce DeWitt, an American physicist. They have four daughters.

DeWitt-Morette entered the University of Caen in 1940 and transferred to the University of Paris in 1943. She spent the wartime years as a student. By 1945, she was working as a theoretical physicist in the laboratory of Frédéric Joliot and Irène Joliot-Curie, two of the leading French physicists of that era.

After the war, DeWitt-Morette traveled to England and Ireland where she met Nobel laureate Paul Dirac and worked with Walter Heitler at the Institute for Advanced Studies in Dublin. She worked in Dublin for several years before accepting an invitation to work at the Institute for Theoretical Physics in Copenhagen, headed by Niels Bohr. DeWitt-Morette received her Ph.D. from the University of Paris in 1947. In 1948, she traveled to the United States where she took a position at Princeton University's Institute for Advanced Study, then headed by J. Robert Oppenheimer. She remained at Princeton until 1950. Here she met and later married fellow physicist Bryce DeWitt.

After her marriage to DeWitt, DeWitt-Morette turned down several offers of tenured professorships from French universities. She persuaded the French minister of education to give her money to open a summer institute for the study of theoretical physics at Les Houches, a small town in the French Alps near Mont Blanc. This institute, known as the Les Houches School, admitted 30 students each summer for advanced courses on standard topics in theoretical physics. Courses lasted for eight weeks, and many of the teachers and students were past or future Nobel Prize winners. DeWitt-Morette directed this summer program until 1972.

Since 1948, most of DeWitt-Morette's life has been spent in the United States. In 1952, she moved to Berkeley where her husband had accepted a position at the Lawrence Livermore National Laboratory. DeWitt-Morette took a position as lecturer in physics at the University of California at Berkeley. When Bryce DeWitt became director of the Institute for Field Physics at the University of North Carolina in 1956, DeWitt-Morette took a nontenured position as visiting research professor at that institution. However, the university would not give her a full position, citing nepotism rules as the reason.

In 1971, DeWitt-Morette and her family moved again, this time to Austin, Texas, where she and DeWitt accepted positions on the astronomy and physics faculties at the University of Texas, Austin. There, she finally became a full professor, although she had to wait until 1987 to become a full-time faculty member.

Beginning in the 1950s at Princeton and continuing through her whole career, DeWitt-Morette has concentrated her attention on path integration, an approach to understanding quantum physics first explored by Richard Feynman in the late 1940s and 1950s. DeWitt-Morette expanded the concept of path integration by new mathematical concepts, especially introducing the semiclassical expansion of the functional integral. She and her students developed theoretical models to explain such phenomena as glory scattering, orbiting, rainbow scattering, and scattering of waves by black holes.

For her work as an educator and theorist, DeWitt-Morette was awarded the Chevalier de l'Ordre National du Mérite and the Chevalier dans l'Ordre de la Légion d'Honneur by the French government in 1981. She has also served on numerous scientific committees and has supervised 22 graduate students during their doctoral work. She continues to teach at the University of Texas at Austin.

Diacumakos, Elaine
(1930–1984)
American
Biologist

Although cell biologist Elaine Diacumakos developed the first technique for inserting and removing material to and from cells, her accomplishments were not recognized for most of her career. With geneticist French Anderson, she later perfected her cell insertion technique so that it could be applied in gene therapy. She is now credited with pioneering one of the early breakthroughs in that field. Diacumakos also explored cancer cells' resistance to drugs.

Born on August 11, 1930, in Chester, Pennsylvania, Diacumakos was one of Gregoris and Olga (Dezes) Diacumakos's two children. Diacumakos received her bachelor's degree in zoology from the University of Maryland at College Park in 1951. She then began doctoral studies at New York University, earning her master's degree in cell physiology and embryology in 1955 and her Ph.D. in 1958. While still a graduate student, she also worked as a research associate at New York University. She married James Chimondies in 1958, the same year she won the prestigious Founders Day Award from New York University.

After Diacumakos completed her Ph.D., she held a number of concurrent positions. She remained at New York University as a research associate until 1964. From 1958 to 1960, she also held a two-year fellowship at Rockefeller University, where she began a fruitful collaboration with Nobel Prize–winning geneticist Edward Tatum. In addition, Diacumakos worked as a research associate at the Sloan-Kettering Division of the Graduate School of Medical Studies at Cornell University in New York City from 1959 to 1963, when she was promoted to the rank of instructor. At Sloan-Kettering, she investigated cancer cells and explored the phenomenon of a cell's resistance to drugs.

Diacumakos returned to Rockefeller University in 1971 to become a senior research associate in biochemistry and genetics. While teaming up again with Edward Tatum at Rockefeller, she began to develop her cell insertion techniques. Since their research at that time focused on discovering whether the organelles of an individual cell contained genetic material, it was necessary for the duo to invent a method for extracting material from and inserting it into the microscopic entities. Diacumakos experimented with minuscule glass needles, which she made herself by bending glass she heated with her Bunsen burner. This method allowed her to create glass needles that were thinner than a strand of human hair and could be used with incredible precision.

After Tatum died unexpectedly in 1975, Diacumakos was promoted to head the Cytobiology Laboratory at Rockefeller. She was unable to raise funding for her projects, though, and her work on genetic material stalled. In the mid-1970s, French Anderson (a medical doctor and genetic researcher) learned of Diacumakos's success with her cell extraction technique. The two worked together for several years to adapt the technique to the fledgling field of gene therapy. In 1979, Diacumakos's and Anderson's efforts were rewarded when they repaired a mouse's genetic defect by placing a copy of a functioning gene (with one of Diacumakos's glass needles) into the damaged cell. A mere 11 years later, scientists conducted the first successful gene therapy on a human.

Diacumakos's role in the development of gene therapy was acknowledged only a week before her death, when her research was commended by the Metropolitan Chapter of the Association for Women in Science. During her lifetime, she was a member of the American Genetic Society, the American Society for Cell Biology, and the Cell Cycle Society. Diacumakos died of a heart attack in her home in Manhattan on June 11, 1984. Her cell extraction techniques are now widely used by other researchers.

Dicciani, Nance K.
(1947–   )
American
Chemical Engineer

Nance K. Dicciani made perhaps her most significant contribution to the sciences with her doctoral dissertation, a research project drawing on the resources of the University of Pennsylvania, the National Science Foundation, and the government of the Soviet Union. In it, she applied chemical engineering to medical imaging, resulting in the later development of ultrasonic scanners that gained widespread use in examining pregnant women. She continued to contribute to scientific research and marketing throughout her career in the chemicals industry.

Dicciani was born in 1947 in Philadelphia, Pennsylvania. Her mother, who was a homemaker, and her father, who was a industrial engineer, both supported Dicciani's interest in the sciences, which began in the fifth grade. With such encouragement, Dicciani never considered herself a lesser scientist than men. She remained in Philadelphia for college, studying chemical engineering at Villanova University. Her penchant for combining pure science with practical applications took hold in her undergraduate years. Dicciani graduated with a B.S. in 1969.

Dicciani continued her education at the University of Virginia, where she earned her M.S. in chemical engineering in 1970. She returned to her home city to work as the superintendent of water treatment and reservoirs in Philadelphia's Department of Public Works. After four years, she continued her education with doctoral study at the University of Pennsylvania, where she wrote her influential dissertation, "Ultrasonically-Enhanced Diffusion of Macro Molecules in Gels." Dicciani earned her Ph.D. in chemical engineering in 1977, and in the 1980s she returned to the University of Pennsylvania to pursue studies in the other component of her career, business, to earn her M.B.A. from the prestigious Wharton Business School in 1986.

Dicciani commenced her career in industry in 1977 as a research engineer with Air Products and Chemicals, Inc. She rose through the ranks there, becoming a research manager the next year, and then receiving promotions to become the director of research for the process systems group in 1981, the director of research and development for her division in 1984, the general manager of her division in 1986, and, finally, the director of commercial development in 1988. During her 14 years with the company, she helped develop technologies for the company's first noncryogenic process for separating nitrogen and oxygen from air as well as identifying a new catalyst for the production of benzene from coke.

In 1991, Rohm and Haas, one of the largest chemical companies in the world, hired her as the business director of its Petroleum Chemicals Division. In 1993, the company elected her vice president, and from 1996 through 1998, she headed the company's worldwide Monomers business. In late 1998 and early 1999, the company named her senior vice president and appointed her to the executive council.

Dicciani has received numerous honors in her career, including the 1986 Professional Achievement Award as well as the 1994 J. Stanley Morehouse Memorial Award, both from her alma mater, the University of Villanova, where she is a member of the Board of Trustees. The Society of Women Engineers also honored her with its 1987 Achievement Award. Dicciani is a senior member of the society and has served on its National Board of Advisors.

Dick, Gladys Rowena Henry
(1881–1963)
American
Physician and Microbiologist

An early woman physician in a field that was dominated and controlled by men, Gladys Dick conducted important research on scarlet fever, a disease that was especially deadly to infants at the beginning of the 20th century. Dick collaborated with her husband to isolate the organism responsible for the disease. She later developed a vaccine for scarlet fever, and at the end of her career, she conducted research on polio.

Gladys Rowena Henry was born on December 18, 1881, in Pawnee City, Nebraska. Her father was William Chester Henry, a prosperous banker and grain dealer, and her mother was Azelia Henrietta Edson Henry. She had two siblings. In 1914, Gladys Henry married George Frederick Dick, a physician. They raised two adopted children.

When she was just five years old, Gladys Dick accompanied her mother to visit a sick child in their neighborhood. During their visit, the child had a seizure. This prompted an intense desire in Dick to know more about the treatment of diseases. She attended public schools in Lincoln, Nebraska, where her family moved when she was a child. After graduation from high school, she attended the University of Nebraska, graduating from there with a B.S. in 1900. Because she was a woman, Dick had trouble finding a medical school that would admit her. She spent three years teaching and taking graduate biology courses until she was finally admitted to the Johns Hopkins Medical School in Baltimore, Maryland, in 1903. Dick completed her M.D. at Johns Hopkins in 1907. She then worked on an internship at Johns Hopkins and studied for a year at the University of Berlin.

In 1911, Dick took her first job in medicine, as the director of the laboratory at the Children's Memorial Hospital in Chicago, Illinois. Here she met her future husband and collaborative investigator, George Dick. By 1914, Dick

Gladys Dick, an early-20th-century physician who conducted breakthrough research into the causes of scarlet fever and eventually developed a vaccine for the disease. *(The Alan Mason Chesney Medical Archives of the Johns Hopkins Medical Institutions, courtesy of National Library of Medicine)*

and her husband had moved to the John R. McCormick Memorial Institute for Infectious Diseases in Chicago. They now turned completely to the study of scarlet fever. Using Koch's postulate (which gives rules for isolating a disease-causing organism) as their guide, they began trials to determine the cause of scarlet fever. At that time, the disease killed 30 percent of children five years old and under who contracted it. Those who did not die often had to live with deafness or heart and kidney disease.

Dick began trying to isolate hemolytic streptococci, an organism that was usually found in scarlet fever patients. They began trying to introduce hemolytic streptococci into animals, but this did not work. Animals appeared resistant to the disease. They then were forced to introduce it into a trial batch of humans, including themselves and many of their friends. Their work was interrupted by World War I. George Dick had to go to Europe to serve as an army doctor, and during a severe bout of flu in 1918, Gladys lost the strain of hemolytic streptococci they had been working on. After the war, they began again. By 1923, they had isolated not only a batch of hemolytic strep-

tococci that they could prove caused scarlet fever, but they had narrowed down the culprit to a specific toxin that was produced by this bacteria. Knowing that the disease was caused by the toxin, they developed a test, called the Dick test, that would tell if a patient had a susceptibility for scarlet fever. They then developed an antitoxin for those who had already contracted scarlet fever, and they later developed a vaccine for the disease.

For their work, they were nominated for, but did not receive, the Nobel Prize in medicine in 1925. In 1926, they won the University of Toronto's Mickle Prize, and in 1933, they were given the University of Edinburgh's Cameron Prize for work in the field of therapeutic medicine. Gladys Dick died on August 21, 1963, in Palo Alto, California.

Diggs, Irene
(1906–1998)
American
Anthropologist

Irene Diggs is best known for her comparative ethnohistorical studies of the descendants of Africans in the Americas. A prolific author, Diggs conducted research in Cuba as well as Central and South America, documenting the history and sociology of these African descendants. Her most important work is *Black Chronology,* and she cofounded the influential journal *Phylon: A Journal of Race and Culture.*

On April 13, 1906, Ellen Irene Diggs was born in Monmouth, Illinois, to Charles Henry and Alice Diggs. Although her parents were of the working class, they encouraged her early ambitions to travel. From 1923 to 1924, Irene Diggs attended Monmouth College on scholarship. She then transferred to the University of Minnesota, where she experienced overt racism for the first time. Rather than deterring her from finishing her education, though, the prejudice she faced only convinced her of the need for African-American role models in U.S. schools. She earned her bachelor's degree in 1928 and began graduate work at Atlanta University in 1932. While a student, she worked under W. E. B. DuBois, the famed black scholar, educator, and activist. In this capacity, she conducted research for some of DuBois's most influential work, including his seminal book *The Dusk of Dawn.* DuBois provided her with freedom and support to explore her own research. While at Atlanta University, Diggs cofounded *Phylon,* a scholarly journal that published articles on race and culture.

Diggs began to fulfill her childhood goal of traveling around the world in the early 1940s. After vacationing in Cuba, she sought to return to the island to study. To this end, she applied for and received a Roosevelt Fellowship from the Institute for International Education at the University of Havana. In Cuba, she collaborated with noted ethnographer Fernando Ortiz, who helped her learn the fieldwork skills necessary to gather information on local cultures. She was awarded her Ph.D. in anthropology from the University of Havana in 1945.

Upon completing her formal education, Diggs traveled throughout Central and South America. She financed her way primarily by writing travel articles for a popular audience. (These essays were syndicated by the Associated Negro Press.) In 1946, Diggs was an exchange scholar in Montevideo, Uruguay. She returned to the United States in 1947 when she joined the faculty at Morgan State University in Baltimore, Maryland. As a professor in the university's anthropology and sociology departments, Diggs continued her research. In the 1950s, she wrote a series of articles on *quilombos*—Brazilian communities founded by runaway slaves. Her most widely acclaimed work is *Black Chronology,* an ambitious book that detailed the accomplishments of black people between 4000 B.C. and the abolition of slavery. She also wrote biographical essays on DuBois (the most famous of these was titled "DuBois and Marcus Garvey"), which were informed by her personal experience with him.

Diggs remained at Morgan State until she retired in 1976, yet she remained active in her field. In addition to conducting research, she held an emeritus faculty appointment at the University of Maryland. She is credited with contributing significantly to the study of comparative ethnohistorical sociology, and she was incomparable in documenting the history of the descendants of African slaves. A member of the American Anthropological Association and the American Association for the Advancement of Science, Diggs was honored in 1978 by the Association of Black Anthropologists for her tireless career as a teacher, role model, and anthropologist. She died in 1998.

Dolan, Louise Ann
(1950–   )
American
Physicist

Louise Dolan, a professor of physics at the University of North Carolina at Chapel Hill since 1990, won the 1987 MARIA GOEPPERT-MAYER Award from the American Physical Society for her work on the theory of elementary particles by applying Kac-Moody algebras to the study of particle physics. Dolan also did important work in string theory, and she was responsible for the reemergence of interest in a specific branch of this study, superstring theory. Dolan's main field of research was theoretical high-energy physics.

Dolan was born on April 5, 1950, in Wilmington, Delaware. She attended Wellesley College, earning her B.A. there in 1971. She then proceeded to the Massachusetts Institute of Technology (MIT) for her doctoral work. After earning her Ph.D. there in 1976, she remained in Boston at

the city's other prestigious institution, Harvard University, as a junior fellow.

Dolan retained this fellowship until 1979, but she did not remain at Harvard throughout this entire period. In 1977, she traveled overseas to Paris, France, where she conducted research as a visiting scientist at the Laboratoire de Physique Théorique de l'École Supérieure. She returned to the United States in 1978, not to Harvard, but to Princeton University, where she was a visiting fellow in the department of physics.

In 1979, Dolan commenced a relationship with Rockefeller University that would last more than a decade. The university hired her as a research associate, but within one year, it had promoted her to the position of assistant professor. She remained at that rank for only two years, becoming an associate professor in 1983.

The next year, Dolan again headed overseas, this time to London, England, where she was a visiting fellow at Imperial College. After she returned, she took up a position concurrent with her Rockefeller University appointment, working as a consultant for the Los Alamos National Laboratory. In 1990, Rockefeller named her a laboratory head. Dolan did not remain at this position long, though, as the University of North Carolina at Chapel Hill hired her as a professor of physics that same year.

Dolan published her first significant paper, coauthored by R. Jackiw and entitled "Symmetry Behavior at Finite Temperature," while still a doctoral student at MIT in 1974. The paper had far-reaching significance, serving as the foundation for new cosmological theories of the early universe. Her 1981 paper, "Kac-Moody Algebra in Hidden Symmetry of Chiral Models," introduced the use of Kac-Moody algebras to particle physics. Her winning of the Maria Goeppert-Mayer Award largely resulted from this work. Dolan published several other significant papers, mostly involving string theory. Dolan's other areas of interest for her research include nonabelian gauge theories, loop space, transfer matrix, exactly integrable systems and spin systems, as well as nonperturbative theory.

Besides the Goeppert Award, Dolan has received several other honors for her work in physics, most notably a John Simon Guggenheim Fellowship from 1988 through 1989. Dolan was also a fellow of the American Physical Society, a very prestigious appointment.

Doubleday, Neltje Blanchan De Graff
(1865–1918)
American
Naturalist and Writer

Writer and wife of publishing tycoon Frank Doubleday, Neltje Doubleday helped to popularize ideas emerging from the new conservation movement of the late 19th and early 20th centuries. Doubleday wrote primarily about birds and plants, although she occasionally wrote about American Indians as well. Light on scientific study, her books and numerous magazine articles nonetheless prompted an increased awareness of humankind's relationship to nature in an era in which nature was generally seen as the raw material to be exploited for the benefit of America's growing industrial base.

Born on October 23, 1865, in Chicago, Illinois, Neltje Blanchan De Graff was the daughter of Liverius De Graff, the owner of a men's clothing store, and Alice Fair De Graff. The De Graff family was prosperous enough to send Neltje to private boarding schools in New York State. Sometime in the late 1870s or early 1880s, she attended the Misses Masters' School in Dobbs Ferry, New York, in the Hudson River valley north of New York City. She also attended St. John's school in New York City proper. In 1886, when she was 20, De Graff married Frank Doubleday, an up-and-coming book publishing baron. They had three children, Felix Doty, Nelson, and Dorothy.

Neltje Doubleday's studies seem to have stopped with her graduation from secondary school. This lack of university training was the norm for women at that time in the United States. Most women were discouraged from pursuing higher education, as it was considered a man's domain. The place for women's activities was the home or, if the woman was from a wealthy family, as a volunteer in a charitable cause. Even though Neltje's family in Chicago does not seem to have possessed wealth on the scale of the Carnegies, Rockefellers, or even Frank Doubleday, they probably had enough money to send their daughter out into the world in the traditional fashion for their class, as a debutante in New York society. It was undoubtedly at one of these events that De Graff met her future husband.

As noted in personal accounts and memoirs of her friends, Neltje Doubleday was an engaging and personable woman. She possessed charm, energy, and warmth and sought out an area in which to make a name for herself. Learning about magazine and book publishing from her husband, she realized that there was a niche in those businesses for her talents as a writer. In her first book, *The Piegan Indians,* published in 1889, Doubleday laid out her interests and style. This was a popular work about an American subject, a Plains Indian tribe. She borrowed from the writings of ethnographers and other writers who looked at the Piegans with greater scientific objectivity, yet she herself was not interested in writing an ethnographic study. This was to be Doubleday's only book on American Indians.

In the 1890s, Doubleday, who always wrote under the pen name of Neltje Blanchan, turned to writing about birds and plants of the United States. Her book, *Bird Neighbors* (1897), was published by her husband's firm of Doubleday, McClure. It discussed bird habitats and sea-

sonal migration, but it did not group the birds by the scientifically accepted system of classification. Instead, it listed them by size and color. *Bird Neighbors* was a huge commercial hit, selling more than 250,000 copies. Doubleday followed this up with a string of books in the early years of the 20th century about birds and plants: *Nature's Garden* (1900), about American wildflowers; *How to Attract Birds* (1902), a book about birds and the plants that draw them; *The American Flower Garden* (1909), about how to create a lush flower garden in a home with large grounds.

Doubleday, who also did volunteer work for the American Red Cross, died suddenly on February 21, 1918, at the age of 52, during an official Red Cross trip to Canton, China. Her work, although of limited scientific value, helped make the conservation movement a respectable force in American life.

Douglas, Allie Vibert
(1894–1988)
Canadian
Astrophysicist

The first Canadian woman to become a professional astrophysicist, Allie Douglas was involved in early studies about the stellar velocity and absolute magnitude of stars. She later did research on the absolute magnitude and parallaxes of Class A stars. Douglas also taught astronomy and physics, and for 20 years, she held an administrative position at Canada's Queen's University. In her roles of teacher and administrator, she encouraged women's participation in engineering and the sciences.

Douglas was born in Montreal, Canada, in 1894. The year of her birth, both of her parents died, and she was sent to live with her maternal grandmother in London. Douglas's grandmother, an Irish woman living in England, took good care of her and her brother, George, but when she was 10, Douglas and her brother were sent back to live with relatives in Canada. Until her 18th birthday, she was cared for by two aunts. These aunts encouraged her interest in science by taking her to lectures held at Montreal's Natural History Society and the McGill University Physics Building. Douglas never married nor had children.

An excellent student at Westmount High School in Montreal, Douglas received a scholarship to attend McGill University. She began her undergraduate studies in 1912 with the intention of majoring in mathematics and physics. However, the beginning of World War I in 1914 caused her to suspend her studies. Following her brother, George, who had enlisted in the army and was stationed in England, Douglas ventured to London and took a job as chief of women clerks at the British War Office's recruiting department. She kept this job throughout the war and won a silver cross as Member of the Order of the British Empire for her efforts. At the end of the war in 1918, she stayed in London to work as registrar of the Khaki University, a correspondence school organized by the Canadian government for soldiers leaving the army. She returned to Montreal in 1919 and, in 1920, finished her undergraduate degree with honors. She immediately began working on a master's degree in physics from McGill, which she won in 1921 for her study of the radiation emanating from radium isotopes.

In 1922, with the help of a postgraduate fellowship, Douglas returned to England to work on a Ph.D. in physics. She had won a place in Ernest Rutherford's Cavendish Laboratory at Cambridge University. She soon became aware of the odds that were stacked against her as a woman at Cambridge. It was possible for her to complete all of her academic requirements and still not be awarded a Ph.D. Abandoning Rutherford's lab, she started working with Arthur S. Eddington at the Cambridge Observatory. Here she began a statistical compilation of the relationship between the stellar velocity and the absolute magnitude of a group of stars. Douglas returned to Canada in 1923, but it took her until 1926 to finish her Ph.D., which was based on a study of the spectrographs of stars in the collection of the Yerkes Observatory in Wisconsin.

Because the male-dominated faculty at McGill refused to grant her tenure, Douglas taught astrophysics at McGill as an instructor, not a professor, from 1927 to 1939. In 1939, she accepted a position as full professor and dean of women at Queen's College, the women's university that was affiliated with McGill. For the next 20 years, until her retirement in 1959, she worked tirelessly to teach and encourage several generations of young women to enter the sciences.

For her work as an astrophysicist, teacher, and administrator, Douglas was awarded an honorary doctorate from the University of Queensland in 1965. She was active in the International Federation of University Women, and in 1967 the National Council of Jewish Women named her its "Woman of the Century." Allie Douglas died on July 2, 1988, at the age of 94.

Dresselhaus, Mildred Spiewak
(1930–)
American
Physicist

Renowned for her work on solid-state physics, Mildred Dresselhaus has opened up new fields of knowledge with her research from the late 1950s through the 1990s. She has done valuable pure-science research on superconductors, materials that have the ability to conduct electricity with very little resistance, and carbon analysis.

Born into a poor family in New York City in 1930, Mildred Spiewak was forced to work in factories as a child to help support her family. However, in spite of the hours she had to work, Spiewak was a good and dedicated student. After she graduated from high school in 1947, she was admitted to Hunter College in New York City. In 1958, she married Gene Dresselhaus, a fellow physics student she met while studying at the University of Chicago. They have four children.

At Hunter, Dresselhaus had intended to study for a degree that would allow her to become an elementary-school teacher, but her math and science professors, noticing that she had made a perfect score in the mathematics part of her admission exam, urged her to concentrate on sciences instead. She realized that she liked science, especially physics, and worked on a major in physics, which she received in 1951. Dresselhaus did so well at Hunter that in 1951–52 she was offered a Fulbright scholarship to study at the Cavendish Laboratory at Cambridge University in England. Cavendish was the premier research facility in physics in the United Kingdom at that time. After completing her postgraduate studies at Cavendish, Dresselhaus worked on her master's degree at Radcliffe College, which she received in 1953. She then went to the University of Chicago to work on a Ph.D. in physics. For her doctorate, she began work in a new field that had attracted few researchers: a form of solid-state physics that centered on superconductors in a magnetic field. Dresselhaus wanted to know how these types of solids reacted inside magnetic fields and what properties they possessed that made them different from ordinary conducting metals.

After finishing her doctorate at Chicago in 1958, Dresselhaus and her husband looked for jobs at a university that would let them both work in the same department. They rejected Cornell University because it would accept only Gene Dresselhaus and not Mildred. However, in 1960, the Lincoln Laboratory at the Massachusetts Institute of Technology (MIT) accepted them both. At MIT, Dresselhaus continued her superconductor research, focusing her work on the metals lead and tin. She was not interested in using her work to invent new technologies. She was content to leave this work of applied science to others. Instead, she enjoyed uncovering new information about the behavior of certain materials under experimental conditions.

Throughout the 1960s and 1970s, Dresselhaus continued her research in solid-state physics: investigating the electrical and optical properties of solids, semimetals, and magnetic semiconductors. In the 1980s, she began looking at carbon as a superconductor. In carbon, she discovered what were called buckyballs, hollow clusters of sixty atoms. (They were given their name because they resemble the geodesic dome created by engineer Buckminster

Fuller.) This phenomenon has proved valuable in speeding drugs into the human body and as an extrastrong form of wire tubing.

Dresselhaus has also pushed MIT and the higher education establishment in general to attract greater numbers of women students to science and engineering. Throughout, she has also been a teacher. She became a full professor in MIT's Department of Engineering in 1968. By 1983, she was a professor in both the engineering and physics departments. In 1985, she was made Institute Professor at MIT, a rare lifetime honor. For her work in physics, she has been elected to the National Academy of Engineering (1973) and the National Academy of Sciences (1985). She has also received the National Medal of Science in 1990.

Dunham, Katherine Mary

(1909–)
American
Anthropologist

Katherine Mary Dunham distinguished herself as a dancer and choreographer, but her expertise as an anthropologist informed her art, as she incorporated dance traditions from various cultures that she studied into her dance innovations. She published three books on her anthropological travels, and she established several foundations to promote the understanding of various cultures, most notably a training center and museum in East St. Louis that educated inner-city youths about African culture.

Dunham was born of mixed heritage (African, Madagascan, French-Canadian, and Native American) on June 22, 1909, in Glen Ellyn, Illinois. Her father, Albert Dunham, was a traveling salesman, and her mother, Fanny June Taylor, died when she was four. Her father then married Annette Poindexter, a former schoolteacher, and moved the family to Joliet, Illinois, where Dunham attended Joliet Township Junior College. She then transferred to the University of Chicago, where she studied anthropology while working her way through school by dancing professionally with troupes she formed—first the Ballets Negre, then the Negro Dance Group, which performed at the 1934 Chicago's World Fair.

Dunham combined dance and anthropology when she received a Rosenwald Foundation travel fellowship, which she used to conduct anthropological field studies on West Indian and African dance in Haiti, Martinique, Jamaica, and Trinidad. She developed choreographic theories based on traditional African dance movements. Dunham returned to the United States for her bachelor's degree in anthropology in 1936, and she subsequently earned a master of science degree from the University of Chicago and a Ph.D. in anthropology from Northwestern University.

In the late 1930s, Dunham moved to New York City, where she established a dance school and touring company, the Katherine Dunham Troupe, which she directed, choreographed, and danced in. She continued to draw on her anthropological background in her staging of dances; for example, she based "L'Ag'Ya," one of her early successes, on a Martinican fight dance. The popularity of her dancing gave her the opportunity to promote African-American culture as well as other diverse indigenous cultures as she toured and performed with her dance company and with the likes of the Duke Ellington Orchestra.

In 1939, Dunham married the renowned Canadian-American stage and costume designer John Pratt, and in 1952, they adopted a daughter, Marie Christine, a five-year-old French-Martinique girl at the time. The couple collaborated throughout their careers, until Pratt's death in 1986.

Dunham published an autobiography, *A Touch of Innocence* (1959), as well as three books about her anthropological travels: *Katherine Dunham's Journey to Accompong* (1946), *The Dances of Haiti* (1947), and *Island Possessed* (1969). In 1965, she disbanded her dance company in order to move to Senegal, where she served as the technical cultural advisor to President Leopold Sedar Senghor and trained the Senegalese National Ballet for the First World Festival of Negro Arts in Dakar.

In 1967, Southern Illinois University in Edwardsville offered her a position as an artist in residence; there, she institutionalized her approach of fusing the sciences and the arts by establishing a dance anthropology program. In 1969, she extended this notion into social politics by establishing the Katherine Dunham Center for the Arts and Humanities in East St. Louis, providing opportunities for residents of this impoverished region to benefit from the study and practice of arts from diverse cultures. She later founded the Children's Workshop and the Dunham Technique Seminar and Institute for Intercultural Communication, as well as establishing the Foundation for the Development and Preservation of Cultural Arts and the Dunham Fund for Research and Development of Cultural Arts.

Dunham remained politically active throughout her career. In 1993, at the age of 83, she undertook a 47-day hunger strike to draw attention to the political plight of Haiti; in the wake of this action, Haiti conducted its first democratic election. Dunham garnered honors galore throughout her lifetime, including the Presidential Medal of Honor, the 1968 *Dance Magazine* Award, the 1983 Kennedy Center Honors Award, and the NAACP Lifetime Achievement Award.

Duplaix, Nicole
(1942–   )
French/American
Zoologist and Ecologist

Noted for her pioneering studies of river otters, Nicole Duplaix is a zoologist who has made the extinction of wildlife by human encroachment one of her primary concerns. She authored a study of a colony of river otters in Suriname and has worked with the World Wildlife Fund to try to save this and other endangered species.

Nicole Duplaix was born in 1942 to George Duplaix, an artist, and Lily Duplaix, a writer. Her parents were wealthy and had homes in the United States and France. After attending primary and secondary school in Paris and Manhattan, Duplaix did undergraduate work at Manhattanville College, in Westchester County just outside of New York City. During her last two years at Manhattanville, Duplaix arranged for a volunteer job as an assistant attendant at New York's Bronx Zoo. Her boss, the zoo's curator of mammals, was in Duplaix's words, "an otter freak." His enthusiasm got Duplaix hooked on this playful species. She also began to realize, even at this relatively early date when consciousness about ecological disaster and animal extinction was just emerging, that river otters and other mammals were being threatened by the encroachment of human habitation. After graduating from Manhattanville in 1965, Duplaix earned a master's degree in animal ecology from the University of Paris in 1966. She left academia for a number of years in the 1970s, but by 1976, she had returned to doctoral studies at the University of Paris. Her dissertation, which she completed in 1980, was about a colony of freshwater river otters on Kapoeri Creek in the South American country of Suriname.

From 1966 to 1974, Duplaix left the field of animal ecology while she lived in Europe and the United States. From 1970 to 1974, she was in London as the wife of a wealthy commodities broker. Following a divorce, she decided to return to work at the Bronx Zoo. At the zoo, she became familiar with a species of large freshwater otter that lived in Surinam. Known to Surinam locals as the *river dog*, these otters can grow to six feet in length and weigh 60 pounds. Encouraged by her mentor at the zoo, Duplaix began to plan for a trip to Suriname to study these animals in their habitat.

Suriname, now an independent country, is a former Dutch colony at the remote northeastern corner of South America. The Dutch arrived here early in the 17th century but only firmly established a presence in the 1680s. Carib Indians on the coast quickly died of imported diseases and slavery. The Dutch, like their English and French neighbors on this coast, then brought in black slaves. Many of these slaves escaped and fled to the jungle backcountry where they intermarried with Indian groups. The descendants of these escaped slaves are known as Bush Negroes. It was these people who helped Duplaix with the logistical support to carry out her studies.

The river dog otter once was common in the Suriname region but because of hunting and human encroachment into their habitat now exist only in the backcountry. Duplaix found that they live in extended family units and that each unit marks a territory that takes up about a mile of river. The families live in small "campsitelike" areas cleared of underbrush along the river and forage from there.

With these findings, Duplaix helped the Suriname government begin planning to try to save this and other endangered otter colonies. She also worked as a speaker for the World Wildlife Fund during the 1980s. She lived in Florida in the early 1990s but moved back to Paris as of 1998.

Dyer, Helen M.
(1895–1998)
American
Biochemist

An early researcher in the causes and cures for cancer, Helen Dyer made many contributions to the understanding of the biochemistry of the body. She discovered compounds that had antimetabolic effects on mammalian bodies and found ways to reverse these effects. She also conducted considerable research on nutrition and its biochemical effects on the stomach and on possible viral causes of cancer.

Born in Washington, D.C., on May 16, 1895, Helen Dyer was the daughter of Joseph Edwin Dyer, the owner of a wholesale grocery business, and Mathilda Robertson Dyer. She attended Western High School, a public secondary school in Washington, graduating from there in 1913. Dyer never married.

Dyer did undergraduate work at Goucher College, in Towson, Maryland, just outside of Washington, D.C. She earned a B.A. from Goucher in 1917. After her graduation, Dyer spent a year teaching physiology at Mount Holyoke in Massachusetts, but she returned to Washington to work in the Hygienic Lab of the U.S. Public Health Service (which later was merged into the National Institutes of Health). From 1921 to 1933, she worked as a research assistant for Carl Voegtlin, biochemistry and pharmacology department head at the Public Health Service. In the late 1920s, she enrolled in the George Washington University Medical School where she worked on a master's degree in biochemistry. Working under future Nobel laureate Vincent du Vigneaud, Dyer won her M.A. in 1929 and continued on a program of doctoral studies, completing this in 1935.

After finishing her doctorate, Dyer worked as an assistant professor of biochemistry at George Washington

University. Besides teaching courses, she continued her work in the laboratory. During this time, she conducted research that aimed to find a dietary replacement for the enzyme methionine. Dyer synthesized a replacement compound called *ethionine,* which, when fed to rats, proved to be both toxic and a growth inhibitor. The antigrowth effects of ethionine were reversed when the rats were switched back to methionine. This work showed that antimetabolites could be manufactured in the laboratory and opened up this area as a field for further study. At this same time, Dyer was working on other research problems. With Vincent du Vigneaud and others, she synthesized and tested amino acid compounds that contained sulfurs. She demonstrated that these compounds could replace cystine in a diet to aid growth. Also with du Vigneaud, she experimented on extracting the active compounds of the posterior pituitary gland.

Because she was a woman, Dyer was not promoted from her beginning rank of assistant professor at George Washington University. So in 1942, she quit her position there and took a job with her friend and mentor, Carl Voegtlin, who had become director of the National Cancer Institute (NCI). She was to remain at the NCI until she retired in 1965. In her early days at the NCI, Dyer conducted research on gastric cancers by examining the cancer-causing effects of compounds such as acetyl-beta-methycholine chloride and histamine diphosphate. Later, she looked at the process by which cancers grow, with special emphasis on an examination of liver cancer.

During her career, Dyer was the author of 60 scientific papers. She won the American Chemical Society's Garvan Medal in 1961. In 1965, she was granted an honorary doctorate from her alma mater, Goucher College. Dyer was a fellow of the American Association for the Advancement of Sciences. She served as a consultant to the NCI for years after her retirement. She died at the age of 103 in 1998.

_E_

Earle, Sylvia Alice
(1935–   )
American
Marine Biologist

A specialist in phycology, the study of seaweeds and other algae, Sylvia Earle has in a varied and interesting career observed sea life in its own habitat by living in a submerged sea station and worked on technologies that allow divers to probe ever deeper into the sea. She has also served as a scientific administrator and worked with conservation organizations to gather information about threatened ocean environments and educate the public to the dangers humans pose to sea species.

Born on April 30, 1935, in Gibbstown, New Jersey, Sylvia Earle is the daughter of Alice Earle, a nurse, and Lewis Earle, an electrical engineer. Earle connected with nature when she was a child living on a farm in New Jersey. Her mother taught Earle and her two siblings to cherish nature. Earle credits this environment with instilling a curiosity about the natural world. Earle made her connection with the ocean when her family moved to Florida in 1948. The town she lived in, Dunedin, just outside of Tampa, was directly on the Gulf of Mexico, and Earle began scuba diving regularly. "I practically had to be pulled out of the water," she remembers. Earle has been married three times and has three children.

Earle enrolled in Florida State University where she studied botany. She graduated with a B.S. from Florida State in 1955, then went to Duke University in North Carolina to get an M.S. in botany in 1956. In 1957, Earle married a fellow science graduate student, John Taylor. For seven years, until 1964, she temporarily withdrew from science to raise her two young children. In 1964, she began working as a sea researcher in conjunction with the beginning of her work on a Ph.D. in marine biology from Duke. She finished her doctoral work and was awarded a Ph.D. in 1966.

Beginning in the mid-1960s, Earle began making numerous dives to examine marine life in its own environment. By the late 1960s, the National Aeronautics and Space Administration (NASA) had devised a plan to study human behavior for long durations in closed and isolated quarters, the equivalent of a long spaceflight, only this experimental look would be done on the ocean floor. Several all-male crews had stayed in an undersea "habitat" that NASA had constructed in 50 feet of water on the floor of the Caribbean Sea. In 1970, NASA decided to put an all-female crew of four women scientists in this habitat, which was called Tektite. For Earle, the psychological and physiological studies that NASA was conducting were secondary to the chance to study marine life for an extended time underwater. This all-women crew unexpectedly grabbed a huge amount of media attention, and Earle became a celebrity. Along with her crewmates, she received the Conservation Service Award, the highest civilian award given by the Department of Interior.

During the 1970s, Earle began collaborating with a British engineer named Graham Hawkes to devise a diving suit that would allow scientists to descend much deeper into the ocean than had previously been possible. In September 1970, Earle dove 1,250 feet into the ocean near Hawaii in a *Jim suit*. Free-floating and unattached to cables, she studied deep-sea creatures while her suit withstood 600 pounds of pressure per square inch. During this time, Earle also worked with marine biologist Katherine Payne on a study of humpback whales, and she was curator of phycology at the California Academy of Sciences. For two years (1990–92), she served as director of the U.S. National Oceanic and Atmospheric Administration. She currently runs an undersea exploration company called Deep Ocean Exploration and Research and is working with the National Geographic Society on Sustainable Sea Expeditions, a study of U.S. national marine sanctuaries.

Earle is a fellow of the American Association for the Advancement of Sciences. She has won the Society of Women

Geographers Gold Medal (1990), the Kilby Award (1997), and the John M. Olguin Marine Environment Award (1998).

Eastwood, Alice
(1859–1953)
Canadian/American
Botanist

Overcoming an early life of hardship and deprivation, Alice Eastwood became the most renowned expert on the flora of the American West. After working informally as a botanist in Colorado for 12 years while also a full-time high-school teacher in Denver, she was invited to become the curator of botany at the California Academy of Sciences in San Francisco. She would stay with the academy for 57 years, expanding its collection, then rebuilding it and her life after the disastrous earthquake of 1906.

Born in Toronto, Canada, on January 19, 1859, Alice Eastwood was the daughter of Colin Skinner Eastwood, an unsuccessful businessman, and Eliza Jane Gowdey Eastwood. Eastwood spent the first five years of her life on the grounds of the Toronto Asylum for the Insane where her father had a job as a caretaker. When she was about age five, Colin Eastwood tried to start a grocery business, but this soon failed. Eastwood's mother, who had been ill, died when Alice was six. Colin Eastwood then left with Alice's brother for Denver, Colorado. Alice Eastwood and her sister were given over to an Ontario convent.

Eastwood first developed an interest in botany while visiting the house of an uncle, Dr. William Eastwood, who grew hybrid plants on the grounds of his home. Alice would take seeds from these plants and grow them back at the convent. In 1873, when Eastwood was 14, she was summoned with her sister to live with her father in Denver. She worked at her father's new store and took frequent trips into the Rocky Mountains. In 1875, Eastwood enrolled in East Denver High School, quickly becoming the best student in her class. However, in her junior year, she was forced to drop out when her father's business again failed. For more than a year, she worked full-time to help support her family. In 1878, she returned to finish high school.

After graduation in 1879, Eastwood began teaching Latin, natural science, and other subjects at East Denver High. She also began to study botany formally and make frequent trips out into the Rocky Mountains to collect alpine plants. Corresponding with other botanists, she became known as an expert of the botany of Colorado and the Rocky Mountains. By 1881, she was able to make a trip back east to Boston to visit the Gray Herbarium in Cambridge. There she met Asa Gray, the most renowned American botanist of the time.

In 1890, Eastwood traveled to California on a collecting trip. In San Francisco, she visited KATHERINE BRANDEGEE, curator of botany of the California Academy of Sciences (CAS). Brandegee and her husband, Townshend Stith Brandegee, invited Eastwood to submit articles to their journal *Zoe*. Impressed with Eastwood's energy and knowledge, in 1892, the Brandegees offered Eastwood their jobs so that they could retire to San Diego. Eastwood accepted and, in 1893, became curator of botany at CAS and editor of *Zoe*.

Eastwood held the CAS job until she retired in 1949. During these 57 years, she traveled to all parts of California, Oregon, Nevada, Utah, New Mexico, Baja California, and the Yukon Territory in Alaska collecting plants. She systematically rearranged and updated the classification of the collection left to her by the Brandegees. After the earthquake of 1906, which toppled and burned down the academy, she single-handedly rebuilt the majority of the herbarium collection that was lost and did this while living in a makeshift apartment. From 1908 to 1911, she was frequently on the rails, traveling back east and to Europe to inspect other herbariums. Bit by bit, the CAS's herbarium was restored and is now housed in Golden Gate Park. As if that were not enough, Alice Eastwood became a driving force in the save-the-redwoods movement in California.

For her efforts, Eastwood became the only woman given a star for distinction in every edition of *American Men of Science* published during her lifetime. In addition, she was a fellow of the American Association for the Advancement of Sciences and an honorary member of the Royal English Forestry Society. In 1950, at the age of 91, she was invited to Stockholm, Sweden, as honorary chairwoman of the Seventh International Botanical Congress. She opened the meeting sitting in the chair once occupied by Linnaeus. Alice Eastwood died on October 30, 1953, at the age of 93.

Eckerson, Sophia Hennion
(c. 1880–1954)
American
Botanist and Microchemist

A botanist whose specialty was the study of the chemistry of living plants, Sophia Eckerson worked as a teacher and researcher. During the early part of her career, she taught at Smith College in Massachusetts and at the University of Chicago. She also took on research projects in which she examined plant seed germination, plant root growth, protein synthesis in plants, and the formation of plant cellulose membranes.

The exact date of Eckerson's birth is unknown, but she was probably born sometime around 1880 in Old Tappan, New Jersey. Her parents were Albert Bogert Eckerson and Ann Hennion Eckerson. A woman who valued her independence and moved around the country to advance her career, she never married.

After graduation from secondary school, Eckerson delayed the beginning of her university education to help put her brothers through college. By 1901, she had enrolled in Smith College in Northampton, Massachusetts. At Smith, Eckerson was inspired by William F. Ganong, who became her friend and mentor, to study botany and plant physiology. She won her B.A. in botany at Smith in 1905 and stayed on to work on her M.A., which she completed in 1907. During these years, she decided to make her life's work the study of plant microchemistry and physiology. In 1909, Eckerson went to the University of Chicago to work on her Ph.D., which she was awarded in 1911.

During and after her work on her master's degree at Smith, Eckerson taught botany and plant microchemistry. From 1905 to 1909, she served as demonstrator and assistant in the botany department. Here she began putting together notes that became an unpublished text she entitled "Outlines of Plant Microchemistry." Eckerson used this introductory text, which outlined the main topics of plant physiology, heavily in her classroom teaching at Smith and later at Chicago. After finishing her doctorate at the University of Chicago, Eckerson was given the post of assistant plant physiologist at that institution, a job she held until 1915. From 1915 to 1920, she held the post of instructor of plant microchemistry. As happened with many other women in academia at that time, Eckerson found that advancement to a position of full professor was almost impossible. These jobs were almost always held open for men only. During the school year 1921–22, Eckerson taught and did research at the University of Wisconsin. She also shuttled back and forth between Wisconsin and Washington, D.C., where she worked for the U.S. Department of Agriculture's Bureau of Plant Industry and later its Cereal Division. In 1924, she took a job as plant microchemist at the newly formed Boyce Thompson Institute for Plant Research in Yonkers, New York. She remained here for the rest of her career and eventually became chair of the institute's Department of Microchemistry.

Eckerson began her in-depth research on plant chemistry with her Ph.D. dissertation at the University of Chicago. This work focused on the "after ripening" of the seed embryo, or exactly how the embryo chemically changes during germination. After examining the microchemistry of seeds during various stages of germination, Eckerson concluded that germination was dependent on the embryo's acidity, its ability to produce certain enzymes, and how well it absorbed water. At the Boyce Thompson Institute, Eckerson conducted detailed experiments that looked at how plants synthesize proteins found in the soil in which they grow. Her research concentrated specifically on how certain plants absorb and synthesize nitrates. During the later stage of her career at the Thompson Institute, she examined the processes by which cotton and other kinds of plants produce cellulose particles.

For her work in plant microchemistry, Eckerson was given a star by her name (indicating that she was a leader in her field) in the 1938 edition of *American Men of Science*. She also served as vice chair (1934) and chair (1935) of the physiological section of the Botanical Society of America. Eckerson retired from active work in 1940 and died on July 19, 1954, in Connecticut.

Eddy, Bernice
(1903–   )
American
Medical Microbiologist

Dr. Bernice Eddy's experience demonstrates the danger posed by institutional sexism, not only to the victim of the prejudice but also to the greater society she is trying to serve. Eddy and her collaborator, Dr. Sarah Stewart, were the first to describe polyoma virus (later renamed SE polyoma virus to honor them) in the late 1950s and warn of its potential hazards in polio vaccines. Because Eddy's superiors had not authorized her research, and because of the expensive implications of her results (a recall of all of the polio vaccines developed by Dr. Jonas Salk), they suppressed her findings until independent researchers verified the same results, forcing an acknowledgment.

Eddy was born in 1903 in Glendale, West Virginia. Her father, Nathan Eddy, was a physician. She graduated from college in Marietta, Ohio, in 1924, and proceeded to the University of Cincinnati, where she received her master's degree in bacteriology in 1925 and her Ph.D. in 1927. In 1938, she married her fellow Public Health Service worker, the physician Jerald G. Wooley, who died when their two daughters, Bernice and Sarah, were still young.

Eddy continued to work at the University of Cincinnati after receiving her doctorate until 1930, when she went to work for the Public Health Service. Then, in 1935, she transferred to the Biologics Control Division of the National Institutes of Health (NIH) in Bethesda, Maryland, where she checked vaccine quality. For 16 years following World War II, Eddy checked the Army's influenza vaccines. In 1944, she was named chief of the flu virus vaccine testing unit. In 1952, she commenced attempts to find polio treatments, and the next year she received the NIH Superior Accomplishment Award for this work.

In 1954, she became a controversial figure when testing the polio vaccine recently developed by Salk. She discovered a correlation between vaccine administration and paralyzation in the test monkeys; she immediately reported her findings to her superiors, which they ignored, and subsequently 200 children developed polio from the live vaccine. Eddy was taken off polio vaccine testing.

Working in collusion with Stewart, Eddy developed a method of growing tissue infected with tumor viruses in isolation from other cells, leading to their discovery of SE polyoma virus, which apparently caused leukemia. Eddy traced this virus as well as the infected polio vaccines back to a common source: the kidney cells of rhesus monkeys from which the vaccine samples were grown. Eddy confirmed her hypothesis when she injected ground-up monkey cells into newborn hamsters, who developed tumors, suggesting that the monkeys were indeed the source of the virus. On July 6, 1959, Eddy reported her findings to her supervisor, Joseph Smadel, who dismissed the cancerous tumors as "lumps." Eddy circumvented his authority by announcing her results at the fall 1960 meeting of the New York Cancer Society, further enraging Smadel, who forbade her from publishing and speaking about her research in public.

Maurice Hilleman's discovery of SV40, the same cancerous virus that Eddy had described, vindicated her personally if not professionally. Hilleman echoed Eddy's warnings when he advised against the use of rhesus monkeys to prepare vaccines, as their wild nature created unpredictability. The NIH, instead of acknowledging her achievements, railroaded Eddy into less and less significant research roles until her 1973 retirement at the age of 70. She then received recognition of her contributions, as the secretary of the Department of Health, Education and Welfare awarded her a special citation that year followed in 1977 by the NIH Directors Award.

Edinger, Johanna Gabrielle Ottelie ("Tilly")
(1897–1967)
German/American
Paleontologist

Tilly Edinger is credited with inventing the field of paleoneurology, as none before her had studied the brains of fossilized animals in her systematic way. She solved the problem of the deterioration of brains by casting the space inside skull fossils to recreate brains. Tilly not only overcame the obstacles barring women from the scientific fields (her father opposed her scientific career), but also she worked in secrecy in the midst of the oppressive Nazi regime and continued to do research even after her hearing impairment forced her to retire from teaching.

Edinger was born on November 13, 1897, in Frankfurt, Germany. Her mother, Anna, supported numerous social causes, and her father, Ludwig, was a wealthy and famous comparative neurologist. His research undoubtedly influenced Edinger's decision to study neurology, yet she differentiated herself by focusing on extinct and fossilized brains. Edinger studied at the Universities of Heidelberg and Munich from 1916 through 1918 before receiving her doctorate in zoology from the University of Frankfurt in 1921 with a dissertation on Mesozoic marine reptiles. She

stayed on at the university for the next six years as a research assistant in paleontology.

In 1927, the Senckenberg Museum in Frankfurt hired Edinger as the curator of its vertebrate collection. When the Nazi regime took power in 1933, her Jewish ancestry forced her underground, working at the museum in secret by using the back entrance and removing her nameplate from her office. Nazi officials discovered her ruse in 1938 and expelled her from her position, so she emigrated from Germany in May 1939, arriving in the United States in 1940. She landed a position with the Harvard Museum of Comparative Zoology that year, and she taught comparative anatomy at Wellesley College until her hearing disabilities became overwhelming. She became a U.S. citizen in 1945.

In 1929, Edinger published her first book, *Fossil Brains,* in which she described her method of studying brain endocasts created by the fossilization of sediment filling the cranial cavity and thus leaving a mold of the space formerly filled by the brain. Since the brain fills the skull to capacity, fossilized molds or even plaster casts of the internal space inside a skull fossil can be used to study the brains of long-extinct animals. In 1948, she published a monograph, "The Evolution of the Horse Brain," tracing the progression of fossil horse brains from *Hyracotherium* to *Equus.* Edinger described this orthogenetic progression not as a linear development, as it would seem, but more like the growth of a many-branched tree. This study remains a classic, even though its conclusions have been modified and expanded upon since its introduction.

Edinger served as the president of the Society of Vertebrate Paleontology from 1963 to 1964, indicating her high stature in her field. One of her discoveries bears her name: The thick bony deposits lining the vertebral column of certain species of fish are called *Tilly bones.* Edinger died when a car struck her while she was crossing the street on May 27, 1967.

Edwards, Cecile Hoover
(1926–   )
American
Nutrition Researcher

Cecile Hoover Edwards devoted her life to the study of nutrition with the specific goal of improving the health of impoverished people. She conducted in-depth research on amino acids, the essential nutritional components of proteins, in order to understand the chemistry and physiology of nutrition. She also educated people about the basics of fortifying the body with proper nutrition. She was on the forefront of the movement that transformed home economics into a more scientific study, enhancing the "how-to" curricula with hard scientific data and advanced educational approaches.

Edwards was born on October 26, 1926, in East St. Louis, Illinois. Her mother was Annie Jordan, a former schoolteacher, and her father was Ernest Hoover, an insurance manager. At the age of 15, Edwards matriculated at Tuskegee Institute, majoring in home economics and minoring in nutrition and chemistry. She graduated with honors in 1946 and continued on with graduate research at Tuskegee, conducting chemical analyses of an animal source protein, supported by a fellowship from Swift and Co. She earned her master's degree in chemistry in 1947.

Edwards then pursued her doctorate at Iowa State University, where she worked as a research associate in nutrition conducting investigations on methionine, one of the essential amino acids. In her dissertation, she concluded that methionine was beneficial not only as a catalyst for protein synthesis but also as a carrier of sulfur, which could be easily released and transferred to other compounds. After earning her Ph.D. in nutrition in 1950, she continued to experiment on methionine, authoring some 20 more papers on the amino acid. In 1951, Edwards married Gerald Alonzo, and together the couple had three children.

In 1950, Edwards returned to Tuskegee as an assistant professor and research associate in foods and nutrition. Within two years, the institute had promoted her to head the department of foods and nutrition, a position she retained from 1952 through 1956. Concurrently, she collaborated with the Bureau of Human Nutrition and Home Economics, working with the Agricultural Research Service of the U.S. Department of Agriculture. She then moved to North Carolina A&T State University as a professor of nutrition, remaining at this post for more than a decade. In 1968, the university promoted her to head the department of home economics until Howard University hired her for the same post in 1971.

In 1974, Howard University promoted Edwards to be the dean of its School of Human Ecology, charging her with the duty of developing a new curriculum. Edwards approached the task with zeal, expanding her focus from nutrition education to a more holistic approach to educating people. She specifically sought to counter the perception that African Americans were not intelligent enough to thrive as well as other ethnicities by instructing the disadvantaged on life's rudimentaries, such as parenting, child care, nutrition, financial budgeting, and job skills.

In 1985, the National Institute of Child Health and Human Development hired Edwards to direct a five-year program studying the effects of nutritional, medical, psychological, and socioeconomic factors on the pregnancies of women with low incomes. While conducting this research, she served as dean of Howard University's School of Continuing Education for one year, from 1986 through 1987. She remained at Howard thereafter as a professor of nutrition.

Edwards, Helen T.
(1936–)
American
Accelerator Physicist

In helping with every aspect of building particle accelerators, Helen T. Edwards created the instrumentation by which to explore quantum mechanical theory through nuclear physics by colliding atomic protons into antiprotons to see what would happen. For this work, the United States Department of Energy granted her the Ernest Orlando Lawrence Award in recognition of her contributions to the understanding of nuclear energy. Edwards was also granted a John D. and Catherine T. MacArthur fellowship.

Edwards was born on May 27, 1936, in Detroit, Michigan. During grade school and in junior high school, she was a self-admitted lousy speller and reader, but she had a natural inclination toward the natural sciences and mechanics. By her high school years, her family had moved to Washington, D.C., where she attended Madeira, an all-girls high school. Edwards then studied physics under Dr. B.D. McDaniels at Cornell University, where she earned her B.A. in 1957. She remained at Cornell first as a researcher and then as a doctoral candidate. McDaniels acted as Edwards's doctoral thesis adviser. Cornell awarded Edwards her Ph.D in physics in 1966. She remained at Cornell's Laboratory for Nuclear Studies for the next four years as a research associate helping to run the 12 GeV electron synchrotron under Dr. Robert Wilson (one GeV equals 1 billion electron volts).

Wilson took Edwards along with him when he departed Ithaca for Chicago to head the staff at the Fermi National Accelerator Laboratory. When she first moved to Fermi, she headed the Switchyard Extraction. Edwards also headed the accelerator division of the lab, where she oversaw the commissioning of the 400-GeV main accelerator. This experience acted as a precursor for her later overseeing of the construction of the highest energy superconducting particle accelerator, the Tevatron (one TeV equals one thousand GeVs, or one trillion electron volts.)

Edwards had a hand in absolutely every aspect of this project, as noted by her supervisor, Dr. Leon Lederman, the director of Fermilab: "superconducting magnet production, magnet testing, cryogenics (in turn divided into Central Helium Liquifier and 24 satellite refrigerators at 600W each), radio frequency, vacuum systems, beam diagnostics, controls, lattice design and orbit theory, magnet parameter acceptance criteria, magnetic correction elements, power supplies, reporting, documentation and cost control, quench protection, extraction system, injection systems and beam transport, beam-abort systems, radiation protection and installation."

In 1986, the Department of Energy granted Edwards the E. O. Lawrence Award in physics for, according to an official statement, "her leadership in the construction and commissioning of the Tevatron at the Fermi lab . . . the first successful superconducting proton synchrotron ever built." For this gold medal award she received $25,000. In 1988, she received a coveted MacArthur fellowship, which gave her financial freedom for one year. From 1989 through 1992, Edwards served as the head and associate director, Superconduction Division, Superconducting Supercollider Laboratory, Dallas, Texas. From 1992 on she returned to Fermi, where she has remained as a guest scientist. She is a fellow of the American Physical Society and a member of the National Academy of Engineering.

Ehrenfest-Afanaseva, Tatiana
(1876–1964)
Russian/Dutch
Theoretical Physicist

Tatiana Ehrenfest-Afanaseva, in collaboration with her husband, Paul Ehrenfest, helped to overturn the existing understanding of statistical thermodynamics by replacing it with a new model based on the ergodic hypothesis that the microscopic mirrored the macroscopic. She also published her own papers on the random status of probability in physics as well as on the teaching of mathematics.

Ehrenfest-Afanaseva was born on November 19, 1876, in the city of Kiev in the Ukraine. Her father, a Russian-Orthodox civil servant, died when she was young, leaving her to the upbringing of an uncle who was taught at the Polytechnical Institute in St. Petersburg, Russia. As a woman, Ehrenfest-Afanaseva could not attend university in St. Petersburg, so she studied education at the women's pedagogical school and later at the Women's Curriculum, which offered university-level programs for women.

Ehrenfest-Afanaseva continued her education in 1902 by traveling with her aunt Sonya to Germany, where the University of Göttingen allowed women students. There, she met Ehrenfest, a physics student from the University of Vienna, and the couple married in 1904. Together, they had two daughters and then two sons. They lived first in Vienna, then in Göttingen, before returning to St. Petersburg, where Ehrenfest had to renounce his Judaism and Ehrenfest-Afanaseva her Russian Orthodoxy in order to live together in compliance with local laws that prevented religious intermarriage.

This stance came back to haunt Ehrenfest later, disqualifying him from consideration for succeeding Albert Einstein at the University of Prague Institute of Physics due to its ban on professors without religious affiliations. However, Ehrenfest did succeed Hendrik Antoon Lorentz as a professor of theoretical physics at the University of Leiden in the Netherlands in 1912. There, Ehrenfest-Afanaseva set up their vegetarian, nondrinking, and nonsmoking household, where they homeschooled their children.

The husband-and-wife team worked effectively in collaboration, as Ehrenfest-Afanaseva's logical mind both grounded and spurred Ehrenfest's creative imagination. In dialogue, the couple developed the underpinnings of their most important idea, published in Felix Klein's 1956 mathematical encyclopedia and translated in 1959 as "The Conceptual Foundations of the Statistical Approach in Mechanics." In this article, they argued against Boltzmann's H-theorem, in which he statistically proved the second law of thermodynamics on the conservation of entropy; Ehrenfest and Ehrenfest-Afanaseva relied on new statistical analyses as well as the ergodic hypothesis, whereby a part reflects the whole, to prove their theory.

This paper was published more than two decades after the suicide of Ehrenfest in 1933. Ehrenfest-Afanaseva's output slowed after her husband's death but accelerated again in the 1950s. Besides this important encyclopedia entry, Ehrenfest-Afanaseva published a paper, based on ideas they both used in lectures as early as 1911, entitled "On the Use of the Notion 'Probability' in Physics," which appeared in the *American Journal of Physics* in 1958. Ehrenfest-Afanaseva also published papers in English, German, Russian, and Dutch on topics as various as axiomatization, randomness and entropy, and teaching pedagogy. She published her last piece, a 164-page monograph on mathematical pedagogy, in 1960, at the age of 84. Ehrenfest-Afanaseva died on April 14, 1964, in Leiden.

Eigenmann, Rosa Smith
(1858–1947)
American
Ichthyologist

The famous marine biologist Carl L. Hubbs described Rosa Smith as "the first woman ichthyologist of any accomplishment." Together with her husband, the zoologist Carl H. Eigenmann, she was the first to describe some 150 species of fish. Eigenmann's career was productive but short-lived. The mother of five children, she devoted herself to their upbringing and never returned to full-time scientific research.

Rosa Smith was born on October 7, 1858, in Monmouth, Illinois, the youngest of nine children. Her mother was Lucretia Gray and her father, Charles Kendall Smith, was a printer who moved his family to San Diego, where he served as the clerk of the San Diego school board. Rosa's tubercular condition prompted this move to a warmer climate.

Smith attended business college in San Francisco in 1880 and simultaneously published her first scientific paper on Pacific coast fish. That year, she read a paper to the San Diego Society of Natural History as its first female member. Most likely, the paper reported her discovery, description, and classification of the blind goby, or *Typhologobius californiensis* (now *Othonops eos*). In the audience was David Starr Jordan, a renowned ichthyologist from Indiana University; Eigenmann's research impressed him sufficiently to invite her to study in Bloomington and on a natural history tour of Europe with 33 other students the next summer. An illness in her family summoned her home to San Diego, and she never graduated from Indiana, though she did work as a reporter in 1886 for the *San Diego Union* newspaper that her brother and brother-in-law owned.

While at Indiana University, Rosa met Carl Eigenmann, a doctoral student in ichthyology, and the couple continued their relationship even after she left Indiana. They married on August 20, 1887. Before marrying, she published 20 scientific papers in which she addressed issues of the embryology, evolution, and classification of fish species, and after marrying, they coauthored 15 scientific papers on fish species of South America and western North America. They parented five children: Margaret, Charlotte, Theodore, Adele, and Thora, two of whom suffered from mental disorders.

Eigenmann accompanied her husband to Harvard University in 1887, where she was the first woman to attend graduate classes in cryptogamic botany under W. G. Farlow. She and her husband also studied Harvard's ichthyological collection of freshwater fish gathered on the Thayer Expedition to Brazil and the Hassler Expedition throughout South America. Eigenmann returned to San Diego when the San Diego Society of Natural History appointed her husband its curator. They conducted independent research until 1891, when Indiana University appointed Carl Eigenmann professor of zoology. He later headed the department and then became dean of the graduate school. Eigenmann ended her scientific career in 1893, as child-rearing responsibilities fell on her exclusively, though she continued to collaborate with her husband by editing many of his scientific writings.

After her husband's death from a stroke on April 24, 1927, Eigenmann returned to San Diego, where she remained, near her children, until her death on January 12, 1947. She was not scientifically active during these last 20 years. Although she lived most of her life in a traditional role as wife, mother, and homemaker, during her brief scientific career she distinguished herself as an important ichthyologist.

Elgood, Cornelia Bonté Sheldon Amos
(1874–1960)
English
Physician

Sheldon Elgood was the first woman physician that the Egyptian government ever appointed, and after her retire-

ment, she was the first woman to be honored for public service by the Egyptian government, which granted her the Decoration of the Nile (third class). In the intervening 23 years of service, Elgood established multiple programs in medicine and education that started on a local level but proved so effective that they were implemented on the national level.

Sheldon Amos was born in 1874 and learned to speak Arabic growing up in Egypt. Both her father and her brother served in the Egyptian judicial system, the former as a judge and the latter as a judicial advisor. She attended medical school at London University, where she earned her M.D. in 1900.

In 1901, Amos returned to Egypt, where the government appointed her to the International Quarantine Board of Egypt. She spent the next two years at the Quarantine Hospitals at El-Tor in Suez, mostly caring for pilgrims returning from Mecca, who were required to stay for monitoring. Based on the multiple cases of dysentery she treated, Amos wrote several papers on this disease. The quality of her work convinced officials to replace her with another woman doctor when she was transferred in 1902 to Alexandria, where she established a private practice and volunteered at the government hospital to establish an outpatient clinic for women and children.

In 1907, Amos married Major Percy Elgood, a year after the Egyptian government transferred her to Cairo. She was also transferred to the Ministry of Education, which charged her with the administration of three girls' schools with 600 students. In typical fashion, Elgood implemented her systems with such efficiency and efficacy that her program grew exponentially, eventually incorporating 600 schools and 20,000 students throughout the entire nation.

Concurrently, Elgood oversaw the building of the first free children's dispensaries in the country as the medical member of a commission established by the Countess of Cromer. Again, an initiative that started locally, in Cairo and Alexandria, proved so successful that dispensaries were built nationwide. Elgood also helped to found a school for training midwives as a member of a board set up by Field Marshal Viscount Kitchener. As per usual, her success was replicated, as schools modeled on this example sprouted up throughout the provinces.

In addition to these works, Elgood also sponsored six Egyptian women to study medicine in England. She held a longtime appointment on the board of Victoria Hospital, and she testified as to hygienic conditions in the country before the Balfour Commission on Public Health in Egypt. During World War I, Elgood contributed to the war effort by establishing the Cairo Voluntary Aid Detachment, which she served as a commandant and medical officer, and working with the Cairo Red Cross Committee.

Elgood's wartime efforts earned her numerous decorations, including the Order of the British Empire and the

French Médaille de la Reconnaissance Française. In 1923, she retired to Heliopolis with her husband, who died in 1941. Two years earlier, she had been promoted to the station of commander with the Order of the British Empire. The advent of the Suez Crisis forced her from Egypt to Cyprus, and she eventually settled in London, where she died on November 21, 1960, at the age of 86.

Elion, Gertrude Belle ("Trudy")
(1918–1999)
American
Pharmacologist

Gertrude Elion and her professional partner George Hitchings approached pharmacological research in a novel way in the 1940s. At the time, researchers used the trial and error method of inventing new drugs. In contrast, Elion and Hitchings compared normal human cell properties with those of abnormal or infectious cells, such as cancer cells, attempting to identify differences so they could attack these undesirable cells without harming the normal cells. This new paradigm of research led to the development of new drugs for treating leukemia, malaria, gout, and autoimmune disorders such as AIDS. For this work, Elion and Hitchings shared the 1988 Nobel Prize in physiology or medicine with James Black.

Elion was born to immigrant parents on January 23, 1918, in New York City. Her father, Robert, was a dentist. In 1933, her grandfather died of stomach cancer, prompting Elion to vow to find a cure for this disease. However, her family had lost their money in the stock market crash of 1929, so they could not afford college for her. Luckily, she received a free scholarship earmarked for women to Hunter College, where she graduated summa cum laude in 1937. Fifteen doctoral programs rejected Elion's applications because she was a woman, though she managed to earn a master's of science from New York University in 1941 while working in the industry and as a chemistry and physics teacher in New York. Elion then worked other industry

Recipient of the 1988 Nobel Prize in medicine, Gertrude B. Elion knew from the age of 15 that she wanted to do cancer research. *(Courtesy of Burroughs Wellcome Company)*

jobs for short stints with Quaker Maid Co. as an analyst of food chemistry and with Johnson & Johnson as an organic chemistry researcher. However, it was not until the exodus of male workers to fight in World War II that the doors opened for Elion as a woman scientist to land a job commensurate with her intelligence and skills. In 1944, Burroughs Wellcome Research Laboratories hired her as Hitchings's assistant. Burroughs Wellcome quickly promoted her to the position of senior research biochemist. She then became assistant to the head of the chemotherapy division in 1963, and in 1967, she headed the experimental therapy division. In this position, she enjoyed the autonomy to make her own decisions without having to answer to male supervisors. She maintained this position until her retirement in 1983, when she became an emeritus scientist with the company.

Hitchings's and Elion's pragmatic research methods led to the development of many new drugs. In 1950, she invented 6-mercaptopurine (6-MP), which proved especially effective against childhood leukemia, with an 80 percent success rate of curing the disease when combined with other anticancer drugs. She also invented allopurinol, which broke down uric acid deposits in the joints of those with gout, offering them much-needed relief. In 1974, Burroughs Wellcome scientists invented the antiviral drug acyclovir based on research performed by Elion. Similarly, after Elion's retirement, her team of researchers developed the anti-AIDS drug AZT based on the foundation of research done by Elion.

Burroughs Wellcome encouraged its scientists to publish their findings after patents had been secured, so Elion published more than 225 scientific papers throughout her career. She also received numerous honorary doctorates (to make up for the fact that she never earned one on her own, as she was forced to choose her career over her education early on).

Elion gained recognition of the significance of her contributions during her active career, receiving the Garvan Medal from the American Chemical Society in 1968. After she retired, though, honors besides the Nobel Prize were heaped upon her: In 1985, she received the Distinguished Chemist Award; in 1991, she received the National Medal of Science and was the first woman inducted into the Inventors Hall of Fame; and in 1997, she received the Lemelson/MIT Lifetime Achievement Award. Elion died on February 21, 1999.

Emeagwali, Dale Brown
(1954–   )
American
Microbiologist, Cancer Researcher

Dale Brown Emeagwali was honored as the 1996 Scientist of the Year by the National Technical Association in

recognition of her discovery of the existence of kynurenine formamidase in the bacteria *Strepomyces parvulus*. Before this discovery, scientists believed that these isozymes existed only in higher organisms; since this discovery, researchers have used this knowledge to better understand the causes of cancer found in the blood, such as leukemia. Emeagwali also actively promoted the study of the sciences and mathematics by African-American students, founding the annual African-American Science Day at the University of Minnesota in collaboration with her husband, the Nigerian aeronautical and astronautical engineer Philip Emeagwali.

Emeagwali was born as Dale Donita Brown on December 24, 1954, in Baltimore, Maryland. Her father, Leon Robert Brown, worked as the superintendent of the production department for the *AFRO-American* newspaper, and her mother, Doris Brown, was a public-school teacher. Dale attended Baltimore city public schools through her 1972 graduation, when she matriculated at Coppin State College. She commenced her study of medicine while there with a summer internship at Meharry Medical School and graduated from Coppin State in 1976.

Brown experienced culture shock when she moved to Washington, D.C., to commence her graduate study in microbiology at Georgetown University, though she assimilated easily, regardless of her race and class. She received support from the Southern Fellowship Fund from 1979 through 1981, her final years at Georgetown. At that time, she met Philip Emeagwali, a doctoral student in civil engineering, on a bus ride back to Georgetown. In 1981, Brown received her Ph.D. in microbiology from Georgetown's School of Medicine. That year, she also married Emeagwali. They have one child together, Ijeoma.

Upon receiving her doctorate, Emeagwali commenced working as a fellow with the National Institutes of Health (NIH) while holding concurrent fellowships with the National Science Foundation, the American Cancer Society, and the Damon Runyon–Walter Winchell Cancer Fund. In 1983, the NIH promoted her to its staff, and two years later, she took up a fellowship with the Uniformed Services University of Health Sciences.

In 1986, Emeagwali moved to the University of Wyoming as a research associate, and then she spent the years from 1987 through 1992 at the University of Michigan as an assistant research scientist. During that time period, she held a concurrent position as a principal investigator for the National Science Foundation and the American Cancer Society. In 1992, the University of Minnesota hired her as a research associate and lecturer, and she remained there until 1996, when she returned to her native city as a member of the biology faculty at Ball State University.

Emeagwali's work focuses on cancer research, and she has proven that the antisense methodology could inhibit cancer gene expression, a discovery that holds important implications for cancer treatments. She also investigates in the fields of fermentation, enzymology, virology, and protein structure and function. In addition to her research, Emeagwali has promoted African-American students to study the sciences and mathematics. When she was growing up, she was often discouraged from these disciplines and was told that they were too difficult for her as a black and as a woman. Emeagwali strives to shatter these stereotypes, both with her own work and with the achievements of the upcoming generations of African-American scientists and mathematicians.

Emerson, Gladys Anderson
(1903–1984)
American
Biochemist and Nutritionist

Gladys Anderson Emerson was the first scientist to isolate vitamin E in its purest form as an antioxidant. This discovery led to the addition of vitamin E to many packaged foods, extending their shelf life significantly. Emerson then performed important research on the vitamin B complex, linking their deficiency to the development of arteriosclerosis, or hardening of the arteries. She also investigated the links between diet and cancer, and late in her career, she played an instrumental role in the defining of the official stances on diet and nutrition of the United States Department of Agriculture and the Food and Drug Administration (FDA).

Emerson was born on July 1, 1903, in Caldwell, Kansas. She was the only child of Louise Williams and Otis Emerson, who moved several times in her youth. She attended elementary school in Fort Worth, Texas, and high school in El Reno, Oklahoma. She double majored at the Oklahoma College for Women, graduating in 1925 with a B.S. in chemistry and an A.B. in English.

Stanford University offered her assistantships in both chemistry and history; Emerson chose the latter, earning her M.A. in history in 1926. She headed the department of history, geography, and citizenship at a junior high school in Oklahoma City before returning to her own education to conduct doctoral research in biochemistry and animal nutrition at the University of California at Berkeley, where she earned her Ph.D. in 1932. She traveled to the University of Göttingen in Germany as a postdoctoral fellow in chemistry under the Nobel laureate Adolf Windaus as well as Adolf Butenandt, a hormone research specialist.

The next year, Emerson returned to the University of California at Berkeley, where she worked in the Institute of Experimental Biology as a research associate. It was here that she conducted research on wheat germ oil, isolating alpha-tocopherol, or vitamin E in its purest form, in 1935. Emerson remained with the institute for a decade until she

joined the pharmaceutical firm Merck and Company in 1942, where she remained for more than a dozen years.

Emerson headed Merck's department of animal nutrition and also conducted research on the vitamin B complex. In one experiment, she deprived rhesus monkeys of vitamin B6, resulting in a development of arteriosclerosis. While still with Merck, Emerson contributed to the war effort during World War II working for the Office of Scientific Research and Development, and after the war, Emerson worked from 1950 through 1953 at the Sloan Kettering Institute. There, she investigated the linkage between diet and cancer.

The University of California at Los Angeles appointed Emerson as a professor of nutrition in 1956. In 1961, she moved along with a handful of other professors from the home economics department in the College of Letters and Sciences into the division of nutritional sciences in the School of Public Health, and the next year, the school appointed her as its vice chairman, a position she retained until her 1970 retirement.

A year before her retirement, Emerson was named by President Richard M. Nixon as the vice president of the Panel on the Provision of Food as It Affects the Consumer, an offshoot of the White House Conference on Food, Nutrition, and Health. In 1970, she testified before the FDA about vitamin and mineral supplements and food additives. Earlier in her career, she had received recognition as well, as the American Chemical Society bestowed on her its 1952 Garvan Award. Emerson died in 1984.

Esau, Katherine
(1898–1997)
Russian/American
Botanist

Katherine Esau focused her interest on the relationship between healthy and diseased plants. Other scientists recognized her as an authority on the anatomical structure and ontogenic development of both kinds of plants.

Esau was born on April 3, 1898 in Ekaterinoslav, Russia. Her father, Johann, was a mechanical engineer and the city engineer for Ekaterinoslav before its citizens elected him mayor. She attended the Golitsin Women's Agricultural College from 1916 through 1917. After the Bolshevik Revolution led to the onset of xenophobia, Esau's German Mennonite family was forced to emigrate, first to Germany with the German Army returning home from their occupation of the Ukraine in World War II. The Esaus landed in Berlin, where Katherine studied at the College of Agriculture from 1919 to 1922. The Esaus then chose to continue on to Reedley, California, in 1922. There, she commenced her dual careers as a plant breeder in practice and an academic botanist. While working first at Sloans Seed Company in Oxnard, California,

Botanist Katherine Esau was recognized as an authority on the anatomical structure and ontogenic development of both healthy and diseased plants. *(Davidson Library, Special Collections, University of California at Santa Barbara)*

on the effects of phloem-limited viruses on the development and structure of plants. She also examined how plants utilized the sieve tube as a conduit for food transferal.

Esau's first two books, *Plant Anatomy,* published in 1953, and *Anatomy of Seed Plants,* published in 1960, saw second editions, the former reprinted in 1965 and the latter reprinted in 1977. She published three other important texts on botany: *Plants, Viruses and Insects* in 1961, *Vascular Differentiation in Plants* in 1965, and *Viruses in Plant Hosts* in 1968. That year she retired.

In 1951, Esau served as the president of the Botanical Society of America. In 1957, when the National Academy of Sciences elected her a member, the university still housed her office in a garage. In 1971, the Swedish Royal Academy of Sciences invited her to join as a member. Esau saw almost an entire century in her lifetime; she died at the age of 99 on June 4, 1997.

Estrin, Thelma
(1924–   )
American
Computer Scientist

A pioneer in the rapidly developing area of biomedical computing, Thelma Estrin has invented and applied cutting-edge computer technology to neurophysiological research. As the founder and director of the Data Processing Laboratory for the Brain Research Institute of the University of California at Los Angeles (UCLA), Estrin introduced digital techniques for recording the impulse firing patterns of neurons, and she was one of the first scientists to translate brain signals into digital data. She also played a key role in utilizing interactive graphics as a brain research tool and has promoted the role of women in science and engineering.

Thelma Estrin was born on February 21, 1924, in New York City, to Billy and Mary Ginsburg Austern. She married Gerald Estrin in December of 1941, when she was only 17. (The couple would eventually have three daughters.) The newlyweds began studying history together, but World War II changed the course of their lives. Gerald enlisted in the army and was sent overseas to serve in the Signal Corps. Due to the labor shortage caused by the war, Thelma was recruited to work in American industry. She was sent to an intensive engineering assistant course at the Stevens Institute of Technology and then worked at Radio Receptor Company. If not for the war, it is unlikely Estrin would have received training and advancement in a field usually closed to women.

Because both Thelma and Gerald Estrin thoroughly enjoyed their work, they remained in engineering after the war. Both enrolled at the University of Wisconsin, where they studied electrical engineering. Thelma earned her bachelor's degree in 1948, her master's in 1949, and her

and then at Spreckels Sugar Company, she pursued her doctorate as an assistant in the botany department at the University of California at Davis. When she received her Ph.D. in 1931, she immediately commenced working as a junior botanist at the university's agricultural experiment station and as an instructor in the botany department.

Slowly and steadily, Esau worked her way up each ladder, climbing the first rung as an assistant professor and botanist from 1937 to 1943, then she climbed to the next rung as an associate professor and botanist from 1943 to 1949. Esau spent the rest of her career on the top rung as a full professor and botanist. The university promoted her in exact six-year increments, not a year sooner, even though she distinguished herself outside the university as a Guggenheim fellow in 1940 and then two decades later as the Prather lecturer at Harvard University. In 1963, she transferred from Davis to Santa Barbara to study the structure and ontogeny of phloem. She focused her attention

Ph.D. in 1951. After graduating, Gerald was invited to join a prestigious group at Princeton University that was creating the first digital and electronic computing machine. Thelma's prospects of finding work in an academic setting were slim. The market was flooded with male engineers who had received training during the war or who had gone to college on the GI Bill. Her luck was not much better in industry. When she applied to work at RCA, they refused to hire her because the company lacked a woman's bathroom.

Thelma Estrin finally found a research position in the early 1950s in the Electroencephalography Department of the Columbia Presbyterian Hospital in New York. In 1953, the couple was offered research positions at the Weizmann Institute of Science in Israel, where they played a key role in building the first computer in the Middle East.

After returning to the United States in 1955, Gerald received a faculty appointment at UCLA. The university's nepotism rules prevented Thelma from being hired in the engineering department, so she became associated with the medical school instead. In 1961, she inaugurated the UCLA Brain Research Institute's first general-purpose computer facility. Called the Data Processing Laboratory, Estrin's brainchild was the first integrated computer laboratory that was specifically intended to create computer technology for neurological research. That same year, Estrin invented digital methods for recording the impulse firing patterns of neurons. In 1963, she embarked on a Fulbright grant in Israel, where she developed techniques to identify EEG patterns that preceded brain seizures and to use feedback to prevent such seizures.

In 1970, Estrin was appointed director of the Data Processing Laboratory, a post she held until 1980. In this capacity, she oversaw efforts to incorporate interactive graphics as a brain research tool. In 1980, after the university's nepotism rules were lifted, Estrin became a full professor at UCLA's School of Engineering and Applied Science (SEAS). She took a leave of absence from 1982 to 1984 to serve as the director of the National Science Foundation's division of Electrical, Computing, and System Engineering. Upon returning to UCLA in 1984, she was named both assistant dean of the SEAS and director of the Engineering Science Extension. In addition to her teaching, researching, and administrative responsibilities, Estrin became actively involved in promoting women's careers in engineering and science.

Estrin retired in 1991, becoming an emeritus professor of computer science at UCLA. Nevertheless, she has continued to contribute to her field. For example, in 1996, she wrote a well-received article on women and computing for the *IEEE Annals of the History of Computing*. Her role in applying technology to health care and neuroscience has been recognized. She was awarded the 1981 Society of Women Engineers Achievement Award and is a fellow of several professional societies, including the American Institute of Medical and Biological Engineering.

Evans, Alice Catherine
(1881–1975)
American
Microbiologist

Alice Catherine Evans was the first woman appointed to a permanent position by the Dairy Division of the Bureau of

Alice Evans, whose research on bacteria found in fresh cows' milk led to the understanding that humans could contact brucellosis by drinking fresh cows' milk. *(National Library of Medicine, National Institutes of Health)*

Animal Industry of the United States Department of Agriculture and the first woman president of the American Society for Microbiology. Her discovery of brucellosis led to the institution of pasteurization in the 1930s as a necessary process in dairy production to prevent the transmission of disease-causing bacteria and viruses.

Evans was born on January 29, 1881, in Neath, Pennsylvania. Her father, William, was a farmer, surveyor, and teacher. From 1898 to 1901, she attended Susquehanna Collegiate Institute in Towanda, Pennsylvania. For the next four years she served as a public-school teacher. During this time, she took advantage of a two-year nature study course for rural teachers offered by Cornell University, and she went on to earn her B.S. in bacteriology from the Cornell College of Agriculture. The University of Wisconsin at Madison awarded her a scholarship, which she used to earn her master's degree in 1910.

That year, the Dairy Division hired her as a dairy bacteriologist. When the division relocated to Washington, D.C., in 1913, they retained her services by offering her a permanent position, thus representing the first permanent appointment of a woman by the division. At that time, the dairy industry and the public believed raw milk to be perfectly safe for drinking. In 1917, while studying contaminated milk, Evans decided to include some uncontaminated milk in her study, and she found evidence of the bacteria *Bacillus abortus* (which causes pregnant cows to miscarry) in her sample. Evans traced a relationship between *Bacillus abortus* and *Micrococcus melitensis* (which causes undulant fever in humans, a disease Evans

contracted and carried for seven years). Evans presented her findings at the 1917 meeting of the Society of American Bacteriologists (now called the American Society of Microbiology); the all-male audience received her report with condescension and unfounded skepticism.

In 1918, the U.S. Public Health Service hired Evans to study epidemic meningitis, influenza, and streptococcal infections. Meanwhile, the scientific community began to reassess its dismissal of Evans's work when, in 1920, other bacteriologists confirmed her results and reclassified the bacteria she identified as a new genus, *Brucella,* and renamed undulant fever as brucellosis. Doctors around the world identified more and more cases of brucellosis, which had remained undetected by mimicking other diseases. This prompted Evans to lobby even harder for the dairy industry to adopt a policy of pasteurization (the heat-treatment process invented by French bacteriologist Louis Pasteur in the 1860s), which the industry finally adopted in the 1930s.

In 1928, a little more than a decade after members of the American Society for Microbiology received this woman's research with incredulity, the society elected Evans as its first woman president. In 1934, she received an honorary M.D. from the Women's Medical College and an honorary Sc.D. from Wilson College in 1936. In 1945, Evans retired and became honorary president of the Inter-American Committee on Brucellosis, a position she held until 1957. She died on September 5, 1975, in Alexandria, Virginia, the same year she was elected an honorary member of the American Society for Microbiology.

_F_

Faber, Sandra Moore
(1944–   )
American
Astrophysicist and Cosmologist

Sandra Faber bore the bad news to the National Aeronautical and Space Administration in June 1990 that its Hubble Space Telescope, in which the agency had invested 220 human-years of planning and testing, had an optical error corrupting all data it gathered. Faber then worked on the solution to the problem, which amounted to putting prescription glasses on the telescope. Faber specializes in the study of the formation, structure, and evolution of galaxies, and she was one of the first astronomers to theorize the existence of *dark matter*, a controversial assertion at the time but now a commonly accepted theory. Faber considers herself more of an observer than a theoretician, and she considers the telescope her time machine, transporting her back to when the light she observes was first formed thousands or millions of light-years ago.

Faber was born on December 28, 1944, in Boston, Massachusetts, and grew up in the Pittsburgh suburb of Mt. Lebanon, Pennsylvania. She attended Swarthmore College in Philadelphia, where she served on the student council budget committee (hardly a harbinger of her career to come), on which she met Andy Faber. The couple married in 1967. She pursued her doctorate in physics at Harvard University but followed her husband to Washington, D.C., where she finished writing her dissertation (on the correlation between galaxy size and spectral absorption-line strength) as a guest of Carnegie Institution's Department of Terrestrial Magnetism under the guidance of her de facto thesis advisor there, VERA RUBIN.

Upon receipt of her Ph.D. in 1972, Faber secured an assistant professorship at the University of California at Santa Cruz, where she has remained ever since, climbing to the position of University Professor (one of only seven in

the UC system). In 1979, she collaborated with graduate student John Gallagher to author the paper "The Masses and Mass-to-Light Ratios in Galaxies," in which they proposed that the universe is made up of 90 percent dark matter. The next year, she joined six other astronomers in a group dubbed the Seven Samurai who noted the gravitational pull of the Centaurus constellation and identified what they called the Great Attractor galaxy there.

Faber played an important role in identifying the problem with the Hubble Space Telescope, which turned out to be a drilling error in the corrector that focused the telescopic image on its electronic eye off-center by a mere 1.3 millimeters. After finding the problem, Faber then worked to fix it, helping to design the corrective lenses used to compensate for the optical error. Other important work Faber has performed includes her assertion of the

One of the first astronomers to theorize the existence of dark matter, Sandra Moore Faber studies the formation, structure, and evolution of galaxies. *(UCSC, courtesy of AIP Emilio Segrè Visual Archives)*

Faber-Jackson relation, which she devised in conjunction with graduate student Robert Jackson whereby they discovered a method of calculating distances to galaxies. Faber specializes in studying mass-to-light ratios and stellar populations in elliptical galaxies, and her long-term goal is to measure the total mass-density of the universe.

One key to this project will be the installation of the Keck II telescope on top of Mauna Loa in Hawaii, where Faber spends several weeks each year making telescopic observations. This new spectrographic telescope, called DEIMOS (for Deep-Imaging Multiobject Spectrograph), will increase the power of the existing Keck telescope tenfold when used in conjunction. (Faber hopes DEIMOS will be up and running by early 2001.)

The American Astronomical Society awarded Faber its Dannie Heineman Prize in 1985 in recognition of her sustained body of influential research. Faber is a member of the National Academy of Sciences and the American Academy of Arts and Sciences.

Falconer, Etta Zuber
(1933–   )
American
Mathematician

Etta Zuber Falconer has dedicated her career to encouraging African-American women to study mathematics and the sciences. A professor of mathematics at Spelman College in Atlanta, Falconer has tirelessly advocated for her students. Her efforts were recognized in 1995 when she received the LOUISE HAY Award for outstanding achievements in mathematics education. Falconer has also exerted a powerful influence through her leadership of several unique programs at Spelman, including the college's NASA Women in Science and Engineering Scholars program (which she directs). In addition, she cofounded the National Association of Mathematicians.

Born in Tupelo, Mississippi, on November 21, 1933, Falconer was the youngest child of Dr. Walter A. and Zadie Montgomery Zuber. Etta's father, a physician, supported her early interest in mathematics. She eventually married Dolan Falconer, with whom she had three children: Dolan, Jr., Alice (Falconer) Wilson, and Walter.

After completing high school, Etta Falconer attended Fisk University in Nashville, Tennessee, where she met one of her lifelong mentors, EVELYN BOYD GRANVILLE. Falconer excelled at Fisk, graduating summa cum laude with a B.S. degree in mathematics in 1953. Although she had planned to teach high-school mathematics, the faculty at Fisk urged her to attend graduate school. Following this advice, Falconer enrolled at the University of Wisconsin, from which she received a master's degree in mathematics in 1954.

Upon the conclusion of her studies at Wisconsin, Falconer embarked on her teaching career, accepting a post at Okolona Junior College in Okolona, Mississippi. She remained there until 1963, at which point she left to teach at Chattanooga (Tennessee) Public School. In 1965, she joined the faculty at Spelman College as an assistant professor. While fulfilling her duties at there, Falconer also began to work toward her Ph.D. at Emory University. In 1969, she was awarded her doctorate in mathematics for her thesis on quasigroups and loops. Soon after completing her Ph.D., Falconer left Atlanta for an assistant professorship of mathematics at Norfolk State College.

In 1972, Falconer returned to Spelman College to serve as a professor of mathematics and the chair of Spelman's mathematics department. She relinquished the latter post in 1982 when she was named chair of the college's division of natural sciences. From that new position, she helped launch—and subsequently directed—Spelman's NASA Women in Science and Engineering Scholars Program (WISE), which encourages promising women to pursue graduate work in mathematics and the sciences. Falconer also introduced the NASA Undergraduate Science Research Program and the College Honors Program to Spelman. In 1990, Falconer was promoted to the directorship of science programs and policy at Spelman, and she became the associate provost for science programs and policy at the college the following year.

Falconer remains at Spelman, where she has been the Fuller E. Callaway Professor of Mathematics since 1990. In addition, her work in the NASA WISE program has proved to be an unequivocal success. From the program's first class of graduates alone, five students began graduate work in the sciences. Falconer's accomplishments have been recognized by her peers as well. Prior to winning the Louise Hay Award, Falconer had received the Spelman College Presidential Faculty Award for Excellence in Teaching in 1994 and the United Negro College Fund Distinguished Faculty Award in 1986. She also founded the National Association of Mathematicians, a professional organization that promotes the concerns of African-American students and mathematicians.

Farquhar, Marilyn Gist
(1928–   )
American
Cell Biologist

Marilyn Farquhar's research spans a diversity of areas: electron microscopy, cell secretion, intracellular membrane traffic, and glomerular permeability and pathology, to name just a few. As a medical school professor, though, she advises her students to focus first on the basic biological sciences, setting a firm foundation from which they can specialize in accordance with whatever direction developments in medical technology head. She calls this belief "intellectual medicine," or physicians' reliance on basic intelligence as a problem-solving and diagnostic skill, instead of relying on overspecialization or dependence on drug manuals.

Farquhar was born on July 11, 1928, in Tulare, California. Growing up in a small town in California's Central Valley, Farquhar's family did not have access to a major medical center, but her mother's internist made a strong impression on Farquhar, as he managed to stay abreast of advances in medicine by diligently reading medical journals. She attended the University of California, where she received an M.A. in 1953, then worked as a junior research pathologist while pursuing her doctorate in experimental pathology, which she received in 1955. She stayed on there as an assistant research pathologist until 1958.

That year, Rockefeller University offered Farquhar a research associateship, which she kept until 1962, when she moved back to California to take up an associate research pathologist position at UC San Francisco. Within two years, the university promoted her to an associate professorship in pathology. In 1966, working with Dorothy Bainton, her first postdoctoral fellow, she discovered that different types of blood cells express different types of enzymes. Eight years later, a biotechnology firm contacted her about practical applications of this discovery, which proved instrumental in the development of diagnoses and treatments of diseases such as AIDS and cancer. From 1968 to 1970, she served as a professor in residence there. In 1970, Farquhar married Nobel laureate George Emil Palade, a fellow cell biologist and collaborator on renal capillary research in the 1950s. Farquhar was married once before, in 1951, and she has two children.

In 1970, Farquhar returned to Rockefeller University, where Palade had a long-standing appointment, for a three-year stint as a full professor. In 1973, she returned again to UC San Francisco as a professor of cell biology and pathology, a position she retained for the next 15 years. In 1987, she joined Palade on the faculty of the Yale University School of Medicine as the Sterling Professor of Cell Biology and Pathology. In 1990, Farquhar left Yale to initiate the Cellular and Molecular Medicine Program at the University of California at San Diego to explore the inner life of single cells. As of 2000, she continues her work at UC San Diego.

From 1981 through 1982, Farquhar served as the president of the American Society of Cell Biology, which awarded her the 1987 E. B. Wilson Medal. In 1988, she received the Homer Smith Award from the American Society of Nephrology. In 1999, the Histochemical Society granted her the Gomori Award in recognition of her outstanding contributions to the fields of histochemistry and cytochemistry.

Farr, Wanda K.
(1895–1983)
American
Biochemist

Wanda K. Farr is recognized for her discovery that cellulose is made up of tiny cellular structures called plastids. A very significant compound found in all plants, cellulose had eluded scientists who had studied it for years because the process of cellulose synthesis, in which the scientists hoped to see the components of cellulose, was always obscured by their research tactics. Farr discovered the reason for this, and in the process she revealed the plastids.

Born to C. Fred and Clara M. Kirkbide on January 9, 1895, in New Matamoros, Ohio, Wanda Margarite Kirkbide had a special relationship with her great-grandfather, Samuel Richardson, a local physician, who nurtured her early interest in plants. Farr's father died of tuberculosis when she was just a child, so her family discouraged her interest in becoming a physician for fear that she would be exposed to tuberculosis, too. Heeding their wishes, Farr enrolled at Ohio University in Athens, where she received her B.S. in botany in 1915. She continued her education at Columbia University, receiving her M.S. in botany in 1918.

While at Columbia, she met her future husband, Clifford Harrison Farr, who was completing his Ph.D. in botany. When Clifford graduated, he accepted a position teaching plant physiology at the Agricultural and Mechanical College of Texas. Farr took a job as instructor of botany at Kansas State College in 1917 in order to be closer to him. The couple soon married, and Clifford took a job at the Department of Agriculture in Washington, D.C., where their son Robert was born. Farr raised her young son and completed the research for her master's degree from Columbia.

Clifford's work took them back to Texas following World War I and then to the University of Iowa, where Clifford held a faculty position in the botany department. In Iowa, Farr worked with her husband, researching the tiny hairs found on plant roots that absorb water and nutrients from the soil. The couple moved to St. Louis in 1925, where Clifford had accepted a position at Washington University, but he died three years later from a long-term heart condition. Farr later married E. C. Faulwetter.

Although Farr had intended to pursue her doctorate at Columbia University, her husband's death changed her plans, and she instead took over his classes at Washington University. From 1926 to 1927, she also worked as a research assistant at the Barnard Skin and Cancer Clinic in St. Louis, where she studied new methods for growing animal cells in culture dishes. She was awarded a grant from the National Academy of Sciences to continue her studies of root hairs, and this work led in 1928 to a position with the U.S. Department of Agriculture at their Boyce Thompson Institute in Yonkers, New York.

Farr's new position was as a cotton technologist, and it was while doing this work that she unraveled the mystery of cellulose. Cellulose makes up the majority of the cell walls of cotton fibrils and provides its structure and stiffness. Farr had studied the chemical content and other characteristics of the fibrils, but the origin of cellulose had yet to reveal itself. The problem, Farr realized, was that plastids seem to disappear into cytoplasm when placed in water for study under the microscope. When the light of the microscope passes through the cellulose plastids, they are rendered invisible. But when the plastids fill with cellulose, they explode and spill the cellulose into the cytoplasm, where they are then visible. Thus, the cellulose molecules seem to arise out of nowhere, fully formed.

Having solved the cellulose mystery, Farr went on to various interesting assignments, including a job for the New York Metropolitan Museum, analyzing the sheets from a 3,500-year-old Egyptian tomb to determine if they were made of cotton or linen. She was appointed director of the Cellulose Laboratory of the Chemical Foundation at the Boyce Thompson Institute in 1936. From 1940 to 1943, she worked for the American Cyanamide Company, and from 1943 to 1954, she worked for the Celanese Corporation of America. Other jobs included serving as the 12th annual Marie Curie lecturer at Pennsylvania State University in 1954; opening her own laboratory, the Farr Cytochemistry Lab in Nyack, New York, in 1956; and teaching botany and cytochemistry at the University of Maine from 1957 to 1960. She also belonged to numerous organizations, including the American Association for the Advancement of Science, the American Chemical Society, the New York Academy of Science, the Royal Microscope Club (London), and the American Institute of Chemists.

Fawcett, Stella Grace Maisie
(1902–1988)
Australian
Botanist

Stella Grace Maisie Fawcett played a central role in demonstrating the effects of overgrazing on Australia's high plains. Fawcett headed a study by the Soil Conservation Board to measure the toll that cattle and sheep were having on the landscape. Her report proved essential to the decision to conserve land in the Bogong High Plains that had been earmarked for hydroelectricity development.

Fawcett was born in 1902 in Footscray, a suburb of Melbourne. Biographical details about her upbringing are scant, but it is known that she spent several years teaching school in the Melbourne area. In 1922, when she was 20 years old, she won a scholarship to attend the University of Melbourne. She planned to study geology but was told by professors that this field was "not for women." As a result, she opted to major in botany instead and received her master's degree in 1936.

Upon completing her education, Fawcett remained at the University of Melbourne as a researcher and a teacher. Although she initially conducted microscopic research, a medical condition prevented her from pursuing this project. Fawcett's new direction was set in the early 1940s when the head of the University of Melbourne's botany department, John Turner, recommended her for the post of first field officer for the state of Victoria's newly created Soil Conservation Board. (Gender discrimination had not vanished, however—the labor shortage caused by World War II simply meant that few men were available to oversee the study.) The alpine and subalpine Australian high plains had been used by ranchers to graze their cattle and sheep since the mid-19th century. By the early 1940s, it was evident that the livestock had caused serious soil erosion, a condition that had only been worsened by the proliferation of rabbits in the area. The Soil Conservation Board wanted to determine and measure the effect of this grazing on plant life in the plains.

Based in Omeo, Victoria, Fawcett traveled the Bogong High Plains by foot, horse, and car. She fenced off two study areas to keep out livestock in order to observe which plants returned to the trammeled ground. She returned at frequent intervals to these areas to assess any changes in the soil condition. Her work was not easy. The terrain was daunting and exhausting to navigate, and the weather was exceptionally inclement (often raining for days at a time). Moreover, ranchers opposed her efforts fiercely.

At the conclusion of her study, Fawcett produced a report detailing her findings. Of perhaps greatest significance, the report proved incontrovertibly that grazing was the primary cause of erosion. Fawcett recommended that fewer cattle be allowed on the land and that sheep be banned from grazing altogether. She also concluded that if ranchers waited until later in the season to bring their cattle to the high plains, much of the erosion would be avoided (since the plants would have a chance to reach maturity). The Soil Conservation Board backed her recommendations, and the situation rapidly improved. Fawcett's report had other far-reaching effects as well. Turner used her findings to stall a hydroelectrical project slated for the Bogong High Plains.

Fawcett returned to the University of Melbourne in 1949, working as a temporary lecturer in ecology. In 1952, she was promoted to senior lecturer. She married a professor, Dennis Carr, in the early 1950s. Together with Turner, she wrote several papers on the progress of the High Plains research effort in 1959. Funded by the University of Melbourne, she returned to the Bogong Plains in 1964 to measure extensively the condition of the fragile soil and vegetation. In 1967, Fawcett moved to the Australian National University in Canberra for a post as a visiting fellow. She remained there until her death in 1988.

Federoff, Nina V.
(1942–   )
American
Molecular Biologist

An accomplished flutist, who once played for the Syracuse Symphony Orchestra and expected to become a professional musician, Nina V. Federoff is now best known for her successful cloning and molecular genetic analysis of the transposable elements in corn, which were first identified by BARBARA MCCLINTOCK in work that earned her the Nobel Prize.

Born in Cleveland, Ohio, on April 9, 1942, Federoff was the daughter of Russian immigrants Olga S. Snegireff Federoff and Vsevold N. Federoff. Federoff's mother, Olga, worked as a Russian language instructor and translator for the United States Air Force Language School. Vsevold, her father, was an engineer for the Carrier Corporation. The family descended from nobles, scientists, clerics, and diplomats under the Russian czars. As a teenager, Federoff was highly motivated in her academic pursuits. She took college history courses and taught music lessons when she was just 16 years old. She graduated from high school first in her class.

In 1959, shortly after graduating from high school, Federoff married Joseph Hacker. Their daughter, Natasha, was born one year later, and the couple divorced in 1962. During this time, Federoff studied at Syracuse University on a scholarship. She worked on the side as a freelance flutist, a music teacher, and, following in her mother's footsteps, a Russian–English translator for the U.S. Air Force Language School. She eventually became the assistant manager of the Translation Bureau at Biological Abstracts. In 1966, she graduated summa cum laude from Syracuse, was voted salutatorian, and was elected to Phi Beta Kappa. In the same year, Federoff married Patrick Gagandize. Their son, Kyr, was born in 1972, and the couple divorced in 1978.

Following her college graduation, Federoff considered working as a professional flutist, but she changed her mind after meeting James D. Watson, who had won the 1962 Nobel Prize with Francis Crick for their work on the structure of deoxyribonucleic acid (DNA). Federoff went back to school, determined to become a physician. Her studies introduced her to laboratory work, and she found that doing experiments was more interesting than anything she had ever done before. While an undergraduate, Federoff was awarded a Na-

tional Science Foundation Undergraduate Summer Research Award to spend three months at Woods Hole Marine Biological Laboratory in Massachusetts, which she has called the turning point for her in terms of committing herself to a life of science.

Federoff continued her education at Rockefeller University after receiving another National Science Foundation Grant for graduate work. In 1972, she graduated with a Ph.D. and was elected to Sigma Xi, the scientific honor society. Her first employment was with the University of California at Los Angeles (UCLA) as an acting assistant professor. During her two years at UCLA, she taught and conducted research on nuclear ribonucleic acid (RNA). She left when she was awarded postdoctoral fellowships from the Damon Runyon–Walter Winchell Cancer Research Fund and the National Institutes of Health.

It was about this time, in the early 1970s, that Federoff first met Barbara McClintock at the Cold Spring Harbor Laboratory. McClintock's discovery of mobile genes in the chromosomes of corn fascinated Federoff and inspired her to study McClintock's complex writings on the subject. By 1978, Federoff was working as a staff scientist for the Department of Embryology at the Carnegie Institution of Washington and conducting research on corn genetics that was directly based on McClintock's research. She had done significant work on bacteriophages and was responsible for the first complete sequencing of a 5S ribosomal gene, which was important because it allowed her to create a map of the subelements of a genetic material found in all cells and in many viruses. Still, it was Federoff's work with transposable elements, or *jumping genes* as they are also known, that she has named her most important contribution. These jumping genes are notable because they can move from one position to another on a chromosome. Federoff found that there were such elements in plants other than corn, which greatly facilitated cloning of genes in other plants, as well.

The controversies surrounding genetic cloning have captured Federoff's interest, and she has spent much time speaking and writing on the issues, even appearing on television and radio talk shows to share her ideas with the general population. She founded the International Science Foundation to foster scientific exchange between scientists from the former Soviet Union and Western scientists. A member of the National Academy of Sciences and the American Academy of Arts and Sciences, she has published nearly 100 articles for both scientific and general audiences. In 1990, Federoff married Michael Broyles, and the following year, she rejoined the Syracuse Symphony Orchestra as a guest flutist. She was named an outstanding contemporary American woman scientist by the New York Academy of Sciences in 1992.

Fell, Honor Bridget
(1900–1986)
English
Cell Biologist

Dame Honor Bridget Fell loved her work in cell biology, because it brought her closer to understanding the incomprehensible mystery of life. She pioneered the organ culture technique, though it amazed her that she could propagate living tissue outside its home environment. Tissue culturing became an important tool in biological research, but Fell always maintained a healthy skepticism toward her own technique, realizing that what tissue cells do in vitro may or may not replicate what they do in the body. Fell was born on May 22, 1900, at Fowthorpe near Filey in Yorkshire, England, the last of nine children. Her mother was Alice Picksgill-Cunliffe, and her father, Colonel William Edwin Fell, was a minor landowner who contributed to the British efforts in the Boer War by purchasing horses in the United States and shipping them to the British army in South Africa.

The secondary school Fell attended, Wychwood School in Oxford, specialized in the instruction of science, especially biology. At the age of 16, Fell matriculated at Madras College of St. Andrews, and at the age of 18, she transferred to Edinburgh University to study zoology under Professor of Animal Genetics Frank Crew, who studied microscopic tissue from the reproductive organs of fowls. There, she earned a B.Sc. in 1923, and later she added a Ph.D. in 1924 and a D.Sc. in 1932.

Crew encouraged Fell to study at Cambridge under Dr. T. S. P. Strangeways, who pioneered tissue culturing. Fell ended up staying at Cambridge for the remainder of her career, first as Strangeways' scientific assistant from 1923 through 1928 through a grant from the Medical Research Council and later, from 1928 through 1970, as the director of the Strangeways Research Laboratory. Fell then spent nine years as a Research Associate in the Department of Pathology at Cambridge University before returning to the Strangeways Research Laboratory in 1979 as a Research Associate.

Fell further advanced Strangeways's tissue culture techniques, using them to uncover the histogenesis and differentiation of bone, cartilage, and other tissues. Fell humbly saw her own role in the advancement of science as just one link in the "chain of discovery," as she built upon the work of those who preceded her and assumed that subsequent generations of scientists would continue to advance her theories.

Fell received multiple honors in her lifetime, including the 1948 Trail Medal from the Linnaean Society, the 1956 Prix Albert Brachet of the Belgian Royal Academy, and the 1965 Prix Charles-Leopold Mayer from the French Academy of Sciences. Honor Bridget also became a fellow of the Royal Society, an unusual allowance by the all-male institution. She received honorary doctorates from Edinburgh University in 1959, Smith College in 1962, Harvard University in 1964, and Kings College in London in 1967. Fell continued to work until her death in 1986. Throughout her career, she demonstrated a reverence for and fascination with the biological processes of life, and she was thankful that she could spend her professional time studying what it was that created life.

Fenselau, Catherine Clarke
(1939–   )
American
Biochemist

Catherine Fenselau is an expert in biomedical applications of mass spectrometry. In the late 1990s, Fenselau oversaw the installation of one of the most powerful mass spectrometers in her laboratory and then its removal and reinstallation in her new lab when she transferred campuses

Catherine Fenselau, an expert in biomedical applications of mass spectrometry. *(Courtesy of Catherine Fenselau)*

within the University of Maryland system. She utilized this spectrometer to study chemical reactions of drugs with proteins. She also conducted research on the chemistry of gaseous ions as well as on posttranslational modifications of proteins.

Fenselau was born on April 15, 1939, in York, Nebraska. She attended Bryn Mawr College, graduating with a B.A. in 1961. She married the next year and had two children. She then pursued a doctorate in organic chemistry at Stanford University under Dr. Carl Djerassi, earning her Ph.D. in 1965. The next year, the University of California at Berkeley hosted her as an American Association of University Women fellow. She remained at Berkeley the subsequent year, 1966 through 1967, as a research chemist for NASA. Then, in 1967, Johns Hopkins University School of Medicine hired her as an instructor; over the next two decades, she rose in the ranks to the position of professor of pharmacology.

Throughout this time, Fenselau pursued a variety of diverse interests and opportunities. The United States Public Health Service granted her a five-year career development award from 1972 until 1977. She has served as the editor in chief of the journal *Biomedical Mass Spectrometry* since 1973. She acted as a consultant for the Medical Chemical Study Section of the National Institutes of Health from 1975 through 1979. The University of Warwick extended a visiting professorship to her in 1980, as did Kansas University in 1986. In 1987, Fenselau accepted a professorship in the Department of Chemistry at the University of Maryland in Baltimore County.

In June 1987, Fenselau coordinated the installation of a new HiResMALDI Fourier mass spectrometer by IonSpec Corporation in her University of Maryland, Baltimore County laboratory. MALDI (matrix-assisted laser desorption/ionization) technology became available in 1987 using the time-of-flight analysis method. Technological advancements led to the development of the Fourier transform mass spectrometry analysis method, which employed a strong magnetic field to take ionic measurements based on the strength of electrical signals emitted during the process. When the University of Maryland appointed her as chairperson of the Department of Chemistry at its College Park campus in the summer of 1998, Fenselau supervised IonSpec's disassembly, transport, and reinstallation of the MALDI Fourier transform mass spectrometer in her new laboratory.

Fenselau is a member of several professional organizations, including the American Association for the Advancement of Science, the American Chemical Society, and the American Society of Mass Spectrometry. Fenselau has received numerous awards throughout her career, including the American Chemical Society's 1985 Garvan Medal and its 1989 Maryland Chemist Award. In 1993, she received a medal from the Spectroscopy Society of Pittsburgh.

For the decade between 1991 and 2001, she received the Merit Award from the National Institutes of Health.

Ferguson, Margaret Clay
(1863–1951)
American
Botanist

Margaret Clay Ferguson advanced the study of botany in the United States, particularly through her groundbreaking studies of native pine trees and plant genetics. She also exerted a considerable influence over the field in her role as the chair of the botany department at Wellesley College. Renowned there as a teacher, Ferguson was also an adept administrator, building the department into one of the leading centers of plant science in the United States. She is also credited with opening the field of botany to women by training more female botanists than anyone before her.

Born in Orleans, New York, on August 20, 1863, Mary Ferguson was one of Robert Bell and Hannah Warner Ferguson's six children. Her parents operated a farm in the township of Phelps, New York, primarily producing cabbage, wheat, and potato crops. Not surprisingly, young Margaret's experiences on the family farm instilled an early interest in agriculture and plant science. Ferguson herself would never marry, choosing instead to dedicate her full energy to her work.

Ferguson embarked on her career at a young age. In 1877, when she was 14, she began teaching botany in the public schools in the Phelps region, and she was appointed assistant principal in 1887. Despite her hard work as a teacher, Ferguson did not abandon her own education, graduating from Genesee Wesleyan Seminary in Lima, New York, in 1885.

In 1888, Ferguson enrolled at Wellesley College, where she studied botany and chemistry and received her bachelor's degree in 1891. For the next two years, she served as the head of the science department at an Ohio seminary before returning to Wellesley as an instructor in 1893. While she continued to teach at Wellesley, Ferguson began a Ph.D. program at Cornell University in 1899, and she was awarded a doctorate in science in 1901. Her thesis on a species of native pine was seminal, as it was one of the first studies to examine thoroughly the functional morphology and cytology of a pine native to North America.

Ferguson remained at Wellesley for the duration of her career. After being named an associate professor in botany in 1904, she was promoted to full professor and chair of the botany department, a position she would retain until 1930. Her tenure as chairperson was marked by successes in diverse areas. She was a highly respected teacher, winning the accolades of both her students and colleagues for the enthusiasm she brought to the classroom. She also continued her own research, taking up a genetic study of *Petunia* in the early 1920s and publishing a total of 27 papers.

As the administrative head of the department, Ferguson made Wellesley's botany program one of the strongest in the nation. She emphasized the importance of chemistry and physics to the botany curriculum. Moreover, she raised funds to build new greenhouses at Wellesley as well as a new botany building. The greenhouses were later named in her honor.

Ferguson retired in 1932, although she remained actively involved in research until 1938. Her many contributions to botanical studies were acknowledged by her peers throughout her career. She was elected president of the Botanical Society of America in 1929, becoming the first woman to hold this distinguished post. In 1943, she was named a fellow of the New York Academy of Sciences. Ferguson spent most of her final years in Florida with family. After moving to San Diego, California, Ferguson died of a heart attack in 1951.

Fieser, Mary Peters
(1909–   )
American
Organic Chemist

Mary Fieser and her husband, Louis Fieser, are renowned in the field of organic chemistry for the textbooks and reference books they wrote during their many years of collaboration and for the research they pursued at Harvard University, particularly their work on the synthesis of lapinone, an antimalarial drug; cortisone, a steroid hormone; and vitamin K.

Born in Atchison, Kansas, in 1909 to Robert J. and Julia Clutz Peters, Mary Peters Fieser was raised in a home where hard work, education, and service to others were highly valued. Her mother, a bookstore owner and manager, had graduated from Goucher College with a degree in English. She was one of seven children, all of whom had been educated at home until they were college age by their mother, Mary Fieser's maternal grandmother. Mary's father was an English professor, first at Midland College, where his father was president, and later at Carnegie Mellon University in Pennsylvania. He also served briefly as the Pennsylvania secretary of labor. Mary and her sister, Ruth, grew up in Harrisburg, Pennsylvania, where they attended a private girls' school. Both Mary and Ruth continued their educations at Bryn Mawr College, with Mary majoring in premedicine and Ruth in mathematics.

After taking a class with Louis Fieser, who taught chemistry at Bryn Mawr from 1925 to 1930, Mary changed her major from premed to chemistry. Louis Fieser's work on the synthesis of quinones, which are present in some plants and berries, drew much attention during the years he taught at Bryn Mawr, and in 1930, he was offered a position at Harvard. In 1937, he became a full professor, and then the Sheldon Emery Professor of Organic Chemistry in 1939.

When she graduated from Bryn Mawr in 1930, Mary accompanied Fieser to Harvard and began work on her master's degree in chemistry at Radcliffe College while pursuing research in his laboratory. In 1931, she was awarded her master's degree in organic chemistry, and on June 21, 1932, she and Louis Fieser were married. One year later, they published their first professional paper together. Over the years, they would coauthor 15 research papers and 17 books. This prolific work was not enough, however, to earn Fieser the professional recognition she deserved. It took the institution 29 years to give her the title of Research Fellow in Chemistry, and she never received a salary.

Although Fieser never pursued a Ph.D., her professional association with her husband and his research team at Harvard benefited her tremendously. She was free to work and conduct her research in an excellent facility and in the company of accomplished scientists without battling the discrimination against women that existed among many of her peers. Her fellow Harvard chemists held her in very high regard. One of these men, William von Doering, was quoted in the *Journal of Chemical Education* as saying that she was "a very gifted experimentalist" and an "active, influential part of the team."

As part of her husband's research team, Fieser helped develop a method for synthesizing large amounts of vitamin K in a short period of time. Useful for its blood-clotting characteristics, vitamin K was previously available only in small quantities from green plants, especially dried alfalfa. The team also worked on the use of napthoquinones in the prevention of malaria, an area closely related to the work Fieser had done for her master's degree. Quinine had been the standard antimalarial drug, but during World War II, when Japan invaded the East Indies, most of the world's supply of quinine became unavailable to the Allies. Focused on finding an alternative to quinine, the Fiesers eventually synthesized lapinone for use in treating malaria. In addition to these significant developments, Fieser also studied the chemical causes of cancer and contributed to the team's synthesis of cortisone, which is used to treat rheumatoid arthritis and other conditions.

In addition to her research, Fieser and her husband coauthored many articles and textbooks. Their first book, *Organic Chemistry,* was published in 1944 and became a best-seller. *Reagents for Organic Synthesis,* the first volume of which was published in 1967, was the first reference work of its kind and is still considered a "bible" for organic chemists. Fieser was an excellent writer, as was her husband, and they both concerned themselves with stylistic details, such as the placement of commas, insuring a level of excellence in the writing that was not common among scientists. Their writing was also notable for being both personal and professional. Pictures of their Siamese cats, included with the preface of most of their books, became a Fieser trademark. There was one book for which Mary claimed sole authorship, and that was the *Style Guide for Chemists,* which she wrote based on a set of grammar, style, and rhetoric notes she had written for contributors to the serial *Organic Reactions.*

Mary was awarded the Garvan Medal by the American Chemical Society in 1971 for her outstanding contributions to research, her writing, and her efforts to teach chemistry students how to write. Throughout her life, Fieser has enjoyed competitive games and used to organize her husband's research team for contests in Ping-Pong, badminton, and horseshoes. Since her husband's death, she has continued to publish updated editions of *Reagents.* In 1994, at the age of 85, she published volume 17.

Fisher, Elizabeth F.
(1873–1941)
American
Geologist

Elizabeth F. Fisher became one of the first women field geologists in the United States, and she also holds the distinction of being the first female geologist hired by an oil company. She was also a professor at Wellesley College for 32 years, teaching courses in geology, geography, mineralogy, and resource conservation. In addition, she authored a textbook on natural resources for junior-high-school students.

Fisher was born in Boston, Massachusetts, on November 26, 1873, to Charles and Sarah Cushing Fisher. Elizabeth Fisher never married. After graduating from high school, she enrolled at the Massachusetts Institute of Technology (MIT). While still a student, she began teaching classes in geology and geography at Wellesley College. She earned her B.S. degree from MIT in 1896 and thereupon accepted a permanent position as an instructor at Wellesley. In 1897, Fisher embarked on a voyage to Russia with the International Geological Congress. For four months, she studied oil wells in Baku, located near the Caspian Sea (in present-day Azerbaijan).

Upon returning to Wellesley after her stint abroad, Fisher resumed her teaching responsibilities. In 1906, she was promoted to the rank of associate professor of geology and mineralogy. Three years later, she advanced to full professor and was also named head of the department of geology and geography at Wellesley. At the same time, Fisher taught extension classes at Harvard and Radcliffe. Her lectures at these schools ranged across an array of topics, but she concentrated most particularly on water supply and water power as well as on agricultural issues, such as soil erosion and the impact of fertilization. In her research, Fisher focused on the conservation of natural resources, river terraces, and shorelines.

In 1918, Fisher charted a new path when she became the first woman to be hired by an oil company. In the midst of a nationwide oil shortage, Fisher worked as a field geologist to locate oil wells in north-central Texas. Simultaneously, she continued to work at Wellesley, where she strove to interest her students in the topics she taught. For instance, in 1919, she brought 12 students on a horseback tour of Glacier National Park in Montana to study geological formations. That same year, in an effort to bring her expertise to younger students, Fisher wrote an influential textbook for junior-high students entitled *Resources and Industries of the United States.* The book stressed her fervent belief in the need for conservation and also described American commerce and industry in relation to that of other nations.

Fisher retired in 1926 when she became a professor emeritus at Wellesley. But even then, she remained active in her field. She took part in a geographical survey of the Florida coast as well as a study of that state's Lake Okeechobee. After discovering that drainage canals had lowered the lake's water level by four feet, Fisher noted that the newly exposed soil was rich in nutrients. Consequently, she advocated draining Lake Okeechobee in order to create nearly one million acres of arable farmland. (Like many other geologists of her generation, she believed that otherwise "unproductive" land should be reclaimed for agricultural use.)

Fisher died on April 25, 1941, following a long illness. A fellow of the American Association for the Advancement of Science and the American Geographical Society, Fisher was respected by her peers and students alike. She had also been an active member of the Appalachian Mountain Club and the Boston Society of Natural History.

Fleming, Williamina Paton Stevens ("Mina")
(1857–1911)
Scottish/American
Astronomer

"Mina" Fleming worked her way up from being a human computer of mathematical equations to overseeing the astronomical photography collection at the Harvard College Observatory. Her keen eye and steady hand aided her in discovering 10 novas and 222 variable stars. Fleming is credited with ushering women into the field of astronomy, opening the way for them by first proving herself both acute and astute. Although the work was not glamorous, it offered one of the few choices for women at the time period.

Fleming was born Williamina Paton Stevens on May 15, 1857, in Dundee, Scotland. Her father, Robert, who

A group of women computers at the Harvard College Observatory, directed by Mrs. Williamina Fleming (standing), who became the foremost American woman astronomer of her time. Edward C. Pickering, director of the observatory, is looking on. *(Harvard College Observatory)*

carved and gilded picture frames as well as furniture, died when she was seven years old. At the age of 14, Fleming started teaching school. In 1877, she married James Fleming and the next year they immigrated to Boston. In 1879, Fleming found herself pregnant and then husbandless within a year. She raised their one son, Edward, on her own.

Fleming secured work as a maid to the director of the Harvard College Observatory, Edward Pickering. At that very time, Pickering was embarking on a huge project classifying stars by their spectral readings. He discovered that women made particularly good workers on this project, which required patient observation of photographs recording the spectral readings of hundreds of stars per photographic plate. Pickering hired Fleming (purportedly to replace an incompetent man who could not record the information properly) in 1879 as a copyist and computer.

By 1881, Fleming had proven herself more than competent, earning a permanent position on Pickering's staff. In 1886, she took charge of the star classification project, devising with Pickering a classification system of 17 classes from A to Q based on the intensity of the hydrogen spectral lines, which revealed the surface temperature and chemical makeup of the stars. Between 1885 and 1900, Fleming oversaw 21 women computers and photoplate inspectors in Pickering's lab, acting as mentor for future luminaries in astronomical research such as HENRIETTA LEAVITT, ANTONIA MAURY, CECILIA PAYNE-GAPOSCHKIN and ANNIE JUMP CANNON. Cannon went on to improve Fleming's classification system. In 1890, Pickering published the *Draper Catalogue of Stellar Space,* crediting Fleming for performing most of the analyses for the complete spectral classifications of 10,351 stars.

Fleming went on to edit the *Annals of the Harvard College Observatory*.

Fleming's corporate appointment to the curatorship of the astronomical photographs in 1898 represented the first time Harvard University extended a corporate appointment to a woman. From then until her death, she acted as the curator at the Harvard College Observatory.

In 1907, she published her own book, *A Photographic Study of Variable Stars Forming a Part of the Henry Draper Memorial*. In this book, she published for the first time 222 variable stars, based on her discovery of a way to identify variable stars, whose light undulated back and forth, by the spikes of brightness on their spectral photographs. In 1906, Fleming became one of only four women and the only American elected to the Royal Astronomical Society. By 1910, it was estimated that she had personally examined 200,000 photographic plates. Fleming died of pneumonia at the age of 54 on May 21, 1911, in Boston, Massachusetts.

Flügge-Lotz, Irmgard
(1903–1974)
German
Aerodynamics Engineer

Irmgard Flügge-Lotz helped advance the age of flight as an aerodynamics engineer. She pioneered the theoretical and technological foundations for automatic control. She was the first woman appointed to a full professorship in engineering by Stanford University. She opened the door for women to enter the field of aerodynamics, which had been dominated by men.

Lotz was born on July 16, 1903, in Hameln, Germany. Her father, Osark Lotz, was a traveling journalist, and her mother, Dora Grupe, assisted in her family's construction engineering business. Visiting building sites as a child peaked Lotz's interest in engineering. Lotz attended schools in Frankenthal, Mönchen-Gladbach, and Hanover, graduating from high school in 1923. She then studied applied mathematics at the Technical University in Hanover, where she was one of very few women students. She earned the degree of Diplom-Ingenieur in 1927, then wrote her thesis on the mathematical theory of heat conduction in circular cylinders to earn the degree of Doktor-Ingenieur in 1929. All the while, she worked full time as a teaching assistant in practical mathematics and descriptive geometry.

Lotz accepted an offer to work for the Aerodynamische Versuchsanstalt in Göttingen, where she solved a problem that had been plaguing Germany's preeminent aerodynamicist, Ludwig Prandlt, involving the application of his lifting line theory to the integro-differential equation governing airplane wings' spanwise lift distribution. Prandlt rewarded her with a promotion to head the aerodynam-

ics group. Upon its publication in 1931, this solution became known as the *Lotz method*. In Göttingen, Lotz met civil engineer Wilhelm Flügge, a native of the city, and the couple married in 1938.

The Deutsche Versuchsanstalt fur Luftfahrt (DVL) had just hired Flügge as a department head, so the newlyweds moved to Berlin. DVL soon hired Flügge-Lotz as an aerodynamics consultant to take advantage of her expertise in flight dynamics, automatic control theory, and discontinuous (or on-off) control systems. In 1944, the company moved them to Saulgau, which came under French occupation after World War II. The French government secured their services in 1947 to work for the French National Office for Aeronautical Research in Paris. For the next year, Flügge-Lotz headed the aerodynamics research group confronting the problems posed by increasing airspeeds, until her husband accepted an offer for a full professorship at Stanford University in the United States.

Stanford's bylaws against nepotism prevented Flügge-Lotz from receiving a full academic appointment, so she lectured in engineering mechanics and supervised doctoral dissertation research in aerodynamic theory. Over the next decade, she expanded her activities to include teaching an introductory course in mathematical hydro- and aerodynamics spanning graduate students' first year while also conducting research in fluid mechanics, numerical methods, and automatic controls. She published *Discontinuous Automatic Control* in 1953 and *Discontinuous and Optimal Control* in 1958.

In 1960, she represented the United States as its only female delegate to the first Congress of the International Federation of Automatic Control in Moscow. Upon her return, Stanford finally recognized her extensive contributions to the school with a dual appointment of full professorship in both engineering mechanics and in aeronautics and astronautics, making her Stanford's first woman engineering professor. Even after her 1968 retirement, Flügge-Lotz continued to research satellite control, heat transfer, and the drag of high-speed vehicles. In 1970, the Society of Women Engineers granted her its Achievement Award. That same year, the American Institute of Aeronautics and Astronautics inducted her as its first woman fellow, and in 1971 she was the first woman to give its von Karman Lecture. Flügge-Lotz died in 1974 at the Stanford Hospital.

Foot, Katharine
(1852–1944)
American
Cytologist

Although little information about Katharine Foot's life and career has been preserved, she was regarded as one of the leading scientists of her time. Foot received no formal education and was never officially employed. Nonetheless, she conducted research that won her the admiration of her peers. Together with her laboratory partner, Ella Church Strobell, Foot developed new techniques for preparing samples to view under the microscope and pioneered the use of photography in cytological studies.

Almost no biographical data about Foot is extant, beyond the fact that she was born in Geneva, New York, in 1852. The first edition of *American Men of Science* (1906) described her education simply as "private schools," but it has been established that she received instruction at the Marine Biological Laboratory in Woods Hole, Massachusetts in 1892—when she was 40 years old.

Because Foot was not employed, it is assumed that she and Strobell privately funded their own research. Foot's work was primarily comprised of microscopic observations of the developing eggs of *Allobophora fetida*. In particular, she and Strobell analyzed the maturation and fertilization of *Allobophora fetida's* eggs. In the process of exploring these reproductive processes, Foot and Strobell hit on a number of innovative techniques. The duo was one of the first to photograph their research samples, which added an element of clarity to their papers. Prior to this development, most researchers simply made drawings of what they viewed through the microscope, a situation that often led to the omission of key details. Many other scientists began to incorporate photomicrographs after Foot and Strobell. Furthermore, Foot and Strobell invented a way of making extremely thin samples of material at low temperatures for viewing under the microscope. In addition, they collected reprints of their various papers into a book, *Cytological Studies, 1894–1917,* which discussed their theories on chromosomes.

In 1892, Foot became a regular member of the Marine Biological Laboratory, a position she held until 1921. However, it is unclear where she resided during this period. Various editions of the *Biological Bulletin* (the Marine Biological Laboratory's annual report) list her address as Denver, Colorado; Evanston, Illinois; and New York City. In 1921, Foot was made a life member of the laboratory. Bulletins from this period cite her as living in France, England, and South Carolina.

Despite the dearth of personal data about Foot's life, it is evident that her scientific work was respected. In his 1915 book, *The Mechanism of Mendelian Heredity,* Thomas Hunt Morgan declared Foot to be one of seven women who were significant contributors to the fledgling field of genetics. Foot's work on cellular morphology was also favorably reviewed in H. J. Mozans' *Woman in Science.* Her listing in *American Men of Science* was accompanied by a star, which indicated her inclusion among the editor's roster of the 1,000 most important scientists in the United States. Foot was also a member of the American Society of Natu-

ralists and the American Society of Zoologists. She died in 1944. Although the exact location of her death is not known, her last address was given as Camden, South Carolina.

Fossey, Dian
(1932–1985)
American
Zoologist

Dian Fossey transformed human understanding of wild gorillas by living among them for a number of years, integrating into their community and coming to recognize their gentle nature, dispelling the myth of their violent nature. Fossey actively crusaded against poaching after one of her favorite gorillas, Digit, was killed by a poacher's bullet. Many believe that Fossey herself fell at a poacher's hand as well, as she was discovered murdered in her cabin in 1985. Within a few years, her 1983 book, *Gorillas in the Mist,* was transformed into a motion picture of the same name starring Sigourney Weaver, who subsequently served as the honorary chairperson of the Dian Fossey Gorilla Fund.

Fossey was born in 1932 in San Francisco, California. She developed a love for animals early in life that her father, George, encouraged, but her mother, Kitty, squelched by disallowing her from raising pets. She entered the University of California in 1950 as a preveterinary student, but she got sidetracked and transferred to San Jose State College, graduating in 1954 with a degree in occupational therapy. She used this degree when she moved to Louisville, Kentucky, in 1956, where she ended up heading the occupational therapy department at Kosair Crippled Children's Hospital.

In 1963, her life turned back in on itself during a six-week trip to Africa; at the Olduvai Gorge archaeological dig site, she met the famed British paleoanthropologist Louis Leakey. When Leakey visited Louisville three years later, he was keeping an eye out for a woman recruit to study mountain gorillas as his student JANE GOODALL was studying chimpanzees: Fossey volunteered, going so far as to have her appendix removed beforehand lest it rupture in the wilderness far from medical help, as Leakey suggested. Inspired, and funded by the National Geographic Society and the Wilkie Foundation, Fossey returned to Africa in 1967, first to the Democratic Republic of the Congo before political unrest forced her out. Within six months, invading soldiers captured and imprisoned her; she escaped to Mounts Visoke and Karisimbi in Rwanda's Parc National des Volcans. There, she set up the Karisoke observational outpost, which eventually grew into the Karisoke Research Center. Known to the locals as Nyirmachabelli, "the woman who lives alone in the forest," Fossey lived not alone but among nine groups of gorillas numbering five to 19, whom she named Digit, Uncle Bert, and Beethoven, among other playful names. She developed mutual trust with the gorilla community she was studying that culminated in 1970 with the first friendly gorilla-human contact: One of her favorite gorillas, Peanuts, an adult male, reached his hand out to touch hers.

That year, Fossey counted only 375 gorillas in her census, making them an endangered species. Because they were losing population to human encroachment on their lands and poaching of their numbers, Fossey began to fear for the survival of the gorillas. When poachers killed her other favorite gorilla, Digit, Fossey retaliated by establishing the Digit Fund and campaigning against poaching. *National Geographic* magazine supported her cause by publishing a cover story on the plight of the gorilla; the story of Digit's tragic murder prompted many readers to donate to her fund, which stemmed the tide of the disappearance of gorillas and indeed led to an increase in population.

Fossey applied her field research to doctoral work by writing it up as a dissertation at Cambridge University, where she earned her Ph.D. in zoology in 1974. Six years later, Fossey took a break from her field study to take up a professorship at Cornell University. This position supported her for three years while she chronicled her experiences at Karisoke in the book *Gorillas in the Mist,* which became a best-seller and a major motion picture. Proceeds from the book's sales funded her return to Karisoke, but a machete cut short her lifework, as she was found slashed to death in her own cabin on December 26, 1985. Her gravestone, placed over a grave Fossey had dug for slain gorillas, reads, "No one loved gorillas more. Rest in peace, dear friend, eternally protected in this sacred ground, for you are home where you belong."

Franklin, Rosalind
(1920–1958)
British
X-ray Crystallographer

Rosalind Franklin's mastery of X-ray crystallography photographic techniques played an extremely important but underappreciated role in the discovery of the structure of DNA, the so-called building blocks of life. One particular DNA photograph of Franklin's proved to be the key that unlocked the mystery, acting as the physical evidence substantiating the theoretical hypothesizing of male scientists who failed to acknowledge Franklin for providing this stepping-stone.

Franklin was born on July 25, 1920, in London, England. Her mother, Muriel, volunteered to social causes what spare time she had from raising five children, and her father, Ellis, was a successful banker who opposed his daughter's desire to pursue a scientific career. However, Rosalind's

insistence prevailed, and she attended Newnham, Cambridge University's women's college, graduating in 1941 with a degree in chemistry. She went on to earn her Ph.D. from Cambridge in 1945 with a dissertation based on research she performed on the structure of carbon molecules at a wartime job as an assistant research officer at the Coal Utilization Research Association.

After the war, Franklin migrated to Paris to conduct chemical research at the Laboratoire Centrale des Services Chimique de l'État. Here she studied colloidal carbon with X-ray crystallography, a method she adopted as her specialty for the rest of her career. She returned to England in 1950, and Kings College appointed her to a research position in 1951. She applied her X-ray crystallography technique to the study of deoxyribonucleic acid (DNA), the little-understood building blocks of life. Although German scientist Max von Laue invented the X-ray crystallography method in 1912 to study the structure of crystalline forms of elements, Franklin experimented with taking X-ray crystallography photographs of elements in their noncrystalline forms. In May 1952, she photographed the structure of "wet" DNA but put the picture aside to focus her attention on her study of "dry" DNA in its crystalline form.

Scientists raced to uncover the secret structure of DNA in order to better understand its function. Franklin's colleague at King's College, Maurice Wilkins, also studied DNA, but friction between the two prevented them from collaborating fruitfully. Franklin's excellence as a woman scientist threatened the proprietary status of the dominant male community of scientists, who were both derisive and jealous of women's scientific abilities. While Franklin focused on physical observation of DNA, rivals James Watson, a brilliant young American working at Cambridge, and his British partner Francis Crick took a more theoretical angle, hypothesizing what structures would make most sense. Despite the King's College/Cambridge rivalry, Wilkins befriended Watson, who convinced Wilkins to show him some of Franklin's X-ray crystallography photographs (without her knowledge or permission), including her study of "wet" DNA. The pattern in the photograph clearly revealed a helical shape like that of a spiral staircase, confirming Watson and Crick's working hypothesis.

Watson and Crick neglected to credit Franklin for the role her work played in confirming their theoretical structure of DNA when they published their famous "double-helix" theory in Britain's major scientific journal, *Nature,* on April 25, 1953. Ironically, Franklin herself probably did not even realize that her photograph had served as the missing puzzle piece that fit Watson and Crick's hypothetical picture together, as she had moved on from studying DNA at King's College to studying tobacco mosaic virus (TMV) at Birkbeck College.

Again, Franklin distinguished herself by preparing the best X-ray crystallography photographs of TMV, revealing its hollow tubular structure to dispel the belief that the compound was solid. She also revealed that RNA (ribonucleic acid, a cousin of DNA) formed the foundation of TMV. In 1956, she learned that she had ovarian cancer, but she still continued to conduct research on her next project, the polio virus. The cancer took her life at the age of 37 on April 16, 1958, in London. When the Nobel Prize was awarded to Watson, Crick, and Wilkins four years later, Franklin's death prevented this official recognition of her contribution to the discovery of the structure of DNA, though many doubt that she would have been included had she lived, considering the degree to which her work had been elided.

Free, Helen Murray
(1923–   )
American
Chemist

Helen Murray Free liberated diabetics with her dip-and-read self-test kit with which they could independently monitor their own blood glucose level instantly at home. She also helped revolutionize clinical diagnostics with her development of dry reagents for urinalysis testing, one of the most standard medical diagnoses. Later in her career, she focused on promoting the importance of chemistry to the mainstream public, serving first as the president of the American Association for Clinical Chemistry in 1990 and three years later as the president of the American Chemical Society (ACS).

Helen Mae Murray was born in 1923 in Pittsburgh, Pennsylvania. She studied chemistry at the College of Wooster in Ohio, graduating with honors in 1944. She entered the pharmaceutical industry as a researcher for Miles Laboratories, Inc., of Elkhart, Indiana, famous for their Alka Seltzer brand. In 1947, Murray married fellow chemist Albert Free, and together the couple had nine children and 12 grandchildren. Free and her husband often collaborated, especially on improvements to the urinalysis process. For these innovations, Free obtained seven patents by 1975. That year, she and her husband reported their combined wisdom on the subject in their second book, *Urinalysis in Laboratory Practice,* an influential text.

Free climbed the ranks at Miles, where she served as the new products manager and the director of Clinical Laboratory Reagents. As her career progressed, she became more focused on professional management. She pursued this interest to attain a Master's in Management with a concentration in Health Care Administration from Central Michigan University in 1978. Thereafter, Free taught at Indiana University in South Bend as an adjunct professor of Management. When Bayer, Inc. bought out Miles Labs, Free became a professional relations consultant in the diagnostics division.

In 1993, the ACS named Free its president. She immediately embarked on several initiatives to raise the public's awareness of the importance of chemistry in modern life. She helped establish the ACS Volunteers in Public Outreach program, which grew to number 15,000 volunteer scientists. She helped found "Kids & Chemistry," a program joining adult scientists with elementary-school children for hands-on experimentation. As well, she chaired the National Chemistry Week Task Force from 1987 through 1992 and carried that momentum into the International Chemistry Celebration.

For these initiatives, Free received the 1995 Laboratory Public Service National Leadership Award, along with her husband. That year, the ACS created the Helen M. Free Public Outreach Award in her honor. Free has received numerous other awards, including ACS's 1980 Garvan Medal for Distinguished Service to Chemistry by a Woman, the American Society for Medical Technology's Professional Achievement Award in Nuclear Medicine, and the 1996 Kilby Award. Additional presidencies included the American Institute of Chemists, the American Association for the Advancement of Science, and the Royal Society of Chemistry. Free has led a very diverse career, first focusing on research and chemical innovation, then focusing on professional management, and finally devoting her energies to outreach and public service as a means of promoting the field of chemistry.

Freedman, Wendy Laurel
(1957–   )
Canadian
Astrophysicist

Wendy Freedman's ultimate goal has been to ascertain a reliable value for the Hubble Constant, or the expansion rate of the universe, which determines a multitude of other astronomical calculations. In recognition of her expertise in this area, she was named one of three principal investigators on an international panel advising the Hubble Space Telescope project, which prioritized work on the Extragalactic Distance Scale as one of the most important uses of the telescope, allotting five years (the longest time allocation of any project employing the space telescope) with the telescope to calculate this figure.

Freedman was born in Toronto, Ontario, on July 17, 1957. During her undergraduate study, she received the University of Toronto's Robertson Award in 1977. Two years later, she graduated with a B.Sc. in astronomy and astrophysics. She stayed on at the university to pursue doctoral work with the assistance of an Ontario Graduate Scholarship in 1979. The next two years, she served as the Amelia Earhart fellow. The subsequent two years (1982 and 1983), she received support under the auspices of the

Wendy Freedman, whose career goal is to ascertain a reliable value for the Hubble Constant, or the expansion rate of the universe. *(Photo by Raleigh Souther, Courtesy of Wendy Freeman)*

Natural Sciences and Engineering Research Council Scholarship and then received her Ph.D. in astronomy and astrophysics in 1984.

Upon receiving her doctorate, Freedman was named a Carnegie fellow of the Carnegie Institution of Washington, D.C. She installed herself at the institution's Palomar Observatory in Pasadena, California, where she commenced her two principal lines of investigation: the determination of extragalactic distance scale and the stellar populations of nearby galaxies to better understand their evolution. Within three years, the institute recognized her excellence as an astronomical investigator and fixed her in a permanent position on the observatory staff; she now serves as a senior research staff astronomer. Along with Shoko Sakai and Barry F. Madore, Freedman formed the core group of the Hubble Space Telescope Key Project on the Extragalactic Distance Scale. These three principals devised a new means of estimating distances to galaxies according to the peak brightness of certain stars. The previous method relied on observations of Cepheid variable stars, whose rarity and difficulty of observation prevented standardized calculations of galaxy distance. The new method used red giant stars as its standard candle for distance calculations.

Freedman and her colleagues specifically focused their attention on the phenomenon called "tip of red giant branch," when these stars burned brightest while their helium core imploded. This phenomenon represents a much more predictable variable by which to measure the distance of the galaxy containing the red giant, and it yielded results within 10 percent of calculations through the Cepheid method.

Freedman has contributed to the advancement of the field of astronomy in diverse ways. She served as a visiting committee member at the Harvard Center for Astrophysics from 1993 through 1997 as well as on the National Research Council's Committee on Astronomy and Astrophysics from 1995 through 1998. In 1997, she chaired the Executive Board of the Center for Particle Astrophysics in Berkeley, California, and she served as a member of NASA's Scientific Oversight Committee for the Next Generation Space Telescope from 1996 through 1998. In 1994, Freedman received the Marc Aaronson Lectureship and prize in recognition of a decade of important contributions to the fields of astronomy and astrophysics. In 1999, the American Physical Society awarded her the Centennial Lectureship.

Friend, Charlotte
(1921–1987)
American
Oncologist, Microbiologist

Charlotte Friend made discoveries that changed the face of cancer research by identifying a retrovirus that produced leukemia in laboratory mice and later identifying cell lines that suggested the possibility of malignant phenotype reversal. Friend further contributed to her field by distributing samples of her discoveries to other researchers, facilitating their research and establishing her discoveries for standard use in cancer research.

Friend was born on March 11, 1921, in New York City, where her Jewish parents had immigrated from Russia. She remained in the city for her undergraduate work, graduating from Hunter College in 1943. During World War II, she served in the United States Navy as an officer in the hematology laboratories. After the war, she pursued her doctorate in immunology at Yale University, which granted her a Ph.D. in 1950. She returned to New York City to conduct research at the Sloan-Kettering Institute for Cancer Research, where she remained until 1966, having attained the rank of associate professor of microbiology by then.

Friend created "a violent storm of controversy" at the 1956 meeting of the American Association for Cancer Research (AACR) when she presented a mouse leukemia virus she had discovered by injecting ultrafiltered tissue from leukemic laboratory mice into adult Swiss mice, which developed leukemia. The conventional wisdom at that time doubted the possibility of cancer-causing viruses. In 1958, in collaboration with the electron microscopist Etienne de Harven, Friend dispelled doubt by publishing the first electron micrographs of the virus, which was later determined to be a retrovirus and was named Friend leukemia virus after its discoverer. Friend and de Harven continued to conduct research on this virus: De Harven coined the term *budding* to describe the process by which cell hosts release retroviruses, and Friend became renowned for her generosity as she freely distributed her virus to researchers, who adapted it as a standard for viral research. The Friend–de Harven collaboration resulted in several important scientific papers in the late 1950s and 1960s: "Electron Microscope Study of a Cell-Free Induced Leukemia of the Mouse: A Preliminary Report" (1958); "Further Electron Microscope Studies of a Mouse Leukemia Induced by Cell-Free Filtrates" (1960); "Structure of Virus Particles Partially Purified from the Blood of Leukemic Mice" (1964); and "A New Method for Purifying a Murine Leukemia Virus" (1965).

In 1966, the Mt. Sinai Hospital appointed Friend as the director of its newly established Center for Experimental Cell Biology, where she remained for the rest of her career as a professor. Here, she made yet another important discovery: Friend erythroleukemia cells, which she also freely distributed to interested researchers. These cells established a precedent that malignant phenotypes could be reversed.

Friend received numerous awards for her pioneering research, including the 1962 Alfred P. Sloan Award for Cancer Research, the 1975 William Dameshek Prize from the American Society of Hematology, the 1979 Prix Griffuel, and the 1982 Papanicolau Award. The National Academy of Sciences elected her a member in 1976, and she served as president of the AACR, the New York Academy of Sciences, and the Harvey Society. Friend died of lymphoma on January 7, 1987. The Charlotte Friend Memorial Lecture honors Friend's contributions to the field of cancer research.

Frith, Uta Auernhammer
(1941–   )
German/British
Psychologist

Brain researcher Uta Auernhammer Frith's ongoing studies of brain disorders have yielded new understandings about conditions such as autism, Asperger's syndrome, and dyslexia. In the process, Frith has uncovered much information about the workings of the normal brain as well. Her work is characterized by its attention to the multiple facets—biological, cognitive, and behavioral—of developmental brain disorders. She has also written and edited several influential books on brain disorders.

Born in Rockenhausen, Germany, on May 25, 1941, Frith grew up in a creative family. Her father was an artist, and her mother a writer. Frith was a precocious child. Although she began her elementary education at an all-girls' school, she demanded to attend the more academically challenging boys' school when she was 12. Later in life, she married Christopher Frith, with whom she would collaborate on some projects.

Frith studied psychology and the history of art at the University of the Saarland in Saarbrucken, Germany. After receiving her degree in 1964, she moved to Britain. In 1968, she was awarded her Ph.D. in psychology from the Institute of Psychiatry at London University.

For most of her career, Frith conducted research at the Developmental Psychology Unit of the Medical Research Council (her division was later renamed the Cognitive Development Unit). She joined the organization in 1968 and remained there until the Unit closed in 1998. While there, Frith studied autism, a brain disorder characterized by impaired language and social development. Most likely caused by brain damage or failure of development after birth, autism often results in total emotional isolation for its victims. Frith has proposed that the primary defect caused by autism is the inability to form "theory of mind." In other words, autistic people cannot comprehend that other people's perceptions, beliefs, and emotions differ from their own.

To test this hypothesis, Frith constructed an experiment with her colleagues Alan Leslie and Simon Baron-Cohen. In 1986, this trio told both autistic and nonautistic children a story about two girls, Sally and Anne, who were playing with a marble. Frith told the children that Sally put the marble in a basket and left the room. However, Anne put the marble in a box when Sally was gone. The key question Frith put to the children was where Sally would look for the marble when she returned. The nonautistic children almost universally stated that Sally would first look in the basket, because she did not know that her friend had moved the marble. But the autistic children (even if they were teenagers or of above-normal intelligence) were convinced that Sally would look in the box. The autistic children were unable to grasp that although they knew where the marble was, Sally did not. In the 1990s, Frith and her collaborators employed imaging techniques to correlate the absence of "theory of mind" to a specific area in the brain. Frith later wrote a book on the disorder, *Autism: Explaining the Enigma*. She also edited a book, published in 1992, called *Autism and Asperger's Syndrome*. (Asperger's syndrome is a disorder much like autism, though Asperger's does not typically impede language development. The main effect of Asperger's is intense social impairment.)

Frith explored dyslexia while at the Developmental Psychology Unit as well. Although dyslexia is typically thought to affect reading and spelling abilities, Frith posited that it can be detected long before a child reaches reading age. For Frith, dyslexia is a basic disorder of speech processing that can even be inherited. She teamed up with her husband in 1995 to apply positron emission tomography (PET) technology to dyslexia research. The PET scans indicated that when nondyslexics took language tests, three brain areas were active, including a connective area called the insula. But when dyslexics took the same test, the insula did not become active. Given her expertise on this topic, Frith was selected to edit a book on dyslexia—*Cognitive Processes in Spelling.*

In 1998, Frith moved to the Institute of Cognitive Neuroscience at University College, London, where she remains active in researching brain disorders. Her work has received numerous honors, including the British Psychological Society's President's Award in 1990. She was also elected a fellow of the British Psychological Society in 1991 and a member of the Academia Europa in 1992.

_G_

Gadgil, Sulochana
(1944–)
Indian
Atmospheric Scientist

Sulochana Gadgil is a leading authority on monsoon climate variability and dynamics. Her study of the monsoon—a seasonal wind that brings copious quantities of rainfall to Asia—includes research into the monsoon's rainfall patterns, the regional climate changes it brings, and its effects on India's agriculture. An applied mathematician by training, Gadgil is a professor at the Indian Institute of Science in Bangalore, India.

Born on June 7, 1944, in Pune, India (near Bombay), Gadgil was the third of Veshwant and Indumati (Kanhere) Phatak's four children. Gadgil's interest in science was encouraged by her father, who was a physician. In 1965, Gadgil married ecologist Madhav Gadgil, a childhood friend and fellow college student. The couple had two children—a daughter, Gauri, and a son, Siddhartha.

Gadgil attended Poona University, where she received her bachelor's degree in chemistry in 1963 and her master's degree in applied mathematics in 1965. She and her husband then traveled to the United States to pursue their doctorates at Harvard University—Sulochana studied applied mathematics, while Madhav focused on biology. She was awarded her Ph.D. in 1970 for her dissertation on the dynamics of the Gulf Stream. The research into fluid dynamics she conducted at Harvard provided a solid foundation for her future work in atmospheric science.

From 1970 to 1971, Gadgil was a research fellow at the Massachusetts Institute of Technology, but she returned to her childhood home of Pune in 1971 when she accepted a position as a pool officer in the Council of Scientific and Industrial Research at the Indian Institute of Tropical Meteorology. In 1973, Gadgil took a faculty position at the Indian Institute of Science in Bangalore, where she remains. She was promoted to associate professor in 1981 and named a full professor in 1986. While at the Indian Institute of Science, Gadgil has authored more than 40 papers on a vast array of topics. Her research has centered on tropical atmospheric and oceanic circulations, rainfall patterns and climate variability during the Asian monsoon, and the effects of the weather phenomenon on the region's agriculture. Some of her more influential papers were published in the *Journal of Climatology* and in *Nature*. Since 1989, Gadgil

Sulochana Gadgil, a leading authority on monsoon climate variability and dynamics. *(Courtesy of Sulochana Gadgil)*

has served as chairperson of the Centre for Atmospheric Sciences at the Indian Institute of Science while simultaneously discharging her professorial duties.

Gadgil's study of the monsoon has provided much-needed information about this climactic phenomenon that is central to Asia's agriculture. She has had a number of honors bestowed on her for her contributions to her field. After winning the India Career Award from the University Grants Commission in 1980, she received the B. N. Desai Award of the India Meteorological Society in 1982. She was also invited to serve on several committees, including the advisory committee of the National Centre for Medium Range Weather Forecasts, the committee on Climate Change and Oceans, and a joint Scientific Committee of the World Climate Research Programme. She is a fellow of the Indian Academy of Science and a member of the Indian Meteorological Society.

Gage, Susanna Phelps
(1857–1915)
American
Embryologist

Although Susanna Phelps Gage was a respected embryologist and comparative anatomist, her work was often ignored. Like most other women scientists of the late 19th and early 20th centuries who were married to scientists, Gage's research was often viewed as a mere adjunct to her husband's projects. Moreover, her considerable scientific accomplishments—including publications on the structure of muscles and the comparative morphology of the brain—were deemed less important than her role as her husband's wife. She was never formally employed. Nevertheless, she was one of only 25 women to be highlighted as particularly significant contributors to their fields in the second edition of *American Men and Women of Science.*

Gage was born on December 26, 1857, in Morrisville, New York, to Henry Samuel and Mary Austin Phelps. Her father was a businessmen, and her mother had worked as a schoolteacher before her marriage. Susanna received her primary education at the Morrisville Union School and the Cazenovia Seminary in Cazenovia, New York. She attended Cornell University, where she earned a Ph.D. in 1880. (Her degree was later described as a "doctor of philosophy.") While at Cornell, Gage earned the distinction of being the first woman in the history of the university to take a laboratory physics class. On December 15, 1881, she married Simon Henry Gage, a professor of histology and embryology at Cornell. The couple's only child, Henry Phelps Gage, was born on October 4, 1886. (He would later attend Cornell and receive a Ph.D. in 1911.)

After graduating from Cornell, Gage undertook independent research in embryology and comparative anatomy. In 1904, she joined the research team at the Bermuda Biological Station. The following year, she began to study neurology, first at Johns Hopkins Medical School and, in 1905, at Harvard University. She incorporated her education in neurology into her research on the comparative morphology of the brain. She also explored the development of the human brain, the comparative anatomy of the nervous system, and the structure of muscle. She published several papers on these topics. One paper—"The Intramuscular Endings of Fibers in the Skeletal Muscles of Domestic and Laboratory Animals"—was included in the 1890 *Proceedings* of the 13th annual meeting of the American Society of Microscopists. An adept artist, Gage illustrated her papers (as well as those of her husband) with meticulous detail.

Considering the lowly status accorded women scientists in this era, Gage was well recognized by her peers. In addition to being elected a fellow of the American Association for the Advancement of Science, she was also a member of the Association of American Anatomists. Nevertheless, her entry in the 1910 edition of *American Men and Women of Science* refers to her as Mrs. S. H. Gage. She died on October 15, 1915. Her husband and son donated $10,000 to create the Susanna Phelps Gage Fund for Research in Physics at Cornell University in 1917. They chose the discipline of physics in honor of her pioneering decision to take physics as a student.

Gaillard, Mary Katharine
(1939–   )
American
Physicist

Mary Katharine Gaillard's research in theoretical physics has brought her international acclaim. She is especially renowned for her work in the field of particle physics, having written groundbreaking papers on gauge theories, supergravity, and superstring theories. Gaillard was able to calculate the mass of the charmed quark before its existence had been proven, and she predicted what the mass of the b-quark would have to be as well. A prolific author, Gaillard has written more than 140 articles and edited two books. Her work has received a number of awards, including the Prix Thibaud.

Gaillard was born on April 1, 1939, in Brunswick, New Jersey, to Marion Catherine (Wiedemayer) and Philip Lee Ralph. After her marriage to Jean Marc Gaillard ended, she wed Bruno Zumino. She had three children with Gaillard—Alain, Dominique, and Bruno.

Gaillard earned her bachelor's degree from Hollins College in 1960 and then enrolled at Columbia University, completing her master's degree in 1961. She opted to pursue her doctorate in France and studied physics at the

University of Paris. In 1964, she was awarded her Dr. du Troisième Cycle and, in 1968, her Dr. Sci. in theoretical physics.

Gaillard launched her career in France, serving as a research assistant at the Centre National de Recherche Scientifique in Paris from 1964 to 1968. Concurrently, she was named a visiting scientist at the European Organization for Nuclear Research in Geneva in 1964, a post she would hold until 1984. She returned to the United States in 1968 when she accepted a research associate position at the University of California at Berkeley. She remains at Berkeley, where she has risen steadily through the academic ranks; she was promoted to head of research in 1973, director of research in 1980, and professor of physics in 1981. Also in 1981, she was appointed faculty senior scientist at the Lawrence Berkeley Laboratory.

Much of Gaillard's innovative research in particle physics was conducted in the 1970s. With B. W. Lee, she wrote an influential paper in 1974 on K mesons in gauge theories. Gauge theories are mathematical theories incorporating both quantum mechanics (the study of the behavior of matter and light on the atomic and subatomic scale) and Einstein's special theory of relativity (which posited that the speed of light is a constant). Gauge theories are most commonly used to describe subatomic particles and their associated wave fields. Gaillard also investigated quarks and short-lived particles such as K mesons. (Quarks are the subatomic particles believed to be the fundamental constituents of matter—composing protons and neutrons. Quarks are thought to exist in different types or "flavors," including "up," "down," and "strange." *Charmed quarks* are a subcategory of *strange quarks*. Strange quarks typically make up K mesons and other transitory particles, while *up* and *down* quarks compose protons and neutrons.)

Gaillard's theoretical acumen was on full display in 1975 when she predicted the mass of the charmed quark before it had even been discovered. (That is, scientists at the time believed that a particle fitting the description of what is now known as the charmed quark must exist, but no one had yet been able to prove that it in fact did so.)

Subsequently, Gaillard edited two books—*Weak Interactions* (1977) and *Gauge Theories in High Energy Physics* (1983) that explored this area of physics in great detail. She also pursued her research, which enabled her to predict the mass of the b-quark. Gaillard continued to be based at Berkeley during the 1980s and 1990s, but she also held a number of distinguished concurrent positions. From 1979 to 1981, she served as the team leader of the Theory Group in France and she worked at the Institute for Theory of Physics at the University of California at Santa Barbara in 1985. She was named a Guggenheim fellow from 1989 to 1990. She remains at the University of California at Berkeley as both a professor of physics and the faculty senior scientist.

In recognition of Gaillard's contribution to the field of theoretical physics, she has been awarded numerous honors. In addition to winning the Prix Thibaud from the University of Lyons in 1977, she received the E. O. Lawrence Award from the Department of Energy in 1988 and the J. J. Sakurai Prize of the American Physical Society in 1993. She was elected to the National Academy of Science in 1991 and is a fellow of the American Academy of Arts and Sciences.

Galdikas, Biruté
(1946–   )
Canadian
Primatologist

Biruté Galdikas ranks with JANE GOODALL and DIAN FOSSEY as one of the most important primatologists. All three conducted long-term observations of species genetically related to humans at the behest of the famous British paleoanthropologist Louis Leakey. While Galdikas's early research brought her praise from the scientific community, her later initiatives to reintroduce orphaned orangutans back into the wild created controversy and met with much criticism.

Galdikas was born on May 10, 1946, at a stopover in Germany while her parents were moving from their native Lithuania to Canada. Her mother, Filomena, was a nurse and her father, Antana, was a contractor. She grew up wandering the wilderness of High Park in Toronto, observing wildlife and wondering about the evolution of humans. These interests led her to study psychology and biology at the University of British Columbia before she transferred to the University of California at Los Angeles, where she received her B.A. in 1966. She remained at UCLA to pursue her master's degree in anthropology, which she received in 1969, the same year she married Rod Brindamour.

At the age of 22, Galdikas met Leakey, who encouraged her to follow in the footsteps of his two previous protégées—Goodall, who studied chimpanzees in their natural habitat in Tanzania, and Fossey, who studied mountain gorillas in their natural habitat in Rwanda—by studying orangutans in their natural habitat in the interior rain forests of the island of Borneo. Galdikas followed up on his suggestion to become the third of "Leakey's Angels." On November 6, 1971, Galdikas set up her observational outpost, which she later dubbed Camp Leakey, in the Tanjung Puting rain forest, which the Indonesian government later established as a national park in part due to Galdikas's urging.

Observing orangutans, who roam the treetop canopies ranging over 40 square kilometers, proved more challenging than Goodall's and Fossey's more fixed observations. Galdikas persisted, eventually familiarizing herself with many of the estimated 1,000 orangutans inhabiting the region. With patience, Galdikas became the first scientist to observe many orangutan behaviors: She witnessed a wild orangutan birth, she documented orangutans utilizing sticks as tools, and she recorded their long calls, a haunting mating ritual. She wrote up these observations into what one academic reviewer called a "monumental" doctoral dissertation, earning her Ph.D. in anthropology from UCLA in 1978.

The next year, her 10-year marriage fell apart under the strain of Galdikas's constant mothering of orphaned orangutans, which she raised alongside her own son Binti in hopes of reintroducing the domesticated animals into the wild. Galdikas married Pak Bohap, a native Bornean rice farmer and Dayak tribal traditionalist, in 1981, and together they parented two children—Fred and Jane, named after her godmother Goodall. As her studies continued, Galdikas shifted her role from observer of wild orangutans to champion of the reintroduction of orphaned orangutans into the wild, a move that many criticized as unscientific.

Indeed, confused by their close association with humans, orphaned orangutans proved difficult to repatriate into the wild. In her 1995 book, *Reflections of Eden: My Years with the Orangutans in Borneo*, Galdikas exhibited extreme detachment when she described an orangutan rape of her cook, characterizing the orangutan's behavior as "worrisome" but altogether natural for a male of his species. In 1990, Galdikas intervened in two high-profile cases of orphan orangutan transport: the so-called Bangkok Six episode, in which orangutans being smuggled through Singapore to Belgrade were discovered but fell ill and died under Galdikas's supervision; and the so-called Taiwan Ten episode, in which orangutans being repatriated to Borneo fell into a bureaucratic morass that Galdikas tried to save them from unsuccessfully. While neither episode represented a simple victory or defeat for Galdikas, they both exposed the complexity of taking a political stance as opposed to conducting pure research.

Despite the controversies surrounding Galdikas's methodologies, she has received much recognition for her pioneering work, including the 1990 PETA Humanitarian Award, the 1992 Sierra Club Chico Mendes Award, the 1993 United Nations Global 500 Award, and the 1997 Tyler Prize for environmental achievement from the University of Southern California. In 1995, the government of Canada (where she teaches at Simon Fraser University) appointed her as an officer of the Order of Canada. Perhaps her most influential role is as the president of the organization she established in 1986, the Orangutan Foundation International.

Gardner, Julia Anna
(1882–1960)
American
Geologist and Paleontologist

Julia Gardner's work as a geologist and paleontologist gained international recognition because she specialized in identifying mollusk fossils that indicated sedimentary strata with petroleum deposits.

Gardner was born on January 26, 1882, in Chamberlain, South Dakota. Four months later, her father, a physician named Charles, died, leaving her mother, a schoolteacher who was also named Julia, to raise their daughter on her own. Bryn Mawr professor and pioneer geologist FLORENCE BASCOM took Gardner under her wing, mentoring her during her undergraduate and graduate studies. Gardner earned her bachelor's degree in 1905 and then returned to Bryn Mawr after a year of teaching to earn her master's degree in 1907. With an enthusiastic reference from Bascom, Johns Hopkins University awarded Gardner a scholarship to pursue her doctorate in paleontology, which she earned in 1911. After working as an assistant paleontologist at the university—sometimes paid and sometimes unpaid—until 1915, she joined the war effort in 1917 by volunteering with the Red Cross and American Friends Service Committee in France until she was injured.

Gardner had worked for the U.S. Geological Survey (USGS) on contract before departing to the fronts of World War I. Upon her return in 1920, the USGS rehired her as a full-time assistant geologist. Gardner distinguished herself as one of the first women employed by the agency. Working on the coastal plain survey in Texas, where she identified oil deposits, she steadily rose through the ranks, becoming an associate geologist in 1924 before being promoted to a geologist in 1928. Her work during this period was later published in 1943 in the text, *Correlation of the Cenozoic Formations of the Atlantic and Gulf Coastal Plain and the Caribbean Region*. In 1936, she transferred to the paleontology and stratigraphy section, where she just used different methods to pursue further her quest for petroleum-bearing rock.

Gardner again contributed to the war effort during World War II, this time by examining shells recovered from the sand ballast of exploded bombs. Upon analysis, Gardner could identify the beaches where the sand originated, thus most likely locating as well the beaches from which the Japanese were launching their incendiary balloons. Gardner's research during the war peaked her interest in Pacific geology, where she focused her attention in the remaining years of her career with the USGS. Immediately upon retiring from governmental work in 1952, she was offered private-sector work applying her petroleum-tracking skills to the islands of the western Pacific. She probably could have pursued a lucrative second career circling the globe in search of hidden oil, but this work discontinued when she suffered a stroke in 1954.

In 1952, Gardner served as president of the Paleontological Society, and that year she received the United States Interior Department's Distinguished Service Award. The next year she served as the vice president of the Geological Society of America. However, in 1954, she retired completely, debilitated by the stroke. She died on November 15, 1960, six years after her stroke.

Garmany, Catharine Doremus
(1946–   )
American
Astronomer

Astronomer Catharine Doremus Garmany specialized in the study of the most massive and hottest of stars, classified as O- and B-type stars. Using satellite telescopes, Garmany observed these stars, the radiation emitted by them, and the elemental properties of the stars, such as temperature, mass, and luminosity, to further the understanding of the evolution and formation of massive stars. In addition to her research studies, Garmany taught astronomy at the university level.

Garmany was born on March 6, 1946, in New York City as Catharine Doremus. Known as Katy, Garmany was the oldest of three children. She developed an interest in astronomy while attending elementary school and was influenced by astronomy books written by Franklyn M. Branley, a children's author. Garmany's father worked as a copy editor and thus was able to bring home a variety of books for the children, including those by Branley. Growing up in an urban environment provided Garmany with ample resources, such as the Hayden Planetarium, which she visited with her mother.

Hoping to further her studies in astronomy, Garmany applied and was accepted into the Bronx High School of Science, a prestigious establishment that focused on teaching its highly motivated students the sciences and mathematics. After completing high school, Garmany attended Indiana University and studied astrophysics. She received her bachelor's degree in 1966 and then continued with graduate studies at the University of Virginia. Garmany earned her master's degree in 1968 and her Ph.D. in astronomy in 1971. Her dissertation dealt with "OB associations," groupings of thousands of O- and B-type stars. Garmany focused on a particular OB association known as III Cepheus and outlined three years of observations conducted at the Kitt Peak National Observatory in Arizona. A year prior to receiving her doctorate, Garmany married George P. Garmany Jr.

After earning her Ph.D., Garmany worked as a research associate at the University of Virginia for several years.

Struggling to find academic positions in astronomy, Garmany had a breakthrough in 1976 when she was awarded the ANNIE JUMP CANNON Award in astronomy by the American Association of University Women. The prominent award opened many doors for Garmany, and she secured a position as a postdoctoral research associate at the Joint Institute for Laboratory Astrophysics (JILA), in Colorado in 1976. Garmany also began teaching astronomy classes at the University of Colorado in Boulder.

At JILA, Garmany continued her research on the O- and B-type stars. The significance of these stars was that they had a shorter life in comparison to other stars. O- and B-type stars were known to explode as supernovas, very bright, fleeting objects that discharged vast amounts of energy into space. The matter emitted by these stars supplied critical elements that provided the means for the development of life on Earth. Garmany hoped the data gathered by her observations would assist in the evaluation of observations from other, faraway galaxies.

In 1981, Garmany joined the Department of Astrophysical, Planetary, and Atmospheric Sciences at the University of Colorado as an associate professor. Garmany became a fellow at the University of Colorado's Center for Astrophysics and Space Astronomy in 1985 and a fellow at JILA in 1990. In 1991, Garmany became the director of the Sommers-Bausch Observatory and Fiske Planetarium. Her duties included teaching, providing shows for the public, directing graduate students, and creating programs for public school groups. Garmany has worked to improve the facilities at the planetarium and the observatory and has succeeded in securing new computer and undergraduate laboratory equipment. In 1997, an astronomy computer laboratory was established in the observatory. Garmany continues to carry on her studies of hot, massive stars while educating students and the public about astronomy.

Geller, Margaret Joan
(1947–   )
American
Astrophysicist and Cosmologist

Margaret Geller's maps of the universe revealed not organized galaxy distribution but rather an intricate patchwork of clustered patterns. Some of these patterns drew clear pictures in the sky, such as *stickman,* the simple figure she and her colleagues observed in their maps, or the Great Wall, an expanse of clustered galaxies. Geller aspired to map more and more of the universe to get a clearer picture of how it developed and expanded.

Geller was born on December 8, 1947, in Ithaca, New York. Her mother encouraged her interest in art and languages, which led to her own interest in nature's patterns, and her father, a crystallographer at Bell Laboratories,

Margaret Geller, whose work as an astrophysicist involves mapping the universe to get a clearer picture of how it developed and expanded. *(Courtesy of Margaret Geller)*

ing specifically at the distribution of galaxies. Gazing through the 60-inch telescope at Mount Hopkins in Arizona, Huchra and Valerie de Lapparent, a French graduate student, photographed this segment of the sky, and Geller interpreted the resulting data. They all expected to find the galaxies evenly distributed through this section of the universe, as if some orderly laws reigned, but to their surprise, they found the galaxies clustered in strange configurations. After reacting in amazement to the interesting patterns in the night sky, she realized that the clustering must be due to gravitational forces, which exist out in space as they do on earth.

The map yielded a count of about 1,000 galaxies clustered together, separated by vast expanses of space devoid of light. The clustering created patterns; one cluster appears to be a child's stick-figure drawing, so Geller and her team dubbed it "stickman." They also identified a cluster that spread about half a billion light-years across, so they named it the Great Wall. Geller undertook a subsequent study in the late 1990s with Harvard-Smithsonian colleague Dan Fabricant with the intention of mapping an even deeper section of space to determine the evolution of the distribution of galaxies. They hoped to survey more than 50,000 galaxies extending more than five billion light-years from Earth.

Geller has won several important awards, including the 1990 MacArthur "genius" award and the 1991 Newcomb-Cleveland Award from the American Association for the Advancement of Science. Though Geller values these awards, she appreciates even more the beauty of the patterns created by the evolution of the universe.

Germain, Marie Sophie
(1776–1831)
French
Mathematician

Marie Sophie Germain was considered the "HYPATIA of the 18th century." With no formal mathematical training, she devised a proof, under certain conditions, for Fermat's Last Theorem, an infamous equation that Fermat scribbled in the margins of one of his books and left unsolved, confounding mathematicians until 1993 when it was finally solved with the help of computers. Germain distinguished herself in both pure and applied mathematics. Her limited recognition stemmed from the hegemonic male control of the sciences in her lifetime and from limitations in her mathematical skills due not to lack of intellect but to lack of formal training, barred as she was from advanced education.

Germain was born on April 1, 1776, in Paris. Her mother was Marie-Madeleine Gruguelin and her father, Ambrose-François, was a silk merchant. When she was 13

gifted her with toys that encouraged her ability to visualize three-dimensionally, and he brought her into his work to interest her in science. She entered the University of California at Berkeley, intending to study mathematics, but physics drew away her attention. When she graduated in 1970, she shifted her attention again, focusing more specifically on astrophysics for her graduate work at Princeton University, where she earned her M.A. in 1972 and her Ph.D. in physics in 1975.

The Harvard-Smithsonian Center for Astrophysics hosted Geller's postdoctoral work as a theoretical astrophysics fellow between 1974 and 1976, followed by a year and a half at Cambridge University in England. She returned to the Center for Astrophysics as a research associate until 1980, when she was named an assistant professor there. Three years later, she earned the title of astrophysicist, and in 1988, Harvard University appointed her as a professor of astronomy.

In 1985, Geller teamed up with Harvard-Smithsonian colleague John Huchra to map a slice of the universe, look-

years old, Germain encountered in her voracious readings an account of Archimedes dying at the hands of Roman soldiers because he was too engrossed in mathematical contemplation to flee the troops. Feeling similarly inspired in spite of oppressive opposition—her parents forbade their daughter from reading by sequestering her candles and leaving her fire unlit at night—Germain defied her parents' attempts to quell her academic interests by stealing candles and studying in conditions cold enough to freeze her ink. As the Reign of Terror raged outside her doors, Germain read Newton and Euler and taught herself Latin, Greek, and differential calculus.

Germain yearned to attend the mathematical lectures of Joseph Lagrange at the newly founded École Polytechnique, which disallowed women students, so Germain coaxed male students into lending her their notes. Under the pseudonym M. Leblanc, she composed a term paper so lucid that Lagrange sought out the author, only to discover it to be a woman. Lagrange became her mentor and also arranged for her to correspond with the German mathematician Karl Friedrich Gauss, which she did as M. Leblanc. Only later, when she came to Gauss's aid politically to save him from French troops occupying his German village, was her true identity revealed, much to Gauss's amazement.

Germain's Theorem, as her partial solution to Fermat's Last Theorem is known, asserted that Fermat's conjecture was true if x, y, and z were prime to one another and to n, if n was a prime number less than 100. Germain devised this solution at the age of 25. However, it did not see the light of day until much later, as it languished unread in a letter to Gauss.

Germain also distinguished herself in applied mathematics, as she was one of the few who even took up Napoleon's challenge to define mathematically the phenomenon first described by Ernest Chladni in 1808 whereby sand placed on a metal or glass plate would vibrate when a violin bow was drawn across the plate's edge. Napoleon called for the establishment of a "prix extraordinaire First Class" from the Institut de France. The prize was offered three times, in 1809, 1813, and 1815, and each time Germain submitted the best solution, though her theory of elasticity retained imperfections due to her lack of formal training. She was finally awarded the prize on January 8, 1816.

In 1822, the Academy of Sciences admitted Germain to its meetings—an unprecedented occurrence—under the auspices of Fourier. However, in 1830, the University of Göttingen denied Gauss's entreaties to award Germain an honorary doctorate. In June 1831, Germain died of breast cancer. The suppression of Germain's mathematical abilities robbed the history of the sciences of her potential contributions, had she been properly trained or appropriately recognized in her lifetime.

Giblett, Eloise Rosalie
(1921–)
American
Hematologist

Eloise Giblett, who has devoted her career to researching human genetics and blood, is perhaps best known for her groundbreaking finding that an inadequate supply of two enzymes—adenosine deaminase and nucleoside phosphorylase—causes inherited deficiencies in the body's immune system. Giblett also discovered a number of new inherited characteristics, called genetic markers, including blood groups and serum proteins. The author of more than 200 scholarly papers, Giblett has written a basic text on genetic markers and is credited with collecting the earliest known HIV-positive blood sample.

On January 17, 1921, Eloise Giblett was born in Tacoma, Washington, to William and Rose Giblett. Young Eloise expressed an early interest in music and planned to

Eloise Giblett, a hematologist who has discovered a number of new genetic markers, including blood groups and serum proteins. *(Photo by Wallace Ackerman Studio, Courtesy of Eloise Giblett)*

seek a career in the field. It was not until she was a student at Mills College in Oakland, California, that her passion for science was kindled.

To pursue this newfound interest, Giblett transferred from Mills to the University of Washington. She earned her bachelor's degree in science in 1942 and intended to continue her scientific education in graduate school. However, World War II altered these plans. As her contribution to the war effort, Giblett served as a medical technician in the WAVES, the women's branch of the U.S. Navy. She then used funds from the GI Bill to study medicine at the University of Washington. She obtained a master's degree in science in microbiology in 1947 and was awarded her medical degree in 1951. Upon graduation, Giblett remained at the University of Washington for her internship and residency. She then completed a postdoctoral fellowship at the University of London, from 1953–55, where she specialized in hematology and human genetics.

Giblett returned to Seattle in 1955 and accepted a joint appointment—a full professorship in medicine at the University of Washington and the associate directorship of the Puget Sound Blood Center. (She was named executive director of the blood center in 1979.) She would spend the next 30 years immersed in her duties as a researcher, educator, and administrator. In the mid-1970s, she made one of her most significant discoveries, determining that insufficient amounts of the enzymes adenosine deaminase and nucleoside phosphorylase—which both play key roles in the purine cycle—cause inherited deficiencies in the body's immune system. (Purines are the constituent elements of nucleic acids, which in turn are the building blocks of DNA and RNA.) Later, Giblett explored the potential of gene therapy to treat these deficiencies.

Giblett has made other important breakthroughs as well. She discovered new genetic markers, such as polymorphisms—variant genetic forms in one population—in blood cell enzymes, as well as serum proteins. (Serum is the fluid portion of the blood that remains after clotting.) Giblett was also one of the first researchers to determine that the proteins in serum are, in fact, genetically determined. Her numerous journal articles on genetic markers and her 1969 textbook, *Genetic Markers in Human Blood*, established her as an authority in the field of immunohematology.

In 1987, Giblett retired from her full-time duties as professor and executive director of the blood center. Thus she was free to return to her early interest in music, and she now plays violin in Seattle area chamber ensembles. Her scientific accomplishments have been amply recognized. In addition to receiving the Emily Cooly Award in 1975, she won the Karl Landsteiner Award in 1976 and the Philip Levine Award from the American Association of Clinical Pathologists in 1978. She was elected to the National Academy of Sciences in 1980 and to the presidency of the American Society of Human Genetics in 1973. Giblett never married or had children. Interestingly, it came to light in 1998 that a blood sample she had collected while in central Africa in 1959 (where she was investigating genetic resistance to malaria with Dr. Arlo Motulsky) is the earliest known HIV-positive blood sample. The existence of the blood sample contradicts the prior assumption that HIV and AIDS did not come into existence in their present forms until the 1970s.

Gilbreth, Lillian Evelyn Moller
(1878–1972)
American
Psychologist and Efficiency Engineer

Lillian Gilbreth pioneered the study of the psychological and physical effects of industrial and domestic work, and she offered solutions to these problems by redesigning industrial and domestic work spaces to increase efficiency and decrease psychological and physical damage. She advanced the study of time and motion analysis and helped usher scientific management into the modern study of management theory.

Gilbreth was born on May 24, 1878, in Oakland, California. Her father, William Moller, ran a successful hardware business while her mother, Annie, raised eight children with the help of the eldest, Lillian. She earned a bachelor's degree in English literature in 1900, becoming the first woman to speak at the commencement of the University of California at Berkeley, where she remained to earn a master's degree in 1902. Her plans to continue toward a doctorate got sidetracked on a trip to Boston in 1903 when she met Frank Bunker Gilbreth, a self-taught construction engineer specializing in increasing efficiencies in workers. The couple married in her home on October 19, 1904.

They had 12 children, whom they raised together according to the principles of efficiency they both espoused. While Frank championed physical efficiency as a means of increasing worker comfort and output, Lillian stressed the importance of psychological comfort to workers' production. Together, they shifted their business, Gilbreth, Inc., from construction to consulting, as demand for better understanding of workers' labor and mind frame increased. Gilbreth pursued a doctorate in psychology at this time; her dissertation, *Psychology of Management*, was published in 1914, before she even received her Ph.D. from Brown University in 1915. This text helped establish the discipline of scientific management, which led to modern management theory.

When Frank died of a heart attack on June 14, 1924, Gilbreth, Inc. suffered a decrease in business as companies doubted Lillian's ability to consult as effectively as her husband. Gilbreth supported her family by taking over her husband's vacant visiting lecturer position at Purdue University, where she was the first female to become a full professor at an engineering school when Purdue's School of Mechanical Engineering appointed her to a chair of management in 1935. In 1939, she added the duty of consultant on careers for women at the university. Two years later, the Newark School of Engineering hired her to head its new Department of Personnel Relations.

At Purdue, Gilbreth was relegated to teaching home economics as opposed to engineering; instead of considering this a defeat, she rose to the occasion by elevating the study of domestic economies to a science. She applied the same time and motion analysis to kitchen tasks and designs as she had to industrial tasks and factory designs as a business consultant. She disseminated her ideas on domestic efficiencies in *The Home-Maker and Her Job,* published in 1927, and *Management in the Home,* published in 1954, as well as in women's magazines such as *Good Housekeeping* and *Better Homes and Gardens.* She demonstrated her design with a model efficiency kitchen for the Brooklyn Gas Company. She also designed a model kitchen for the Institute of Rehabilitation Medicine at the New York University Medical Center, demonstrating increased efficiencies for handicapped homemakers. She collaborated with Edna Yost in 1944 to write up her ideas on helping the disabled improve their domestic skills in *Normal Lives of the Disabled.*

Gilbreth remained active academically well into her eighties. The Society of Industrial Engineers, which admitted no women, made her an honorary member in 1921. In 1966, she became the first woman to receive the Hoover Medal for distinguished public service by an engineer. Her children Frank Gilbreth Jr. and Ernestine G. Carey collaborated on two memoirs recounting the rollicking adventures of Gilbreth family life, *Cheaper by the Dozen,* which was published in 1924 and later made into a movie, and *Belles on Their Toes,* published in 1950. Gilbreth died on January 2, 1972, at the age of 93.

Giliani, Alessandra
(1307–1326)
Italian
Anatomist

A university student at a time when it was extremely rare for women to be admitted into an institution of higher learning, Alessandra Giliani studied not only philosophy but also the practical science of anatomy. She is the first woman to be recorded by historical documents as practicing anatomy (or what would today be called pathology).

Born in 1307 in Periceto, a small town in northern Italy, Alessandra Giliani was almost certainly from a wealthy family. Little is known about her childhood or her family background, but her admission to the University of Bologna around the year 1323 indicates that, at the very least, she studied with a tutor during her childhood and teen years. Only a tiny fraction of men at that time, and almost no women, were given any education, much less one provided by private tutors. Giliani never married.

During the late Middle Ages, the time in which Giliani lived, university studies centered on the study of philosophy. Philosophy and religion were intertwined to the point that philosophy was understood to be a means by which religion could be better understood. Unfortunately for the devout Christian, the most admired philosophers were ancient pagan Greeks such as Plato and Aristotle, whose views often did not align with Christian dogma.

The greatest scholars of the era immediately before Giliani's birth were theologian-philosophers such as Boethius, Thomas Aquinas, and Bonaventure. These three and other philosophers engaged in a struggle about the teachings of the ancient Greek philosophers. By the time Giliani entered the University of Bologna, the ideas of the Greek philosophers had been attacked and were temporarily in disrepute.

Even though Giliani's main interest seems to have been anatomy, she would have been required to study philosophy as a foundation for further learning. The main universities of Europe at that time—a group that included not only the University of Bologna but the University of Paris and Oxford University in England—required seven years of philosophy studies before a student could begin to study for one of the professions—law, medicine, or theology. However, this requirement was not always strictly enforced, and it appears that Giliani studied philosophy for only a few years before beginning her medical training.

The school that Giliani attended—the University of Bologna—coalesced in the ancient market town of Bologna, Italy, in the later 12th century. According to a professor who taught there at that time, in 1200 the university had around 10,000 students. The "university" at Bologna was not like a modern institution of higher learning. There was no central administrative authority; rather, it was organized as a collective of student guilds, which hired the most eminent authorities in law, philosophy, and medicine as teachers. Gradually, these informal guilds took on a more cohesive structure and became colleges that were loosely affiliated in a "university."

Giliani's most important teacher at Bologna was the medical doctor and anatomist Mondino dei Luzzi. Mondino, one of the leading authorities on human anatomy of his day, believed that a thorough medical education had to include hands-on knowledge of the human body, expertise that could be attained only through dissection of the human body. Alessandra Giliani soon became Mondino's most skilled assistant and a deft prosecutor, as an anatomist who prepared a corpse for dissection is called. She even developed a technique to drain a corpse of blood and replace the blood with fluid dyes. These dyes then marked even small veins for the medical students to see, a valuable teaching tool.

Unfortunately, this talented woman would not live to achieve a full life or career. She died suddenly, probably of some kind of infectious fever, on March 26, 1326, at the age of 19.

Gleditsch, Ellen
(1879–1968)
Norwegian
Nuclear Chemist

A pioneer in the field of radiochemistry, Ellen Gleditsch conducted early studies on radium, barium, and uranium to determine the chemical composition of these elements and understand their process of radioactive decay. Gleditsch's career was not easy. She encountered numerous obstacles to her advancement because of her gender. Determined not to be pushed aside, Gleditsch persevered in her quest to do science and ended her life with a full career as a researcher, teacher, educator, and mentor to several generations of women scientists.

Born on December 29, 1879, in Norway, Gleditsch was the oldest of 10 siblings. She was a good student and was especially adept in the sciences. Because of her proven ability as measured by her grades in secondary school, Gleditsch should have had no problem being admitted to university. However, when she graduated from high school around the year 1897, Gleditsch was not allowed to take the Norwegian university entrance examination, which was given to men only. These restrictions were not unique to Norway. At that time, they were in place in most of the major universities in the Western world, including Yale and Harvard in the United States.

Denied admission to the University of Oslo, Gleditsch instead studied pharmacology at a less prestigious school in Norway. She was awarded a degree in pharmacology in 1902, then, as a result of her degree, was grudgingly admitted into the University of Oslo where she began studying chemistry. She studied chemistry at that institution from 1902 to 1907. In 1907, seeking a friendlier intellectual climate, Gleditsch applied for, and was admitted to, MARIE SKLODOWSKA CURIE's Institute of Radium as a laboratory worker and research scientist.

Gleditsch remained at Curie's lab in Paris until 1912. At first, she was given the task of separating and purifying crude radium minerals into purer forms of radium. She worked on radium separation for a half day, for which work she was able to avoid paying a fee to use Curie's lab, then for the rest of her working day Gleditsch was allowed to work on her own projects under the supervision of Marie Curie. Gleditsch used this opportunity to begin a lifetime's study of the chemical composition of radioactive substances. Not much was known about these variants of radium and uranium, which were later termed isotopes, when Gleditsch began her work. Gleditsch was especially interested in the rate of radioactive decay of these various isotopes. In 1890, the physicist Ernest Rutherford coined the term *half-life* to describe the process of this decay. Gleditsch concentrated on trying to determine the half-life of these substances, especially radium. She published her first papers on this subject in 1909.

In 1912, Gleditsch returned to the University of Oslo where she had been given a job as instructor and research chemist. She remained there for two years before she won a scholarship to go to the United States to continue her research at Yale University, which had more advanced equipment than the lab in which she worked in Norway. At Yale, where, again because of her gender, she was tolerated rather than embraced, Gleditsch finished her study of the half-life of radium, which she calculated to be 1,686 years. This extremely accurate estimation was to stand for 35 years until researchers with more sophisticated equipment established a definitive half-life of radium at 1,620 years. Despite having used much cruder measuring devices, Gleditsch had been only 66 years off.

Around 1916, Gleditsch returned to Norway where she would spend the rest of her career as a teacher at the University of Oslo. Because of the circumstances of her own struggle to become a scientist, she devoted a lot of her time to encouraging young women to enter the sciences. From 1926 to 1929, she served as president of the International Federation of Women Scientists, and in 1929, she was appointed full professor of chemistry at the University of Oslo, the institution that had denied her an education when she had begun her studies following her graduation from high school.

During her career, Gleditsch authored 150 papers and books, many of them popular science books aimed at a broad market of nonscientists. She was awarded honorary doctorates from the University of Strasbourg in France in 1948 and the Sorbonne in Paris in 1962. Gleditsch died of complications from a stroke in Norway on June 5, 1968. She was 88 years old.

Glusker, Jenny Pickworth
(1931–   )
British/American
Crystallographer, Chemist

Researcher Jenny Pickworth Glusker studied crystallography, the science of crystal structure at the atomic level. Glusker's work on cancer-causing agents has greatly furthered cancer research. An active member of the crystallography community, Glusker has worked to enhance scientists', as well as the public's, understanding and appreciation of crystallography. Glusker has published numerous works on crystallography, including two textbooks, and she has been involved with the American Crystallographic Association.

Born Jenny Pickworth on June 28, 1931, in Birmingham, England, Glusker was the oldest of three children. Her father, Frederick Alfred Pickworth, and her mother, Jane Wylie Stocks, were both doctors. Because Glusker's parents hoped she would carry on the family tradition and enter the medical profession, they were not initially enthusiastic about Glusker's interest in chemistry.

Glusker was determined to pursue a career in chemistry, and she thus attended Somerville College at Oxford University, majoring in chemistry. She worked in the laboratory of Sir Harold Thompson—who did not approve of women being in the laboratory—and carried out research using infrared spectroscopy. During her undergraduate years, Glusker met her future husband, Donald L. Glusker, an American attending Oxford University as a Rhodes scholar. After earning her bachelor's degree in chemistry in 1953, Glusker began working toward her D.Phil. in chemistry in the laboratory of DOROTHY CROWFOOT HODGKIN, a chemist who in 1964 won the Nobel Prize. Glusker studied the structure of vitamin B_{12}, a complex compound that presented many challenges to the crystallographers of the time, and she was awarded her doctorate in 1957.

After completing her research work at Somerville College, Glusker and her fiancé made the choice to move to the United States to get married. Though her parents opposed the decision, the couple journeyed to the United States and married in late 1955. Both succeeded in gaining postdoctoral research fellow positions at the California Institute of Technology in 1955, and Jenny Glusker worked with chemist Linus Pauling. After their one-year postdoctoral experience ended, the pair moved to Philadelphia, hoping the area would provide the best professional opportunities for two married chemists. Glusker secured a position as a research fellow at the Institute for Cancer Research at the Fox Chase Cancer Center in 1956, where she worked in the laboratory headed by A. Lindo Patterson. With Patterson's approval and support, Glusker was able to work on a part-time basis while raising her three children.

Glusker became a research associate at the Institute for Cancer Research in 1957 and studied the crystallography of such small molecules as citrates. Her work on citrates furthered the understanding of the citric acid cycle, a series of reactions involving the breakdown of metabolic products, on a molecular level. After Patterson died in 1966, Glusker assumed control of the laboratory. In 1967, she was named associate member of the institute, and in 1979, she became

a senior member. She also became increasingly involved in teaching, accepting a position as adjunct professor of biochemistry and biophysics at the University of Pennsylvania in 1969. Glusker returned to the study of vitamin B_{12} and researched the structure of cancer-causing substances through the 1990s. She also became increasingly interested in the structure of macromolecular compounds.

Glusker has been extremely active in the scientific community. She was a member of the U.S. National Committee for Crystallography from 1974 to 1990 and served as the organization's chairperson from 1982 to 1984. Glusker was elected president of the American Crystallographic Association in 1979 and founded the association's newsletter, which she also edited for 15 years. In 1978, Glusker received the American Chemical Society's Philadelphia Section Award. Glusker has also been awarded the American Chemical Society's Garvan Medal (1979), the American Crystallographic Association's Fankuchen Award (1995), and an honorary doctorate from the College of Wooster. A prolific writer and editor, Glusker has published numerous works, including the classic textbook *Crystal Structure Analysis: A Primer.*

Goeppert-Mayer, Maria
(1906–1972)
German/American
Physicist

Maria Goeppert-Mayer has been called the "Marie Curie of the atom," as she contributed greatly to the understanding of atomic structure with her nuclear shell theory, in which she accounted for the existence of the "magical numbers" of protons and neutrons in particularly stable elements. Goeppert-Mayer also contributed to the Manhattan Project by calculating the separation properties for the fission isotope of uranium, a key component in the building of the atomic bomb.

Goeppert was born on June 28, 1906, in Katowice, in the Upper Silesia region of Germany (now Poland). Her father was a pediatrics professor who moved the family to Göttingen in 1910 to teach at the university (the seventh generation of academic scholars in the family). Before Goeppert even enrolled at the University of Göttingen in 1924, she was already in the bosom of the intelligentsia, as her father entertained the likes of Niels Bohr and Max Born. Goeppert intended to study mathematics, but she switched to physics under the influence of her teacher, Born. She wrote her doctoral thesis on the phenomenon of an orbiting electron jumping to an orbit closer to the atom's nucleus, calculating the probability that the electron would emit two photons of light in the process. She received her Ph.D. in 1930, but her complex calculations were not confirmed experimentally until some 30 years later.

Maria Goeppert-Mayer, who won the 1963 Nobel Prize in physics for formulation of the shell nuclear model of the atom. *(AIP Emilio Segrè Visual Archives, Physics Today Collection)*

In 1930, she married the physical chemist Joseph Edward Mayer, and the couple had two children, Maria Ann and John. The two scientists immigrated to the United States, where Mayer had accepted a professorship at Johns Hopkins University. Goeppert-Mayer received an assistantship in the physics department, but she did not receive any pay, commencing a trend that plagued her early career, whereby she essentially volunteered her scientific skills to universities that refused to compensate her, purportedly due to antinepotism regulations. When Mayer moved to Columbia University in 1939, Goeppert-Mayer again donated her time and expertise to the university as a lecturer. In 1940, Goeppert-Mayer and her husband coauthored the text *Statistical Mechanics,* and in 1941, Sarah Lawrence College offered her a half-time teaching position, her first paid job.

The male exodus to World War II opened the door for Goeppert-Mayer, who worked on the atomic bomb project under Harold Urey at the Strategic Alloy Metals Laboratory. With 20 scientists working under her supervision, she performed the calculations that proved integral to the separation method of uranium-235, the fission isotope. She continued this research at Los Alamos in 1946, then she and her husband followed the postwar scientific migration to Chicago. Goeppert split her time between the Argonne National Laboratory, where she was a senior physicist at a consultant's salary, and the Enrico Fermi Institute of Nuclear Studies at the University of Chicago, where she was an unpaid professor. Not until the University of California at San Diego was wooing her did the University of Chicago offer her a paid professorship, which she took for one year before moving to San Diego in 1960 as a full professor in the school of science and engineering.

Goeppert-Mayer performed her most important work in 1948, studying the series of seven "magic numbers" (2, 8, 20, 28, 50, 82, and 126) associated with the proton- or neutron-count in the atomic nuclei of particularly stable elements. Following a suggestion from Fermi in 1950, she hypothesized that these magic numbers corresponded to the shell numbers in nuclei, much as the closed shell theory postulated the same about atomic shells. Upon calculation, she confirmed her nuclear shell model, with spin orbit couplings corresponding exactly with the magic numbers. Working independently in Heidelberg, Germany, Johannes Hans Daniel Jensen arrived at similar results at about the same time, so the two collaborated on a book in 1955, *Elementary Theory of Nuclear Shell Structure.*

In 1963, Goeppert-Mayer and Jensen shared the Nobel Prize for physics with Eugene P. Wigner. Goeppert-Mayer was only the third woman to win a Nobel Prize for physics, the second American woman to win a Nobel Prize, and the first American woman to win the prize for physics. After a protracted illness, she died on February 20, 1972, in San Diego.

Goldberg, Adele
(1945–)
American
Computer Scientist

As part of a research team at Xerox's famed Palo Alto Research Center (PARC), Adele Goldberg played a significant part in developing and popularizing the object-oriented computer language Smalltalk. A precursor to contemporary computer languages such as Java, Smalltalk revolutionized computer programming by enabling users more easily to program and customize applications. Goldberg later launched her own company, ParcPlace Systems (now ParcPlace-Digitalk), which adapted Smalltalk for corporate uses and marketed the new version to businesses who could benefit from it. In 1999, Goldberg founded another business—Neometron, Inc.—an Internet support provider. The recipient of the prestigious Lifetime Achievement Award from *PC Magazine,* Goldberg is also dedicated to improving computer education in colleges.

Adele Goldberg was born in Cleveland, Ohio, on July 7, 1945, to Morris and Lillian Goldberg. Goldberg grew up in Chicago before attending the University of Michigan at Ann Arbor for her undergraduate education. She would later have two children of her own—daughters Rebecca and Rachel.

In 1973, Goldberg earned her Ph.D. in information science from the University of Chicago. She completed her dissertation, "Computer-Assisted Instruction: The Application of Theorem-Proving to Adaptive Response Analysis," while working as a research associate at Stanford University.

As she pursued her study of education technologies at Stanford, Goldberg was recruited by Xerox, which offered her a position at PARC in 1973. She accepted and remained there for the next 15 years as a laboratory and research assistant. In 1979, Goldberg was promoted to manage the Systems Concepts Laboratory (SCL) at PARC. The SCL team developed Smalltalk-80, an object-oriented programming language that differed greatly from other programming languages of that time, such as BASIC, FORTRAN, and COBOL, by virtue of the fact that Smalltalk was not a procedure-oriented language. Procedure-oriented programs restrict their users to keyboard commands, while Smalltalk—an object-oriented language—utilizes overlapping windows on graphic display screens. Not only was Smalltalk's innovative format simpler to use (thereby improving productivity among computer programmers), it was also customizable and could be transferred among applications with minimal effort.

Goldberg's role in the SCL group was vital. Not only did she assist in developing Smalltalk's user interface, but she also parlayed her background in education technology to create understandable demonstrations of Smalltalk. She worked at schools in Palo Alto, California, to test the technology and to gain understanding into how people interact with and use technology. She also urged Xerox to market Smalltalk. However, the company was notorious for allowing brilliant inventions to languish, and it resisted Goldberg's entreaties as well. Eventually, Goldberg persuaded Xerox to allow her and three colleagues to form a subsidiary that would market Smalltalk. Goldberg launched this venture—dubbed ParcPlace Systems—in 1988 and became the fledgling company's president and chief executive officer.

With Goldberg at its helm, ParcPlace enjoyed early success. She popularized Smalltalk, making the language well known among programmers and corporate employees alike. She wrote or cowrote the first books on Smalltalk, including *Smalltalk 80: The Language* in 1989 and *Smalltalk Developers Guide to Visual Work* in 1995. In 1992, she scaled back her duties at ParcPlace, limiting her titles to chief strategist and chairman of the board. The company went public in 1994, and she resigned in 1995 (though she still owned a stake in ParcPlace).

In 1999, Goldberg again started her own company when she formed Neometron, Inc. At the same time, she continues to pursue her interest in education, formulating computer science courses at community colleges in the United States and at schools abroad. In the same vein, she is a board member and adviser at Cognito Learning Media, a provider of multimedia software for science education. Her role in revolutionizing computer programming has been widely recognized. In addition to winning *PC Magazine*'s Lifetime Achievement Award in 1996, she was also elected president of the Association for Computing Machinery (ACM) from 1984 to 1986. She was also included in *Forbes*'s "Twenty Who Matter" list and was the recipient of the 1987 ACM Software Systems Award.

Goldhaber, Gertrude Scharff
(1911–1998)
German/American
Physicist

Gertrude Goldhaber discovered that spontaneous fission emits neutrons in 1942, but the U.S. government deemed this information classified, so publication of her paper announcing this discovery was delayed until after the war, in 1946. She made other contributions to both theoretical and experimental physics, but she was perhaps more influential as a voice supporting the role of women in the sciences. During her tenure at Brookhaven National Laboratory, she established several initiatives geared toward women in the sciences: In 1958, she began a training course for precollege science teachers to promote girls' interest in the sciences; and in 1979, she cofounded Brookhaven Women in Science. "The vicious cycle which was originally created by the overt exclusion of women from mathematics and science must be broken," she stated, adding that "it is of the utmost importance to give a girl at a very early age the conviction that girls are capable of becoming scientists."

Goldhaber was born on July 14, 1911, in Mannheim, Germany. She studied at the Physics Institute of the University of Munich, writing her doctoral dissertation under Walther Gerlach to earn her Ph.D. in 1935. She left Hitler's Germany for London, England, to conduct postdoctoral work as a research associate at Imperial College under George P. Thomson for the next five years.

In 1939, she married fellow physicist Maurice Goldhaber, and together the couple had two sons, Alfred and Michael. That year, they moved to the University of Illinois, where Goldhaber spent a decade as a research physicist before the university appointed her special research assistant professor of physics in 1948. Besides her discovery of neutron-emission in spontaneous fission during this time, she also collaborated with her husband to identify beta rays with atomic electrons.

The Argonne National Laboratory secured Goldhaber's services as a consultant from 1948 until 1950, when she moved to the Brookhaven National Laboratory as an associate physicist, the first woman with a Ph.D. on its scientific staff. By 1962, Goldhaber had worked her way up to senior physicist, a title she retained until 1979. After retiring, she took on adjunct professor positions at Cornell University from 1980 to 1982 and at Johns Hopkins University from 1982 until 1986.

The American Physical Society admitted Goldhaber as a member in 1947, and she became one of the few women to read a paper before the organization in the late 1950s. In 1960, she was a member of the National Research Council's Committee on the Education and Employment of Women in Science and Engineering. In 1972, she became the third woman physicist elected to the National Academy of Sciences.

Goldhaber's name graces several honors, including the Gertrude S. Goldhaber Prize, established in 1992 by Brookhaven Women in Science, to be awarded to a female graduate students in physics; Boston University's Gertrude and Maurice Goldhaber Prize for outstanding new graduate students in physics; and Harvard University's Maurice and Gertrude Goldhaber Prize for outstanding graduate students in physics. Goldhaber died in 1998.

Goldhaber, Sulamith
(1923–1965)
Austrian/American
Physicist

Sulamith Goldhaber made significant contributions to the understanding of elementary particle physics. Together with her husband, Goldhaber conducted groundbreaking research using nuclear emulsions, particle accelerators, and hydrogen bubble chambers. Her achievements include observing the first nuclear reactions of antiprotons, becoming the first to measure the spin of the K meson, and codiscovering the A meson.

Sulamith Goldhaber was born on November 4, 1923, in Vienna, Austria. Her parents—Abraham and Toni (Reinisch) Low—emigrated to Palestine when she was a small child. While attending Hebrew University, she met her husband, Gerson Goldhaber, whom she married on May 8, 1947. The couple had one child together—Amos Nathaniel.

After receiving her bachelor's and master's degrees from Hebrew University in 1947, Sulamith Goldhaber moved to the United States with her husband. The newly-wed couple attended the University of Wisconsin, where Sulamith earned her Ph.D. in 1951.

The Goldhabers then migrated to New York, as Gerson was offered a faculty position at Columbia University. Su-

lamith found a post at Columbia as a research associate in radiochemistry. This work introduced her to the theories and techniques of high-energy physics—the field she would pursue for the rest of her career. Goldhaber moved again in 1953, this time to California, where Gerson had accepted an assistant professorship at the University of California at Berkeley.

Sulamith Goldhaber was hired as a research physicist at the university's radiation lab in 1954. She and her husband worked as a team, using nuclear emulsion technology to analyze particles. Nuclear emulsion, which shares many attributes with photography, utilizes a glass plate thinly coated with a silver halide compound to detect a single particle. In effect, the emulsion "stops" the particle, and special developing procedures reveal its "tracks." By measuring these tracks, physicists can determine the particle's range as well as an estimate of its initial energy. Sulamith Goldhaber was renowned for her skills in using this technology and was able to produce a number of interesting findings. She was one of the first to observe the interactions of negative K mesons with protons. (Mesons are subatomic particles composed of quarks and antiquarks. Various types of mesons have been discovered, including Pi mesons and K mesons). She also reported the first nuclear interaction of antiprotons and was the first to witness the mass splitting of the charged E hyperons (another class of subatomic particle).

In the mid-1950s, Berkeley's particle accelerator (then called the Bevatron) became central to the work of both Goldhabers. A particle accelerator is a device that produces a beam of fast-moving, electrically charged atomic or subatomic particles and can reveal fundamental information about subatomic matter. In 1956, Sulamith Goldhaber presented a paper on heavy mesons and hyperons at the Rochester Conference. Her discussion revolutionized the field, making particle accelerators thereafter the most important tool for the study of subatomic particles. In 1959, Sulamith Goldhaber was named a physicist at Berkeley's Lawrence Radiation Laboratory. She branched out to incorporate hydrogen bubble chambers in her research in the early 1960s. Using this technology, Goldhaber became the leading authority on the interaction of the K+ meson with nucleons. She and Gerson together discovered the A meson, and they were also the first to measure the spin of a K meson. The "triangle diagram" for studying resonant states is attributed to the Goldhabers.

Unfortunately, Sulamith Goldhaber's brilliant career was cut short. In 1965, she departed on a sabbatical tour around the world with her husband. In Madras, India, she suffered a brain hemorrhage. Surgery revealed that she had a brain tumor. She died on December 8, 1965, without ever regaining consciousness. A member of the American Physical Society and Sigma Xi, Goldhaber is considered a pioneering researcher in her field.

Good, Mary Lowe

(1931–)
American
Chemist

In a long career in academia, government, and business, Mary Lowe Good has contributed to science by her research, teaching skills, and administrative talents. Through her investigations into the Mössbauer technique, Good was able to discover new paths in inorganic chemical research. Later, Good took her research skills into private business when she worked for the oil, materials-production, and biotechnology industries. Good later served stints as an adviser and administrator with the National Science Foundation, the president's Council of Advisors on Science and Technology, and the U.S. Department of Commerce.

Born on June 20, 1931, in Grapevine, Texas, a suburb of Dallas, Mary Lowe Good is the daughter of John Wallace Lowe and Winnie Mercer Lowe. Her father was variously a teacher, principal, and superintendent of the local schools, and her mother taught English and mathematics in the Grapevine high school. Lowe was one of four siblings, all of whom attended universities with the encouragement of their parents. In 1952, while a student at the University of Arkansas, Lowe married Bill J. Good, a physicist. They have two children.

After graduation from high school, Good attended Arkansas State Teachers College, from which she earned a B.S. in chemistry in 1950. In 1951, she enrolled in the graduate chemistry program at the University of Arkansas in Fayetteville. She earned a master's degree from the University of Arkansas in 1953 and began teaching chemistry at Louisiana State University (LSU) in Baton Rouge the following year. While teaching at LSU, Good continued research in inorganic chemistry, specializing in solvent extraction of metal complexes. This work resulted in a Ph.D., which was awarded her by the University of Arkansas in 1955.

In 1958, Good was appointed associate professor of chemistry at the newly created LSU New Orleans campus. Here she continued research on the extraction of metals from solvents and also began a new phase of research using Mössbauer spectroscopy to understand the chemical structure of solids such as iron and tin halides. Good expanded the use of Mössbauer spectroscopy to examine the chemical structure of ruthenium compounds. By submitting ruthenium to Mössbauer spectroscopy, she uncovered the ways that this metal oxidizes and forms metal-to-metal bonds. Good also discovered much data about ruthenium's electronic properties.

Good became a full professor of chemistry at LSU New Orleans in 1963, and by 1974, as a result of her work with ruthenium, she was appointed Boyd Professor at LSU, a position endowed with nearly $200,000 a year in research support funds. In 1978, Good moved from LSU New Orleans back to the main campus in Baton Rogue where she continued her work in material science. She remained at LSU until 1980 when she resigned to take a position as vice president of research and director of the laboratory for Universal Oil Products, Inc., a subsidiary of the Allied Signal Company. She remained with Allied Signal until 1993, working on the chemistry of oil refining and helping to develop long-lasting antifouling coatings for U.S. Naval ships.

In 1993, President Bill Clinton appointed Good undersecretary for technology in the Department of Commerce. This was merely the latest of a string of governmental appointments that had begun with President Jimmy Carter, who had appointed Good to an oversight committee of the National Science Foundation. In 1991, President George Bush appointed Good to serve on his Council of Advisers on Science and Technology. For her work in chemistry, Good was awarded the American Chemical Society's (ACS) Garvan Medal in 1973 and the ACS's Charles Lathrop Parsons Award (for outstanding public service) in 1991. She has been presented honorary doctorates from LSU, the University of Arkansas, and Duke University. Currently, she is a dean of the College of Information Science and Systems Engineering at the University of Arkansas, Little Rock.

Goodall, Jane

(1934–)
English
Zoologist

Jane Goodall revealed to the world a firsthand account of chimpanzee culture, which turned out to be as fascinating as human culture. She dispelled several myths about chimpanzees, such as the belief that they ate only plants and insects and the notion that humans alone used tools. However, Goodall also observed the dark side of chimpanzees, behavior that mirrored the human will to violence. Goodall set up a foundation to help protect chimpanzees, and she educated people the world over about the realities of chimpanzee behavior and culture.

Goodall was born on April 3, 1934, in London, England. Her mother, Vanne Morris, was a homemaker and writer, and her father, Mortimer, was an engineer. As a child, Goodall's most beloved toy was a stuffed chimpanzee named Jubilee, a precursor to her later companions. Jane grew up at the Birches, her grandmother's estate in Bournemouth, on the sea. Her parents divorced, leaving her mother with no money to educate Goodall except by sending her to secretarial school, after which she worked as a secretary at Oxford University and for documentary filmmakers.

When an old friend invited Goodall to visit her in Kenya, she jumped at the chance to fulfill this childhood dream. There, she met the famous British paleoanthropologist Louis Leakey, who hired her as an assistant secretary before offering her a position studying chimpanzees in their natural habitat over a sustained period of time to familiarize us with our closest genetic relatives, which Goodall readily accepted. On July 16, 1960, she set out for Tanzania, where she set up camp at Gombe Stream on the banks of Lake Tanganyika. In her first year there, she observed chimpanzees eating meat (they were believed to be herbivores); she even saw chimpanzees hunt and kill young baboons to eat them. She also observed one chimpanzee in particular (named David Graybeard by Goodall) stick a stem of grass into a termite hole and then retract it covered with delicious termites. Chimpanzees use tools, Goodall concluded, dispelling the myth that only humans know to use tools.

Without having earned even a bachelor's degree, much less a master's, Goodall applied to Cambridge University's doctoral program in zoology, which accepted her in 1961. She wrote up her ethological observations as a thesis and earned her Ph.D. in 1965. In 1962, the *National Geographic* started preparing an article on Goodall by sending Dutch photographer Baron Hugo van Lawick to Gombe; the two fell in love and married on March 28, 1964, and three years later Goodall mothered a son officially named after his father but unofficially called Grub, or "bush baby" in Swahili. Goodall and Lawick divorced a decade after they married. She remarried in 1975, but her second husband, Derek Bryceson, head of the Kenyan national park system, died of cancer in 1980.

The longer Goodall observed chimp culture, the more secrets they revealed. She was disturbed to discover chimpanzees waging war against each other, though it demonstrated a closer affinity with their human counterparts. In 1975, human conflict visited Gombe, as Zairian rebel soldiers attacked the Gombe compound, kidnapping four students who they ransomed two months later. In 1977, Goodall established the Jane Goodall Institute to promote conservation of chimp habitat and protection of chimps. Goodall eventually left full-time field study to travel around the globe informing people about chimpanzees. For her contribution to scientific understanding, the British Empire named her a commander, the National Geographic Society awarded her the Hubbard Medal, and the Japanese crowned her with its equivalent of a Nobel, the Kyoto Prize. The ongoing research at Gombe represents the longest continuous animal research project ever, an accomplishment Stephen Jay Gould called "one of the Western world's great scientific achievements."

Goodenough, Florence Laura
(1886–1959)
American
Psychologist

Florence Laura Goodenough developed several methods of measuring and interpreting children's intelligence, including the Minnesota Preschool Scale, which included verbal and nonverbal scores in its calculations. She also employed time and event sampling in her work, which would later prove useful to other fields. Goodenough's most significant contribution to child psychology was her advocacy against the notion that IQ remained constant throughout a person's life. Instead, she contended that intelligence needed to be measured over the course of an individual's life. In 1937, *American Men and Women of Science* listed her among their 1,000 most important American scientists.

Born on August 6, 1886, in Honesdale, Pennsylvania, Goodenough grew up with eight siblings on a small family farm. She attended the Millersville, Pennsylvania, Normal School, where she received her B.Pd. (bachelor of pedagogy) in 1908. Upon graduating, she embarked on a teaching career at various schools in Pennsylvania and New Jersey. She was particularly captivated by her work at the Vineland (New Jersey) Training School, where she was able to work closely with developmentally disabled children.

At the same time that Goodenough worked as a teacher, she also continued her education. In 1920, she earned her B.S. from Columbia University. The following year, she served as director of research at public schools in Rutherford and Perth Amboy, New Jersey (her position was akin to that of a school psychologist today), and she studied for her master's under Leta Hollingworth at Columbia. After completing her M.A. in psychology in 1921, Goodenough moved to Stanford University for doctoral studies. She worked with Lewis Terman in his groundbreaking research on gifted children. Terman, who helped formulate the Stanford-Binet IQ test for children, included Goodenough in many aspects of his research and even listed her as a contributor in his book on the project. She was awarded her Ph.D. in psychology in 1924. Her dissertation, *Measurement of Intelligence By Drawings,* was published as a book in 1925. In this text, she outlined her "Draw the Man" test for preschoolers. By defining strict criteria for evaluating the drawings, Goodenough's test provided an accurate indication of the child's intelligence.

After receiving her Ph.D., Goodenough accepted a position as chief psychologist at the Minneapolis Child Guidance Clinic in Minneapolis, Minnesota. In 1925, she became an assistant professor at the University of Minnesota's Institute of Child Welfare, and she was made a full professor there in 1931. Goodenough's tenure at the university was a fruitful one. In 1931, she published her second book, *Anger in Young Children,* in which she interpreted a seven-year study of 41 children. She also cowrote the *Handbook of Child Psychology* in 1933. Perhaps her most important work, the handbook put forth the first professional critique of "ratio IQ." Goodenough was adamant that the practice of viewing IQ as a fixed constant was detrimental to children. For her, children should not be categorized at a certain intelligence at an early age. Instead, she advocated measuring intelligence over the total life span—not just in children and adolescents.

While at the University of Minnesota, Goodenough also devised the Minnesota Preschool Scale, which revised the Stanford-Binet test to include younger children. Comprised of both verbal and nonverbal sections, the Minnesota Preschool Scale gave a fairly accurate estimate of early mental ability. She was also a pioneer in using observational techniques such as time sampling (studying all of a subject's behavior during a set amount of time) and event sampling (observing how often certain behaviors are performed). These methods would later become central in studying the natural behavior of both humans and animals.

Goodenough remained at Minnesota for the remainder of her career. In 1947, she was forced to retire early because of a degenerative disease, and she went blind a few years later. She died on April 5, 1959, in her sister's home in Florida of a stroke. Her legacy was profound. She was the first psychologist to argue against pigeonholing young children according to their intellectual abilities, and she also improved existing intelligence tests. Her contributions were recognized by her peers. In addition to serving as president of the National Council of Women Psychologists in 1941, she was elected president of the Society for Research in Child Development in 1946 and 1947. Totally devoted to her work, Goodenough never married or had children.

Granville, Evelyn Boyd
(1924–   )
American
Mathematician

Evelyn Boyd Granville holds the distinction of being one of the first two African-American women to earn a doctorate in mathematics. (The other is MARJORIE LEE BROWNE.) In addition to teaching at three universities, Granville was a major contributor to the Vanguard and Mercury space programs. She also played an active role in formulating mathematics programs for elementary and secondary students. Her textbook, *Theory and Application of Mathematics for Teachers,* has become a standard source for educators.

On May 1, 1924, Evelyn Boyd was born in Washington, D.C., to William and Julia Walker Boyd. Her father was a custodian at the family's apartment building, while her mother was an examiner for the U.S. Bureau of Engraving and Printing. After her parents separated when she was young, Boyd was raised primarily by her mother. When it came time for Boyd to begin high school, Washington's public schools were still racially segregated. However, she was fortunate to attend Dunbar High School, which offered high-level courses and had excellent teachers.

After graduating from Dunbar as valedictorian, Boyd enrolled at Smith College in 1941, where she majored in mathematics and physics. She had originally planned to become a high-school math teacher, but her stellar performance led her professors to advise her to pursue a college teaching position. In 1945, Boyd graduated summa cum laude from Smith and began graduate studies at Yale University. She received her Ph.D. in mathematics in 1949 for her dissertation, *On Laguerre Series in the Complex Domain.*

Boyd accepted a postdoctoral fellowship at New York University's Institute of Mathematics and Science. Despite her high-powered degree and academic success, Boyd was unable to find a faculty position at a university with a doctoral program—most likely because she was an African-American woman. In 1950, Boyd was offered an associate professorship at Fisk University, a prominent all-black college in Nashville, Tennessee. Although she enjoyed teaching, Boyd opted to return to Washington in 1952, and she took a job as an applied mathematician at the U.S. Army's Diamond Ordnance Fuze Laboratories, developing missile fuses.

In 1956, Boyd changed positions again and began working for the International Business Machines (IBM) Company. At IBM, she played a significant role in formulating orbit computations and computer procedures for the Project Vanguard and Project Mercury space programs. Boyd left IBM in 1960, when she married the Reverend Gamaliel Collins and moved to Los Angeles, California, with him. From 1960 to 1962, she was a research specialist at the Computation and Data Reduction Center of the U.S. Space Technology Laboratories, analyzing rocket trajectories and methods of orbit computation. In 1962, she moved to the North American Aviation Space and Information Systems Division, where she worked as a research specialist studying celestial mechanics, trajectory, and orbit computation as well as digital computer techniques for the Apollo space program. She returned to IBM (this time in Los Angeles) as a senior mathematician in 1963.

Boyd's marriage to Collins ended in divorce in 1967, the same year she left IBM for a faculty position in the mathematics department at California State University (Cal State) in Los Angeles. Struck by her students' low level of mathematics preparedness, Boyd became involved in various programs aimed at addressing the problems in elementary math education. From 1968 to 1969, she taught an elementary school supplemental mathematics program through the state of California's Miller Mathematics Improvement Program. In 1970, she headed an after-school mathematics enrichment program for elementary students, and in 1975, she cowrote *Theory and Application of Mathematics for Teachers* with Jason Frand. The book quickly became an integral part of education programs and was used at more than 50 colleges nationwide. Boyd married Edward Granville in 1970, taking his last name. She remained at Cal State until 1984, when the couple moved to Texas. From 1985 until 1988, Granville taught mathematics and computer science at the University of Texas at Tyler. In 1990, she was named to the Sam A. Lindsey Chair there, and she remains a visiting professor.

Granville's achievements were recognized in 1989 when Smith College awarded her with an honorary doctorate. With this honor, she again made history, becoming the first African-American woman mathematician to receive such an award. She is a member of the American Mathematical Society and the Mathematical Association of America. Granville's career has served as an inspiration to both women and African Americans, proving that the field—long dominated by white men—is open to all. Several of her former students would later credit her with motivating them to earn doctorate degrees in mathematics.

Grasselli Brown, Jeanette G.

(1929–)
American
Analytical Chemist

Analytical chemist Jeanette Grasselli Brown worked in the private sector for nearly four decades and did much to establish the emerging fields of infrared and Raman spectrometry. Grasselli's development of problem-solving methods in analytical chemistry had many practical applications and were used to study environmental pollution issues and identify various industrial contaminants, including those found in gasoline. Grasselli sought to make the uses of spectroscopy understandable to the general public and to private industry.

Born Jeanette Gecsy on August 4, 1929, in Cleveland, Ohio, Grasselli was the eldest of two children born to Nicholas W. Gecsy and Veronica Varga. Grasselli's parents were Hungarian immigrants, and she was raised in a largely blue-collar, Hungarian neighborhood. Though Grasselli's parents were not wealthy, they valued education and hoped that she would pursue a college education.

Grasselli became interested in chemistry when she took a beginning chemistry course in high school. She was in a college preparatory program, and she had been planning to major in English. Her attraction to chemistry and the support of her chemistry instructor motivated Grasselli to change her proposed major. She received a full scholarship to Ohio University and graduated summa cum laude in 1950.

After completing her undergraduate chemistry studies, Grasselli accepted a position as a project leader and infrared spectroscopist with Standard Oil of Ohio (since 1985 BP Amoco, Inc.), where she spent her entire professional career. The infrared spectrometer was a relatively new instrument at the time, and Grasselli's task was to master it. Spectroscopy dealt with the measurement of the interactions of matter and electromagnetic radiation. Scientists were able to analyze these measurements and create a graphic representation of a physical system or phenomenon, known as a spectrum. The infrared spectrometer allowed Grasselli to study substances at the atomic and molecular levels in a nondestructive manner. Because the instrument could analyze any substance in almost any quantity, even a very small amount, its potential in practical applications was promising. The spectrometer, for instance, made it possible to test soil, air, and water to study the effects of pollution and to identify unknown pollutants and chemicals. Grasselli, aware of the possibilities of infrared spectroscopic analysis, collaborated with the Cleveland coroner's office to help detect and classify unidentified crime substances.

While working at Standard Oil, Grasselli earned her master's degree from Western Reserve University in 1958. She worked with Standard Oil until 1988, eventually working her way up to director of corporate research of environmental and analytical sciences. After her retirement, Grasselli remained active, acting as a consultant and lecturer on spectroscopic analysis and its numerous applications. She was elected to the Ohio Board of Regents in 1995, was a board member for a number of corporations, and chaired the board of the Cleveland Scholarship Programs beginning in 1995. Grasselli also found time for marriage—she married Glenn R. Brown in 1987 but retained the name Grasselli from a former marriage as her professional name.

Grasselli has received numerous honors and awards for her work, including the American Chemical Society's Garvan Medal in 1986, its Fisher Award in Analytical Chemistry in 1993, and its Encouraging Women into Careers in Science Award in 1999. She also received the Distinguished Service Award from the Society for Applied Spectroscopy in 1983 and the William Wright Award from the Coblentz Society in 1980. Grasselli has received honorary doctorate degrees from eight universities, including Ohio University (1978), Clarkson University (1986), Notre Dame College (1995), and Kenyon College (1995). She was elected to the Ohio Women's Hall of Fame in 1989 and the Ohio Science and Technology Hall of Fame in 1991. She has written numerous journal articles and books, including *Analytical Raman Spectroscopy*.

Green, Arda Alden

(1899–1958)
American
Biochemist

Educated in medicine, Arda Alden Green devoted her career to researching enzymes and proteins. Green developed a number of protein isolation and purification methods that were adopted by the scientific community. Her work on enzymes contributed to the work of Carl and GERTY THERESA RADNITZ CORI, the husband and wife team who won the Nobel Prize in 1947 for uncovering the intermediate steps in the conversion of glycogen to glucose. Green also studied bioluminescence and was responsible for discovering how and why fireflies produce a glowing effect.

Arda Green was born on May 7, 1899, in Prospect, Pennsylvania. She attended the University of California at Berkeley and graduated in 1921, earning highest honors in chemistry and honors in philosophy. Green entered the graduate program in philosophy at Berkeley but switched to medicine after one year. After studying medicine for two years at Berkeley, Green took a break from her studies to work in laboratory research at Harvard University Medical School for one year. When she returned to Berkeley, she received a Leconte Memorial Fellowship. Green then transferred to Johns Hopkins University School of Medicine, where she published her first article, on electrolyte conductivity. In 1927, Green was awarded her medical degree.

After earning her M.D., Green received a National Research Council Fellowship and returned to Harvard to work in Cohn's laboratory. There she succeeded in developing a method for evaluating the balance between oxygen and hemoglobin as well as the outcome of hydrogen ion activity. Green's process, which involved figuring out the solubility of hemoglobin, became a standard method. Green stayed on at Harvard for a number of years following her fellowship, working in Cohn's laboratory and as a pediatrics research associate. During these years, Green developed methods to isolate and purify proteins. She also carried out immunological research, particularly with regard to measles prevention, and continued to investigate hemoglobin. Green also tutored biochemistry students at Radcliffe College.

In 1941, Green left Harvard to become a research associate in pharmacology at Washington University in St. Louis, Missouri. Working in the laboratory of Carl and Gerty Cori, Green successfully isolated the enzyme phosphorylase. After one year at Washington University, Green became an assistant professor of biological chemistry. She remained in St. Louis for an additional three years, studying muscle proteins and developing other purification processes, including one for the enzyme aldolase, which was responsible for breaking down fructose.

In 1945, Green became a staff member of the research team at the Cleveland Clinic. There Green made a significant discovery when she isolated an organic compound that her research team called serotonin. Green found this new substance in blood serum, and further studies indicated serotonin's importance in a number of biochemical processes, including the stimulation of muscles and circulation. Green's discovery significantly advanced the understanding of the functioning of the central nervous system. In 1953, Green returned to Johns Hopkins University to work at the McCollum-Pratt Institute. She studied the enzyme luciferase, which was responsible for making fireflies glow, and discovered the enzymatic reaction that caused the glowing effect.

Expanding on her research on luciferase, Green began working on the isolation and crystallization of enzymes involved in the bioluminescence of bacteria. Green's health began to deteriorate, however, and in 1955, she was diagnosed with breast cancer. Green died on January 22, 1958, just a few months before she was to be presented with the American Chemical Society's Garvan Medal.

Gross, Carol A.

(1941–   )
American
Bacteriologist

Noted bacteriologist Carol Gross has focused her research on the production of cell proteins in response to heat. She has also studied the ribonucleic acid (RNA) polymerase enzyme that regulates various functions in both RNA and deoxyribonucleic acid (DNA). The editor of the *Journal of Bacteriology,* Gross is a respected professor and researcher.

Carol Gross was born on October 27, 1941, in Brooklyn, New York. Her father, Samuel Polinsky, was an attorney, while her mother, Mollie Hausman Polinsky, was employed as a school guidance counselor.

After earning her bachelor's degree from Cornell University in 1962, Gross attended Brooklyn College, where she received her master's degree in 1965. While attending Brooklyn College, Gross met and married her husband. She later gave birth to two children—Steven (in 1965) and Miriam (in 1969). Gross moved to the University of Oregon and was awarded her Ph.D. in bacteriology in 1968 for her dissertation on the regulation of lactose production in the *Escherichia coli* bacteria.

Gross began a postdoctoral fellowship at the University of Oregon in 1968. She remained there until 1973, when she accepted a post as a project assistant at the University of Wisconsin. While there, she began to study RNA—the single-stranded molecules that are transcribed from DNA in a cell. During this period, Gross also developed an interest in cancer research, and she consequently transferred to the McArdle Laboratory for Cancer Research at the University of Wisconsin in 1976. Over the course of the next three years, she advanced to the rank of associate scientist. In 1979, she joined the Wisconsin faculty as an associate professor in the department of bacteriology. Gross left Wisconsin briefly in 1985 for a visiting professorship in Nanjing, China, where she lectured on gene therapy and recombinant DNA. That same year, she was appointed to the scientific advisory board of the Damon Runyon–Walter Winchell Cancer Research Fund. Gross was promoted to a full professorship at the University of Wisconsin in 1988.

In 1993, Gross joined the faculty of the University of California at San Francisco (UCSF) as a professor in the department of stomatology (diseases of the mouth) and microbiology. While fulfilling her teaching obligations, Gross has concentrated on two areas of research—the response of cells to high heat, and RNA polymerase (which is an enzyme that binds compounds and transcribes DNA, thereby regulating how DNA interacts with a cell). Gross's high-heat studies stem from the observation that when cells are subjected to intense temperatures, they begin to produce copious quantities of certain proteins, which are capable of growing and thriving at ordinarily lethal temperatures.

Gross remains at UCSF, and her contributions to her field have been widely recognized. In 1992, she was elected to both the American Academy of Arts and Sciences and the National Academy of Sciences. In addition, she was named editor of the *Journal of Bacteriology* in 1990 and became a member of the editorial board of *Genes and Development* that same year.

Guthrie, Mary Jane
(1895–1975)
American
Zoologist

Mary Jane Guthrie's most significant scientific contributions were made in the field of cytology (which is the branch of biology concerned with the structure, function, pathology, and life history of cells). Specifically, Guthrie researched the cytoplasm (which is the protoplasmic substance surrounding the cell's nucleus) of the female reproductive system and the endocrine glands (such as the thyroid, adrenal, and pituitary glands, which release enzymes directly into the body). She also did work concerning experimental pathology and organ culture.

On December 13, 1895, Mary Jane Guthrie was born in Bloomfield, Missouri, to George Robert Guthrie and his wife, Lula Ella Boyd. Guthrie herself would never marry or have children. She received her bachelor's degree in 1916 and her master's in 1918 from the University of Missouri. She pursued her doctorate at Bryn Mawr College in Pennsylvania, earning her Ph.D. in zoology in 1922. While at Bryn Mawr, Guthrie also worked as a demonstrator in the biology department from 1918 until 1920 and as a zoology instructor from 1920 to 1921.

Upon completing her formal education, Guthrie returned to the University of Missouri, where she would remain as a faculty member for nearly 30 years. After being appointed an assistant professor in 1922, she subsequently rose through the academic ranks, becoming an associate professor in 1927 and a full professor in 1937. Like many other women scientists of her generation, though, Guthrie often experienced blatant discrimination in finding funding for her research projects. For instance, in 1934, she was told by an official at the Rockefeller Foundation that a female scientist such as herself would need to provide more credentials to receive a grant than her male counterparts. In essence, she was required to submit proof of her scientific prowess above and beyond anything that would be asked from a man.

Despite the obstacles she encountered, Guthrie was highly respected in her field. A prolific writer, she coauthored several influential textbooks. Of particular note were her contributions to the *Textbook of General Zoology* in 1938. (She also collaborated with John M. Anderson to produce *General Zoology* in 1957 and *Laboratory Directions in General Zoology* in 1958.) Guthrie left the University of Missouri in 1951 for Detroit, Michigan, where she accepted a position as a research associate at the Detroit Institute of Cancer Research. She concurrently conducted independent research at Wayne State University (also in Detroit). In 1955, Guthrie became a full professor at Wayne State while retaining her post at the Institute of Cancer Research. She retired from both positions in 1960.

During her long career, Guthrie became a member of numerous professional societies. In addition to the American Society of Zoologists and the American Association of Anatomists, she joined the Genetics Society, the American Association of Mammologists, and the Tissue Culture Association. She was also elected to the American Association for the Advancement of Science and the American Society of Naturalists. Despite her focus on research, however, Guthrie had many interests outside of science. An avid stamp and furniture collector, she also enjoyed reading, theater, golf, and horseback riding. Mary Jane Guthrie died in 1975 at the age of 80.

H

Hahn, Dorothy Anna
(1876–1950)
American
Organic Chemist

A lifelong educator, Dorothy Anna Hahn built a strong chemistry department at Mount Holyoke College that produced many professional women chemists. Hahn was also a prominent researcher, and her work in ultraviolet spectroscopy led to significant advances in chemistry. Hahn collaborated with many of her students on research projects and published numerous scientific articles with her students.

Born on April 9, 1876, in Philadelphia, Pennsylvania, Hahn was the younger of two daughters born to Carl S. Hahn and the former Mary Beaver. Hahn's mother was a Philadelphia native, but her father was a German immigrant who dabbled in many professions. Though he worked primarily as a bookkeeper or clerk, he also worked as a linguist and vendor of artificial flowers.

After graduating from Miss Florence Baldwin's School, which later came to be known as the Baldwin School, Hahn attended Bryn Mawr College in Bryn Mawr, Pennsylvania. In 1899 she graduated, earning bachelor's degrees in both chemistry and biology. Hahn secured a position as professor of chemistry at the Pennsylvania College for Women in Pittsburgh, where she remained until 1906. She also simultaneously taught at Kindergarten College from 1904 to 1906. At Kindergarten, Hahn taught biology. Hahn decided to return to her studies in 1906, and she spent a year at the University of Leipzig in Germany. In 1907, Hahn returned to Bryn Mawr to continue with her graduate studies in organic chemistry, and in 1908, Hahn found a faculty position at Mount Holyoke College in South Hadley, Massachusetts, where she remained for the duration of her career.

In 1913, Hahn published her first article, on which she worked with a graduate student. The paper outlined Hahn's confirmation of the ring structure of organic compounds known as hydantoins. Vitamin B was an example of a hydantoin. Hahn and her student completed their research using the ultraviolet spectroscopic method developed by Hahn's colleague, EMMA PERRY CARR. A year after publishing the paper, Hahn was promoted to associate professor of chemistry.

Hahn began collaborating with organic chemist Treat B. Johnson when she attended Yale University as an American Association of University Women fellow from 1915 to 1916. In 1916, she was granted her Ph.D. At Yale, Hahn studied and wrote about the relationship between electrons and chemical valence. She also began to investigate cyclic polypeptide hydantoins and continued this research after returning to Mount Holyoke. In fact, Hahn and her students worked on the synthesis of organic molecules and their analysis through spectroscopic methods for more than a decade. In 1918, Hahn was made a full professor, a position she held until her retirement in 1941. Hahn devoted her time to training her students using a hands-on, collaborative approach, and she inspired many to pursue graduate studies in chemistry, particularly at Yale.

Hahn published nearly two dozen articles from 1913 to 1940 and collaborated with Treat Johnson on several works, including *Theories of Organic Chemistry,* which was published in 1922. She and Johnson also authored *Pyrimidines: Their Amino and Aminoxy Derivatives,* published in 1933. In addition to teaching, Hahn also worked in the private sector researching coal tar products. Hahn enjoyed traveling and spent many summers on the coast of Connecticut, where she became an avid sailor. After Hahn's death in 1950, a seminar room in a new chemistry building at Mount Holyoke was furnished in her honor.

Hamerstrom, Frances
(1907–1998)
American
Wildlife Biologist

Frances Hamerstrom studied wildlife biology under the famous ecologist Aldo Leopold, and she became the first woman in this country to earn a master's degree in wildlife management. She studied the greater prairie chicken population, and she reversed population decreases to promote the continuation of the species.

Hamerstrom was born as Frances Flint on December 17, 1907, in Needham, Massachusetts. She led a privileged life as a child, traveling to Europe and posing as a fashion model. She married Frederick Hamerstrom in Orlando, Florida, in 1931. She flunked out of Smith College but finished up her undergraduate work at Iowa State College, earning her bachelor of science degree in 1935. She then proceeded to the University of Wisconsin at Madison, where she studied wildlife biology under Aldo Leopold to become the first woman in the nation to receive a master's degree in wildlife management in 1940.

She and her husband and their two children remained in Wisconsin, where they moved into an antebellum house overlooking a prairie. Hamerstrom studied greater prairie chickens, which would rise in the morning on their land and do a ritualistic mating dance. As word spread, increasing numbers came to their land to watch this daily dance. The Hamerstroms turned their house into a sanctuary, as owls nested in the rafters. In 1949, Hamerstrom started working as a game biologist for the Conservation Department, which later became the Wisconsin Department of Natural Resources (WDNR). Through her efforts, the dwindling number of greater prairie chickens turned around and gained population. Hamerstrom spearheaded initiatives to buy and protect grasslands habitats: The Dane County Conservation League and the Society of *Tympanuchus cupido pinnatus* bought thousands of acres in the Buena Vista Marsh in southern Portage County and the Paul Olson Wildlife Area west of Stevens Point. She remained with the WDNR until 1972.

Hamerstrom then directed the Raptor Research Foundation from 1974 through 1976. She focused much of her research on golden eagles, publishing *An Eagle to the Sky* in 1970; in it she described her rehabilitating and raising of golden eagles. In her next book, *Strictly for Chickens,* she focused on the greater prairie chickens. Later books included *The Wild Foods Cookbook* and an autobiography, *My Double Life: Memoirs of a Naturalist.* Interestingly, both she and her husband were avid hunters, sacrificing the lives of individual animals while preserving whole species.

In 1982, Hamerstrom became an adjunct professor at the College of Natural Resources at the University of Wisconsin at Stevens Point. She continued her study of kestrels, or sparrow hawks, an investigation she sustained for 20 years. She also acted as an activist in support of rain forest protection, traveling to Africa and South America to visit rain forests. She and her husband traditionally hosted apprentices in wildlife management at the Plainfield home, which became somewhat of a menagerie. She continued to take in these students even after her husband's 1990 death.

Hamerstrom received numerous distinctions in her career, including awards from the Wildlife Society in 1940 and 1957, the 1960 Josselyn Van Tyne Award from the American Ornithologist's Union, the 1964 Chapman Award from the American Museum of Natural History, the 1980 United Peregrine Society Conservation Award, and the 1985 Edwards Prize from the Wilson Ornithological Society. Hamerstrom died on August 31, 1998.

Hamilton, Alice
(1869–1970)
American
Physician and Industrial Toxicologist

Alice Hamilton commenced her career as a physician specializing in bacteriology and pathology, but her experience at Chicago's Hull House (founded by Jane Addams) caring for immigrants opened her eyes to the dangerous conditions suffered by industrial workers. She devoted the rest of her career to improving working conditions, especially lobbying against the toxic exposure commonplace at that time.

Hamilton was born on February 27, 1869, in New York City. Her father, Montgomery, co-owned a wholesale grocery firm, while her mother, Gertrude, homeschooled her four children. Alice received not only an academic education, learning languages, literature, and history, but also a moral education, learning to value personal liberty and its incumbent responsibilities. At the age of 17, Hamilton followed the family tradition of attending Miss Porter's School for Girls in Farmington, Vermont, where she decided to become a doctor.

Hamilton first attended the uncertified Fort Wayne College of Medicine before matriculating at the University of Michigan School of Medicine in 1892 and receiving her M.D. in 1893. Hamilton then practiced as an intern in Minneapolis and Boston, specializing in bacteriology and pathology. She spent the next two years in Germany, studying first at the University of Leipzig in 1895 and then at the universities of Munich and Frankfurt in 1896. Upon returning to the United States, she continued her study and research at Johns Hopkins University.

In 1897, Hamilton took up a professorship of pathology at the Women's Medical College of Chicago, simultaneously continuing to study and conduct research at the University of Chicago. With the closure of the women's school in 1902, Hamilton migrated briefly to Paris to study at the Pasteur Institute before returning to Chicago to work at the McCormick Institute for Infectious Diseases, where she remained until 1909.

In Chicago, Hamilton became a member of Hull House and donated her time and expertise helping immigrants, many of whom had contracted diseases from breathing toxic fumes created by one of the many industrial processes they were constantly exposed to, such as smelting lead. In 1908, in an unprecedented move, the governor of Illinois created a commission to survey work-site hazards, especially persistent ones such as long-term exposure to lead. In 1910, the governor appointed Hamilton to head the survey, which ended up documenting 578 cases of work-related lead poisoning in its publication of *A Survey of Occupational Diseases*. The state of Illinois followed up on these findings by passing laws mandating safer workplaces, regular medical examinations, and worker's compensation for work-related accidents and illnesses.

At an international conference on occupational accidents and illnesses in Belgium, Hamilton ran into the U.S. Commerce Department's commissioner of labor, Charles O'Neil, who asked her to take up the same project she was conducting on the state level and apply it to the national scope. She spent the next decade investigating occupational poisons for the U.S. Bureau of Labor Statistics. In 1919, Harvard University appointed her as its first woman professor, teaching industrial medicine in its School of Public Health. Though this represented a major step for Harvard, it was a small step at that, as the university neglected to promote her from the lowest rank of assistant professor throughout her 16 years on the faculty. However, she did carry enough sway to stipulate a half-time appointment, allowing her time to continue with the survey. After she published the national results, other states followed the example set by Illinois in passing occupational safety laws.

Hamilton published three influential books in her career: *Industrial Poisons in the United States* in 1925, *Industrial Toxicology* in 1934, and her autobiography, *Exploring the Dangerous Trades* in 1943. Her political opinion held increasing sway as she established herself as an important figure in America, and she used her bully pulpit to support birth control and oppose capital punishment and war. Hamilton died on September 22, 1970, at the age of 101.

Alice Hamilton, whose work caring for immigrants at Chicago's Hull House encouraged her to focus her career on improving the working conditions of industrial laborers. *(National Library of Medicine, National Institutes of Health)*

Hardy, Harriet
(1905–1993)
American
Pathologist

Harriet Hardy was a pioneer in the field of occupational medicine. Through her investigation of a respiratory disease among Massachusetts factory workers, Hardy discovered berylliosis—a human-made illness caused by the inhalation of particles of the metal beryllium. Hardy then applied her clinical expertise to the prevention and treatment of occupational diseases. In addition to establishing the National Beryllium Registry at Massachusetts General Hospital (MGH), she also founded (and for 24 years directed) the Occupational Medicine Clinic at MGH. Moreover, she was one of the first scientists to correlate the development of certain kinds of cancer with exposure to asbestos. In 1971, Hardy became the first woman to be granted a full professorship at Harvard Medical School.

On September 23, 1905, Harriet Hardy was born in Arlington, Massachusetts. After graduating from Wellesley College in 1928, she enrolled at Cornell University and received her medical degree in 1932. Intending to become a general practitioner, Hardy launched her career at Northfield Seminary in Massachusetts upon completing her residency at Philadelphia General Hospital.

Hardy's simple, small-town medical practice ended in 1939, however, when she accepted a position as a college doctor and researcher at Radcliffe College in Cambridge, Massachusetts. In 1940, she joined the staff at MGH and explored with Joseph Aubt the effects of lead poisoning. After she was named to the Massachusetts Division of Occupational Medicine in 1945, her first assignment was to investigate a respiratory disease that plagued factory workers in the Sylvania and General Electric fluorescent lamp factories in Lynn and Salem, Massachusetts. The workers at these plants were suffering from shortness of breath, rapid weight loss, and coughing. Hardy determined that the disease was berylliosis—caused by the workers' exposure to beryllium, which was used in the manufacture of fluorescent lights. She subsequently became a leading expert on beryllium poisoning.

To move her findings from the laboratory into the realm of public health, Hardy established the National Beryllium Registry at MGH. The new unit quickly became a model for tracking occupational hazards and developing guidelines for their control. In 1947, Hardy founded the Occupational Medicine Clinic at MGH, which she would direct until she retired in 1971. In this new capacity, she continued her research, studying mercury poisoning and anthrax exposure among farm workers. Hardy also worked with the Atomic Energy Commission in Los Alamos, New Mexico, to examine the effects of radiation on the human body, and she served as a safety consultant for the nation's first nuclear reactor. In 1949, she formed the Occupational Medical Service (now called the Environmental Medical Service) at the Massachusetts Institute of Technology (MIT) to establish further connections between industrial hazards and illness. In 1954, she became one of the first scientists to demonstrate a link between asbestos and cancer. Hardy broke new ground again in 1958 when she joined the faculty at Harvard Medical School as the first female associate clinical professor at the school. In 1971, she became a full professor, again a first for a woman at Harvard Medical School.

Hardy's numerous accomplishments were lauded by her peers. She was named Woman of the Year by the American Medical Women's Association in 1955. The author of more than 100 scientific articles, Hardy also cowrote the influential book *Industrial Toxicology* with ALICE HAMILTON in 1949. An outspoken critic of hazardous working conditions, Hardy's findings led to changes in workplace environments. After her discovery of berylliosis, for example, fluorescent light manufacturers found a harmless calcium compound to use in place of beryllium. In addition, her research on the harmful effect of benzene caused the U.S. government to reduce by half the amount of benzene used in industry. Hardy died of lymphoma on October 13, 1993. In 1999, MIT named a new building the Harriet L. Hardy Library in her honor.

Harris, Mary Styles
(1949–)
American
Geneticist

As director of the Sickle Cell Foundation of Georgia and as that state's director of genetics services, Mary Styles Harris

played an essential role in educating the public about sickle-cell anemia and the importance of early testing for the disease. A highly regarded geneticist, Harris launched her own consulting business—Harris & Associates—to advise companies involved in genetic engineering. She also used two scientific grants to produce innovative television programs on the relationship between science and medicine and on health issues affecting minorities.

Mary Styles was born in Nashville, Tennessee, on June 26, 1949. Her father was a physician who completed his medical degree shortly after her birth. When Styles was a few months old, the family moved to Miami, Florida, where her father established his practice. Her mother, Margaret, held a degree in business but stayed at home to care for the family. Styles's father died when she was only nine, and the family suffered financial hardship. Despite these difficulties, Styles excelled in high school.

Upon completing her secondary education, Styles enrolled at Lincoln University in Pennsylvania. One of the first women to enter this college, Styles graduated in 1971. Although friends and family believed she would pursue a medical degree, Styles instead opted to enter a graduate program, preferring to conduct research rather than treat patients. In 1972, she married Sidney Harris, an engineering student. Together, the newlyweds attended Cornell University, where Mary earned her Ph.D. in genetics in 1975 and Sidney received a doctorate in engineering.

In 1975, Mary Harris was awarded a postdoctoral fellowship from the National Cancer Institute to conduct research at the New Jersey University of Medicine and Dentistry on the chemical structure of viruses. After two years, Harris realized that she no longer wanted to carve out a career in grant-based research. Consequently, she accepted the position of executive director of the Sickle Cell Foundation of Georgia. Sickle-cell anemia is a chronic hereditary blood disease that disproportionately affects Africans and people of African descent, and causes red blood cells to become sickle shaped and dysfunctional. As director of the foundation, Harris was responsible not only for fund-raising and other administrative duties but also for implementing measures to prevent the spread of the disease. She gained particular recognition for encouraging genetic testing to detect the disease at birth. (If the illness is identified in an infant, it can be treated more successfully than if it is allowed to progress.) Beginning in 1979, while continuing to head the foundation, Harris also taught genetics at Morehouse College's medical school.

While working in these two capacities, Harris also sought new ways to shape public health policy and to influence popular opinion about health matters. For example, in 1979, she took an innovative approach to raise awareness about medical issues. After receiving a National Science Foundation grant, she used the funding to produce television documentaries about health and medicine.

Harris added to her responsibilities in 1982 when she was named director of Georgia's genetics services (she held this post concurrently with those at the Sickle Cell Foundation and Morehouse). Her work at the genetics services department enabled her to address the importance of genetic testing in Georgia and other states. Her most impressive accomplishment was to coordinate a seven-state genetic screening program for infants. During this hectic period, Harris had her only child.

In 1986, Harris left Georgia when her husband was offered a job at Claremont College in California. She resigned from her myriad positions and founded her own company in California. Harris & Associates advised companies that used genetic engineering on public relations matters. In the early 1990s, she again parlayed science grants into media projects in an effort to draw attention to health issues affecting African Americans and other minorities. One such project was *To My Sisters . . . A Gift For Life*—a 40-minute documentary about breast cancer in the African-American community. In 1987, she was named chief executive officer of BioTech Publications, which creates audiovisual educational materials on health topics. She remains at BioTech.

Harris's diverse career has done much to raise public awareness about health and disease—especially about the health problems facing African Americans, women, and other minorities. Her work at the Sickle Cell Foundation and Georgia genetics services helped advance the cause of genetic testing for sickle-cell anemia. A member of the American Public Health Association and the American Society of Human Genetics, Harris was also named one of *Glamour* magazine's Outstanding Young Women in 1980.

Harrison, Anna Jane
(1912–1998)
American
Chemist

Anna Harrison distinguished herself as the first woman president of the American Chemical Society and as the fourth woman president of the American Association for the Advancement of Science. She also rose through the ranks of academia much faster than most women, who were held back by discriminatory promotion policies. Harrison joined the Mount Holyoke faculty in 1945 as an assistant professor and attained a professorship within five years, whereas most women scholars waited up to 10 years between promotions. In testament to her scientific excellence, she received 20 honorary degrees.

Harrison was born on December 23, 1912, on a farm in Benton City, Missouri. Her parents were Mary Jones and Albert Harrison. She commenced her education in a one-room country school, where she returned as a teacher after graduating from the University of Missouri in 1933. After

two years of teaching, she returned to the University of Missouri to earn her master's degree in 1937 and her Ph.D. in physical chemistry in 1940.

The Sophie Newcomb College of Tulane University hired Harrison fresh out of graduate school as an instructor in chemistry, and within two years, she was promoted to an assistant professorship. She made a parallel move to the Holyoke faculty, remaining an assistant professor, but she quickly climbed to an associate professorship by 1947. She became a full professor in 1950, and a decade later, she chaired the department of chemistry for six years, until 1966. A decade after this, she was named the William R. Kenan Jr. Professor of Chemistry, an endowed chair she inhabited until her 1979 retirement into emerita status at Holyoke. The next year, she served as a distinguished visiting professor at the United States Naval Academy.

Harrison's field of research was vacuum ultraviolet spectroscopy. She focused more of her attention on the classroom than on the laboratory, though. She distilled her understanding of approaches to teaching chemistry in the 1989 textbook *Chemistry: A Search to Understand,* which she coauthored with her Holyoke colleague Edwin Weaver. They addressed as their audience students who were genuinely interested in chemistry but did not intend to enter the field professionally. Outside the realm of academia, Harrison sought to educate the general populace as well as lawmakers about scientific issues—to inform the former in their voting decisions and the latter in their legislative decisions.

Harrison received multiple recognitions of her scientific excellence, including the 1949 Frank Forrest Award from the American Chemical Society, a 1960 Citation of Merit from the University of Missouri, the 1969 Manufacturing Chemists Association Award in College Chemistry Teaching, the 1977 James Flack Norris Award for Outstanding Achievement in the Teaching of Chemistry, and the 1982 Chemical Education Award from the American Chemical Society. More significantly, alumnae from Mount Holyoke's class of 1968 voted her as one of the people who had had the greatest impact on their lives. Harrison died on August 8, 1998, at the age of 85. In honor of her life, Mount Holyoke College set up the Anna Harrison Fund for Faculty Research.

Harvey, Ethel Browne
(1885–1965)
American
Embryologist

Ethel Browne Harvey won renown when she discovered that sea urchin eggs, when separated into nucleated and enucleated parts by centrifuge, could reproduce, independent of both male and female components. This discovery, along with the publication of her classic text, *The American Arbacia and Other Sea Urchins,* secured a place for her in the history of science, though scholarship from the early 1990s suggested that she deserved to win a Nobel Prize for experiments conducted during her years of graduate research.

Browne was born on December 14, 1885, in Baltimore, Maryland. Her mother was Jennie Nicholson, and her father, Bennet Browne, was a physician. She attended the Bryn Mawr School in Baltimore and then Goucher College, where she earned her bachelor's degree in 1906. She proceeded to Columbia University, where she studied zoology to earn her master's degree in 1907. While pursuing her doctorate at Columbia, she taught science in private school and then worked as an assistant in biology at Princeton University. She also started "summering" at the Woods Hole Marine Biological Laboratory, which elected her into its corporation in 1909 while she was still a graduate student. At Columbia, she conducted her dissertation research on a cytological study of male gametes of the aquatic hemipteran *Noctonecta* under E. B. Wilson. She earned her Ph.D. in 1913.

Browne first spent a year as the Sarah Berliner Fellow at the University of California, then the next year as an assistant in histology at the Medical College of Cornell University. In 1916, she married Edmund Newton Harvey, a fellow biologist who specialized in bioluminescence. They spent their honeymoon in Japan, and she spent some of her time researching at the Misaki Marine Biological Station. While her husband held a steady position at Princeton University, Ethel Harvey conducted itinerant research around the world, researching at the Oceanographic Institute of Monaco in the early 1920s and then at the Naples Zoological Station in the mid-1920s. In the late 1920s, she landed back in the United States at Washington Square College of New York University as an instructor in biology.

Starting in 1931, she conducted independent research at Princeton University, which offered her laboratory space but not a salaried position (the only money she ever received was a 1937 grant from the American Philosophical Society). In 1950, Harvey became the first woman in half a century to be elected a trustee of the Marine Biological Laboratory and only the second woman ever. In 1956, she published her masterwork (and only book publication), *The American Arbacia and Other Sea Urchins.*

Harvey did her most famous work on sea urchins, whose eggs she spun in a centrifuge until they separated into nucleated and enucleated parts, which spontaneously regenerated. The scientific term for conception in this sexual void is parthenogenetic merogonyl, or lacking both male and female nuclei, thus lacking chromosomes and genes: an immaculate conception, so to speak.

Some of Harvey's work would have been even more famous had it been acknowledged. In 1991, cell biologist Howard M. Lenhoff published a discussion of Harvey's 1909 graduate paper on induction in hydra, whereby trans- plantation "induced" a secondary axis of polarity in the host. Hans Spemann and HILDE MANGOLD performed almost the identical experiment in 1924, announcing their results as the discovery of the "organizer" principle, whereby grafted tissue altered the growth of its host tissue to match its own growth. Spemann won the Nobel Prize for this discovery (Mangold had died before the presentation of the Nobel). Lenhoff presented compelling evidence that Spemann knew of Browne's 1909 experiment but did not acknowledge it. Harvey herself acknowledged her discovery of the organizer, as the phenomenon was called, according to anecdotal evidence from a Woods Hole colleague. Lenhoff interpreted the history of this episode as an indicator of the depth of prejudice exercised against women in the field.

Harvey spent her last years of research at the Woods Hole Marine Biological Laboratories, from 1959 to 1965. In 1956, Goucher College awarded her an honorary degree of Doctor of Science. She was also honored as a fellow of the Institut International d'Embryologie and an honorary membership in the Societa Italiana de Biologia Sperimentale. Harvey died on September 2, 1965, in Falmouth, Massachusetts.

Harwood, Margaret
(1885–1969)
American
Astronomer

A working scientist for more than 50 years, Margaret Harwood spent most of her career as the director of a major observatory. She logged thousands of hours in astronomical observation and supervised thousands more hours of observation made by students and colleagues. A pioneering woman scientist in a field dominated by men, Harwood did not always get the jobs she would have liked or the credit she deserved for her discoveries.

Born on March 19, 1885, in Littleton, Massachusetts, Margaret Harwood was the daughter of Herbert John Harwood and Emelie Augusta Green Harwood. She never married nor had children.

Harwood was an outstanding student in high school. After completing her secondary education, she enrolled in 1903 in Radcliffe College, the women's college of Harvard University. Harwood studied sciences as an undergraduate at Radcliffe, earning a B.A. from that college in 1907. Immediately following her graduation, she began to work doing astronomical computation at the Harvard College Observatory.

Harwood was one of a number of women scientists whom the Harvard Observatory's director, Edward C. Pickering, hired to do this exhaustive and unrewarding job. Other well-known women astronomers who worked at the Harvard Observatory included ANNIE JUMP CANNON and

WILLIAMINA FLEMING. This task often consisted of laborious examinations of photographic plates on which Harwood and other women astronomers would compile star positions and look for astronomical anomalies. They also examined photographs of star spectra. Pickering liked to boast that no male astronomer could do this work as well as the women because the males did not have the patience for numbing, detailed star indexing. Few of the women at the Harvard Observatory during this era made rapid career advances, and none of them became full members of the Department of Astronomy at Harvard.

Perhaps sensing that her career at the Harvard Observatory would be laden with frustrations, Harwood took a position as a research fellow at the MARIA MITCHELL Observatory in Nantucket, Massachusetts, in 1912. Named after the woman who is considered the first American woman astronomer, the Maria Mitchell Observatory was a place that suited Harwood. Here she was freed from an unvarying diet of drudgery and had considerably greater choice in her research topics. Around this same time, she began work on a master's degree in astronomy from the University of California at Berkeley, most of the actual study of which was done at the Maria Mitchell Observatory.

By 1916, Harwood had completed her M.S. in astronomy from the University of California. That same year, she was promoted to director of the Maria Mitchell Observatory. At the observatory, Harwood studied variable stars, that is, those stars whose brightness varies over time. She also kept watch for random objects such as asteroids that entered our solar system. In 1917, she discovered a new asteroid that had entered the solar system, named Asteroid No. 886. In an incident that shows the small regard in which women astronomers were held by their mostly male peers, Harwood was not credited with the discovery of this object. Instead, another astronomer, George H. Peters, was given credit for the discovery, even though he observed Asteroid No. 886 (later named Washingtonia) four days after Harwood had.

In spite of this professional discourtesy, Harwood went on to have an outstanding career. She was elected to the American Association for the Advancement of Science and the Royal Astronomical Union. In 1962, several years after she retired, Harwood received the Annie J. Cannon Prize in Astronomy from the American Association of University Women and the American Astronomical Society. Harwood died in Massachusetts in 1969 at the age of 84.

Hawes, Harriet Ann Boyd
(1871–1945)
American
Archaeologist

Hawes was the first woman to lead a large archaeological dig, organizing the 1901 excavation of a 3,000-year-old Cretan town. This find represented the first well-preserved urban site from the Minoan Age to be uncovered in Crete.

Boyd was born on October 11, 1871, in Boston, Massachusetts. Her mother, who was named Harriet as well, died when she was a baby, leaving her father, Alexander Boyd, who owned and ran a leather business, to raise her along with her four brothers. Boyd attended Smith College to study classics, receiving her bachelor's degree in 1892. She spent the next four years teaching classics in a private school before returning to school herself for four years of graduate study at the American School of Classical Studies in Athens, Greece, from 1896 through 1900. During this time, she also practiced her social advocacy by volunteering as a nurse in Thessaly during the Greco-Turkish War in 1897 and in Florida in the Spanish-American War in 1898.

This school disallowed women from participating in field research, ostensibly according to social conventions—this kind of protocol usually conspired to bar women from meaningful professional work at this time period. So Boyd boldly applied the remainder of her grant money, combined with her personal savings, to finance a trip to the island of

Harriet Boyd Hawes, whose 1901 excavation of a 3,000-year-old Cretan town was the first major archaeological dig led by a woman. *(Courtesy of Wellesley College Archives)*

Crete, which she undertook with a woman friend, traveling in Victorian dresses on mules and unaccompanied by a male. As the first American archaeologist to dig in Crete, she uncovered several Iron Age tombs at Kavousi, on the eastern side of the island. This fieldwork served as the data for her thesis, which she submitted to Smith College to earn her master's degree in 1901.

The American Exploration Society recognized the importance of Boyd's work by offering her a grant to return to Crete that same year. Boyd moved her dig to a nearby site called Gournia, where she unearthed evidence of a village predating the tombs she had discovered earlier that year at Kavousi. During this time, Boyd was teaching Greek archaeology and modern Greek at Smith. She secured sufficient financing to return to Gournia in 1903 and 1904 to finish uncovering the entire village. She supervised more than a hundred local workers who excavated their ancestors, also workers, thus offering for the first time a glimpse of everyday life from the Bronze Age in Crete. Boyd wrote up her findings in a book, *Gournia, Vasiliki and Other Prehistoric Sites on the Isthmus of Hierapetra, Crete,* which the American Exploration Society published in 1908.

While on this expedition in Crete, Boyd met Charles Henry Hawes, a British anthropologist. The couple married on March 3, 1906, and after Boyd's book appeared, they collaborated on another book, *Crete, the Forerunner of Greece,* published in 1909. They had two children together, Mary and Alexander, whom Boyd Hawes raised during a hiatus in her professional career. During World War I, she returned to her social activism by working in a Serbian army hospital camp on Corfu and in hospitals in France. She returned to her teaching career in 1920, when Wellesley College appointed her to a lectureship in pre-Christian art, a position she retained until 1936. In the 1930s, she supported the union movement in Massachusetts. She died on March 31, 1945.

Hay, Elizabeth Dexter

(1927–   )
American
Cell Biologist

Elizabeth Hay discovered in the early 1960s that epithelial cells could create collagen, a radical realization that was immediately challenged. She went on to help establish the field of extracellular matrix in the realm of cell biology in the 1970s. Besides contributing to her field through research, Hay has also contributed to it as the chair of Harvard Medical School's Department of Anatomy and Cell Biology for almost two decades, inspiring students and junior faculty as their mentor.

Hay was born on April 2, 1927, in St. Augustine, Florida. Her father, a doctor, became an army physician

Elizabeth Hay, who in the 1970s helped to establish the field of extracellular matrix in the realm of cell biology. *(Courtesy of Elizabeth Hay)*

during World War II, at which time the family moved to Mississippi and then to Kansas. Hay commenced her undergraduate career in 1944 at Smith College, where she studied biology under S. Meryl Rose, who became her mentor. Together, they conducted research on amphibian limb regeneration at the Marine Biological Laboratory in Woods Hole, Massachusetts, after her graduation summa cum laude from Smith in 1948. Rose advised Hay against pure research, because, for women at that time, more opportunities were open to an M.D. than a Ph.D.

Hay went straight to Johns Hopkins Medical School, graduating in 1952 and pursuing her internship at the Osler Service Hospital of Johns Hopkins. She remained there as an anatomy instructor before being promoted to an assistant professorship in 1956. The next year, she joined the Anatomy Department at the Medical College of Cornell University as an assistant professor. In the mid-1950s, Hay began applying electron microscopy to the study of cells, and at this same time, the *Journal of Cell Biology* and the American Society of Cell Biology (ASCB) were born in New York City, where she was working. Hay became involved in the ASCB immediately, and she later served as the organization's president.

In 1960, Hay moved to the Harvard Medical School as an assistant professor. Within four years, she was named the Louise Foote Pfeiffer Associate Professor of Embryology, and in 1969, she retained the title when she was granted a full professorship. She served as the chair of Harvard Medical School's Department of Anatomy and Cellular Biology from 1975 through 1993.

Hay's research focuses on cell migration, which she likens to Rome at traffic hour, with cells traversing the complex landscape of the developing embryo. What she finds fascinating is how cells know where to head, which depends not only on genetically programmed instructions but also on their reading of the surroundings. Hay's research interests spanned the multidisciplinary spectrum of topics and approaches to cell biology, including autoradiographic studies of nucleic acid and protein synthesis in embryos and regenerates; fine structure development of the muscle, skin, and eye; tissue interaction in the developing cornea; and immunohistochemistry of collagen. Another main interest of Hay's concerns embryonic epithelial-mesenchymal transformations, which she hypothesized could help to explain how cancer cells migrate similarly to embryonic cells. Hay edited the book *Cell Biology of Extracellular Matrix* in 1981.

In 1988, she received the first of several important awards: the Alcon Award for Vision Research. The next year, the ASCB honored her with its highest prize, the E. B. Wilson Award. In 1992, the American Association of Anatomists granted her the Henry Gray Award. Besides being a devoted scientist, Hay is also an avid mycophile, gathering up to 40 pounds of wild mushrooms for dinner parties.

Hay, Louise Schmir
(1935–1989)
French/American
Mathematician

Louise Schmir Hay was renowned for her research in mathematical logic, recursive function theory, and theoretical computer science. She also held the distinction of being the first woman to head a mathematics department at a research-oriented university in the United States. During her career, Hay won the admiration of both colleagues and students for her teaching ability.

Hay was born to Samuel and Marjem Szafran Szmir on June 14, 1935, in Metz, France. Hay's parents had emigrated to France from Poland. After her mother died in 1938, her father married Eva Sieradska Szmir within a few months. Because they were Jewish, the Szmirs were forced to spend World War II fleeing from the Nazis. They found refuge in Switzerland in 1944, remaining there for one year. In 1946, they moved to the United States and changed the family name to Schmir. Hay's father owned a delicatessen in New York City.

Although Hay evinced no affinity for mathematics as a child, a high school geometry teacher sparked her interest in the subject. During her senior year at William Taft High School in the Bronx, Hay won a prize in the Westinghouse Science Talent Search for a project on non-Euclidean geometry. After graduating as class valedictorian, Hay enrolled at Swarthmore College. She married John Hay at the end of her junior year. In 1956, she received her B.A. in mathematics, and then she followed her husband to Cornell University, where she studied mathematical logic. In 1958, she transferred to Oberlin College, where her husband had been offered a teaching position. She completed her master's thesis on infinite valued predicate calculus at Oberlin and was awarded her M.A. from Cornell in 1959.

After a one-year teaching stint at Oberlin, Hay worked at the Cornell Aeronautical Laboratory in Buffalo, New York. John Hay then took a faculty position in Boston, and Louise Hay again followed, teaching at a junior college for one year and at Mount Holyoke College for three. In 1963, she gave birth to her first son. A discussion with mathematician Hannah Neumann about the possibility of having both a meaningful career and children prompted Hay to pursue her doctorate. While her husband remained in Boston, Hay moved to Cornell University in 1963. She wrote her dissertation on recursion theory in 1964 and also had twin boys. She received her Ph.D. in 1965.

Hay stayed at home in 1964 to take care of her three small children, but by 1965, she had returned to work, teaching part-time at Mount Holyoke. From 1966 to 1967, she conducted research at the Massachusetts Institute of Technology on a National Science Foundation fellowship. Her marriage to John Hay ended in divorce in 1968, and that same year, she accepted an associate professorship at the University of Illinois at Chicago. In 1970, she married Richard Larson, a colleague in the mathematics department. She was promoted to full professor in 1975. Her work in the 1970s was focused on the classification of index sets of recursively enumerable sets. In addition to publishing several highly regarded papers on this topic, she also introduced the concept of the "weak jump." Cognizant of her own difficulties balancing family and career, Hay cofounded and was actively involved in the Association for Women Mathematicians, becoming the first speaker invited to address the group on a mathematical topic in 1974. Hay and Larson spent 1978 in the Philippines on Fulbright scholarships. In 1979, Hay was named acting head of Illinois's mathematics department, becoming the only female head of a research-oriented university mathematics department during this period. She served in this capacity until 1989. In addition to her research, Hay was noted for her teaching and commitment to her students.

Hay's colleagues recognized her contributions to her field. She was named secretary of the Association for Symbolic Logic in 1982 and was also appointed to the executive board of the Association for Women in Mathematics (AWM) from 1980 to 1987. After being diagnosed with cancer in 1974, Hay suffered a relapse in 1988. She died in Oak Park, Illinois, on October 28, 1989. In honor of her stellar career, the AWM established an annual award given in Hay's name to a woman who has made a significant contribution to mathematics education. The award is intended to pay tribute to Hay's life and work as a teacher, researcher, and administrator.

Hazen, Elizabeth Lee
(1885–1975)
American
Microbiologist

Elizabeth Hazen collaborated with RACHEL BROWN to discover a new drug, harvested from soil microbes, that cured not only human diseases but also proved effective in a host of other applications, including acting as a mold inhibitor to protect artwork and other important historical artifacts.

Hazen was born on August 24, 1885, in Rich, Mississippi. Her parents, William and Maggie Hazen, farmed cotton there until their untimely deaths during Hazen's childhood. An aunt and uncle raised her and her sister in Lula, a nearby town. Hazen attended the State College for Women (later named Mississippi University for Women) in Columbus, graduating in 1910. Hazen taught high school science until 1916, when she returned to her own education to study bacteriology at Columbia University, where she obtained her master's degree in 1917.

Hazen spent the next decade researching in the U.S. Army diagnostic laboratories in Alabama and in New York. She also returned to Columbia to pursue her doctorate in microbiology, which she received in 1927 at the age of 42. She proceeded to teach at the university's College of Physicians and Surgeons for the next four years, until the New York State Department of Health offered her a research position in 1931 in its Division of Laboratories and Research. There, Hazen specialized in identifying disease-causing fungi, becoming an expert in this underinvestigated field.

Hazen blanketed the country collecting soil samples in search of microbes that might prove useful in antibiotics. She would culture these samples in nutrient-rich petri dishes, which she sent to Albany, where fellow Department of Health researcher Brown worked, extracting the active agents from the microbes. She sent these back to Hazen at her New York City laboratory for further analysis. In 1948, Hazen collected a sample near Warrenton, Virginia, that turned out to contain a microbe with two fungus-killing agents, one of which was previously undiscovered.

In late 1950, Brown and Hazen announced their discovery of fungicidin, which they renamed *nystatin* in honor of their place of work, the New York State Department of Health. They patented this new drug, which ended up having a diversity of effective uses: to cure vaginal yeast infections; to prevent Dutch elm disease in trees; to kill molds on produce; and, most interestingly, to protect priceless artwork and historical documents from mold in the wake of the 1966 flooding of Florence. Instead of pocketing the proceeds, the two women established the Brown-Hazen Fund to administer the profits from their patented drug, funneling the money into grants and back into their own research.

In 1955, Hazen coauthored *Laboratory Identification of Pathogenic Fungi Simplified*. That year, she and Brown shared the Squibb Award in Chemotherapy, and they later shared the Sara Benham Award from the Mycological Society of America. Hazen retired from active research in 1960, but she continued to contribute to her field as a permanent guest investigator in the Mycology Laboratory at Columbia University. In 1968, the New York State Department of Health honored Hazen with its Distinguished Service Award.

Two decades after Hazen and Brown shared their first award, they shared another, the first Chemical Pioneer Award from the American Institute of Chemists. Hazen died on June 24 of that year, 1975.

Hazlett, Olive Clio
(1890–1974)
American
Mathematician

Olive Clio Hazlett was the most prolific woman mathematician before 1940, writing 17 research papers on an array of topics. Her area of expertise was linear algebra, and she focused on such concerns as division algebras, modular invariants, nilpotent algebras, and the arithmetic of algebras. Hazlett was frustrated by the dearth of advancement opportunities available to her in academics despite the high quality of her work. Like many other women mathematicians, she never reached the level of full professor and was often relegated to teaching large, introductory classes.

Hazlett was born in Cincinnati, Ohio, on October 27, 1890, though she grew up in Boston, where she attended the area's public schools. She never married or had children, perhaps because periods of her adulthood were marked by mental illness.

After earning her bachelor's degree from Radcliffe College in 1913, Hazlett pursued graduate work at the University of Chicago. She was awarded her master's degree in 1913 and her Ph.D. in mathematics in 1915. She wrote her dissertation—*On the Classification and Invariantive Characterization of Nilpotent Algebras*—under the guidance of renowned mathematician Leonard Dickson. Hazlett was only his second female doctoral student.

After receiving her Ph.D., Hazlett was named an Alice Freeman Palmer Fellow of Wellesley College, and she spent from 1915 to 1916 researching invariants of nilpotent algebras at Harvard University. In 1916, she accepted an associate position at Bryn Mawr College. She was hired as an assistant professor in the mathematics department of Mount Holyoke College in 1918. Although she was eventually promoted to an associate professor in 1924, she was dissatisfied with the environment at Holyoke, finding that her algebraic research was stymied by mediocre library facilities and a lack of time.

Hazlett left Holyoke in favor of an assistant professorship at the University of Illinois in 1925. By this time, her reputation as an outstanding mathematician was well established. When she applied for the job at Illinois, her former mentor Dickson called her "one of the two most noted women in America in the field of mathematics." In 1928, she was awarded a Guggenheim Fellowship. With this support, she spent two years studying in Italy, Switzerland, and Germany, producing a paper, "Integers as Matrices," from her research. She returned to Illinois in 1930 and was promoted to associate professor.

Although Hazlett found that her research opportunities had improved at Illinois, she remained unhappy with her teaching assignments. She wrote a letter to the chair of the department in 1935, complaining that she was forced to teach large, introductory level classes, while her younger, less experienced male counterparts were given smaller graduate seminars. Her unhappiness spiraled into a series of mental breakdowns in the 1930s and 1940s from which she never recovered. She was on administrative leave from the university from 1936 to 1938, returning to teach in the early 1940s. (During World War II, she was given work by the U.S. Signal Corps.) In 1946, her condition worsened. She remained on administrative leave until 1959, when she officially retired from the university.

Hazlett was a prolific writer and researcher, having written 17 important papers—all before 1930. That her peers respected her work is evident in her active involvement in professional societies. She was a member of the Council of the American Mathematical Society from 1926 to 1928 and served as the associate editor of its journal—*Transactions of the American Mathematical Society*—from 1923 to 1935. She was also elected to the American Association for the Advancement of Science, the American Mathematical Society, and the New York Academy of Sci-

ences. She died on March 8, 1974, at her home in Peterborough, New Hampshire.

Healy, Bernadine
(1944–)
American
Cardiologist

The first woman to head the National Institutes of Health (NIH), Bernadine Healy has been at the forefront of shaping medical policy and research in the United States. In addition to a stellar academic career and her post at NIH, she served as president of the American Heart Association (AHA) and as deputy science advisor to President Ronald Reagan. Healy has concentrated particularly on women's health issues, notably by instituting groundbreaking research into women's heart disease and changing NIH policy to require drug trials to include both men and women (previously, most tests had focused exclusively on men, even if the drugs in question were intended for broader use).

Born on August 2, 1944, in New York City, Healy was the second daughter of Michael and Violet (McGrath) Healy. She spent her childhood in Queens, New York, where her parents ran a small, home-based perfume business. She expressed an early interest in medicine, and she was encouraged in this direction by her father, a strong proponent of equal opportunities for women. He enrolled her at the academically competitive Hunter College High School, from which she graduated first in her class.

Healy attended Vassar College and earned her B.A. in chemistry in 1965. She entered Harvard Medical School as one of only 10 women in a class of 120, and she experienced intense sexism from classmates who berated her for "using up" a spot better filled by a man. She persevered and received her M.D. cum laude in 1970, completing her internship and residency at the Johns Hopkins Hospital in Baltimore, Maryland.

After a two-year stint at NIH's National Heart, Lung, and Blood Institute, Healy returned to Johns Hopkins in 1976. She served there as a professor of medicine and cardiology and also undertook her first administrative duties—becoming the director of the coronary care unit in 1977 and the assistant dean for postdoctoral programs and faculty development in 1979. During this period, Healy married George Bulkley, with whom she had a daughter. The couple divorced in 1981. In 1985, she wed cardiologist Floyd Loop and had another daughter with him. Healy left Johns Hopkins in 1984 when President Reagan appointed her deputy director of his Office of Science and Technology. In 1985, she accepted the directorship of the Cleveland Clinic Foundation, overseeing a research budget of $36 million. While retaining that post, Healy also served as president of the American Heart Association from 1988 to

1989, where she spearheaded a major effort to research heart disease in women.

In 1991, President George Bush named Healy director of the sprawling NIH, an agency encompassing 13 research institutes, 16,000 employees, and a research budget exceeding $9 billion. At the time, NIH was plagued by a mass exodus of researchers and was also embroiled in political controversies (such as whether to use fetal tissue for research). In response, Healy instituted the Shannon Awards—grants designed to keep researchers active during funding lapses—and "town meetings" where NIH scientists could pinpoint problems. Healy also remained focused on women's health issues. She launched the $625 million Women's Health Initiative, which studied diseases afflicting women. Moreover, she dictated that NIH would only fund clinical drug trials that included both men and women (if the drug was intended to treat both genders). Healy also oversaw research concerning the effects of vitamin supplements, diet, and hormone replacement therapy on women between the ages of 49 and 79.

Healy left NIH in 1993 and was named dean of the College of Medicine and Public Health and professor of medicine at Ohio State University in 1995. Under her leadership, the college expanded its cancer research, established its Heart and Lung Institute, and became officially recognized as a Center of Excellence in Women's Health. In 1997, Healy also accepted the chair of the Ohio State University Research Commission—a task force reviewing the entire university's research. She remains at Ohio State.

Healy's impact on medical research in the United States has been profound. She has received numerous awards, including the 1992 Dana Foundation's Distinguished Achievement Award and two AHA special awards for service. She is a member of the Institute of Medicine of the National Academy of Sciences, serves on the editorial boards of several leading medical journals, and has authored or coauthored more than 220 articles, as well as two books intended for a lay audience—*Staying Strong and Healthy from 9 to 99* (1995) and *A New Prescription for Women's Health* (1996). She has continued to treat patients throughout her career.

Heloise
(c. 1098–1164)
French
Philosopher and Physician

A legendary figure in medieval France, Heloise is credited with being a knowledgeable herbalist/physician as well as a philosopher. However, there are no historical documents that verify that she actually practiced medicine, and the little we know of her rests mostly on the records of her relationship with the scholar and philosopher Peter Abelard and the few surviving letters she wrote to him.

Born in or near the year 1098 somewhere in northern France, Heloise was sent as a teenager to live with her uncle, Fulbert, the canon of the Notre Dame Cathedral in Paris. Around 1116, when Heloise was 17, her powerful uncle arranged to hire as her private tutor the respected scholar Peter Abelard, who then was in his late thirties. Judging from the position held by her uncle and the fact that she was offered an education, Heloise's family must have been wealthy and influential. She must have also possessed considerable intellectual skills, because it was unusual for a woman, even one from a wealthy family, to be given an education at that time.

By 1116, Abelard had made an international reputation for himself as one of the most brilliant philosophers of his day. He had earlier studied in Paris with William of Champeaux and at the cathedral school of Laon with the respected theologian Anselm. He had taught at several centers of learning in France, and in 1113, he returned to Paris to teach at the cathedral school of Notre Dame. It was at this school that he met Fulbert, who hired him to be Heloise's tutor.

Peter and Heloise quickly became infatuated with one another and soon began an affair that resulted in Heloise's pregnancy. She was sent by her family to a convent in Brittany to give birth to a son, named Peter after his father but usually referred to as Astralabe. After the birth of Astralabe, Abelard offered to marry Heloise. But according to his memoirs, *Historia calamitatum,* written later in his life, Abelard was at first refused by Heloise. He persisted, and the couple eventually married in a secret ceremony. In 1118, finding out about the secret marriage, suspecting it was not legitimate and also suspecting that Heloise was about to be abandoned, Fulbert hired a pair of thugs to attack and castrate Abelard.

Publicly humiliated by this act, Abelard retired to the monastery of St. Denis. He arranged for Heloise to become a nun at the nearby convent of Argenteuil. Heloise remained a nun at Argenteuil for 11 years, while Abelard, a contentious man who possessed a knack for alienating his protectors, wandered from monastery to monastery. In 1129, when all of the nuns of the Argenteuil were expelled from their nunnery, Heloise, with Abelard's help, set up a nunnery on property Abelard owned in Paraclete, near the town of Nogent-sur-Seine.

At Paraclete, Heloise became first prioress, then abbess of the nunnery. Her superior education, as well as shrewd understanding of human motivations, certainly must have made her an able administrator. It was at Paraclete that Heloise acquired her reputation as a physician, for she is reputed to have healed sick nuns there many times.

Not much is heard of Astralabe after his birth, and it is unclear if he too lived for a time at Paraclete. In the 1140s, Heloise asked for help in getting him an ecclesiastical position. Peter Abelard died around the year 1142, and

Heloise lived for 22 more years, dying on May 15, 1164. At her death, by her request, she was buried next to Abelard in the cemetery at Paraclete. Their bodies were reinterred next to each other in 1817 in Père Lachaise Cemetery in Paris.

Herrad of Landsberg
(c. 1135–1195)
German
Herbalist, Astronomer, and Philosopher

A noted herbalist and physician, Herrad of Landsberg was also a scholar who possessed broad and deep learning. She was the capable administrator of a large convent of nuns, but she is best known for the encyclopedia she wrote, *Hortus deliciarum* (The Garden of Delights), which summarized a lifetime of study and thought about medicine, botany, religion, history, astronomy, and philosophy.

Probably born in the French-German border territory of Alsace sometime around the year 1135, Herrad was sent at an early age to the convent of Hohenburg on Mount St. Odile in Alsace. Nothing is known of her family, but the young age at which she found herself in a monastery indicates that she was not a primary member of a wealthy family. She may have been an orphan, or it is also possible that she was the illegitimate child of a well-connected man.

Whatever the circumstances of her life, Herrad attracted the attention of the nuns at Hohenburg. The sisters must have detected considerable intelligence in Herrad, because she was given a thorough education at the monastery. By 1860, she had become a teacher herself and was giving instruction in a number of subjects at the monastery. She began writing the *Hortus deliciarum* in about 1160 and probably finished it about 1170. During this period (in 1167), Herrad became abbess of Hohenburg.

Herrad's *Hortus deliciarum* was the first encyclopedia ever compiled by a woman. The encyclopedia form was popular in the classical period and the Middle Ages. Three encyclopedias were compiled during Roman times—those of Varro, Celsus, and Pliny. Of these, only copies of Pliny's work survived into the Middle Ages.

Roman encyclopedias were originally meant as texts to train youth and included the subject matter—the seven "liberal arts" of grammar, dialectic, rhetoric, geometry, arithmetic, astronomy, and music—that were thought appropriate for a well-rounded education. Later, Pliny expanded the realm of subject matter included in encyclopedias by adding bits and pieces of information outside the liberal arts that interested him.

With the conversion of the Roman Empire to Christianity, early Christian scholars such as St. Augustine urged writers to compile new encyclopedias that would include discussion of symbols and parables of the Bible. In around 630, Isidore of Seville wrote an uncompleted Christian en-

cyclopedia, but it was not until the time of Herrad that writers returned to this task again. Slightly before Herrad began her project, Lambert of St. Omer wrote his *Liber floridus* (1120), and around 1130, Hugh of St. Victor wrote his *Didascalion*.

Herrad's *Hortus deliciarum* covers strongly religious themes. It has sections that are devoted to the rituals and daily rhythms of the Hohenburg convent, and it also offers spiritual advice for the nuns and novices who lived there. Herrad was also very concerned with feast days. She put together a table that laid out the exact day for Easter and Christmas week 532 years into the future (this table of feast days ends in 1706).

Also important in Herrad's work was a listing and explanation of herbal medicines. She probably borrowed sections that had been developed by the famous medical school at the University of Salerno in Italy, but she also included much information that was specific to the region in which she lived. The work is richly illustrated with drawings that show the plants being discussed.

In 1187, Herrad oversaw the construction of a large hospital that was built on the grounds of the Hohenburg convent. She ran this institution, which ministered to sick nuns and lay people, until her death at Hohenburg on May 15, 1195.

Herschel, Caroline Lucretia
(1750–1848)
English
Astronomer

Caroline Herschel lived in the shadow of her brother, William, King George III's royal astronomer. While not assisting him, she managed to distinguish herself as an important astronomer in her own right, discovering eight comets and compiling catalogs that brought order to the night sky.

Herschel was born on March 16, 1750, in the city of Hanover in what is now Germany. Her mother, Anna Ilse Moritzen, opposed education for her daughters except in domestic arts, while her father, Isaac, an oboist in the Hanovarian Foot Guards, surreptitiously encouraged her musical abilities and thirst for knowledge, introducing her to astronomy by showing her a visible comet when she was young. During her teenage years, her mother treated her as a kind of household slave. She escaped from under her mother's yoke in 1772 when William, by now an organist in Bath, England, summoned her and provided their mother with a replacement servant.

William trained his sister as a singer and, more importantly, as an astronomer. She helped him make his own telescopes, pounding and sieving horse manure for molds to hold the mirrors that he hand polished, once for 16 hours straight while Caroline fed him. In 1779, they commenced an ambitious project of mapping all the lights in the night sky, with William making the methodical telescopic "sweeps" of the sky and Caroline recording his observations. The next day, she made clean copies of her notes and calculated the positions of the observed objects, leaving her scant time for sleep. This hard work paid off in 1781 when William discovered a new planet, which he named Uranus in honor of King George, who rewarded him with a yearly salary of £200 as the royal astronomer.

In April 1786, the Herschels moved to a residence in Slough that could house all their telescopic equipment, appropriately called the Observatory House. Four months later, on August 1, 1786, while William was away and Caroline had independent access to the telescope, she discovered a new comet, later dubbed the "first lady's comet." The Royal Society published her letter of announcement, bringing her some notoriety, most notably from King George, who granted her a yearly income of £50, the first money she felt her own to dispose of at her will. At about this same time, William married Mary Pitt, and Caroline repaired from the residence to her own quarters, which she resented at first, but later she grew to appreciate as it offered her independence. Over the next decade, she discovered seven more comets.

Herschel used her cataloging skills, gained from noting her brother's observations, to repair the discrepancies between the catalog Flamsted published in 1725 and this first royal astronomer's own unpublished notes. In 1798, the Royal Society published her *Index to the Catalogue of 860 Stars Observed by Flamsted but Not Included in the British Catalogue,* including 560 previously uncataloged stars. She spent the remainder of her career mentoring her nephew, William's son John, who followed in his father's and aunt's footsteps to become an astronomer. Together, John and Caroline observed 2,500 nebulae that she cataloged on her own and sent back to him in 1825. Three years later, he submitted this list to the Royal Astronomical Society, which responded by awarding her a gold medal. In 1835, the Royal Society inducted her as an honorary member, one of the first women so honored. A decade later, on her 96th birthday, she got word from the King of Prussia that he had awarded her his gold medal for science. Perhaps the most appropriate recognition of her life occurred posthumously, after her January 9, 1848 death: In 1889, a small planet was named Lucretia in honor of her contributions to astronomy.

Herzenberg, Caroline L.
(1932–)
American
Physicist

During her varied and distinguished career, Caroline L. Herzenberg has gained recognition for her research into the Mössbauer effect. She also analyzed the first lunar samples from the Apollo missions and developed instruments for fossil fuel studies. In addition, Herzenberg has dedicated herself to raising awareness about the accomplishments of women scientists. Her critically acclaimed book, *Women Scientists from Antiquity to the Present,* provides biographical and bibliographic information about more than 2,500 women.

Born on March 25, 1932, in East Orange, New Jersey, Caroline Stuart Littlejohn was the daughter of Charles and Caroline Littlejohn. In 1961, she married Leonardo Herzenberg, with whom she had two children.

Herzenberg attended the Massachusetts Institute of Technology after she won the Westinghouse Talent Search in high school. She earned her bachelor's degree in 1953 and then pursued graduate work at the University of Chicago, receiving her master's in 1955 and her Ph.D. in physics in 1958.

Upon completing her studies, Herzenberg remained at the University of Chicago as a postdoctoral fellow from 1958 to 1959. During this year, her work focused on measuring the products of nuclear reactions between lithium isotopes and isotopes of beryllium and boron, research that laid the foundation for future investigations into heavy ions. From 1959 to 1961, Herzenberg was a postdoctoral fellow at the Argonne National Laboratory in Argonne, Illinois. It was here that she launched her pioneering work concerning Mossbauer spectroscopy. Of particular note, she and her colleagues at Argonne succeeded in verifying the existence of the Mossbauer effect. The Mossbauer effect is the name given to the phenomenon whereby an atom in a crystal does not recoil when it emits a gamma ray but instead transfers all the emitted energy to the gamma ray (resulting in a sharply defined wavelength).

In 1961, Herzenberg was named assistant professor of physics at the Illinois Institute of Technology (IIT). She continued to explore the Mossbauer effect as well as experimental low-energy physics using a Van de Graaff accelerator. After establishing a Mossbauer-effect research facility at IIT, she turned her attention to the geological applications of the effect. Her work on the spectra of different rock types led her to conclude that it would be possible to analyze rocks and minerals from lunar and planetary surfaces using Mossbauer spectroscopy. In light of this discovery, she was awarded a grant from the National Aeronautics and Space Administration to examine the lunar samples from the Apollo missions.

Herzenberg's academic career stalled in 1967, however, when she was denied tenure at the University of Chicago. For the next 10 years, she held a number of consultant and temporary positions. From 1967 to 1970, she was a research physicist at the IIT Research Institute. In 1970, she was promoted to senior physicist at the Research Institute but left in 1972 for a visiting associate professorship at the

University of Illinois Medical Center. After a stint as a visiting professor at California State University at Fresno from 1975 to 1976, she returned to the Argonne National Laboratory as a physicist in 1977. Her work at Argonne has ranged across a number of topics, including a fossil-energy instrumentation program that sought to develop instrumentation for process control of coal conversion and combustion plants. During the 1980s, she also studied radioactive waste disposal, technology for arms control verification, and fossil energy utilization.

Another facet of Herzenberg's career has been her ongoing efforts to collect information about women scientists. In addition to publishing several articles on the historic role of women in science, she authored *Women Scientists from Antiquity to the Present* in 1986. The book made available a wealth of hitherto scattered bibliographic information on many women scientists. Herzenberg's numerous accomplishments have won her the respect of her peers. She was elected a fellow of the American Association for the Advancement of Science and the American Physical Society. She also served as president of the American Association of Women in Science from 1988 to 1990. In 1989, she became the first scientist to be inducted into the Chicago Women's Hall of Fame.

Hewitt, Jacqueline N.
(1958–   )
American
Astrophysicist

Jacqueline Hewitt holds the distinction of being the first person to discover Einstein rings. Although Albert Einstein's theories predicted the existence of these ring-shaped objects produced by distant galaxies, they remained undetected until Hewitt's breakthrough. Einstein rings are an example of gravitational lensing—the phenomenon in which electromagnetic radiation (including light) is affected by its passage through a gravitational field. Hewitt's finding is significant because Einstein rings can be used to measure the size of galaxies and the volume and percentage of dark matter in the universe.

Born on September 4, 1958, in Washington, D.C., Hewitt is the daughter of Warren and Grace (Graedel) Hewitt. She married nuclear physicist Robert P. Redwine in 1988, with whom she had two children—Keith (born in 1988) and Jonathon (born in 1993). She attended Bryn Mawr, graduating magna cum laude with a degree in economics in 1980.

Hewitt chose to pursue graduate work in science because of an astronomy class she took during her sophomore year at Bryn Mawr. After enrolling at the Massachusetts Institute of Technology (MIT), she found that a command of physics was key to understanding astronomy. With her professor Frank Ockenfels, she launched her investigation of gravitational lensing and collected data using the Very Large Array (VLA) radio telescope (located near Socorro, New Mexico). She opted to use a radio telescope rather than an optical one because it is extremely difficult to detect the optical images of gravitational lenses. By contrast, gravitational lenses emit large amounts of energy at radio wavelengths and are thus easier to discern through that method. Hewitt was awarded her Ph.D. in physics in 1988.

After completing her doctorate, Hewitt accepted a postdoctoral fellowship with the Very Long Basineline Interferometry unit at MIT from 1986 to 1988. While analyzing the data she collected with the VLA in New Mexico, she observed a distinct ring on her computer screen. It was an Einstein ring, which she was not expecting to find in her more general search for gravitational lenses. Since Hewitt's groundbreaking discovery, several other Einstein rings have also been found.

In 1988, Hewitt joined the research staff at the Department of Astrophysical Sciences at Princeton University. She kept this position for only a year and then returned to MIT as an assistant professor of physics. She was promoted to associate professor there in 1994. While fulfilling her teaching and research obligations, Hewitt also serves as the principal investigator for the Radio Astronomy Group of the Research Laboratory of Electronics at MIT.

Hewitt remains on the faculty of MIT. She has received a number of awards for her unique accomplishments in the field of astrophysics, and she won the David and Lucile Packard Fellowship in 1990. For her contributions to the study of gravitational lenses, her colleagues at MIT nominated her for the 1995–96 Harold E. Edgerton Award. In 1995, she was the recipient of the MARIA GOEPPERT-MAYER Award. The significance of Hewitt's work is profound. Einstein rings have the potential to allow physicists to determine the size of the universe and to answer questions such as whether the universe will keep expanding or will eventually collapse on itself.

Hibbard, Hope
(1893–1988)
American
Zoologist

Hope Hibbard specialized in cytology as a professor of zoology at Oberlin College for some 34 years. She also focused her efforts on advocating the role of women in academia and especially of women in research. It was her belief that women could rise to the top of their chosen field and that women compounded the discrimination suffered at the hands of men with their own self-imposed limitations, choosing marriage over their own careers. She urged women to try to combine work with marriage.

Hibbard was born in 1893. For her undergraduate study, she attended the University of Missouri, where she earned her bachelor's degree in 1916. She started her graduate work there, receiving her master's degree in 1918, but then transferred to Bryn Mawr College to continue on her doctoral work. She simultaneously worked there as a demonstrator in biology at the college and earned her Ph.D. in zoology in 1921. Upon receiving her doctorate, she became an associate professor at Elmira College from 1921 through 1925. She then received the Sarah Berliner Fellowship until 1926, when she traveled to France, where she worked as a preparateur in the comparative anatomy techniques laboratory of the University of Paris. She remained there as an International Education Board fellow and earned her D.Sc. degree in zoology from the Sorbonne in 1928.

She then commenced her long-standing relationship with Oberlin College, a small, private liberal arts school in Ohio with a reputation for innovative curricula. She started there as an assistant professor of zoology in 1928, and within three years, the college had promoted her to an associate professorship. Within another three years, she had reached the rank of full professor. Hibbard rose through the ranks of academia exceedingly fast; during this time period, women usually languished at the same rank for years, receiving a promotion only at the longest possible intervals allowed by institutional statutes, if at all. However, Hibbard not only rose quickly to the rank of full professor, but also she served as the department chair in zoology from 1954 through 1958. In 1952, she was appointed to an endowed chair as the Adelia A. Field Johnston Professor of Zoology. She also served as a trustee at the Woods Hole Marine Biological Laboratory.

Hibbard conducted her research on marine biology, invertebrate animals, and the structure of cells. Her publications addressed the specific topics of the tissues and organs of limpets, squid, earthworms, and silkworms, as well as on the Golgi apparatus. Hibbard also focused her attention on the status of women in academia as well as in society at large. She was an active member of the American Association of University Women (AAUW) and wrote a paper on the organization in 1933. She wrote numerous other papers on women's issues, including "Vocations for Women and How College Can Prepare Them," in 1935, and "The Life of Oberlin Women Today," in 1937. She was also a charter member of the Oberlin chapter of the League of Women Voters.

Hibbard presented a paper entitled "Women in Research" at the Ohio State University's Vocational Conference in November 1937. In the paper, she urged women not to abandon research careers in favor of marriage, which damaged the credibility of all women researchers by

Renowned zoologist Hope Hibbard at work in her Oberlin College Laboratory. *(Oberlin College Archives, Oberlin, Ohio)*

promoting the belief that women are not reliable as long-term researchers, but to attempt to combine the two. She pointed out that women could make it to the top of their fields, citing several examples of women who were sufficiently intelligent and hardworking. She suggested that women in many ways created their own barriers to professional upward mobility and that it was time to get outside the paradigm whereby women self-imposed limitations.

Hibbard became an honorary life member of the AAUW in 1987. She died in 1988, and Oberlin College established the Hope Hibbard Memorial Award in her honor.

Hildegard of Bingen
(1098–1179)
German
Healer

Hildegard was born in 1098 in Germany into a noble family, but as the 10th child, in accordance with custom, her life was tithed to the church. Appropriately, she started having visions at the age of three, but she kept them secret in self-defense. At the age of eight, she commenced her religious education at the hand of Jutta, an anchoress who devoted her life to solitary existence in contemplation of God. Jutta's only human contact came in the form of her teaching of rudimentary skills, such as reading the Psalter in Latin, to a handful of girls from noble families. Hildegard remained under Jutta's tutelage until the death of her mentor when Hildegard was 38 years old. The Benedictine monastery at Disibodenberg chose to transform the anchorage, attached as it was to the church, into a convent, and the monks elected Hildegard head of the convent. In 1141, Hildegard had a vision of God, and she dedicated her life thereafter to the recording of her visions as a means of honoring God. She even received a papal imprimatur sanctioning her writings as truly inspired by God. In 1150, she moved the convent to Bingen and later established a sister convent, named Eibingen, across the Rhine River.

That year, Hildegard wrote two books, *Physica* and *causae et curae,* together known as *Liber subtilatum,* or "The Book of Subtleties of the Diverse Nature of Things." In these texts, which did not bear the same mark of divine inspiration as her other works but rather seemed inspired by Hildegard's experience, she discussed natural history and the curative effects of natural substances. Hildegard based her medicinal philosophy on ancient Greek tenets dividing the world into four elements (fire, air, water, and earth) that corresponded to the four bodily humors (choler, or yellow bile; blood; phlegm; and melancholy, or black bile). Hildegard's cures sought to restore balance between the humors by introducing natural substances that would promote a return to equilibrium. For example, Hildegard characterized the flower tansy as hot and slightly damp, thus corresponding to the elements of fire and water; she prescribed tansy to cure catarrh and coughs by curbing the flow of the phlegmatic humors.

Hildegard's writings spanned a wide variety of topics, touching on issues as diverse as the place of humans in the divine cosmology as well as the physiological functions of human anatomy. It is believed that Hildegard was the first to describe a female orgasm, which she accorded a very positive view in breaking with the Christian tradition of demonizing female sexuality.

Modern scrutiny of Hildegard's writings revealed a correlation between her descriptions of her visions and the conditions endemic to migraine sufferers. Hildegard experienced visions of intense light, seeming like "extinguished stars," followed by periods of sickness, paralysis, and blindness, which then gave way to a rebounding sense of euphoria. These conditions correspond exactly to the experience of migraine headaches. What distinguishes Hildegard from most migraine sufferers is how she interpreted her experience, taking a positivistic view of her visions as gifts from the heavens instead of as hellish torture.

The 900-year anniversary of Hildegard of Bingen's birth was 1998, and there were celebrations commemorating this event at Wellesley College and the University of California at Los Angeles as well as conferences and symposia in Burlington, Vermont, and in Mainz, Germany.

Hildegard died in 1179, and though she has been beatified, she has not yet been canonized.

Hill, Dorothy
(1907–1997)
Australian
Geologist

Dorothy Hill's long and varied career was significant in several respects. Her research on fossil corals contributed greatly to knowledge about their structure and morphology. She also recognized the need for Australian geologists to have a broader understanding of their continent's stratigraphy (the geological study of the stratified rocks that compose the earth's crust), and she was instrumental in mapping the geology of the Australian state of Queensland. She used this data to provide an updated interpretation of the continent's geology. A prolific author, Hill wrote numerous articles on topics ranging from Paleozoic corals to the origins of the Great Barrier Reef to the history of the University of Queensland's geology department.

On September 10, 1907, Dorothy Hill was born in Brisbane, Australia, to Robert Sampson and Sarah Jane (Kingston) Hill, the third of her parents' seven children. Hill's father worked in a large store. Hill's early education, at the Brisbane Girls' Grammar School from 1920 to 1924, introduced her to science. She would never marry nor have children.

Hill's initial plans to study medicine in Sydney were thwarted by her family's lack of funds. Fortunately, she won one of 20 entrance scholarships to the University of Queensland. She intended to major in chemistry but switched to geology after a class with Professor H. C. Richards. Hill graduated in 1928 with a first class honor's degree in geology. She also became the first woman to win the university's Gold Medal, given annually to the most outstanding student. She then won a Foundation Traveling Scholarship to the Sedgwick Museum (geology department) at the University of Cambridge in Britain, where she received her Ph.D. in geology in 1932.

Upon completing her doctorate, Hill spent two more years at Cambridge as a research fellow, and she published two papers on various coral features. She returned to Australia and the University of Queensland as a research fellow in 1937, where she would remain for the duration of her career. Her first project in Australia involved basic taxonomic work, as she labored to classify the known coral fauna (fauna are the animals of a specified region or period) to gain a greater understanding of Queensland's wide-ranging stratigraphy. In 1943, she published a paper reinterpreting the Australian Paleozoic record. During this period, she also collaborated with oil companies that were mapping and drilling in central Queensland. Hill believed that by gaining a fuller understanding of the region's geology—and thus being able to determine whether oil was present—she could help her geology students find future work in industry.

World War II interrupted Hill's work. From 1942 to 1945, she served in the Women's Royal Australian Naval Service ciphering and coding transmissions. In 1946, she was named a lecturer in geology at the University of Queensland, followed by promotions to reader in 1955 and research professor in 1959. During this period, Hill continued to make headway in collecting and systematizing paleontological and geological information. After much effort, Hill completed a volume on coelenterates for the *Treatise on Invertebrate Paleontology* in 1956. (She wrote a second volume on Archaeocyathida for the treatise in 1972.) She also devoted herself to her overarching goal of creating a comprehensive geology of Queensland that would provide detailed information about the fossilized strata of the region. In keeping with this goal, she edited the *State Geology* monograph published by the Geological Society of Australia in 1960. Hill officially retired in 1972, though she continued to make daily trips to the university for the next 15 years. In 1978, she completed her massive study, *Bibliography and Index of Australian Paleozoic Coral.*

Hill received numerous awards and honors for her diverse contributions. In 1956, she became the first woman to be elected a fellow of the Australian Academy of Science, and in 1970, she became the first woman to serve as the organization's president. She won the Lydell Medal from the Geological Society of London in 1964 as well as the W. R. Browne Medal and the Mueller Medal. She is credited not only with conducting pioneering research into fossil coral and fauna but also with elevating the status of the study of geology in Australia. A constant supporter of her students, Hill sought to improve science education in Australia and to keep the country's finest scientists at home. She died in Brisbane on April 23, 1997. The chair in paleontology and stratigraphy was named in her honor as was the Dorothy Hill Library at the University of Queensland.

Hobby, Gladys Lounsbury
(1910–1993)
American
Microbiologist

Gladys Lounsbury Hobby played a central role in the development of antibiotics. Although Andrew Fleming discovered penicillin in 1928, he was unable to produce sufficient quantities of the antibiotic for it to be of use to patients. It was left to Hobby, as part of a research team at Columbia Medical School, to bring penicillin from the laboratory to the wider world. Her team was the first to cure a patient with a penicillin injection, and she saved countless lives by helping to solve the problem of how to make penicillin in large quantities. Later employed at Pfizer Pharmaceuticals, she aided in the discovery of other antibiotics, including Terramycin, Biomycin, and streptomycin. Her book, *Penicillin: Meeting the Challenge,* detailed the intense and collaborative efforts that ultimately led to the mass production of penicillin.

On November 19, 1910, Hobby was born in New York City. After receiving her bachelor's degree from Vassar College in 1931, she attended Columbia University, earning her Ph.D. in bacteriology in 1935. From 1934 to 1943, Hobby was engaged in the effort to perfect penicillin as an antibiotic for widespread use. She was a member of a research team at Columbia Medical School that included Karl Meyer and Martin Henry. During the same period, she also worked at Presbyterian Hospital in New York City.

The bacteria-fighting properties of penicillin had been recognized since 1928 when Andrew Fleming discovered that the mold *Penicillium* excreted a liquid that killed microbes. Unfortunately, Fleming could not manufacture enough of the substance to explore his finding further. His groundbreaking work was dismissed as a laboratory curiosity until the 1930s when Howard Florey and Ernst Chain created the first version of penicillin after Fleming. However, they too could not produce a sufficient quantity of the antibiotic to use effectively on humans. Hobby's team at Columbia set out to accomplish this goal. After becoming the first to cure a patient with a penicillin injection, the group experimented with deep-tank fermentation. This technique proved successful and made it possible to produce the drug on a grand scale—enabling, among other things, the immediate treatment of thousands of U.S. soldiers fighting in World War II.

In 1944, Hobby joined Pfizer Pharmaceuticals, where she continued her research into antibiotics, contributing greatly to the discovery of several new ones. Her group found that bacteria-fighting organisms often lived in soil, and the antibiotic Terramycin resulted from these efforts. She also helped develop streptomycin and Biomycin, and elaborated on how antimicrobial drugs work. Hobby left Pfizer in 1959 to become chief of research at the Veteran's Administration Hospital in East Orange, New Jersey. There she focused on chronic infectious diseases. She concurrently served as assistant research clinical professor in public health at Cornell Medical College.

Hobby retired in 1977, though she continued to work, dedicating herself to science writing. Having authored more than 200 articles during her career, she wrote *Penicillin: Meeting the Challenge* in 1985. This book described the development of penicillin in meticulous detail. She also founded and edited the journal *Antimicrobial Agents and Chemotherapy.* A member of the American Association of Science, the American Academy of Microbiology, and the American Society of Microbiology, Hobby was well respected by her fellow researchers. She died suddenly of a heart attack on July 4, 1993, in her home in a retirement community in Pennsylvania. The effects of her contributions cannot be overestimated. It is reckoned that between 1945 and 1955 alone, more than 1.5 million Americans survived what could otherwise have been fatal infections because of penicillin.

Hodgkin, Dorothy Crowfoot
(1910–1994)
English
X-ray Crystallographer

Dorothy Crowfoot Hodgkin is considered the founder of protein crystallography for her indefatigable work mapping the molecular structure of important compounds such as penicillin. She won the 1964 Nobel Prize in chemistry for her structural analysis of vitamin B_{12}, but the crowning achievement of her long career was her deduction of the structure of insulin, which she commenced in the 1930s and completed in the late 1960s.

Hodgkin was born on May 12, 1910, in Cairo, Egypt, where her father, John Crowfoot, worked as an archaeologist for the British Ministry of Education. Her mother, Grace, was an artist and an expert in Coptic textiles. Hodgkin grew up separated from her parents by World War I, as she remained in the safe haven of England while her mother accompanied her father to his new post in Sudan as the director of the Ministries of Education and Antiquities. For her 16th birthday, her mother gave her a book on X-ray crystallography by its pioneer, William Henry Bragg, a gift that helped determine her future.

In 1928, Hodgkin entered Somerville, the women's college of Oxford University, and graduated with a bachelor's degree in chemistry in 1932. She proceeded to Cambridge University, where she conducted postgraduate research under J. D. Bernal, who was working with X-ray crystallography. Hodgkin and Bernal applied X-ray crystallography first on pepsin, which they fully analyzed by 1934. That year, Hodgkin began to suffer from rheumatoid arthritis, a condition that afflicted her the rest of her career.

Also in 1934, Hodgkin returned to Somerville as a researcher and teacher with her own lab space, though it was submerged in the basement of the Oxford Museum. She pursued her doctorate, writing her dissertation on her X-ray crystallography study of cholesterol, which she completed in 1937 to earn her Ph.D. On December 16 of that year, she married African studies specialist Thomas L. Hodgkin. Together the couple had three children, raised by family and nannies while their father taught in the north of England and their mother worked at Oxford. He finally joined her on the Oxford faculty in 1945.

Dorothy Crowfoot Hodgkin *(AIP Emilio Segrè Visual Archives, Physics Today Collection)*

Hodgkin spent the years during World War II studying penicillin, one of the first antibiotics, which Ernest Chain had brought to her for analysis. From 1942 through 1946, Hodgkin and her graduate student Barbara Rogers-Low deciphered penicillin's molecular structure. In honor of this significant achievement, the Royal Society inducted Hodgkin as a fellow, only the third woman so honored by 1947. The next year, she commenced analysis of vitamin B_{12}, an even more complex molecule necessary to prevent pernicious anemia in humans. Hodgkin and her team worked for six years collecting data; as they were preparing to interpret this data, Hodgkin met Kenneth Trueblood of the University of California at Los Angeles, where he had programmed a computer to calculate crystallographic readings. Trueblood and Hodgkin collaborated cross-continentally to arrive at the molecular structure of vitamin B_{12} in 1956. A year later, Oxford finally appointed Hodgkin as a reader (the equivalent of a full professorship in the United States), and in 1958, they finally equipped her with adequate laboratory facilities.

In 1964, Hodgkin became the first British woman to win a Nobel Prize in the sciences. The next year, she received Britain's Order of Merit, the second woman so honored, after Florence Nightingale. Hodgkin spent the remainder of the 1960s deciphering the 777 atoms of insulin by analyzing 70,000 X-ray spots, and in 1969, she announced her results (which were refined in 1988 with the help of advanced computers). Hodgkin retired in 1977. She suffered a stroke and died on July 30, 1994.

Hoffleit, Ellen Dorrit
(1907–   )
American
Astronomer

Dorrit Hoffleit distinguished herself as an astronomer by discovering more than 1,000 new variable stars. She worked in the observatories of both Harvard and Yale, and later in her career, she became interested in the history of astronomy, especially women's contributions to the field. She also directed the Maria Mitchell Observatory on Nantucket for more than two decades.

Hoffleit was born on March 12, 1907, in Florence, Alabama. She attended Radcliffe College for her bachelor's degree, which she received in 1928. She remained in Cambridge working at the Harvard Observatory as an assistant while continuing with graduate work at Radcliffe. She earned her master's degree in 1932 and then spent the next six years pursuing her doctorate, finally earning her Ph.D. in astronomy in 1938. That year, Harvard Observatory promoted her to the position of research associate. In 1943, Hoffleit contributed to the war effort as a mathematician with the Ballistic Research Laboratories of Aberdeen Proving Ground in Maryland. Unlike many women, who lost their wartime positions to returning men, Hoffleit retained her position until 1948 and remained a consultant with the labs until 1961. In 1948, she returned to Harvard Observatory as an astronomer, and during this time period, she published her first important book, *Some Firsts in Astronomical Photography* (1950). During the 1955 through 1956 academic year, she lectured at Wellesley College.

In 1956, Hoffleit moved from the Harvard Observatory to the Yale Observatory, first as a research associate until she reached the rank of senior research astronomer later in her career. She held a concurrent appointment as the director of the Maria Mitchell Observatory stationed on Nantucket Island in Massachusetts, a post she held for more than two decades. From 1958 until 1960, she served as the vice president of the American Association of Variable Star Observers and then as its president from 1961 through 1963. She published her second important book, *Bright Star Catalogue,* in 1964. In 1976, she com-

Dorrit Hoffleit, an astronomer who has discovered more than 1,000 new variable stars, shown in June 1938 after receiving her Ph.D. from Radcliffe College. *(Courtesy of Dorrit Hoffleit)*

menced a relationship with the Astronomical Observatory of the Pasteur Institute, located in Strasbourg, France.

Later in Hoffleit's career, she grew interested in the history of astronomy and specifically the role of women in promoting the advancement of the field, prompting her to write biographies of MARIA MITCHELL, WILLIAMINA FLEMING, and ANNIE JUMP CANNON. In 1992, she published a historical account of the field specific to her place of employment, *Astronomy at Yale, 1701–1968.* She also served as the editor of the journal *Meteoritics* and contributed more than 400 articles to scientific journals throughout her career, discussing her research topics of variable stars, stellar spectra, proper motions, meteors, galactic structure, and bright stars, among other topics.

Hoffleit has been a member of numerous professional organizations, including the American Association for the Advancement of Science, the Meteoritical Society, the International Astronomical Union, and the American Astronomical Society. In 1984, Smith College granted her an honorary doctorate of science. Though retired, she continued to work eight-hour days into her nineties.

Hogg, Helen Battles Sawyer
(1905–1993)
American
Astronomer

Helen Sawyer Hogg helped to popularize astronomy with her weekly column in the *Toronto Star,* "The Stars Belong to Everyone," which ran for three decades. She also contributed to academic astronomy, focusing her research on variable stars in globular clusters and writing more than 200 papers throughout her career. She also served as the president of several professional organizations, including the Canadian Astronomical Society, which she helped found in 1971.

Hogg was born on August 1, 1905, in Lowell, Massachusetts. She majored in chemistry at Mount Holyoke College until 1925, when she decided to switch to astronomy. In 1926, the Harvard astronomer ANNIE JUMP CANNON visited Mount Holyoke, solidifying Hogg's decision to pursue astronomy. Hogg graduated that year and went straight to the Harvard Observatory to work with Cannon and Harlow Shapley, who were investigating star clusters. While working at the observatory, she taught at Smith College in 1927 and at Mount Holyoke in 1930 through 1931, and she pursued her doctorate from Radcliffe College (since Harvard University did not offer women graduate degrees then). She obtained her Ph.D. in 1931.

The year before, she had married Frank Hogg, who was studying astronomy at Harvard, and together they had three children. He landed a job at the Dominion Astrophysical Observatory in Victoria, British Columbia, so the couple moved there after their marriage and after Hogg finished her doctorate. As was the case so often with academic couples at that time, the institution offered the man a job but expected the wife, no matter how qualified, to volunteer her time if she wanted to work in her field there. While at Dominion, Hogg devised her method for measuring the distance of galaxies beyond the Milky Way by tracking their brightness cycles.

In 1935, the University of Toronto offered Frank a position, and Hogg worked as an assistant at the David Dunlop Observatory, which promoted her to a research associateship in 1938. In 1940, she returned to Mount Holyoke as a visiting professor and acting chair of the department of astronomy. When she returned to Toronto in 1941, the university offered her a research associateship, which she retained for the next decade. In 1947, she published the *Bibliography of Individual Globular Clusters,* and in 1955, she published the *Second Catalogue of Variable Stars in Globular Clusters.* In 1950, Hogg's husband died, and within one year, the University of Toronto offered her an assistant professorship (she worked her way up to full professor by 1957). Also in 1951, she commenced writing her astronomy column for the *Toronto Star.* Between 1955 and 1956, she served as the program director in astronomy for the National Science Foundation in Washington, D.C.

The American Astronomical Society presented Hogg with the Annie Jump Cannon Prize in 1950. She served as the first woman president of the physical sciences section of the Royal Society of Canada in 1960 and of the Royal Canadian Institute from 1964 through 1965. In 1967, she received the Centennial Medal of Canada, and in 1976 she was made a Companion of the Order of Canada, one of the country's highest honors. Two observatories bear her name: the

Helen Sawyer Hogg, whose weekly column in the *Toronto Star,* "The Stars Belong to Everyone," helped to popularize astronomy. *(The Mount Holyoke College Archives and Special Collections)*

Canadian National Museum of Science and Technology's observatory in Ottawa and the University of Toronto's southern observatory in Chile. Hogg retired in 1976 and died of a heart attack in Richmond Hill, Ontario, on January 28, 1993.

Hollinshead, Ariel Cahill
(1929–   )
American
Pharmacologist and Virologist

Ariel Hollinshead developed important vaccines, first for all four types of lung cancer and later an experimental cocktail for AIDS patients that tested very well in preliminary trials. Due to its combination of an antiretrovirus and a vaccine, it was called chemoimmunotherapy. Hollinshead also promoted the role of women in the sciences and lobbied for more recognition of women in cancer research.

Hollinshead was born on August 24, 1929, in Allentown, Pennsylvania. She attended Ohio University in Athens, Ohio, where she earned her bachelor's degree in 1951. She then pursued graduate work at George Washington University (GWU) in Washington, D.C., earning her master's degree in 1955 and her Ph.D. in pharmacology in 1957. The next year she married, and she has two children.

She remained at the GWU Medical School for the rest of her career, starting out as a research fellow in pharmacology from 1957 through 1959 while simultaneously filling a research fellowship in virology at the College of Medicine at Baylor University. In 1959 GWU appointed her an assistant professor of pharmacology. She continued to rise up the ranks, from associate professor in 1961 through 1973 to full professor since 1974. In 1964, she also added duties as the director of the Laboratory for Virus and Cancer Research.

In the 1970s, Hollinshead discovered a new approach to treating cancer by using a unique vaccine that covered against all four major types of lung cancer. During that time, she also invented a method for isolating antigens from cell membranes by means of a low-frequency sound. In the late 1990s, she studied two patients suffering from AIDS, treating them using thiophenyl urea (TUR) as an antiretroviral agent in combination with an antiretroviral vaccine. Three weeks of intravenous TUR doses were followed by nonvirion antigen vaccine intradermally once a month for three months. This treatment dramatically lowered HIV DNA reading in one patient within six months and in the other within nine months, and measurements in both patients of infection and immunosuppression levels normalized. This chemoimmunotherapy treatment showed great promise for further study.

Hollinshead conducted research in many different areas, including chemotherapy of animal virus diseases, such as cancer; nucleoprotein chemistry of viruses; cancer immunogenetics and immunoprophylaxis; environmental carcinogens; isolation, purification, and identification of animal and human tumor-associated antigens; and human-to-human hybridoma research.

Hollinshead has supported women studying the sciences by serving as a coordinator for Graduate Women in Science at her regional chapter of Sigma Delta Epsilon. She has also belonged to several other organizations, including the Society for Experimental Biology and Medicine, the American Society for Clinical Oncology, and the American Association for Cancer Research. In 1992, she served as a mentor for Women in Cancer Research, an offspring of the American Association for Cancer Research born from an informal gathering at the 1988 national meeting in New Orleans and launched on May 25, 1988, to recognize women's issues and contributions to cancer research.

Hoobler, Icie Gertrude Macy
(1892–1984)
American
Biochemist

Trained as a physiologist and chemist, Icie Macy Hoobler combined these two disciplines to study the biochemistry of nutrition in humans, especially women, infants, and children. Her studies helped establish the minimum requirements for many vitamins and minerals in the human diet at a time when this information had been inadequate. During years of detailed work, she also examined metabolism during a woman's reproductive cycle and illuminated the relationship between diet, nutrition, and healthy pregnancies in women.

Born on July 23, 1892, in Daviess County, Missouri, Icie Gertrude Macy was the daughter of Perry Macy and Ollevia Elvaree Critten Macy. Macy was a ninth-generation descendant of Thomas Macy, a Baptist minister who was one of the first settlers of Nantucket, Massachusetts. She grew up on the family farm. From 1926 to the mid-1930s, Macy raised her two nieces after the death of their mother, Macy's sister Ina. In 1938, when she was 44, she married B. Raymond Hoobler, a physician.

Even though neither of Hoobler's parents was well educated, they encouraged their children to pursue an education. She received her primary education in a one-room Missouri schoolhouse and her secondary education at a boarding school. Hoobler's interest in science became clear during her secondary studies, but to satisfy her father, who had decided that she was going to be a musician, she studied music in high school and at the teacher training college she attended in Missouri. After receiving an A.B. from Central College for Women in 1914, Macy enrolled in the University of Chicago in 1916 where, with the encouragement of chemist MARY SHERRILL, she began to study chemistry. Hoobler earned a B.S. in chemistry from the University of Chicago in 1916.

For her graduate education, Hoobler attended the University of Colorado, where she studied biochemistry under the direction of Robert C. Lewis. She received her M.S. in chemistry from the University of Colorado in 1918 and began her Ph.D. studies in that field at Yale the same year. At Yale, Hoobler studied with Lafayette B. Mendel of Yale's Sheffield Scientific School. Mendel encouraged Hoobler in her studies of the nutritional value of cottonseed flour, which was being substituted for wheat flour during World War I. Government authorities were concerned about using cottonseed flour as a food staple because of so-called cottonseed meal injury, the physical deterioration of animals who consumed this product. Hoobler determined that cottonseed was not deficient in vitamins or nutritional value, but it contained the poisonous compound gossypol, which caused harmful side effects. For her work on this and other studies, Hoobler was awarded her Ph.D. in 1920.

After finishing her Ph.D., Hoobler worked for a while at a hospital in Pittsburgh and at the Department of Household Science at the University of California at Berkeley. In 1923, she took a job at the Merrill-Palmer School for Motherhood and Child Development in Detroit, Michigan. She also worked as a researcher at the Children's Hospital of Michigan in Detroit. She remained at Merrill-Palmer until her retirement in 1959.

At Merrill-Palmer, Hoobler worked on a 25-year study of nutrition and human physiochemistry that was underwritten by the Children's Fund. As the leader of this research team, Hoobler focused research on six topics: the metabolism of women during their reproductive cycles, the composition of human milk, infant growth and development, childhood growth and development, the nutrition of meals given to children in Michigan's child-care institutions, and blood studies related to health and disease.

Perhaps Hoobler's greatest contribution to the health of American mothers and their children was the discovery that vitamin D and vitamin B needed to be added to milk to prevent rickets, a bone disease that was then common among children.

For her work, Hoobler was awarded the University of Colorado's Norlin Achievement Award in 1938, the American Chemical Society's Garvin Award in 1946, and the Modern Medicine Award in 1954. She returned to her native Missouri in 1982 and died in Gallatin, Missouri, the town nearest her childhood farm, on January 6, 1984. She was 92 years old.

Hopper, Grace Murray
(1906–1992)
American
Computer Scientist

Grace Hopper was a pioneer in the field of computer science, helping to invent the first compiler (A-O) as well as the first computer languages (Flow-Matic and COBOL). She got her start with computers in the U.S. Navy during World War II, and she remained in the service until her retirement, after which the navy promoted her to rear admiral, the first woman to hold this rank.

Grace Brewster Murray was born on December 9, 1906, in New York City to Mary, a homemaker, and Walter Murray, an insurance broker. She followed in the footsteps of her great-grandfather, a navy admiral, and her grandfather, a civil engineer who showed her how to survey on field trips together. Murray commenced her long relationship with Vassar College as an undergraduate studying mathematics and physics for her bachelor's degree in 1928. She proceeded to Yale University for her master's degree in 1930, then returned to teach at Vassar while continuing on doctoral work at Yale. During that time, she married Vincent Foster Hopper, an English literature teacher. She received her Ph.D. in 1934 when Vassar promoted her to an instructorship. By 1939, the college had raised her to the rank of assistant professor.

After World War II broke out, Hopper enlisted in the U.S. Navy's Women Accepted for Voluntary Emergency Services, which assigned her to the Board of Ordnance computing project at Harvard University working under Howard Aiken. When she reported to duty on July 2, 1944, he assigned her to the Mark I computer, the prototype of the modern personal computer, although it measured 51 feet long and eight feet high. Hopper and her colleagues were responsible for fixing the machine when it broke, including when a moth flew into its gears; from then on, they called the process of fixing breakdowns *debugging*. In 1945, Hopper and her husband divorced, and in 1946, Hopper resigned from Vassar to continue on at Harvard even after the war effort ceased, working with the Mark I and its successors, the Marks II and III.

Realizing the vast potential of the computer, she also realized the need for people to transform this potential into understandable programs, so she joined the Univac division of the Remington Rand Corporation in 1949 as a systems engineer. One of her first innovations was a compiler, which allowed the computer to generate its own programs. In 1957, she essentially translated computer languages into plainer English (both for the machine and for the human operating it), resulting in first the Flow-Matic language and later in the COBOL (common business-oriented language) in 1959. She climbed the ranks in the company from director of automatic programming to chief engineer to staff scientist by the time she retired in 1971. She continued to climb the ranks simultaneously in the navy, rising to the post of commander by the time she retired in 1979.

That year, the Department of Defense awarded Hopper its highest mark of distinction, the Distinguished Service Medal. Six years later, the navy further recognized Hopper by promoting her one final time, to rear admiral,

Grace Murray Hopper, a computer science pioneer who helped to invent the first compiler and the first computer languages (Flow-Matic and COBOL). *(Special Collections, Vassar College Libraries)*

the highest rank ever attained by a woman. She received a host of other honors for her achievements, including the 1946 Naval Ordnance Development Award, the 1973 Legion of Merit, induction into the Engineering and Science Hall of Fame in 1984, and the National Medal of Technology in 1991. She died at the age of 85 on January 1, 1992.

Horney, Karen
(1885–1952)
German/American
Psychoanalyst

Karen Horney created controversy early in her career when she broke from Freudian psychoanalytical theory, conceiving of female psychology on its own terms instead of in (subordinate) relation to male psychology. As her own theoretical paradigms matured, she further distanced herself from Freud by focusing attention on the present instead of fixating on the past. Horney left her permanent mark on her field as the founding dean of the American Institute of Psychoanalysis and the founding editor of the *American Journal of Psychoanalysis*.

Horney was born in 1885 in Hamburg, Germany. She studied medicine at the Universities of Freiburg, Göttingen, and Berlin, earning her M.D. from the latter in 1913. She had married Oskar Horney in 1909, and she entered analysis with Karl Abraham the next year. In 1920, she helped to found the Berlin Psychoanalytic Institute. By 1926, she had

separated from her husband, and in 1932 she emigrated to the United States.

That year, Frank Alexander hired Horney as the associate director of the newly founded Chicago Psychoanalytic Institute, where she remained for two years. In 1934, she moved to New York City, where she established a practice, taught, and became a supervisor of the New York Psychoanalytic Institute. The year 1941 was significant for Horney as she cofounded three mainstays in the field of psychoanalysis: the American Institute of Psychoanalysis (acting as its dean until her death); the *American Journal of Psychoanalysis* (serving as its editor until her death); and the Association for the Advancement of Psychoanalysis.

Horney's theoretical thought can be divided into three distinct phases: In the 1920s and early 1930s, she attempted to redefine feminine psychology within a Freudian paradigm; in the late 1930s and early 1940s, she revised Freudianism by replacing his biological determinism with societal and interpersonal influences; and in the late 1940s and early 1950s, she emphasized how people could attain self-understanding by recognizing and abandoning self-destructive defense mechanisms. A fourth phase developed posthumously when a collection of her essays was published under the title *Feminine Psychology* in 1967. This anthology revisited her controversial break with Freudian theory in the 1930s, when she replaced his notion of penis envy with her notion of womb envy as well as rejecting Freud's notions of female masochism and feminine development. The timing of this reprinting coincided with the rise of second-wave feminism, which acknowledged Horney as the first great psychoanalytic feminist.

In addition to writing numerous journal articles, Horney published many important books, including *The Neurotic Personality of Our Time* (1937), *New Ways in Psychoanalysis* (1939), *Our Inner Conflicts* (1945), and *Neurosis and Human Growth* (1950). She was instrumental in the progression of psychoanalysis from its Freudian infancy to its post-Freudian maturity. She retained many Freudian tenets key to psychoanalysis, such as free association, but she rejected dubious Freudian concepts, such as the libido theory and the primacy of infantile sexuality. Most importantly, Horney identified Freudian psychoanalysis as the product of a male genius within a male-dominated society. One of Horney's main projects, therefore, was to transform psychoanalysis into a practice that took women's perspectives and problems into account. Horney died in 1952. In 1990, the International Karen Horney Society was founded to continue the dissemination of her psychoanalytical ideas and practices.

Horstmann, Dorothy Millicent
(1911–   )
American
Epidemiologist

Although Dorothy Millicent Horstmann played a central role in developing the polio vaccine, her accomplishments have often been overlooked. Her principal scientific achievement was her 1952 discovery that polio reached the nervous system through the bloodstream. Horstmann spent almost her entire career at the Yale University School of Medicine.

On July 2, 1911, Horstmann was born in Spokane, Washington, to Henry and Anna (Humold) Horstmann. After receiving her bachelor's degree from the University of California in 1936, Horstmann remained at that institution, earning her medical degree in 1940. She served her internship at the San Francisco City and County Hospital from 1939 to 1940, and she completed her residency at Vanderbilt University by 1942.

Horstmann accepted a position at the Yale University School of Medicine in 1942. She would remain associated with Yale for the duration of her career. After being named an associate professor in 1945, she held a National Institutes of Health postdoctoral fellowship from 1947 to 1948. Soon thereafter, she embarked on the research that would aid in the battle against polio (a devastating viral disease that frequently caused paralysis and deformity in children). Scientists were then debating the question of how the virus reached the central nervous system (which is where it wreaked its havoc). After examining the findings of her fellow researcher, William McDowell Hammon (whose

Dorothy Horstmann, who played an essential role in the development of the polio vaccine. *(National Library of Medicine, National Institutes of Health)*

work had revealed that injections of gamma globulin—an antibody-rich serum extracted from plasma—produced temporary immunity to polio), Horstmann became convinced that polio was an infectious malady that reached the nervous system through the blood.

To test her hypothesis, Horstmann conceived of an experiment in 1952 to determine whether polio did indeed first appear in the bloodstream and then travel to the nervous system. After she fed monkeys and chimpanzees minute quantities of polio, she carefully examined their blood for traces of the virus. Although none of the primates immediately developed polio symptoms, Horstmann observed traces of the virus in their blood. Many of the animals were later afflicted with polio's debilitating paralysis. Parallel work conducted by David Bodian at Johns Hopkins University confirmed Horstmann's research.

During the late 1950s and early 1960s, Horstmann participated in field trials to ensure that polio vaccines were safe and effective. She also advanced to the rank of professor of epidemiology and pediatrics in 1961. In 1969, she was named the John Rodman Paul Professor of Epidemiology and Pediatrics at Yale. She continued her innovative work, investigating maternal rubella and the rubella syndrome in infants.

After retiring from her full academic duties in 1982, Horstmann remained at Yale, first as an emeritus professor and later as a senior research scientist. She received four honorary doctorates and a number of other commendations. Notably, she won the James D. Bruce Award from the American College of Physicians in 1975, Denmark's Thorvold Madsen Award in 1977, and the Maxwell Finland Award of the Infectious Diseases Society of America in 1978. In addition, she was elected to the National Academy of Sciences, the American College of Physicians, and the Royal Society of Medicine. She remained concerned with scientific issues into the late 1990s, joining with other scientists to warn about the perils of global warming.

Hrdy, Sarah Blaffer
(1946–   )
American
Anthropologist

Sarah Hrdy overturned the notion that the males of certain monkeys, the langurs, kill babies of their own species in response to overpopulation. She discovered instead that they kill the babies of rival males to make the females more receptive to mating, thus passing along their own genes and destroying the genetic pool of rivals. She then shifted her focus from male behavior to female behavior, realizing that female langurs maximized their sexual contact by mating with males other than the dominant male, who would not harm her babies lest they be his own offspring.

Sarah Blaffer was born on July 11, 1946, in Dallas, Texas, and she grew up in Houston. Her family lived very comfortably on money made in the oil industry, but the social conventions of her milieu discouraged women from pursuing an education or career. Blaffer's mother, however, encouraged her daughter's academic interest, supporting her move away from home at 16 to attend a girls' boarding school in Maryland. From there, Blaffer matriculated at Wellesley College, where she studied philosophy while also taking creative writing courses. While conducting research for a novel on Mayan folklore, Blaffer decided that studying cultures held more interest for her than fiction writing. In the wake of this realization, she transferred to Radcliffe to major in anthropology.

Blaffer wrote her undergraduate thesis on the demon H'ik'al, a spirit who policed sexuality, earning her B.A. in 1969. Three years later, she published her thesis as a book, *The Black-man of Zinacantan*. That year, 1972, she married fellow anthropologist Daniel Hrdy, with whom she had three children. By the time of her marriage, she was taking film courses at Stanford University, where she also took a course with Paul Ehrlich, who introduced her to the idea of overpopulation. This prompted her interest in solving the langur mystery of infanticide. The Hrdys departed for Mount Abu in Rajasthan to study the langurs for her doctoral dissertation.

After observing the langurs, she realized that the infanticide occurred independently of overpopulation, which prompted her to consider other root causes. She noticed the cyclical nature of male domination: One male would dominate small troops of females and rival males, with a new male gaining domination every 28 months or so. Male ascendancy meant determination of the gene pool, so dominant males killed the young of rivals. She also noticed that the females were not necessarily faithful to the dominant male but rather spread the gene pool among all the males without the dominant male being any wiser, as he could not afford to commit infanticide on what might be his own genes.

This research earned Hrdy her Ph.D. from Harvard in 1975. As with her undergraduate thesis, she published her dissertation two years later as a book, *The Langurs of Abu*, introducing the notion of sociobiology, a Darwinian theory of animal evolution where gene transfer rules behavior. Hrdy later applied her learning on sex and gender differentiations in primates to humans, resulting in the 1981 book, *The Woman That Never Evolved*. In it, she argued that human women had escaped the "evolutionary trap" of rivalry, instead choosing to join together. In 1984, Hrdy joined the anthropology department at the University of California at Davis, and she still worked there in the late 1990s.

Huang, Alice Shih-Hou
(1939–   )
Chinese/American
Microbiologist

Alice Shih-Hou Huang's pioneering research into viruses led to major breakthroughs in the understanding of how viruses function. Of particular note, Huang played a significant role in the discovery of reverse transcriptase, an enzyme that allows viruses to convert their genetic material into deoxyribonucleic acid (DNA). In addition, her work with defective interfering particles—abnormal viruses that interfere with the reproduction of normal viruses—won her the Eli Lilly Award in Microbiology and Immunology in 1977. She also holds the distinction of being the first Asian-American to head a national scientific society in the United States.

Huang was born in Nanchang, China, on March 22, 1939. When the communists took control of the country in 1949, her parents—Quentin and Grace Betty Soong Huang—sent Alice and her siblings to the United States. Alice Huang became an American citizen during her senior year of high school.

From 1957 to 1959, Huang attended Wellesley College in Massachusetts before transferring to the School of Medicine at the Johns Hopkins University in Baltimore,

Maryland. Initially interested in becoming a practicing physician, Huang soon decided to commit herself to a career in medical research. She earned her bachelor's degree in 1961, her master's degree in 1963, and her Ph.D. in microbiology in 1966.

Immediately after graduating, Huang spent a year at the National Taiwan University in Taipei, Taiwan, as a visiting professor. She returned to the United States in 1967 for a postdoctoral fellowship at the Salk Institute for Biological Studies in San Diego, California. The following year, she married a fellow virologist at the Salk Institute, David Baltimore, with whom she had one daughter.

Huang and Baltimore took their work to the Massachusetts Institute of Technology (MIT) in 1968. Huang was a postdoctoral fellow at MIT from 1968 to 1969 and a research associate there from 1969 to 1970. At this time, scientists understood little about how viruses replicated, but the young couple immersed themselves in the subject. Huang's study of viral genetics facilitated Baltimore's discovery of the reverse transcriptase enzyme, for which Baltimore shared the 1975 Nobel Prize in medicine. Reverse transcriptase enables viruses to convert ribonucleic acid (RNA) into DNA. This conversion process is of particular concern where it is used by a certain class of virus, the retrovirus, which includes the virus that causes acquired immune deficiency syndrome—AIDS. Retroviruses carry their genetic blueprints in the form of RNA, but with the help of reverse transcriptase, these viruses are able to synthesize DNA from RNA. This reversal of the usual cellular process makes it possible for a retrovirus to incorporate itself permanently into the DNA genome of an infected cell, thereby effectively taking it over and allowing the retrovirus to copy itself further.

In 1971, Huang was named an assistant professor of microbiology and molecular genetics at Harvard Medical School. She was promoted to associate professor in 1973 and to full professor in 1979. At the same time, she served as an associate at Boston City Hospital from 1971 to 1973 and as the director of the infectious diseases laboratory at Children's Hospital in Boston from 1979 to 1989. While at Harvard, Huang explored defective interfering particles, which are abnormal viruses that stymie a normal virus's ability to reproduce. Specifically, Huang studied a rabies-like virus that produces mutant strains that interfere with the virus's future growth. Her hope was that in gaining an understanding of where the mutant strains came from and how they worked, scientists would one day be able to employ defective interfering particles to prevent viruses from spreading in humans. Huang's work in this area was recognized in 1977 when she received the Eli Lilly Award.

Huang remained at Harvard until 1991. She was then appointed dean for science at New York University. She is credited with having advanced the field's understanding of viruses and has been repeatedly recognized for her achieve-ments. From 1988 to 1989, she served as president of the American Society of Microbiology, and she was named to the editorial boards of several scientific journals, including *Intervirology, Journal of Virology,* and *Review of Infectious Diseases.*

Hubbard, Ruth
(1924–   )
Austrian/American
Biologist

Ruth Hubbard spent her long career at Harvard University, focusing first on the biochemistry of vision and the synthesis of visual pigmentation before expanding the scope of her investigations to include the sociology of science. Most importantly, she documented how the social status of scientists determines how they gather and interpret scientific data.

Hubbard was born on March 3, 1924, in Vienna, Austria. She emigrated along with her family to the United States in 1938, settling in Brookline, Massachusetts, in time for her to attend high school there. She proceeded to Radcliffe College to study biochemical sciences, earning her bachelor's degree in 1944. Hubbard conducted research in clinical laboratories for several years before returning to Radcliffe to pursue her doctorate. She spent 1948 through 1949 in London, England, at University College Hospital Medical School as a U.S. Public Health Service predoctoral fellow. She received her Ph.D. in biology in 1950.

Hubbard then commenced her long-standing relationship with Harvard University, starting as a research fellow in biology in 1950. From 1952 to 1953, she traveled to Copenhagen, Denmark, where she conducted research at the Carlsberg Laboratory under a Guggenheim Fellowship. Upon her return to Harvard in 1953, the university promoted her to the position of research fellow. She continued to climb the ranks throughout her career there, becoming a research associate in 1958, the same year she entered a marriage that produced two children. She became a lecturer a decade later. She served as a visiting professor at the Massachusetts Institute of Technology in 1972, and in 1973, she commenced a sustained relationship with the Marine Biological Laboratory at Woods Hole, Massachusetts, where she also served as a trustee. After she had worked at Harvard for two and a half decades, the university finally promoted her to the rank of professor of biology in 1974.

Hubbard conducted her primary research on the biochemistry and photochemistry of vision in vertebrates and invertebrates. She widened the scope of her research to include the politics of health care, focusing on women's health issues. Stepping further back for perspective, she considered the sociology of science; specifically, she investigated the degree to which the socioeconomic status of researching scientists—namely their sex, race, class, and institutional affiliations—determines what questions they ask in their research, and what answers they entertain (and what answers they might reject).

Hubbard has published extensively in her field, with more than 150 articles appearing in books, journals, and magazines. She has coedited several books, including *Genes and Gender II: Pitfalls in Research on Sex and Gender* (1979), *Women Look at Biology Looking at Women* (1989), and *Reinventing Biology: Respect for Life and the Creation of Knowledge* (1995). She has also authored several books, including *The Politics of Women's Biology* (1990), *Exploding the Gene Myth* (with Elijah Wald, 1993), and *Profitable Promises: Essays on Women, Science and Health*.

In 1967, Hubbard received the Paul Karrer Medal from the Swiss Chemical Society. Though she retired to emerita status in 1990, she continued to exert influence in her field and in the culture at large. In 1997, for example, she wrote an editorial reaction to sheep cloning, published in *The Nation,* in which she cautioned against the control of cloning decisions by market forces and urged public forums to discuss the implications of cloning, lest it lead to another Nazi regime attempting to create a master race.

Hudson, Mary K.
(1949–   )
American
Space Physicist

Mary K. Hudson is the head of the department of physics and astronomy at Dartmouth College, where she was also a founding member of the Women in Science Program. She focuses her research on space physics, conducting radiation belt studies and space weather modeling. As a young physicist, she won the James B. Macelwane Medal from the American Geophysical Union (AGU).

Hudson was born on January 6, 1949, in Santa Monica, California. In 1967, she worked as a mathematician for the McDonnell-Douglas Corporation while studying physics at the University of California at Los Angeles (UCLA). The next year, she served as a National Science Foundation trainee at UCLA. In 1969, she graduated summa cum laude with a bachelor of science degree in physics.

Hudson remained at UCLA for her graduate work, simultaneously serving as a research physicist at the Space Sciences Laboratory of the Aerospace Corporation from 1969 through 1971. That year, she received her master of science degree. She conducted her doctoral studies as a research assistant under her adviser, Charles F. Kennel, to earn her Ph.D. in physics in 1974.

Upon receiving her doctorate, Hudson moved from UCLA to the University of California at Berkeley, which hired her as an assistant research physicist, a position she held from 1974 through 1981. She was then promoted to the rank of associate research physicist through 1984 while concurrently holding a senior fellowship at Berkeley. In 1984, Dartmouth College hired Hudson as an associate professor of physics, though she simultaneously served as a research physicist at Berkeley from 1984 through 1986. Also in 1984, she married and eventually had two children. In 1990, she served as a visiting associate professor in the electrical engineering department at Stanford University. That same year, Dartmouth promoted her to the title of professor. In 1996, Dartmouth appointed her head of the department of physics and astronomy.

Hudson has focused her research on space physics, studying the effects of geomagnetic storms and advancing space plasma theory through ionosphere plasma simulation. She has also studied auroral particle acceleration and heating. Her team has published its research regularly in the *Journal of Geophysical Research,* which she has served as an associate editor. Some articles that have appeared in the journal's pages include the following: "Evidence for a Global Disturbance with Monochromatic Pulsations and Energetic Electron Bunching" (1999), "Simultaneous Satellite and Ground-Based Observations of a Discretely Driven Field Line Resonance" (1999), "A Statistical Study of Pc 3-5 Magnetic Pulsations Observed by the AMPTE/Ion Release Module Satellite" (1999), and "Simulations of Radiation Belt Formation During Storm Sudden Commencements" (1998).

In 1984, Hudson received the Macelwane Medal, granted to the outstanding young geophysical scientist by the American Geophysical Union. She then served as the secretary of the AGU's Magnetospheric Physics section from 1986 through 1988. After becoming the head of Dartmouth's department of physics and astronomy, she received the first Celebration of Excellence Award from the Burlington, Vermont-based Pizzagalli Construction Company. The award not only recognized her excellence in research but also honored her mentoring of students and advocacy of women in science as a positive role model.

Hyde, Ida Henrietta
(1857–1945)
American
Physiologist

Ida Hyde broke down many barriers in the male domination of the sciences throughout her career: She was the first woman to receive a doctorate from the University of Heidelberg, the first woman to conduct research at the Harvard Medical School, and the first woman elected to membership in the American Physiological Society. However, her main contribution to her field, her invention of a stimulating microelectrode small enough to insert into a cell, has been obscured, as the scientific establishment has not officially acknowledged this as her innovation.

Hyde was born in 1857. Her mother was an astute businesswoman who supported the family until the Chicago Fire of 1870 burned their home and business. Thereafter, responsibility for supporting the family fell partly on Hyde, who taught public school for several years before commencing college at the age of 24.

She began her undergraduate study at the University of Illinois in 1881 while continuing to teach; it was not until 1888 that she devoted her full attention to her college studies, transferring to Cornell University to earn her bachelor's degree in 1891. She continued her education at Bryn Mawr College, where she also worked as an assistant in biology. At this time as well, she commenced a long-standing relationship with the Marine Biological Laboratory at Woods Hole, Massachusetts, where she worked for many of its summer programs.

Hyde traveled to Europe to continue her doctoral studies as a Collegiate Alumnae European Fellow, first at the University of Strasbourg and later earning her Ph.D. from the University of Heidelberg in 1896. That year, she conducted postdoctoral work at the University of Bern and at the Naples Zoological Station, where she later helped establish the Naples Table Association for Promoting Scientific Research by Women, an initiative to provide fellowships to American women scientists. Hyde returned to the United States in 1897 to continue her postdoctoral studies at Radcliffe College and to conduct research at the Harvard Medical School. She then taught prep-school histology and anatomy for four years.

In 1899, the University of Kansas offered Hyde an associate professorship in physiology. When the university established an independent department of physiology in 1905, it promoted her to the rank of full professor. The year before, she continued her own education at the University of Liverpool and thereafter spent summers enrolled at the Rush Medical College, where she earned her M.D. in 1911. She published two important textbooks at this time, *Outlines of Experimental Physiology* in 1905 and *Laboratory Outlines of Physiology* in 1910. Throughout her academic career, she focused her research on both vertebrate and invertebrate animals, investigating physiological phenomena such as circulation, respiration, nervous functions, and even the development of embryos.

Continuing her support of women in the sciences after her 1925 retirement, Hyde endowed scholarships earmarked for women studying the sciences at the University of Kansas and at Cornell University. In 1945, the year she died, Hyde established the Ida H. Hyde Women's International Fellowship of the American Association of University Women.

Hyman, Libbie Henrietta
(1888–1969)
American
Zoologist

Libbie Hyman's major contribution to the sciences was her publishing of *The Invertebrates,* six volumes of which appeared between 1940 and 1968, a year before her death. The monumental task of compiling this comprehensive reference is even more impressive for the fact that Hyman supported herself on sales of her earlier books throughout its preparation, as she held no paid appointment for the majority of her career.

Hyman was born on December 6, 1888, in Des Moines, Iowa. She attended the University of Chicago, where she earned her bachelor of science degree in 1910. She continued on at the university, pursuing a doctorate in zoology, which she received in 1915. Charles Manning Child retained her in the department of zoology, where she worked as an assistant until 1928, when she was promoted to an associateship. Hyman conducted her research on the physiology and morphology of the lower invertebrates, specifically focusing on the taxonomy of planarians, or free-living flatworms.

During this time, Hyman published several books on invertebrate and vertebrate zoology, anatomy, physiology, and embryology that proved successful enough to provide her some income for the remainder of her career. In 1919, she published *A Laboratory Manual for Elementary Zoology,* which was reprinted in 1926. In 1922, she published *A Laboratory Manual for Comparative Vertebrate Anatomy.* She also published *Comparative Vertebrate Anatomy* in 1942 after she started composing her masterwork, *The Invertebrates.*

In 1931, Hyman's position at the University of Chicago ended when the department chair retired, leaving her without a supporter in the department. According to Margaret W. Rossiter in *Women Scientists in America,* other members of the department found her abrasive, and she may also have been ostracized due to her Jewish heritage. This abandonment occurred at the beginning of the depression, increasing the difficulty for a woman to find a job in the sciences; Hyman spent the next six years unaffiliated with any institution, though she continued her research independently at the American Museum of Natural History. Recognizing her need and the worthiness of her research, the museum offered her office space as well as a laboratory from 1937 on, but not a paid position. Despite these obstacles, Hyman persisted in her preparation of the manuscripts for the six volumes of *The Invertebrates.*

Though her work went unpaid, it did not go unrecognized. In 1954, the National Academy of Sciences awarded her the Elliot Gold Medal, and in 1961, the academy elected her as a member. In 1960, the Linnaean Society of London awarded her its gold medal. She received honorary degrees from the University of Chicago in 1941, Goucher College in 1958, Coe College in 1959, and Upsala College in 1963. She also served as the president of the Society of Systematic Zoology from 1959 through 1963. She maintained memberships in several professional organizations, including the American Society of Zoologists, the American Society of Naturalists, and the American Society of Limnology and Oceanography. Hyman died on August 3, 1969, in New York City.

Hypatia of Alexandria
(c. 370–415)
Greek
Mathematician and Philosopher

Regarded as the first woman in history to teach advanced mathematics, Hypatia of Alexandria was a mathematician, scientist, and philosopher. A dedicated educator, Hypatia lectured at the Neoplatonic School as well as publicly to the masses. She wrote a number of books and critiques on mathematics and philosophy, but none survived. Hypatia was considered to be a brilliant scholar and the only famous female scholar in ancient Egypt. She strongly influenced her students, who came from near and far to hear her famed lectures.

Little is known about the details of Hypatia's life. She was born sometime around A.D. 370. Her father, Theon of Alexandria, was a well-known mathematics scholar and philosopher who most likely acted as her educator. Theon taught at the Museum of Alexandria, a cultural and intellectual gathering place that included a number of schools, a celebrated library, and public lecture halls. Theon encouraged Hypatia, in her quest for knowledge and taught her mathematics; sciences, including astronomy; and philosophy. Theon also instructed Hypatia about world religions, rhetoric, and the fundamentals of teaching. With this well-rounded education, Hypatia learned how to captivate audiences with her knowledge and speaking abilities. It was believed that Hypatia's intelligence exceeded that of her father's.

Hypatia taught at the Neoplatonic School in Alexandria, becoming the school's director around A.D. 400. At the Neoplatonic School, Hypatia was a mathematics and philosophy instructor, and her lectures were popular and well attended. In her philosophical teachings, Hypatia was influenced by Neoplatonism, a theory of philosophy developed in the third century A.D. by Plotinus and Iamblichus. Neoplatonism was a combination of Platonism and mysticism, and Plotinus believed that there was an ultimate reality that was beyond the intellectual grasp of humans. In addition to her teachings, Hypatia was also known for donning a philosopher's cloak and walking through Alexandria to give impromptu philosophical lectures.

Hypatia wrote a number of books, including the *Astronomical Canon* and commentaries on *Arithmetica* by Diophantus and the astronomical works by Ptolemy. Hypatia also edited *On the Conics of Apollonius,* which defined the conic sections that came to be known as the parabola, hyperbola, and ellipse. Hypatia's contribution to the mathematical concept of conics made the ideas easier to understand, and her work was well received. In cooperation with her father, Hypatia also worked on an 11-part commentary on *Almagest* by Ptolemy as well as a new edition of Euclid's *Elements.*

Largely from surviving letters by Synesius of Cyrene, one of Hypatia's best-known students, it was thought that Hypatia might have been involved with the invention of the plane astrolabe, a device used by Greek astronomers to ascertain the position of the sun and stars. Hypatia may have also worked on the graduated brass hydrometer, an instrument used to measure the gravity of liquids, and the hydroscope, a device designed to allow for the study of objects under water. Synesius may have assisted Hypatia with these inventions.

Hypatia met with a brutal death in 415 when she was killed by a violent mob. Though the specific reasons for her murder are unclear, it was believed she may have been killed because of her seemingly pagan beliefs and her independent nature. A rivalry between Hypatia's friend Orestes, Alexandria's pagan governor, and Cyril, the bishop of Alexandria, also may have resulted in her death. A learned and talented educator and scientist, Hypatia was an accomplished mathematician and philosopher admired by many.

_ I _

Ildstad, Suzanne T.
(1952–   )
American
Immunologist and Surgeon

Transplant surgeon and immunologist Suzanne Ildstad's pioneering research on bone marrow cells paved the way for the development of new treatments for deadly diseases. Ildstad's discovery of a type of bone marrow cell that appeared to facilitate the success of bone marrow transplants offered hope to the medical community and to those facing these transplants, which were conducted primarily as a last resort. The potential of Ildstad's research also applied to organ transplants. Ildstad applied her research to the treatment of a variety of autoimmune diseases.

Born on May 20, 1952, in Minneapolis, Minnesota, Suzanne Ildstad may have been destined to enter medicine. Her grandmother worked as a scrub nurse for brothers William and Charles Mayo, who established the renowned Mayo Clinic. Ildstad's mother was also a nurse, and her training had been completed at the Mayo Clinic. Ildstad was also influenced by a psychiatrist neighbor who knew of Ildstad's interest in medicine. When Ildstad was in high school, her neighbor aided her in securing a summer position at an adolescent psychiatric center.

During her freshman year of college at the University of Minnesota, Ildstad met David J. Tollerud, a fellow student whom she married a year later, after she completed her first year of college and Tollerud his third. Ildstad graduated summa cum laude in 1974 with a bachelor's degree in biology. She then entered medical school at Mayo Medical School in Rochester, Minnesota, planning to become a surgeon. After earning her medical degree in 1978, Ildstad completed her surgical residency at Massachusetts General Hospital in Boston, Massachusetts. During her four years in Boston, Ildstad developed an interest in research, particularly in regard to transplants.

After her residency, Ildstad worked as a research fellow at the National Institutes of Health from 1982 to 1985.

Suzanne T. Ildstad, a surgeon and immunologist whose discovery of a type of bone marrow cell that facilitates bone marrow transplants has offered hope to many people suffering from terminal diseases. *(Courtesy of Suzanne T. Ildstad)*

There she investigated the role of marrow transplants in terms of organ transplant rejection. It was during this period that Ildstad developed her progressive theory regarding T cells. A bone marrow transplant was a dangerous procedure with a high fatality rate. This was because the patient's bone marrow, and thus the immune system, was destroyed then replaced with the replacement marrow, which had to come from a matching donor. Many factors could lead to the failure of the transplant and the death of the patient. One threat was graft versus host disease, in which the T cells, or immune cells, from the healthy donor marrow attacked the new host. In the late 1970s and early 1980s, research aimed to remove T cells from donor marrow to prevent graft versus host disease but engraftment failed in 70 percent of the transplants. Researchers concluded that T cells were a critical part of engraftment, but Ildstad had a different idea. She believed another, as-yet-undiscovered cell, one that facilitated engraftment, was being removed with the T cells.

Ildstad began to explore her idea after joining the University of Pittsburgh in 1988. With a research team, Ildstad conducted extensive experiments to discover and isolate the cells that helped bone marrow to take, or engraft. Ildstad found the facilitating cells in 1994. The cells possessed protein markers similar to those of T cells, which was most likely why they had been removed with the T cells during the earlier research studies. Ildstad's discovery meant graft versus host disease could be avoided, and it also had implications in other areas of medicine, including organ transplantation. The development of chimeric immune systems, which combined the characteristics of the donor and the recipient, became a viable possibility with Ildstad's research.

In 1996, Ildstad left the University of Pittsburgh and joined the Allegheny University of the Health Sciences. In 1999, Ildstad became the director of the Institute for Cellular Therapeutics and professor of surgery at the University of Louisville. There she continued her research of the chimeric immune system and has applied the technique in clinical trials. Ildstad is also exploring the possibilities of using induced chimerism to treat and cure bone marrow disorders, sickle-cell anemia, and autoimmune diseases, such as diabetes and rheumatoid arthritis.

Suzanne Ildstad's breakthrough discovery of the facilitating cells of the bone marrow have opened up a host of possibilities for the treatment of countless disorders and diseases. In 1997, Ildstad was elected to the National Academy of Science's Institute of Medicine. She received the James A. Shannon Director's Award in 1991 and has been recognized by numerous organizations for her work. Though a devoted physician and researcher, Ildstad makes her family, including her two children and her husband, who also became a physician, a priority.

- J -

Jackson, Shirley Ann
(1946–   )
American
Physicist

The first African-American woman to receive a doctorate from the Massachusetts Institute of Technology (MIT), physicist Shirley Jackson was also one of the first two African-American women to attain a doctorate in physics in the United States. After a stellar career in private industry and academics, Jackson was appointed to head the U.S. Nuclear Regulatory Commission (NRC) and was named president of Rensselaer Polytechnic Institute. Her groundbreaking research on the interaction of electrons on liquid helium films earned her election to the American Physical Society.

The second daughter of George and Beatrice Jackson, Shirley Jackson was born on August 5, 1946, in Washington, D.C. Her father cultivated her early interest in science. After obtaining her B.S. from MIT in 1968, Jackson continued to study physics, earning her Ph.D. from MIT in 1973 for her dissertation, "The Study of a Multiperipheral Model with Continued Cross Channel Unitarity." While at MIT, Jackson was also actively involved in the university's Black Student Union. Dedicated to increasing the number of African Americans matriculating at MIT, she helped boost African-American student enrollment at the elite university from two to 57 in one year.

From 1973 to 1974, Jackson was employed as a research associate at the Fermi National Accelerator Laboratory. After accepting the post of visiting science associate at the European Organization for Nuclear Research in 1974, Jackson returned to Fermi in 1975 as a research associate in theoretical physics. From 1976 to 1977, Jackson worked at the Stanford Linear Accelerator and the Aspen Center for Physics. She married physicist Morris Washington, with whom she had one son, Alan.

Jackson entered the private sector in 1978 when she accepted a position at Bell Telephone Laboratories. Initially hired to work in the theoretical physics division at Bell, Jackson later transferred to the company's scattering and low-energy physics research laboratory. She would remain at Bell until 1995, exploring an array of topics ranging from Landau theories of charge density waves in one and two dimensions to the electronic, optical, magnetic, and transport properties of novel semiconductor systems. In 1991, Jackson became a professor of physics at Rutgers University, where she remained until 1995, working as a consultant to Bell labs all the while.

In 1995, President Bill Clinton appointed Jackson to be the commissioner of the NRC, the body responsible for ensuring the safety of the U.S. nuclear industry. Jackson thus became the first African-American woman to sit on the NRC as well as the first woman and the first African American to lead it. Jackson's duties at the NRC were myriad. In addition to preventing accidents at nuclear power plants, she was to oversee the safe disposal of nuclear waste. Soon after she took the helm of the NRC, allegations surfaced that the organization was lax in its enforcement of safety standards. Jackson resolved to bolster the agency's authority, and she won the respect of consumer rights activists, including Ralph Nader.

Jackson's diverse accomplishments have been duly recognized by her peers. Honored with MIT's Taylor Compton Award, she was also elected a fellow of the American Physical Society and the American Academy of Arts and Scientists. She has won the New Jersey Governor's Award in Science as well as the Candace Award. In 1999, she was named the 18th president of Rensselaer Polytechnic Institute. She has continued to serve as chief of the NRC and is active in education, science, and public policy organizations.

Jacobi, Mary
(1842–1906)
American
Pediatrician

Mary Jacobi paved the way for women to enter the field of medicine, first by practicing medicine herself and next as the founder of the Association for the Advancement of Medical Education for Women. She headed the children's department at Mount Sinai Hospital, which became renowned as the model for pediatric facilities at that time.

Mary Putnam was born in 1842. In 1864, she attended the Female Medical College in Pennsylvania, but this institution kept her for only one year, as she surpassed its ability to teach her after only a "brief course." She graduated from the New York College of Pharmacy, the first woman to do so, and then became the first woman to apply to the École de Médecin in Paris, France. The prestigious institution accepted her but advised her to dress in menswear to disguise herself, though she specifically ignored this advice and dressed as she always had, in women's attire. She studied there until her 1871 graduation, when she returned to the United States as a fully accredited medical doctor.

Putnam ensconced herself in the New York City medical community upon her return to the United States, gaining entrance into the New York Academy of Medicine. In 1872, she founded the Association for the Advancement of Medical Education for Women to redress some of the problems she experienced trying to gain a medical education as a woman in the United States at that time. In New York, she acquainted herself with Abraham Jacobi, an influential figure in the medical establishment in the city. He gained an excellent reputation not only as an adept physician but also as a man of integrity who championed the cause of raising health standards, especially of the urban poor. Putnam was taken with Jacobi, and the couple married in 1875.

Mary Jacobi presided over the dispensary at Mount Sinai Hospital for its first year of existence in 1874, when she hosted 4,592 consultations and filled 13,004 prescriptions in the two basement rooms accorded to her. Within two years, she had expanded the dispensary to four rooms, each corresponding to one of the hospital's departments: internal medicine, surgery, gynecology, and pediatrics. The dispensary transformed into the Children's Department in 1878, the first such pediatric service in a New York hospital and indeed the first in the country. The department received a $25,000 endowment from Michael Reese of California to sustain its activities.

In 1904, the new "Children's Pavillion" at Mount Sinai Hospital housed 58 beds in rooms with views of Central Park. The roof was converted into a nursery-solarium and a playground with a protecting balustrade. Besides working for the Mount Sinai Children's Department, Jacobi also maintained a large private practice, and she taught at the New York Infirmary for Women & Children. Outside of her professional life, she supported the suffrage movement and encouraged women to enter the field of medicine despite the inherent obstacles that she had to navigate herself. After several years of declining health, Jacobi died in New York City on June 10, 1906.

Jacobs, Aletta Henriette
(1854–1929)
Dutch
Physician

Aletta Jacobs was born in February 1854, in the Groningen province of the Netherlands. From the age of six, she wanted to follow in the footsteps of her father and older brothers (she was the eighth child in a family of 11) by becoming a doctor. At finishing school, instead of studying how to be a demure female, she scrutinized Dutch law to find that no law expressly prohibited women from attending university. At the age of 16, she took her first step in that direction by passing the examination to become an apothecary and apprenticed under one of her brothers. Then, in 1871, she petitioned the Dutch liberal Prime Minister Thorbeck to allow her to attend lectures in medicine at the University of Groningen. With the blessing of both her father and the prime minister, she entered the University of Groningen on April 29, 1871, at the age of 17. After Jacobs set this precedent, universities nationwide opened their doors to women students, a result that brought Jacobs a sense of accomplishment.

Jacobs struggled against the opposition of male students (including her brother Johan) and against chronic malaria and a serious bout of typhoid fever. In April 1878, she passed her examinations allowing her to practice medicine. Intent on earning her M.D., though, she wrote a thesis entitled "Localization of Physiological and Pathological Phenomena in the Brain" to earn her medical degree in April 1879, making her the first female physician in the Netherlands.

Jacobs opened the office of her practice on January 17, 1882. Male doctors tried to convince her to limit her practice to midwifery, but she heeded her own instinct instead. She did devote most of her care to women and children, though. She noticed a correlation between the social circumstances of her patients and their medical conditions, prompting her to commence her lifelong devotion to social causes. She first lobbied for legislation to shorten shopkeepers' work shifts and to allow them to get off their feet when not serving customers. She later rallied against prostitution as a health hazard to the prostitutes as well as to the wives who contracted venereal diseases from their wayward husbands. Dutch legislators reluctantly agreed to legally abolish prostitution.

More controversial was Jacobs's support of birth control, which she viewed as a medical necessity. In 1882, she obtained a *pessarium occlusivium,* and thereafter she prescribed diaphragms. She became notorious internationally as the first outspoken proponent of contraception. In the late 1880s, she met the radical journalist Carel Victor Gerritsen. The couple established a "free union" in protest against marital law, but in 1892, they acquiesced to social pressure to marry.

Jacobs retired from medicine in 1904, and her husband died in 1905. Thereafter, she devoted herself to the advancement of women's issues, namely suffrage and peace. She chaired the Dutch Committee of the International Suffrage Alliance, and from this position, she organized a conference of the International Congress of Women at the Hague on April 28, 1915. Remarkably, she managed to convince delegates from countries that were engaged in war against each other to negotiate together. In 1919, her suffragette activism paid off as women won the right to vote. Jacobs wrote up her life experiences in the book *Memories: My Life as an International Leader in Health, Suffrage, and Peace,* which finally saw publication in English translation in 1996. Jacobs died in 1929.

Jemison, Mae C.
(1956–   )
American
Physician and Astronaut

In September 1992, Mae Jemison became the first African-American woman in space when the space shuttle *Endeavor* lifted off beyond the earth's atmosphere. Jemison resigned from NASA the very next year to commence work on applying issues from the social sciences in the technological realm of the hard sciences, specifically tailoring existing and emerging technologies for use in developing countries.

Jemison was born on October 17, 1956, in Decatur, Alabama, the youngest of three children born to Dorothy and Charlie Jemison, a schoolteacher and a maintenance worker, respectively. When Jemison was three, the family moved to Chicago, where an uncle planted the seeds of her interest in the sciences, specifically anthropology, archaeology, and astronomy. She graduated from Morgan Park High School in 1973, and at the age of 16, she received a National Achievement Scholarship to Stanford University, where she earned degrees in both chemical engineering and Afro-American studies in 1977.

While in medical school at Cornell University, she traveled to Cuba, Kenya, and Thailand to practice medicine. She earned her M.D. in 1981 and fulfilled her

Astronaut Mae Jemison, the first African-American woman in space. *(Courtesy Mae Jemison)*

medical internship at the Los Angeles/Southern California Medical Center in 1982. She remained in Los Angeles to practice general medicine at the INA/Ross Loos Medical Center until January 1983, when she joined the Peace Corps as a medical officer assigned to Sierra Leone and Liberia. She returned to Los Angeles in June 1985 to work for CIGNA Health Plans.

In 1987, NASA accepted Jemison from a pool of almost 2,000 applicants to enter its astronaut training program as one of only 15 candidates. On September 12, 1992, the space shuttle *Endeavor* launched into space with Jemison aboard as a science mission specialist. As the first woman of color to travel through space, Jemison gained much notoriety. That year, Detroit's public school system named one of its alternative schools the Mae C. Jemison Academy, and the next year *People* magazine named her one of the Fifty Most Beautiful People in the World.

Instead of basking in this limelight, though, she used it as a springboard to advance social causes by founding The Jemison Group, Inc. in 1993, after resigning from NASA. This foundation supports the research and development of advanced technologies for implementation in developing countries, administering programs such as Alafiya, a satellite telecommunications system established to improve health care in West Africa, and The Earth We Share, an experiential-based science camp for young teens from around the world. Also in 1993, Jemison served as the Montgomery Fellow at Dartmouth College, teaching a course on space-age technology and developing countries. The Ivy League college not only retained her as a faculty member in its Environmental Studies Program but also hosted the Jemison Institute for Advancing Technology in Developing Countries, an extension of The Jemison Group based in Houston, Texas.

Jemison has received much recognition in her career: In 1991, *McCall's* magazine named her one of the Ten Outstanding Women of the '90s, and in 1993, she was named one of the Fifty Most Influential Women by *Ebony* magazine. She won the Essence Award in 1988, the Johnson Publications Black Achievement Trailblazers Award in 1992, and in 1993, she received both the Turner Trumpet Award and the Kilby Science Award. Both the National Women's Hall of Fame and the National Medical Association Hall of Fame inducted her as a member. Perhaps her most appropriate distinction, though, was appearing in an episode of *Star Trek: The Next Generation.*

Jex-Blake, Sophia Louisa
(1840–1912)
English
Physician

Sophia Jex-Blake confronted opposition at every step in her pursuit to practice medicine, but she defied all expectations by pushing her will through. She attended classes at the Harvard Medical School and at Edinburgh University against protests from male students and faculty alike. She went on to establish first the London School of Medicine for Women and then the Edinburgh School of Medicine for Women.

Jex-Blake was born in 1840 in Hastings, England. Her mother was Mary Cubitt, and her father, the prominent physician Thomas Jex-Blake, had retired from practice by the time she was born. Against the Anglican beliefs of her parents, she attended classes at Queen's College in London starting in 1858. Much to their vexation, the next year the college offered Jex-Blake a tutorship that her parents allowed her to accept only as an unpaid position, as work did not befit a middle-class woman at the time. She retained the position until 1861.

After teaching in Germany, Jex-Blake traveled to the United States, where her encounter with Dr. Lucy Sewell, the resident physician at the New England Hospital for Women, convinced her to pursue medicine. She studied medicine in New York City under ELIZABETH BLACKWELL from 1865 until 1868. In 1867, she and fellow New England Hospital student Susan Dimmock applied to the Har-

vard Medical School, which soundly rejected them by a vote of seven opposed and only one in favor; undeterred, they pushed their hand by reapplying for the next year, and in the meanwhile, they attended lectures by Dr. Hasket Derby, much to the consternation of the medical school faculty.

In 1868, the death of her father beckoned her back home to comfort her mother in grieving. While her mother recovered from her grief, Jex-Blake wrote up her experiences in the United States in the book *A Visit to Some American Schools and Colleges,* relating her encounters with American experiments in coeducation. She also investigated the possibility of continuing her medical training. Parliamentary law prohibited women from studying medicine. Jex-Blake circumvented this statute by migrating to Scotland in 1869, where she repeated her Harvard tactic of attending medical lectures. Against the protestations of the male students and faculty, Jex-Blake and her cohort, Edith Pechy, took and passed their examinations in 1873, but university statutes prevented them from receiving the medical degree. Back in England, Jex-Blake and three other women pursued diplomas in midwifery; rather than grant Jex-Blake and the other women degrees, the Board of Examiners of the Royal College of Surgeons resigned, as they lacked any other means of preventing the women from gaining what they sought. Jex-Blake publicly called their decision into question, first by cofounding the London School of Medicine for Women in 1874 with ELIZABETH GARRETT ANDERSON. Second, she waged a political campaign against the barring of women from the practice of medicine, a movement that found support from Russell Gurney, who pushed legislation through Parliament in 1876 calling for equal educational opportunity for women. The Irish College of Physicians followed suit by qualifying and licensing Jex-Blake in 1877 as its first woman doctor.

Jex-Blake suffered a bitter disappointment when Anderson passed her over in tapping Isabel Thorpe to head the London School of Medicine. Instead of wallowing in her sense of betrayal, Jex-Blake turned around and founded the Edinburgh School of Medicine for Women in 1886. She also established a successful practice there while rallying for women's suffrage and entrance into the professions. She retired in 1899, retreating to Tunbridge Wells in Kent, England. Jex-Blake died in 1912.

Johnson, Katherine Coleman Goble
(1918–   )
American
Aerospace Technologist

Katherine G. Johnson, as she is known, played an instrumental role in the National Aeronautics and Space

Administration's (NASA) space program by making the necessary but complicated computations predicting the flight paths of satellites and spacecraft. Computers now handle these calculations, but Johnson performed them by hand before these kinds of computer programming and technology were available. Johnson had to account for not only the airspeed of the craft but also the interplanetary trajectories affecting the gravitational pull on the crafts. Johnson's work pioneered a new approach to navigation, taking into account variables exclusive to space travel.

Katherine Coleman was born on August 26, 1918, in White Sulphur Springs, West Virginia. Her father, a laborer, was so committed to the education of his children that he moved the family at the beginning of each school year to the town of Institute, where the influence of West Virginia State College created educational opportunities for African Americans that were not available elsewhere. This experience must have made a strong impression on Coleman, as she studied education there and graduated summa cum laude in 1937 with a bachelor of science degree in mathematics. She continued with graduate study in mathematics and physics at West Virginia University.

While in graduate school, Coleman met James Goble, and the couple married. They had three daughters, all of whom followed in their mother's footsteps as mathematicians and teachers. Goble taught math and French (her other college major) in Marion, Virginia, and in Newport News, Virginia, where she also directed the programs of the local USO. In 1953, NASA hired her as a mathematician at its Langley Research Center in Hampton, Virginia. There, she worked first on the B-17 bomber before applying her computational skills on tracking the orbits of unmanned satellites. One such satellite was the *Earth Resources Satellite,* which circled above the world in search of buried minerals and other natural resources invisible to the eye.

After the death of her first husband, Goble married James A. Johnson and took the name Katherine G. Johnson. As the space program progressed from unmanned to manned flights, Johnson continued to make the necessary calculations for their safe passage into space. It was imperative that these spacecraft exit and reenter the earth's atmosphere at precise times, as the passage of seconds and minutes changed the trajectory calculations radically. Johnson and her team worked up the computations governing astronaut Alan B. Shepard Jr.'s first spaceflight. She later worked on the Apollo moon missions, first setting their itinerary and afterward analyzing the data gathered in the Apollo flights.

NASA recognized Johnson's contributions to the space program with its Group Achievement Award twice, first in 1967 and next in 1970. She received other special achievement awards in 1970, 1980, and 1985. She continued to work for NASA until her 1992 retirement. She then devoted her time to breaking down cultural stereotyping of mathematics as a difficult or boring topic. To promote the study of mathematics, she participated in a U.S. Department of Education-sponsored television documentary promoting mathematics. She has also contributed her time to the Girl Scouts and the YWCA.

Johnson, Virginia E.
(1925–)
American
Sex Therapist

With William Howell Masters, Virginia E. Johnson studied human sexuality under laboratory conditions. After the duo published their findings in a book entitled *Human Sexual Response,* the uproar that ensued heightened public interest in sex therapy. It also boosted the popularity of the pair's Reproductive Biology Research Foundation in St. Louis, Missouri, which treated couples with sexual problems. In addition to counseling clients, Johnson taught sex therapy to professional practitioners and coauthored several more books with Masters.

The elder of Herschel and Edna (Evans) Eshelman's two children, Johnson was born on February 11, 1925, in St. Louis, Missouri. Her family moved to Palo Alto, California, in 1930, returning to St. Louis three years later. Johnson's personal life was more turbulent. After two brief marriages in the 1940s, she wed George Johnson, an engineering student and dance bandleader, in 1950. After having two children, Scott Forstall and Lisa Evans, the couple divorced in 1956. Johnson next married her professional collaborator, William Masters, in 1971. They were divorced in 1992.

Johnson never received a college diploma. She enrolled at Drury College in Springfield, Missouri, in 1941, but she left school after her freshman year. For the next 16 years, Johnson held a number of odd jobs. In 1957, though, she was offered a position as a research staff associate for William Masters—then an associate professor of clinical obstetrics and gynecology at Washington University in St. Louis. Masters was just embarking on the first clinical study of human sexuality. Over the next seven years, the study gathered scientific data from 694 volunteers, using electroencephalography, electrocardiography, and color monitors. Johnson and Masters studied their subjects in various modes of sexual stimulation and described the four stages of sexual arousal, demonstrated the inadequacies of certain types of contraceptives, discovered that vaginal secretions in some women prevent conception, and noted that sexual satisfaction does not decline with age.

In 1964, Johnson and Masters formed the Reproductive Biology Research Foundation in St. Louis to treat couples with sexual problems. Johnson served as the foundation's research associate. (She was promoted to assistant director in 1969 and codirector in 1973). In 1966, the duo published their seminal work—*Human Sexual Response*. Although written in an academic style, the book became a popular hit, selling more than 300,000 copies. As a result, their practice at the foundation boomed. In 1970, they published their second book, *Human Sexual Inadequacy*, which proposed that sexual dysfunction stems from cultural attitudes rather than physiological or psychological issues.

After marrying in 1971, Johnson and Masters continued their professional alliance, cofounding the Masters and Johnson Institute in 1973. Johnson oversaw daily operations, while Masters concentrated on scientific research. The couple coauthored their third book—*The Pleasure Bond: A New Look at Sexuality and Commitment*—which described commitment and fidelity as the basis for enduring sexual bonds. Masters and Johnson subsequently began training dual-sex therapy teams and leading regular workshops for college professors and marriage counselors. Masters and Johnson's fourth book, *Homosexuality in Practice* (1981), asserted that homosexuality was a learned behavior. The work engendered significant controversy over both its conclusions and its methodology. Masters and Johnson's final collaborative publication was released to even more criticism. Published in 1988, and coauthored by Robert Kolodny, *Crisis: Heterosexual Behavior in the Age of AIDS* inaccurately predicted an epidemic of AIDS cases among heterosexuals and suggested that it might be possible to contract the disease from toilet seats.

Masters and Johnson's personal and professional bonds frayed. The board of their institute was dissolved, and Johnson's son-in-law, William Young, became acting director. In 1992, Masters and Johnson divorced. Johnson took most of the institute's records and continued to pursue her research. She formed the Virginia Johnson Masters Learning Center in St. Louis in 1994, where she remains director. Her new organization produces instructional materials for couples with sexual dysfunction.

Johnson is credited with helping to launch the scientific study of human sexuality. With Masters, she formed a new field of scientific inquiry as well as a new branch of counseling and psychology. She trained countless professionals in the practice of sex therapy, and with Masters, she brought sexual dysfunction out of the bedroom and into the medical community. She and Masters jointly received the Sex Education and Therapists Award in 1978 and the Biomedical Research Award of the World Sexology Association in the following year. Johnson is a member of the American Association for the Advancement of Science and the Society for the Study of Reproduction.

Joliot-Curie, Irène
(1897–1956)
French
Nuclear Physicist

Irène Joliot-Curie collaborated with her husband, Jean Frédéric Joliot-Curie, in the discovery of artificial synthesis for radioactive material and the discovery of the neutron. For these advances, the wife and husband couple was awarded the 1935 Nobel Prize in chemistry.

Curie was born on September 12, 1897, in Paris, the daughter of the Nobel laureate physicists Pierre and Marie Curie. She followed in their footsteps by studying under the Faculty of Science at the University of Paris. She served in World War I as a nurse radiographer. After the war, she joined her mother as an assistant at the Institute of Radium. While working in that laboratory, she conducted her research on alpha rays of polonium that served as the locus of

Irène Joliot-Curie, with husband Frédéric. The two were jointly awarded the 1935 Nobel Prize in chemistry for their discovery of artificial radioactivity. *(Société Française de Physique, Paris, courtesy AIP Emilio Segrè Visual Archives)*

her doctoral dissertation. She was hooded as a doctor of science in 1925.

Irène met Frédéric Joliot when he was also an assistant at the Institute of Radium, and the couple married in 1926. They had two children together, one daughter named Hélène and one son named Pierre. They fused not only their personal lives but also their professional lives, conducting much of their research in tandem. Together, they discovered that stable elements could be destabilized to create radioactive material artificially. In 1933, they bombarded alpha particles at the stable element boron, which created a radioactive compound of nitrogen. They jointly published their results in the 1934 paper, "Production artificielle d'éléments radioactifs. Preuve chimique de la transmutation des éléments." The Nobel committee recognized the profound implications of their discovery by awarding both Joliot-Curies the Nobel Prize in chemistry for 1935.

The Faculty of Science in Paris had appointed her as a lecturer in 1932, and in 1937, it conferred on her the title of professor. The year before, the French government had named her undersecretary of state for scientific research. In 1938, Joliot-Curie's research on the behavior of heavy element neutrons opened up one door on the way to the discovery of nuclear fission. The next year, the Legion of Honour inducted her as an officer. During World War II, Frédéric led the underground resistance movement as the president of the Front National.

After the war, Irène succeeded her mother as the director of the Institute of Radium. Also in 1946, she was named a commissioner for atomic energy, a position she retained for six years during which France amassed its first atomic stockpile. After the war, she worked to promote peace as a member of the World Peace Council. She also contributed her energy to women's rights as a member of the Comité National de l'Union des Femmes Françaises.

As with her mother, Marie Curie, who died from sustained overexposure to radiation, Irène Joliot-Curie also died from lifelong overexposure to radiation. She contracted leukemia and died in Paris on March 17, 1956. Her husband succeeded her as director of the Institute of Radium. He also continued to oversee the construction of the 160 MeV synchrocyclotron at the new center for nuclear physics in Orsay, a project she had commenced before dying.

Jones, Mary Ellen
(1922–1996)
American
Biochemist

Mary Ellen Jones established her reputation as a biochemist early in her career when she isolated carbamyl phosphate—one of many molecules that form the building blocks of biosynthesis. Her discovery paved the way for important work on deoxyribonucleic acid (DNA) and ribonucleic acid (RNA) and helped to reveal how cells produce the molecules that comprise DNA. She is also renowned for her research on enzyme action, specifically on how metabolites (the products of metabolism) control enzyme activity. In addition, she proved that enzymes can be involved in more than one task. Her studies of metabolic pathways were essential to later work on fetus development and cancer.

Born on December 25, 1922, in La Grange, Illinois, Mary Ellen Jones was one of the four children of Elmer Enold and Laura (Klein) Jones. After receiving her bachelor's degree from the University of Chicago in 1944, Mary Ellen Jones began doctoral studies at Yale University, where she earned her Ph.D. in biochemistry in 1951.

From 1951 to 1957, Jones worked with future Nobel Prize winner Fritz Lipmann at the Biochemical Research Laboratory at Massachusetts General Hospital. She made her first important discovery while collaborating with Lipmann and his team. Lipmann had already noted a group of molecules that he believed to lie at the heart of biosynthesis, but Jones became the first to isolate one of these molecules—carbamyl phosphate, which is one compound in the sequence by which nucleotides are made. (Nucleotides are the links in the helical chain that is DNA.)

Jones remained at the Biochemical Research Lab until 1957. With Lipmann, she was one of the first researchers to recognize that adenosine triphosphate (ATP) is split to produce a mononucleotide and inorganic pyrophosphate during biosynthetic reactions. (The ATP division releases energy.) Jones and Lipmann hypothesized that the synthesis of DNA and RNA takes place when inorganic pyrophosphate is released from ATP and other trinucleotides. Biochemist Arthur Kornberg later proved this theory.

In 1957, Jones took a faculty position in the biochemistry department at Brandeis University. She left Brandeis in 1966 to become an associate professor of biochemistry at the University of North Carolina at Chapel Hill (UNC). She advanced to the rank of professor in 1968, the same year she was appointed to be a professor in the university's zoology department. In the late 1960s, she collaborated with Thomas Shoaf and discovered that five human enzymes necessary for life occur in only two—rather than five—distinct proteins. With this finding, Jones established that enzymes can be involved in more than one task. In 1971, she moved to the University of Southern California, where she served as a professor of biochemistry until 1978. She then returned to UNC and was named chair of the biochemistry department, becoming the first female department head on the university's medical school faculty. In 1980, Jones once again broke barriers at UNC when she became the first woman to be named a Kenan Professor—one of the highest honors conferred by the school.

Although Jones stepped down as department head in 1989, she remained active in her field until she retired in 1995. She won international recognition for her work. In addition to being elected to the Institute of Medicine in 1981, she was inducted into the National Academy of Sciences in 1984. Moreover, she was honored as the North Carolina American Chemical Society's distinguished chemist and was named president of the American Society of Biological Chemists in 1986. She died of cancer on August 23, 1996.

Joshee, Anandibai
(1865–1887)
Indian
Physician

Anandibai Joshee broke precedent by becoming the first Hindu Brahmin woman to receive a medical degree in the Western world. Though she died before she was able to practice medicine back in India, she made an indelible mark on the culture by inspiring Indian women to pursue medical education.

Yamuna, as she was named before her marriage, was born on March 31, 1865, in Poona (now Pune), India. Her parents, Ganpatrao and Gungubai Joshee, were well-to-do landlords in Poona. Her father supported her love of learning by transforming one of the rooms of their home into a school, where she took Sanskrit lessons from Gopal Vinayak Joshee, a widowed postal worker and Sanskrit scholar. At the age of nine in 1874, she married Gopal, who was 20 years her senior. At the age of 14, she gave birth to their only child, who lived a mere 10 days. Convinced that the child would have survived with proper medical attention, Anandibai (a name meaning *joy* bestowed on her in respect of her marriage) devoted her life to becoming a doctor.

There was no precedent for women doctors in India at the time, and Joshee met vehement opposition to her studies. However, her intentions found support in the United States after Gopal wrote a letter to the Reverend Doctor R. G. Wilder that appeared in the journal he published through Princeton University, *Missionary Review*. A Mrs. Carpenter of Roselle, New Jersey, was so moved by this letter to the editor that she pledged her support for Joshee by offering her housing in the United States. Before leaving India, Joshee became the first Indian woman to deliver a public address, which she entitled "The Courage of Her Conviction." Anandibai presented the address at Serampore College Hall in Calcutta on February 24, 1883, and posited the need for women doctors in India and furthermore pledged not to convert from Hinduism to Christianity to achieve her goals.

Joshee sold her gold wedding bangles for passage to the United States. She also received financial support from

H. E. M. Jones, director-general of the Indian Post Office, who established the Jones Fund for her with his own contribution of 100 rupees in addition to 750 rupees donated by her supporters and 200 rupees from the governor-general of India. Joshee arrived on Ellis Island on June 4, 1883, where she was received by the Carpenter family, with whom she stayed until October 1, 1883.

She immediately applied to the Woman's College of Pennsylvania (later called the Medical College of Pennsylvania and Hahnemann University School of Medicine), and Dean RACHEL BODLEY recognized her commitment to learning medicine by admitting her with a generous financial aid package. Joshee wrote a 50-page thesis (the longest in her class) entitled "Obstetrics Among the Aryan Hindoos," which earned her a doctor of medicine degree on March 11, 1886. Queen Victoria wrote her a message of congratulations, signaling the significance of Joshee being the first Indian woman to receive a medical degree.

Dewan Meherjee Cooverjee of the Albert Edward Hospital in Kolhapur, India, offered Joshee a position as the physician-in-charge of the Female Ward for a salary of 300 rupees, free lodging and board, and a one-way ticket from New York to India. However, Joshee never got the chance to capitalize on this position to implement her goals of improving health conditions for women and children in India, as she had contracted tuberculosis on a speaking tour of the eastern United States. Joshee died on February 26, 1887, soon after her return to India. Though she got scant opportunity to practice medicine before she died, her act of earning a medical degree proved significant in inspiring other Indian women to pursue the study and practice of medicine.

Joullié, Madeleine M.
(1927–   )
American
Chemist

An organic chemist, Madeleine Joullié is renowned for her work synthesizing organic compounds such as tilorone, furanomycin, and cyclopeptides, which have been valuable as antibiotic and antiviral drugs. She was also a hardworking and highly respected teacher who mentored many students into their first professional experiences as research chemists.

Born on March 19, 1927, in Paris, Madeleine Joullié soon left her native country to travel with her mother and father, an international businessman, to various parts of the world. Eventually, the family settled in Brazil where Mr. Joullié set up a permanent business. In 1959, Madeleine Joullié married Richard Prange, an American physicist. She has two children, Hélène and Pierre.

Joullié was given her secondary education at a private school in Rio de Janeiro. In Brazil, Joullié adhered to a strict schedule of school and family. Mr. Joullié felt that for her university education, Madeleine needed a more open, free-wheeling environment. Accordingly, in 1945, he sent her to Simmons College in Boston, Massachusetts. Joullié lived at Simmons's International House but did not spend much time socializing. Instead, she preferred to concentrate on her studies, especially the sciences, which she discovered engaged her imagination. She graduated with a B.S. in chemistry in 1949, then immediately enrolled in the chemistry master's degree program at the University of Pennsylvania in Philadelphia. In her graduate studies, Joullié decided to specialize in organic chemistry. She won her M.S. in 1950 and her Ph.D. in 1953, both from the University of Pennsylvania.

After getting her Ph.D., Joullié decided that she wanted to remain in academia as a teacher and researcher. The University of Pennsylvania offered her a job as the instructor of undergraduate organic chemistry courses. She held this position until 1957 when she was promoted to research associate and began working much more on experimental projects in the laboratory. There was no money then for her to hire graduate students or postdoctoral fellows to assist her. Instead, she had to do most of her work herself with the help of undergraduates. By 1968, she was appointed to the rank of associate professor; she was made a full professor at the University of Pennsylvania in 1974.

In the early and mid-1970s, Joullié, now with the support of a funded group of postdoctoral assistants, began to study the reactions of simple compounds such as ketene and sulphur dioxide. By studying the chemical interplay between these compounds with relatively simple structures, she and her team developed knowledge of how more complex compounds would react chemically. They then put this knowledge to work by synthesizing the drug tilorone, which is an interferon inducer, that is, a protein that increases the resistance of cells to attacks by viruses. Later, she and her team studied how carbon–carbon bonds convert into carbon–oxygen bonds. She also synthesized furanomycin, an antibiotic that had been isolated from a natural source. In doing this, her team also better explained furanomycin's complex molecular structure. In the 1980s, she synthesized certain kinds of highly complex compounds called cyclopeptides. She and her team also examined the molecular arrangement of a depsipeptide group called the didemnins, which are derived from a marine animal found in the Caribbean called the Trididemnum.

Joullié has written three books and has had more than 180 articles and reviews published in professional journals. She has received a number of awards for her work, including the American Chemical Society's (ACS) Garvan Award in 1978 and the ACS's Henry Hill Award in 1994. She has continued to teach and conduct research at the University of Pennsylvania.

K

Karle, Isabella L.
(1921–   )
American
Chemist and Crystallographer

A chemist who has specialized in the study of the crystal structure of solids, Isabella Karle has devised new ways to determine the molecular makeup of complex organic substances. Together with her husband, Nobel Prize–winning chemist Jerome Karle, she has worked out theoretical and practical solutions that show this molecular structure. From this, a three-dimensional picture of molecular structure can be constructed. This has aided in the analysis of complex organic structures and production in the laboratory of substances that previously could be obtained only in nature.

Born in Detroit, Michigan, on December 2, 1921, Isabella Lugoski is the daughter of Zygmunt Lugoski and Elizabeth Graczyk Lugoski. Her father was a house painter, and her mother worked as a seamstress. Both of Lugoski's parents spoke little English. As a result, Isabella Lugoski began to speak English only when she attended public schools in Detroit. In 1942, she married Jerome Karle. They have three children.

Karle got hooked on chemistry while in high school. In 1938, she enrolled at Wayne State University in Detroit. Later that year, Karle transferred to the University of Michigan. She received her B.S. in chemistry from there in 1941. After graduation, she found that teaching assistantships, which helped fund the education of graduate students, were being given exclusively to males. Only after she had received a grant from the American Association of University Women was she able to enter the graduate chemistry program at Michigan. She won her M.S. in chemistry there in 1942 and her Ph.D. in 1944.

Karle's professional work at Michigan centered on the study of gaseous molecules through the technique of electron diffraction. Photographs of the interaction of electrons with gas molecules were taken as electrons were shot through a gas. These images of diffracted, concentric rings allowed scientists to learn about the angles of gas molecules and the distance that lay between them—a three-dimensional picture of the molecular arrangement.

In 1946, after the end of World War II, Karle took the knowledge she had gained from her work at Michigan to the U.S. Naval Research Laboratory (NRL) in Washington, D.C., the only lab that offered jobs to both Karle and her husband. At the NRL, Isabella and Jerome Karle began working together to solve a problem that had not yet been mastered in chemistry—figuring out a way to determine the three-dimensional molecular structure of solids by way of X-ray diffraction. This technique, which worked so well with gases, was not effective with most organic compounds. X-ray diffraction could tell the intensity of organic crystals but not its phase or molecular pattern. During the late 1940s, both Karles worked on this problem. Jerome Karle, with his partner, Herbert A. Hauptman, devised a theoretical solution, while Isabella Karle designed and built a machine that would measure the crystal phase through X-ray diffraction. For their part of the work, Jerome Karle and Herbert Hauptman won the Nobel Prize in chemistry in 1985.

Isabella Karle went on to use her technique to analyze the crystal structures of the venom found in the skin of certain South American frogs; a substance found in a Panamanian wood that repels worms and termites; and peptides, clusters of amino acids, some of which are useful as drugs. She is a strong defender of the value of basic scientific research. "Often," she has said, "it takes 10 or 50 or even more years for the knowledge derived from basic research to be applied to high technology, but the pool containing basic information must be replenished, otherwise it will run dry."

For her research in the crystal structure of molecules, in 1993 Karle was given the Bower Award, the highest honor bestowed by the Franklin Institute. She was the first woman to receive this prize. She was awarded the American Chemical Society's Garvan Award in 1976 and elected to the National Academy of Sciences in 1978. In 2000, Karle continued to work at the Naval Research Laboratory.

Kaufman, Joyce Jacobson
(1929–   )
American
Chemist

Trained as a physical chemist, Joyce Kaufman specialized in quantum chemistry, the application of the theories of quantum mechanics to the study of complex molecules. Employed by Johns Hopkins University's chemistry department and its School of Medicine, Kaufman devised theoretical approaches that helped in the design of drugs that affected the central nervous system and brain chemistry. She also worked on computer modeling of organic compounds.

Born on June 21, 1929, in New York City, Joyce Jacobson is the daughter of Robert Jacobson and Sara Seldin Jacobson. Her father was a salesman. Jacobson's parents separated in 1935, at which time Sara Jacobson took Joyce to live with her at Sara's parents' house in Baltimore, Maryland. In 1940, Sara Jacobson married Abraham Deutch. Joyce Jacobson was largely raised by her stepfather, who encouraged her in her studies. In 1948, Jacobson married Stanley Kaufman. They have one child.

Joyce Kaufman proved to be a gifted student in Baltimore's public schools. She began reading at age two and was especially inspired by the biographies of scientists, notably Marie Curie. When she was nine, she attended a summer program for students who excelled in math and science. She graduated from an accelerated science program at a Baltimore high school and enrolled in Johns Hopkins University in 1945. Kaufman graduated from Johns Hopkins with a B.S. in chemistry in 1949.

Kaufman switched back and forth between career and education in the 1950s. From 1949 to 1952, she worked at the U.S. Army's Chemical Center in Baltimore, beginning as a technical librarian and finishing up there as a research chemist. In 1952, she returned to Johns Hopkins as a research chemist for one of her undergraduate professors. She worked in the chemistry labs, studying the isotope exchange reactions of boron hydrides. By 1958, she had reenrolled at Johns Hopkins as a student. She won an M.S. in physical chemistry in 1959 and, in 1960, was granted a Ph.D. In 1960, she attended a summer workshop at the University of Uppsala in Sweden where she studied theoretical approaches to quantum chemistry.

After getting her doctorate, Kaufman joined the Martin Company's quantum chemistry group in 1960. She became head of the group in 1962 and remained with Martin until 1969, working on various applications of quantum mechanics to chemistry. In 1969, Kaufman began her long affiliation with Johns Hopkins University. That year, she was appointed an associate professor in the Johns Hopkins School of Medicine's anesthesiology department. She was also named as the principal research scientist in Johns Hopkins's chemistry department.

At Johns Hopkins, Kaufman studied drug toxicity, comparing the standard approaches that used animals to determine toxicity with theoretical models that she developed via quantum chemical analysis. Her approach focused on drugs used as anesthetics; on neuroleptics, drugs that treat mental disorders; and on painkilling narcotics. These compounds are complex and had not been subject to quantum chemical analysis before. Her studies not only gave an accurate reading of the molecular structure but also aided in the design of drugs and made the prediction of toxicity easier without costly and time-consuming animal trials. Throughout her research work, Kaufman has also served as a teacher and has trained numerous graduate students in the techniques of quantum chemistry.

For her efforts, Kaufman was awarded the Dame Chevalier by France's Centre National de la Recherche Scientifique in 1969. She received the American Chemical Society's Garvan Medal in 1974 and was elected a fellow of the American Physical Society and the American Institute of Chemists. She has continued to teach and conduct research at Johns Hopkins University.

Keller, Evelyn Fox
(1936–   )
American
Physicist, Molecular Biologist, and Historian

Trained as a physicist, Evelyn Keller later shifted her interests to molecular biology and the history of science. As a research scientist, Keller worked to construct mathematical models that would explain chemotaxis—how and why cells move in response to chemical stimulation. She also worked to explain mathematically certain kinds of cellular pattern formations. In her later career, she devoted herself to examining the role of women in science and the history of science.

Born on March 20, 1936, in New York City, Keller is the child of Albert and Rachel Fox, Russian Jewish immigrants to the United States. Evelyn Fox married Joseph Keller in 1964. She has two children.

Keller did well in high school, but in spite of the efforts of her older brother, Maurice, a biologist, to pique her curiosity about science, she remained uninterested in science as a career. Fox graduated from a public high school in New

Evelyn Fox Keller, who has worked to construct mathematical models to explain how and why cells move in response to chemical stimulation, a process known as chemotaxis. *(Photo by Donna Coveney, Massachusetts Institute of Technology)*

York City in 1953 and enrolled in Queens College. There she drifted into science by writing papers on science fiction literature and physics for her English class. At the end of the year, she transferred to Brandeis University in Massachusetts. There she majored in physics but only as an entry into a medical school. However, by her senior year, she had realized that she had come to love physics for itself and for what she termed the "discipline of pure, precise, definitive thought. . . . I fell in love with the life of the mind. . . . and the image of myself succeeding in an area where women had rarely ventured." Keller graduated at the top of her class with a B.S. in physics from Brandeis in 1957.

Upon her graduation from Brandeis, Keller received a National Science Foundation grant to pursue her graduate education in physics. She decided to attend Harvard for her graduate work and began to study for her master's degree while a student at Harvard's women college, Radcliffe, be-

ginning in 1957. Her first years at Harvard were, in her words, a time of "unmitigated provocation, insult, and denial." She felt that she was singled out for abuse because she was one of the few women in the program and because she did not view physics as an exercise in computational skill, but rather as "a vehicle for the deepest inquiry into nature." In the summer of 1959, Keller accepted her brother's invitation to study with marine biologists at the Woods Hole lab. Here she experienced an epiphany. She realized that she could combine work in physics with molecular biology, which suited her intellectual style better than her previous work. At Harvard, she studied with Walter Gilbert, a future Nobel laureate and physics professor who had taken a similar turn toward biology. Keller was awarded a Ph.D. in physics from Harvard in 1963.

Keller began her professional career as an assistant professor at the Cornell University Medical School in New York City (1963–69). She then moved to the State University of New York in Purchase where she was associate professor of natural sciences from 1972 to 1982. During the 1960s, her research focused on mathematical modeling of biological processes, but by the 1970s, she had begun to devote more and more time to a critique of the relationship between the scientific establishment and women scientists. She opened up the then-taboo subject of the dearth of women in the sciences in a lecture she delivered at the University of Maryland in 1974. In this lecture, entitled "The Anomaly of a Woman in Physics," Keller asserted her hope that "the political awareness generated by the women's movement . . . will support young women who today attempt to challenge the dogma, still very much alive, that certain kinds of thought are the prerogative of men."

Keller followed this with a biography of the geneticist BARBARA MCCLINTOCK, *A Feeling for the Organism* (1983), and a book of essays about women and the sciences, *Reflections on Gender and Science* (1985). From 1988 to 1993, Keller held down a position at the University of California at Berkeley as professor of the history of sciences. Since 1993, she has been a professor in the Program in Science, Technology, and Society at the Massachusetts Institute of Technology.

For her work, Keller has won the Distinguished Publication Award of the Association of Women in Psychology (1986). She also won the MacArthur Foundation Fellowship Award in 1993.

Kelsey, Frances Oldham
(1914–   )
Canadian/American
Physician and Pharmacologist

A pharmacist who later became a physician, Frances Kelsey is renowned for her courageous stand against a large drug company in the early 1960s. While an officer at the U.S. Food and Drug Administration (FDA), Kelsey rejected attempts by an American drug company to market thalidomide in the United States. Under intense pressure to reverse her decision, Kelsey continued to reject the use of the drug.

Frances Oldham was born in Vancouver, British Columbia, on July 24, 1914, to Frank and Katherine Oldham. In 1943, Oldham married Fremont Ellis Kelsey, a pharmacologist she met while both were students at the University of Chicago. She has two daughters.

From her youth, Kelsey knew that she wanted to be a scientist. In 1930, she moved from Vancouver to Montreal to study for an undergraduate degree at McGill University. She completed a B.Sc. and M.S. in pharmacology in 1934 and 1935, respectively. After finishing her studies in Canada, Kelsey enrolled in the University of Chicago where she studied for her doctorate in pharmacology, which she attained in 1938. Kelsey remained at the University of Chicago as a researcher and teacher in her field. During her early work at Chicago, Kelsey and her husband discovered that certain drugs that produced no side effects in adult rabbits were toxic to the fetuses of female rabbits if the drugs were ingested by the adult female during her pregnancy. This was an important study showing the differences in reactions in adult mammals and their fetuses to the same drug.

After her marriage to Ellis Kelsey in 1943, Frances Kelsey was forced to relinquish her job at the University of Chicago's pharmacy department. The university did not allow husbands and wives to work together in the same department. Instead, Kelsey spent the next seven years raising her daughters and attending the university's medical school. By 1950, she had completed the medical courses and become an M.D. From 1954 to 1960, Kelsey worked and interned at the University of South Dakota where her family had moved after her husband was offered a job there. In 1960, the Kelseys moved to Washington, D.C., and Frances Kelsey got a job reviewing applications from drug companies for licenses for their new drugs.

Kelsey had come to the FDA just as Congress had launched an investigation about the close ties between the drug industry and the industry's regulators at the FDA. Kelsey's predecessor, Dr. BARBARA MOULTON BROWNE, had resigned her job because her decisions against the new products of the drug companies were so often reversed by her superiors.

The first application that Kelsey found on her desk in November 1960 was from Merrell, a large manufacturer that wanted approval for thalidomide, a drug that was being used in Europe as a sleeping aid and to ease "morning sickness" in pregnant women. Kelsey rejected the request because in her estimation, Merrell had not provided enough clinical testing data about the drug. Early in 1961,

Kelsey read an article about thalidomide in a British medical journal that noted that the drug had caused nervous system damage to some patients who had taken it. Kelsey also noticed that there was no information about whether thalidomide might cause side effects to patients with liver or digestive problems or about its effects on fetuses.

For these reasons, she rejected the drug in January 1961 and again in March. Kelsey was put under heavy pressure by the drug companies to reverse her decision, but she refused. She was vindicated by studies released in November 1961 that linked thalidomide with severe birth defects in children born to mothers who had taken the drug in the first three months of their pregnancies. Some of these babies were born with flipperlike arms and legs, others without ears, and others with no limbs at all.

For her courageous stand against this drug, Kelsey was given the Distinguished Federal Civilian Service Award, the U.S. government's highest honor for a civilian employee, by President John Kennedy in 1962. Kelsey remained with the FDA for the rest of her career. Because of her proven integrity and abilities, she was made chief of the FDA's Investigative Drugs Branch in 1963, its Division of Oncology and Radiopharmaceutical Drugs in 1966, and the Division of Scientific Research in 1971. By 2000, Kelsey was still working as head of the FDA's Office of Compliance.

Kessel, Mona
(1956–)
American
Physicist

Throughout her career in universities and at the National Aeronautics and Space Administration (NASA), Mona Kessel has studied the physical phenomena that occur in the region between the Sun and the Earth. Kessel has been especially interested in the bow shocks of Earth and of passing comets. During the late 1980s and through the 1990s, she has been able to contribute to an understanding of these phenomena because of fresh information from a series of solar probes and Earth-orbiting satellites.

Born on April 20, 1956, in Junction City, Kansas, Mona Kessel is the daughter of Richard A. Rusk and Louise S. Gallion. Her father was an army officer who later taught systems management at the University of Southern California, and her mother was a quality control manager for the General Services Administration. In 1983, she married Robert A. Kessel, also a physicist. They have two children, Sarah and Ellen.

Early in her schooling, Kessel found that she had an affinity for math and science. However, she had no idea that one day she would work as a space scientist. "When I was growing up, I didn't even consider [space physics] as an option because I didn't know about it," Kessel has said.

"Looking back, I realize that what I enjoyed most was problem solving." She took every math course that was offered but did not take a physics course until her senior year in high school. In physics, she found a scientific discipline that appealed to her. However, her high school counselors recommended that she should study engineering in college.

Kessel enrolled as an undergraduate at Baker University in Baldwin City, Kansas, in 1974. She anticipated moving on to engineering at the University of Southern California after two years, but she enjoyed physics so much that she stayed. She was awarded a B.S. in physics from Baker in 1978. In 1979–80, Kessel worked as an engineer at the Southwestern Bell Telephone company. This proved to be a lucrative but unfulfilling job. Realizing that her true passion was for scientific research, Kessel enrolled in the graduate physics program at the University of Kansas in 1981. Here she concentrated on the subdiscipline of space physics, the study of physical phenomena encountered in the region that lies between the Earth and the Sun. Kessel completed her Ph.D. in space physics from the University of Kansas in 1986.

After receiving her doctorate, Kessel and her husband moved to England, where they worked at the Mullard Space Science Laboratory from 1986 to 1990. At Mullard, Kessel continued the work on shocks that she had begun during her graduate studies at Kansas, but now she concentrated on bow shocks. She also got involved with flight projects, notably the European Space Agency's mission Cluster. By 1990, she was back in the United States teaching undergraduate astronomy at DeKalb College in Atlanta. In 1991, she took a position at NASA's Greenbelt Center in Greenbelt, Maryland.

At Greenbelt, Kessel intensified her study of bow shock, concentrating on Earth's bow shock. Like a boulder in a river, the interaction of Earth's magnetosphere (ending at the magnetopause) with the solar wind creates a bow shock upstream. At the bow shock, the solar wind is slowed, heated, and partially deflected around the Earth's magnetosphere. At the bow shock and just upstream, waves are created that, under certain conditions, can penetrate into the magnetosphere. Kessel finds these interactions fascinating.

Kessel has been fortunate to have benefited from the International Solar Terrestrial Physics (ISTP) program, funded and administered by NASA and the European, Russian, and Japanese space agencies. Together these agencies have launched probes at the Sun and maintain a number of Earth-orbiting satellites. These sources have provided a huge amount of data about the Sun, the solar wind, Earth's bow shock, the magnetosphere, and the intimate relationship that exists between them.

In almost real time, Kessel and ISTP tracked the effects of a solar "storm" that was ejected from the Sun on January 7, 1997. They saw the solar images, which indicated that a magnetic cloud was being shot out of the Sun, then measured the arrival of this magnetic surge at Earth's bow shock two days later. This magnetic burst compressed and bent Earth's magnetosphere, knocking out a Telstar satellite.

Kessel has been principal investigator on a number of NASA studies. She is a life member of the American Geophysical Union and is also one of the Women of NASA, participating in chat sessions with schools all over the United States. She recently won a NASA National Resource Award.

Kil, Chung-Hee

(1899–1990)
Korean
Physician

An early Korean woman doctor and political activist, Chung-Hee Kil joined the struggle for Korean independence at the same time that she was fighting for respect for women doctors within Korean society. To help train more women doctors in Korea, she cofounded the first medical school for women in Korea.

Born on February 3, 1899, Chung-Hee Kil was the daughter of Mr. and Mrs. Hyun-Suk Kil. When Mr. Kil died while Chung-Hee was in her early teens, Kil, her mother, and Kil's two siblings moved in with her paternal grandfather. The grandfather, who was from the Korean nobility, encouraged Kil to get an education, which was unusual for that time. Most Korean girls then were raised to be wives and mothers, and few were offered formal education. In 1925, Kil married Tak-Won Kim, a fellow physician. They had three children.

Kil lived through some of the most turbulent times in recent Korean history. In 1910, the Japanese government invaded and occupied Korea. The Japanese made Korea a part of its imperial system, and it was ruled by a Japanese governor-general. Most of the top posts in the new Korean administration were likewise taken by Japanese. Speaking Korean was officially forbidden, although this was an almost impossible law to enforce. In March 1919, Kil signed a petition demanding Korean independence from the Japanese. She also took part in large demonstrations against the Japanese that were held in Seoul and other Korean cities. In May 1919, the Japanese cracked down on this movement, attacking demonstrators and jailing leaders. More than 500 Koreans were killed, and thousands were imprisoned. Kil probably escaped arrest because of her young age and the fact that she was a woman.

To get her medical degree after the Japanese occupation, Kil was forced to travel to Japan, where she studied at the Tokyo Women's College from 1919 to 1923. Here she encountered Japanese hostility and suspicion but, determined to get her degree, she kept largely to herself. At the end of her medical school training, Kil helped with emergency efforts to save victims of the Tokyo earthquake of September 1, 1923, an event that affirmed her belief that she had entered the right profession.

Kil began her medical career with an internship at the Government General Hospital in Seoul. While at the General Hospital, she married Tak-Won Kim, a doctor specializing in internal medicine. Soon after their marriage, Kim and Kil had to part to continue their medical education. Kil returned to Tokyo where she met an American doctor, Rosetta Hall, who ran Tong-Dae-Moon Women's Hospital in Seoul. Hall persuaded Kil to work at Tong-Dae-Moon after she returned to Korea. There Kil performed an important task of treating Korean women, many of whom were reluctant to visit male physicians.

To train more women physicians, Kil and her husband founded the Chosen Women's Medical Training Institute in Seoul in 1928, the first school to train Korean women to become doctors. Because the Japanese were suspicious of Kil and Kim for their political activities, and because the idea of training Korean women to become doctors was frowned on by many prominent Koreans, Kil had a hard time finding the money to keep the school running. Somehow, though, she and her husband scraped by; by 1938, the year it was reorganized as the Seoul Women's Medical College, the institute had an enrollment of 64 students.

On the institute's change to a college in 1938, the Japanese took over the institution and fired Kil. The next year, Kil's husband died after having been imprisoned by the Japanese for his political activities. For the next 25 years, Kil worked as a private physician in Seoul. She retired from medicine in 1964, and in 1979, she emigrated to the United States, where her two daughters had come to establish their own medical practices.

Later in her career, Kil was recognized for her pioneering work. In 1959, she was given an award by the Korean public health ministry for her contribution to medical education in Korea. The city of Seoul recognized her work publicly in 1960, and in 1961, she was honored by Ewha Woman's University for her role in training the first generation of Korean women doctors. Chung-Hee Kil died in 1990 at the age of 92 in Cheltenham, Pennsylvania.

King, Helen Dean
(1869–1955)
American
Zoologist

A research scientist more than a teacher, Helen Dean King began her career studying the embryology of toads. However, her greatest contribution to science was her experiments on mammal inbreeding and the breeding and domestication of wild animals. She was briefly infamous when the popular press of her day noticed her work on inbreeding and reported incorrectly that she considered the taboo against incest a superstitious relic.

Born on September 27, 1869, King was the daughter of William and Lenora King. Both her mother and father came from wealthy families, and education was valued in the King household. King had one sister; she never married or had children.

For her secondary education, King studied at the local Oswego Free School, where she excelled in the sciences. In 1887, she enrolled as an undergraduate at Vassar College at nearby Poughkeepsie, New York. Vassar, an all-female institution, at that time led all other women's colleges in the amount of money it put into its science programs. King earned her B.A. from Vassar in 1892. After taking several years away from studies, King decided to enroll as a graduate student in zoology at Bryn Mawr, a women's college in Pennsylvania near Philadelphia.

At Bryn Mawr, King came under the sway of Thomas Hunt Morgan, a gifted scientist and teacher of zoology. Although King studied physiology and paleontology, Morgan directed King's studies toward morphology, the subdiscipline of zoology that was his specialty and that King had decided to pursue as well. She received her degree in zoology from Bryn Mawr in 1899 with a dissertation entitled "The Maturation and Fertilization of the Egg of *Bufo lentiginosus.*" After completing her doctorate, King stayed at Bryn Mawr as a lab assistant. She continued her work on toads and other amphibians, with an emphasis on developmental anatomy.

King moved to the University of Pennsylvania, located in Philadelphia, in 1906. She served as a research fellow in zoology there for two years, and in 1908, she began a long association with the Wistar Institute of Anatomy and Biology, also located in Philadelphia. By 1927, King had become a full professor at Wistar. She served for many years on the institute's advisory board.

At Wistar, King expanded her areas of research and came into her own as a researcher. She worked on breeding pure strains of rats that were used as experimental animals in the institute's labs. While doing this, King became interested in the effects of inbreeding. She conducted an experiment in which brother–sister albino rats were inbred and compared their growth and activity to noninbred albino rats. Kings published numerous papers about her results, which concluded that there were no significant differences between the inbred offspring and the offspring of noninbred pairs.

King also conducted domestication experiments with wild Norway rats captured on the streets of Philadelphia. Beginning with six pairs of the wild rats, King bred this group down 28 generations. She noted the mutations that resulted from this process, such as different hair type and colors, and concluded that domestication in captivity leads to a diversity in wild species rather than the anticipated homogenous effects.

King was recognized for her scientific contributions. She was listed and starred in *American Men of Science* (the star signifying that she was one of the top thousand scientists in the country). In 1932, she was given the Ellen Richards Prize by the Association to Aid Scientific Research for Women. She was editor of *The Journal of Morphology and Physiology* from 1924 to 1927 and served as vice president of the American Society of Zoologists in 1937. She was also a member of the American Association for the Advancement of Science and a fellow of the New York Academy of Sciences. Helen King died in Philadelphia at the age of 85 on March 9, 1955.

King, Louisa Boyd Yeomans
(1862–1948)
American
Horticulturist

A self-educated gardener and horticulturist, Louisa King devoted her life to studying horticultural techniques and educating the public about the care and selection of plants, shrubs, and trees. She was one of the founders of the Garden Club of America, an organization that encouraged gardening enthusiasts to pursue their avocation. King was also well known as a magazine and book writer on gardening.

Born on October 17, 1862, in Washington, New Jersey, Louisa Yeomans was the daughter of Mr. and Mrs. Alfred Yeomans. Her father was a Presbyterian minister, and her mother was a housewife. The third of five siblings, Louisa Yeomans described her life as a child as intellectually challenging and carefree. In 1890, when she was 26, she married Francis King and moved with him to Illinois. They had three children.

King received her secondary education at private schools in New Jersey, and she does not seem to have attended a college or university. Her education in botany and horticulture began when she married Francis King. The newlyweds moved into the house of her mother-in-law, Mrs. Henry W. King, in Elmhurst, Illinois, a suburb of Chicago, and remained there for the first 12 years of their marriage.

The elder Mrs. King was a knowledgeable gardener with a fine garden on the land that surrounded her house. She had cultivated more than 200 varieties of herbs and numerous flowers, plants, and fruit trees. Mrs. King also had a well-stocked library of books on horticulture and botany, and she subscribed to numerous gardening magazines.

Louisa King, who had not been exposed to such gardening lore before, found that she was taken with both an academic study of plants and plant species, and with the practical, dirty-hands work of preparing soils, pruning, spraying, and weeding. She accompanied Mrs. King on walks through the garden during which the older woman explained to her daughter-in-law the peculiarities of various plant species as she instructed her full-time gardener about garden chores that Mrs. King had planned. Through her mother-in-law, Louisa King learned the Latin names of plants and trees and became acquainted with new ideas about gardening that were being advocated by horticulturists such as Gertrude Jekyll and William Robinson.

In 1902, King and her husband moved into their own home, Orchard House, in Alma, Michigan. There, King began to plan and plant her own garden. She also wrote various experts in horticulture, striking up friendships with a number of them, especially Gertrude Jekyll. As King developed her own style and tastes, she decided to write about gardening for national magazines such as *Garden Life, House Beautiful,* and *Country Life.* Following the lead of her friend, Gertrude Jekyll, and other advocates of "modern" gardening, King promoted gardens that utilized designs and plants that fit more naturally into the surrounding landscapes. She also championed solid fields of color for flower plantings rather than the hodgepodge explosions favored by traditional Victorian gardeners.

King's first gardening book, *The Well-Considered Garden,* was published in 1915. She followed this with nine more books about garden planning, soil management, and even tool care. King was also active in gardening groups. In 1913, she was one of the founders of the Garden Club of America. In 1914, she helped found the Women's National Farm and Garden Association, which encouraged women to study horticulture and botany in colleges and helped women manage farms during the absence of their husbands in World War I.

King won numerous honors for her work, among them the Massachusetts Horticulture Society's Robert White Medal (1921) and the Medal of Honor from the Garden Club of America (1923). She was also elected a fellow of the Royal Horticulture Society in the United Kingdom. Louisa King died on January 16, 1948, at the age of 84 in New York State.

King, Mary-Claire
(1946–   )
American
Geneticist

As a geneticist, Mary-Claire King has taken on a variety of important projects. Early in her career, she proved that the genetic makeups of humans and chimpanzees are almost identical. Later, she was successful in identifying marker genes that indicated whether women in certain families carry a high genetic risk of developing breast cancer. King has also worked with families in Argentina, Bosnia, Rwanda, and Ethiopia who are trying to track children and grandchildren who have disappeared in guerrilla wars in those countries.

Mary-Claire King was born just outside of Chicago, in Wilmette, Illinois, on February 27, 1946, to Harvey and Clarice King. Her father was an executive for Standard Oil of Indiana, and her mother was a housewife. In 1972, she married Robert Colwell, whom she divorced in 1980. She had one child.

King loved math and sciences in high school. She enrolled in Carleton College in Minnesota and earned a B.A. in mathematics in 1966. For her graduate work, King decided to attend the University of California at Berkeley (UC Berkeley). She began work at Berkeley in biostatistics but also took courses in genetics, which she liked because to her it was akin to working out a puzzle. Upset by the war in Vietnam, King dropped out of the university in 1969 and for several years, worked for consumer advocate and social activist Ralph Nader.

In the early 1970s, she returned to Berkeley and began work toward her Ph.D. in genetics in the molecular biology lab of Allan C. Wilson, her mentor. King won her Ph.D. in 1973 with a dissertation about a study that she had done in Wilson's lab comparing the genetic makeups of humans and chimpanzees. She surprised the scientific world by finding that there was almost no genetic difference between the two species, indicating how close has been the evolutionary history of humans and chimpanzees.

By 1974, King had been hired as an assistant professor of epidemiology at UC Berkeley's School of Public Health. She became a professor of epidemiology at Berkeley in 1984 and a professor of genetics at Berkeley's Department of Genetics and Molecular Biology in 1989.

In 1975, King began looking at the genetic makeup of certain women, trying to figure out if particular families carried a gene that made them likely to develop breast cancer. This approach was at odds with the prevailing wisdom of the time: that cancer was the result of damage to genes caused by chance or by such environmental factors as chemical ingestion or radiation exposure.

At that time, scientists were just beginning their painstaking assembly of the human genetic profile—the 100,000 or so genes carried by each human. Most of these genes were unknown; however, some could be identified by the genetic proximity to so-called marker genes. By 1990, King had succeeded in identifying a marker gene for breast cancer, which she called BRCA1. Even though the actual breast cancer gene itself was found by another team in 1994, King's work had gone a long way toward narrowing down the search.

Beginning in 1984, King got involved with families in Argentina who were searching for the grandchildren they had lost during Argentina's "dirty war," which was the gov-

ernment-sponsored kidnapping and murder campaign against leftist activists that went on from 1976 to 1983. There were many women among these activists, and a number of them were murdered and buried at sea or in unmarked graves. Children captured with their mothers, or babies born to imprisoned mothers, were then anonymously put up for adoption. King put her expertise at the service of the grandparents of these children, who wanted to track down their grandchildren.

Because mitochondria DNA is passed only through the mother, King used the DNA of mitochondria, a cellular component that aids cells to convert energy, to track the children. In this way, she could conclusively prove whether particular children were the grandchildren of individual women. More than 50 Argentine children have been reunited with their grandparents as a result of this work.

In 1995, King moved to Seattle to work as professor of genetics and to head a genetics lab at the University of Washington. She is also the codirector of the Human Genome Diversity Project, a part of the larger Human Genome Project. She has served on committees of the National Institutes of Health, the National Cancer Institute, and the National Institute of Medicine.

King, Reatha Clark
(1938–)
American
Chemist

The child of an illiterate farmworker and a domestic servant, Reatha Clark King's early research in the field of fluoride chemistry proved vital to the rocketry of the National Aeronautics and Space Administration's (NASA) space program. After 25 years of academic work, including stints as a researcher, professor, dean, and university president, King became the president and executive director of the General Mills Foundation in 1988. She remains in that position and also sits on the board of several major companies, including the Exxon Corporation.

Reatha Belle Clark was born to Willie and Ola Watts Campbell Clark on April 11, 1938, in Pavo, Georgia. The second of the couple's three daughters, Clark and her mother and sisters moved to Moultree, Georgia, after Willie and Ola divorced when Clark was still young. In Moultree, she and her sisters picked cotton and tobacco after school and during summers to earn extra money for the family. Clark later married N. Judge King, with whom she had two children, N. Judge III and Scott.

After graduating as valedictorian of her high-school class in 1954, King received a scholarship to study in Atlanta, Georgia, at Clark College. Initially intending to major in home economics, with the long-term goal of becoming a teacher in that field, King's life changed

A chemist and academic administrator, Reatha Clark King has focused her career on championing educational opportunity for all people. *(Courtesy of Reatha Clark King)*

dramatically when she enrolled in the introductory chemistry class that was required of all home economics majors. King was captivated by both the subject matter and by the inspiring personality of the instructor, Alfred Springs. Springs became a mentor to King, reinforcing her new-found love of chemistry and encouraging to pursue her studies at a higher level. Upon receiving her undergraduate degree, King took Springs's advice and entered the University of Chicago's graduate program in chemistry. Winning a Woodrow Wilson Scholarship facilitated this pursuit. At Chicago, King focused on physical chemistry and took a particular interest in thermochemistry. She earned her Ph.D. in chemistry in 1963.

King took a post as a researcher with the National Bureau of Standards in Washington, D.C., after completing her graduate studies. In that capacity, she strove to design materials that could safely handle oxygen difluoride, a highly corrosive substance. She also investigated other types of intermetallic compounds (substances that are generally hard and brittle and do not follow the ordinary rules of valency). Her work ultimately assisted NASA in building more serviceable rockets for the space program. King was a valued employee at the Bureau of Standards, consistently earning high performance ratings.

King moved from the Bureau of Standards to New York City's York College in 1968, where she began teaching. A paper she wrote in 1969 on fluoride flame calorimetry garnered a Meritorious Publication Award, and in 1970, the school promoted her to Assistant Dean of Natural Sciences and Mathematics. King also earned an M.B.A. from Columbia University during the early 1970s after taking a leave of absence from York. She left the college entirely in 1977 when she was named president of Metropolitan State University in St. Paul, Minnesota. She held that position for 11 years, overseeing dramatic expansions in the school's graduate programs and an overhaul of its general curriculum. After resigning from Metropolitan's presidency in 1988, King became the president and executive director of the General Mills Foundation.

King's accomplishments have been widely recognized. In 1988, she was honored in Minneapolis–St. Paul as Twin Citizen of the year. Moreover, she holds honorary doctorates from a slew of schools, including Alverno College, Carleton College, Empire State College, Marymount Manhattan College, Nazareth College of Rochester, Rhode Island College, Seattle University, Smith College, and the William Mitchell College of Law.

Kirch, Maria Winkelmann
(1670–1720)
German
Astronomer

A major contributor to the science of astronomy, Maria Kirch was given scant credit for her more than 25 years of work in the field. In spite of the resistance she met from the German academic establishment of her time, Kirch persevered with her astronomical studies. She is credited with being the first woman to discover a comet, and she also compiled data for calendars sold by a major German scientific organization.

Maria Winkelmann was born in 1670 in the small town of Panitsch, which is located near Leipzig, at that time in the German principality of Saxony. She was the daughter of an educated man, a Lutheran minister, who insisted that she be given an education equal to that received by boys. This desire on the part of Herr Winkelmann to educate his daughter was unusual for the time; he hired a private tutor to teach her the standard subjects of philosophy, mathematics, and literature. She later married and had at least one son.

During Maria Winkelmann's schooling, it became clear that she had an aptitude for science and a particular interest in astronomy. To satisfy her curiosity about astronomy, Herr Winkelmann sent Maria to study with a local skywatcher and self-educated astronomer named

Christoph Arnold. Dubbed "the astronomical peasant," Arnold studied the skies every clear night, a discipline by which he memorized an amazing amount of data about stellar and planetary movements. He taught Winkelmann everything he knew and gave her a good basic education in star lore.

While studying with Arnold, Winkelmann met the distinguished astronomer, Gottfried Kirch, who would visit the astronomical peasant to compare notes on stellar and planetary observations. Kirch and Winkelmann became friends, then sweethearts. In 1692, when Winkelmann was 22, they married. By 1700, Maria Kirch had moved with her husband from Saxony to Berlin, the capital of the state of Prussia and an increasingly important German city. Gottfried Kirch had taken a job as astronomer with the Prussian Royal Academy of Sciences, and Maria Kirch worked with him as a partner in his observations and calculations.

Maria and Gottfried Kirch took turns observing the skies. They alternated nights, one night with Maria making observations and notes, the next with Gottfried handling this duty. Sometimes, both stayed up to observe the sky, but they would each take a different section of sky to look at. It was during one of these sessions at the observatory in Berlin that Maria Kirch spotted and identified a new comet. The identification of new comets was an important discovery at that time. The next evening, Kirch showed Gottfried her discovery in the night sky. The prejudice against women scientists in Europe at that time was so intense that the Kirches decided to report the discovery as Gottfried's. This announcement was made to the Prussian Royal Academy of Sciences in 1702. It was not until eight years later, in 1710, that Gottfried felt free to admit that the actual discoverer was his wife, Maria.

Besides conducting astronomical observations for many years, Maria Kirch also compiled data for the annual calendar that the Academy of Sciences published, and she served as the general editor of this calendar. In addition, Kirch authored several astrological pamphlets in German. At that time, astrology, the prediction of individual life circumstances tied to astronomical data at a person's birth, was still loosely linked with astronomy; however, soon astronomers would dissociate themselves completely from astrology, which they would view as a set of superstitions, not as a scientific discipline.

In 1712, after her husband's death, Kirch was informally banned from the Academy of Sciences observatory and its calendar business. Kirch then went to work at the private observatory of Baron Frederick von Krosigk. However, she returned to the academy's observatory in 1716 as an "assistant" to her son, Christfried, who had also become an astronomer. She died at the age of 50 in 1720.

Kistiakowsky, Vera E.
(1928–)
American
Physicist

An activist for women's participation in the sciences, Vera Kistiakowsky has also worked as a research physicist, teacher, and arms control activist. She began her professional career as a nuclear chemist but switched the focus of her research efforts to nuclear physics, then particle physics, and finally astrophysics.

Born on September 9, 1929, in Princeton, New Jersey, Vera Kistiakowsky is the daughter of Mr. and Mrs. George Kistiakowsky. Vera's father was a physical chemist who taught at Harvard and served as President Dwight D. Eisenhower's science advisor. Because she grew up in an environment in which science was a part of family life and discussion, Kistiakowsky never assumed that it was a vocation for men only. In 1951, Kistiakowsky married a fellow student at Berkeley; she has three children.

Kistiakowsky graduated from high school in 1944 when she was 15 and turned 16 at the beginning of her first year at Mount Holyoke, a women's college in Massachusetts. At first, Kistiakowsky decided on a premedical-school major, but soon she decided that she was attracted to chemistry and physics. Because Mount Holyoke did not have a good physics department at that time, and because it seemed more likely that she would be able to find a teaching job in chemistry than physics, she switched her major to chemistry, although she preferred courses and research topics that straddled the line between chemistry and physics. She graduated from Mount Holyoke with an A.B. in chemistry in 1948.

For her graduate studies, Kistiakowsky enrolled in the doctoral program for nuclear chemistry at the University of California at Berkeley. The faculty for that program was one of the best in the world at that time. The star professor was Glenn Seaborg, discoverer of plutonium and Nobel Prize winner for his work on transuranic elements. Kistiakowsky worked with her professors on studies of the energy states of promethium, an element that does not occur in nature because all of its isotopes are subject to extreme radioactive decay. These studies were more problems in physics than chemistry; nonetheless, Kistiakowsky was awarded a Ph.D. in chemistry from Berkeley in 1952.

From 1952 to 1954, while she was waiting for her husband to finish his Ph.D., Kistiakowsky worked at the Naval Radiological Defense Laboratory, located in the San Francisco Bay area. She also began sending out letters trying to find a job at a university as a teacher. Out of the approximately 100 letters that she sent, she received one reply, and even that one was a rejection. In 1957, when her husband was given a job as an instructor of physics at Columbia University, Kistiakowsky managed to get a research associate's job there too. She remained at Columbia, working with CHIEN-SHIUNG WU, a prominent woman physicist, until 1959.

From 1959 to 1962, Kistiakowsky worked as an assistant professor of nuclear physics at Brandeis University; then in 1963, she moved to the Massachusetts Institute of Technology (MIT) where she would remain for the rest of her career. She became a professor of physics at MIT in 1972. By then, her research interests had shifted from nuclear to particle physics.

During the late 1960s and through the 1970s and beyond, Kistiakowsky has been active in promoting the participation of women in the sciences. She helped to organize a seminar about women in the sciences at a National Organization for Women conference in 1969, and in the early 1970s, she organized a group in Boston called Women in Science and Engineering, which was later merged with the Association for Women in Science.

In the late 1970s and into the 1980s, Kistiakowsky led efforts to protest the U.S. arms buildup, especially President Reagan's Strategic Defense Initiative (SDI), also known as Star Wars. She organized a petition campaign to get science professors to agree not to accept government grants for studies that were part of the SDI program, an effort that spread to many other campuses beyond MIT. Kistiakowsky also gave dozens of lectures outlining her reservations about U.S. military expenditures.

By the time she retired from teaching and research at MIT, Kistiakowsky had authored more than 80 papers on nuclear physics, particle physics, and astrophysics. She was presented with Mount Holyoke's Centennial Alumnae Award in 1972 and was elected a fellow of the American Association for the Advancement of Science in 1984. In her own eyes, however, her greatest achievement was the nurturing of her own children while trying to leave a saner world for them to live in than the one into which she was born.

Kittrell, Flemmie Pansy
(1904–1980)
American
Nutritionist

The first African-American woman to earn a Ph.D. in nutrition, Flemmie Pansy Kittrell was an internationally respected nutritionist and educator. During her 40-year career, Kittrell raised awareness about child development and family welfare and emphasized the importance of a child's early home environment to his or her development. In addition to traveling abroad extensively, Kittrell also founded Howard University's school of human ecology.

Born on December 25, 1904, in Henderson, North Carolina, Kittrell was the youngest daughter of James and Alice (Mills) Kittrell. Both parents traced their lineage back to African and Cherokee ancestors, and they conveyed a deep respect of learning and literature to each of their children. Kittrell herself would not marry or have children.

After receiving her bachelor's degree from the Hampton Institute in North Carolina in 1928, Kittrell taught home economics at Bennett College in Greensboro, North Carolina. With the support of her Hampton professors, she then enrolled at Cornell University to pursue graduate studies. She earned her master's degree in 1930 and her Ph.D. in nutrition in 1938.

Kittrell returned to Bennett College in 1938. Two years later, she accepted a position as the head of the home economics department at Hampton College. That institution also named her dean of women. After receiving a personal invitation from Howard University president Mordecai Johnson, Kittrell left Hampton in 1944 to preside over Howard's home economics department. She immediately began to develop a curriculum that broadened the department's focus to include such issues as child development and nutrition.

In 1947, Kittrell made her first international journey when—sponsored by the U.S. government—she conducted a nutritional survey of Liberia. During the course of this work, she discovered that nearly 90 percent of the citizens of that West African nation suffered from a type of malnutrition known as hidden hunger. In response to the recommendations that she made, the Liberian government overhauled its agricultural and fishing industries. After returning to Howard, Kittrell soon pursued her international activism again. In 1950, she won a Fulbright Award, which enabled her to collaborate with Baroda University in India. As a result of her stay in that country, she formulated a new educational plan for nutritional research. Kittrell returned to India in 1953 to teach home economics classes and nutritional seminars. In 1957, she led a team to Japan and Hawaii to research home economics educational programs in those places. She then headed three tours to western and central Africa between 1957 and 1961.

Even with her busy travel schedule, Kittrell fulfilled her academic duties at Howard University. Indeed, she achieved a significant goal in 1963, when she convinced the university to dedicate a new building to her innovative school of human ecology. The building also hosted one of the first Head Start programs.

Kittrell retired from Howard University in 1972, although she remained active in her field, with projects including a stint as a visiting senior fellow at Cornell from 1974 to 1976. After serving as a Moton Center senior research fellow in 1977, she was named a Fulbright lecturer in India in 1978. During the course of her career, Kittrell received numerous honors. Selected as Hampton University's outstanding alumna in 1955, she received the Scroll of Honor from the National Council of Negro Women in

1961. The Home Economics Association established a scholarship fund in her name. She is now recognized as a key figure in the development of nutritional education. Kittrell died of cardiac arrest on October 3, 1980.

Kivelson, Margaret Galland
(1928–)
American
Physicist

A professor of space physics, Margaret Kivelson has made important contributions to the understanding of the magnetic fields in the areas surrounding Earth and Jupiter. She has also studied the magnetic fields of Jupiter's moons and of asteroids.

Born on October 21, 1928, in New York City, Margaret Galland knew in high school that she wanted to be a scientist but was unsure how well she could do with such a career. "I had low expectations for my career and so did others," she has said. "While still in high school, I had been advised to become a dietitian by an uncle who was a professor. He was well aware of where women with a scientific bent fit readily into the academic institution, and it wasn't in physical sciences!" Ignoring her uncle's advice, she began to study physics. In 1949, she married fellow physicist and chemist Daniel Kivelson. They have two children.

A good student in high school, Kivelson was accepted into Radcliffe College, Harvard's women's college, in 1946. She received her A.B. from Radcliffe in 1950, then immediately began work on a master's degree, which she completed in 1952. She was awarded a Ph.D. in physics from Harvard in 1957.

As she was completing her doctorate, Kivelson moved in 1955 with her husband to Los Angeles, California, where she took a job as a consultant in physics for the Rand Corporation, a major think tank with close ties to the U.S. government. She held the Rand position until 1971. In 1967, Kivelson took a research position at the University of California at Los Angeles (UCLA), where she worked as a research geophysicist at the Institute of Geophysics and Planetary Physics (IGPP). In 1971, she was appointed to the faculty of UCLA as an assistant professor; that same year, she was promoted to associate research geophysicist at IGPP. Since then, she has steadily climbed the ranks within the geophysics and space physics community at UCLA. By 1977, she had become professor at UCLA's Department of Earth and Space Sciences, and in 1983, she also assumed a position as professor at IGPP. She chaired the Department of Earth and Space Sciences from 1984 to 1987, and in 1999–2000, she was interim director of IGPP.

Besides working in her field at UCLA, Kivelson held down a number of advisory positions in both academe and

Space physicist Margaret Kivelson has made important contributions to the understanding of the magnetic fields in the areas surrounding Earth and Jupiter. *(Courtesy of Margaret Kivelson)*

government. From 1977 to 1983, she served on the Board of Overseers at Harvard College. She also served on the National Aeronautics and Space Administration's (NASA) Advisory Council from 1987 to 1993, the National Research Council's Committee on Solar-Terrestrial Research from 1989 to 1992, the National Science Foundation's Advisory Committee on Geosciences from 1993 to 1997, and from 1998 to 2000, she cochaired the UCLA Academic Faculty Senate's Committee on Gender Equity Issues.

Kivelson has focused her research interests on the magnetic fields of Earth, Jupiter, and Jupiter's moons, Ganymede and Io. From 1995 to 2000, she was principal investigator of the magnetometer data generated by the *Galileo Orbiter*, which was placed in orbit around Jupiter. From this data, she and her team discovered and measured the intrinsic magnetic field of Ganymede and found a magnetic signature for Io. She also found a magnetic signature that indicates an intrinsic magnetic field for the asteroid Gaspra.

For her work, Kivelson was honored by the Guggenheim Foundation, which awarded her a grant in 1973–74. She was also presented Harvard University's 350th Anniversary Alumni Medal in 1986. For her work on the *Galileo Orbiter*, she was given NASA Group Achievement Awards in 1995 and 1996, and she was elected to the National Academy of Sciences in 1999.

Klein, Melanie Reizes
(1882–1960)
German/British
Psychologist

One of the pioneering women in the field of psychology and psychoanalysis, Melanie Klein opened up important avenues of psychoanalytic thought by her analysis of children and of adults with severe mental illnesses such as schizophrenia and depression. She played a major role in the psychoanalysis movement begun by Sigmund Freud.

Born on March 30, 1882, in Vienna, Austria, Melanie Reizes was the youngest child of Moriz and Libussa Reizes. With her three siblings, Reizes grew up in poverty in Vienna. Because her family did not have enough money, she never attended university. She married a cousin, Arthur Klein, in 1903 and had three children.

Melanie Klein's marriage took her out of poverty. Her husband was an engineer who earned a decent income from his profession. Within a few years of her marriage, Klein moved with her husband and family to Budapest, then a major city of the Austro-Hungarian Empire.

Because her marriage did not bring her happiness, Klein began a lengthy series of psychoanalytic sessions in Budapest. Psychoanalysis, as elaborated by its main proponent, Sigmund Freud, was then in its infancy, but the number of analysts multiplied rapidly. Klein found one, Sandor Ferenczi, in Budapest to work with her. Impressed with the ideas that guided psychoanalysis, Klein began to study Freud's works. With the encouragement of her own analyst, Ferenczi, she then began to work as an analyst.

Her first patient was her son, Erich. Rejecting Freud's notion that children could not be analyzed, Klein developed a theory about neuroses in children that emphasized a child's relationships with other people in his life, especially his mother. Klein believed that neuroses in children sprang out of the child's perception that he or she was not receiving emotional support and warmth from the mother. This perception caused a child to fantasize acts of hurt and revenge on the mother, which in turn caused shame in the child. The conflict between the desire to love the mother and the wish to hurt her caused neurotic behavior.

Klein published a first attempt at working out this mother–child relationship as a scientific paper around 1918. As a result of this paper, she was granted membership into the Hungarian Psychoanalytic Society and began practicing analysis and developing a psychoanalytic practice. Even though Klein differed with Freud on whether children could be successfully analyzed and on the causes of childhood mental problems (Freud stressed the role of fathers; Klein of mothers), she did agree with Freud on the crucial role of the unconscious as the place where distorted mental impulses were incubated. Following Freud's central idea, Klein felt that neurotic behavior could be cured only by talk therapy that uncovered these hidden unconscious impulses.

In 1921, Klein finally broke off her unhappy marriage and moved to Berlin where she became a member of the Berlin Psychoanalytic Society. She continued analyzing children in Berlin, but her stay there was not as fruitful as she had hoped it would be. Analysts in Berlin never accepted her because she had no formal university training. In 1926, Klein moved to England and joined the British Psychoanalytic Society. She remained in Britain for the rest of her life. Klein eventually wrote a book, *The Psychoanalysis of Children* (1960), that became one of the standard Freudian texts about the analysis of children. In the latter part of her career, she also wrote about her belief that extremely disturbed patients, such as schizophrenics and people with serious depression, could be successfully treated by analysis. Klein died at age 78 on September 22, 1960.

Klieneberger-Nobel, Emmy
(1892–1985)
German/British
Microbiologist

An early pioneer in the field of microbiology, Emmy Klieneberger-Nobel laid the foundation for a scientific understanding of *Mycoplasma*, a genus of nonmotile microorganisms that are in many ways intermediate between bacteria and viruses. Klieneberger-Nobel discovered the L-form of mycoplasmas such as *Mycoplasma arthritides, Mycoplasma neurolyticum,* and *Mycoplasma pneumoniae,* and she was able to explain their cellular and colonial morphology.

Born on February 15, 1892, in Frankfurt, Germany, Emmy Klieneberger was the daughter of Abraham Adolph Klieneberger and Sophie Hamberger Klieneberger. Emmy was the youngest of four siblings, her two brothers and sister being much older than she was. From a poor family, Klieneberger's father established a successful business as a wine merchant in Frankfurt. Her mother was a housewife. In 1944, Klieneberger married Edmund Nobel, a Viennese pediatrician and researcher living in London. She was married to Nobel for only two years before his death in 1946. They had no children.

Klieneberger-Nobel was from a family of assimilated Jews who baptized Emmy when she was seven (although the family did not regularly attend church). She was an average student in secondary school and attended a teacher's training college. After receiving her teaching certificate in 1911, she decided to continue her university education by studying mathematics and sciences at the University of Göttingen. At the beginning of World War I in 1914, Klieneberger-Nobel returned to Frankfurt to study botany and zoology at Goethe University in that city. She earned her Ph.D. in botanical zoology in 1917.

From 1919 to 1922, Klieneberger-Nobel taught physics, chemistry, and biology at a girl's school in Dresden. However, missing laboratory work, she took a job in 1922 at the Hygiene Institute in Frankfurt. Here she developed her specialty of bacteriology and became a member of the German Society for Hygiene and Bacteriology. Hitler's rise to power in 1933 prompted Klieneberger-Nobel to emigrate to England where she began work at the Lister Institute. She would remain at Lister for the rest of her professional career, retiring from there in 1962.

Klieneberger-Nobel's most important work came after her move to the Lister Institute. From the beginning at Lister, she focused her research efforts on mycoplasmas, a type of microorganism whose workings and variations were then largely a mystery. In 1934, she identified for the first time differences within mycoplasmas and began to establish differences between them and other bacteria.

Klieneberger-Nobel worked to develop a culture in which the organisms of a bronchopneumonia found in rats and mice could be grown and thus examined, something that no researcher had managed to do. Developing a special nutrient agar of her own, Klieneberger-Nobel successfully grew this microorganism and identified it as an L-form of mycoplasma. She later also used this technique to isolate, identify, and reproduce *Mycoplasma arthritides*, a microorganism that causes a severe form of arthritis in rats, and *Mycoplasma pneumoniae*, which causes a form of bronchopneumonia in rats. Her basic research on these mycoplasmas gave scientists a greater understanding of how these microorganisms cause infectious diseases and helped in the understanding of mycoplasmas that cause human diseases.

For her work, Klieneberger-Nobel won the Jenner Memorial Scholarship from the Lister Institute in 1935. In 1967, she was made honorary member of the Robert Koch Institute and won the Robert Koch Medal for achievements in microbiology. She was also an honorary member of the International Organization for Mycoplasmology. She died at the age of 93 in 1985 in England.

Knopf, Eleanora Bliss
(1883–1974)
American
Geologist

During her long tenure with the United States Geological Survey (USGS), Eleanora Bliss Knopf made significant contributions to the understanding of the geological history of mountainous regions in New England and Pennsylvania. She was persistent in her efforts to incorporate the most current techniques and concepts into her work. For instance, she was one of the first geologists to use stereoptically viewed airplane photographs for field mapping. She also collaborated with her geologist husband, Adolph Knopf, in analyzing the Boulder batholith formation in the Rocky Mountains.

Eleanora Bliss Knopf was born on July 15, 1883, in Rosemont, Pennsylvania, to Tasker Howard and Mary Anderson Bliss. Her father was the chief of staff of the United States Army during World War I.

After enrolling at Bryn Mawr College in 1900, Knopf earned both her bachelor's degree in chemistry and a master's degree in geology by 1904. Her interest in geology was cultivated by Bryn Mawr professor FLORENCE BASCOM, who was the first American woman to receive a Ph.D. in the subject. Upon graduating, Knopf remained at Bryn Mawr as an assistant curator in the college's geological museum and as a demonstrator in a geological laboratory there. In 1909, she began her challenging doctoral dissertation on the geology of the Doe Run-Avondale district, near Bryn Mawr. She was awarded her Ph.D. in 1912.

Immediately after completing her doctorate, Knopf accepted a position with the USGS. In this capacity, she continued to explore areas near the site of her dissertation research. While at the USGS, she met esteemed geologist Adolph Knopf, who was a widower with three children. They wed in 1920 and soon thereafter moved to New Haven, Connecticut, where Adolph Knopf had been offered a prestigious faculty post at Yale University. However, Yale refused to hire women, so Eleanora Knopf was forced to work out of her husband's office and teach private courses, although she did continue to receive USGS assignments. Much of her work for the USGS involved interpreting geologic faults and folds in the Taconic region—a mountainous area along the border between New England and New York.

Knopf was adamant about incorporating the latest tools and methods in her research for the USGS. She sought advice from European geologists about how best to analyze accurately the effects of high temperatures and pressures on rock. From her thorough reviews of international geological literature (among other endeavors, she translated a number of German scholarly papers into English), she introduced new concepts and approaches to other

American geologists. For example, she was an early advocate of using laboratory work to examine the deformation of rocks. She also embraced new technology, becoming one of the earliest geologists to use stereoptically viewed airplane photos (thereby having what appeared to be a three-dimensional view of an area).

Adolph and Eleanora Knopf left New England in 1951 when Adolph opted to take a faculty position at Stanford University in California. Eleanora Knopf was hired as a research associate at the school, and she continued to complete projects for the USGS until 1955. She also teamed up with her husband to explore the Boulder batholith in the Rocky Mountains. (A batholith is a deep-seated rock intrusion that often forms at the base of a mountain and is uncovered only by erosion.) Adolph died in 1966, but she continued their joint project until her own death of arteriosclerosis on January 21, 1974. By incorporating new ideas and methodologies from European geologists, Knopf introduced important innovations to her field and improved the techniques of her colleagues.

The recipient of a presidential Young Investigator Award, a Guggenheim Fellowship, and a MacArthur Fellowship, Mimi Koehl's research involves the application of fluid dynamics and solid mechanics in the study of biological structure. *(Photo by Joy Schaber, Courtesy of Mimi Koehl)*

Koehl, Mimi A. R.
(1948–   )
American
Biologist and Biomechanist

Active in the interdisciplinary field of biomechanics, Mimi Koehl has studied a variety of marine and insect creatures to understand how their morphology, that is, the design of their bodies, allows them to interact with the environments in which they live. To understand this relationship she mixes engineering with biology, field investigation with laboratory studies.

Mimi Koehl was born on October 1, 1948, in Washington, D.C., to George and Alma Koehl. Her father was a physics professor, and her mother was an artist. Koehl was influenced by both her parents. From her father, she learned that she could be involved in the sciences, and through her mother, she became fascinated with visual arts. As a teenager, Koehl studied art on her own time, and she began a formal study of art during her first years in college. In 1993, she married Thomas Powell.

In 1966, Koehl enrolled in Gettysburg College, located in Gettysburg, Pennsylvania. Here she decided to switch from visual arts to biology after enjoying a biology course that she was required to take. Koehl graduated magna cum laude from Gettysburg with a B.S. in biology in 1970.

From 1970 to 1976, Koehl studied zoology at Duke University in North Carolina. Her work in biomechanics began at Duke with a study of the mechanical design of soft-bodied animals such as sea anemones. She used this information to explain how the bodies of sea anemones both withstand and utilize the water currents and waves

flowing around them. Even though she had encountered teachers and fellow students who did not take her seriously as an undergraduate, Koehl discovered that her gender did not seem to be an obstacle to her work nor her career advancement in graduate school and beyond.

After Koehl received her Ph.D. from Duke in 1976, she worked as a postdoctoral researcher at the University of Washington in 1976–77 and at the University of York in the United Kingdom in 1977–78. She was appointed assistant professor of biology at Brown University in 1978–79, and in 1979, she began a long association with the University of California at Berkeley, where she is still a professor in the Department of Integrative Biology.

At Berkeley, Koehl continued to delve deeply into biomechanics research. Through field observation and laboratory analysis, she sought to understand how, for instance, marine plants and animals deal with the cyclical water motion of waves, how other aquatic animals feed on unicellular plants suspended in the water around them, and how insect wings evolved.

Central to her quest "are the basic physical rules about how organisms interact with the air and water around them." Occasionally, she has encountered results that are in opposition to accepted explanations about particular creatures. This occurred with the comblike appendages of many planktonic animals, which marine biologists had assumed to be leaky sieves. After a detailed study of the fluid dynamics of such appendages, Koehl concluded that the holes in these appendages were too small to allow seawater to pass through them and that they catch food particles by creating water currents toward the mouth rather than by sieving.

Koehl's work has been discussed in *Scientific American* and *Discover* magazines as well as on television programs such as *Nova,* and she has published dozens of papers in scientific journals. For her research, in 1983 she was one of the first scientists to receive a presidential Young Investigator Award. She also won a Guggenheim Foundation fellowship (1988) and a MacArthur Foundation fellowship (1990). In 1993, she was selected as a fellow in the California Academy of Science and in 1998 as a Phi Beta Kappa Visiting Scholar.

Koller, Noemie Benczer
(1933–   )
Austrian/American
Physicist

An Austrian by birth but later a naturalized U.S. citizen, Noemie Benczer Koller has pioneered in several areas of nuclear and condensed matter physics. Koller is best known for her studies of the hyperfine interactions of nuclei in solids and ionized atoms. She also has conducted investigations in nuclear spectroscopy and nuclear magnetic moments.

Born in Vienna, Austria, on August 21, 1933, Noemie Benczer is the daughter of Dr. Maurice Benczer and Rica

Noemie Koller, a pioneer in nuclear and condensed matter physics. *(Courtesy of Noemie Koller)*

Benczer. Her father worked as a chemist and businessman, and her mother was a bookbinder. The family fled France during the Nazi occupation and found asylum in Cuba in 1942 and, subsequently, in Mexico in 1943. Benczer completed her high school education at the Lycée Français in Mexico City. When she was 17 years old, she emigrated to the United States. Benczer married Earl Leonard Koller, also a physicist, in 1956. She has two children—Daniel, a physicist, and David, a geologist.

Koller enrolled in Barnard College, the women's college of Columbia University, in 1951, and she earned a B.A. in 1953. She received a master's degree in nuclear structure physics at Columbia University in 1955. She stayed at Columbia for her doctorate, which she completed in 1958 under the direction of Professor CHIEN-SHIUNG WU. Professor Wu is known for her demonstration of the breakdown of parity conservation in beta decay. Koller's work at Columbia involved studies of beta decay and parity nonconservation in weak interactions.

In 1958, Koller began her career as a research associate in nuclear physics at Columbia University, a position she held for two years. In 1960, she moved to Rutgers University in New Brunswick, New Jersey, where she was hired as an assistant professor, the first woman ever to be appointed to a faculty position there. By 1965, she had become an associate professor and was promoted to the rank of professor of physics in 1970.

At the beginning of her career, Koller continued her studies in general nuclear physics, but by the late 1960s and early 1970s, she became interested in using the Mössbauer effect to investigate magnetic materials. The Mössbauer effect, named after R. L. Mössbauer, who discovered the phenomenon in 1957, is the emission without recoil of a gamma ray from an atomic nucleus embedded in the lattice of a solid. Koller used Mössbauer effect spectroscopy coupled with nuclear physics techniques to measure electron spin densities and surface magnetism of iron-based materials.

Koller also studied decay rates in the atomic nuclei of calcium and gold. In her study of the decay rate of calcium (40 Ca), she was the first person to identify the double gamma decay of this isotope to the ground state. In another important series of studies, Koller measured the interplay of single particles and collective motions in nuclei, and she demonstrated how a simple relation based on constant gyromagnetic ratios for nucleon pairs could describe a broader range of nuclear electromagnetic transitions. More recently, she has developed new techniques leading to very high-precision measurements of magnetic moments of short-lived nuclear states of unstable, exotic nuclei at very high spin, energy, or deformation. She was the first to make a direct measurement of the magnetic moments of superdeformed nuclear states. She then developed new techniques to study magnetic properties of nuclei far from the valley of stability. The properties of these nuclei provided the main impetus for the construction of the new "rare isotope accelerator," which is being planned in the United States.

Koller has been active in university administration and professional organizations as well as research. In 1992, she became associate dean in the Faculty of Arts and Sciences at Rutgers, a post she held until 1996. In 1993–94, she was chair of the Division of Nuclear Physics of the American Physical Society, and from 1993 to 1996, she was a member of the Department of Energy's and National Science Foundation's Nuclear Science Advisory Committee.

Koller was given a New Jersey Women of Achievement Award in 1997. She has also been selected as a fellow of the American Physical Society and the American Association for the Advancement of Science. She continues to teach physics and conduct research at Rutgers.

Kovalevskaia, Sofia Vasilyevna
(1850–1891)
Russian
Mathematician

A math prodigy, Sofia Kovalevskaia overcame intense discrimination because of her gender to become one of the leading mathematicians of the 19th century. In an era when women were shut out of university study, especially in the sciences, Kovalevskaia not only managed to earn a doctorate but also became the first tenured woman professor in any European university. Her career reached its pinnacle when she won a coveted prize from the French Academy of Sciences for the solution of an extremely difficult math problem.

Born in Moscow on January 15, 1850, Sofia Vasilyevna Korvin-Krukovsky was the daughter of General Vasily Korvin-Krukovsky and Yelizaveta Korvin-Krukovsky. Korvin-Krukovsky's family was wealthy and belonged to the Russian nobility. When her father retired from the army in 1858, he moved his family to their country estate of Palobino in western Russia near St. Petersburg. Here Sofia had a wonderful childhood, which was enlivened by her discovery that she loved to solve mathematical problems. According to her memoirs, *Recollections of Childhood,* her first math puzzles were presented to her when her family ran out of wallpaper and covered the walls of her room with her father's student notes from a calculus class. "I passed whole hours before that mysterious wall," Sofia wrote, "trying to decipher even a single phrase, and to discover the order in which the sheets ought to follow each other."

Realizing that his daughter possessed unusual talent in math, Vasily Korvin-Krukovsky hired private tutors for her. At 19, when her years of tutoring were finished, Sofia knew

that she would not be admitted into any Russian university. To get around this obstacle, she arranged a marriage of convenience with Vladimir Kovalevsky. The two of them then traveled to Heidelberg, Germany, where Sofia began informally studying university-level mathematics. The Kovalevskys's marriage lasted until 1880; Sofia Kovalevskaia had one daughter from this match.

In 1871, Kovalevskaia moved to Berlin to study with Karl T. Weierstrass, the father of mathematical analysis. At first, Weierstrass wanted nothing to do with her. However, when she handily solved a difficult mathematical problem he had given her as a way to get rid of her, Weierstrass took her as a private student and became her mentor. Under Weierstrass's tutelage, Kovalevskaia wrote several papers that solved math problems. One was an analysis of differential equations; another concerned the shape of the rings of Saturn. Her third paper, which was an analysis of the mathematics of ellipses, won her a doctorate from Germany's University of Göttingen in 1874.

At 24, Kovalevskaia tried to find work as a mathematician. This proved even harder than getting an education. When no university would offer her a job, she returned to Moscow to be with Vladimir Kovalevsky, who had returned several years earlier. Her daughter, Sofia, was born in Moscow in 1878. For six years, Kovalevskaia tried to be a conventional wife, but without her work in mathematics, she grew bored. Finally, in 1880, she separated from her husband, temporarily left her daughter with a friend, and moved to Paris where she was elected to the French Mathematical Society.

In 1883, Kovalevskaia finally got the break she had been waiting for. One of her former students, Gösta Mittag-Leffler, now a professor at the University of Stockholm, offered her a teaching job. Kovalevskaia moved to Stockholm, sent for her daughter, and a year later was offered a full professorship. In 1886, at the urging of Mittag-Leffler, Kovalevskaia submitted an anonymous paper to the French Academy of Sciences analyzing a topic assigned by the academy: the rotation of a solid body around a fixed point. Citing her "precise and elegant form" in solving the problem, the academy awarded her its prestigious Bordin Prize in 1888. When her identity was revealed, there was considerable surprise that the paper's author was a woman. Kovalevskaia died unexpectedly of pneumonia at the young age of 41 on February 10, 1891.

Krieger, Cecelia
(1894–1974)
Austrian/Canadian
Mathematician

The first woman to earn a doctorate in mathematics from a Canadian university, Cecelia Krieger was known for her inspired teaching, her devotion to her students, and her translation of two works, *Introduction to General Topology* (1934) and *General Topology* (1952), by the mathematician Sierpinski.

Other than the fact that she was born on April 9, 1894, in Jaslo, Poland (then a part of the Austro-Hungarian Empire), not much is known about Krieger's childhood and youth. Krieger was from a Jewish family and grew up in the Austro-Hungarian Empire, which at that time was the site of a thriving Jewish cultural community. Yet, in spite of the accomplishments of Austro-Hungarian Jews such as Sigmund Freud, Franz Kafka, and others, there was much discrimination against Jews in universities and professions. This caused Krieger, her mother, and sisters to emigrate to Canada in 1920. In 1953, she married Dr. Zygmunt Dunaij.

Krieger had studied mathematics and physics at the University of Vienna for a year before she moved to Canada. On her arrival, she enrolled as an undergraduate in the University of Toronto in 1920. Even though she at first spoke little English, Krieger completed her B.A. at the University of Toronto by 1924.

She began her graduate work immediately and was supervised by W. J. Webber. She won her master's degree in mathematics in 1925 and her Ph.D. in 1930. In her graduate studies and beyond, Krieger worked with modular elliptic functions, principles of mechanics, the theories of numbers and functions, and the theory of sets. Her dissertation, which was published in two parts in the *Transactions of the Royal Society of Canada* (1928 and 1930), was entitled "On the Summability of Trigonometric Series with Localized Properties—On Fourier Constants and Convergence Factors of Double Fourier Series."

Even before she had completed her Ph.D., Krieger was appointed an instructor in mathematics at the University of Toronto. She was promoted to lecturer in the math department in 1931, a position she held for 12 years. In 1943, at the age of 49, she became assistant professor of mathematics, a position she was to hold until her retirement in 1962. That her career ended at an assistant professor's rank is due to the discrimination that she faced as a woman in a scientific discipline.

Krieger is best known for her translation of two works that advanced the mathematical subdiscipline of topology. Also known as analysis situs, topology is a branch of mathematics begun in the 19th century by Bernhard Riemann and Henri Poincaré. However, topology was relatively neglected by mathematicians until the early part of the 20th century when it was revisited and advanced conceptually. The idea of a *geometrical configuration*, developed by earlier mathematicians, was changed to a *point set*. The notion of topology as being composed of triangles, circles, polygons, and other geometrical abstractions was dropped in favor of an entity composed of typically infinite points. Any group of things, such as a set of numbers or algebraic entities, was conceived of as a topological space. Krieger's translation and teaching helped English-speaking students understand these new conceptual advances.

Krieger was also a gifted teacher. One of her students, the mathematician CATHLEEN SYNGE MORAWETZ, credits Krieger with keeping her on course in the sciences. "For my career, she played a fundamental role," Morawetz has said. "I was in my final year at the University of Toronto. World War II had almost ended, and there was a call for teachers to go to India. The idea of living in an exotic country appealed to me. . . . But I ran into Miss Krieger one day [and told her my plans]. She was horrified. 'Why aren't you going to graduate school?' 'I haven't the money,' I said. 'Easily fixed,' said Miss Krieger. She [arranged for a scholarship from the Canadian Association of University Women]. Miss Krieger had delivered."

After her "official" retirement in 1962, Krieger continued to teach part-time at the University of Toronto and Upper Canada College until her death at the age of 80 in 1974.

Krim, Mathilde
(1926–   )
Italian/American
Virologist

Mathilde Krim is the cofounder and cochair of the American Foundation for AIDS Research (AmFAR), the preeminent national nonprofit organization devoted to combating acquired immunodeficiency syndrome (AIDS). Krim was a highly respected virologist researching the use of the interferon protein to treat cancer when she first encountered patients with AIDS. Troubled by the lack of organized efforts to fight this new disease, Krim founded the AIDS Medical Foundation to raise funds for AIDS research and to boost public awareness of the malady. In 1985, that organization merged with another nonprofit group to form AmFAR.

Krim was born Mathilde Galland on July 9, 1926, in Como, Italy. Her mother was Czechoslovakian and her father a zoologist from Switzerland. In 1932, the family moved to Switzerland to escape Italy's economic depression.

Krim attended the University of Geneva in Zurich, Switzerland, receiving her bachelor's degree in genetics in 1948 and her Ph.D. (also in genetics) in 1953. Soon after completing her education, Krim became involved with Zionist movements. She married a Bulgarian medical student (whose name and fate are unknown), converted to Judaism, and moved with her new husband to Israel.

In 1953, Krim joined the Weizmann Institute in Rehovot, Israel, as a junior scientist, later becoming a research associate. With her research team, Krim pioneered the use of amniocentesis to determine the gender of a fetus. The procedure involves analyzing fetal chromosomes found in the amniotic fluid. While in Israel, Krim gave birth to her daughter, Daphne. In 1958, she met and married Arthur Krim, who would later launch Orion Pictures. The following year, she left the Weizmann Institute and returned to the United States with Arthur. She found a position as a research associate at Cornell University's Division of Virus Research. Krim then began to investigate interferon—a protein that is produced naturally in most animal species and that appears to inhibit tumors and boost the immune system.

Krim left Cornell in 1962 to become an associate at the prestigious Sloan-Kettering Institute of Cancer Research in New York. Hoping to investigate interferon's cancer-fighting potential, she lobbied for the center to form an interferon laboratory. These efforts were initially stymied, as the costs of interferon treatment were prohibitive. In 1975, however, after new cloning techniques made the mass production of interferon financially feasible, Krim was tapped to found and head Sloan-Kettering's interferon evaluation program. From 1981 to 1985, she concentrated on using interferon to treat leukemia.

While conducting her interferon research, Krim came in contact with AIDS patients, as she was testing the effectiveness of interferon to treat Karposi's sarcoma (a cancer that often afflicts people with AIDS). Krim became outraged by the lack of public funding for fighting the disease and was even more incensed by the dearth of efforts to raise public awareness about it. In response, she founded the AIDS Medical Foundation in 1983, which awarded grants for AIDS research. She merged her group with another nonprofit in 1985, creating AmFAR. In 1986, Krim left Sloan-Kettering in order to devote herself wholly to fundraising for AIDS awareness and education. In her role as cochair of AmFAR, Krim has deployed her stellar reputation as a scientist and her husband's powerful connections in the entertainment industry to make the organization a world leader in the fight against AIDS. She has publicly promoted the use of condoms and safe needle exchanges, and she lobbied Congress to pass the 1990 Disabilities Act to protect people with AIDS from discrimination. She has also sought to inform scientists and citizens alike of a dangerous new disease—multidrug-resistant tuberculosis, which disproportionately afflicts those with human immunodeficiency virus (HIV) and AIDS as well as the poor and the homeless. In addition to her numerous duties with AmFAR, Krim has worked as an associate research scientist at both St. Luke's Roosevelt Hospital Center and the College of Physicians and Surgeons at Columbia University since 1986.

Krim is acknowledged as one of the foremost forces in the global battle against AIDS, especially in raising the public's awareness about AIDS, its causes, modes of transmission, and epidemiological patterns. AmFAR has raised an unprecedented $40 million for research and education. The organization has also brought together government officials, drug company executives, drug treatment experts, gay activists, and minority leaders to fight AIDS. The author of more than 70 scientific papers, Krim remains cochair of AmFAR and is active at St. Luke's and Columbia. She has also received several awards, most notably the 1993 John W. Gardner Leadership Award.

Kübler-Ross, Elisabeth
(1926–   )
Swiss/American
Psychiatrist

During her tenure as assistant professor of psychiatry at a Chicago hospital, Elisabeth Kübler-Ross began to experiment with ways that terminally ill patients could cope with the emotional burdens of imminent death. She organized group therapy sessions in which patients were encouraged to talk about the feelings they held toward their condition. From this work, Kübler-Ross developed her specialty of treating terminally ill patients and surviving family members. In medical terms, it is called *thanatology;* however, outside of medicine it is usually known as death and dying. Kübler-Ross's work in this field has offered a new way for doctors and patients to think about the experience of dying.

Born on July 8, 1926, in Zurich, Switzerland, Elisabeth Kübler was one of three identical triplets. Her father, Ernest Kübler, was an executive in an office supply company, and her mother, Emma Kübler, was a housewife. Because Kübler and her sisters looked exactly alike, it was hard for other people to distinguish them. Even Kübler's parents had trouble telling them apart. In 1957, Kübler married Emmanuel Ross, an American physician. They have two children.

During high school, Kübler-Ross found a way to separate herself from her two sisters by doing volunteer work in a hospital for refugees from the war zones that surrounded Switzerland during World War II. Because of its neutrality, Switzerland had remained relatively unaffected by the war. Her work with these wounded refugees gave Kübler-Ross her first experience of deep human suffering and influenced her to begin undergraduate work that would lead toward a medical degree.

After graduating from secondary school in 1944, Kübler-Ross took several years off to do humanitarian relief work for war refugees. She eventually began working at a medical laboratory, an experience that led to medical school at the University of Zurich. She began working on her medical degree in 1950 and became a medical doctor in 1957. After marrying Emmanuel Ross, Kübler-Ross went to New York where she enrolled as a resident in the pediatrics program at Columbia Presbyterian Medical Center. However, her pregnancy kept her out of that position, and she took up a residency in psychiatry at the Manhattan State Hospital from 1959 to 1962.

As a practicing psychiatrist at the Colorado General Hospital in Denver from 1962 to 1965, Kübler-Ross worked with the most difficult to treat patients such as schizophrenics. Breaking with tradition, she used a new approach on these patients: Mixed in with drug and talk therapy, she asked the patients what activities would be most helpful for them. Kübler-Ross felt that this approach was moderately successful with her patients, so when she moved to Chicago to work at the University of Chicago's Billings Hospital, she continued this technique.

At Billings, Kübler-Ross initiated a group therapy seminar in which dying patients spoke to each other and to doctors about their feelings toward dying and death. From this encounter, and from her one-on-one work, Kübler-Ross identified five emotional stages a dying person typically goes through from the time he or she is informed about the fatal illness until the time of death. These are denial, anger, bargaining, depression, and acceptance.

Kübler-Ross pointed out that dying used to occur in a patient's home. By the 1960s, however, most people with terminal illnesses were put into hospitals in which they spent their dying days. Kübler-Ross made the then-revolutionary suggestion that hospices be set up for dying patients, either as a separate center or in a patient's home. The hospices would be places where the patients' pain could be managed and where a warmer emotional setting could afford an easier and more meaningful emotional passage to death.

Kübler-Ross left Billings Hospital in 1970. Since then, she has maintained a private practice and has written and lectured about death and dying. She wrote a number of books, the most well-known of which are *Death and Dying* (1969), *AIDS: The Ultimate Challenge* (1987), and her memoir, *The Wheel of Life* (1997). She was a member of the American Association for the Advancement of Science, the American Psychiatric Association, and the American Holistic Medical Association.

Kuhlmann-Wilsdorf, Doris
(1922–)
German/American
Metallurgist

Renowned metallurgist and materials scientist Doris Kuhlmann-Wilsdorf is an expert in tribology, the study of both the effects of friction on moving machine parts and the methods of lubrication. She has made several significant contributions to her field, including her innovative design for electrical metal-fiber brushes. She also developed

a model for surface deformation that takes into account erosion as well as friction and wear. The author of more than 250 articles, Kuhlmann-Wilsdorf has been on the faculty of the University of Virginia since 1963 and has served as a consultant to leading corporations and the National Institute for Standards and Technology.

Born in Bremen, Germany, on February 15, 1922, Kuhlmann-Wilsdorf was the daughter of Adolph Friedrich and Elsa (Dreyer) Kuhlmann. Prior to entering Göttingen University in 1942, she worked as an apprentice metallographer and materials tester from 1940 to 1942. She completed her undergraduate and graduate studies at Göttingen, receiving her Ph.D. in materials science in 1947. She then continued her research as a postdoctoral fellow at Göttingen. In 1949, she accepted a fellowship at Bristol University in England, where she worked with Nobel laureate Nevill Mott. She married Heinz G. F. Wilsdorf in 1950.

With her new husband, Kuhlmann-Wilsdorf became a lecturer in the physics department of the University of Witwatersrand in Johannesburg, South Africa, in 1950. The couple moved to the United States in 1956. Her husband was appointed director of laboratories at the Franklin Institute in Philadelphia, Pennsylvania, and Kuhlmann-Wilsdorf took a position as an associate professor at the University of Pennsylvania (Penn) in the same city. She was promoted to the rank of full professor in 1961 and remained at Penn until 1963, when she left to join the faculty of the University of Virginia in Charlottesville as a professor in the department of engineering physics. In 1966, she was named university professor of applied science.

Kuhlmann-Wilsdorf's tenure at the University of Virginia has been marked by her myriad contributions stemming from her investigation into the behavior and properties of various metals. For example, she determined why aluminum sheets only crinkle under pressure while those made from other metals break. She also developed the model for surface deformation. But one of her most important endeavors has been her design for electrical metal-fiber brushes. If ship drives are modified in the future, her invention could well lead to the widespread conversion to electric motor (rather than diesel) engines on naval vessels. If this occurs, ships will be lighter as well as more maneuverable and efficient.

In addition to pursuing her research and fulfilling her teaching obligations, Kuhlmann-Wilsdorf has concurrently served as a consultant to a number of top corporations, including General Motors Technical Center (1960–70), Chemstrand Research Laboratories (1964–66), General Dynamics (1985–87), and Maxwell Laboratories (1987–89). She also worked with the National Institute for Standards and Technology from 1981 to 1982.

Kuhlmann-Wilsdorf's achievements in her field have been well recognized. In 1989, the Society of Women

Engineers gave her its Annual Achievement Award, citing her "foundational and preeminent contributions to our understanding of the mechanical behavior of solids." She also won two Medals for Excellence in Research from the South Eastern Section of the American Society for Engineering Education in 1965 and 1966. Furthermore, she received the Heyn Medal from the German Society for Materials Science. She remains a full professor at the University of Virginia, where she continues to refine the electrical brush technology. Her research interests include work hardening, tribology, melting, and electrical contacts. She also enjoys lecturing on the reconciliation of science and religion.

Kuperberg, Krystyna
(1944–   )
Polish/American
Mathematician

Born in Poland, Kuperberg came to the United States to work on her Ph.D. and decided to stay in America for her career. She is best known for her work in topology and discrete geometry, and she has also contributed as an educator and participant in professional societies.

Born on July 17, 1944, in Tarnów, a city near Kraków, Krystyna Trybulec is the daughter of Barbara and Jan Trybulec, both college-educated pharmacists who managed the local pharmacy in her town. In 1959, when she was 15, her family moved to northern Poland, first to the city of Malbork, then to Gdansk, a seaport on the Baltic that later was the birthplace of the Solidarity movement that swept the Polish communist regime from power in the 1980s. In 1964, she married Wlodzimierz Kuperberg, a fellow mathematician she met at the University of Warsaw. They have two children, Greg and Anna.

Kuperberg received her secondary education from state schools in Tarnów, Malbork, and Gdansk. As early as primary school, she knew that she had an aptitude and affinity for mathematics. However, the difference between the math she studied in high school and what she encountered in 1962 when she entered the University of Warsaw was profound. Kuperberg was excited by what she has termed "the wonderful world of abstraction" that she found at the university. Her first math professor's first lecture was in algebra—an axiomatic definition of the determinant—but much more advanced than the concepts she had studied in high school.

After she graduated with an M.S. in mathematics in 1966, Kuperberg enrolled as a graduate student at the University of Warsaw under Karol Borsuk, a teacher who had entranced her as an undergraduate with his lectures on topology. She studied mathematics with Borsuk at the University of Warsaw until 1969. She left Poland in the fall of 1969 to live in Sweden where she worked as an instructor at the University of Stockholm from 1970 to 1972. In 1972, she moved to Houston, Texas, where she enrolled in Rice University and began work on her Ph.D. At Rice, Kuperberg studied with W. H. Jaco and concentrated on problems relating to topology. She received her Ph.D. from Rice in 1974.

In 1974, Kuperberg's career took her to Auburn University in Auburn, Alabama, where she was appointed to the position of assistant professor. By 1984, she had become a full professor of mathematics at Auburn. During her career, Kuperberg has tackled a number of interesting conceptual problems. In 1984, she solved a problem left by the mathematician Knaster about the bi-homogeneity of continua. Later she studied fixed points and the topology of dynamical systems. She found in 1993 a smooth counterexample to the Seifert conjecture that every nonzero vector field on the three-dimensional field has at least one closed orbit. This construction was soon applied to real analytic and piecewise linear settings. In the mid-1990s, she has worked, often with her son, Greg, who is also a mathematician, on problems related to aperiodic flows. She has given more than 50 lectures on this subject, including talks to the American Mathematical Society (AMS) and the International Congress of Mathematics.

For her work, Kuperberg received the Kosciuszko Foundation's Alfred Jurzykowski Award in 1995 and the Auburn University Creative Research Award in 1999. She has served on the AMS's Council and other committees of the AMS and the Mathematical Association of America. From 1994 to 1999, Kuperberg held the position of Alumna Professor of Mathematics at Auburn. Currently, she is once again a full professor.

Kwolek, Stephanie L.
(1923–   )
American
Chemist

An award-winning chemist, Stephanie Kwolek is known for her work on low-temperature polymers with the DuPont Company, the organization she worked with for her entire career. Kwolek's best-known discovery is the tough, rigid material known by the trade name Kevlar. Used in radial tires and for brake pads, racing sails, fiber-optic cables, and spacecraft shells, Kevlar is most famous as the lightweight, protective material used in bulletproof vests.

Kwolek was born in 1923 in New Kensington, Pennsylvania, a town on the Allegheny River near Pittsburgh in the western part of the state. After attending public school in New Kensington, she enrolled in the women's college of Carnegie Mellon University in Pittsburgh in 1942.

Kwolek earned a B.S. in chemistry from Carnegie Mellon in 1946 and intended to continue her university studies by getting a medical degree. Because she needed to earn more money to continue her education, she took a job as a chemistry researcher at the DuPont Company's textile fiber laboratory in Buffalo, New York.

Once she began to work at DuPont, Kwolek discovered that she liked the job of research chemist. Accordingly, she dropped her plans to attend medical school. She got her job just as millions of men began to return to the United States after having served in the U.S. armed forces during World War II. Initially, she and other newly hired women were told that they might be moved to other work once male chemists began applying for jobs at DuPont. However, because she had begun work at the beginning of research on low-temperature polymers, she was able to hold on to her job.

In 1950, Kwolek transferred to DuPont's Pioneering Research Laboratory in Wilmington, Delaware. Here she became deeply involved with experiments in petroleum-based synthetic fibers condensed at low temperatures. From these experiments, Kwolek and the DuPont lab invented a process to create and manufacture polymer substances that would be trademarked as Kapton polyimide film and Momex aramid fibers.

During the 1960s, Kwolek was able to create pure monomers from which polybenzamides were synthesized. In 1964, she worked with liquid crystalline polymers, discovered an ideal solvent, and determined the exact conditions under which these polymers could be spun.

If these low-temperature polymers were worked with under conditions too moist or too hot, they would be subject to hydrolysis and self-polymerization. By repeated and painstaking experimentation, Kwolek found out which conditions were suitable for the creation of these polymers. Even though the end result was a polymer that was fluid and cloudy (rather than the preferred viscous and clear polymer), Kwolek continued with her work. After the solution was spun out, she discovered that her synthetic material was much stiffer and stronger than other polymers.

By 1971, DuPont had begun to market Kwolek's invention as Kevlar. This synthetic material proved to have half the density of fiberglass, while at the same time it was a remarkable five times stronger than steel.

Kwolek worked as a research associate at DuPont until her retirement in 1986. During her career, she won many awards for her work, including the Howard Potts Medal from the Franklin Institute in 1969, the Chemical Pioneer Award from the American Institute of Chemists in 1980, induction into the Inventors Hall of Fame in 1995, and the Perkin Medal from the Society of Chemical Industry in 1997. Kwolek was only the second woman to win the Perkin Medal during its 97-year history. She continues to act as a consultant in polymer chemistry and frequently talks at colleges and secondary schools about science and careers in science.

L

Lachapelle, Marie-Louise
(1769–1821)
French
Midwife

A midwife at a time when women were not allowed to become doctors, Marie-Louise Lachapelle became one of the most knowledgeable practitioners of obstetrics in the late 18th and early 19th centuries. She performed around 40,000 deliveries and compiled notes about obstetrics and childbirth that were published as a massive compendium on the subject after her death.

Born in Paris on January 1, 1769, Marie-Louise Dugès was the daughter of Louis Dugès, a health official, and Marie Jonet Dugès, a midwife at the Hôtel Dieu hospital in Paris. In 1792, at the age of 23, Marie-Louise Dugès married a doctor named Lachapelle. She had one daughter and had to return to work as a midwife in 1795 after the death of her husband.

Marie Jonet Dugès came from a multigenerational family of midwives and taught her daughter her profession. By the time she was 15, Marie-Louise Lachapelle had delivered her first baby, according to her account a difficult birth in which she was able to save mother and child.

In 1795, Lachapelle's husband and mother died, and Lachapelle took over her mother's position as head of the maternity ward at Hôtel Dieu. At that time in France and other European countries, public hospitals were charitable institutions, many funded by the church, for poor patients. Wealthy and middle-class patients generally received medical care at private clinics or in their homes. The Hôtel Dieu was funded by Cathedral Notre Dame in Paris.

Lachapelle worked at Hôtel Dieu for a year before taking a leave of absence to study with a renowned German obstetrician, Franz Carl Naegele, in Heidelberg. By 1797, she was back in Paris where she was asked to head the maternity ward at a new hospital, the Hospice de la Maternité, which was set up as a teaching as well as a working hospital.

At La Maternité, Lachapelle, along with the obstetrician Jean-Louis Baudelocque, trained hundreds of midwives in the techniques that she was developing. One of the first things she did was to deal with the problem of hygiene that beset all public hospitals during that era. During Lachapelle's life, medical doctors and staff still did not adequately understand the role of infectious agents such as bacteria and viruses in causing death at hospitals. Basic hygiene conditions at many hospitals were appalling. Thousands of mothers and their babies died of childbed fever while staying at hospitals. Lachapelle began trying to get hospital staff to make La Maternité a cleaner place. She also limited the number of attendants and visitors who were allowed in the maternity wards. Both measures helped to reduce child mortality.

Lachapelle taught her midwife students via lectures about birthing techniques, during some of which she used a manikin to illustrate problems that students were likely to encounter in births. She also supervised her students during their deliveries. Whenever a mother died, students were required to attend autopsies so that they might better understand what went wrong and how to correct mistakes in the future.

Lachapelle had begun writing what would become a three-volume work about obstetrics and midwifery when she fell ill with stomach cancer in 1821. She died of that disease on October 4, 1821, but a nephew who was a doctor completed her book from her notes. *Pratique des accouchements; ou mémoires et observations choisies, sur les pointes les plus importants de l'art* was published in 1825. Translated into German, it was republished throughout the 19th century and became an indispensable work for those who aided in the birth of children.

Ladd-Franklin, Christine
(1847–1930)
American
Mathematician and Psychologist

A scientist with wide-ranging interests, Christine Ladd-Franklin made significant contributions to the fields of mathematical logic and psychology. Her research in mathematical logic was conducted primarily in a discipline called symbolic logic. Ladd-Franklin's advancements in psychology were in the field of color theory, in which she offered theories about how humans perceive color.

Born in Windsor, Connecticut, on December 1, 1847, Christine Ladd was the daughter of Eliphalet Ladd and Augusta Niles Ladd. Her father was a prosperous New York merchant, and her mother came from a politically well-connected family. Until she was six, Ladd lived in New York City. After that, she lived in Windsor, where her family moved after her father retired. When her mother died in 1860, Ladd was sent to live with a grandmother in Portsmouth, Vermont. She received her secondary education at a private school in Massachusetts. In 1882, at the

Christine Ladd-Franklin, Vassar College 1869, whose accomplishments in the field of psychology included significant theories about how humans perceive color. *(Special Collections, Vassar College Libraries)*

age of 34, she married Fabian Franklin, a fellow mathematician (she changed her surname to Ladd-Franklin). She had two children.

Ladd-Franklin began her university education at Vassar College, a newly established women's college in Poughkeepsie, N.Y., in 1867. She completed her B.A. in just two years and then taught science and mathematics in high schools for nine years. By 1878, she had returned to university-level studies. She had requested to be admitted to Johns Hopkins University in Baltimore as a "special student" (the university did not yet officially admit women students). James J. Sylvester, a mathematics teacher there who knew of Ladd-Franklin from math papers she had published, got her admitted and obtained a $500 yearly grant for her. By 1882, she had completed work for her Ph.D. Even though she had finished her work and her dissertation, "The Algebra of Logic," was published in a Johns Hopkins mathematics department periodical in 1883, Ladd-Franklin was not formally granted a Ph.D. at that time. The university, which would not award any woman a doctorate until 1893, finally gave Ladd-Franklin her doctorate in 1926 when she was 78 years old.

After her marriage, Ladd-Franklin remained in Baltimore, and even though she held no formal position at the university, she continued to work on mathematical–logical problems in association with Sylvester and Charles Sanders Peirce, another Johns Hopkins professor. By 1886, she had turned her attention to understanding the way the human mind perceives color. Combining mathematics with psychology, Ladd-Franklin concentrated on how the mind takes separate images from the two eyes and unifies them into a single perception. In 1891, she traveled with her husband to Germany where she worked with G. E. Müller at the University of Göttingen and Hermann von Helmholtz in Berlin on color-vision problems. After studying color perception, Ladd-Franklin presented a paper at the International Congress of Psychology in London in 1892 in which she advocated the idea that color differences were distinguished by the mind's perception of three kinds of light—white, yellow, and blue. Two other colors—red and green—arise in the yellow light spectrum. Thus the mind perceives the basic colors of white, yellow, blue, red, and green, from which all other colors arise. This theory eventually became for many years the standard way psychologists explained the phenomenon.

Ladd-Franklin returned to Baltimore where she served as an unpaid lecturer at Johns Hopkins until 1909. In 1914, she and her husband moved to New York City, and she became an unpaid lecturer in psychology at Columbia University until 1927. She was active in the Association of Collegiate Alumnae, an organization that pushed universities to open doors fully to women. Ladd-Franklin was the author of 64 scientific papers and a book, *Colours and Colour Theories*, published in 1929. A year before her death on March 5, 1930, she was praised as "a merciless and indomitable logician . . . whose grasp of the phenomenon of color vision is truly remarkable."

Laird, Elizabeth Rebecca
(1874–1969)
Canadian
Physicist

Elizabeth Rebecca Laird chaired the physics department at Mount Holyoke College, transforming it into a graduate-quality program despite the fact that it enrolled only undergraduates. She was the first woman physicist accepted by Sir J. J. Thompson to conduct research at the Cavendish Laboratory of Cambridge University.

Laird was born on December 6, 1874, in Owen Sound, Ontario. Her mother was Rebecca Laird, and her father, the Reverend John Laird, was a Methodist minister. After graduating from the London Collegiate Institute in 1893, she matriculated at University College of the University of Toronto. There she majored in mathematics and physics, placing first in both divisions in each of her last three years. In 1896, she earned her bachelor of arts degree and received the university's Gold Medal. Although her nomination for an 1851 exhibition scholarship after her graduation came with strong support, nevertheless her gender prevented her from receiving this postgraduate funding. Laird therefore spent the next year as a mathematics instructor at Ontario Ladies College.

In 1898, she received a postgraduate fellowship in physics from Bryn Mawr College, and she spent one of the years abroad as a fellow at the Max Planck laboratories in Berlin. She returned to Bryn Mawr to study under L. Stanley McKenzie, earning her Ph.D. in physics and mathematics in 1901. Mount Holyoke College hired her in 1901 as an assistant in physics and retained her for the next 40 years. The college promoted her to the rank of instructor in 1902, and in 1903, it appointed her as the head of the physics department. The next year, she became a full professor.

In 1909, Laird became the first woman to conduct research at Cambridge University's Cavendish Laboratory under Sir J. J. Thompson on a paid sabbatical leave. Mount Holyoke supported its faculty's pursuit of continuing research in their fields through these salaried sabbaticals, and Laird took advantage of this to develop an undergraduate physics program staffed by professors who rivaled the expertise of graduate faculties.

Laird took subsequent leaves to study at the University of Würzburg under a Sarah Berliner Research Fellowship from 1913 through 1914 and at the University of Chicago in 1919. In 1925, Laird served as an honorary research fellow at Yale University. She focused her research on "soft" X-rays, spectroscopy, thermal conductivity, spark radiation, and the electrical properties of biological material in the microwave region.

In 1940, Laird retired to emerita status at Mount Holyoke, although she continued to conduct research. She volunteered her expertise working on radar development for the National Research Council of Canada at the University of Western Ontario, where she became an honorary professor of physics from 1945 until 1953. During this time, she conducted research on the absorption of ultra-high frequency radiation by tissue under grants from the Canadian Cancer Foundation.

Laird received honorary degrees from the University of Toronto in 1927 and from the University of Western Ontario in 1954. She died in 1969, and the next year, the University of Western Ontario established the Laird Memorial Lecture Series.

Lancefield, Rebecca Craighill
(1895–1981)
American
Bacteriologist

Working at the Rockefeller Institute for Medical Research, Rebecca Craighill Lancefield made her scientific reputation in the study of streptococcus, a multifaceted bacterium whose workings and groupings were little understood when she began her career. During six decades, Lancefield identified more than 50 types of streptococcus and discovered how various strains reacted with the human body. Her research helped doctors to better understand diseases caused by this agent and to develop treatments to heal patients suffering from it.

Born on January 5, 1895, at Fort Wadsworth in Staten Island, New York, Rebecca Craighill was the daughter of William Edward Craighill, an officer in the U.S. Army, and Mary Byram Craighill, a homemaker. During her childhood and youth, Craighill traveled around the United States as her father was based at different army stations. She was encouraged to excel in her studies by both parents but especially by her mother, who had been influenced by Julia Tutwiler, an advocate for education for women. The third of six daughters, Rebecca Craighill was the first one to attend college. In 1918, she married Donald Lancefield, a biologist she met while they attended Columbia University. She had one child.

In 1912, Rebecca Lancefield traveled to Massachusetts to enroll as an undergraduate in Wellesley College, a college for women. She first intended to study French and English but found herself instead attracted to science when she took her first zoology course. She graduated from Wellesley with a B.A. in zoology in 1916. Lancefield briefly taught science at a girls' boarding school in Vermont, but wanting to pursue a career in science, she enrolled in

Rebecca C. Lancefield, who identified more than 50 types of the streptococcus bacterium. *(Photo by RI Illustration Service, Don Young, Courtesy of Wellesley College Archives)*

Columbia University in New York City. Working intensively with Hans Zinsser, a noted bacteriologist, she earned a master's degree in bacteriology from Columbia in 1918.

Lancefield wanted to begin work on a Ph.D. in bacteriology immediately, but the entry of the United States into World War I delayed her plans. Her husband was drafted into the army and assigned to work at the Rockefeller Institute. Lancefield also applied for a job at the Rockefeller Institute and was hired to assist Oswald Avery and Alphonse Dochez in their early studies of the streptococcus bacterium. After the war, she followed her husband for a one-year teaching stint in Oregon and then returned with him to New York. In 1922, she began working again at the Rockefeller Institute, this time with Homer Swift, while also working on her Ph.D. dissertation in bacteriology at Columbia.

Lancefield won a Ph.D. from Columbia in 1925. Her dissertation, "The Immunological Relationships of *Streptococcus viridans* and Certain of Its Chemical Fractions," summarized the studies she had made at the Rockefeller Institute that showed there was no link between rheumatic fever and a class of streptococcus called *Streptococcus viridans.* Lancefield continued her work on streptococcus, and by 1933, she had devised a precipitin test by which she found five different strains, or subvarieties, of this bac-

terium. Lancefield later isolated the cause of rheumatic fever, which was by group A streptococcus, and determined that the virulence of a particular group A streptococcus was caused by a protein that protects that strain from being enveloped by human white blood cells, which normally kill invasive organisms.

For her contributions to the understanding of bacteriology, Lancefield won many awards, including the American Heart Association's Achievement Award (1960) and the Medal of the New York Academy of Sciences (1970). She served as president of the Society of American Bacteriologists in the late 1940s and, in 1970, was elected to the National Academy of Sciences. Rebecca Craighill Lancefield died in New York City of complications from a broken hip on March 3, 1981. She was 86 years old.

Lavoisier, Marie Anne Pierrette Paulze
(1758–1836)
French
Chemist and Translator

The daughter of wealthy aristocrats, Marie Lavoisier was an experimental chemist, organizer of scientific salons, and translator of the works of English chemists into French. She lived through some of the most difficult and exciting times in French history—the flowering of the scientific revolution during the late 1700s as well as the horrors of the French Revolution, which claimed her husband and father among its victims.

Born on January 20, 1758, in Montbrison, a town in the Loire River valley of central France, Marie Anne Pierrette Paulze was the daughter of Jacques Paulze, a politician and financier, and Claudine Thoynet Paulze, an aristocratic woman whose uncle was the controller-general of finance for the regime of Louis XVI. During her childhood and youth, Marie Paulze's father was director of the French East India Company, then a partner of the Ferme-Générale, a group of private financiers who had been given the job of collecting taxes for the government. She was one of three siblings and the only girl in her family. In 1771, at the age of 13, she married Antoine Laurent Lavoisier, also an aristocrat and a man who worked for her father. Marie Lavoisier's mother died in 1761 when Marie was three years old. Afterward, Lavoisier was sent to live and get an education in a convent. Hers must have been a fairly good education, although it was not the equal of what a boy would have received then in France. After she married, she continued her education with her husband who was interested in a career as a philosopher and scientist.

Philosophy and science were seen as two branches of the same desire for universal knowledge. Both Lavoisier and her husband read books about philosophy and science; they also frequently invited scientists and philosophers to dine with them and spend long evenings in conversation. These informal seminars about science and philosophy proved popular among intellectuals of their day, and the Lavoisiers's house was seen as the place to go when a visiting scientist came to Paris.

In 1775, Antoine Lavoisier was appointed royal gunpowder administrator, a position that required the couple to live in Paris. The couple moved to the Arsenal, the royal administration's gun and gunpowder storage center in Paris. This site also served as an experimental laboratory for weapons and armaments and included numerous rooms for administrators.

According to Marie Lavoisier, the period from 1775 (when she was 17) to 1793, when Antoine Lavoisier and Jacques Paulze were imprisoned, were the happiest of her life. She and her husband got along wonderfully and shared a mutual interest in science. During this time, Lavoisier, with her husband, conducted most of her scientific experiments.

One of the Lavoisiers's long-running experimental programs was the study of plant respiration and transpiration, that is, how the organism exchanges oxygen for carbon dioxide, which involves the loss of water vapor into the atmosphere. Marie Lavoisier kept extensive notes and made many drawing about these experiments. She did the same for experiments that she and her husband conducted with potassium chlorate gunpowder. She also helped her husband prepare his book, *Traité élémentaire de chemie,* published in 1789, and translated the works of chemists Joseph Priestley and Richard Kirwan.

Because they belonged to the aristocracy and had worked as high officials in the royal government, Antoine Lavoisier and Jacques Paulze were viewed as traitors to the revolutionaries who took over the French government in 1789. Both were guillotined on May 8, 1794.

Marie Lavoisier eventually regained some of her properties and lived a genteel life in Paris. She did not conduct further scientific experimentation but continued to host salons for well-known scientists of her day. She died suddenly, perhaps of a heart attack, in Paris on February 10, 1836, at the age of 78.

Leakey, Mary Douglas Nicol
(1913–1996)
British
Paleontologist and Anthropologist

Known for many years as the junior partner in the Leakey–Leakey team, Mary Leakey was in fact the scientific equal of her more famous husband, Louis Leakey. She was the one who actually made several of the famous finds of early humanoid bones and footprints that altered our understanding of the evolution of man from apes during the past few million years. With her husband, and on her own, she authored numerous books and articles that explain her research into the paleontological record of early humans and apes.

Born in London, England, on February 6, 1913, Mary Nicol was the daughter of Erskine Nicol and Cecilia Frere Nicol. Her mother was from a family that had excavated Stone Age sites in England in the late 1700s, and her father was a painter who frequently took the family to southwestern France, where Mary first encountered ancient human drawings of animals found in caves. Her experience with cave paintings in France made her realize that she would like to study and work as an archaeologist. She would marry Louis Leakey in 1936 and with him would have three sons, one of whom, Richard, would carry on her work as an early primate paleontologist.

Encouraged by Dorothy Liddell, an archaeologist she met in the late 1920s, Leakey began studies in geology and archaeology at the University of London in 1930. From 1930 to 1934, she did her first fieldwork excavating early Stone Age sites around Henbury, Devon, in southwestern England. She developed a specialty at analyzing stone tools and discovered that, like her father, she had a facility for drawing these tools and the sites in which they were uncovered.

Mary Nicol first met Louis Leakey in 1933, and by 1934, they were working together on a dig in Clacton, England, where the skull of a hominid, a humanlike ape that walked on two feet like humans, had been found. That same year, she illustrated Leakey's book, *Adam's Ancestors.* By 1935, Mary and Louis Leakey had fallen in love, and Leakey asked her to join him for an excavation at his favorite site, the Olduvai Gorge in the British colony of Tanganyika (later to be the independent nation of Tanzania). Leakey later said that this spot "has . . . come to mean more to me than any other in the world." Here and in Kenya, she and her husband began a long series of digs at sites where human and ape remains were known to be numerous. It was not until 1948 that their first really big find was made. On Rusinga Island in Lake Victoria, Kenya, they found the skull of a prehuman ape species called *Proconsul africanus.* Louis Leakey speculated that this ape was the "missing link" between present-day humans and apes, the species that would show how humans diverged from apes about 16 million years ago. (Later evidence would show that this guess was not accurate.)

Other finds followed for Mary Leakey. In 1959, she discovered the jaw of an early hominid she named *Zinjanthropus.* By using carbon-dating techniques, the Leakeys determined that this creature lived 1.75 million years ago, the earliest hominid then found in Africa. A few years later, Mary Leakey's son Richard discovered the jaw of a true human ancestor, which was given the name *Homo habilis.*

Homo habilis and the much more apelike *Zinjanthropus* had coexisted for a while in Africa.

After Louis Leakey's death in 1972, Mary Leakey continued her work alone. One of the finds that she was most proud of was her discovery in 1978 of hominid footprints in a layer of sediment beneath some lava beds in Laetoli, Tanzania. These prints clearly show that hominids were walking upright like humans at around 3.5 million years ago, a time before which any humanlike tools have been found. As Leakey explained in a *National Geographic* article, "This new freedom of forelimbs posed a challenge [for the hominids]. The brain expanded to meet it [by later figuring out how to use tools]. And mankind was formed."

For her work, Leakey was awarded the Hubbard Medal by the National Geographic Society, the Boston Museum of Science's Bradford Washburn Award, and the Society of Women Geographers Gold Medal. She was also granted an honorary doctorate from Oxford University in 1981. She died at age 83 on December 9, 1996.

Leavitt, Henrietta Swan
(1868–1921)
American
Astronomer

During a career spent exclusively at the Harvard Observatory, Leavitt studied the brightness of stars and compiled charts of variable stars, suns whose brightness varies in regular intervals. Using her work on the brightness of variable stars, other astronomers were able to calculate the distance of faraway objects such as galaxies, thereby exponentially extending the size of the known universe.

Born on July 4, 1868, in Lancaster, Massachusetts, Henrietta Swan Leavitt was one of seven children of George Leavitt and Henrietta Kendrick Leavitt. Her father was a Congregational minister who preached at a church in Cambridge, Massachusetts. Leavitt attended public elementary and secondary schools in Cambridge. She never married nor had children.

When she was 17, Leavitt moved with her family to Cleveland, Ohio, where her father had taken another preaching job. From 1885 to 1888, Leavitt attended Oberlin College. She transferred to the Society for the Collegiate Instruction of Women (which later became Radcliffe) in 1888 and received an A.B. from there in 1892. Leavitt concentrated on astronomy in college, but an illness soon after her graduation left her deaf. She did not return to her studies or the workplace until 1895, when she began to work as an unpaid volunteer at the Harvard Observatory. For seven years, Leavitt performed the most tedious calculation and measuring chores at the observatory. Finally in 1902, she was hired as a paid member of the staff, which was supervised by Edward Pickering, the observatory's head.

Henrietta Leavitt, who established the first method of measuring large-scale distances in the universe. *(Harvard College*

By 1907, Leavitt had become chief of the section of the observatory that measured star brightness. The work of establishing an accurate measure of a star's true brightness was one of the most important practical tasks of turn-of-the-century astronomy. This was not a matter of theoretical speculation but of nuts-and-bolts astronomical comparison and calculation. The apparent brightness of stars varied from observations made with the eye and the star image that appeared on film. Also, photographs made at different observatories gave different brightnesses for the same stars, a clear impossibility unless the star itself changed brightness over time. Leavitt sat down with almost 300 photographs from 13 observatories and systematically calculated the true brightness of 46 stars that lay near the North Pole star. This reckoning, which standardized the brightnesses for stars in this region, was adopted by the International Committee on Photographic Magnitudes in 1913.

While studying star magnitudes, or brightness, Leavitt became interested in another astronomical problem that related to brightness. She began trying to understand the relationship of brightness to time in Cephid stars in a star group called the Magellanic Clouds. Cephids are stars whose brightness waxes and wanes over a regular period of time. Leavitt established that the brightness of each Cephid star in the Magellanic Clouds was linked to the length of its pulsation cycle: the longer it took to wax and wane, the brighter the star.

In 1912, she published a table that linked pulsation to brightness. The longest pulsation period was 127 days, the shortest 1.25 days, and the average was about five days. This table allowed other astronomers to compare these stars' true brightness (as calculated by Leavitt) with their apparent brightness. From this comparison, they could accurately calculate the distance in light-years of these stars from Earth. Before Leavitt's work, distances could be calculated out only to 100 light-years. Now, astronomers were astonished to find that the Cephids in the Magellanic Clouds were 100,000 light-years away, which meant that this star group was outside our own galaxy; indeed, they were totally separate galaxies from our own.

Having established this extraordinarily important way of calculating distances in the universe, Leavitt then returned to her work of measuring star brightness in various parts of the sky, a chore that would occupy the rest of her career. She died of cancer at the age of 52 on December 12, 1921.

Le Beau, Désirée
(1907–1993)
Austro-Hungarian/American
Chemist

Désirée Le Beau developed methods for recycling rubber (primarily from old tires) to be used in new products. She was a colloidal chemist (the branch of chemistry dealing with those substances composed of small, insoluble, non-diffusable particles suspended in a medium of different matter), and she worked in industry for most of her career. Le Beau held several patents for the rubber recycling processes she invented, and she was the first woman to chair the American Chemical Society's Division of Colloid Chemistry.

On February 14, 1907, Désirée Le Beau was born in Teschen, Austria-Hungary (now part of Poland) to Phillip and Lucy Le Beau. She would later marry Henry Meyer on August 6, 1955. Although she planned to become a pharmacist, she inadvertently attended a chemistry class at the University of Vienna, and this brief exposure to the subject captivated her interest. She went on to earn her undergraduate degree in chemistry from the University of Vienna. She then attended the University of Graz in Austria, receiving her Ph.D. in chemistry in 1931.

After completing her formal education, Le Beau accepted a position with the Austro-American Rubber Works in Vienna in 1932. She moved to Paris in 1935 to work as a consultant at the Société de Progrès. Because of Hitler's escalating aggression in Europe, however, she emigrated to the United States in 1936, joining the Dewey & Almy Chemical Company in Massachusetts as a research chemist upon her arrival. In 1940, Le Beau was named a research associate at the Massachusetts Institute of Technology's (MIT) Department of Chemical Engineering and Division of Industrial Cooperation.

Le Beau remained at MIT until 1945, when she was appointed the director of research at the Midwest Rubber Reclaiming Company in Illinois. The company was the world's largest independent reclaiming company, and it afforded Le Beau the opportunity to examine the structures of natural and synthetic rubbers and clays. Her research led her to a number of breakthroughs. In 1958, she formulated and later patented a process to use reclaimed rubber in a tie pad for railroads. She also obtained patents for producing reclaimed rubber in particulate form and for developing methods of reclaiming amines (derivatives of ammonia in which hydrogen atoms have been replaced by radicals containing hydrogen and carbon atoms) and acids. Le Beau was named a Currie lecturer at Pennsylvania State College in 1950.

Le Beau's influence on the field of reclamation chemistry was profound. Her research was especially crucial during World War II, when rubber was needed for both the war effort and domestic production. Her insights helped create rubber that would otherwise not have been available. These accomplishments won her a number of prizes and honors, including the 1959 Society of Women Engineers Achievement Award for her contributions to the field of rubber reclamation. In addition to becoming the first woman to chair the American Chemical Society's Division of Colloid Chemistry, she was also the first woman to chair the society's St. Louis division. Elected a fellow of the American Institute of Chemists, she authored many papers on the topic of reclamation.

Lehmann, Inge
(1888–1993)
Danish
Mathematician and Geologist

Trained as a mathematician but later seduced by the science of geodesy, a field of geology whose study is the shape, size, and composition of Earth's surface and subsurface, Inge Lehmann proved that there was a solid core at Earth's center. This idea upset the prevailing wisdom that Earth's subsurface was completely molten.

Lehmann was born in Copenhagen, Denmark, on May 13, 1888, the daughter of Alfred Georg Ludvig Lehmann and Ida Torsleff Lehmann. Her father was a professor of psychology at the University of Copenhagen who believed that his two daughters should be as well educated as any man. Accordingly, he sent Inge to the first school in Denmark in which boys and girls were taught together rather than being segregated by gender. Lehmann never married nor had children.

In 1907, Lehmann enrolled at the University of Copenhagen as a math student. She completed her undergraduate degree there in 1910, then spent a year studying math at Cambridge University in England. For six years (1912–18), Lehmann put her math skills to use by working as an actuary at a Danish insurance company. She then returned to the University of Copenhagen to get a master's degree in mathematics, which she finished in 1920. She followed this up by studying math for a while at the University of Hamburg.

In the early 1920s, Lehmann became acquainted with the new techniques of seismology that were being employed to study Earth's subsurface. Intrigued with this branch of science, she began work in 1925 at the Royal Danish Geodetic Institute. She helped install the first seismographs in the institute's offices that year. Lehmann returned to the University of Copenhagen yet again, this time to get a master's degree in geology with a specialty in geodesy, which she finished in 1928. That year, she was appointed head of the Royal Danish Geodetic Institute's department of seismology. In this capacity, she supervised the institute's subsurface seismology work in Denmark and Greenland, and she edited the institute's bulletins.

Lehmann also began studying readings at her lab in Copenhagen of earthquake shock waves that traveled through Earth's core from other regions. Of special interest were so-called P-waves, shock waves that reached Lehmann from quakes in the Pacific Ocean region, on the other side of Earth from Denmark. These waves had the possibility of opening up the core of Earth for scientific inspection. At the time, almost all geologists believed that the huge area of Earth below the solid outer mantle was entirely liquid, composed of molten rock. This molten rock is the material that spews up from volcanoes and lava flows. However, by analyzing the P-waves received at her office in Denmark, Lehmann offered proof that a sphere of solid rock exists at the core of Earth. This core is surrounded by hot, liquid material, which is, in turn, wrapped by the solid mantle at the edge of Earth's surface. She announced this finding in 1936, and it immediately caused a furor in the world of geology. A few years later, other studies confirmed her work, and this theory is now universally accepted.

Lehmann officially retired in 1953 but continued her work for years after that at various universities around the world. For her achievements, the Royal Danish Society of Science awarded her its Gold Medal in 1965. She was also awarded the American Geophysical Union's Bowie Medal in 1971, and in 1997, this same group named a medal after her—the first award by this organization to be named after a woman. Lehmann died in 1993, at the age of 105.

Leopold, Estella Bergere
(1927–   )
American
Paleoecologist

Estella Bergere Leopold is recognized as a leading authority on paleoecology—the study of prehistoric organisms and their environments. Her research technique generally involves comparing the pollen and spores of ancient plants with those that exist today. During her 21-year career with the United States Geological Survey, she made important findings about the effects of climate on the evolution and extinction of prehistoric plant species. She later headed the Quaternary Research Center at the University of Washington, Seattle, and has been active in several crucial conservation efforts.

One of Aldo and Estella Bergere Leopold's five children, Leopold was born in Madison, Wisconsin, on January 8, 1927. Her father, the conservationist Aldo Leopold, authored *A Sand County Almanac* and instilled in his children a love of both science and nature. Estella Leopold would never marry or have children.

Leopold attended the University of Wisconsin at Madison, earning her bachelor's degree in botany in 1948. After receiving her master's degree from the University of California at Berkeley, she contemplated pursuing Ph.D. work at the University of California at Los Angeles, but she was troubled by tales of gender discrimination at the institution. Instead, she enrolled at Yale University, where she developed her interest in paleoecology under the tutelage of George Evelyn Hutchinson. She was awarded her Ph.D. in botany in 1955.

Although she longed for an academic career, Leopold was intimidated by the misogynistic attitudes prevalent at the time. Instead, she opted to join the Paleontology and Stratigraphic Branch of the United States Geological Survey (USGS) in Denver, Colorado. Her distinguished career there spanned from 1955 to 1976. In her work in the Rocky Mountains, Leopold made a number of important discoveries. Foremost among these was her finding that the extinction and evolution of ancient species into newer variants was more prevalent in the middle of the continent because of the wider degree of seasonal changes. By contrast, coastal areas (with their more moderate climates) supported older species, such as the giant redwood.

Like her father, Leopold was an ardent conservationist. She led the Defenders of Florissant, Inc., in a successful campaign to save Colorado's Florissant fossil beds. Although the National Park Service had recognized that the area needed protection in 1962, no substantive action was taken. After a protracted lobbying campaign, Leopold's group succeeded in having the U.S. Congress designate the area as a national monument in 1969. During this period, Leopold also served on the boards of the Environmental

Defense Fund, the National Audubon Society, and the Nature Conservancy.

In 1967, Leopold at last fulfilled her goal to teach when she became an adjunct professor at the University of Colorado. In 1976, she left both the USGS and the university to direct the Quaternary Research Center at the University of Washington. (The Quaternary period is the current era of the earth's history, which began roughly 2 million years ago). She left that post in 1982 but remained a professor of botany at the University of Washington.

Leopold's contributions to the field of paleoecology are vast, and her achievements have been amply recognized. She was elected to the National Academy of Sciences in 1974 and to the Geological Society of America a decade later. In addition to serving as the president of the American Association for the Advancement of Science in 1995, she was the associate editor of *Quaternary Research* from 1976 to 1983, and she continues to work on the journal's editorial board as well as on that of *Quaternary International*. In addition, her conservation efforts won her the 1969 Conservationist of the Year Award from the Colorado Wildlife Federation. She remains at the University of Washington and hopes to write a book on paleoecology.

Lepeshinskaia, Ol'ga Borisovna Protopova
(1871–1963)
Russian
Biologist and Physician

Educated as a physician in Russia and Switzerland, and an active participant in revolutionary politics before and after the Russian Revolution of 1917, Ol'ga Lepeshinskaia quickly rose through the ranks of the Soviet Union's science establishment, eventually becoming one of its authorities in the field of cellular biology. Because of her association with Trofim Lysenko, the discredited Soviet-era geneticist, Lepeshinskaia's reputation plummeted even before her death in 1963.

Born in August 1871 in Perm, a city in central Russia just west of the Ural Mountains, Ol'ga Borisovna Protopova was the daughter of a farmer who owned a modest estate in that region. There is little information about her childhood and youth. However, she must have received a good education; otherwise, she would not have been able to pursue an advanced medical education in Russia and abroad later in her life. In the mid-1890s, she met and married Panteleimon N. Lepeshinskii, a revolutionary activist in St. Petersburg. Lepeshinskaia had at least one child, a daughter, from this union.

The first record of Lepeshinskaia's appearance in the historical record in Russia appears in 1894 when she joined the St. Petersburg Union of Struggle for the Emancipation of the Working Class. This was an outlawed dissident group that was working to create revolutionary change in Russia. Their goals were the removal of the czar, replacement of the czarist political system with a more progressive government, and passage of laws that would guarantee rights for industrial workers and peasants. Lepeshinskaia was active in the St. Petersburg Union of Struggle while apparently at the same time attending a medical school in St. Petersburg. In 1897, she graduated with a degree in surgery from the Rozhdestvensky School. The year after that, she joined the Russian Social Democratic party, and later she followed Vladimir Lenin when he broke from the Social Democrats to form the Russian Communist party. For her organizing activities, she was sent into exile around 1900. She ended up in Switzerland where she continued her medical studies at the University of Lausanne. She graduated from there with a medical degree in 1902.

Lepeshinskaia returned to Russia in 1903 and began participating in revolutionary politics in the town of Pskov, which is in the St. Petersburg region. By 1915, she was in Moscow, where she was awarded a medical degree from the University of Moscow. However, between 1897 and 1913, owing to frequent arrests and Siberian and European exiles, she never had a settled life in Russia. Lepeshinskaia had developed close ties to Lenin by the time he returned to St. Petersburg in 1917 and was a high-level Communist operative during the revolution that swept the communists to power.

Lepeshinskaia's scientific career began in earnest only after the revolution, by which time she was in her late forties. From 1920 to 1926, she was a staff researcher at a Moscow hospital. In 1926, she moved to the Timiriazev Institute's Histological Laboratory, a position she held until her death. She also worked simultaneously in the All-Union Institute for Experimental Medicine's Cytology Lab and became head of this lab from 1957 to 1963.

From about the mid to late 1930s until her death in 1963, Lepeshinskaia was a central figure of Soviet science in the fields of cytology, histology, and plant biology. She worked hard to maintain her ties with Soviet leaders such as Joseph Stalin and his successor, Nikita Khrushchev. During Stalin's time, especially, it was impossible for other scientists to disagree with the party line about scientific knowledge. To do so could mean death or exile to a Siberian prison camp. Unfortunately, Lepeshinskaia established the party line in her fields. Because a free give-and-take of ideas was forbidden, she was not challenged to produce the work she was probably capable of. She allied herself with Trofim Lysenko, whose ideas about plant genetics were completely discredited in the 1960s. Indeed, Lepeshinskaia's own papers about the healthful effects of enemas, the formation of blood vessels from egg yoke, and other concepts were ridiculed as "rubbish" by a later generation of dissident Russian scientists. By the time of her death in 1963 at the age of 92, Lepeshinskaia's reputation had been undermined and her work demolished.

L'Esperance, Elise Depew Strang
(c. 1879–1959)
American
Physician

From a beginning in pediatric medicine, Elise L'Esperance later switched her focus to the study and prevention of cancer. She believed that the key to preventing cancer was an early detection of the disease. To accomplish this, she founded an institution, the Strang Tumor Clinic, that specialized in the early detection of cancer in women and children. Later, she expanded the clientele base of her business to include men and adolescents.

The details of Elise Strang's early life are sketchy, but it is thought that she was born sometime around 1879, probably in New York City. Her father, Albert Strang, was a physician, and her mother, Elise Depew Strang, was a housewife who came from a wealthy and politically connected New York family. While a medical student, she married David L'Esperance, a lawyer.

She completed her secondary schooling early and, following her father's career path, in 1896 at the age of 16 enrolled in medical school at Woman's Medical College, the teaching branch of the New York Infirmary for Women and Children, founded in 1868 by ELIZABETH BLACKWELL. Between 1900 and 1902, she completed her residency in pediatric medicine at Babies Hospital in New York City.

For eight years, L'Esperance worked at a private practice in pediatrics in New York. By 1910, she had become interested in pathology, the medical specialty in which human organs and tissues are examined to detect and understand the workings of disease at the cellular level. In 1910, L'Esperance began study under James Ewing, the pathologist in residence at the Cornell Medical School in New York City. She studied with Ewing until 1912 when she took a job as pathologist and lab director at the New York Infirmary for Women and Children, a job she would hold until 1944. Beginning in 1919, L'Esperance also served as an instructor in surgical pathology at Bellevue Hospital in New York City.

Originally, L'Esperance had become a pathologist because she wanted to contribute to the fight against tuberculosis, a deadly disease that especially afflicted the poor during the early part of the 20th century. However, by 1930, her attention had shifted to cancer, a disease whose workings were still poorly understood. Her desire to contribute to an understanding of cancer became personal in 1930 when her mother died of the disease. With money she had inherited from her uncle, former senator Chauncey

Depew, L'Esperance and her sister, May, founded the Strang Tumor Clinic in 1930.

As a pathologist, L'Esperance was well aware of how cancer could slowly grow in one organ of the human body without being detected. Often cancer was found only after it had spread to other parts of the body, at which point the chances of a successful cure were usually naught. Accordingly, L'Esperance encouraged women and children to come into the Strang Clinic for cancer examinations even if they felt well and healthy. In the clinic's first year of operation, L'Esperance and her colleagues found early cancers in 71 women who had seemed to be in good health.

By 1940, the Strang Clinic had become so popular that L'Esperance added another clinic at the Memorial Center for Cancer and Allied Diseases. By 1950, her clinics had examined 35,000 patients. Actual cancers were found in about 2 percent of the patients, but precancerous conditions were discovered in 18 percent of the patients.

During her career, L'Esperance authored 30 papers about pathology and cancer in medical journals. She also served for a time as editor of the *Journal of the American Medical Association.* For her contributions to medicine, L'Esperance was elected a fellow of the New York Academy of Medicine. She was also a member of the American Cancer Society, the Woman's American Medical Association, and the American Radiologists Society. She died on January 21, 1959 in Westchester, New York.

Levi-Montalcini, Rita
(1909–   )
Italian/American
Physician and Embryologist

Trained as a physician, Rita Levi-Montalcini was attracted to bioscientific research early in her career. Just before World War II, she began a study of the growth of nerve cells in chick embryos. A Jew, Levi-Montalcini was prohibited from working in Italian medical laboratories by the fascist government of Benito Mussolini in 1938. After the war, she was invited to join the faculty of the Department of Zoology at Washington University in St. Louis where she continued her research. Her discovery with coworker Stanley Cohen of a substance that promoted the growth of nerve cells in embryos was a milestone in neuroscience in the 1950s. For this work, Levi-Montalcini was awarded the Nobel Prize for medicine and physiology in 1986.

Born on April 22, 1909, in Turin, Italy, Rita Levi-Montalcini was the daughter of Adamo Levi, an electrical engineer, and Adele Montalcini, a painter. She had two older siblings, but throughout her life, Rita was especially close to her twin sister, Paola, who, like her mother, became an artist. In 1938, Levi-Montalcini became engaged to

Germano Raising, a fellow medical student. However, Raising died before they were wed. She never married.

The Levi-Montalcini household was warm and supportive of its children. Because Mr. Levi wanted them to get to know children of all groups, Rita and Paola attended public elementary schools. For middle and secondary school they went to a private girl's academy, which did not teach all of the courses one would need to attend university.

Between 1927 and 1930, Levi-Montalcini remained at home and occupied herself with reading, especially trying to fill in the gaps of her education. By 1930, she decided that she wanted to enter the School of Medicine at the University of Turin. After studying with a friend for six months, she passed the entrance exam and soon began work with Guiseppe Levi (no relation), a professor of pathology and histology.

After earning her M.D. in 1936, Levi-Montalcini worked at the University of Turin's Institute of Anatomy, but in 1938, she was fired, along with Guiseppe Levi and all other Jews, on orders from the Mussolini government. Be-

fore she was fired, Levi-Montalcini had been studying the development of nerve growth in chick embryos. During the war, Levi-Montalcini continued her work unofficially in a makeshift lab she set up in her house. Because of the entry of the German army into Italy in 1943, she had to stop her research as she and her family fled south to Florence where they spent the remainder of the war in hiding.

After the war, on the strength of a paper she had published about her chick embryo research, Levi-Montalcini was invited by Viktor Hamburger to come to Washington University in St. Louis to work with his team on embryological research. In 1949, she published a paper that reported her findings that neurons in the nerve cells of chick embryos and other observed animals grow in regular patterns, thus indicating a genetic programming of nerve cell growth. Levi-Montalcini, with fellow researcher Stanley Cohen, documented environmental factors that affected this nerve cell growth. By the mid-1950s, she and Cohen had identified, extracted, and purified a substance, named nerve growth factor (NGF), that caused nerve cells to reproduce.

From 1969 to 1979, Levi-Montalcini served as director of the newly created Institute of Cell Biology in Rome, which was jointly funded by Washington University and the Italian government. For her work, Levi-Montalcini received the Lasker Award for medical research in 1986, the same year she won the Nobel Prize. She is also a member of the Association for the Advancement of Science (U.S.); the European Academy of Science, Arts, and Letters; and the National Academy of Science of Italy. Although she officially retired in 1979, she lives in Rome and continues to do research at the Institute of Cell Biology.

Levy, Jerre
(1938–)
American
Psychologist and Neurophysiologist

A curiosity about how people perceive objects and process information led Jerre Levy to study psychology and later brain physiology. By observation and experimentation on a wide variety of people, Levy helped to show that the two halves of the brain (each called a brain hemisphere) analyze information differently. Levy's findings were widely, and often inaccurately, reported in the popular press during the 1970s and after, with accounts usually stating that the left hemisphere is the linear, detail-processing part of the brain, while the right exclusively analyzes information holistically and is the center of emotion.

Born in Birmingham, Alabama, Jerre Levy grew up in the small town of Demopolis, Alabama. Levy is the daughter of Mr. and Mrs. Jerome Levy. Her mother was a

Rita Levi-Montalcini, who discovered nerve growth factor (NGF), a substance that makes nerves grow and may someday be used in treatment for such conditions as Alzheimer's disease. *(Bernard Becker Medical Library, Washington University School of Medicine)*

housewife, and her father owned and ran a clothing store in Demopolis. On her graduation from high school in 1956, she enrolled in the University of Miami.

Levy's high-school counselors had advised her that science was a man's occupation and that it would be hard for a woman to make a career for herself. Taking this advice to heart, Levy as an undergraduate majored in psychology, the one science in which women seemed to be able to work.

After graduating from the University of Miami with a B.A. in psychology in 1962, Levy enrolled in the university's graduate program in biological psychology. She worked hard in that program for three years before being told by the department chairman that he would not give her a grant to finish her doctoral studies. His reasoning was that the money would be wasted on her because no university would hire a woman as a professor on a tenure track. Levy finally managed to get a grant from the marine science department, but she stayed at the university for only another year and a half. In 1967, she was admitted to the doctoral program in biology at the California Institute of Technology (Caltech) in Pasadena.

At Caltech, Levy found a sympathetic mentor in Roger Sperry, a neurobiologist who had begun a study of people with severe epilepsy whose corpus callosum, the bundled cord of nerve fibers that link left and right brain hemispheres, had been surgically cut. This surgery reduced the frequency and severity of epilepsy in these patients, and it also made them ideal candidates for the study of the roles of these separate halves of the brain in processing visual, verbal, and spatial information.

Previous neurophysiological studies of people with damage to one or another side of the brain had already proven that, in general, the left hemisphere processes most verbal information while the right hemisphere is involved with processing faces and spatial information. In her studies of people with a severed corpus collosum, Levy established that there is an interplay between right and left brains. In processing visual information, the left hemisphere decodes specific details of a picture or sight; the right hemisphere puts these details together to form a whole.

Levy received her Ph.D. from Caltech in 1970 and continued her research at the University of Pennsylvania, which hired her as an assistant professor in 1972. She stayed at the University of Pennsylvania until 1977, when she was hired by the University of Chicago where she currently is a full professor.

Levy's continuing experiments in neurophysiology have further refined our understanding of brain function. She has shown that, depending on whether the function is visual or verbal, both hemispheres to some degree can perform detailed analytical processing, just as each can also perform holistic processing for different functions.

Jerre Levy, whose research has helped to reveal how the two hemispheres of the brain analyze information differently. *(Courtesy of Jerre Levy)*

For her work, she won a National Science Foundation Grant in 1976 and a Spencer Foundation Grant in 1979. She is a member of the Society of Experimental Psychology and the International Neuropsychological Symposium.

Levy, Julia
(1934–   )
Canadian
Pathologist

Trained as an experimental pathologist, Julia Levy has explored new approaches to treating cancers and autoimmune diseases such as arthritis. To attack specific cancer sites and avoid therapies that attack a broad spectrum of cells, many of which are healthy, she has developed anticancer drugs that are sensitive to light rays that can be directed to highly defined areas inside the body.

Born on May 15, 1934, in Singapore, then a British island colony at the tip of the Malay Peninsula, Julia Levy is the daughter of a Dutch banker and English mother. The family lived in Singapore because her father's business was centered in the British Malaysian and the Dutch East Indies colonies. In 1941, just before the outbreak of World War II, Levy's father sent her, her sister, and mother to Vancouver, British Columbia, where they remained for the duration of the war. At war's end, Mr. Levy joined them there.

Levy attended public schools in Canada and, on her graduation from high school in 1951, enrolled in the University of British Columbia. As an undergraduate, she studied immunology and bacteriology. She graduated with honors with a B.A. from the University of British Columbia in 1955. After completing her undergraduate degree, she married and, with her husband, enrolled in the University of London. In England, she worked on a Ph.D. in experimental pathology, which she completed in 1958.

In 1959, Levy returned to Canada and was hired as an assistant professor at the University of British Columbia. Splitting her time between research and teaching, she eventually became a full professor.

Beginning in the 1960s, Levy began to conduct research on new approaches to treating cancer. After much experimentation, she focused on a combined approach that mixed drugs that would strengthen the body's own immune system with light-sensitive chemotherapies. In 1981, she founded a Canadian drug company, Quadra Logic Technologies, with four fellow scientists to market her ideas. The company has since changed its name to QLT PhotoTherapeutics.

For the university and Quadra Logic, Levy refined her light-sensitive approach. She has perfected a chemotherapeutic drug that, unlike most chemotherapy drugs, is activated only when it is exposed to light. After administering the light-sensitive biochemical agent, Levy, or the attending doctor, then threads a thin fiber-optic cable into the patient's body and directs it to the exact spot on the lung, liver, pancreas, or other organ where the tumor is growing. Laser light is then beamed through the fiber-optic cable to the cancer site. When the laser hits the spot of the organ or tissue where the cancer is to be found, it activates the light-sensitive biochemical drug, which then attacks the cancer at only that one place. In this way, the treatment does not weaken the patient by attacking healthy cells like other broad-based chemical therapy treatments.

Levy has extended her light-sensitive biochemical approach to treatments of arthritis, AIDS, and age-related vision loss. For her work, she was elected to a prestigious organization of scientists, the Royal Society of Canada, in 1981. In 1987, she was appointed to Canada's National Advisory Board of Science and Technology. She also continues to teach at the University of British Columbia.

Lewis, Margaret Adaline Reed
(1881–1970)
American
Cellular Biologist

An innovator in cellular biology, Margaret Lewis was the first researcher to successfully grow mammalian tissue culture in vitro, a milestone in her field. She also conducted research on the kinds of media that were best suited for growing cell cultures, and later in her career, she studied white blood cells and the characteristics of malignant cancer cells.

Born on November 8, 1881, in Kittanning, Pennsylvania, a small town in the western part of that state, Margaret Adaline Reed was the daughter of Joseph Cable Reed and Martha Adaline Walker Reed. In 1910, she married Warren Harmon Lewis, a fellow cellular biologist with whom she frequently collaborated on scientific investigations. They had three children.

In 1897, Lewis enrolled in Goucher College in Maryland from which she received an A.B. in biology in 1901. Between 1901 and 1908, she pursued graduate studies in biology and zoology at Bryn Mawr in Pennsylvania as well as at the Universities of Paris, Zurich, and Berlin.

Although she did not earn a postgraduate degree, she developed considerable theoretical and experimental expertise in her field and was able to land a job in 1904 as lecturer in physiology at the New York Medical College for Women. Between 1904 and 1907, Lewis not only worked at the Medical College for Women but as a lecturer in zoology at Barnard College and as a lab assistant for Columbia University biologist T. H. Morgan. Between 1907 and 1909, she worked as an instructor of biology at Columbia and Columbia's women's college, Barnard.

Lewis stayed out of the workforce for five years after her marriage to Warren Lewis in 1910, but by 1915, she had returned to laboratory research at the Carnegie Institute's Department of Embryology located in Baltimore. She remained as a research associate at the Carnegie Institute until 1946, and from 1946 to 1964, she was a member of the Wistar Institute of Anatomy and Biology in Philadelphia.

Lewis's early work centered on investigations into culturing mammalian tissue, something that no scientist had successfully done. Lewis knew that these failures probably were the result of not knowing which ingredients to mix for an agar solution in which the tissues would be grown. After much experimentation, she came up with a salt solution in which she was able to culture bone marrow and spleen cells from a guinea pig.

In 1911 and 1912, Lewis and her husband worked on further formulations of media in which different kinds of cells could be grown. Because their interest lay in the structure of cells, they needed a clear medium that would enable them to examine the cells under a microscope. They developed such a culture, known as the Locke-Lewis solution, in which they grew cells in one drop on the underside of a glass. This was then placed in the hollow of a microscope slide for examination.

During the 1920s and 1930s, Lewis examined the workings of a kind of white blood cell called monocytes. She determined that monocytes could transform themselves into a cell type known as macrophages, which ate and destroyed microorganisms in the bloodstream. Previous to this research, scientists had assumed that monocytes and macrophages were distinct cell types.

At this same time, Lewis began looking at chromosomes in malignant cancer cells to discover if abnormal chromosomes could be a cause of these cancers. They discovered that these cancer cells were equally malignant whether they had normal or abnormal chromosomes and concluded that abnormal chromosomes were not the cause of the malignancy. Later, Lewis followed up this cancer research with studies of substances that might be able to eliminate or reduce the size of cancer cells.

Lewis won a number of awards for her work in cell biology, including honorary life membership in the Tissue Culture Society, the William Wood Gerhard Gold Medal of the Pathological Society of Philadelphia (1938), and a listing with a star (signifying the highest achievement) in the 1938 edition of *American Men of Science*. Lewis died at age 88 on July 20, 1970.

Libby, Leonora Woods Marshall
(1919–1986)
American
Physicist

The only woman to work with Enrico Fermi's team that created the world's first nuclear chain reaction, Leonora Libby had a remarkable career as a nuclear physicist. Besides participating in the Manhattan Project, the secret nuclear weapons work during World War II, she was later involved with theoretical and experimental work in particle physics, the push to understand the nature of the atomic nucleus that was one of the major subjects of physics research in the 1950s and 1960s. She ended her career with studies in the use of physics to aid in archaeological research and in the physics of cosmology.

Born on August 9, 1919, in LaGrange, Illinois, a suburb of Chicago, Leonora Woods was the daughter of Wreightstill Woods and Mary Holderness Woods. Her father was a lawyer in LaGrange, and the family also owned a small farm near the town. One of five siblings, Leonora Woods occasionally helped with farmwork such as apple picking. In 1943, she married John Marshall, a physicist, whom she divorced in 1966, the same year she married

Williard Frank Libby, a chemist. She had two children with John Marshall.

Exceptionally bright, Leonora Libby earned a B.S. in chemistry from the University of Chicago in 1938 at the age of 19. She immediately went on to graduate school at the University of Chicago, first studying with James Franck, a refugee from Nazi Germany. Franck taught Libby and his other students about recent discoveries that had been made in Europe regarding the application of quantum mechanics to chemistry. However, after Franck gave Libby a warning that she might have a difficult time finding a job in chemistry because she was a woman (he also mentioned that he had had the same problem in Germany because he was a Jew), she decided to switch to the study of physics under Robert Mulliken.

Mulliken, the 1966 Nobel Prize winner for physics, guided Libby to her dissertation in nuclear spectroscopy; she received her Ph.D. from Chicago in 1943, just as her laboratory was being turned into an experimental nuclear reactor under the direction of famed Italian physicist Enrico Fermi. Libby was soon drafted into the Fermi team, and from 1943 to 1946, she worked on the Manhattan Project, first with Fermi in Chicago, then later at Hanford, Washington, where she and her husband, John Marshall, were part of the team that planned and constructed a plutonium-processing plant.

From 1946 to 1960, Libby worked at the University of Chicago's Institute for Nuclear Studies, which, until his death in 1954, was headed by Enrico Fermi. At Chicago, and at the nearby Argonne Nuclear Lab, Libby began to participate in the new high-energy physics experiments that were conducted with the aid of recently constructed particle accelerators. The machines sped electrons and other nuclear particles to tremendous rates of speed and then sent them smashing into the nuclei of atoms. The resulting demolition of the nuclei and splatter of ejected nuclear particles enabled physicists to better understand the composition of atomic nuclei. As a result of these studies, a much more complicated picture of atomic matter emerged.

Libby continued her high-energy physics research at several universities in New York State during the early 1960s and at the University of Colorado where she remained until 1973. After her marriage to Williard Libby, she moved to Los Angeles where she served as adjunct professor at the University of California's School of Engineering and participated with her husband on problems related to carbon-dating archaeological remains.

During her career, Libby was the author of more than 60 technical papers and three books. In November 1992, at the 50th anniversary of the first nuclear chain reaction in Chicago, Libby was one of four women—the others were LISE MEITNER, MARIE CURIE, and IRÈNE JOLIOT-CURIE—honored as a pioneer in nuclear science. Libby died on November 10, 1986, in Santa Monica, California.

Lin, Ch'iao-chih
(1901–1983)
Chinese
Physician

One of the first western-trained women medical doctors in China, Ch'iao-chih Lin served the people of her country as an obstetrician/gynecologist, researcher, and community leader. She worked most of her life as a doctor and professor of obstetrics/gynecology in the Chinese capital of Peking (now called Beijing), during which time she witnessed extraordinary chaos and political change. By the end of her life, she had become a legendary and beloved figure in China, a woman who seemed to embody the suffering and endurance of the Chinese in the 20th century.

Ch'iao-chih Lin was born on December 23, 1901, probably in the southern province of Fukien. There is little available information about her childhood and youth. It is likely that she came from a relatively affluent family. Later in her life, she described herself as "an old intellectual, coming from the old society," thus her father may have been a university-trained government official or could have come from a local family with some property and income. It is know that by age 16, Lin was living in the Fukien port city of Amoy, where in 1917, she witnessed the destruction wrought by a typhoon, an event that inspired her to start thinking about becoming a doctor. She never married and had no children.

Because so little is known about Lin's youth, there is no information about her early schooling. Obviously, she had a good elementary and secondary education, which was in itself extremely unusual for a woman in China during the early part of the 20th century. Probably she was educated by private tutors. Her first known schooling was as a student at the medical school of Peking Union Medical College (PUMC), where she enrolled in 1921 at the age of 20. Lin was fortunate to have been of university age at that time. The PUMC was only six years old in the year she enrolled. It was the creature of the western-oriented Chinese Medical Board and the Rockefeller Foundation, the latter of which supported it with grants. The aim of the PUMC was to create a core group of western-trained doctors in China. It was hoped that these physicians would, in turn, train other doctors. Through this process, China would eventually have enough doctors to meet the medical needs of its people. Lin was a member of the first class at PUMC that included women. She graduated in 1929, joined the PUMC faculty as a professor of obstetrics and gynecology, and in 1932–33, studied the latest techniques in her field at Manchester University and the University of London. She also studied at the University of Chicago in 1939–40.

Lin witnessed tumultuous change during her time at the PUMC. In 1928, the year before she graduated, Nationalist leader Chiang Kai-shek took over the capital. His forces held Peking for only nine years before the Japanese army occupied the city in 1937. The Japanese shut the PUMC down in 1942, and Lin moved to another hospital in the city. After World War II, the city was seized by the Red Army in 1949. Relatively apolitical, Lin remained in the city and at the hospital. In 1948, Lin became head of the obstetrics/gynecology department at PUMC (which was later renamed Capital Hospital). From the 1950s through the 1970s, she served as a delegate in a number of People's Congresses, the Chinese Communist legislature. She was elected a fellow of the Chinese Academy of Biological Sciences in 1955.

Lin also played an important role in the famous Barefoot Doctors program that began in the late 1950s. This was a government-sponsored effort to fill China's medical needs temporarily by training local people in basic medical care. By 1968, more than 29,000 Chinese had been given this instruction.

She was praised for her work during the 1950s but for a while fell victim to the excesses of the cultural revolution in the late 1960s, when she, along with many other professionals, was singled out for punishment and shipped to the countryside to work as a manual laborer on farms. By the early 1970s, she seems to have been "rehabilitated," because she was included in a group of dignitaries that visited Canada and France.

In 1980, at the age of 79, she helped found, and became vice president of, the Family Planning Association of China. She died in 1983.

Lloyd, Ruth Smith
(1917–1995)
American
Anatomist

Ruth Smith Lloyd holds the distinction of being the first African-American woman to earn a Ph.D. in anatomy. A highly respected researcher, Lloyd spent most of her career as a professor at Howard University in Washington, D.C. Her work focused predominantly on fertility issues, including the female sex cycle and the relationship of sex hormones to growth.

On January 25, 1917, Ruth Smith Lloyd was born in Washington, D.C. She was the youngest of Mary Elizabeth and Bradley Donald Smith's three daughters. After graduating from Dunbar High School in Washington, D.C., Lloyd enrolled at Mount Holyoke College in Massachusetts, where she received her bachelor's degree in 1937. Although she planned ultimately to pursue a career as a high school teacher, Lloyd opted to obtain a master's degree first. She began graduate work at Howard University and was awarded her master's in zoology. During the course of her studies at Howard, her professors recognized her abilities and encouraged her to continue on and work toward her doctorate. One of Lloyd's mentors at Howard recommended that she explore the anatomy program at Western Reserve University in Cleveland, Ohio. Lloyd followed this advice, and in 1941, became the first African-American woman to receive a Ph.D. in anatomy. Her time at Western also sparked her interest in fertility issues, to which she was exposed by working with the monkey colonies in the school's fertility laboratories.

In 1941, immediately after completing her education, Lloyd accepted a position as an assistant in physiology at Howard University's college of medicine. She then transferred briefly to Hampton Institute, where she taught zoology, but her main focus for the next year was on raising her young family. She had married a physician, Sterling M. Lloyd, in 1939, and the couple had three children together—Marilyn, Sterling Junior, and David. Lloyd returned to Howard University as a lecturer in 1942, and she would remain at that institution for the rest of her career. She was promoted to the rank of assistant professor of anatomy in 1958 and advanced to associate professor in the university's college of medicine shortly thereafter. Both her research and teaching concentrated on the field of endocrinology and medical genetics. Endocrinology is the study of the endocrine glands, which are those glands that produce internal secretions that are introduced directly into the bloodstream and are then carried to the organs they affect. Endocrine glands include the thyroid, pituitary, and adrenal glands.

Lloyd finished her career at Howard's medical college and then retired to her home in Washington, D.C. She remained modest about her groundbreaking accomplishments, typically describing herself as an average person with a normal life. Her influence at Howard, however, was profound. Along with her research and teaching duties, she directed the school's academic reinforcement program and became known as "Mama Lloyd" to an entire generation of Howard students. Her peers recognized her achievements as well. She was named a member both of Sigma Xi (an honorary scientific society) and the American Association of Anatomists. She died of cancer at the age of 78, on February 5, 1995.

Logan, Martha Daniell
(1704–1779)
American
Botanist

An early self-trained botanist in the British North American colonies, Martha Logan was an important collector of plant species from the Carolinas. She is known from her correspondence with famed botanist John Bartram and through the serial publication of her gardening advice column, "The Gardener's Kalendar," in the *South Carolina Gazette.*

Born on December 29, 1704, in St. Thomas Parish, South Carolina, Martha Daniell was the daughter of Robert and Martha Daniell. Her father was a wealthy and prominent citizen of the colony who served as deputy governor and owned ships, several plantations, and slaves. In 1718, at age 14, Martha Daniell married George Logan Jr. She had eight children, six of whom lived to be adults.

There are no records of the sort of schooling that Martha Logan received, but it is clear from existing documents that she learned to read and write and was taught about plants and horticulture from an early age. Owing to her father's wealth and status, she was likely taught by a private tutor. It is equally likely that because she was a girl her formal education ended after she had learned reading and writing.

After their marriage, Martha and George Logan moved to a plantation—given to them by Martha's father—that was 10 miles from Charleston, the principal city of colonial South Carolina. For the next 20 years, nothing is heard of Martha Logan, almost certainly because she was consumed with caring for her large family. Judging from her later activities, she probably oversaw the growing of a plantation vegetable garden and the planting of decorative trees and shrubs as well as fruit trees and herbs. She must also have made excursions into the Carolina woods to collect native plants and trees.

In spite of her inherited wealth, Martha and George Logan seem to have fallen into financial trouble by the time Martha was 38, in 1742. Either because of financial mismanagement, or as a possible result of her husband's death, Martha Logan took out an advertisement that year in the colony's leading newspaper, the *South Carolina Gazette,* offering her services as a teacher and setting up her house as a boarding school for children. She also advertised for work as an embroiderer and seamstress.

In 1751, Logan's first anonymous "Gardener's Kalendar" piece appeared in the *South Carolina Gazette.* Even though the piece was not signed, it seems that the identity of its author was widely known in the Charleston area, indeed in most of South Carolina. By 1753, she had been forced to sell her plantation estate and had moved into a house in Charleston where she dealt in hard-to-find seeds, flower roots, and fruit stones. It is at about this time, when she was in her late forties, that she embarked on botany as a business and serious avocation.

Within the next seven years, Logan had begun correspondence with John Bartram, one of the leading botanists in colonial North America. Bartram had been appointed a royal botanist by King George III and traveled extensively throughout the colonies. Aided by an industrious Captain

North, who regularly ferried his ship between the various colonies, Bartram and Logan sent each other samples back and forth between the Carolinas and Philadelphia, Bartram's home. In a letter to the English botanist, Peter Collinson, Bartram wrote, "I received a lovely packet [of seeds and cuttings] . . . from Mistress Logan, my fascinated widow. Her garden is a delight and she has a fine one."

Such correspondence and plant and seed exchanges were to continue until Martha Logan's death, in Charleston, South Carolina, in 1779 at the age of 75.

Logan, Myra Adele
(1908–1977)
American
Physician

Myra Adele Logan is the first woman to perform open heart surgery as well as the first African-American woman elected to the American College of Surgeons. Known as a selfless doctor interested in serving the community, Logan worked tirelessly at Harlem Hospital for most of her career. She also conducted important research on antibiotics and breast cancer.

Born in 1908, in Tuskegee, Alabama, Myra Logan was the eighth child of Warren and Adella Hunt Logan. Logan's childhood was relatively privileged since her father was a trustee and treasurer of the renowned Tuskegee Institute. (Activist Booker T. Washington was a neighbor and friend of the family.) In 1943, Logan married well-known artist Charles Alston. The couple had no children, preferring to dedicate themselves to their professional careers.

Logan was valedictorian of her class at Atlanta University, from which she earned her bachelor's degree in 1927. She then received her master's degree in psychology from Columbia University in New York before working at a YWCA office in Connecticut. Logan subsequently chose to study medicine and enrolled at New York Medical College. Aided by a four-year Walter Gray Crump Scholarship, she was awarded her medical degree in 1933. She then completed her internship and residency at Harlem Hospital in New York. Her medical training there was diverse, as she was called upon to do everything from delivering babies to repairing stab wounds.

Logan remained at Harlem Hospital as an associate surgeon after finishing her residency. She also acted as a visiting surgeon at Sydenham Hospital. In 1943, Logan made history when she became the first women ever to perform open-heart surgery—in only the ninth operation of the kind. Despite her hospital duties and her busy private practice, she nevertheless found time to conduct vital research. As antibiotics became available for the first time, Logan investigated aureomycin and published her findings in the *Archives of Surgery* and the *Journal of American*

Medical Surgery. She turned her attention to breast cancer in the 1960s. By developing a slower X-ray process, she was able to detect minute differences in the density of breast cancer tissues and thus could detect tumors earlier than had previously been possible. In addition, Logan is remembered for participating in one of the initial medical group practices in the nation. She was a charter member of the Upper Manhattan Medical Group of the Health Insurance Plan—an early version of today's common practices that house multiple specialties under one roof.

Logan was also an activist committed to social issues. She served as a member of the New York State Committee on Discrimination, from which she resigned in protest when New York's Governor Dewey willfully ignored the legislation proposed by the committee in 1944. In addition to working with Planned Parenthood, Logan was a member of the National Association for the Advancement of Colored People and sat on the New York State Workman's Compensation Board.

After retiring in 1970, Logan took the time to pursue her interests in theater and music. Her professional achievements were recognized when she was elected to the American College of Surgeons. She died on January 13, 1977, at Mount Sinai Hospital in New York, succumbing to lung cancer at the age of 68. Her husband died only four months later.

Long, Irene Duhart
(1951–)
American
Aerospace Medicine Physician

As director of the Biomedical Operations and Research Office at the Kennedy Space Center (KSC), Irene Duhart Long is one of the highest-ranking African-American women at the National Aeronautics and Space Administration (NASA). At KSC, Long is responsible for an array of programs and facilities, managing all aspects of the center's aerospace and occupational medicine activities. In addition to conducting life-sciences research, including projects such as the Controlled Ecological Life Support System, Long's team provides support for key aspects of the Space Shuttle program. Long also played an important role in founding the Space Life Sciences Training Program, which brings college students to KSC to learn about the space program.

Long was born on November 16, 1951, in Cleveland, Ohio, to Andrew and Heloweise Davis Duhart. Long became interested in the space program at a young age. After watching reports of space missions when she was nine, she decided to have a career in aerospace medicine.

Long attended Northwestern University, receiving her bachelor's degree in biology in 1973. She then enrolled at

St. Louis University's School of Medicine. After earning her medical degree in 1977, she completed a two-year general surgery residency at the Cleveland Clinic and Mount Sinai Hospital in Cleveland. Following her residency, Long returned to school and was awarded a master's degree in aerospace medicine from Wright State University in Dayton, Ohio. This program required an additional three-year residency, which she undertook at Ames Research Center in Mountain View, California.

In 1982, Long fulfilled her childhood dream when she joined the staff at KSC as chief of the Occupational Medicine and Environmental Health Office. Her responsibilities were myriad. In addition to heading a team of physicians providing medical services to astronauts in emergencies, she was charged with protecting the health and safety of nearly 18,000 workers, civil servants, and contractors at the Space Center. To this end, her office performed employee physicals and oversaw the inspections of work spaces to protect employees from hazardous materials. She also coordinated efforts among multidisciplinary teams from the Department of Defense, environmental health agencies, and astronaut offices to insure successful launches and to prepare for potential disasters. In 1984, Long helped create the immensely successful Space Life Sciences Training Program. The program developed a curriculum to inspire students to study science and mathematics and to bring college students to NASA to become involved in the space program. Participants learn about topics such as space physiology in plants, humans, and animals.

After 12 years as chief of the Occupational Medicine and Environmental Health Office, Long was promoted in 1994 to director of the Biomedical Operations and Research Office at KSC. Now responsible for the center's aerospace and occupational medicine activities, Long oversees employee medical clinics at KSC and the adjacent Cape Canaveral Air Station. Her office supervises KSC's environmental and ecological monitoring program, including the testing of groundwater and the safe removal of asbestos. Moreover, Long manages all of KSC's life-sciences research, such as the Controlled Ecological Life Support System project—a long-term NASA effort to develop a self-sufficient food chain capable of supporting long-term human stays in space. Long's employees also provide scientific support to workers processing space shuttle payloads that are biological in nature and offer support to the space shuttle program.

Long remains director of the Biomedical Operations and Research Office, where her tireless efforts have been acknowledged. In 1995, she won the president's special award of the Society of NASA Flight Surgeons. A member of the Aerospace Medicine Association, the Society of NASA Flight Surgeons, and the National Medical Association, Long also continues to pursue research outside of NASA. In

a landmark paper in 1982, she determined that it was safe for people with the sickle-cell trait—a condition not the same as sickle-cell anemia—to fly.

Lonsdale, Kathleen Yardley
(1903–1971)
British
Crystallographer

Using the recently invented X-ray spectrometer, Kathleen Lonsdale determined the crystal pattern of molecules in numerous carbon-based compounds. She did important work on six-membered rings, demonstrating that there was variability between them, and began a lifelong project of deciphering the structure of thousands of the crystals, which she published in her *International Tables*, the standard handbook of crystallographers to this day.

Born on January 28, 1903, Kathleen Yardley was the daughter of Henry Frederick Yardley and Jessie Cameron Yardley. Yardley's father worked for the British Post Office and, during the Boer War, served in the British army. His alcoholic misbehavior caused Yardley's mother to leave him in 1908, when Kathleen was five. She was raised by her mother in a small town just outside of London. In 1927, Yardley married Thomas Lonsdale, also a crystallographer.

In 1919, Kathleen Lonsdale enrolled in Bedford College, the women's college of the University of London. She completed her B.S. in physics from Bedford in 1922, and because of the high score she got on her final written exams, she was hired to work in the X-ray crystallography laboratory of Nobel Prize–winner William Bragg. Within a year, Bragg and Lonsdale moved to the Royal Institution while Lonsdale worked on her master's thesis, which she received from University College of the University of London in 1924.

Yardley worked at the Royal Institution for four years before moving to the University of Leeds with her new husband. She received her Ph.D. from the University of Leeds in 1929, and in 1930, she and her family returned to London where she rejoined the Royal Institution in 1934. She remained at the Royal Institution until 1946, when she began doing teaching and research at University College's crystallography department, which she founded. In 1949, Lonsdale was made a full professor; she retired from full-time work in 1968.

Lonsdale began her scientific career learning how to use the X-ray spectrometer, which was an invention of her mentor William Bragg. The X-ray spectrometer shot X rays through molecules of substances such as diamonds, benzene, and other compounds to determine their molecular patterns and their three-dimensional structure. Lonsdale's first tasks were the analyses of crystals in the molecules of succinic acid and ethane compounds.

In the late 1920s at the University of Leeds, Lonsdale studied the structure of six-membered rings of benzene, called benzene rings. Earlier work by William Bragg and others of six-membered rings of diamonds showed that these rings were puckered in their structure. Thus it was assumed that all six-membered rings had this same structure. By careful mathematical analysis of her spectroscopic results, Lonsdale was able to prove that the structure of benzene rings was flat and hexagonal. Because the benzene ring was found in many organic compounds, this new information opened up many new avenues of research for organic chemists.

During the 1930s and 1940s, Lonsdale continued her massive project of analyzing the structure of crystal compounds. She began shooting X rays through compounds such as diamonds from different angles to determine the spacing between carbon atoms, and she did studies of diamonds and other crystals at a variety of temperatures.

During World War II, Lonsdale, who was a Quaker pacifist, refused to serve as a fire watcher on the grounds that even that kind of duty would be contributing to the war effort. For this decision, she was fined and briefly jailed. Lonsdale continued her antiwar efforts through old age when she actively participated in Ban the Bomb rallies and organizing in England.

For her work in science, Lonsdale was voted a fellow of the Royal Society in 1945, becoming the first of two women to be admitted to that group. In 1956, she was made a Dame Commander of the British Empire, the equivalent of knighthood for women. She won the Royal Society's Davy Medal in 1957, and in 1967, she became the first woman president of the British Association for the Advancement of Science. Lonsdale died of leukemia on April 1, 1971, in London.

Love, Susan
(1948–)
American
Physician

Trained as a surgeon, Susan Love has devoted her career to the treatment of breast cancer. By the late 1980s, concerned that the usual avenues for breast cancer treatment were inadequate, Love began to write articles and a book for the general public on this issue. She also founded the National Breast Cancer Coalition, a lobbying group that pressed for greater funding for breast cancer research.

Born on February 9, 1948, in Little Silver, New Jersey, Susan Love is the daughter of Mr. and Mrs. James Love. Her father was a machinery salesman who was frequently transferred to new sales territories by the company he worked for. For this reason, Love went to middle and secondary schools in Puerto Rico and Mexico City. Love lives with her partner, Helen Cooksey, and has a daughter.

In 1966, Love enrolled in Notre Dame of Maryland, a small women's college run by the Sisters of Notre Dame. Love took premed courses at Notre Dame, and for a short while became a novitiate nun. However, after six months, she quit and eventually transferred to Fordham University in New York City, where she finished her premed courses and earned a B.A. in 1970. After finishing her undergraduate degree, Love was admitted to the State University of New York's Downstate Medical College, located in Brooklyn, and graduated from there in 1974. She later completed a residency in surgery.

In 1980, Love entered private practice as a surgeon. Many of her cases were breast cancer patients, and Love soon realized that there was a role for her in this field. For several years, she worked as director of Beth Israel Hospital's breast clinic in New York City. In 1982, she became a surgeon specializing in cancer cases at the Dana Farber Cancer Institute's Breast Evaluation Center. In 1988, she moved to Boston to found the Faulkner Breast Center at

Kathleen Lonsdale, whose *International Tables* is a standard handbook for cystallographers. *(AIP Emilio Segrè Visual Archives)*

A surgeon renowned for her innovative treatments of breast cancer, Susan Love also founded the National Breast Cancer Coalition. *(Courtesy of Susan M. Love)*

Faulkner Hospital. The breast center was the first medical organization of its kind to have a cross-disciplinary, all-female staff. While at Faulkner, Love also worked as an assistant professor at the Harvard Medical School.

Love had developed tremendous experience with treating and counseling breast cancer patients, and she was unhappy with a number of approaches that were considered standard in this field. For instance, she felt that the typical treatment of surgery, radiation, and chemotherapy were used too often and that doctors did not do a good job laying out alternatives to this treatment to their patients. Also, she felt that there was far too little emotional support being given breast cancer patients by mostly male doctors.

To redress these problems, Love wrote a hugely popular book, aimed at a general audience of women, entitled *Susan Love's Breast Book* (1990). In her book, Love laid out alternative paths women could take if they discovered they had breast cancer. She suggested that women whose cancer had been detected early could opt for a removal of the cancerous node in their breast rather than a complete mastectomy, or complete removal of the breast, which was typically recommended by surgeons. The survival chances, Love noted, were about equal for both methods. Love has

also questioned the need for mammograms, which are used to detect early breast cancers, for most women under the age of 50.

In 1992, Love moved to Los Angeles to direct the University of California at Los Angeles's (UCLA) Breast Center. She quit this job and briefly retired from medicine in 1996. However, she has returned to surgery as adjunct professor of surgery at the UCLA School of Medicine where she is working on a fiber-optic cable that can be threaded through the milk ducts in a woman's breast to look for early cancers.

Lovelace, Augusta Ada Byron
(1815–1852)
British
Mathematician

The daughter of one of England's most notorious and talented poets, Ada Lovelace was a mathematical prodigy and one of the most talented mathematicians of her age. Because women, and especially women of her position, were discouraged from participating in science, she wrote few papers in her field. However, her most famous paper, signed with her initials only, was a lucid explanation of the concept of the computer, written almost a hundred years before the first computer was built.

Born on December 10, 1815, in London, Augusta Ada Byron was the daughter of George Gordon, known to most as Lord Byron, and Annabella Milbanke Byron. Her father was one of the founders of the English romantic literary movement and one of the most famous and accomplished poets of his generation. Byron's parents lived together only briefly; after their separation, she lived with her mother. In 1835, Ada Byron married William King, a nobleman who later became the earl of Lovelace. She had three children.

At an early age, Ada Lovelace displayed a remarkable talent for mathematics. This was not entirely surprising, as her mother, whom her father had dismissively called the "princess of parallelograms," also had a strong interest in math. Because Annabella Byron was wealthy, she could afford to hire tutors to educate her daughter, which was itself unusual for young girls at the time. Her mathematics tutor was Augustus DeMorgan, who confided to Annabella Byron that "[Ada's] powers in [mathematics] has been something so utterly out of the common way for any beginner, man or woman." He further added that if Ada had been a man, she would have gone to Cambridge or Oxford and become a mathematician "of first-rate eminence."

In 1833, when she was 17, Ada met the mathematician Charles Babbage at a coming-out ball given for young ladies in London. She and Babbage were instantly drawn into conversation with one another because of their shared interest in mathematics. Babbage described to her an invention he was working on, a difference engine, a machine that would compute differential equations automatically. Insurance companies and shippers had a great need for such machines, which could compute actuarial tables and ocean navigation quickly and accurately. Babbage had built a prototype of this device and received a grant from the government to construct a full-size machine.

Ada Lovelace threw herself into the work of understanding Babbage's machine. She taught herself differential equations and engaged in extensive correspondence with Babbage. Babbage, who was disorganized and inept at business, blew the grant he had been given without ever building the full-scale machine, but within a few years he had embarked, with Lovelace's help, on the conceptual work for a more complicated device, which he called an analytic engine. The logical structure of this device was remarkably similar to that of the digital computers of the 20th century.

An Italian, L. F. Menabrea, wrote an article in French about Babbage's work. Lovelace translated and greatly expanded Menabrea's paper, essentially remaking it as an original exposition of Babbage's ideas. In her work, "Sketch of the Analytical Engine Invented by Charles Babbage . . ." (1843), Lovelace compared the workings of the analytic engine to the weaving looms that had been newly invented by Frenchman J. M. Jacquard. Babbage had proposed using punch cards to feed data into his machine much as Jacquard used punch cards to tell his automatic looms what patterns to weave. "We may see most aptly that the Analytic Engine weaves algebraic patterns just as the Jacquard loom weaves flowers and leaves," Lovelace wrote." She added that "the Analytical Engine has no pretensions whatever to originate anything. . . . The machines . . . must be programmed to think and cannot do so for themselves."

Lovelace was only to live for nine more years, during which time she largely gave up formal mathematical studies. Instead, she devised what she thought to be a foolproof mathematical scheme to win at gambling. Falling deeply in debt because of her gambling losses, she also became addicted to morphine. In 1852, she contracted cancer of the uterus. She died in England of this disease at age 36, on November 17, 1852.

Lubchenco, Jane
(1947–   )
American
Marine Ecologist

Trained as a zoologist and ecologist, Jane Lubchenco has devoted her career to understanding how marine creatures fit into their environments. She has also spent considerable time studying how human-made pollution is affecting Earth environments. Because she believes that Earth's environment is threatened by human industrial activity, she has participated in a number of groups that have urged industries and governments to develop new techniques that will lessen the threat to the interlocking environments of the world.

Born on December 4, 1947, in Denver, Colorado, Lubchenco is one of five siblings. Her mother, a physician, showed her that it was possible for a woman to work and raise a family. In the early 1970s, Lubchenco married Bruce Menge, who is also an ecologist. They have two sons.

In 1965, Lubchenco enrolled in Colorado College, which is located in Colorado Springs. She majored in biology and received her B.S. in 1969. She immediately enrolled in the master's program in ecology at the University of Washington and earned her M.S. from that institution in 1971. Lubchenco then began work on a Ph.D. in ecology at Harvard University, which she completed in 1975. She worked as an assistant professor at Harvard until 1977 while at the same time serving as a research investigator in ecology for the National Science Foundation and a visiting professor of ecology at the Discovery Bay Marine Lab.

In 1978, Lubchenco and her husband moved to Corvallis, Oregon, to begin an unusual joint career at Oregon State University. Because both Lubchenco and her husband wanted to work while sharing the duties of raising a family, they sought a working environment that would let them share teaching and research duties. One spouse would always be working, while another would be looking after the children. However, these two roles would be alternated about every six months so that each could continue his or her career. Also, both would remain on a tenure track. When the administration at Oregon State agreed to this arrangement, Lubchenco and her husband began teaching and conducting research there.

Lubchenco has always liked to have several irons in the research fire at once. At the same time that she moved to Oregon, she also began doing field research in Panama. This research lasted for six years and resulted in a number of published papers about the ecology of plant–herbivore interactions, predator–prey interactions, and algal ecology. In 1979, she won the Mercer Prize from the Ecological Society of America for the best paper on ecology published in 1978.

In the 1980s, Lubchenco built on her career successes by working as a visiting professor in ecology at the University of Antofagasta in Chile and the Institute of Oceanology in Quingdao, China. She also served as a council member of the Ecological Society of America from 1982 to 1984 and on that organization's awards committee from 1983 to 1986. In 1988, she was promoted to full professor of zoology at Oregon State University. After 1988, when her children were older, she was able to resume full-time work. She was chairperson of the department from 1989 to 1992.

One of Lubchenco's most important contributions to scientific activism was her help in authoring the "Sustainable Biosphere Initiative," a report about the precarious state of Earth's environment that was published in 1991. This report urged that world governments begin to make efforts to promote biodiversity by saving endangered species and by initiating programs to sustain fragile ecological systems.

For her work in ecology and for her activism, Lubchenco has won numerous awards, including a grant to work as a Pew Scholar in Conservation and the Environment (1992) and a MacArthur Fellowship in 1993. She was elected president of the Ecological Society of America in 1992, and in 1995, she was elected president of the American Association for the Advancement of Science. She continues to teach and do research at Oregon State University.

Lucid, Shannon W.

(1943–)
American
Astronaut

Astronaut Shannon Lucid has set the record for the most flight hours in orbit—5,354 hours, or 223 days—of any woman in the world. This record made Lucid the most experienced American astronaut, male or female. In recognition of these feats, President Bill Clinton awarded Lucid with the Congressional Space Medal of Honor in December 1996, the first woman to receive it.

Shannon Wells was born on January 14, 1943, in Shanghai, China, where her parents were serving as missionaries. The family was imprisoned in a Japanese internment camp until a prisoner exchange freed them. They returned to the United States and settled in Bethany, Oklahoma, where she graduated from high school in 1960. Fascinated by Robert Goddard, the inventor of modern rocketry, she earned her pilot's license after high school in

Astronaut Shannon Lucid, shown reading a book in the Spektr module aboard the *Mir* space station. *(Courtesy of NASA and the Russian Space Agency)*

her first step toward her eventual career. She studied chemistry at the University of Oklahoma to earn her bachelor of science degree in 1963.

She remained at the university as a teaching assistant in the chemistry department until 1964, when she became a senior laboratory technician at the Oklahoma Medical Research Foundation. Next she served as a chemist at Kerr-McGee in Oklahoma City from 1966 until 1968. She then worked as a graduate associate in the Department of Biochemistry and Molecular Biology from 1969 until 1973. She conducted graduate study in biochemistry during this period, earning her master's degree in 1970 and her Ph.D. in 1973. She married Michael F. Lucid while still in school, and she delivered the second of her three children the day before an important examination, which she passed. After receiving her doctorate, she returned to the Oklahoma Medical Research Foundation as a research associate.

In January 1978, Lucid entered the first class of NASA's astronaut corps to admit women, and she became an astronaut in August 1979. She conducted numerous duties on ground before entering space. She first entered space aboard the space shuttle *Discovery* on a seven-day voyage in June 1985. She flew as a mission specialist twice aboard the shuttle *Atlantis,* in 1989 and in 1991, and once aboard the *Columbia,* in 1993. During these missions, she performed important tasks and conducted numerous experiments: She used the remote manipulator system to retrieve satellites; she activated the automated directional solidification furnace; she operated the shuttle solar backscatter ultraviolet instrument to map atmospheric ozone; she deployed the fifth tracking and data relay satellite; and she conducted neurovestibular, cardiovascular, cardiopulmonary, metabolic, and musculoskeletal medical experiments on herself as well as experiments on radiation measurements, polymer morphology, lightning, microgravity effects on plants, and ice crystal growth in space.

In March 1996, after a yearlong training regimen in Star City, Russia, she lifted off in the *Atlantis,* bound for the space station *Mir.* Onboard, she served as a board engineer 2, conducting experiments in the life sciences and the physical sciences as well as in diplomacy, as she coexisted with two Russian cosmonauts for 188 days. On this six-month mission, she traveled a total of 75.2 million miles, most of them at 17,300 miles per hour while 250 miles above the Earth's surface.

Upon her return aboard the *Atlantis,* which was twice delayed, Lucid's own body served experimental purposes as scientists studied the effects on the human body of prolonged weightlessness from living in space. Amazingly, Lucid walked on her own two feet upon disembarking into the Earth's gravitational pull. While in space, Lucid maintained her connection to Earth by eating M&M candies, a story that made for interesting media coverage and probably increased sales of the chocolate confections.

Lyon, Mary Frances
(1925–)
English
Geneticist

Working for the Medical Research Council (MRC) of the United Kingdom, Mary Francis Lyon spent her career investigating the effects of radiation on genetic mutation. She also concerned herself with investigations of other agents that might cause genetic mutations, and she studied the process of genetic mutation itself and how this mutation might be applied to medicine.

Born on May 15, 1925, in Norwich, England, Lyon is the daughter of Clifford James Lyon and Louise Frances Kirby Lyon. Both of her parents came from working-class families. Her father, who was a tax inspector for the British government, was the son of a barrel maker. Her mother, who worked as a schoolteacher before she was married, was the daughter of a shoemaker. Both parents encouraged Lyon in her education. She never married and has no children.

Because her father's job required that he move to a new locale every several years, Mary Lyon attended a number of elementary and secondary schools. She first developed an interest in biology when she was 10 and was attending the King Edward VI High School in Birmingham. There she won an essay contest in which the reward was a set of books about flora and fauna. In 1939, when she was 14, she began studies at her final secondary school in Surrey. By now, she knew that she liked the sciences. Despite the disruptions in her education caused by World War II, she studied physics, chemistry, and mathematics and passed an entrance exam into Cambridge University.

At Cambridge's Girton College, Lyon formally studied zoology but informally took as many courses as she could in genetics, which was not offered as a major. She graduated from Cambridge in 1946 with a B.A. in zoology and remained at Cambridge for her master's degree and doctorate in genetics. Her advisor was R. A. Fisher, for whom she conducted research on the genetic mutation of a mouse colony. Later, for her dissertation, she studied how a particular defect in balance was genetically passed down through successive generations of mice. Lyon was awarded her M.A. and Ph.D. in 1950.

After winning her doctorate, Lyon was hired by Conrad Waddington to be a researcher at the Institute of Animal Genetics (IAG) in Edinburgh, Scotland. One of the primary grants for the research being done at the IAG came from the MRC. At that time, there was a great deal of concern about possible genetic mutation that might result from exposure to radiation. This radiation came from atmospheric nuclear testing and nuclear power plants. There was also concern about genetic mutations that might result from nuclear war.

For the next 40 years, Lyon worked on genetic mutation and radiation, first at the IAG, then at the MRC Radiobiology Unit in Oxfordshire. One of her major discoveries resulted from work she did on nonradiation genetic mutations, the so-called Lyon hypothesis, which explains how X chromosomes can sometimes be inactive in mammals. She also studied mutant genes in mouse *t*-complexes and showed that distorted genes can mutate a set of genes called responder genes. Finally, in the case of germ-cell mutation, she concluded that only a small percentage of mutation was the result of environmental radiation.

Lyon won numerous awards for her work, including the Francis Amory Prize of the American Academy of Arts and Sciences (1977), the San Remo International Prize for Genetics (1985), and the William Allan Award of the American Society of Human Genetics (1986). In 1973, she was elected a fellow of the U.K. Institute of Biology and the Royal Society. From 1962 to 1986, she headed the MRC's Genetics Division, and from 1986 to 1990, she was deputy director of the MRC. She retired from active scientific research in 1990.

M

Maathai, Wangari Muta

(1940–)

Kenyan

Biologist, Anatomist, Ecologist

A scientist and environmental and social activist, Wangari Muta Maathai is best known for the Green Belt Movement, which she founded in Kenya in 1977. The Green Belt idea, which is based on reforestation and maintaining a sustainable environment, has now spread to more than 30 countries in Africa as well as the United States.

Born in Nyere, Kenya, in 1940, Wangari Maathai excelled in her secondary education. In the late 1950s, she went to Nairobi, capital of what was then the British colony of Kenya, to study biology at the University of Nairobi. Her university undergraduate years coincided with a period of great upheaval in Kenya. An independence movement led by Jomo Kenyatta had sputtered throughout the 1950s. By the early 1960s, Kenyatta had gained enough international support for independence that the British were forced to negotiate with him. Independence was granted in 1963, and Kenyatta ruled as president of Kenya from 1964 to 1978, the year of his death.

After completing her undergraduate degree, Maathai remained at the University of Nairobi where she entered the graduate school. She received a doctorate in veterinary anatomy from that institution in the mid-1960s, becoming Kenya's first woman Ph.D. Maathai stayed at the University of Nairobi to teach veterinary anatomy. In the late 1960s, she married a young politician who would eventually lead her into her calling as a social activist. In 1976, she became chair of the Department of Veterinary Anatomy, and in 1977, she was promoted to full professor in the department, both firsts for a woman in Kenya.

Maathai first used the idea of tree planting as a way to engage the imagination of Kenyans during the election bid of her husband in 1974. Acting on a campaign promise her husband had made to a constituency in a poor part of Nairobi, Maathai organized a planting campaign in the neighborhood. Seeing that similar campaigns under the leadership of government officials almost always failed, Maathai enlisted the help of the local women from the neighborhood who faithfully tended the saplings.

In 1977, as an active member of the National Council of Women in Kenya, Maathai expanded her tree planting idea onto a national level. An educational, environmental, economic, and social program, the Green Belt idea began to solve several problems at once. Because Kenya was largely deforested by the late 1970s, firewood was unavailable as a source of heat and cooking. Ordinary Kenyans instead had to rely on expensive imported coal and gasoline for these needs. Reforestation enabled poor Kenyans to partially harvest these new forests for fuel while at the same time preventing the erosion that damaged their croplands. Replantings made the forests ecologically and economically sustainable.

Green Belt also carves out a role for Kenyan women, who generally are the ones who lead and tend these reforestation efforts. By putting ordinary women in leadership roles, Green Belt gives them a role in decision making in their communities and the nation at large.

Under Maathai's direction, Green Belt has planted 7 million trees in Kenya, many of them native species. Income is provided to poor families not just through gathering and selling firewood but also by tending figs, citrus, and nut trees, which are among the species planted.

Maathai continued her activist work into the 1990s. In 1989, her group stopped the construction of a high-rise building that was to have been constructed in a Nairobi park. In 1991, Maathai participated in a new political movement called the Forum for the Restoration of Democracy, which opposes the dictatorship of President Daniel Arap Moi. In 1999, Maathai was again involved in controversy when she and a group of her followers organized protests to stop a real estate development in Karura Forest, which is located near Nairobi.

For her work, Maathai has won the Hunger Project's Africa Prize (1991) and Sweden's Right Livelihood Award (1995). She has also been imprisoned several times and beaten by police and government party thugs. As of 2000, she continues to agitate for environmentally friendly government and corporate policies and to fight for democracy and women's rights in her home country.

MacGill, Elsie Gregory
(1908–1980)
Canadian
Aeronautical Engineer

An innovative engineer and astute businesswoman, Elsie MacGill was the first woman to design and build an airplane. In 1937, her appointment to chief aeronautical engineer of the Canadian Car and Foundry Company made her the first woman chief engineer of a North American airplane company. She put her skills to good use during World War II when she oversaw the production of British aircraft manufactured in Canada.

Born on March 27, 1908, in Vancouver, British Columbia, Elsie MacGill was the daughter of parents who were both involved in Canada's legal system. Her father was a lawyer, and her mother, Helen Gregory MacGill, served for many years as a juvenile court judge. She was the first woman in Canada appointed to this position. Both of MacGill's parents encouraged her to pursue higher education, and the example of her mother and grandmother, both suffragettes, must have inspired her to believe that she could enter the mostly male field of engineering. In 1943, at age 35, MacGill married E. J. Soulsby, a businessman.

After completing her secondary education in Vancouver, MacGill enrolled in the University of Toronto, where she soon decided to take a degree in electrical engineering. A good student, MacGill graduated with an engineering degree in 1927 when she was 19, becoming the first woman to have earned an electrical engineering degree from that institution. After graduation, MacGill was hired to work for the Austin Aircraft Company in Pontiac, Michigan. At the same time, she applied to, and was accepted by, the graduate school in engineering at the University of Michigan where she studied aeronautical engineering.

From 1927 to 1929, MacGill worked on the practical details of airplane design and construction at the Austin company while studying the latest theories and applications of aeronautical engineering at the University of Michigan. Shortly before her final examinations, MacGill fell ill with a fever that marked the onset of polio. Determined to complete her degree, MacGill took her exams in her hospital bed. She passed and earned a master's degree.

Again she achieved a first, becoming the first woman to have earned such an academic degree from the University of Michigan.

After a period of physical recovery in Vancouver, MacGill returned to finish her formal engineering studies. She enrolled in the doctoral program in aeronautical engineering at the Massachusetts Institute of Technology, earning her doctorate in the study of airflows in two years. She then took a job with the Fairchild Aircraft Company in Montreal where she supervised stress analysis studies on aircraft wings and fuselages. By 1937, she moved up to the position of chief aeronautical engineer at the Canadian Car and Foundry Company. Here she designed a fighter trainer plane, the Maple Leaf Trainer II, most of which were sold to the Mexican Air Force.

MacGill's most prodigious feat was the conversion of a former railroad boxcar factory into an aircraft plant in 1939 at the beginning of World War II. She had to supervise the retooling of thousands of machine tools and coordinate the duplication of the 25,000 parts that went into the Hawker Hurricane fighter plane in its factories in England. The factory geared up in record time and began producing 40 planes a month in January 1940. By March 1941, the factory employed 4,500 people and was producing 23 planes a week.

MacGill formed her own aeronautical engineering consulting firm in 1943, and for the rest of her life, worked as an independent consultant. She served as a technical advisor to the United Nations' Civil Aviation Organization. She aided this group in the drafting of international regulations for aircraft design.

For her work, MacGill was elected a member of the Canadian Aeronautics and Space Institute and the Royal Aeronautical Society of the United Kingdom. She received the Order of Canada (1971) and the Engineering Institute of Canada's Julian C. Smith Award (1973). She died in 1980 in Canada.

Mack, Pauline Beery
(1891–1974)
American
Chemist

An inspiring teacher and devoted research scientist, Pauline Mack taught chemistry to several generations of students at Pennsylvania State University. She is also known for the long-term nutrition studies she conducted on a cross section of American families. Later in her career, she perfected a device that measured calcium-content bone density in humans.

Born on December 19, 1891, in Norborne, Missouri, Pauline Beery was the daughter of John Beery and Dora Woodford Beery. Her parents owned and ran a general store in Norborne. While teaching at Penn State, Beery met Warren B. Mack. They married in 1923. She had no children.

Always a good student, Mack attended public schools in Norborne. Her secondary curriculum included Latin, German, and mathematics. After graduation from high school in 1910, Mack attended Missouri State University. Intending to major in Latin, Mack instead was diverted by a charismatic chemistry teacher, Dr. Herman Schlundt, and soon decided to study chemistry instead.

Mack completed her B.A. in chemistry at Missouri State in three years and then began a six-year stint of high-school teaching in Missouri. Mack also applied, and was accepted, to Columbia University's master's program in chemistry. During her summer vacations, Mack often traveled to New York City to work on her M.S. degree, which she earned in 1919.

Immediately after completing her master's degree, Mack landed a job as an instructor of chemistry and applied chemistry at Penn State. Although employed by the Home Economics Department, she also taught freshman chemistry in the Department of Chemistry. She estimated that during her 30 years at Penn State, she taught about 12,000 students.

Early on, Mack developed her own projects and research topics. In 1927, she began publishing a magazine called *Chemistry Leaflet,* which was distributed to beginning students of chemistry in high school and college. A lively periodical full of practical applications of chemistry, *Chemistry Leaflet* was written to encourage students to become chemists. Later it proved so successful that the American Chemical Society published it under the title *Chemistry.*

During her early years at Penn State, Mack worked on her Ph.D., which she completed in 1932. Her dissertation was a study of the calcium chemistry of bone in mammals. In 1935, Mack began a 15-year study of nutrition known as the Pennsylvania Mass Studies in Human Nutrition. Selecting 100 urban families and 100 rural families from different income and social class groups, Mack examined each family's diet, income level, and education and cross-referenced this information with measurements of the children's skeletal maturation and the amount of minerals in the bones of all family members. The study was a pioneering effort to quantify what constituted a proper diet and how best to educate the public about changing dietary habits for the better. With Mack as project director, the Pennsylvania Mass Studies in Human Nutrition began with five part-time researchers and ended with a staff of 65.

In 1952, Mack moved to Texas to take the job of dean of the College of Household Arts and Sciences at Texas Women's University. She held this job until 1962 when she became director of the Research Institute at Texas Women's University. Here she perfected an X-ray technique to measure the mineral content of human bones. She also worked

with the National Aeronautics and Space Administration in a study of bone loss among astronauts in space.

For her work in the field of applied chemistry, Mack was awarded the American Chemical Society's Garvan Medal in 1950 and was the first woman to win the Silver Snoopy Award from the American astronauts. She also was awarded honorary doctorates from Moravian College for Women in Pennsylvania and Western College for Women in Ohio. Mack died of a stroke in Texas on October 22, 1974, at the age of 83.

Macklin, Madge Thurlow
(1893–1962)
American
Geneticist

An early researcher in the links between inherited genetic traits and cancer, Madge Thurlow Macklin was also a strong proponent of teaching genetics in medical schools. A medical doctor by training, Macklin taught herself genetics in order to understand the links between genetic inheritance and disease. She conducted numerous statistical studies on human populations to establish that there was a genetic as well as an environmental factor to susceptibility to disease.

Born on February 6, 1893, in Baltimore, Maryland, Madge Thurlow was educated in public schools in her hometown. In 1919, she married Charles C. Macklin, a medical doctor and later a professor of medicine. They had three children.

She entered Goucher College in Maryland in 1910 and received an A.B. degree from that institution in 1914. A diligent student, Macklin won several scholarships at Goucher and later at the Johns Hopkins School of Medicine, which she began attending in 1915. In 1919, she was awarded an M.D. degree from Johns Hopkins. She taught physiology there for one year before moving with her husband to the University of Western Ontario in London, Ontario, in 1921.

Macklin had a difficult time at Western Ontario University. The university, which had hired Charles Macklin as a professor of histology, did not accept the idea of husband and wife working at the same level on the same faculty. For this reason, Macklin was offered only an instructor's position in the Department of Histology and Embryology. In 1930, after teaching at Western Ontario for eight years, she was grudgingly promoted to adjunct assistant professor.

Although she was a teacher of histology and embryology at the medical school of Western Ontario University, Macklin pushed hard to introduce genetics as a part of the curriculum. "Medical genetics should be taught in the medical school in the final year of medicine," she wrote in

1933. "[The course should be taught] by a medically trained person conversant with the broader aspects of the science of genetics, . . . not . . . by a geneticist who is not medically trained . . . [and is not acquainted] with the phenomena of disease."

In 1945, Macklin, who was outspoken and aggressive in her views, was fired by Western Ontario University. She had made a name for herself by then as a medical geneticist and quickly found another job at the U.S. National Research Council (NRC). In 1946, she began conducting cancer research at Ohio State University for the NRC. She was also given research positions in the Ohio State zoology department and medical school. She retired from active teaching and research in 1959.

During her career at Western Ontario University and at the NRC, Macklin wrote more than 200 papers about medical genetics and cancer. By submitting reams of family case histories to statistical analysis, she was able to show that the likelihood of developing certain kinds of cancer—breast cancers and gastric cancers, for instance—were increased by genetic inheritance.

For her work, Macklin was presented the Elizabeth Blackwell Award in 1957. In 1959, she was elected president of the American Society of Human Genetics. At the age of 69, she died of coronary thrombosis on March 14, 1962, in Toronto, Ontario, Canada, where she had moved to be near her children.

Makhubu, Lydia Phindile
(1937–   )
Swazi
Chemist

As a researcher, Lydia Phindile Makhubu has studied the plants used in traditional African medicine. In addition to looking for new remedies, Makhubu has sought to preserve ancient knowledge about cures. Her work on *Phytolacca dodecandra* has been especially promising for its ability to eradicate a waterborne parasite that causes schistosomiasis in humans. Makhubu has also dedicated herself to improving science education in Africa and to increasing opportunities for women scientists in developing nations. A professor at the University of Swaziland since 1973, Makhubu has also proved to be an adept administrator. She holds the distinction of being the first Swazi woman to earn a Ph.D.

Makhubu was born on July 1, 1937, at the Usuthu Mission in Swaziland (a nation on the eastern border of South Africa). Her father was a teacher who trained to become a medical technician while working in South Africa. Upon his return to Swaziland, he worked at various health clinics. As a result, Lydia Makhubu grew up around doctors and developed an early interest in medicine.

In 1963, Makhubu received her bachelor's degree from Pius XII College in the country of Lesotho. While an undergraduate, she became fascinated with organic chemistry. After winning a Canadian Commonwealth Scholarship, she left Africa to study at the University of Alberta in Edmonton, where she earned her master's degree in 1967. She then began doctoral studies in medicinal chemistry at the University of Toronto, receiving her Ph.D. in 1973.

After completing her education, Makhubu immediately returned home and accepted a post at the University of Swaziland, where she remains. She rose through the academic ranks, becoming a senior lecturer in 1979 and a full professor in 1980. She immersed herself in the study of medicinal plants in the 1970s and 1980s, collecting plants used by traditional Swazi healers and analyzing them in her university laboratory. Her efforts underscored her commitment to unifying her traditional culture with modern medicine. Since many Swazis still used healers as their primary care providers, Makhubu hoped that her analysis could help the healers determine the proper doses to administer. Ultimately, she believed that healers and doctors could work side by side. Her research in the 1980s was centered on *Phytolacca dodecandra* (also known as endod). This African plant kills water-dwelling snails that carry a worm parasite that causes schistosomiasis—a condition in which the worms burrow into the human body and damage the liver and lungs.

Makhubu has also embraced administrative responsibilities at the University of Swaziland. She was named dean of the science faculty in 1976 and pro-vice-chancellor in 1978. In 1988, she was appointed as the university's vice-chancellor, a post she still holds. No other woman in southern Africa has held a comparable position at a university. From 1989 to 1990, she led the Association of Commonwealth Universities—becoming the first woman to hold this position as well.

Makhubu's career has also had another facet. Troubled by the lack of science education in Swaziland and other African nations, she founded the Royal Swaziland Society of Science and Technology in 1977, which she has headed ever since. The organization facilitates cooperation among scientists and also attempts to inform the general public about the importance of science. The Society has produced radio programs discussing science's relevance to everyday concerns. Makhubu was also aware of the impediments women in developing nations encounter in their quest to become scientists. Makhubu again created a new organization to attack the underlying causes of this inequity. In 1989, she helped found the Third World Organization of Women in Science (TWOWS), which sought to promote women in science and technology and to bolster women's research efforts. One of TWOWS's most effective programs is in giving travel grants to women from developing nations to attend overseas conferences and seminars.

Makhubu's numerous accomplishments in the realms of both research and strengthening science education have been widely recognized. In addition to winning grants from the European Economic Community, USAID, and the MacArthur Foundation, she has served on several international councils, such as the World Health Organization's Medical Research Council. She is also the vice president of the African Academy of Sciences. Along with her myriad interests and responsibilities, Makhubu enjoys a full family life. She is married to surgeon Daniel Mbatha, with whom she has two children.

Maltby, Margaret Eliza
(1860–1944)
American
Physicist

Margaret Maltby set a number of firsts in her career as a scientist. She was one of the first women to receive an undergraduate degree in science from the Massachusetts Institute of Technology (MIT). She was the first American woman to receive a doctorate from the University of Göttingen in Germany and the first woman to receive a Ph.D. in physics from any German university. She conducted research on conductivities of different aqueous compounds and, for many years, taught physics at Barnard College, the women's branch of Columbia University.

Born on December 10, 1860, in Bristolville, Ohio, Margaret Eliza Maltby was the daughter of Edmund and Lydia Maltby. Maltby never married, but in 1901, she adopted Philip Randolph Meyer, the orphaned son of one of Maltby's close friends. She raised Philip Meyer as a son and later kept in close contact with his three children, whom she considered her grandchildren.

Margaret Maltby attended primary and secondary schools in Bristolville, but because of deficiencies in the education of young women at that time, she had to take a year of preparatory courses at Oberlin College in Ohio before she could enter that institution's freshman class. She earned a B.A. from Oberlin in 1882. Thinking she would like to become an artist, Maltby moved to New York City for a year in 1882–83. However, she soon returned to Ohio where she taught high school for four years.

Maltby had taken courses in mathematics, chemistry, and civil engineering at Oberlin, so she knew that she had an aptitude for the sciences. In 1887, she decided to enroll as a special student in physics at MIT, a status reserved for female students at that all-male institution. Maltby finished her degree four years later in 1891. She spent two more years at MIT as a graduate student.

In 1893, Maltby was accepted at the University of Göttingen as a doctoral student in physical chemistry. Studying under Walther Nernst, a Nobel Prize–winning chemist,

Maltby conducted research in the measurement of substances with high electrolytic resistances. Using alternating currents to measure these conductivities, Maltby applied the Wheatstone bridge to measure the wavelength of these currents. This amounted to a new application of the Wheatstone bridge, a device for measuring electric resistance. By careful measurement to eliminate difficult experimental errors, Maltby contributed to science by offering data for this problem. Her dissertation on this subject won her a Ph.D. from the University of Göttingen in 1895.

After completing her doctoral work, Maltby returned to the United States where she taught physics in Massachusetts and Ohio for three years. In 1898, she was invited to spend a year as a research assistant at the Physikalisch-Technische Reichsanstalt in Germany. Here she established standards for measuring conductivities of aqueous solutions of alkali chlorides and nitrates.

In 1899, Maltby again returned to the United States. She taught theoretical physics for a year at Clark University in Worchester, Massachusetts, then moved to New York City where she had been hired as a chemistry instructor at Barnard College. A year later, she was appointed adjunct professor in Barnard's physics department. She was promoted to associate professor and chairperson of the physics department at Barnard in 1913. She remained at Barnard as department chair until she retired in 1931.

Maltby did little research after she began teaching at Barnard, but she became heavily involved in the campaign to encourage women to pursue university education. From 1913 to 1924, she was a member of the Committee on Fellowships of the American Association of University Women (AAUW). She was also personally involved with the recruitment and support of women students in Barnard's physics department.

For her research and teaching efforts, Maltby was listed with a star (signifying that she was among the top thousand scientists in America) in the first edition of *American Men of Science* in 1906. In 1926, the AAUW named one of its fellowship grants after her. Gradually incapacitated by arthritis, Maltby died in New York City on May 3, 1944, at the age of 83.

Mandl, Ines Hochmuth
(1917–   )
American
Biochemist

One of the most innovative biochemical researchers of the 20th century, Ines Mandl was forced to leave her native Austria when it was swallowed up by Nazi Germany. She eventually settled in the United States where she began her career in biochemistry. She has studied the solubility of amino acids in enzymes and the roles enzymes play in breaking down collagenase and in the growth of tumors in the female reproductive system. Her later work concentrated on the role of enzymes in breaking down elastic tissues such as lung tissue.

Born in Vienna, Austria, on April 19, 1917, Ines Hochmuth is the daughter of Ernst Hochmuth and Ida Bassan Hochmuth. Her father was a prominent and wealthy industrialist in Austria, and Ines had a comfortable life as a child and young adult. In 1936, at the age of 19, she married Hans Alexander Mandl. They have no children.

Mandl received her primary education at a public school in Austria and her secondary education at a private high school. At the time she was living in Vienna, Mandl had no intention of becoming a scientist. Her life fell within the boundaries of an upper-class woman of her time: She would marry early and focus on her role of wife and mother. The rise of Hitler's National Socialist Party, and Germany's takeover of Austria in 1938, changed Mandl's life.

In 1938, Mandl and her husband fled Austria for the United Kingdom. However, fearing detention in Britain because they were citizens of a hostile nation, the Mandls moved again, this time to Ireland. Financially insecure for the first time in her life, Mandl realized that she ought to seek higher education. In 1940, she enrolled in the University of Cork and received a B.S. in biochemistry from that institution in 1944. In 1945, she and her husband moved yet again, to the United States where they were reunited with her parents.

In the United States, Mandl worked briefly for the Apex Chemical Company in New Jersey before being hired as an assistant to the noted biochemist Carl Neuberg. Neuberg was another refugee from Nazism. Before the rise of Hitler, he had been a world-renowned biochemist in Germany and a professor at the University of Berlin. Mandl worked for Neuberg at the Interchemical Corporation and, with his encouragement, returned to university to get an advanced degree in biochemistry. She enrolled at the Polytechnic Institute of Brooklyn and won an M.S. in biochemistry in 1947 and a Ph.D. in 1949.

After completing her degree, Mandl took a job as research associate at the Columbia University medical school's Department of Surgery. By 1955, she had transferred to the medical school's Department of Microbiology, and in 1959, she became director of the obstetrics and gynecology labs at the medical school's Delafield Hospital.

During the 1950s and early 1960s, Mandl studied a compound found in the human body called collagenase, which is a group of enzymes that breaks down collagen into a soluble liquid. During these studies, she managed to isolate and purify a collagenase found in the bacterium *Clostridium histolyticum,* a process that had not been achieved before. Eventually, this particular collagenase was used to treat third-degree burns, bedsores, and herniated disks.

Later, Mandl switched her attention to the causes of pulmonary emphysema. She pioneered studies in elastin, the flexible tissue the lungs and some other human tissues are made of. She studied the role of enzymes in breaking down the flexibility and elasticity of the elastin tissues and discovered how this process works. Her contributions to this field have opened up possible therapies for certain kinds of pulmonary emphysema.

By 1976, Mandl had become a full professor at the Columbia medical school. She continued to work there until her retirement in 1986. For her achievements, she has won the Carl Neuberg Medal (1977) from the American Society of European Chemists and Pharmacists and the American Chemical Society's Garvan Medal (1982). She has also been elected a member of the American Academy of Arts and Sciences and the New York Academy of Sciences. She and her husband live in New York and Hawaii.

Mangold, Hilde Proescholdt
(1898–1924)
German
Embryologist and Biochemist

Although she was only 26 years old when she died, Hilde Mangold had already begun groundbreaking work to determine how organs and tissues develop in an embryo. With her mentor, Hans Spemann, she had completed a series of experiments that established the presence of a so-called organizer, a particular cell that directs tissue and organ development in embryos. Spemann would win the Nobel Prize in medicine in 1935 for his extension of this work.

Born on October 20, 1898, in Gotha, Thuringia, a province in central Germany, Hilde Proescholdt was the daughter of Ernest Proescholdt and Gertrude Bloedner Proescholdt. She was the middle of three siblings raised in a prosperous family. Her father owned a soap factory in her hometown. In 1921, she married Otto Mangold, who was also an embryologist. They had one child.

After completing her secondary education, Mangold enrolled in the University of Jena for the term 1918–19 and then transferred to the larger University of Frankfurt. Mangold must have known even in high school that she had a talent in the sciences. However, it was not until she was at the University of Frankfurt that she found a specific discipline that appealed to her. At Frankfurt in 1920, she attended lectures given by embryologist Hans Spemann. Through Spemann, Mangold learned of the new experiments that were occurring in the field of embryology. In 1921, Mangold transferred again, this time to the University of Freiburg where she could study under Spemann.

These were difficult times in Germany. World War I had just ended in a devastating defeat of the German nation. The German economy was in a shambles, and millions were unemployed. There were even shortages of food in many cities. Throughout all of this, Mangold managed to stay focused on her studies, an indication of the concentration and seriousness she brought to this task.

Between 1921 and 1923, Mangold studied and conducted research for her Ph.D. at the Department of Zoology at the University of Freiburg. According to Viktor Hamburger, a fellow student who would later become a renowned biochemist in the United States after being forced to leave Germany in the late 1930s, Mangold was not only a gifted student, perhaps the most gifted of those who worked with Hans Spemann, but she was also an extremely likable person—emotionally open, funny, and cheerful.

Her doctoral dissertation, entitled "On the Induction of Embryonic Transplants by Implantation of Organizers from Different Species," was the synthesis of her work of trying to isolate the organizer, the chemical compound responsible for the development of certain organs during the growth of embryos. To find the organizer, Mangold performed difficult transplants of tissue from the embryos of one species of salamander to the embryos of another. She used salamanders because they had short breeding seasons and many experiments could be conducted in a relatively short amount of time.

Mangold transplanted a part of the gastrula of the newt, *Triturus cristatus,* to the gastrula of *Triturus taeniatus.* This transplantation resulted in the development in *Triturus cristatus* of neural tubes, notochords, intestines, and kidney tubules typical of *Triturus taeniatus.* Migratory organizer cells from the transplanted material had penetrated from the surface of these embryos to direct the development of these tissues.

Even though Mangold was working under the direction of Spemann, her meticulous work and theoretical collaboration made this discovery possible. After completing her Ph.D. in 1923, Mangold moved with her husband to Berlin. Otto Mangold had been appointed director of the experimental embryology section at the Kaiser Wilhelm Institute, and Hilde had taken time off to raise their infant child. Mangold died on September 4, 1924, as a result of the explosion of a gas stove in her apartment.

Manton, Sidnie Milana
(1902–1979)
English
Zoologist

Fascinated by animals from an early age, Sidnie Manton grew up to become a renowned zoologist. She made a name for herself in her profession by studying and comparing the anatomy and embryology of arthropods, a biological phylum that includes insects, spiders, and crustaceans. Her most involved project was a 15-year study of Onychophora, a tropical arthropod whose class lies somewhere between annelid worms and the more typical arthropod. She also is the author of *A Manual of Practical Vertebrate Morphology,* a standard zoological textbook.

Born in London on May 4, 1902, Manton was the daughter of George and Milana Manton. She attended a private girl's school before enrolling at Cambridge University. In 1937, she married J. P. Harding. They had two children.

At Cambridge, Manton studied zoology from her undergraduate years. In the mid-1920s, she earned a master's degree, and in 1927, a Ph.D. in zoology from that university. After finishing her doctoral work, Manton stayed at Cambridge. From 1927 to 1935, she taught in the Department of Zoology as a demonstrator in comparative anatomy. In 1935, she was promoted to director of natural sciences at Cambridge's Girton College, a position she held until 1942.

Throughout her years at Cambridge, Manton was active in zoological research. In 1928–29, she participated in an expedition to the Great Barrier Reef of Australia where she studied crustaceans and coral growth. Manton also journeyed to Tasmania to study the terrestrial and marine crustaceans found on and just off that island.

In 1942, Manton decided to move to London to take up a teaching and research position at the University of London. From 1943 to 1946, she was a visiting lecturer at the University of London. In 1946, she secured a full-time position as assistant lecturer, a position she kept until 1949. She was promoted to reader in 1949 and held that position until 1960.

At the University of London, Manton continued the work she had begun earlier on crustaceans. It was her idea that the form and structure of the chitinous and calcareous exoskeletons of these creatures matched the creature's habits, that the skeletal evolutionary history coincided with the foraging and survival habits of this class of creature. Her studies of crustaceans and Onychophora laid much of the groundwork for future studies of these creatures in later years.

In 1950, Manton began a 15-year study of the relationship between the form/structure and foraging/survival habits of Onychophora. During this period of intense research, Manton proved the existence of this relationship and offered a hypothesis that the form/structure-creature habit relationship extended to most other arthropods. She also speculated that this relationship was central to these species' evolutionary development.

Manton's work won her numerous awards and other forms of recognition. In 1948, she became the first woman to be elected to the Royal Society. She was awarded the

Gold Medal by the Linnaean Society of London in 1963, and in 1968, the University of Lund in Sweden awarded her an honorary doctorate. Throughout the years that she devoted countless hours to the study of arthropods, Manton also maintained a private interest in cats. She bred several new breeds of cats and wrote a book about domestic cats. Manton died in London on January 1, 1979.

Manzolini, Anna Morandi
(1716–1774)
Italian
Anatomist

Upon her marriage at the age of 20, Anna Manzolini seemed to be a typical Italian woman of her time. Her education was minimal, and she expected to raise a family and run a household in Bologna, Italy, her hometown. However, after six years of marriage, she was forced to learn anatomy, her husband's occupation. In time, she would be considered one of the foremost anatomists of her era, and she would make discoveries about parts of the human body that had not been observed before.

Born in Bologna in 1716, Anna Morandi was the daughter of Charles and Rose Morandi. Little is known about her parents or siblings. In 1736, she married Giovanni Manzolini, her childhood sweetheart.

During the first six years of her marriage, Manzolini had a typical marriage. She had six children and stayed at home to raise them and direct life at her home. Her husband was an artist and sculptor who also was a professor of anatomy at the Medical Institute of Bologna, a school that was a part of the University of Bologna.

The training and teaching of doctors then was not nearly as professional as it has become today. In all likelihood, Giovanni Manzolini was not a doctor and had never worked as a practicing physician. His main skills were artistic. However, there was a niche for him in the medical profession where a skilled artisan and sculptor could earn a living modeling replicas of human organs and tissue, which were used to teach medical students.

In order, to learn his trade, Giovanni Manzolini had to dissect numerous human bodies. Human dissection enabled him to see firsthand the location and appearance of the organs and tissue that he would make out of wax. In this way, he acquired considerable knowledge about the human body.

In 1742, Giovanni Manzolini fell ill with tuberculosis and found it difficult to continue teaching anatomy and making anatomical models. It was at this point that Anna Manzolini stepped up to take his place at the university. Overcoming her fear of human dissection, she, with her husband's help, began to acquire the detailed knowledge of the body that an anatomist needed to know. She also dis-covered that she was extremely gifted in making anatomical models out of wax.

For several years, Anna Manzolini virtually replaced her husband in the classroom and in the sculpting studio. Gradually, however, his health improved. Rather than sending Anna back to their home, Giovanni split time with her lecturing and model making. They worked as a scientific team until he died in 1755 from the effects of his illness.

After her husband's death, Anna Manzolini was given the position of lecturer of anatomy at the Medical Institute. The next year, she was promoted to professor of anatomy, a position she held for the rest of her life.

Manzolini became famous not just in Italy but also throughout Europe for her medical discoveries. Possessing an extremely sharp eye for detail, she brought previously unobserved parts of the body—such as some of the small muscles of the eye—to the attention of the medical community. She was also able for the first time to reproduce capillary blood vessels and nerves in wax.

For her work, Manzolini was honored by Joseph II of Austria and invited by Czaress Catherine II to come to Russia to lecture. She was also elected to the Italian Royal Society, the Russian Royal Scientific Association, and the British Royal Society. She died in 1774 in Bologna. Her wax models can still be seen in the collection of the Institute of Science in Bologna.

Margulis, Lynn Alexander
(1938–   )
American
Biologist

Although theoretical biologist Lynn Margulis's innovative ideas were once dismissed by her colleagues, she is now credited with having changed the way scientists understand both the cell and the overarching theory of evolution. Margulis's hypothesis that cells containing nuclei (eukaryotes) originated when cells without nuclei (prokaryotes) symbiotically joined together posited that symbiosis was the primary mechanism of evolution. Although this theory ran contrary to the longtime assumption that competition (and thus natural selection) was the driving force of evolution, and therefore was initially scorned, it has come to be widely accepted.

The eldest of four daughters, Margulis was born in Chicago, Illinois, on March 5, 1938, to Morris and Leone Alexander. Her father, an attorney who served a term as the Illinois assistant state attorney, owned a company that produced and marketed road-painting materials. Her mother ran a travel agency and managed the family home.

Lynn Alexander was a precocious child, qualifying for early admission to the University of Chicago at the age of 15. Her course work there sparked her interest in science, as she was introduced to theories of heredity and genetics—concerns that would occupy much of her future career. While in college, Alexander met Carl Sagan, who was then a graduate student in physics. They married in 1957, the same year Alexander received her B.A. (The couple had two children together.) After the wedding, Lynn and Carl Sagan moved to the University of Wisconsin, where she earned a master's degree in zoology and genetics in 1960. She then pursued doctoral studies at the University of California at Berkeley, earning her Ph.D. in 1965. (By this time, she and Sagan had divorced.)

While still in graduate school, Lynn Sagan began overturning accepted genetic theories. Most biologists then believed that all of a cell's genetic information was contained in its nucleus. In the early 1960s, Sagan discovered that other cellular bodies—chloroplasts (nutrient-producing organelles in green plant cells) and mitochondria (organelles that process oxygen to create energy)—also contain deoxyribonucleic acid (DNA—the primary component of genetic material), thereby disproving the nucleus-only belief.

This research led Sagan even further. Because both chloroplasts and mitochondria resemble certain types of independent organisms, Sagan postulated that these organelles once existed as prokaryotes that had symbiotically joined together to form eukaryotes to better their chances for survival. She also hypothesized that cell hairs (which allow a nucleated cell to move) first came into being when several highly mobile small bacteria (called spirochetes) merged. This serial endosymbiotic theory (SET) proposed that symbiotic cooperation was the primary impetus driving genetic mutation and evolution. SET ran contrary to the conventional wisdom of the day that genetic change and evolution were caused by the competitive forces of natural selection.

Sagan's ideas were initially derided—15 journals rejected her first comprehensive article before the *Journal of Theoretical Biology* agreed to print it in 1966. That same year, she became an adjunct assistant of biology at Boston University, where she would eventually gain the rank of full professor and remain until 1988. In 1967, she married Thomas Margulis, whose last name she took and with whom she had two more children. (They divorced in 1980.) While at Boston University, Margulis's revolutionary ideas began to find acceptance. Most biologists eventually concurred with the overarching premise of SET, especially as it pertained to mitochondria and chloroplasts. She also won support for a new classificatory scheme. While living matter was traditionally organized according to two broad categories—the plant and animal kingdoms—Margulis advocated a five-kingdom set: animals, plants, fungi, protists (organisms with cell nuclei that do not conform to the first three groups), and monera (bacteria and other microorganisms lacking cell nuclei). This system is now widely used.

In 1988, Margulis left Boston University for a professorship in the botany department at the University of Massachusetts at Amherst, where she remains. In 1983, she was elected to the National Academy of Sciences, and she has been awarded several honorary degrees. In addition to serving on more than two dozen committees and codirecting NASA's Planetary Biology Internship Program, Margulis is a prolific author. She has published more than 130 articles and 10 books (some of which she cowrote with her son Dorion). Her first book, *Origin of Eukaryotic Cells*, published in 1970 and later reissued under the title *Symbiosis in Cell Evolution*, is now considered a seminal work in theoretical biology. She is acknowledged as one of the few living scientists who have caused paradigm shifts in their fields.

Maria the Jewess
First century
Egyptian
Chemist

Although the art of alchemy is often dismissed as a fool's errand, ancient alchemists actually developed many of the procedures and tools used in modern chemistry. One of the more influential alchemists was an Egyptian woman who wrote under the name Maria (or Mary, or Miriam) the Jewess. None of Maria's complete works have survived, but several fragments of her writings remain extant. She is also referenced in other alchemy texts. Her most significant contributions to chemistry were her inventions, which she described in detail. She is credited with inventing the water bath (a distilling device) and the *kerotakis*, which allowed gases to act on metal.

The ultimate goal of the alchemist was to transmute "lower" metals into gold and silver. They generally believed metals to be living organisms evolving toward the perfection of gold, but they also understood the importance of experimentation. Many, in fact, were astute scientists who did much to lay the foundation for modern chemistry. A number of Egyptian alchemists were women—most likely because the practice originated in Mesopotamia, where women chemists incorporated alchemical techniques to produce perfume and cosmetics.

Maria the Jewess is thought to have lived in Alexandria, Egypt, in the first century of the common era. At this time, Alexandria was an educational center and one of the world's most cosmopolitan cities. Maria is believed to have written many different treatises—none of which survive in their entirety. Her most famous work, the *Maria practica*, was excerpted in several collections of ancient alchemy that still exist.

While Maria conceived theoretical bases for alchemy, her greatest achievement was the invention of laboratory equipment. Foremost among these was her *balneum mariae*—a water bath—which is used in modern laboratories and kitchens to this day. Much like a double boiler, this device consisted of two containers, one suspended within the other. The outer vessel was filled with water and heated, thereby slowly heating the substance held in the inner container. The advantage offered by the water bath was that it made it easier to heat a substance gradually and maintain it at a constant temperature.

Maria also conceived the *tribikos*—a still—which was used to separate substances in a liquid through the process of distillation. Maria's tribikos was comprised of an earthenware closed container into which a liquid was poured and heated. As the liquid evaporated, the gas was pumped into another container and cooled. As it condensed back into a liquid, it traveled through three copper delivery spouts, which brought the liquid into three glass containers. In addition to describing her tribikos, Maria offered instructions for constructing one. She detailed how to make copper tubing from sheet metal and recommended flour paste to seal the open joints.

Another of Maria's inventions was the *kerotakis,* which proved helpful in Maria's investigations of the effect of arsenic, mercury, and sulfur vapors on metals. The kerotakis was specifically designed to enable gases to act on metals. This apparatus consisted of a cylinder with a domed top, which was placed over a fire. Sulfur, mercury, or arsenic solutions were heated in a pan near its base, as a piece of metal was attached to a plate and suspended from the domed cover. Gas from the solutions rose past the plate and reacted with the metal. As in the tribikos, the gas reached the top of the dome and cooled back into liquid. However, the liquid was not carried off but ran back down the sides of the kerotakis to be evaporated again. Initially, the vapors from the solution turned the metal black, as they created a black sulfide then named Mary's Black. Alchemists thought this was the first stage of transmutation. Continued heating eventually produced a goldlike alloy.

Alchemy, as Maria the Jewess practiced it, was a true experimental science. Her inventions represented early laboratory equipment, and her effect on other alchemists was profound. Unfortunately, much of Maria's work was lost in the third century, when the Roman Emperor Diocletian persecuted Alexandrian alchemists and burned their texts. Few advances in laboratory chemistry took place after the fall of Alexandria until the 17th century.

Marrack, Philippa
(1945–   )
British/American
Biochemist

Educated in Britain but working most of her life in the United States, Philippa Marrack has devoted her professional life to understanding how mammalian immune systems work. She has concentrated her research on uncovering the role played by T and B cells in fighting antigens within the body that can cause diseases such as cancer, toxic shock, and rheumatoid arthritis. She has also taught immunology at several universities and has served as an adviser on numerous scientific organizations.

Born in Ewell, England, on June 28, 1945, Philippa Marrack received her secondary education in England. In 1963, she enrolled as an undergraduate at New Hall, a college of Cambridge University. After taking a course in biochemistry, she decided to pursue an undergraduate degree in that field. She graduated with a B.A. from Cambridge in biochemistry in 1967 and immediately began graduate work at Cambridge. In 1970, she completed her Ph.D.

From her graduate days, Marrack was interested in the interplay between genes, T and B cells, and antibodies, specifically how T and B cells work to produce antibodies and fight pathogens in the body, and the interplay between T and B cells and genes such as the H-2 gene.

Marrack pursued this early research as a postdoctoral fellow in Cambridge and, from 1973 to 1979, as a fellow at the University of California at San Diego. She then moved to New York State where she briefly worked as an assistant professor at the University of Rochester. In 1979, she moved to Denver to take a position as associate professor at the University of Colorado Health Sciences Center in Denver and researcher at the National Jewish Hospital and Research Center. By 1988, she was a full professor in the Department of Immunology. That same year, she became head of the Division of Basic Immunology at the National Jewish Hospital.

At the University of Rochester, and later in Denver, Marrack refined her research to explore the varieties of T cells and determine these cells' functions. She also devised experiments to observe H-2 linked genes, the so-called immune response genes, and suggested ways that these genes controlled the response of T cells to antigens such as cancerous cell growth. This work led to further studies in which the influence of T cell hybridomas on B cell antibody response was examined. Marrack also studied how T cell receptors recognize and fight antigens and examined the crystal structure of antigens such as the MMTV superantigen.

By examining how biochemical drugs such as interferon react with the immune system, Marrack's basic research has helped in the fight against cancer. Her work has also added to our understanding of toxic shock syndrome and rheumatoid arthritis.

Marrack has received numerous honors and awards for her work in immunology, including election to the National Academy of Sciences in 1989 and the American Academy of Arts and Sciences in 1991. She has also received the Royal Society's Wellcome Foundation Prize

(1990) and the University of Chicago's Howard Taylor Ricketts Prize (1999). As of 2000, Marrack continues to teach at the Health Sciences Center and conduct research at the National Jewish Medical and Research Center.

Matzinger, Polly Celine Eveline
(1947–   )
American
Immunologist

Polly Matzinger was a relative latecomer to science—she did not begin graduate work until she was 29—but when she finally decided that science could be an interesting career, she made up for lost time in a hurry. Working as a researcher in several labs in Europe and the United States, Matzinger developed a theory that the body's immune system responds to danger rather than what it detects as being foreign. This theory, called the danger model, challenges a long-standing model, the self-nonself model, about how the immune system works and has caused a great deal of controversy in the medical world.

Born on July 21, 1947, in La Seyne, France, Matzinger is the daughter of Simone and Hans Matzinger. Her father, who is Dutch, was a World War II resistance fighter, and her mother, who is French, is a former nun. Matzinger has never married. Her children are her four dogs—Annie, Lilly, Charlie, and Roy.

After finishing her secondary education in California in 1965, Matzinger worked at a series of odd jobs in that state while occasionally attending college. By the early 1970s, she was working as a cocktail waitress near the University of California at Davis. At her job, she met a group of science professors who often came to the bar after work for drinks and began engaging them in arguments about science topics. One of these scientists, Robert "Swampy" Schwab, encouraged Matzinger to return to the university to study science. Matzinger heeded Schwab's call and, in 1976, graduated with a B.S. in biology from the University of California at Irvine. After receiving her undergraduate degree, Matzinger enrolled in the graduate school at the University of California at San Diego. She earned a Ph.D. in biology from that institution in 1979.

For 10 years, Matzinger worked in Europe, first on a fellowship from the U.S. National Institutes of Health that sent her to Cambridge University's Department of Pathology and then as a researcher at the Basel Institute for Immunology in Switzerland. As a doctoral student, Matzinger had been puzzled about the existing theory—the self-nonself theory—that explained how the immune system worked. For instance, she reasoned, if T cells, which are a type of blood cell and a major component of the body's immune system, attack a foreign substance such as a virus or a skin graft, why does it not also attack equally foreign sub-

Polly Matzinger, an immunologist, and her first border collie, Annie. It was while watching Annie guard sheep that Polly made the second major mental breakthrough in the creation of the danger model. She saw that Annie paid little attention to noises in the woods as long as the sheep grazed quietly, but she rushed off to investigate if the sheep became distressed. Polly realized that dendritic cells (the sheepdogs of the immune system) might do the same, reacting only if alerted by alarm signals from distressed bodily tissues. *(Courtesy of Polly Matzinger)*

stances such as food in the digestive tract or a fetus in a mother's womb? The self-nonself theory had no answer for this question, yet it was still considered by most immunologists as the best explanation of the phenomena of immune response. With no encouragement from her peers, Matzinger set aside her doubts and tried to understand how the body made the self-nonself discrimination.

By the early 1990s, and now working at the U.S. National Institutes of Health's (NIH) Laboratory of Cellular and Molecular Immunology, Matzinger decided to revisit this question. Prodded by discussions with Ephraim Fuchs, who was questioning why the immune system did not react against cancers, Matzinger put together the elements of her theory of immune response. T cells are activated only if another type of cell, called dendritic cells, warns them that injury has occurred to other, formerly healthy cells. Whether a cell has originated from inside the body or is from outside (that is, foreign) is not relevant. What matters is that it is damaged.

Matzinger's theories have begun to be tested clinically. The most interesting results so far have been in the field of organ transplants. By treating an experimental group of rats that have received skin grafts with drugs that block the response of dendritic cells, doctors have enabled the rats to accept these grafts without traditional immunosuppressant drugs. By treating a group of monkeys with these same blocking drugs, another group of researchers has found a way to transplant kidneys in monkeys. The most immediately interesting new predictions are in cancer treatment. By injecting cancer with bacteria that contains a heat shock protein, some doctors have noted dramatic improvement in patients' conditions and signs that T cells have been activated to fight the cancer.

As of 2000, Matzinger continues to work at the NIH's Laboratory of Cellular and Molecular Immunology. For her insights about the immune system, she has been elected a lifetime honorary member of the Scandinavian Society of Immunology.

Maury, Antonia Caetana
(1866–1952)
American
Astronomer

Antonia Maury's most notable achievement was her improvement of a star classification system. She also identified and studied spectroscopic binaries, which are pairs of stars so close together that they appear to be a single star. Maury was the first woman at the Harvard Observatory to be acknowledged as the author of a publication. Although many of her ideas were rejected during the early years of her career, Maury's approaches were eventually accepted during her lifetime.

On March 21, 1866, Antonia Maury was born in Cold Spring, New York, to Mytton and Virginia Draper Maury. It is not surprising that Maury would eventually pursue a scientific career, given that both her parents had scientific backgrounds. Her father was an amateur naturalist (and an Episcopalian minister), while her mother was the sister of the astronomer Henry Draper, who held the distinction of being the first person to photograph stars' spectra. (Spectra are the colors and lines that occur when a star's light is passed through a prism.) Her younger sister CARLOTTA JOAQUINA MAURY was a noted paleontologist.

Maury's early education took place at home under the tutelage of her father. She then attended Vassar College. She was inspired by MARIA MITCHELL to study astronomy

and graduated in 1887 with honors in astronomy, physics, and philosophy.

In 1888, Maury joined a team of scientists at the Harvard Observatory who were working under the observatory director, Edward Pickering, to classify stars according to differences in their photographed spectra. Despite her interest in the task, Maury was constantly at odds with Pickering, who ordered her to use a classification system he had devised with WILLIAMINA FLEMING. Maury preferred to take a different approach to the work, paying attention to such details as the thickness and sharpness of dark lines in the spectra. Despite this friction, Maury remained at the Harvard Observatory until the early 1890s, studying and cataloging bright northern stars. She resigned after completing the task. (Her work was eventually published in 1897.)

After leaving the observatory staff, Maury spent the next 20 years lecturing and tutoring. She continued to conduct research at the observatory, focusing on spectroscopic binaries. (She and Pickering had first identified these pairs in 1889, which was a significant feat given that even with the most advanced telescopes of the days, these stars—which orbit one another in close proximity—appear to the eye as only a single star.) Maury focused particularly on one binary pair called Beta Lyrae, closely examining changes in its spectra. In 1918, Maury returned full-time to the Harvard Observatory, where she remained until her retirement in 1935. In 1933, she published an influential book on her observations of Beta Lyrae.

Although Maury's classification system was initially spurned by Pickering, her innovations began to gain acceptance in the early 1900s. In 1905, the Danish astronomer Ejnar Hertzsprung advocated Maury's classification over Pickering's. Hertzsprung went on to apply several other of Maury's ideas. For example, Maury's observation that differences in the appearance of spectral lines confirmed that stars of the same color could differ in size and brightness became the central tenet of Hertzsprung's diagram of star development. By the end of Maury's career, she had received full recognition from her peers. She was awarded the American Astronomical Society's ANNIE JUMP CANNON Prize in 1943 for her star classification system. After retiring in 1935, Maury continued to visit the observatory every year. She died in New York on January 8, 1952. She never married or had children.

Maury, Carlotta Joaquina
(1874–1938)
American
Paleontologist

Carlotta Joaquina Maury made a number of important contributions to paleontology. Her area of expertise was in the stratigraphy (the study of the nature, distribution, and relations of the stratified rocks of the Earth's crust) and fossil fauna of Brazil, Venezuela, and the West Indies. In addition to heading her own paleontological expedition to the Dominican Republic, Maury is credited with describing new species and genera of fossil fauna. A professor, she also served as a consultant to the Venezuelan division of the Royal Dutch Shell Petroleum Company and was named the official paleontologist of Brazil.

On January 6, 1874, Maury was born at Hastings-on-Hudson, New York. She was one of Mytton and Virginia Draper Maury's three children. It is not surprising that she pursued a career in science, given her family's background. Her father—an Episcopalian minister—was also an avid amateur fossil collector, and a number of other relatives devoted themselves to science full time. The hydrographer and meteorologist Matthew Fontaine Maury was her cousin, and her grandfather John William Draper was a renowned physicist. Maury's sister ANTONIA CAETANA MAURY was a well-regarded astronomer. Maury herself never married or had children.

After attending Radcliffe College and developing an interest in zoology and geology, Maury enrolled at Cornell University, where she earned her bachelor's degree in 1896. She received the Schuyler Fellowship for graduate research from Cornell, which she used to study paleontology for two years at the University of Paris. She returned to Cornell in 1898 and was awarded her Ph.D. in 1902 for her thesis, "A Comparison of the Oligocene of Western Europe and the Southern United States." She thereupon continued her research at Columbia University from 1904 to 1906, during which time she also worked as an assistant in the school's department of paleontology.

In 1907, Maury joined the Louisiana Geological Survey and was responsible for investigating and reporting on the state's petroleum and rock salt deposits. She left the survey in 1909, opting instead to lecture at Barnard College (for very little pay), where she remained until 1912. While teaching at Barnard, she participated in a geological expedition to Venezuela headed by Arthur Clifford Veatch. In 1912, she moved to the Huguenot College of the University of the Cape of Good Hope in South Africa, where she served as a professor until 1915.

Maury had the opportunity to head her own expedition in 1916 when she won a Sarah Berliner Fellowship. She organized and launched the Maury expedition to the Dominican Republic that year. Since 1910, Maury had also been consulted on geological and stratigraphic issues by the Venezuelan division of the Royal Dutch Shell Petroleum Company. In 1918, her responsibilities expanded again, as she joined the Brazil Survey, which led to her appointment as the official paleontologist to the Brazilian government. She also wrote detailed reports for the American Museum of Natural History in New York. She continued her work with Shell and the Brazilian government until her death.

Maury's years of research in South America and the West Indies provided significant information about stratigraphy and fossil fauna in these regions. She was elected a fellow of both the Geological Society of America and the American Geographical Society. In addition to participating in the American Association for the Advancement of Science, she was also a corresponding member of the Brazilian Academy of Sciences. She died in her home in Yonkers, New York, on January 3, 1938.

Maxwell, Martha Dartt
(1831–1881)
American
Naturalist

Martha Dartt Maxwell gained international acclaim for her innovative natural history displays. She was a pioneer in displaying animals in habitat groupings rather than simply putting similar species together in glass cases as was then customary. Maxwell is also credited with improving the practice of taxidermy and changing the way museums exhibited natural history collections. She launched her own museum shortly after beginning to collect and preserve Rocky Mountain fauna, and she exhibited her work at the Centennial Exhibition of 1876.

Martha Dartt was born to Spencer and Amy Sanford Dartt on July 21, 1831, in rural Pennsylvania. Amy became an invalid when Martha was young, and when Spencer—who had been a farmer—died in 1833, the family was thrown into chaos. Amy eventually remarried in 1841, to Josiah Dartt, Spencer's first cousin. Martha would later credit her grandmother, Abigail Sanford, for imbuing her with a love of nature by taking the young girl on long walks through the woods, identifying various animals.

In 1851, Martha left for Oberlin College in Ohio with plans to become a teacher. She had to drop out in 1852 (her family could no longer afford tuition) and return to her parents, who were then living in Baraboo, Wisconsin. She was teaching at a local school when James Maxwell, a Baraboo businessman, hired her in 1853 to chaperone two of his children at Lawrence College in Wisconsin. In return for her services, he agreed to cover her tuition. She had been there less than a year when Maxwell proposed to her. Although he was 20 years her senior, with six children of his own, she agreed. They were married in 1854 and had a daughter in 1857.

Martha Maxwell spent the next several years caring for her new home and stepchildren. However, the once financially secure James Maxwell saw his resources dwindle in the wake of the panic of 1857. Newspapers were rife with stories of miners making fortunes in the recently discovered gold mines of Colorado, and James convinced Martha that they should try their luck as well. In 1860, they moved to Pikes Peak, Colorado, leaving their daughter in the care of Martha's family. A chance encounter with a German taxidermist in the early 1860s inspired Martha to take up that discipline. In 1862, lonesome for her daughter and with her marriage troubled, Maxwell returned to Baraboo, where she studied taxidermy. She then moved to a temperance colony in New Jersey. In 1868, James Maxwell persuaded her to return to Colorado.

Back in the Rocky Mountains, Maxwell's career as a taxidermist and naturalist blossomed. Her work was characterized by habitat displays. Instead of merely presenting the animals lifelessly in glass cases, lumped together by species, Maxwell constructed elaborate artificial habitats that showed different kinds of birds and animals coexisting as they did in nature. The effect was astounding. In 1870, she sold one of her collections to Shaw's Garden in St. Louis. Even the Smithsonian Institution bought some of her specimens. In 1874, she opened her own museum—the Rocky Mountain Museum—in Boulder, Colorado. She moved it to Denver in 1876 but was unable to make the venture profitable.

Despite this setback, Maxwell's reputation as a naturalist was solidified in 1876 when she was asked to produce an exhibit for the Philadelphia Centennial Exhibition. Her display was one of the most popular at the internationally attended event. She built a Rocky Mountain landscape with streams, plains, caves, and mountains, and she populated it with bears, mountain sheep, buffaloes, mountain lions, fish, and turtles—all preserved in perfect detail. In front of this scene was a simple placard that read "Woman's Work." People were fascinated with her exhibit and even more so with Maxwell. *Harper's Bazaar* and other magazines ran stories on this "Colorado Huntress." She won the admiration of several curators there, including Robert Ridgeway, who named a species of owls after her in 1877. In 1878, her sister published a biography of her that also drew a great deal of attention. Called *On the Plains and Among the Peaks; or, How Mrs. Maxwell Made Her Natural History Collection,* the book presented her in heroic terms.

After the centennial, Maxwell did not return to Colorado or her husband. Unhappy in the marriage, she remained on the East Coast, where she attended classes in Boston, oversaw several exhibitions of her collections, and worked occasional odd jobs to make ends meet. She died in Massachusetts on May 31, 1881, of an ovarian tumor.

McClintock, Barbara
(1902–1992)
American
Geneticist

Barbara McClintock's pioneering experiments on maize plants formed the cornerstone of modern genetics. She was awarded the 1983 Nobel Prize in physiology or medicine for her discovery of transposable genes. Although this revolutionary finding (that a gene's location on the chromosome is not fixed) occurred in 1950, it was largely ignored until the 1970s because few could understand McClintock's maverick ideas.

Born on June 16, 1902, in Hartford, Connecticut, Barbara McClintock was the third daughter of Thomas and Sara Handy McClintock. Her father was a homeopathic physician who encouraged Barbara's scientific inclinations. Her relationship with her mother, however, was strained, and young Barbara spent large portions of her childhood with relatives in rural Massachusetts.

Although Sara McClintock forbade her daughters to attend college—because education might make them unmarriageable—Barbara McClintock yearned to pursue her studies. After her father intervened on her behalf, McClintock enrolled at Cornell University in 1919. She received her B.S. in 1923 and immediately began graduate studies there. Cornell was a leading genetic research institute, though it differed from many of its counterparts. Researchers at Cornell worked with maize instead of the more conventional fruit fly. Although maize registered genetic changes more slowly than fruit flies, it offered many easily observable genetic traits, such as the color of the kernels on the cob. While still a graduate student, McClintock made maize an even more ideal research tool. She developed new staining techniques that could identify each of the plant's 10 chromosomes, and she proceeded to map the position of

Barbara McClintock, the first woman to win an unshared Nobel Prize in physiology or medicine. *(Cold Spring Harbor Laboratory Archives)*

the genes on the maize chromosomes. In 1927, she earned her Ph.D. in botany.

After graduating, McClintock remained at Cornell as a botany instructor. Despite her sterling reputation, she never received a faculty appointment at Cornell (women were then almost universally denied positions at coeducational colleges). She thrived there, however, publishing more than 10 papers on maize chromosomes. Most significant among these was a 1930 paper published with colleague Harriet Creighton that proved that chromosomes carried and exchanged genetic information. The coauthors had crossed maize crops with opposite traits and found that physical pieces of the ninth chromosome had actually changed places. The experiment was a landmark, considered to be one of the great works of modern biology.

Although several colleagues believed McClintock should have won the Nobel Prize for her maize experiments, she remained unable to find a permanent faculty position. From 1931 to 1936, McClintock used short-term fellowships to continue her research for brief stints at Cornell, the California Institute of Technology, and the University of Missouri, where she was finally named to the full-time faculty in 1936. Although her research won her the accolades of her peers, she left the school in 1941 under acrimonious circumstances.

In 1941, she accepted a temporary position at the Cold Spring Harbor Laboratory in New York. She remained there for the rest of her life, her research financed by the Carnegie Institution. In 1944, she began the work—concentrating on broken maize chromosomes—that would win her the Nobel Prize. Observing kernels on a self-pollinating ear of corn with distinctive pigmentation, McClintock recognized that some genetic information inhibiting pigmentation had been lost (the kernels should have been colorless). In 1948, she discovered that genetic material can be released from its original position on the chromosome and reinserted in another place. She coined the term *transposition* for this phenomenon. In addition to determining that genes need not have fixed positions on chromosomes, McClintock also described and characterized two new kinds of genetic elements: a controlling one that works like a switch to turn off and on genes expressing physical characteristics; and an activator that causes the on-and-off switch to move from one location on the chromosome to another. She published her initial findings in 1950 and a more detailed paper in 1953, both of which were ignored by the scientific community. (Scientists were then convinced that genes were stable, with fixed locations.) McClintock eventually stopped trying to convert her colleagues, choosing to publish the results of her continued research only in the annual reports of the Carnegie Institution rather than in academic journals.

The significance of McClintock's transposition theory was finally comprehended in the 1970s when molecular bi-

ologists discovered transposable elements in bacteria. She won the Albert Lasker Basic Medical Research Award in 1980 and a MacArthur Foundation Fellowship in 1981. She was awarded the Nobel Prize in 1983. She died on September 2, 1992. She never married nor had children. Her contributions to modern genetics were fundamental.

McDuff, Margaret Dusa
(1945–)
English
Mathematician

Dusa McDuff was the first recipient of the American Mathematical Society's Ruth Lyttle Satter Prize in 1991. Her career developed in tandem with the growth of the feminist movement, and she gained confidence in her ability to define her own career goals at the same time that the feminist movement supported this kind of growth for all women.

Margaret Dusa Waddington was born on October 18, 1945, in London. She grew up in Edinburgh, Scotland, where her father was a professor of animal genetics and her mother was an architect with the Scottish Development Office. Although her maternal lineage bristled with intelligent, independent women (both her grandmother and great-grandmother were authors and political activists), she retreated into traditional female roles as a teen, going so far as to decline a scholarship to Cambridge University in order to stay by the side of her boyfriend. She attended the University of Edinburgh instead, earning her bachelor's degree in mathematics in 1967.

By the time she entered graduate school, she had married and taken her husband's name, though he did not follow her to Cambridge. In the middle of her graduate study, she followed him to Moscow for six months, where she studied topology with Israel M. Gel'fand. Back at Cambridge, she attended topology lectures by Frank Adams and studied functional analysis under G. A. Reid. McDuff wrote her doctoral thesis in operator theory, solving a well-known problem in von Neumann algebras that was published in the *Annals of Mathematics*. McDuff earned her Ph.D. in 1971.

McDuff conducted postdoctoral work at Cambridge under a two-year Science Research Council Fellowship. In 1973, she was appointed as a lecturer at the University of York, where she worked with Graeme Segal and essentially wrote a second dissertation. In 1974, she accepted a one-year visiting professorship earmarked for a woman at the Massachusetts Institute of Technology. In 1976, she accepted a lectureship at the University of Warwick, but before starting there, she spent a year at the prestigious Institute for Advanced Study at Princeton University, where she met fellow mathematician Jack Milnor. Upon her return to England, she focused her professional life on the study of groups of diffeomorphisms and foliations, while in her personal life, she divorced her husband, with whom she had had one child.

In 1978, McDuff accepted an untenured assistant professorship with the State University of New York (SUNY) at Stony Brook sight unseen. The university promoted her to an associate professorship and then to a full professorship in 1984. During that time, she married Jack Milnor and had her second child by him. Between 1991 and 1993, McDuff chaired the Department of Mathematics at SUNY Stony Brook. She conducted much of her research on symplectic topology, publishing two monographs in collaboration with Dietmar Salamon on this topic, as well as writing some 30 of her 50 research papers on it.

In 1991, the American Mathematical Society awarded McDuff with the first Ruth Lyttle Satter Prize. In 1994, the Royal Society of London inducted McDuff as one of its very few female members, and in 1995, the American Academy of Arts and Sciences named her as a fellow. In support of the women's movement, McDuff has worked closely with the Women in Science and Engineering program at SUNY Stony Brook, encouraging first-year female students to enter the fields of science, mathematics, or engineering.

McNally, Karen Cook
(1940–   )
American
Geologist

Karen Cook McNally, who heads the Charles F. Richter Seismology Laboratory at the University of California at Santa Cruz, has significantly influenced the study of earthquake source processes. Using the seismic gap theory, McNally has predicted earthquakes with uncanny precision. Her research has also focused on crust and upper mantle structure, earthquake statistics, and fracture mechanics.

Born in 1940, in Clovis, California, McNally experienced earthquakes firsthand during her childhood, but geology did not particularly interest her as a child. She learned ranching from her father and music from her mother. Although McNally's parents encouraged her to attend nearby Fresno State College, she sought independence in an early marriage. While raising her two daughters, she took classes part-time.

Magazine articles on earthquakes piqued McNally's interest in the subject. In 1966, she and her husband divorced. Soon thereafter, McNally took her daughters and moved to Berkeley, California, to study geology. Upon receiving her bachelor's degree from the University of California at Berkeley, she began graduate studies there. McNally was awarded her M.A. in 1973 and her Ph.D. in geophysics in 1976.

The same year she completed her doctorate, McNally accepted a fellowship at the California Institute of Technology (Caltech) in Pasadena. It was here that she first encountered the seismic gap theory, which was then a new approach to predicting earthquakes. In essence, the theory posits that earthquakes are more likely to occur in places where routine earth movement has not taken place. Typically, the monumental plates that compose the earth's crust move constantly, sliding past each other so smoothly that only a seismograph can detect the motion. Plates stop moving when friction locks them together—sometimes for decades or even centuries. Pressure builds up in the stationary area because the plates as a whole continue to slide. Eventually, pressure builds up to the point that it fractures the rock and releases the plates in a sudden burst of energy and movement. The result is a major earthquake. In 1977, McNally and her team from Caltech installed seismographs in a region of Mexico. For several weeks, the geologists found that the machines registered slowly building tremors. However, on the day before the quake, the plates were completely silent, which led the group to believe that a quake was imminent. A few hours later, an earthquake measuring a tremendous 7.8 on the Richter scale erupted within 31 miles of the spot where McNally's group had predicted it would occur.

In 1978, McNally was promoted to senior resident fellow geologist at Caltech, where she remained until 1986. During this period, she predicted five more Mexican earthquakes. Using the data from these earthquakes, she refined seismic gap theory. McNally concluded that as the pressure increases in a gap area, weak spots will collapse first, thereby causing a series of small quakes. These undetectable quakes continue to crack and shake the rock until the strongest rocks give way—and a major earthquake ensues.

McNally moved to the University of California at Santa Cruz in 1986, becoming a professor of geophysics. She also was appointed head of the university's Institute of Tectonics and its Charles F. Richter Seismology Laboratory. In addition to continuing her extensive field research, McNally has applied sophisticated laboratory analyses to her work as well, incorporating synthetic seismograms and forward and inverse modeling approaches. Sine 1976, McNally has held a concurrent position as a consulting seismologist with Woodward-Clyde Consultants in San Francisco.

McNally has greatly aided efforts to predict earthquakes and has also published work on regional tectonics and the structure of the Earth's crust. Her peers have recognized her contributions to the field. In addition to being elected to the Seismological Society of America, she is a member of the American Geophysics Union and the American Academy of Arts and Sciences. She has continued to carry out her work at the University of California at Santa Cruz.

McNutt, Marcia Kemper
(1952–)
American
Geophysicist

Marcia Kemper McNutt researches the Earth's plate tectonics, or the shifting of the sublayers of the Earth's crust. She specifically focuses on mapping the ocean floor, using both sea-bound techniques of echo-soundings and readings taken from satellites. She has hypothesized that the North American continent will someday split in half, though not for a long time in human terms. She is president and chief executive officer of the Monterey Bay Aquarium Research Institute (MBARI) in Moss Landing, California.

McNutt was born on February 19, 1952, in Minneapolis, Minnesota. She attended Northrop Collegiate School (now the Blake Schools), graduating as the class valedictorian in 1970. She studied physics at Colorado College, graduating summa cum laude and Phi Beta Kappa with a bachelor of arts degree in 1973. She then pursued graduate study at the Scripps Institution of Oceanography under a National Science Foundation Graduate Fellowship. She earned her doctorate in earth sciences in 1978. That same year, she married and would have three children. Her first husband died unexpectedly in 1990, and McNutt later married Ian Young.

McNutt's first professional appointment placed her at the University of Minnesota as a visiting assistant professor, filling in for a professor who was on sabbatical. From 1979 through 1982, she worked as a geophysicist predicting earthquakes for the United States Geological Survey, stationed in Menlo Park, California. In 1982, the Massachusetts Institute of Technology (MIT) hired McNutt as an assistant professor. Over the next 15 years, MIT promoted McNutt first to associate professor (in 1986), and later the university named her the Griswold Professor of Geophysics.

During her tenure at MIT, she directed the Joint Program in Oceanography and Applied Ocean Science and Engineering, a collaborative graduate program between MIT and the Woods Hole Oceanographic Institution. She also has served as a standing member of the National Aeronautics and Space Administration Science Steering Group Geopotential Research Mission, as well as serving shorter stints on various committees in her field, such as the Geodynamics Committee.

As president and CEO of MBARI, which is funded by the Packard Foundation, McNutt directs the institute's research, specifically geared toward designing and building both tethered and autonomous underwater vehicles for sampling and observing the ocean, its floor, and its inhabitants. She has participated in almost two dozen oceanographic expeditions herself. Her previous research focused on the southern oceans, which had received little attention

as they are not major commercial traffic lanes nor are they strategic zones militarily. Her research shed light on offshore Africa, which might have oil deposits based on the fact that it separated 160 million years ago from basins off Brazil that are rich in oil.

The American Geophysical Union granted McNutt the James B. MacElwane Medal in 1988. McNutt later became the president-elect of the American Geophysical Union. In 1995, the Public Broadcasting System produced a television series, *Discovering Women,* which featured McNutt, following her on exploration trips to Tahiti to study xenoliths, or rocks hurled from the Earth's interior, and showing her at home raising her three children while performing her duties as an academician.

Mead, Margaret
(1901–1978)
American
Anthropologist

The best known anthropologist of the 20th century, Margaret Mead greatly expanded both the audience for and the topics investigated by anthropological studies. Mead was the first anthropologist to examine child-rearing practices and the role of women in other cultures. Unlike many of her contemporaries (who believed that genes determined human personality and roles), Mead held that culture was the primary shaper of human behavior. The author of hundreds of articles and books, Mead brought her message to a popular audience. Her most famous publications—including *Coming of Age in Samoa* and *And Keep Your Powder Dry*—were national best-sellers. For most of her career, Mead was associated with the American Museum of Natural History.

Born on December 16, 1901, in Philadelphia, Pennsylvania, Margaret Mead was the eldest of the five children of Edward and Emily Mead. Edward was an economics professor at the University of Pennsylvania's Wharton School of Business, while Emily was a sociologist. Mead's mother and grandmother (a child psychologist) cultivated the skills that she would later utilize as an anthropologist, teaching her to observe other children and take notes on their behavior.

Mead began her academic career at DePauw University in Indiana in 1919, but she transferred to Barnard College in 1920. There, she met prominent anthropologists Franz Boas and RUTH BENEDICT. Inspired by their work, she majored in anthropology, graduating in 1923. That same year, she married Luther Cressman. (Their marriage ended quickly.) Mead earned her master's degree in psychology from Columbia in 1924, and in 1925, she traveled to the island of Tau in Samoa to conduct fieldwork for her Ph.D. (which she would receive from

Best-known anthropologist of the 20th century, Margaret Mead was the first anthropologist to examine child-rearing practices and the role of women in other cultures. *(National Library of Medicine, National Institutes of Health)*

Columbia in 1929). Her thesis explored whether adolescent girls in Samoa experienced the same anxieties and concerns as female American teenagers. She concluded that they did not and parlayed her research into her first book (published in 1928)—*Coming of Age in Samoa*—which proposed that culture (not genetic determinants) accounted for these differences.

After returning from Samoa in 1926, Mead was named assistant curator of ethnology at the American Museum of Natural History in New York. She would remain affiliated with this institution for her entire career, becoming associate curator in 1942 and curator in 1956. In 1928, Mead married New Zealand anthropologist Reo Fortune. That same year, the couple researched the Manus people in New Guinea. Their fieldwork provided material not only for several highly regarded scientific papers but also for Mead's books *Growing Up in New Guinea* (1930) and *Sex and Temperament in Three Primitive Societies* (1935). These popular works further explored the notion that social behaviors (including gender roles) were determined by cultural forces.

Mead's marriage to Fortune ended in divorce, and in 1936, she married Gregory Bateson, another anthropologist. Their daughter, Mary Catherine, was born in 1939. (The couple would divorce in 1951.) In the late 1930s and early 1940s, Mead and Bateson studied the Balinese people. Again, Mead transformed her field notes into a book—*Balinese Character*—which was published in 1941. This text pioneered the extensive use of photographs in anthropological works. In 1942, Mead applied her methodology to American culture, producing the best-seller *And Keep Your Powder Dry: An Anthropologist Looks at America,* which compared and contrasted American culture to seven others. From 1947 to 1951, she was a visiting lecturer at Columbia University's Teachers' College.

During the 1950s and 1960s, she lectured and wrote extensively about a range of topics, including education and family life as well as more academic topics. She also returned to the Pacific Islands, New Guinea, and Bali, recording changes in the cultures she had studied previously. In 1964, she officially retired from the Museum of

Natural History, although she continued to maintain an office there until her death on November 15, 1978.

Mead is credited with widening the focus of anthropology. While the discipline had long neglected the study of women and children, Mead made these essential subjects of anthropological inquiry. Mead also popularized the often esoteric discipline with a mainstream audience. Her numerous accomplishments did not go unrecognized. She became the president of the Anthropological Association in 1960 and was elected to the National Academy of Sciences in 1975. After her death, the conclusions of *Coming of Age in Samoa* were called into question by Australian anthropologist Derek Freeman, who accused Mead of leading her subjects to the responses she wanted and of misunderstanding their teasing answers to her questions about sexuality. Whether or not these criticisms are true, they do not diminish Mead's considerable legacy.

Meitner, Lise
(1878–1968)
Austrian/Swedish
Physicist

Lise Meitner played a central role in one of the most significant scientific discoveries of the 20th century. Together with her nephew, Otto Frisch, Meitner determined that atomic nuclei could be split. This theory of nuclear fission revolutionized notions of atomic structure and enabled the invention of both nuclear power and the atomic bomb. Meitner was also the first female full professor of physics in Germany, though she experienced considerable discrimination because of her gender.

Born on November 7, 1878, in Vienna, Austria, Lise Meitner was the third of eight children born to Philipp and Hedwig Meitner. Her father was a wealthy lawyer, and her mother was a member of Vienna's elite social circle. Although Lise expressed an early interest in science, her parents discouraged her ambition. Her father sent her to study to be a French teacher at the Elevated High School for Girls in Vienna.

But Meitner still wished to be a scientist, and she worked intensely with a private tutor to pass the competitive university entrance exams. In 1901, Meitner enrolled at the University of Vienna, where she studied physics. She received her Ph.D. in 1906, becoming only the second woman ever to earn a physics doctorate from that institution. Unable to find a faculty position after graduation, she remained at Vienna as an assistant in her adviser's laboratory.

In 1907, Meitner moved to the University of Berlin's Institute for Experimental Physics to study under Max Planck—one of the progenitors of quantum physics. She persuaded a young chemist, Otto Hahn, to hire her as his

Lise Meitner, who, with her nephew, Otto Frisch, announced in 1939 that the nucleus of an atom could be split. *(AIP Emilio Segrè Visual Archives)*

assistant, and the pair began a fruitful collaboration, focusing initially on the behavior of beta rays (negatively charged particles emitted during the breakdown of radioactive atoms) as they passed through aluminum. In 1912, the duo moved to the nearby Kaiser-Wilhelm Institute after Hahn was hired to work in its new radioactivity department. Although their research was intermittently interrupted by World War I, by 1918 the duo had discovered what was then the second-heaviest element, which they named protoactinium (later shortened to protactinium).

Meitner's career was buoyed by this groundbreaking discovery. In 1918, she was named head of a new department of radioactivity physics at the Kaiser-Wilhelm Institute. In 1926, she was appointed to the faculty of the University of Berlin, becoming the first woman full physics professor in Germany. She continued to investigate beta particles and reunited with Hahn in 1934 to determine what would occur when the heaviest natural elements were bombarded with neutrons. However, political turmoil caused Meitner to abandon her work temporarily. Jewish by birth (although baptized as an infant), Meitner experienced increasing anti-Semitism, and she fled to Sweden and a post at the Nobel Institute of Theoretical Physics in Stockholm in 1938.

In December of 1938, Hahn wrote Meitner a letter explaining a conundrum he had encountered in the laboratory. While bombarding uranium with neutrons, he did not produce a heavier substance, as expected. Instead, he created what appeared to be barium, an element much lighter than uranium. Meitner puzzled over this paradox with her nephew, Otto Frisch, who was visiting her in Sweden. They realized that Hahn's results could be explained if he had split the uranium nucleus—rather than simply adding or subtracting particles from it. This conclusion ran contrary to accepted notions, since most physicists believed it was impossible to split a nucleus. Meitner and Frisch reasoned that the electric charge of the heavy nucleus had offset the forces binding the nucleus together. They also determined that splitting a uranium nucleus would release a tremendous amount of energy. (In the small quantities Hahn had used, the energy output was not apparent.) In 1939, Frisch and Meitner published a paper explaining their discovery. Frisch named the process *fission*, likening the splitting of the atomic nucleus to the process of cell division.

Unwittingly, the two physicists had provided the theoretical framework for the atomic bomb, though they did not appreciate its full consequences until 1945, when the first atomic bomb was dropped. Meitner continued to live and work in Sweden, even after her official retirement in 1947. In 1968, she moved to London to be closer to Frisch. Although she did not share Hahn's Nobel Prize, she received numerous awards for her achievements—including the Max Planck Medal (1949) and the Enrico Fermi Award of the U.S. Atomic Energy Commission (1966). She died on October 27, 1968. When researchers discovered element 109 in 1982, they named it meitnerium in her honor. She never married nor had children.

Mendenhall, Dorothy Reed
(1874–1964)
American
Physician

Dorothy Reed Mendenhall's career as a research physician was significant on a number of fronts. She was the first person to discover that Hodgkin's disease—a cancer characterized by the enlargement of lymph nodes—was not a form of tuberculosis, as had previously been believed. She also identified the specific cell type that indicates Hodgkin's disease, which was subsequently named after her. After her first child died because of poor obstetrics, Mendenhall embarked on a lifelong crusade to reduce infant mortality rates. As a result of her efforts, nationwide standards for evaluating the weight and height of children were established. She also launched programs to educate new moth-

ers and pregnant women about proper nutrition, and she helped form some of the country's first infant welfare clinics.

The youngest of William Pratt and Grace Kimball Reed's three children, Dorothy Reed was born on September 22, 1874, in Columbus, Ohio. (Both her parents were descendants of English colonists who emigrated to the United States in the 17th century.) William Pratt Reed worked as a shoe manufacturer.

After earning her bachelor's degree from Smith College in 1895, Reed became one of the first women to enroll in the Johns Hopkins School of Medicine in Baltimore, Maryland (that university had only recently changed its men-only admission policy). She received her medical degree in 1900.

In 1901, Reed began a fellowship in pathology at Johns Hopkins. In addition to teaching classes on bacteriology during this period, Reed began researching Hodgkin's disease. Although Hodgkin's disease was then thought to be a form of tuberculosis, Reed disproved this erroneous theory. She discovered that Hodgkin's patients all carried a specific type of cell in their blood. These abnormally large cells—now named Reed cells in her honor—became the primary way of identifying the disease. Reed's careful work produced the first detailed descriptions of the tissue changes that characterize Hodgkin's disease. She was also the first to elucidate that the disease progressed through several stages and that a patient's chances for survival decreased with each stage. Reed left Johns Hopkins in 1903 to become a resident physician at Babies Hospital of Columbia University, where she remained until 1906.

Reed married Charles Mendenhall in 1906, and soon thereafter they moved to Wisconsin, where he had a faculty position. Although the couple would have four children together, their first baby died a few hours after birth. This tragic event irrevocably altered Mendenhall's career. She became a tireless advocate for infant welfare, and her efforts led to the creation of Wisconsin's first infant welfare clinic in 1915. In 1914, Mendenhall took a position as a field lecturer in the University of Wisconsin's department of home economics. In this capacity, she crafted a correspondence course in 1918 for new mothers and pregnant women (titled "Nutrition Series for Mothers") and traveled throughout Wisconsin, lecturing about the importance of nutrition. In the 1920s, she launched the school's first ever sex hygiene class. However, her most important contribution at the University of Wisconsin was to institute a nationwide campaign to measure and weigh all children under the age of six, thereby establishing healthy weight and height standards for American children.

Beginning in 1917, Mendenhall also worked intermittently as a medical officer for the United States Children's Bureau, writing bulletins for federal and state agencies about the importance of nutrition and child care. In 1926, she began a study of maternity care and birthing practices in Denmark, which had a low incidence of childbirth complications. After noting that the Danish relied more on specialized midwives and natural methods than on medical intervention, she helped popularize natural childbirth in the United States. She also proposed that obstetrics be elevated to its own medical specialty.

Mendenhall retired from both the University of Wisconsin and the Children's Bureau in 1936. By that time, her constant efforts had succeeded dramatically. By 1937, in fact, Madison, Wisconsin, boasted the lowest infant mortality rate in the nation. Mendenhall withdrew from public life after the death of her husband, and she herself succumbed to heart disease in Chester, Connecticut, on July 31, 1964. In her memory, her family established a scholarship fund for women medical students at Johns Hopkins, and Smith College named Sabin-Reed Hall after her and FLORENCE SABIN.

Merian, Maria Sibylla
(1647–1717)
German
Artist and Naturalist

A talented painter and close observer of the natural world, Sibylla Merian was one of the leading painters of flora and fauna at the end of the 17th and beginning of the 18th centuries. She was especially interested in detailed, realistic paintings of insects, although she frequently painted amphibians, birds, and plants as well. She is best known for the magnificent collection of drawings she made on her trip to Suriname in 1699.

Sibylla Merian was born on April 2, 1647, in Frankfurt am Main, then an independent city-state near the Rhine River in central Germany. She was the daughter of Matthaus Merian, a well-known engraver. Her mother was the daughter of J. T. de Bry, an engraver and book publisher who in 1624 printed a famous book of engravings about the new world, *Collectiones Peregrinationum in Indiam*. Matthaus Merian died in 1650 when Sibylla was three. Her mother married the artist Jacob Marell, who taught Sibylla to draw, paint, and engrave. In 1665, Merian married Andreas Graff, also an artist.

Before she became an accomplished artist, Merian was an astute observer of nature. In 1660, when she was 13, she noted in her journal observations about the life cycle of some silkworms that she kept in her house. In great detail, she described how the silkworm transforms from a caterpillar to a moth. By 1670, after her marriage to Graff, she had moved to Nuremberg where she began making a living painting flowers on tablecloths and embroidering them on textiles. She also executed engravings of flower scenes.

Merian had taken flowers as her subject as an opportune moment. By 1670, flower paintings, especially of tulips, had come in vogue in Europe, and Merian must have done quite well with her paintings, drawings, and engravings. At about this time, she also began to put insects in the flower works. By 1674, Merian had shifted much of her attention from flowers to insects. She put together a large of collection of these creatures and continued her notebook observations of each species. She also began to paint them in a very realistic manner.

Around 1681, Merian's marriage to Graff ended. For a while, she moved back to Frankfurt to help her mother with the family's publishing business. By 1685, she had moved again, this time to join the mystical religious commune that had gathered around John Labadie. The Labadists were a protestant sect who sought ecstatic union with God. They only lasted until 1688, but while Merian was with them, she happened to see a collection of brilliantly colored butterflies that had been brought back from the West Indies by Cornelius van Sommelsdijk. This was a collection she never forgot, and it eventually prompted her to visit the New World.

From about 1690 to 1699, Merian lived in Amsterdam, as always supporting herself painting and doing engravings. By 1699, at the age of 52, Merian had succeeded in persuading the city fathers of Amsterdam to fund a trip to the Americas. She chose as her destination the Dutch colony of Suriname, located on the northern tip of South America. For two years, Merian and her daughter traveled across Suriname, spending countless hours in country campsites under the hot sun or being devoured by mosquitoes at night to collect and paint plants, birds, and insects.

When Merian returned to Amsterdam in 1701, she had accumulated enough material to keep her busy for the rest of her life. For the next 16 years, she drew and cataloged these creatures and plants, many of which had never been described before. Although she never received any official awards from scientific organizations during her lifetime, her collection of drawing, engravings, and paintings were used in the late 1700s by Carl von Linné to develop a binomial nomenclature to name and classify species of living things. Merian died in Amsterdam in 1717 at the age of 70.

Mexia, Ynes Enriquetta Julietta
(1870–1938)
American
Botanist

A woman who worked at a variety of jobs and lived in many places before discovering a love of botany, Ynes Mexia was renowned during the 1920s and 1930s for her energy and zeal as a plant collector. Between 1925 and

1938, she made eight expeditions to collect plant specimens, during which time she traveled to the American Southwest, Mexico, and South America. It is estimated that in the 13-year period in which she was active, Mexia collected more than 137,000 plants.

Born on May 24, 1870, in Washington, D.C., Ynes Mexia was the daughter of Enrique Antonio Mexia and Sarah R. Wilmer. Ynes's father was an employee of the Mexican government and probably worked for the Mexican embassy in Washington. Her mother was divorced from a previous husband and had six children from her first marriage. Enrique Mexia separated from Sarah Wilmer when Ynes was three and returned to Mexico. Ynes Mexia married twice, once in 1897, and after the death of her first husband, again in around 1906. Her second marriage ended in divorce. She had no children.

After Mexia's father left the family, she went with her mother to Texas where she lived until 1885, when her mother sent her to be educated at Saint Joseph's Academy in Emmitsburg, Maryland. Mexia was 15 when she attended St. Joseph's. Not much is known about her education before she attended that institution or about how long she stayed there. She may have remained at St. Joseph's for as little as a year before traveling to Mexico to manage her father's hacienda, which she acquired after his death. Her time at boarding school appears to have been the last formal education she would receive until much later in her life.

Mexia remained in Mexico for about 20 years and began a successful business raising poultry and livestock at her hacienda. In 1909, she fell ill and was forced to go to San Francisco to seek medical treatment. During her recuperation, she learned that her second husband, who was 16 years younger than she, had bankrupted her business. The following year, 1910, the Mexican Revolution began, which made travel around Mexico dangerous. A 10-year-long struggle, it would ravage the countryside and leave a million Mexicans dead.

Depressed and recovering from her illness, Mexia decided to remain in San Francisco. Turning 40, she began a job as a social worker. Gradually, between 1915 and 1925, Mexia discovered natural sciences and botany in particular. She began taking trips into the California countryside with the Sierra Club and began to study natural sciences at the University of California at Berkeley.

In 1925, Mexia took her first botanical collecting trip. Accompanied by Stanford botanist Roxanna Stinchfield Ferris, Mexia ventured into the mountains of western Mexico for two months. Her fluent Spanish and knowledge of the country made her indispensable, and she also learned about the business of botanical collection. Mexia returned on her own to western Mexico in 1926 and came back to the United States with approximately 33,000 specimens, 50 of which were new discoveries.

After a collecting trip to Alaska in 1928, Mexia traveled to Brazil in 1929. She traversed the length of the Amazon River and took a hydroplane from the jungle into the Andes Mountains. From there, she came down to Lima and the Pacific Ocean on a train. Before she left the jungle, she shipped home 65,000 specimens.

Over the next nine years, she would travel to Equador and Colombia, the American Southwest, and back to Mexico. At the age of 67, Mexia experienced chest pains on a collecting trip to Mexico. After returning to San Francisco, she discovered that she had lung cancer. She died of that disease in California on July 12, 1938. Although she had won no awards for her work, her death was a loss to her fellow botanists who praised her as a "person who learned much that was new about the vegetation of the territories in which she collected."

Micheli-Tzanakou, Evangelia
(1942–   )
Greek/American
Physicist

A leader in the field of biomedical engineering, Evangelia Micheli-Tzanakou received her early scientific training in Greece before coming to the United States to pursue her career. She has served as a teacher and a researcher, mostly at Rutgers University in New Jersey. Micheli-Tzanakou has been most interested in devising mathematical models to explain the functioning of the brain and in understanding how the human visual system works.

Born in Athens, Greece, on March 22, 1942, Evangelia Micheli-Tzanakou studied in elementary and secondary schools in her home city. After graduating from secondary school in Athens, she enrolled in Athens University in the early 1960s. As an undergraduate, she studied physics. She earned a B.S. in that discipline from the University of Athens in 1968. For four years, Micheli-Tzanakou taught science at the high school level in her native country. Then, in 1972, she embarked for the United States where she had been accepted into the graduate program in physics at Syracuse University. She earned a master's degree in physics from Syracuse in 1974 and her Ph.D. from that same institution in 1977.

After completion of her Ph.D., Micheli-Tzanakou was hired to teach and conduct research in physics at Rutgers University. She soon began specializing in an exciting, new cross-disciplinary subject, biomedical engineering. Combining rigorous conceptual theory and mathematical speculation typical of physics and engineering with the experimental procedures of medical science, biomedical engineers seek to forge new ways of understanding the working of the human body while at the same time devising machines and analytical tools that will help doctors more effectively treat their patients.

For the first four years of her stay at Rutgers, Micheli-Tzanakou worked as a postdoctoral researcher. In 1981, she was promoted to assistant professor and rose to the rank of associate professor in 1985. In 1990, she was promoted to full professor. That same year she also was named dean of the Department of Bioengineering.

The focus of Micheli-Tzanakou's research has been the human brain. Seeking to understand the interplay of the vast system of interrelated neural networks that constitute our brain, Micheli-Tzanakou has devised mathematical optimization models whose goal is to help her and other researchers understand how a normal brain works compared to a brain that is in some way dysfunctional. For instance, she has compared a population of subjects who have Alzheimer's and Parkinson's diseases with people whose brains have remained normal during their aging. Micheli-Tzanakou has also spent a considerable amount of time trying to understand how the human brain sorts out and makes sense of pattern recognition, a crucial component of our ability to see. In another experiment, Micheli-Tzanakou deduced that educated people maintain brain function better than noneducated people, perhaps because their brains are challenged and stimulated with ongoing learning exercises.

Micheli-Tzanakou's work has been recognized by her peers. She has been elected a fellow of the Institute for Electrical and Electronics Engineers (IEEE), and in 1985, she received the IEEE's Outstanding Advisor Award. She was one of the founding fellows of the American Institute for Medical and Biomedical Engineering and is an honorary member of both the British Brain Research Association and the European Brain and Behavior Society. She has been an associate editor and board member of the magazine *IEEE Transactions*. As a member of the advisory group for the Douglass Project for Women in Math, she has also been involved in encouraging women and especially minority women to enter the sciences. In 1992, she was given the Achievement Award by the Society of Women Engineers. As of 2000, she continues to teach and do research at Rutgers.

Miller, Elizabeth Calvert
(1920–1987)
American
Biochemist

A specialist in carcinogenesis, or the biochemical mechanism whose result is the uncontrolled growth of cells commonly known as cancer, Elizabeth Miller contributed many discoveries that helped researchers understand how cancer occurs. As a researcher, she showed how carcinogenic compounds cause chemical reactions at the molecular level,

especially with DNA, to begin the process of cancer growth. As a teacher, Miller trained more than 40 researchers and taught hundreds of graduate students as a professor at the oncology department of the University of Wisconsin Medical School.

Born on May 2, 1920, in St. Paul, Minnesota, Elizabeth Calvert (known since childhood as Betty) was the daughter of William Lane Calvert and Mary Elizabeth Mead Calvert. The middle of three siblings, Elizabeth was encouraged by both parents to study and to attend college. Her father held a doctorate in agricultural economics and worked for the agriculture department of the University of Minnesota and later the Farm Credit Administration. Her mother was also a college graduate and a homemaker. In 1942, Elizabeth Calvert married James A. Miller. She had two children.

After graduating from high school in Anoka, a small town near Minneapolis, in 1937, Elizabeth Miller enrolled in the University of Minnesota. Because of an inspiring teacher she had had in high school, Miller became fascinated with chemistry and studied agricultural biochemistry at the university, earning her B.S. in 1941. That same year, she enrolled as a graduate student in the biochemistry department of the University of Wisconsin. She earned her M.S. in biochemistry from that institution in 1943 and her Ph.D. from there in 1945.

While still a graduate student, Miller began collaborating on research projects with her husband, a practice the two would continue for the next 42 years. Her first projects were studies of the growth of cancers in rats that had been induced by ultraviolet light and carcinogenic chemicals.

In 1945, after Miller finished her Ph.D., she was invited to join the staff of the McArdle Laboratory for Cancer Research (later the Department of Oncology) at the University of Wisconsin medical school. Here she continued her research on rats and discovered that certain kinds of dyes formed covalent bonds with some liver proteins in the study population. Later, Miller and her research partners found that in order to begin the process of uncontrolled cellular growth, carcinogenic compounds had to metabolize with microsomal enzymes, producing an electron-deficient enzyme that would bond with electron-rich macromolecules such as DNA. They further found that carcinogenisis could be activated and deactivated by controlling the amounts of carcinogens and noncarcinogens that were given to the mice. Miller's findings about the chemical reactions of microsomal enzymes allowed other scientists to probe the way in which hormones and drugs are detoxified by the tissue of humans and other mammals.

Along with her research, Miller also taught and served as an administrator. By 1969, she had been promoted to full professor at the oncology department, and from 1973 to

1987, she was the associate director of the oncology laboratory, which by the time of her death employed 200 people.

Miller won many awards for her work in the field of oncology. In 1976, she and her husband shared Brandeis University's Rosenstiel Award for Basic Medical Research. She also shared with her husband the American Cancer Society's National Award in Basic Sciences (1977) and the University of Toronto's Gairdner Foundation Award (1978). In 1978, she was elected to the National Academy of Science. Miller fell ill from cancer in 1986 and died in Madison, Wisconsin, of liver and bone cancer on October 14, 1987.

Mintz, Beatrice
(1921–)
American
Developmental Biologist

A teacher of biology and researcher in mammalian development, Beatrice Mintz has established a number of firsts in her field. She was the first scientist to produce in the laboratory individual mice that were composed of two genetically different sets of cells. These allophenic mice (the adjective, coined by Mintz, refers to creatures that concurrently carry different sets of cells, each with different genetic traits) allowed her to examine the cellular-genetic origins of such mammalian characteristics as blood production, cellular growth, vertebrae growth, and the growth of photoreceptor cells. Her work was used to study the origins of certain kinds of cancers and to understand the processes of normal and abnormal growth in mammals.

Born on January 24, 1921, in Brooklyn, New York, Beatrice Mintz is the daughter of Samuel Mintz and Janie Stein Mintz. She attended primary and secondary public schools in New York City. After graduation from high school in 1937, Mintz enrolled in Hunter College in New York City. She studied biology and earned an A.B. from that college in 1941. In 1942, she entered graduate school at the University of Iowa. She earned an M.S. from there in 1944 and a Ph.D. in 1946. Both graduate degrees were in zoology.

In 1946, Mintz took a job as instructor of biology at the University of Chicago. She became an associate professor in the biology department, teaching and conducting research. In 1960, she moved to Philadelphia where she took a job as a researcher at Fox Chase Cancer Center's Institute for Cancer Research.

At Iowa and at the beginning of her work in Chicago, Mintz studied how sex hormones affected the development of the reproductive systems of amphibians such as the frog *Rana clamitans* and the salamander *Ambrystoma mexicanum*. In one series of experiments, she removed the pituitary glands in a group of very young amphibians to find out what effect this deficit would have on the sex development hormones in these creatures. She discovered that normal sex development occurred, indicating that the pituitary gland played no role in this hormonal interplay.

By the mid-1950s, Mintz had changed her focus from amphibians to mammals, specifically mice. In one experiment, she examined germ cells for a specific genetic mutation on the eighth and ninth days of development of a population of mice embryos. This particular mutation affected the sex organs of the mice. She found no evidence of the mutation on the eighth day, but on the ninth day, the mutation appeared. This proved that this genetic trait did not originate in the sex organs; rather, it appeared in the germ cells and migrated later to the sexual organs of the mice.

During the early to mid 1960s, Mintz worked to perfect a technique to fuse two different mice embryos into a single embryo, a feat that had never been accomplished before. She succeeded in doing this by removing the *zona pellucida,* the membrane enveloping the egg and early embryo, with a solution of pronase, a protein-destroying enzyme, at an exact temperature and for an exact time. This left her with intact, exposed embryos, which at the eight- to 10-cell phase would stick to each other and continue to grow. Needless to say, this work was extremely delicate and time-consuming. Mintz's lab has produced more than 25,000 successful fusions since the first one in 1967.

Mintz's procedure allowed her to produce allophenic mice, which gave her insight into how creatures develop from the early cellular stage and how genes affect this development. One of the more important phenomena she studied was immune tolerance. She discovered that an allophenic mouse had immune tolerance to both the genetically different cells it carried. Normally, an animal will reject tissue transplanted in it from a genetically dissimilar creature. However, the allophenic mouse accepted tissue from either set of its extended genetic family. She hypothesized that the reason for this was that the mouse's lymphoid cells simply did not react to recognized genetic material. This was a rejection of the theory that there was some kind of blocking agent that prevented immunological reaction.

Mintz continued her experiments into the 1970s, '80s, and '90s. She later studied carcinoma cells in embryonic mice to understand the nature of the abnormality of the carcinoma. She also used allophenic mice to study cancer-casing viruses and human growth hormone.

Mintz won numerous awards for her work, including the New York Academy of Science's Award for Biological and Medical Sciences (1979), the Medal of the Genetics Society of America (1981), and the Amory

Prize of the American Academy of Sciences (1988). She was elected to the National Academy of Sciences in 1973 and the Vatican's Pontifical Academy of Sciences in 1986. As of 2000, nearing 80 years of age, she continues to conduct research at Fox Chase's Institute for Cancer Research.

Mitchell, Maria
(1818–1889)
American
Astronomer

The first woman astronomer in the United States, Maria Mitchell conducted astronomical research for the U.S. Coast Survey, a federal government agency. A native of Nantucket, a coastal town in Massachusetts, and from an early age a student of astronomy, Mitchell was ideally situated for the star-charting work requested of her by the survey. Later, after she had become famous owing to the discovery of a comet, she taught and continued compiling astronomical observations at Vassar College.

Born on August 1, 1818, on Nantucket, an island off Cape Cod, Maria Mitchell was the daughter of William Mitchell and Lydia Coleman Mitchell. Maria was the third of 10 children in a devout Quaker family. Her father was a schoolteacher for much of Mitchell's youth; later he worked as a bank clerk. Mitchell's mother was a librarian. Mitchell remained single for all of her life.

Quaker belief emphasizes equality between the sexes, and Maria Mitchell benefited from this. She was given a good primary and secondary education in the local schools. Furthermore, her father encouraged her love of nature and especially her curiosity about the stars. Besides being the teacher at the local school, William Mitchell also was a local amateur astronomer of note. In his house he kept a sextant, a reflecting telescope, and a Dollard telescope for viewing solar eclipses. Maria

Maria Mitchell, at the age of 29, discovered a new comet. She is shown here with her students at Vassar College. *(Special Collections, Vassar College Libraries)*

learned to use all of these instruments as well as set chronometers for ships' captains. Both father and daughter frequently advised seafarers about how to determine their position from the location of stars.

Following the norm for young women in her era, Mitchell left school at the age of 16. However, she did not stop her studies when she finished her formal schooling. On her own time, she continued to study mathematics and astronomy. She was helped in her studies by the job she held at the Nantucket Atheneum, the local library. Beginning in 1936, when she was 18, she worked there during the afternoons and on Saturday evenings. She had the rest of her time to herself. She worked at the Atheneum, and studied the stars in her ample free time, for the next 20 years.

From a small observatory on the roof of the local bank building, Mitchell began helping her father by sharing duties observing the stars almost every clear night. Using a four-inch telescope lent to her by the U.S. Coast Survey, she did calculations that were used by sailors— the altitude of stars, which helped determine time and latitude; and moon culminations and occulations, which aided in determining longitude. She also looked for nebulae and double stars.

In 1847, while performing her usual work, Mitchell observed a comet through the telescope and used her math skills to calculate its orbit. William Mitchell notified the director of the Harvard Observatory of this find. The discovery was verified and passed along to the king of Denmark who had offered a gold medal to the first person to discover a new comet using a telescope.

This discovery, which won Mitchell the gold medal, made her a celebrity overnight. In 1848, she became the first woman elected to the American Academy of Arts and Sciences, and in 1849, she was hired as a full-time, paid astronomer by the U.S. Coast Survey to continue with the work she had previously done for free—computing the positions of stars to aid mariners in navigation.

In 1865, Mitchell was hired to be the first professor of astronomy at the newly founded Vassar College in Poughkeepsie, New York. Vassar was a women's college, and Mitchell was one of its most popular teachers. During her 25-year career, she mentored more than 20 women who later appeared in *Who's Who in America*. She also became active in the movement to include women fully as students and professors in the nation's universities. From 1873 until her retirement in 1888, she was active on the educational committees of the Association for the Advancement of Women.

For her professional achievements, Mitchell was awarded three honorary doctorates during her lifetime. Posthumously, a crater on the moon was named after her, and the Maria Mitchell Society of Nantucket, which runs the Maria Mitchell Observatory, was formed in her honor.

Mitchell died in 1899 in Lynn, Massachusetts, at the age of 71.

Morawetz, Cathleen Synge
(1923–   )
Canadian
Mathematician

A mathematician from a family of mathematicians and playwrights, Cathleen Synge Morawetz has worked as a researcher, teacher, and administrator. After a brief stint at the Massachusetts Institute of Technology (MIT), Morawetz took a job at New York University's (NYU) Courant Institute, where she has stayed for most of her career. She is best known for her work on the applications of partial differential equations to describe patterns such as airflow and the reflection of various waves off different objects.

Born in 1923, Cathleen Synge is one of three daughters of Irish parents who had immigrated to Canada. Her father, John L. Synge, was a mathematician and a teacher of mathematics. Her mother, Elizabeth Allen Synge, had begun to study math in college but had switched to history on the advice of a family member who told her there would be no jobs for a woman in that field. Elizabeth Synge eventually dropped out of college to support her husband while he finished his degree. In 1945, Cathleen Synge married Howard Morawetz. They have four children and five grandchildren.

Morawetz's family encouraged her to study and attend college. Like her mother, she was interested in history, but it was also apparent to her and her teachers that she had a facility for mathematics. During her last years in high school, she was persuaded by her high-school teacher to apply for a scholarship at the University of Toronto, which she won. She entered the University of Toronto in 1940 and, for a while, studied electrical engineering. Finding that subject difficult, she returned to math and earned a B.A. in mathematics from the University of Toronto in 1944.

In 1944, Morawetz traveled to the United States to continue her education. She enrolled in the master's program in mathematics at MIT. Midway through her studies at MIT, she married Howard Morawetz, whom she had met earlier at the University of Toronto. He had in the meanwhile moved to New Jersey where he had taken a job as a chemist. After completing her M.A. at MIT in 1946, Morawetz moved to New York City to begin working on a Ph.D. at NYU's Courant Institute, one of the leading centers of advanced mathematical studies in the United States. Her first job there was unglamorous; she was put to work to solder connections on a form of early computer that solved linear equations.

During World War II, the mathematicians at the Courant Institute had worked on problems related to the war—for instance, how an artillery shell would react when it hit the steel turret of a tank. These issues had presented the mathematicians with interesting but lower-level problems. Nonetheless, the government funding kept the institute running. After the war, Richard Courant, the institute's founder, realized that the institute would still need government military funding to pay for his staff. Now, however, the research was more theoretical and abstract. Cathleen Morawetz benefited from this new approach.

After receiving her Ph.D. from NYU in 1951, Morawetz spent a few years at MIT before returning to the Courant Institute. From the 1950s, she did intensive theoretical work with partial differential equations on scattering theory, that is, determining the way waves such as airflow, electromagnetic waves, and light and sound waves scatter when they hit and are reflected off different types of objects. Morawetz also taught at the institute; from the mid-1950s, she rose from assistant professor to full professor, the title she attained in 1965. By 1984, she had become director of the Courant Institute.

Morawetz served on the Mayor of New York's Commission for Science and Technology and has been president of the American Mathematical Society. She retired from teaching and research in the late 1980s and continues to live in New York City.

Morgan, Agnes Fay
(1884–1968)
American
Biochemist

A leader in research on the effects of diet on the human body, Agnes Fay Morgan was also a well-respected teacher and administrator. Morgan spent most of her career at the University of California at Berkeley where she conducted a number of pioneering studies about how vitamins are processed by the human body. She also studied the nutritional content of the foods produced by farmers in California and examined how various kinds of food-processing techniques affected the nutritional value of foods.

Born on May 4, 1884, in Peoria, Illinois, Agnes Fay was the daughter of Patrick John Fay and Mary Dooley Fay, both immigrants from Galway, Ireland. Her father was a laborer and builder and her mother a homemaker. The family was not wealthy, but Agnes, always a good student, managed to earn a college scholarship from a local philanthropist because of her excellent grades. She was the only one of four siblings in her family to attend college. In 1908, she married Arthur Morgan, a high school teacher who later worked as a sales manager for a California wheat company. They had one son.

In 1900, at the age of 16, Morgan entered Vassar College in New York State to begin her college studies. She soon transferred to the University of Chicago to be nearer her family. At Chicago, she studied organic chemistry with Julius Stieglitz, a well-known chemist. She had originally wanted to be a research chemist and work for a private company. While at Chicago, she came to realize that it would be very difficult to achieve this goal, so she decided to become a researcher and teacher of chemistry at a university.

Morgan finished her undergraduate studies in 1904, earning a B.S. in chemistry from the University of Chicago. She remained at Chicago for her master's and doctoral programs, earning her M.S. in chemistry in 1905 and her Ph.D. in 1914. Between 1905 and 1914, she taught chemistry at various universities, including the University of Washington and the University of Montana.

In 1915, following completion of her Ph.D., Morgan was hired as assistant professor of nutrition at the University of California at Berkeley, beginning her long stay at that university. By 1919, she had been promoted to associate professor, and in 1923, she was made full professor in the Department of Home Economics. Morgan was a skilled and forceful administrator who introduced much-needed changes to her department. She made students take more science courses, including biochemistry, physiology, anatomy, and medicine. She also established a graduate program and directed the research of many graduate students.

Morgan concentrated her own research on nutrition, especially the interplay of vitamins and the human body. She studied the use of sulfur dioxide as a preservative in vitamin pills and found that it was useful in conserving vitamin C but degraded thiamine. She conducted research about the effects of various vitamins on bodily organs and found that vitamin D and calcium exert an effect on the parathyroid glands, which control bone growth. She warned physicians that they should be careful not to give too much vitamin D and calcium to infants. Morgan looked at how vitamin A affected the thyroid and the effect of riboflavin and pantothenic acid on the adrenal glands. Morgan also discovered that vitamin B affects skin and hair pigmentation; a deficiency of vitamin B causes hair to turn gray. Although many of her findings had been suspected, none had been proved scientifically before Morgan studied them.

For her work, Morgan won the American Chemical Society's Garvan Award in 1949 and the American Institute of Nutrition's Borden Award in 1954. She was elected a fellow of the American Institute of Nutrition in 1962, and in 1963, the *San Francisco Examiner* presented her with the Phoebe Apperson Herst Gold Medal. Morgan retired from active teaching and research in 1954. She died of a heart attack in Berkeley on July 20, 1968.

Morgan, Ann Haven
(1882–1966)
American
Zoologist and Ecologist

A trained zoologist, Ann Haven Morgan also helped to promote the ideas of conservation and ecological awareness among ordinary people in the United States. She wrote several books about ecology, including the ecology of wetlands, which was her specialty, and her book *Kinship of Animals and Man* is one of the first books to discuss the impact of human economic and social development on the natural world. Within her field, Morgan was known for her study of aquatic insects and later hibernating animals.

Born on May 6, 1882, in Waterford, Connecticut, Ann Morgan was the daughter of Stanley Griswold Morgan and Julia Douglass Morgan. She was the oldest of three siblings and often spent her summers and free time exploring the woods around her house, an activity that she later turned into scientific research. She never married.

After completing her secondary education and taking some time off to earn money, Morgan entered Wellesley College in Massachusetts in 1902. She was 20, slightly older than most of her classmates. Morgan spent a year at Wellesley and, not liking it, decided to transfer to Cornell University in New York State. At Cornell she majored in zoology and received a B.S. in that subject in 1906. From 1906 until 1909, she taught at Mount Holyoke College in South Hadley, Massachusetts, before returning to Cornell for graduate work in zoology. Back at Cornell, she studied under James G. Needham, a professor of aquatic zoology and specialist in the ecologies of streams and ponds. Morgan was awarded a Ph.D. from Cornell in 1912.

After earning her doctorate, Morgan returned to Mount Holyoke to teach and conduct research. By 1914, she had been promoted to associate professor of zoology. In 1916, she became department head, and in 1918, she was made full professor. Morgan would remain head of the zoology department until her retirement in 1947.

At Cornell, during her doctoral research, Morgan had begun to study freshwater aquatic insects, especially the mayfly, which was the focus of her dissertation. At Mount Holyoke, she continued with the study of wetlands insects and taught courses in this subject. Her 1930 book, *Field Book of Ponds and Streams*, not only became a standard textbook on the subject, but it also became a favorite of people who enjoyed fishing, especially fly-fishing. Morgan gave detailed descriptions of the kinds of insects one would en-

counter on streams of the northeastern United States and the types of creatures, including fish, that ate these insects. The book was an angler's dream.

In 1926, Morgan landed a grant that allowed her to spend a summer in British Guiana on the tropical northern coast of South America where she studied the ecology of insect and aquatic life along streams in that part of the world. By the 1930s, Morgan, who considered herself a general zoologist, had branched out to study hibernating animals. She published a popular textbook on the subject in 1939 called *Field Book of Animals in Winter*. Many summers between 1918 and 1946, Morgan also taught marine zoology at the Woods Hole Marine Biological Laboratory in Woods Hole, Massachusetts.

For her work, Morgan was listed with a star in the fifth edition of *American Men of Science*. She also won research fellowships from the National Academy of Sciences, the Rockefeller Foundation, and the American Association for the Advancement of Science. It is probably for her concern for preserving the natural world that she is best remembered. In *Kinship of Animals and Man*, she wrote, "Humanity is facing two very old problems, living with itself and living with its natural surroundings. Conservation is one way of working out these problems. . . . It is applied Ecology." Ann Morgan died of stomach cancer on June 5, 1966, in South Hadley, Massachusetts, at the age of 84.

Morgan, Lilian Vaughan Sampson
(1870–1952)
American
Biologist and Geneticist

Trained as a biologist but later turning to genetics, Lilian Morgan made major contributions to our understanding of genetic inheritance. Morgan was also committed to her family, and for almost 27 years, she remained out of the laboratory while she either cared for her sister or raised her own children. When she did get back to scientific work, she studied the ways genetic characteristics are passed from one generation of fruit fly to another. Her most important discovery was of the attached-X chromosome.

Born in 1870 in Hallowell, Maine, Lilian Vaughan Sampson was the daughter of George and Isabella Sampson. When Lilian Sampson was three, both of her parents and her younger sister, Grace, died of tuberculosis. Lilian and her older sister, Edith, were sent to the care of their maternal grandparents, Thomas and Elizabeth Merrick, who lived in Germantown, Pennsylvania, not far from Philadelphia. In 1904, when she was 34, Sampson married Thomas Hunt Morgan. They had four children.

Morgan enrolled in Bryn Mawr in 1887 and majored in biology under the tutelage of Edmund B. Wilson. After graduating with honors in 1891, Morgan won a scholarship to continue her education in Europe, where she studied anatomy at the University of Zurich. In 1892, she returned to the United States to continue her studies in the graduate program at Bryn Mawr. She received her M.S. in biology from that institution in 1894. It was during these two years that she met her future husband, T. H. Morgan, who was her graduate faculty adviser.

For 10 years, between 1894 and 1904, Morgan was not involved in science as a student, teacher, or researcher. She probably spent many of these years tending to her sister, Edith, who died in her twenties of tuberculosis. Morgan may also have taken care of her aging grandparents. In 1904, at the age of 34, after she wed T. H. Morgan, Lilian Morgan returned briefly to science. In the summer of that year, she did lab work on planarian regeneration at Stanford University where she had accompanied her husband for a summer teaching post. When they returned to New York City that fall (T. H. Morgan now worked and did research at Columbia University's zoology department), Morgan did not return to the lab. Soon she began having children. She remained away from the laboratory until her children were old enough to tend for themselves. In 1921, she began work in a lab of her own, and on her own projects, at Columbia University.

During her years away from research, Morgan had kept abreast of developments in biology, especially the turn of some biologists toward genetics. Her husband, who would win a Nobel Prize in physiology and medicine in 1933, was one of the leaders in this quest.

In her lab, Morgan began to study genetic inheritance traits through *Drosophila,* the common fruit fly, which was an ideal vehicle because of its short life cycle. Almost immediately, she discovered something important. In one of her breeding batches, she discovered a female that had characteristics of a male. She crossed this female with normal males and found that the X chromosome sex-linked recessive passed from mother to daughter and father to son, exactly the reverse of what should have happened. By chromosomal analysis, she confirmed what she had suspected: that the mother had an attached-X chromosome strain. The attached-X strain of fruit flies proved to be excellent candidates to explore the crossing-over of chromosome segments at meiosis, which helped to confirm theories about how sex is genetically determined. Thus its discovery was important for the evolving science of genetics.

Morgan later discovered the closed-X chromosome in fruit flies, which also became a vehicle to study various speculations about sex crossover frequencies and gynandromorphs, individuals whose bodies display part female and part male characteristics.

Morgan continued her work at the California Institute of Technology (Caltech) in Pasadena when she and her family moved there in 1928. She never attended meetings of scientific societies or presented papers, and she did not have an official position at Columbia or Caltech until after her husband died in 1945. Finally, in 1946, at the age of 76, she became a research associate at Caltech. Lilian Morgan died in Los Angeles in 1952. She was 82 years old.

Moss, Cynthia
(1940–)
American
Zoologist

Schooled in philosophy but trained as a journalist, Cynthia Moss changed careers in her late twenties when she became "completely hooked on elephants." After traveling to Africa on a sightseeing vacation, Moss began studying elephants with British elephant researcher Iain Douglas-Hamilton. She has since become an authority on elephant behavior and has devoted her life to understanding and protecting African elephants.

Born on July 24, 1940, in Ossining, New York, Cynthia Moss is the daughter of Julian and Lilian Moss. Her father was a journalist and newspaper publisher, and she was one of two siblings.

Even though Moss loved to be in the countryside when she was younger, she had only a passing interest in science when she was in high school. After graduation from secondary school in Ossining, Moss enrolled in Smith College in Northampton, Massachusetts. In college, she chose philosophy rather than any of the sciences as a major. She graduated from Smith with a B.A. in philosophy in 1962.

After completing her undergraduate education, Moss began a career as a journalist in 1962. She went to work in New York City for the national news magazine *Newsweek* and worked her way up through the ranks from a researcher and fact checker to a reporter. Moss worked for *Newsweek* for five years—until what she thought was going to be a routine vacation changed her life. A friend of Moss's had spent time in Africa, and after receiving a number of letters from this person, Moss decided to travel to Africa to see the continent for herself. She found herself so overwhelmed and enchanted with Africa, and Kenya in particular, that she quit her job to remain in East Africa.

One of the reasons Moss decided to remain in Africa was her close encounters with the African elephant. During her trip, she had visited the field encampment of Iain Douglas-Hamilton, a zoologist who was conducting research on elephant behavior. Douglas-Hamilton was studying several herds of elephants in Tanzania, a nation adjoining Kenya's southern border. Moss was so taken with the animals, and with the idea of trying to understand them better, that she joined Douglas-Hamilton's study group and worked with him for three years.

After Douglas-Hamilton's grant money ran out, Moss began writing for American magazines from East Africa. She also worked as a freelance research assistant for several other elephant researchers. However, what she most wanted was funding to conduct her own elephant studies. She was finally able to get financial support from a group called the African Wildlife Foundation (AWF), an American-based environmental organization dedicated to the study and preservation of threatened African species.

Beginning in 1972, with money from the AWF, Moss set up the Amboseli Elephant Research Project and began studying elephants that lived near Lake Amboseli in southeastern Kenya. After more than 25 years' study on 900 elephants, Moss has made many interesting discoveries. She has found that elephants have an intricate and multilayered social organization. At the smallest level is the family group, which is headed by the oldest female, the matriarch, and is composed of other females and their calves. When male elephants reach maturity, they join a loose organization of all-male groups. The females and males come together only during mating season.

Moss further found that the several family groups compose a larger social group called the bond group. In turn, a number of bond groups constitute the largest elephant social group called the clan group. Elephants communicate with one another through at least 25 vocal gestures, some of which are in such a low-frequency range that humans cannot hear them. These low-frequency sounds can carry quite far, over many miles, and allow elephants to stay in touch with one another as they wander looking for forage.

Since the late 1980s, Moss has been involved with efforts to save the African elephant species. She devoted much energy to securing the signing in October 1989 of the Convention on International Trade in Endangered Species, which banned the sale of elephant ivory. She also continues to lobby African governments to set up programs that allow humans and elephants to coexist. For her efforts, Moss has received a conservation award from the Friends of the National Zoo in Washington, D.C., and the Medal for Alumnae Achievement from Smith College (1985). She continues to direct the Amboseli Elephant Research Project.

Moufang, Ruth
(1905–1977)
German
Mathematician

Ruth Moufang made several important contributions to the field of projective geometry. By utilizing algebra to study the forms of projective geometry, she founded a new mathematical specialty. She devised what are now named Mo-

ufang planes and Moufang loops in her honor. Barred from a professorial career by restrictive Nazi policies, Moufang was forced to pursue a career in German industry. Ironically, she never published any scientific papers after she returned to academia in 1946, but she did earn the distinction in 1957 of becoming Germany's first female full professor of mathematics.

On January 10, 1905, Moufang was born in Darmstadt, Germany. Her parents, Eduard and Else Moufang, encouraged both of their daughters to pursue higher education. While Moufang's sister Erica became an artist, Moufang began studying mathematics at the University of Frankfurt. She passed her teacher's examination in 1929 and selected projective geometry as her specialty in 1930. (Projective geometry is the branch of geometry that is concerned with the properties and invariants of geometric figures under projection. Projection is the transformation of points and lines in one plane onto another plane by connecting corresponding points on the two planes with parallel lines.) For her doctoral work, she studied under Max Wilhelm Dehn, who in 1907 wrote one of the first systematic expositions of topology.

Moufang's graduate work would be the most significant of her career. In it, she transported the axioms David Hilbert had postulated for the realm of plane geometry into the field of projective geometry. In her algebraic analysis of projective planes, Moufang devised what is now termed the Moufang plane, a projective plane in which every line is a translation line. After earning her doctoral degree in 1931, Moufang won a yearlong fellowship in Rome. She returned to Germany in 1932 and accepted a teaching position at the University of Konigsberg. In 1933, she went back to the University of Frankfurt. During this second stint at Frankfurt, her work explored nonassociative algebraic structures (later christened Moufang loops in her honor). She also published one paper on group theory in 1937, which proved to be highly influential. In it, she applied the study of a particular type of group to number theory, knot theory, and the foundations of geometry.

Although Moufang became only the third woman in German history to earn a certificate to lecture at the university level (called habilitation), by 1937 she was prohibited from teaching in the university system. Hitler's minister of education refused Moufang permission to teach because she was a woman. Excluded from the familiar environment of academia, Moufang instead found work as an industrial mathematician in private industry, accepting a research post at the Krupps Institute in Essen in 1937. While at Krupps, she concentrated primarily on elasticity theory.

After World War II ended, Moufang was invited to return to the University of Frankfurt, where she was granted the right to teach (*venia legendi*) in 1946. She ascended the ranks of academia, making history in 1957 when she became the first woman in Germany to be appointed a full professor of mathematics. Although she supervised doctoral students and taught, Moufang's research dwindled to a halt upon her return to the university.

Despite the fact that she published a total of only 12 papers, Moufang exerted a powerful influence on mathematical theory. She established a new mathematical specialty, which stimulated future research in the application of algebra to projective geometry. The nonassociative loops she studied were later used to examine properties of cubic hypersurfaces in projective space. Moufang died on November 26, 1977.

_N_

Neufeld, Elizabeth Fondal
(1928–   )
American
Molecular Biologist

A specialist in cellular biochemistry, Elizabeth Neufeld established an international reputation as a researcher for her work tracing the causes of genetic diseases, especially diseases in the mucopolysaccharidoses (MPS) group. Her research has enabled doctors to offer couples prenatal testing to determine whether their child will be affected with diseases such as Hurler, Hunter, and Sanfilippo syndromes and Tay-Sachs and I-cell diseases.

Born on September 17, 1928, in Paris, Elizabeth Fondal Neufeld is the daughter of Russians who left their native country after the revolution in 1917. In the early 1940s, they had to move again to escape the Nazis. This time they moved to the United States.

In New York City, where the family settled, Neufeld was encouraged to study. She developed an interest in biology in high school and made this subject her major when she enrolled as an undergraduate at Queens College. After graduating with a B.S. in biology from Queens College in 1948, Neufeld studied biochemistry at the University of California at Berkeley. She earned a Ph.D. from that institution in 1956.

Neufeld's early work was in the cellular biology of marine animals and plants. She investigated processes such as cell division in sea urchins and the growth of polymers in plant cell walls. In 1963, the National Institutes of Health (NIH), a federal government agency, hired her as a research biochemist. At the NIH, she worked for the National Institute of Arthritis, Metabolism, and Digestive Diseases. Here she began studying MPS disorders such as Hurler, Hunter, and Sanfilippo syndromes.

Hurler and Sanfilippo syndromes are relatively rare diseases characterized by the deterioration and death of nerve tissues in the human body. Both are genetically inherited and in 1963 were little understood. Findings by researchers in France had established that a certain kind of enzyme, called lysosomal enzymes, played an important role in cell metabolism. Researchers speculated that Hurler and Sanfilippo syndromes were caused by an overload of sugars in nerve cells, more sugars that the cells could metabolize, thus causing them to go into a kind of shock and eventually die. These syndromes mainly affected prepubescent children and caused mental deterioration, loss of motor skills, and vision and hearing problems. Most children afflicted with this syndrome died, usually before the onset of puberty.

Neufeld tried to understand these disorders through the prism of her earlier work in plant cell growth but failed. Then she and her colleague, Joseph Fratantoni, accidentally stumbled onto a discovery: They mistakenly mixed cells from a Hunter syndrome patient with cells from a patient with Hurler syndrome. These two cells produced cells that were near normal. With more work, they confirmed that the damage to the nerve cells was the result of too much sugar. They further found that the cause of this sugar overload was the absence of certain enzymes that would consume the excess. Finally, Neufeld and Fratantoni isolated a particular defective gene that caused this condition. Aided by this information, other researchers developed tests for the prenatal diagnosis of this condition.

Because of her dedicated work and administrative skills, Neufeld advanced within the NIH. From 1973 to 1979, she headed the NIH's Human Biochemical Genetics section. In 1984, she moved to Los Angeles where she became the first woman to chair the Department of Biological Chemistry at the medical school of the University of California at Los Angeles (UCLA). In 1982, Neufeld won the Albert Lasker Clinical Medicine Research Award, one of the top awards in her field. She was elected to the American Association for the Advancement of Science in 1988, and in

1994, she won the top award granted by the U.S. government for scientific achievement, the National Medal of Science. As of 2000, she continues to teach at UCLA.

Nice, Margaret Morse
(1883–1974)
American
Ornithologist

Educated in foreign languages and child psychological development, Margaret Nice later turned a childhood passion for birds into a lifelong adult scientific mission to study bird behavior. Nice, who achieved international recognition for her work, was best known for her multiseasonal study of the habits and song of the song sparrow. She also compiled the first complete book about birds in the state of Oklahoma and was instrumental in making European ornithological research available to scientists in the United States.

Born on December 6, 1883, in Amherst, Massachusetts, Margaret Morse was the daughter of a college professor and a homemaker. From an early age, Morse was encouraged to enjoy nature and learn about it. She and her six siblings were given a small garden plot to tend, and she frequently went with her mother into the countryside where she especially liked watching for birds. In 1909, Morse married Leonard Blaine Nice, a professor of physiology and pharmacology. They had five children.

In 1901, Nice enrolled in Mount Holyoke College for her undergraduate education. Gifted at languages, Nice decided to study foreign languages after being disappointed with the way zoology was being taught. She was not interested in animal classification and anatomy, and even less with dissecting animals, especially birds. In 1906, she earned a B.A. in French from Mount Holyoke.

Soon after she graduated, Nice heard a lecture given by Clifton Hodge, a biology professor at Clark University in Worchester, Massachusetts. Hodge specialized in the economic impact of animals on humans, and his observational approach to animals made Nice realize that she did not have to dissect birds to study them. In 1907, she enrolled in Clark as a graduate student in biology, working on a project to study the food preferences and food gathering habits of bobwhite quail. Nice had hoped to win a Ph.D. with her study, but in 1909, she married Leonard Nice and dropped out of college to support him while he finished his Ph.D.

For the next 10 years, Clark was largely frustrated in her desire to study birds. Instead, she studied psychology at Clark University and earned an M.S. in that field in 1915. The size of her family continued to grow, and she had little time to conduct research, although she tried her best to carve out some time to continue her bobwhite studies. In the late teens and early 1920s, she did manage to publish about a paper a year, either on psychology or ornithology.

In 1913, Nice and her family moved to Norman, Oklahoma, where her husband had taken a teaching job at the University of Oklahoma. Here, on numerous camping trips with her family, she began to study the birds of that state. Eleven years later, in 1924, this work resulted in the book *The Birds of Oklahoma*, the first complete, scientific study of birds in Oklahoma.

In 1927, Nice and her family moved again, this time to Columbus, Ohio, when her husband took a teaching job at Ohio State University. Here, on a 60-acre patch of land near her house, she began observing and taking extensive field notes about the behavior of the song sparrow. Between 1928 and 1931, she published numerous articles about her observations of the song sparrows' nest-building habits, mating rituals, chick-raising behaviors, and male singing behaviors. By 1931, she had been elected to the American Ornithologists' Union (AOU), only the fifth woman to be selected for membership in that organization.

Nice continued to do ornithological studies into the 1950s and 1960s. She studied duck behavior in Canada and worked to stop pesticide use and the building of dams that would ruin bird habitats. For her pioneering work, she was elected a fellow of the AOU in 1937. In 1942, she was awarded the Brewster Medal, the highest honor of the AOU. She received an honorary doctorate from Mount Holyoke in 1955. Margaret Nice died at the age of 91 on June 26, 1974, in Chicago, Illinois.

Nichols, Roberta J.
(1931–   )
American
Physicist and Engineer

A physicist and engineer, Roberta Nichols is internationally recognized for her work in alternative fuel research. A researcher and executive at Ford Motor Company since 1979, Nichols supervised a team that conducted fuel and engine tests for cars that would run on liquefied petroleum gas (LPG), compressed natural gas (CNG), liquefied natural gas (LNG), and methanol. She was also part of the management team for the United States Advanced Battery Consortium, which was doing advanced battery technology development for all-electric and hybrid electric (battery/engine) vehicles.

Born on November 29, 1931, in Los Angeles, California, Roberta J. Nichols is the daughter of Robert Fulton Hilts and Winifred Vos Hilts. Her father was an engineer at the Douglas Aircraft Company, and before she married, her mother worked at the Douglas aircraft plant where she met Robert Hilts. After her marriage, Winifred Hilts stayed at home as a housewife. With her first husband, William Mc-

Physicist and engineer Roberta Nichols, who is recognized around the world for her research on alternative fuels, shown here with her 1954 300SL Mercedes-Benz Gullwing Coupe. *(Courtesy of Roberta Nichols)*

Donald, an entomologist, Roberta Nichols had two children, Kathleen and Robert. She is now married to Alfred Yakel, a design engineer.

Nichols began working as a mathematician at the Douglas Aircraft Company in 1957, where she was group leader of a team that collected and analyzed data on the Thor missile. In 1958, she moved to Space Technology Laboratory (STL), which later became TRW Space Division. Then in 1960, when the Aerospace Corporation was formed out of STL to provide technical direction for the U.S. Air Force Space Programs, she was transferred to the new company. As part of the technical staff, she was involved in research on various phenomena related to missiles and spacecraft, including small propulsion engines, techniques for wake quenching, wind-tunnel studies of reentry vehicles, and chemical lasers.

While working full time, Nichols continued her education in her spare time. She graduated with a B.S. in physics from the University of California at Los Angeles in 1968. Continuing her education in graduate school at the University of Southern California, she earned an M.S. in environmental engineering in 1975 and a Ph.D. in engineering in 1979.

From 1979 until her retirement in 1995, Nichols worked at the Ford Motor Company where she had responsibility for their alternative-fuel vehicle development. For 10 years, she was principal research engineer in the Ford Scientific Research Laboratory, where she did ethanol engine design and development work for Ford of Brazil, built experimental methanol vehicles for state, county, and local government fleets in California, and produced prototype vehicles for operation on LPG, LNG, and CNG.

In 1985, development of the Flexible Fuel Vehicle (FFV) began, for which Nichols holds three patents. The FFV can operate on alcohol or gasoline or a combination of both, all in one fuel system. This is done with the help of a fuel sensor that tells the engine control system what mixture is in the tank. In 1989, Nichols left the Research Laboratory to head up the Alternative Fuel Department as part of the Environmental and Safety Engineering Staff. From here, the FFV technology was transferred to the Production Engineering Activity, with first production in 1993, where it continues today. In 1991 Nichols joined the electric vehicle team, where she was manager of the Electric Vehicle Strategic Planning Department. Her primary task was to interact with government and serve as a spokesperson for Ford as they began development of battery-powered vehicles.

Nichols has received a number of awards and honors for her work. In 1974, she was awarded the Institute for the Advancement of Engineering's Outstanding Engineer Merit Award. She received the Society of Women Engineers' National Achievement Award in 1988, was named a fellow of the Society of Automotive Engineers in 1990 (she was their first female fellow), and, more recently, was elected to the National Academy of Engineering. As of 2000, she is regent's professor at the College of Engineering of the University of California at Riverside.

Nightingale, Dorothy Virginia
(1902–   )
American
Chemist

A teacher and researcher in chemistry, Dorothy Nightingale devoted her career to teaching and cultivating a new generation of chemists and conducting experiments in organic synthetic chemical reactions. She is especially known for her experiments in chemiluminescence, light emitted during chemical reactions, and her studies in Friedel-Crafts reactions.

Born on February 21, 1902, in Fort Collins, Colorado, Dorothy Nightingale is the daughter of Jennie Beem Nightingale and William David Nightingale. Before her marriage, Jennie Nightingale had been a teacher and later a secretary in Indianapolis. William Nightingale was a rancher. Dorothy Nightingale never married.

Both of Dorothy's parents, but especially her mother, encouraged her to study. Her first introduction to chemistry came when she was nine years old. A group of students living in a boardinghouse managed by Dorothy's mother showed her experiments at the chemistry lab at nearby Colorado State University.

In 1918, Nightingale enrolled as an undergraduate at the University of Missouri. She was drawn into the study of chemistry by Herman Schlundt, one of her professors, who urged her not only to major in that subject but also to continue her studies in graduate school. Nightingale earned an A.B. from the University of Missouri in 1922 and a master's degree in chemistry from that same institution in 1923. Between 1923 and 1929, she was an instructor in chemistry at the University of Missouri. Around 1924, she began work on her Ph.D. in chemistry from the University of Chicago. She finished her Ph.D. in 1928.

The University of Missouri, and the chemistry establishment in general, was slow in rewarding Nightingale for her excellent work. In spite of the fact that she published many papers in the 1920s through the 1960s (by one count, 40 percent of the publications on organic chemistry from the University of Missouri's chemistry department), promotions were slow in coming. She was only promoted to assistant professor in 1939, a full 10 years after she had earned her Ph.D. and 15 years after she had been hired as an instructor.

Nightingale began her career by studying the chemiluminescence of organomagnesium halides. She also worked to explain the process of Lewis acid-catalyzed organic reactions called Friedel-Crafts reactions. Within this group of reactions, Nightingale specialized in alkylation, the introduction of alkyl group chemicals such as paraffinic hydrocarbon groups into aromatic hydrocarbons. This type of chemical reaction had first been studied as early as 1877, but because it included a huge number of variants, there were many questions still to be answered about these reactions at the time when Nightingale began her research. Her efforts helped create a more thorough scientific understanding of this particular and complex chemical phenomenon. The petroleum industry (which uses this process in producing high-octane gasoline) and various other companies that produce solvents, synthetic rubber, and plastics benefited from her research.

While conducting her research, Nightingale also supervised graduate students, a role she enjoyed. In her career, she oversaw the work of 23 doctoral students and 27 master's students. She also was deeply involved in trying to encourage more women to become scientists, helping in this task through her work with the American Association of University Women and the American Chemical Society.

For her work, Nightingale won the American Chemical Society's Garvan Award in 1959. She retired from the University of Missouri as a full professor in 1972 and moved to Boulder, Colorado, where she still lives.

Nightingale, Florence
(1820–1910)
English
Nurse and Statistician

Florence Nightingale is best known as a nurse and a social reformer, though she did significant work as a statistician

as well. Her statistical analyses of the phenomena she witnessed as a nurse, charting the death rates that she attributed to unsanitary conditions, led to hospital reforms.

Nightingale was born on May 12, 1820. Although her mother opposed a mathematical education as unnecessary for a future wife and mother, her father, William Nightingale, educated her in this subject that he loved. She was also tutored in mathematics by James Sylvester, who instructed her in arithmetic, geometry, and algebra. Nightingale went on to tutor children in these subjects, using the Socratic method of questioning to spark her students' interest in mathematics. She wove stories around her mathematical lessons, creating word problems out of the situations familiar to her young students. For example, she directed her students' attention to the distance they walked to school each day and then prodded them to extrapolate the distance to the equator based on their daily walk.

In 1854, while working as an unpaid superintendent of a London facility for sick women, Nightingale received a

Florence Nightingale, who charted death rates attributed to unsanitary conditions, which led to hospital reforms in England in the mid-1800s. *(National Library of Medicine, National Institutes of Health)*

summons from the British secretary of war, Sidney Herbert, recruiting her to aid in the cause of the Crimean War. Along with 38 other nurses, Nightingale traveled to Scutari to administer medical aid to wounded soldiers. However, she not only tended to the sick but also tracked their progress or decline.

Nightingale firmly believed that social phenomena could be measured objectively and analyzed mathematically, leaving the results open to statistical interpretation. She graphed the statistical data she collected on what she called a "polar-area diagram," a kind of pie chart representing different causes of death. From her information gathered in Scutari, she prepared charts from different time periods throughout 1854 and 1855 as a means of illustrating the inverse relationship between the level of sanitation and the death rate: As the former rose, the latter diminished. She published her findings, including these polar-area diagrams, in her 1858 book, *Notes on Matters Affecting the Health, Efficiency and Hospital Administration of the British Army*. Her results left little room for interpretation, as they clearly displayed the efficacy of improved sanitation. In response, Queen Victoria and the Parliament and the British military all agreed to the necessity of reform in the regulation of hospital hygiene.

Nightingale standardized her mode of statistical compilation in the Model Hospital Statistical Form, which was widely adapted to create consistency in statistics. In 1858, the Royal Statistical Society voted her a fellow. During the American Civil War, Nightingale served as a health-care consultant to the United States government. After the war, in 1874, the American Statistical Society inducted Nightingale as an honorary member. Interestingly, we remember Nightingale in her roles befitting women: as a nurse, as a champion of social causes, as a political activist, and as a supporter of women's equality. We do not remember her in the role that played the key to her significance: as a mathematician and as a statistician, roles more commonly associated with men. After Nightingale died on August 13, 1910, she was remembered by one man, Karl Pearson, as the "prophetess" of applied statistics.

Noddack, Ida Tacke
(1896–1979)
German
Chemist

A chemist who worked in the field of nuclear physics and explored the concept of nuclear decay, Ida Noddack was a crucial part of a small team that in the 1920s claimed to have discovered two new elements. She was given credit for only one of these discoveries, element 75. Noddack later suggested that results of an experiment by Enrico Fermi indicated that he had split an atom. This interpretation seemed so outlandish at the time that it was dismissed out of hand. However, in 1939, scientists confirmed that nuclear fission was possible, as Noddack had speculated.

Born in 1896 in Germany, Ida Tacke began her undergraduate university career at the beginning of one of the bleakest periods in modern German history—the start of World War I in 1914. In 1926, she married fellow chemist Walter Noddack.

Noddack received her undergraduate degree from the Technical University in Berlin in 1919. She remained at Technical University to work on her Ph.D. in chemistry, which she completed in 1921. In 1925, she was hired as a research scientist at Physico-Technical Research Agency in Berlin.

At Berlin, Noddack began the first of the experiments to look for as yet undiscovered elements that she believed lay hidden in certain ores, especially molybdenite and columbite. Noddack was part of a three-person team that included Walter Noddack, the director of the Physico-Technical Research Agency's chemical lab, and Otto Berg, an expert in the analysis of X-ray spectra derived from electrons beamed into the nuclei of different materials.

In her first year at the research agency in Berlin, Noddack and her team managed to isolate and identify a new element, which they identified as element 75. They made this determination by firing electrons into the nuclei of the ores they had refined. The collision of the electrons with the nuclei caused X rays to be emitted from the atomic nucleus. Only a few years before, it had been discovered that the atomic number of elements could be figured from wavelengths of the emitted X rays. It took Noddack a further year to extract a gram of this element, which they named *rhenium* (after the Rhine River), from more than 600 kilograms of molybdenite ore.

At the same time that they were working on element 75, Noddack and her coworkers also used the same electron/X-ray technique on a sample of refined columbite ore, which probably contained about 10 percent uranium. By doing this, they believed that they had discovered another elemental material, element 43, which they named *masurium*. However, this claim was shot down by other scientists who theorized that this as yet imaginary element could not exist naturally because it had a half-life of only 210,000 years, much less time than the age of Earth itself. It was assumed that any naturally occurring bits of element 43 had long ago decayed into other elements. Credit for the discovery of element 43 was later given to Emilio Segre and Carlo Perrier who created it artificially in 1937 by smashing atomic subparticles into molybdenum, element 43.

In the late 1930s and into the mid-1940s, Ida Noddack and her husband worked at the University of Freiburg and at the University of Strasbourg in Alsace, a part of France that the Nazi government had annexed into Germany. They do not seem to have spoken out against the treatment of

German Jewish scientists by the Nazi government, preferring instead to focus only on scientific questions. With the defeat of the Nazis in 1945, the Noddacks fled Germany for Turkey, where they lived in exile for 12 years. They finally returned to Germany in 1956. Ida Noddack worked at the Institute for Geochemical Research in Bamberg, Germany, until her retirement in 1968.

Noddack won a number of prizes and established several firsts for her work during her career. In 1925, she was the first woman to address the German Chemical Society. In 1931, she won the Justus Leibig Medal from the German Chemical Society, and in 1934, she won the Scheele Medal from the Swedish Chemical Society. In 1966, she was given an honorary doctorate by the University of Hamburg, and that same year, she won the High Service Cross from the West German government. She died in Bad Neuenahr, Germany, in 1979.

Noether, Emmy
(1882–1935)
German
Mathematician

One of the most influential and creative mathematicians of the 20th century, Emmy Noether ignored the obstacles placed because she was a woman. By persevering in the career she loved, mathematics, she eventually overcame the prejudices that were stacked against her by academic conservatives. She is credited with proving two theorems that validated the most basic principles of Albert Einstein's general theory of relativity and later with developing a type of algebra, known as abstract algebra, that transformed the discipline of mathematics.

Born on March 23, 1882, in Erlangen, Germany, a small town in the German state of Bavaria, Amalie Emmy Noether (who was known to all as Emmy) was the daughter of Max Noether and Ida Amalia Kaufmann Noether. Her father was a well-respected mathematics professor at the University of Erlangen and her mother was from a well-to-do family in Cologne. Emmy Noether never married.

Dr. Noether encouraged all of his children to study and attend college. An aptitude toward science seems to have run in the family. Not only Emmy but two of her three brothers became scientists. At that time, very few German universities admitted women students, so a woman's education usually ended at the secondary level. The highest academic qualification a woman could hope for was as a secondary teacher.

Emmy Noether was encouraged to pursue the traditional female route to an education. She attended a girl's high school and, in 1900, graduated with a degree that qualified her to teach French or English. However, Noether did not want this to be the end of her education. Instead,

Emmy Noether, a mathematician credited with proving two theorems that validated the basic principles in Albert Einstein's theory of relativity. *(Special Collections Department, Bryn Mawr College Library)*

she won permission to audit courses at the University of Erlangen so that she could prepare for the college entrance exams. Between 1900 and 1902, she took these courses, although she received no credit for them.

In 1903, she passed the entrance examination and for a semester studied math at the University of Göttingen, again as an auditor, not a full student, because neither Göttingen nor Erlangen admitted women as full students. When the faculty of the University of Erlangen finally voted to admit women in 1904, Noether transferred there. She completed her studies by earning a Ph.D. in mathematics from the University of Erlangen in 1907.

Noether's Ph.D. dissertation, entitled "On Complete Systems of Invariants for Ternary Biquadratic Forms," was a work that included more than 300 invariants of ternary biquadratic equations, which were computations about three-dimensional forms that shifted and rotated. Noether described this paper as a "jungle of formulas," and it won her a degree summa cum laude, with highest honors.

Between 1908 and 1915, Noether worked as an unpaid teacher at the University of Erlangen while continuing to work on mathematical problems that interested her. At first, she stuck with rather standard mathematical problems, but gradually she pushed herself into a new area by devising problems and theorems "of extreme generality and abstractness of approach" that were her hallmark.

In 1915, Noether was invited back to the University of Göttingen as an unpaid Privatdozent, a lecturer who received fees from students and not the university itself. She jumped at this chance because, at that time, Göttingen was one of the leading centers of math and physics in the world. Here she worked with such leading mathematicians as Felix Klein and David Hilbert. It was at Göttingen that Noether proved the two theorems that validated Einstein's general theory of relativity. She also developed what are now called Noetherian rings, a mathematical structure that consists of a set and two operations, often involving polynomials and hypercomplex numbers. According to many mathematicians, Noetherian rings and other problems worked out by Noether changed the study of structures into an abstract, theoretical pursuit. This transformation of the purpose of mathematics was Noether's most important contribution.

In 1933, Noether was barred by the new Nazi regime from teaching in Germany. Fortunately, she was able to obtain a teaching job at Bryn Mawr in Pennsylvania where she worked until 1935. Noether died unexpectedly on April 14, 1935, in Pennsylvania, of complications following surgery to remove a uterine tumor. She was 57 years old.

Noguchi, Constance Tom
(1948–   )
American
Physicist and Biochemist

Trained as a physicist, Constance Noguchi early in her career fused physics with biochemistry to try to understand the crippling and often fatal disease of sickle-cell anemia. She has spent 25 years on this effort in her career at the U.S. National Institutes of Health (NIH). Her work has yielded new ways to treat this disease. She hopes in the future to better understand the genetic underpinnings of sickle-cell disorder. Such an understanding may eventually yield a possible gene replacement therapy that could offer a cure for this malady.

Born on December 8, 1948, in Kuangchou, China, Constance Tom is the daughter of James Tom and Irene Cheung Tom. Her father was a Chinese-American engineer who went to China to work before World War II. He remained in China during the war and married Irene Che-

ung, a Chinese woman. Constance was the third child in a family of four and the last to be born in China. The Toms were forced to flee China when Mao Zedong's victorious Red Army overran the country in 1949. In 1969, Constance Tom married Phil Noguchi. They have two children.

From China, the Tom family moved to San Francisco, James Tom's hometown. Constance grew up in a home full of books. It was a place where learning was encouraged, and Constance and her sisters learned from their parents that all of the children would be expected to go to college. By high school, Noguchi knew that she had an aptitude for the sciences. She expected to study a curriculum of courses that would prepare her for admission into medical school.

In 1966, Noguchi enrolled as an undergraduate at the University of California at Berkeley. At first, she took premedical courses but soon became disenchanted with this path. She switched her major to physics because she enjoyed independent theoretical work more than some of the rote learning she had encountered in her biology courses. When Noguchi graduated from the University of California in 1970, she enrolled as a graduate student in physics at George Washington University in Washington, D.C. She

Constance Tom Noguchi, a pioneering research scientist whose work has improved the understanding and treatment of sickle-cell disease. *(Courtesy of Constance Tom Noguchi)*

earned a Ph.D. in theoretical physics from George Washington in 1975.

After winning her Ph.D., Noguchi was hired as a researcher at the NIH's National Institute of Diabetes, Digestive, and Kidney Diseases (NIDDKD). She worked for the institute's laboratory of chemical biology.

Noguchi decided to focus her work on the inherited disorder called sickle-cell anemia. A blood disease most often found among people of African descent, sickle-cell disease is caused by a genetic mutation of the hemoglobin, the red blood cells that carry oxygen to cells throughout the body. People with sickle-cell disease have red blood cells that soften when they have given up their oxygen content. These cells then twist around one another and form into a shape that is similar to the shape of the crescent moon or the farm tool called the sickle (hence the disease's name). Normal red blood cells will remain in a full, circular form, in Noguchi's words, "like a jelly-filled doughnut." The damaged sickle-shaped red blood cells often thicken the blood and block small capillaries throughout the body.

Noguchi began her research by developing a way to measure the extent to which sickle cells have formed in a person's blood. This diagnostic tool can tell a doctor how severe the disease is in a particular patient. She has also conducted experiments with hydroxyurea, a drug that increases the amount of a type of hemoglobin called fetal hemoglobin in the blood. Fetal hemoglobin is especially resistant to the deformations that cause sickle-cell disease. Noguchi and her team have conducted experiments with erythropoietin, a drug that makes it easier for the body to produce new red blood cells. They have found that dosages of erythropoietin allow patients to lower the amount of hydroxyurea they have to take, which is beneficial because of the often harmful side effects that are produced by hydroxyurea.

For her dedicated and pioneering work, Noguchi has won the Public Health Special Recognition Award in 1993 and the NIH EEO Recognition Award in 1995. Currently she heads the NIDDKD's molecular cell biology lab.

Novello, Antonia Coello
(1944–   )
American
Physician and Public Health Administrator

Trained as a pediatrician, Antonia Coello Novello has focused her work on treating children with kidney diseases. Later, she studied public health administration and worked as an adviser on health issues for the U.S. Congress. Her concern for children, for AIDS patients, and her active role in speaking out against tobacco companies and alcohol addiction brought her to the notice of President George Bush who appointed her surgeon general. She was the first woman and Hispanic to hold this position.

Born in 1944 in Fajardo, Puerto Rico, Antonia Coello is the daughter of Antonio Coello and Ana Delia Coello. After her father died when she was eight years old, her mother, a schoolteacher, raised her. In 1970, she married Joseph Novello, a U.S. Navy flight surgeon.

As a child and young adult, Novello suffered from a chronic disease of the colon, which required frequent hospitalization. This condition was cured only by surgery in 1962 when she was 18 years old. Her experience with illness gave Novello the empathy to understand physical suffering in patients and inspired her to become a doctor.

After graduation from high school in Puerto Rico, Novello entered the University of Puerto Rico in 1962. Within three years, she earned a B.S. in premedical studies. Following her graduation in 1965, Novello entered the medical school of the University of Puerto Rico. She graduated in 1965 and, with her new husband, Joseph Novello, moved to Ann Arbor, Michigan, to begin her medical residency at the University of Michigan Medical Center.

In Michigan, Novello decided to specialize in pediatric medicine. She began working in the pediatric nephrology unit at the University of Michigan hospital, caring for children with kidney diseases. Her dedication and skill impressed her supervising doctors. During her time there, she won an Intern of the Year award, the first woman to have been given that honor.

In the mid-1970s, Novello and her husband moved to Washington, D.C., where she worked as a pediatrician. By 1980, she had decided that she wanted to move into the public health part of medicine. She enrolled in Johns Hopkins University's public health program and, in 1982, earned a master's degree in that subject. After getting her M.S. in public health, Novello was hired as a medical and public health adviser by the U.S. Congress. She remained at this job for four years, then took a position as deputy director of the National Institute of Child Health and Human Development (NICHHD). At this job, Novello helped to set government health priorities for the nation's children. One of her greatest efforts was toward getting funding to help children with acquired immune deficiency syndrome (AIDS). While working at NICHHD, Novello also held down the position of professor of pediatrics at George Washington University Hospital in Washington, D.C.

By 1989, Novello's work had come to the attention of the staff of President George H. W. Bush. When the post of surgeon general opened, President Bush nominated Novello to fill this position. Congress quickly voted to confirm this appointment.

During her tenure as surgeon general from 1990 to 1993, Novello made smoking, teenage drinking, and AIDS her major priorities. The surgeon general is the head of the U.S. Public Health Service, the government's major public health agency. Although the surgeon general directs an agency of thousands of employees and oversees doctors who work in some of the poorer areas of the country, the main job is raising awareness of health issues through education and public discussion. During her term, Novello persuaded beer and wine companies to stop targeting youth in their advertisements. She also focused awareness of the problem of AIDS transmission from mothers to their babies. As of 2000, Novello works as director of the New York State Department of Health.

Nüsslein-Volhard, Christiane
(1942–)
German
Molecular Biologist and Geneticist

A leader in the study of the molecular processes that occur at the very beginning of life just after an embryo has been formed, Christiane Nüsslein-Volhard has become probably the best-known woman scientist in Germany of her generation. The creature whose embryos she has examined throughout most of her career is *Drosophila,* the common fruit fly. Nüsslein-Volhard's research has uncovered the mysteries of how organisms begin pattern formation of their unique characteristics within hours of the formation of an embryo.

Born on October 20, 1942, in Magdeburg, Germany, Christiane Nüsslein-Volhard is the daughter of Rolf Volhard and Brigitte Haas Volhard. Her father was an architect, and her mother was from a family of artists and musicians. She has four siblings.

By the time she entered middle school, Nüsslein-Volhard knew that she wanted to be a scientist. An excellent student, she was encouraged in her path by her parents. She delivered the graduation speech at her high school—a speculative address about communication among animals.

After graduation from high school in 1960, Nüsslein-Volhard enrolled as an undergraduate at Johann-Wolfgang Goethe University in Frankfurt, a major city in north-central Germany. Here she began studying biology. After finding biology "dull, flat," Nüsslein-Volhard switched her major to physics. She graduated from Goethe University in 1964 and enrolled in the graduate school of the University of Tübingen. She had decided to study in the new field of biochemistry.

As she had in her undergraduate days, Nüsslein-Volhard again found that she had entered a field that did not entirely appeal to her—this particular biochemistry curriculum emphasized organic chemistry over biochemistry. She was fortunate that the Max Planck Institute, which offered graduate studies in molecular biology and virology, was nearby. She managed to conduct much of her studies at Max Planck where she earned a diploma, an equivalent of a

master's degree, in these subjects in 1968. By 1973, she had earned her Ph.D. in molecular biology and genetics at the University of Tübingen (in affiliation with the Max Planck Institute) as well.

During her six years in graduate school, Nüsslein-Volhard had begun to work with *Drosophila* embryos in an effort to understand more about how genetic traits are passed from one generation to another. According to her, *Drosophila* is an ideal subject for genetic research because it contains "a small number of chromosomes, and the existence of giant chromosomes of the salivary glands provided a unique physical measurement for the numbers of genes and the analysis of chromosomal aberrations."

In graduate school and later in her research, Nüsslein-Volhard focused on finding out how fertilization takes place at the molecular level. She has uncovered how this life-form begins its existence and has identified 120 "pattern" genes that control how the organism will develop as an embryo. The implications of her studies reach far beyond the fruit fly. This basic genetic research has opened avenues of research about human genetics as well.

Much of her work was done as a postdoctoral fellow at labs of Walter Gehring in Zurich and at the European Molecular Biology Laboratory. However, Nüsslein-Volhard always longed to return to the Max Planck Institute, and she did so in 1981 as a group leader of a laboratory there. By 1990, she had risen to the position of director of the institute's Department of Genetics.

Nüsslein-Volhard has won many awards for her work, including the Albert Lasker Award in 1991, the Louis Jeantet Prize in 1992, and the Alfred P. Sloan Jr. Prize in 1993. However, her most important honor was the Nobel Prize in medicine she won in 1995. As of 2000, she continued to work at the Max Planck Institute.

Nuttall, Zelia Maria Magdalena
(1857–1933)
American
Archaeologist

Zelia Nuttall was a pioneering archaeologist in the study of the pre-Columbian cultures of Mexico and Central America. Although not professionally trained in archaeology, she made many important contributions to the field in a career that lasted for more than 50 years. She also was a tenacious researcher who tracked down and reinterpreted historical documents relating to the Aztecs and Maya and the voyages of English sea captains Sir Francis Drake and John Hawkins.

Born in 1857 in California, Nuttall was the daughter of an American father and Mexican mother. Her mother piqued her interest in Mexico, and Zelia educated herself about that country's history and culture by voraciously reading everything she could that touched on those subjects. In 1880, when she was 23, Nuttall married French archaeologist Alphonse Pinart who had come to California to study the Native American cultures of the American Pacific Coast. Nuttall and Pinart were divorced in 1888. They had one daughter, Nadine.

Nuttall attended private schools in the United States but never studied at a college or university. During the late 19th century, women were not admitted to many colleges, and most women never studied in institutions of higher learning. Thus Nuttall's lack of university training fell in line with the norm for women at that time. However, from her mother, and later her own readings, she learned a lot about Mexico and its history. From her husband, Alphonse Pinart, she undoubtedly learned of the evolving techniques of archaeological research. This proved enough to launch her into a field that was itself relatively new and still open to talented enthusiasts who were not professionally trained.

In 1884–85, Nuttall made her first trip to Mexico. She spent a good bit of her time in the central valley of Mexico, the area that surrounds the capital of Mexico City. Here she began collecting terra-cotta heads at the ruins of the ancient city of Teotihuacán, which lies about 30 miles outside present-day Mexico City. She wrote her first paper analyzing the significance of these heads and comparing them with other, similar objects. This paper was published in a professional journal in 1886.

Between 1886 and 1902, Nuttall balanced the duties of raising her daughter and traveling to Mexico to conduct further research. In 1902, after her daughter had reached the age of maturity, Nuttall moved permanently to Mexico. She bought a house in Mexico City and began acquiring a large personal collection of antiquities. She also began collecting native Mexican plants, including many that had medicinal properties. In time, the garden would become as important a collection for her as the antiquities.

Nuttall's abilities were recognized early on by Harvard's Peabody Museum, which named her honorary special assistant in 1887. The National Museum of Mexico appointed her honorary professor of archaeology in 1908. She also was an advisor to the University of California's Department of Archaeology. Nuttall held the positions from Harvard and the National Museum of Mexico until her death in 1933.

Nuttall was also able to travel widely in pursuit of her research. One of her most notable achievements was the proper identification and interpretation of a manuscript that dated from the early Spanish conquest period, later known as the Codex Nuttall. Nuttall found this priceless document in a library in Florence, Italy. In Mexico City, Nuttall found documents that had been ignored for several hundred years that pertained to the raids of the English privateers, Francis Drake and John Hawkins, on Spanish convoys and port towns in parts of the Spanish-American empire.

For her devoted work, Nuttall was elected to the American Anthropological Association, the American Association for the Advancement of Science, and the Geographical Society. She was also listed in the first edition of *American Men of Science*. Zelia Nuttall died in Mexico City in 1933 at the age of 76.

-O-

Ochoa, Ellen
(1958–)
American
Physicist, Electrical Engineer, and Astronaut

Trained as a physicist and electrical engineer, Ellen Ochoa has conducted research on optical systems that automate, and thus speed up, information processing. Since joining the National Aeronautics and Space Administration (NASA) in 1991, Ochoa has worked on numerous scientific and technical problems relating to spaceflight and space shuttle missions. She has served as an astronaut on three spaceflights.

Born in Los Angeles, on May 10, 1958, Ochoa moved with her family to La Mesa, a suburb of San Diego, when she was young. In La Mesa, she attended primary and secondary schools. She is married to Coe Fulmer Miles. They have one son.

After graduating from Grossmont High School in La Mesa in 1975, Ochoa attended San Diego State University where she received a B.S. in physics in 1980. Later in 1980, she enrolled in the graduate program of Stanford University. She received an M.S. and Ph.D. in electrical engineering from Stanford in 1981 and 1985, respectively.

During her work for her Ph.D. at Stanford, Ochoa began research on the use of optical devices that could examine and evaluate the readiness of high-tech systems. After earning her doctorate, she continued this work at the Sandia National Laboratory in Albuquerque, New Mexico, and at the NASA Ames Research Center in the late 1980s. During this time, she invented an optical inspection system, an optical object recognition system, and a method for removing noise in images. She is the co-patent-holder on three inventions.

At NASA Ames, Ochoa served as chief of the Intelligent Systems Technology Division. Here she supervised

35 engineers and scientists on the development of computational systems for aerospace missions. In 1990,

Astronaut Ellen Ochoa, wearing her flight suit before boarding the space shuttle *Discovery* on April 6, 1993. *(Courtesy of NASA)*

Ochoa applied for a slot in the highly competitive astronaut-training program. She was chosen for this training in 1991.

After completing her training, Ochoa worked at the NASA Johnson Space Center in Houston on a number of different technical assignments. She checked flight software and served as a crew representative on a committee that oversaw the development of robotic devices for the shuttle. She also supervised the testing of these devices and training of the crew in their use.

By 1993, Ochoa was ready for her first spaceflight. From April 6 to 12, she served as mission specialist on a flight of the shuttle *Discovery*. On this nine-day mission, the shuttle crew studied the effect of solar activity on Earth's climate and environment through a series of atmospheric and solar tests. On this mission, Ochoa used the remote manipulator system to deploy and capture the *Spartan* satellite, whose job it was to study the solar corona.

Ochoa flew on *Discovery* twice more, in 1994 and 1999. On the 1994 flight, she served as payload commander, again sending out and retrieving a satellite from the shuttle. On the 1999 flight, she was the mission specialist and flight engineer. On this flight, she coordinated the delivery of four tons of supplies to the *International Space Station* during an eight-hour space walk.

Ochoa's honors include NASA's Outstanding Leadership Medal (1995) and Exceptional Service Medal (1997). She has also been awarded the Women in Aerospace's Outstanding Achievement Award and the Hispanic Engineer's Albert Baez Award for Outstanding Technical Contribution to Humanity. As of 2000, she continues to work for NASA.

Ogilvie, Ida H.
(1874–1963)
American
Geologist

The founder of the Department of Geology at Barnard College, which for many years was the women's college of Columbia University, Ida Ogilvie was also one of the pioneering women geologists in the United States. She is best known for the research she conducted on petrology, the study of the origins of rocks, and on glacial geology. Ogilvie was also an inspiring teacher who encouraged young women to pursue careers in the sciences.

Born on February 12, 1874, in New York City, Ida Ogilvie was the daughter of Clinton Ogilvie and Helen Slade Ogilvie. The Ogilvies were a wealthy family and expected their daughter to marry well and raise a family of her own. However, after receiving an excellent education at the Brearley School, a private academy in New York

City, Ogilvie surprised her parents by deciding to postpone marriage so that she could study in Europe. She never married.

In 1896, at the age of 22, Ogilvie returned to the United States to enroll in Bryn Mawr, a women's college in Pennsylvania near Philadelphia. At Bryn Mawr, she discovered the discipline of geology and realized she liked studying about Earth and its origins. She apprenticed herself to FLORENCE BASCOM, Bryn Mawr's professor of geology, and earned a degree in geology in 1900. After taking her undergraduate degree, Ogilvie enrolled in the graduate program in geology at the University of Chicago. Here she studied petrology and published her first geological paper in 1902. Unsure if she would be able to get a teaching job in her specialty of petrology, she also studied the geology of areas that had once been underneath the glaciers of the last Ice Age. Ogilvie then transferred to Columbia University where she earned a Ph.D. in geology in 1903.

After finishing her Ph.D., Ogilvie was hired as a teacher at Columbia University's graduate geology program. She also began lecturing about geology at Barnard College and founded the Barnard Geology Department in 1903. By 1905, she had been promoted to tutor, and in 1910, she became an instructor. Ogilvie was appointed an assistant professor in 1912, an associate professor in 1916, and became a full professor at Barnard in 1938. She was chairperson of Barnard's Geology Department from 1903 until her retirement in 1941.

During the early part of her career, Ogilvie ventured out into different parts of North America on geological field trips. In Maine and New York State, she studied rock formations and deposits left by glacial activity. She also traveled to California and Mexico. In New Mexico, she became an expert of the rock formations and petrology of the Ortiz Mountains.

In about 1910, Ogilvie bought a farm in Bedford, New York. Here she became interested in raising Jersey cattle and breeding horses. During World War I, she threw herself into the Women's Land Army and brought a number of young women to her farm to increase its production of vegetables and beef to help the war effort.

Ogilvie devoted most of the last part of her career to teaching and administration. She was known as an inspiring and devoted teacher. She also organized drives to endow scholarships for women in the sciences at Barnard, Columbia University, and Bryn Mawr.

As recognition for her work, Ogilvie was elected to the Geological Society of America, the American Association for the Advancement of Science, and the New York Academy of Sciences. After her retirement, she lived on a farm she had bought in Germantown, New York. She died there at the age of 89 on October 13, 1963.

Osborn, Mary J.
(1927–)
American
Biochemist

A teacher and researcher for most of her career at the University of Connecticut Health Center School of Medicine, Mary Osborne is best known for her study of the structure of lipopolysaccharide, a molecule found on the surface of certain pathogenic bacteria. She has also conducted research to improve various chemotherapeutic drugs by understanding how they work in the body at the molecular level.

Born on September 24, 1927, in Colorado Springs, Colorado, Mary Jane Merten is the daughter of Arthur and Vivian Merten. Her father worked as a machinist, and her mother taught school. When she was young, the family moved to Los Angeles, California, and Merten grew up in West Los Angeles and Beverly Hills. She is married to Ralph Osborn, a painter, and has no children.

Osborn's parents encouraged her to study and excel in school. Once, when she became interested in nursing as a profession, they asked her why she would not instead consider becoming a doctor. After graduating from a public high school in Los Angeles in 1944, Osborne enrolled in the University of California at Berkeley. As an undergraduate, she began studying premed courses in preparation for medical school. Soon, however, she realized that she did not have the aptitude to consult with patients. She then switched her major to physiology and graduated from UC Berkeley with a B.S. in that subject in 1948.

After graduation, Osborn enrolled in graduate school at the University of Washington. Realizing that "she liked bench research and could do it well," she focused on getting a doctorate in biochemistry. During her graduate years, she studied the workings of vitamins and enzymes that interacted with folic acid to function within the body. Her Ph.D. dissertation centered on methotrexate, a substance that cancels out the action of folic acid in certain enzymes, and which she also found to be a cancer-fighting compound. Osborn was awarded a Ph.D. for this work from the University of Washington in 1958.

After graduation, Osborn moved to Connecticut to begin work as a researcher and teacher at the School of Medicine of the University of Connecticut. Here she switched her research focus to a member of the molecule complex called polysaccharide. The particular molecular structure within the polysaccharide family that she was interested in was called lipopolysaccharide. This molecular cluster appears only on certain kinds of bacteria, including the pathogenic bacteria salmonella, shigella, and cholera bacillus. Lipopolysaccharide is found on the surface of these bacteria and is the agent responsible for their immunological signature and the toxic reactions that these bacteria cause within the body. Osborne labored to understand the structure of this polysaccharide and how it is constructed, that is, its biosynthesis. She uncovered much new information about how this entity was formed. This work has resulted in new antibiotics that can more effectively eliminate these forms of bacteria.

By 1968, Osborn had become a full professor at the University of Connecticut's medical school. She has also been editor of several prestigious journals in her field, including *Biochemistry* and the *Journal of Biochemistry*. As a result of her research, Osborn has been elected to the American Academy of Arts and Sciences (1977) and the National Academy of Sciences (1978). She continues to teach and do research at the University of Connecticut School of Medicine.

P

Panajiotatou, Angeliki
(1875–1954)
Greek
Physician

The first woman to graduate from the medical school of the University of Athens, Angeliki Panajiotatou was so far ahead of her time that for many years she found it impossible to teach or work as a doctor in her native country. Instead, she immigrated to Egypt, which had a sizeable Greek population, and began a successful career there as a teacher and doctor.

Born in Greece in 1875, Angeliki Panajiotatou must have come from a family that encouraged the education of women. Both she and her sister attended the University of Athens at a time when it was extremely rare for women to pursue higher education in that country.

In 1896, at the age of 21, Panajiotatou and her sister enrolled in the University of Athens. Both studied there until about 1900 when they graduated with degrees in medicine. Panajiotatou then went to Germany to continue her medical education. By 1905, she had finished with her advanced education in Germany and had returned to Greece where she was hired as a lecturer on the faculty of her alma mater.

She found a rude reception among the almost exclusively male students at the University of Athens medical school. Many of the male students found her presence among them offensive. They protested having her as a teacher, shouted "Back to the kitchen" at her in classes and eventually boycotted her classes, leaving her with no students. Apparently, the medical school faculty and the university administration did little to support her. Within the year, she was forced to resign.

Undeterred, Panajiotatou immigrated to Alexandria, Egypt, which had a large Greek population. At that time, the British ruled Egypt in a colonial arrangement that fell short of direct colonial rule. Nonetheless, the British kept an army in Egypt and exercised considerable influence over the local governments. Panajiotatou probably received as welcome an invitation as she did because of the British presence.

For a number of years, Panajiotatou stayed in Alexandria where she worked as a doctor in the municipal administration. In around 1908, she passed an examination that enabled her to work also as a quarantine officer. She joined the Quarantine Service of Egypt and remained with that organization for 30 years. A few years before World War I, Panajiotatou was invited to take a job as a professor at Cairo University. She had already developed an interest in infectious tropical diseases, the sort of things she would have seen on her daily rounds as a municipal doctor and quarantine officer. At Cairo University, she lectured and conducted research on the epidemiology of diseases such as cholera and typhus.

In 1918, at the end of World War I, Panajiotatou decided to return to Alexandria. She was offered a job as the chief of the laboratory of the Greek Hospital there. In Alexandria, she continued to study treatments for and schemes to reduce the outbreaks of infectious diseases. She became a well-known and beloved figure in the Greek community in Alexandria. During her stay in that city, she somehow also found time to write a book about how the ancient Greeks handled public health problems in order to prevent disease.

By 1938, Greece was at last ready for a female professor of medicine. That year, Panajiotatou was invited back to teach at the University of Athens. This time, the students received her with enthusiasm. Angeliki Panajiotatou died in Athens in 1954 at the age of 79.

Pardue, Mary Lou
(1933–   )
American
Biologist and Geneticist

Trained as a biologist, Mary Lou Pardue has devoted most of her career to studying genetic structures in insects.

Because much of her work involves insects whose genetic makeup is similar to higher-level creatures such as humans, her research has the potential to aid in the fight against human diseases such as cancer. Although primarily a researcher, Pardue has also taught biology and genetics at the Massachusetts Institute of Technology (MIT) since 1972.

Born on September 15, 1933, in Lexington, Kentucky, Mary Lou Pardue is the daughter of Louis Arthur Pardue and Mary Allie Marshall. Her father, who was a professor of physics at the University of Kentucky and Virginia Polytechnic Institute, encouraged her to study sciences.

After graduating from high school in Virginia in 1951, Pardue enrolled as an undergraduate in the College of William and Mary in Williamsburg, Virginia. She studied biology and received a B.S. in that subject from William

Mary Lou Pardue, a biologist whose career has focused primarily on studying genetic structures in insects. *(Courtesy of Mary Lou Pardue)*

and Mary in 1955. Two years later, in 1957, Pardue enrolled in the graduate program of the University of Tennessee in Knoxville to study radiation biology. She earned her M.S. in 1959. After taking six years off from her studies, Pardue returned to graduate school, this time enrolling in the doctoral biology program at Yale in 1965. At Yale, she studied under Joseph Gall and conducted important work that resulted in new techniques to extract deoxyribonucleic acid (DNA), the basic genetic building blocks for all living organisms.

At Yale, and later at MIT, Pardue decided to focus her research on the nature of chromosomes in what are called eukaryotic organisms. These are organisms whose DNA is found in the nucleus of their cells, as opposed to prokaryotic organisms such as bacteria and viruses, whose DNA is found in the cytoplasm.

Pardue's organism of choice for her biogenetic research was the subspecies of fruit fly called *Drosophila melanogaster.* She chose the fruit fly because it has a short life cycle and many generations can be bred quickly, which is convenient for a researcher into genetic inheritance and traits. Furthermore, *Drosophila*'s genetic processes are similar to higher-level creatures, and therefore lessons learned from them can be extrapolated to mammals.

In order to study the structure and function of DNA and ribonucleic acid (RNA) of the fruit fly, that is, to discover how genes are paired and sequenced, Pardue first had to devise a technique of extracting and examining the genetic material of DNA without causing damage to this minute and sensitive substance. She and her mentor, Joseph Gall, were able to flatten chromosomes, put them on a microscope slide, then stunt their biological activity by adding an alkaline solution to the mix. The alkaline solution loosens the hydrogen bonds holding the DNA and chromosomes together. Then radioactive RNA is put into the mix. The mix is heated to facilitate incubation and bonding of the RNA to the DNA. After hybridization, stray bits of RNA are removed, and the RNA-DNA hybrids are photographed through a device that picks up the radioactivity of the RNA strands.

Later work by Pardue focused on how variations in heat affected genetic activity of fruit flies. She found that changes in environmental temperature of as little as 10 degrees could shut down genetic processes.

In recognition of her work, Pardue has been awarded the Ester Langer Award for Cancer Research in 1977, the Yale Graduate School's Lucius Wilber Cross Medal in 1989, and was chosen to deliver the Katherine D. McCormick Distinguished Lecture at Stanford University in 1997. She was also elected a member of the National Academy of Sciences in 1983 and a fellow of the American Academy of Arts and Sciences in 1985. She continues to teach and conduct research at MIT.

Patrick, Ruth

(1907–)
American
Limnologist

A student of streams, rivers, and wetlands at a time when that sort of investigation was unfashionable in the scientific community, Ruth Patrick has made significant contributions to humankind's understanding of the elements that constitute a healthy river ecosystem. She has documented the effects of water pollution on wetlands and has also studied groundwater systems.

Born on November 26, 1907, in Topeka, Kansas, Ruth Patrick is the daughter of Frank Patrick, a lawyer and banker, and Myrtle Moriah Jetmore Patrick. Myrtle Patrick was college educated but chose to remain at home to raise her family. In 1931, Ruth Patrick married Charles Hodge IV. Acting against custom at that time, Patrick decided to keep her maiden surname. She had one child, Charles, who is a physician.

Frank Patrick, Ruth's father, was a frustrated botanist who had become a banker at his father's insistence. As a result, he often went on outings with his family on Sundays to collect plants, especially diatoms, a multispecies type of one-celled algae found in fresh and saltwater. Frank Patrick encouraged Ruth to study nature. She had her first microscope at age seven and soon began her own collection of specimens.

After graduation from public school in Topeka, Patrick attended the University of Kansas in Lawrence to begin studying biology. Patrick's mother did not want her attending the University of Kansas because the family could not account for the men Patrick might be dating. Therefore, Patrick enrolled in her second undergraduate year in Coker College, a women's institution in South Carolina. She earned a B.S. in botany from Coker in 1929, then enrolled at the University of Virginia where she worked with Ivey F. Lewis, the foremost expert on algae in the United States. She earned her M.S. and Ph.D. degrees from the University of Virginia in 1931 and 1934, respectively.

In 1933, Patrick moved to Philadelphia with her husband when he took a job as an entomologist at Temple University. That year, Patrick began a more than 60-year association with Philadelphia's Academy of Natural Sciences (ANS). She began her career there as an unpaid researcher in 1933. By 1937, she had become a volunteer curator in the Microscopy Department, and at roughly the same time, she began teaching botany at the Pennsylvania School of Horticulture. She was appointed to her first full-time, paid position in 1945, and in 1947, she founded the Limnology Department, of which she was the chairperson for 25 years. From 1950 to 1970, she was an adjunct professor lecturing on botany at the University of Pennsylvania.

Patrick began her career by studying diatoms, a type of algae that has a silica shell and forms a part of phytoplankton, a major component in the marine and freshwater food chain. She began by studying diatoms in river systems in the eastern United States. In the mid-1940s, with the help of the Atlantic Refining Company, Patrick began a study to determine which kinds of diatoms grew in the various streams and rivers of the Conestoga basin, a riverine system in Pennsylvania.

Knowing that different diatoms need different nutrients, she reasoned that diatoms would tell a lot about what compounds were found in the water at different spots. However, she not only measured the occurrence of diatoms but also put together a survey of as many variables as she could to measure the river system's health.

Patrick found that a healthy river system is indicated by a variety of organisms with none dominating over others. She also found that heavy pollution killed off the most sensitive species, opening the way for the dominance of certain species that are not as affected by the pollutants. Her conclusion was that a healthy wetlands ecosystem needed biodiversity—many different kinds of organisms living in the water and along a river's banks.

Patrick has continued her studies of diatoms, water pollution, groundwater, and wetland ecosystems into the 1990s. She has published more than 175 papers and articles and is the coauthor of the definitive text on diatoms, *Diatoms of the United States*. In recognition of her work, she has been elected to the National Academy of Sciences (1970) and the American Academy of Arts and Sciences (1976). She was awarded the Ecological Society of America's Eminent Ecologist Award (1972), the Philadelphia Award (1973), the John and Alice Tyler Ecology Award (1975), and the National Medal of Science (1996). Patrick currently lives in Philadelphia and is curator of the ANS's Limnology Department collection.

Patterson, Francine

(1947–)
American
Animal Psychologist and Endangered Species Activist

A developmental psychologist by training, Francine Patterson has devoted her professional career to studying the behavior of gorillas. She is deeply involved in a multiyear study whose goal is the evaluation of gorillas' ability to communicate thinking and experiences and express emotion. She has written a number of professional papers and popular books about her work with one gorilla in particular, an animal named Koko. In addition, Patterson is deeply involved in efforts to save gorilla habitats in Africa and establish a gorilla refuge on the island of Maui in Hawaii.

Born on February 13, 1947, in Chicago, Illinois, Francine Patterson is the daughter of Dr. Cecil H. Patterson and Frances Spano Patterson. Her father was a psychology professor and her mother was a nutritionist and homemaker.

After graduating from University High School in Urbana, Illinois, in 1965, Patterson enrolled in the University of Illinois at Urbana. As an undergraduate, she studied psychology, earning her B.A. in that discipline in 1970. Following her graduation, Patterson enrolled in the graduate school of Stanford University. There in 1972, she began working to teach nonverbal communication to a gorilla named Koko. Her experiments had several purposes: to establish whether gorillas had the intelligence to learn nonverbal sign language, and if so, to evaluate how large a vocabulary they could acquire.

Patterson's work with Koko and Ndume, another gorilla who eventually joined her group, was the main focus of her doctoral work. After earning her Ph.D. from Stanford in 1979, she received permission to care for Koko and another gorilla named Michael and continues her work with them.

In 1976, Patterson established the Gorilla Foundation with Dr. Ronald Cohn and Barbara Hiller. She is now president and research director of the organization. This nonprofit foundation is dedicated to the preservation of gorillas in the wild, which are endangered. The foundation also works to protect other endangered species and to continue research into gorilla intelligence, behavior, and psychobiology. The foundation hosts a website (www.koko.org), publishes books and scientific papers, films and videotapes, and the journal *Gorilla*. It is also in the process of establishing the Allan G. Sanford Gorilla Preserve in Maui, Hawaii, which is intended to be a place to house and study endangered gorillas.

During her 25 years with Koko and the other gorillas of her study group, Patterson has learned that gorillas are capable of acquiring a fairly extensive vocabulary. She has studied gorilla gestures and interpreted these gestures as a kind of language that gorillas have among themselves. She has also been able to teach American Sign Language (ASL) to several of the gorillas—1,000 words of ASL in Koko's case. Koko can converse in ASL and asks questions in ASL as well. Furthermore, Koko can conduct a "bilingual" conversation by responding in sign to questions asked in English. She has learned to read some printed words and has scored between 85 and 95 on the Stanford-Binet Intelligence Test.

Also during her experiments, Patterson has got Koko to paint and draw representational figures. The most interesting part of Patterson's work has explored Koko's emotional landscape and sense of time. She has learned to talk in ASL about her feelings and uses words such as *happy, sad, afraid, enjoy, mad,* and *love* to express feelings, either in answer to questions from researchers or unprompted on her own. Koko has also learned how to correctly and appropriately use words such as *before, after, later,* and *yesterday* to express her awareness of time.

Patterson is a member of the board of consultants of the Center for Cross-Cultural Communication in Washington, D.C., and editor of *Gorilla.* She received the Rolex Award for Enterprise in 1978, the Award for Outstanding Professional Service from the Preservation of the Animal World Society (PAWS) in 1986, the Kilby International Award in 1997, and was the 14th Annual Genesis Awards Guest of Honor.

Payne-Gaposchkin, Cecilia Helena
(1900–1979)
British/American
Astronomer

An inventive and accomplished astronomer, Cecilia Payne-Gaposchkin began her career as a physicist but became fascinated with stellar objects instead. While still in university, she switched her path to astronomy, which was not then a separate field of study from physics and mathematics. During a career that lasted for more than 50 years, Payne-Gaposchkin studied stellar magnitudes and chemical composition, the behavior of variable stars, and stellar evolution.

Born in Wendover, England, on May 10, 1900, Cecilia Payne was the daughter of Edward John Payne and Emma Leonora Helena Pertz Payne. Her father was a lawyer and also a writer whose specialty was American history. Her mother was a painter. She was one of three siblings. In 1934, she married Russian astronomer Sergei Gaposchkin. They had three children.

Payne-Gaposchkin's parents strongly supported education for all of their children, which was unusual at that time. As a result, Cecilia received a very good primary and secondary education. She attended St. Paul's Girls' School in London, which is where she was first able to study mathematics and science in a thorough way.

After graduating from St. Paul's in 1919, Payne-Gaposchkin enrolled in Newnham College at Cambridge University. At Cambridge, she gravitated to physics, which displaced her earlier interest in botany. However, her first year at Cambridge, she heard a lecture by astronomer Arthur Eddington, who described his expedition to Brazil to observe an eclipse. This encounter made her curious about astronomy and resulted in her eventual concentration on it as her area of study. Payne-Gaposchkin graduated from Cambridge in 1923 with a B.A. in astronomy. At her graduation, she had already published a paper in one of the astronomical journals and was a member of the Royal Astronomical Society.

After graduation, Payne-Gaposchkin applied for and won a National Research Fellowship to attend Radcliffe College, Harvard's women's college, to work on a Ph.D. in astronomy under Harlow Shapley, the director of the Harvard Observatory. At Harvard, she began studying the voluminous collection of stellar spectra photographs that had been made at the observatory since the 1890s. She compared the spectra of these photos with atoms whose properties could be observed in laboratories on Earth and was able to establish a scale to measure stellar temperatures. She was also able to figure out a way to calculate the kind and amount of elements that stars were composed of. This work was put forth in her Ph.D. dissertation and later in a book entitled *Stellar Atmospheres: A Contribution to the Observational Study of Matter at High Temperatures*. She earned her Ph.D. from Harvard in 1925.

After graduation, Payne-Gaposchkin stayed at Harvard where she taught graduate and undergraduate courses, compiled observational data for the observatory, and edited observatory publications. She became a U.S. citizen in 1931. Promotions were slow in coming. For her first 13 years, she worked as a poorly paid researcher and assistant professor. In 1938, she finally won a promotion when she was appointed Phillips Astronomer at the observatory. However, she was not able to become a full professor at Harvard until 1956, at which time she had been at that institution for 31 years.

In the 1930s, after marrying Sergei Gaposchkin, who had moved to Harvard to work, she concentrated her research on variable stars, especially pulsating Cepheid variables and exploding stars called novae. She and her husband continued this work into the 1940s and early 1950s, supervising a team of astronomers who compiled what came to be standard reference charts on variable stars. This work gathered 1.2 million observations on variable stars.

In recognition of her work, Payne-Gaposchkin was awarded honorary degrees from Cambridge University in 1950 and Smith College in 1951. She won the American Astronomical Society's highest award, the Henry Russell Lectureship, in 1976. She died of lung cancer on December 7, 1979, in Cambridge, Massachusetts.

Pearce, Louise
(1885–1959)
American
Pathologist

Louise Pearce distinguished herself both professionally and personally. As one of the early women physicians in America, she developed the drug tryparsamide, which she tested in the Belgian Congo in 1920 as a cure for African sleeping sickness. In her personal life, she challenged convention by

living in a lesbian household with SARA JOSEPHINE BAKER, a physician whose establishment of the Division of Child Hygiene in 1908 decreased the infant mortality rate dramatically in New York City, and the writer Ida Alexa Ross (I. A. R.) Wylie.

Pearce was born in 1885. She attended Stanford University for her undergraduate study, earning her A.B. in 1907. She spent the next year at the Boston University School of Medicine before transferring to Johns Hopkins University School of Medicine, where she earned her M.D. in 1912. Pearce remained at Johns Hopkins Hospital for her internship that same year. Pearce then proceeded to the Rockefeller Institute for Medical Research, where she remained for the rest of her career. For her first decade there, from 1913 through 1923, she was a fellow, and from 1923 through 1951, she served as an associate member.

At Rockefeller Institute, Pearce conducted research on the biology of infectious and inherited diseases. She studied a strain of syphilis in rabbits that was similar to human syphilis. She published a series of articles in the *Journal of Experimental Medicine* tracking the reciprocal effects of concomitant infections. Interestingly, when she combined syphilitic inoculations on rabbits that she had also performed vaccinal inoculations on, the resulting incidence of syphilis proved quite severe. Pearce surmised that the combined inoculations did not increase susceptibility but rather decreased resistance.

Pearce continued to conduct experiments on rabbits, breeding a colony of rabbits over the years that carried more than two dozen strains of hereditary diseases and deformities for her team to research. One particular strain that Pearce focused her attention on was a virus her team isolated that closely resembled human smallpox; this pox almost destroyed her rabbit colony. As well, she and her team identified a rabbit tumor, later known as the Brown-Pearce tumor, which could be transplanted. This discovery proved a boon to cancer research worldwide, as it allowed for the exporting of the tumor from one subject to another to study the spread of cancer.

Pearce gained most of her renown for her work on trypanosomiasis, or African sleeping sickness, a disease spread through infectious organisms carried by the tsetse fly. Pearce identified a drug that combated this disease, according to preliminary studies. In an effort to confirm these studies, Pearce traveled to the Belgian Congo in 1920 to administer the drug on humans at native hospitals. She achieved an 80 percent cure rate for the 77 patients she treated, confirming her discovery as a success.

Pearce gained renown in Princeton, New Jersey, for cohabiting with Baker and Wylie in Trevenna Farm, or the House of the Ladies, an old home they refurbished by installing plumbing, heating, and electricity. Local residents referred to the threesome as "The Girls," a term that carried derogatory connotations on the lips of some and reverential connotations on the lips of others, depending on their attitude toward this group of women living an openly lesbian lifestyle.

Pearce contributed to the advancement of women in the sciences and medicine by serving on the board of the Woman's Medical College of Philadelphia from 1941 through 1946 and then as president from 1946 through 1951. The Belgian government honored her work in its colony, twice decorating her for discovering a cure for African sleeping sickness. Pearce died in 1959. Her contributions to scientific understanding might have been even greater if her files documenting her research had not been destroyed after her death.

Peden, Irene Carswell
(1925–   )
American
Electrical Engineer

Irene Peden conducted groundbreaking research on radio wave propagation through the polar ice pack in Antarctica, where she was the first American woman to live and work on the interior of the continent. She conducted many never-before-attempted experiments, requiring her not only to develop the experimental methods but also to invent mathematical formulas for interpreting the resulting data.

Peden was born on September 25, 1925, in Topeka, Kansas. Her mother was a country schoolteacher, and her father, J. H. Carswell, worked in the automobile business. A required high school chemistry course sparked her interest in science. She later focused her interest on electrical engineering, which she majored in at the University of Colorado for her 1947 bachelor of science degree.

Peden worked for two years as a junior engineer at the Delaware Power and Light Company, from 1947 through 1949. She then obtained the same position at the Aircraft Radio Systems Laboratory at the Stanford Research Institute, which promoted her to the rank of research engineer in 1950; in 1954, she joined its antenna research group. Concurrently, she worked as a research engineer for the Midwest Research Institute from 1953 to 1954.

Peden pursued graduate study at Stanford while working there, receiving her master's degree in 1958. She continued to work there as a research assistant at the Hansen Laboratory and then as an acting instructor in electrical engineering. She also continued to conduct graduate research, writing her doctoral dissertation on measurement techniques for microwave circuits to become the first woman to earn her Ph.D. in electrical engineering from Stanford, in 1962. That year she married Leo J. Peden and mothered his two daughters, Jefri and Jennifer.

The year before, she accepted an assistant professorship at the University of Washington (UW), where she became the sole woman faculty member. She rose through the ranks there, becoming an associate professor in 1964 and a full professor in 1971. From 1973 through 1977, she acted as the associate dean of engineering and as the associate chair of the electrical engineering department from 1983 through 1986. Despite this distinguished career at UW, she continued to receive less pay than her male counterparts, so in the 1990s, she and her electrical engineering colleague Nancy Nihan filed a sexual discrimination suit against the university, seeking compensation for years of underpayment.

In 1970, Peden received a grant from the National Science Foundation (NSF) to study radio waves in Antarctica. The U.S. Navy, which transported researchers to the continent, refused to bring a woman into the harsh arctic conditions, though it would bring her male graduate students there, and women from other countries routinely studied in Antarctica. Finally, the NSF convinced the navy, and Peden became the first American woman to live and work in Antarctica. Not surprisingly, she not only survived but she also conducted pioneering research, measuring both very low frequency and very high frequency radio wave propagation as a means of investigating polar ice packs and the lower ionosphere above the Antarctic.

Numerous organizations recognized the significance of Peden's work by honoring her: She received the 1973 Society of Women Engineers Achievement Award, the 1987 Outstanding Civilian Service Medal from the U.S. Army, and Centennial Medals from the Institute of Electrical and Electronics Engineers in 1984 and from the University of Colorado in 1988. She served as the director of the Division of Electrical and Communications Systems at the NSF for two and a half years. In 1993, the prestigious National Academy of Engineering elected Peden as a member.

Pellier, Laurence Delisle
(19??–   )
French/American
Metallurgist

Laurence Delisle Pellier, a metallurgist, patented a process in 1956 for gold-plating surgical needles as a means of resisting corrosion. This innovation, together with her other significant contributions to the field of metallurgy, earned Pellier the 1962 Annual Achievement Award from the Society of Women Engineers (SWE). After working for two decades in the industry, Pellier founded her own metallurgical consulting firm in Westport, Connecticut, in 1967.

Pellier was born in Paris, France, but she immigrated to the United States while she was still young. There, she studied chemical engineering at the City College of New York. She graduated in 1939 cum laude with a bachelor of science degree. She then proceeded to the Stevens Institute of Technology to conduct graduate study on metallurgy. When she received her master of science degree in 1942, she was the first woman to graduate from Stevens Institute. She then commenced doctoral research in physical metallurgy at Columbia University, but she never completed her Ph.D.

While studying for her master's degree, Pellier worked first as a research associate for the International Nickel Company and then as a research fellow with the General Bronze Company, where she conducted research on alloys using powder metallurgy until 1946. That year, Sylvania Electric Products hired her as a senior metallurgical engineer, a position she retained until 1951. At Sylvania, Pellier applied electron microscopy to the study of physical metallurgy. In 1951, she moved to the American Cyanamid Company, where she studied corrosion, electron metallography, and electroplating (as well as electroless plating.) It was during her tenure at American Cyanamid that she obtained her gold-plating patent. She also received one other patent in her career.

In 1956, the same year she received her gold-plating patent, Pellier joined the Sigmund Cohn Corporation as a research metallurgist, applying her knowledge of metal alloys to the design and manufacture of fine instruments. Two years later, she returned to the International Nickel Company, this time as a senior scientist studying electron metallography on high-temperature alloys.

In 1960, she toured Europe, attending metallurgical meetings in Cambridge, England; Delft, Holland; and her native city of Paris. She published one of her more important papers, entitled "Direct Examination by Electron Transfer of Inconel-X," in the *Fifth International Congress for Electron Microscopy,* a text that came out in 1962. That year, she moved to the Burndy Corporation as a metallurgist before going independent as a consultant in 1964. She institutionalized this consultant's position by founding her own consulting firm, Pellier-Delisle Metallurgical Laboratory, in 1967.

Pellier had established herself as an important metallurgist well before winning the SWE Achievement Award. Pellier also received the Micrography Prize from the American Society for Metals in 1949 and the same prize from the American Society for Testing and Materials in 1952. Pellier's memberships included the American Institute of Mining and Metallurgical Engineers, the French Society of Metallurgy, the Electron Microscope Society of America, French Engineers in the U.S.A., New York Electron Microscopists, and the New York Microscopical Society.

Pendleton, Yvonne
(1957–)
American
Astrophysicist and Astronomer

Yvonne Pendleton has worked most of her career in the Planetary Systems Branch of the National Aeronautics and Space Administration (NASA) at the Ames Research Center in California as an infrared observational astronomer. Her research has focused on the life cycles of distant stars: "I am interested in the whole cycle of stars—where they are born, how they live their lives, and how they die," Pendleton stated. She also has mentored young students interested in the universe as part of the Women of NASA (WON) project, hosting web chats, or informative discussions transpiring via the Internet.

As a young student, Pendleton was discouraged from pursuing the sciences, and she experienced self-doubt about her scientific abilities in comparison to the seemingly self-confident boy students. However, she persevered, taking three years of chemistry and two years of biology in high school, catching up on physics courses, which intimidated her early on, in college. Her sister, Olga, paved the way by earning her Ph.D. in statistics, and she served as Pendleton's primary role model. She even dedicated her doctoral dissertation to her sister. Pendleton earned her Ph.D. in astrophysics from the University of Santa Cruz in 1987.

Pendleton's research originally focused on the interstellar medium, or the expanse of space between stars filled with interstellar dust. She specifically considered how this dust accretes into more coherent objects, such as planets and stars. More recently, she has focused her attention closer to home, studying the dust within our own solar system ever since the discovery in 1992 of the Kuiper Belt Objects, or a ring at the outer reaches of the solar system that represents the cosmic debris left over from the formation of our solar system. In studying this material, Pendleton hoped to gain a better understanding of the organic material that made up the planets in the solar system, specifically Earth.

NASA colleague Dale Cruikshank directed Pendleton's attention to the Kuiper Belt, and the pair collaborated on NASA's inaugural observations through the Keck I and II 10-meter telescopes at the Keck Observatory atop Mauna Kea in Hawaii in October 1996. Pendleton and Cruikshank had extended their collaboration beyond their professional relationship by marrying in January 1996. Pendleton admitted that her commitment to her career took its toll on her first two marriages, but her third marriage seemed destined, as both astronomers had asteroids named after them that followed surprisingly similar orbits. Pendleton had two children from her previous marriages, a daughter born in 1987 and a son born in 1989.

Besides her work at NASA, Pendleton has served as the school astronomer at Peterson Middle School and Laurelwood Elementary School, which her children attend. Through this and her web chats in the WON program, Pendleton has said she hopes to mentor young students and encourage them to pursue the study of astronomy, just as her sister inspired her to persevere in her study of the sciences. As payment in kind, Pendleton has mentored her sister's son in his study of science and astronomy.

Pennington, Mary Engle
(1872–1952)
American
Chemist

Mary Engle Pennington spearheaded many safety advances in the processing, storage, and transportation of perishable foods as head of the U.S. Department of Agriculture's Food Research Laboratory and later as a private consultant. She proved persistent when confronted with obstacles, earning a doctorate, despite the fact that she was denied a bachelor's degree as a woman and applying for a position at the U.S. Department of Agriculture (USDA) under the name M. E. Pennington to avoid rejection based on sexual discrimination.

Pennington was born on October 8, 1872, in Nashville, Tennessee. Her mother was Sarah B. Molony, and her father, Henry Pennington, was a successful label manufacturer. Pennington became interested in foods early on through gardening with her father, and she became interested in chemistry through a library book. She majored in chemistry at the Towne Scientific School of the University of Pennsylvania, but the institution did not grant bachelor's degrees to women, so it offered her a certificate of proficiency in 1892. Undeterred, she pursued a doctorate in chemistry with minors in zoology and botany, earning her Ph.D. in 1895 at the age of 22 through a statutory loophole that allowed women to become doctors of philosophy in "extraordinary cases."

Pennington held postdoctoral fellowships in chemical botany at the University of Pennsylvania for two years and in physical chemistry at Yale University for one year. In 1898, she took up an instructorship in physiological chemistry at the Women's Medical College, a position she retained until 1906. Concurrently, she established the Philadelphia Clinical Laboratory to perform analyses for physicians, and the city of Philadelphia consulted her professional opinion on perishable food storage. Duly impressed with the quality of her work, the city of Philadelphia's Department of Health and Charities appointed her to head its bacteriological laboratory, where she concentrated her efforts on reducing bacterial contamination in ice cream.

At this time, Pennington worked as a consultant for the USDA's Bureau of Chemistry, headed by Harvey W. Wiley, a friend of her family who urged her to apply for a position in the new research laboratory being established to handle prosecutions of the Pure Food and Drug Act passed by Congress in 1906. On his advisement, Pennington veiled her gender by signing her application as "M. E. Pennington." It shocked the Food Research Laboratory when a woman showed up to fill the position of bacteriological chemist in 1907, but by the next year, her quality work earned her an appointment as the head of the lab. In this position, Pennington effected changes in the packaging, warehousing, transportation, and refrigeration of perishable foods.

During World War I, Pennington served as a consultant for the War Shipping Administration, determining that only 3,000 of the country's 40,000 refrigerated cars had adequate air circulation. After the war, President Herbert Hoover presented her with the Notable Service Award, and in 1919, she switched to industry work as manager of research and development at the insulating materials manufacturer American Balsa Company. In 1922, she started her own business as a perishable foods consultant.

Pennington received many honors in her career, most notably the 1940 Garvan Medal from the American Chemical Society. She was the first woman inducted into the Poultry Historical Society's Hall of Fame, to honor her innovation of new methods for slaughtering poultry that preserved freshness longer. She served the associations in her fields, directing the National Association of Ice Industries's Household Refrigeration Bureau from 1923 to 1931, and she continued in these capacities until the end of her career: At the time of her death on December 27, 1952, in New York City, she was still vice president of the American Institute of Refrigeration.

Penry, Deborah L.
(1957–)
American
Biological Oceanographer

The National Science Foundation bestowed its 1993 Alan T. Waterman Award to Deborah L. Penry, the second female to receive this honor as the most outstanding scientist under age 35. Penry's research demonstrated a correspondence between chemical reactions and digestive reactions of benthic organisms, which inhabit the ocean floor.

Penry was born on February 28, 1957, on the Chesapeake Bay, in Maryland. Her interest in oceanic biosystems sprouted as a youngster on fishing trips with her father, as she was more fascinated by the cleaning of the fish, when she viewed the contents of its stomach, than by the catching or eating of the fish. She indulged this interest as an undergraduate studying biology at the University of Delaware, where she earned her B.A. in 1979.

Penry continued her biological studies as a research assistant in invertebrate ecology at the Virginia Institute of Marine Science while conducting graduate study at William and Mary College in Williamsburg, Virginia, for her 1982 master's degree. For the next year, she gained diverse work experience: as a lab chemist for Core Labs, Inc., in Lake Charles; as a research associate for the U.S. Department of Energy Program; and as a brine disposal monitor for McNeese State University. She then proceeded to the University of Washington (UW) for doctoral study under oceanography professor Peter A. Jumars. Her research focused on the intersection between ichthyology, or the study of fish, and limnology, or the study of aquatic ecosystems. She earned her Ph.D. in oceanography in 1988.

The School of Oceanography at UW retained Penry as a postdoctoral researcher for the next two years, when she collaborated with Jumars on the work linking benthic digestive processes to the similar processes of chemical reactions. Penry and Jumars considered this correlation in other comparative studies, testing whether other organisms similarly corresponded. They published their findings in several coauthored articles: "Modeling Animal Guts as Chemical Reactors," in the 1987 *American Naturalist;* "Digestion Theory Applied to Deposit Feeding," in the 1989 *Lecture Notes on Coastal and Estuarine Studies;* and "Gut Architecture, Digestive Constraints and Feeding Ecology in Deposit-Feeding and Carnivorous Polychaetes," in the 1990 *Oecologia.*

Penry continued her postdoctoral research in the Horn Point Laboratory at the University of Maryland, and in 1991, the University of California at Berkeley hired her as an assistant professor in its integrative biology department. Just as Penry's research crossed disciplines, so, too, did this department integrate several fields, including zoology, paleontology, oceanography, and botany, making the appointment particularly appropriate.

Penry also combined her professional career with a family life as a wife and mother. She described the intersection of these spheres in an article entitled "Puppy Peek-a-boo and Cell: Required Reading for Scientist-Parents." In it, she discussed the human tendency to reduce our fellows to the roles in which we immediately perceive them—as a mother, when reading a children's book to her daughter, or as a prominent scientist, when receiving the Alan T. Waterman Award. Penry's experience bespeaks the superficiality of such divisions, as she constantly fills multiple roles simultaneously, sometimes teaching scientific concepts to her young daughter and other times "parenting" her undergraduate students as their mentor.

Perey, Marguerite Catherine
(1909–1975)
French
Physicist

Marguerite Perey was the first scientist to isolate element 87, a naturally occurring element whose existence was predicted by Mendeleyev as early as 1870, though intense scientific experimentation failed to yield conclusive evidence of this element until 1939, when Perey discovered it at the Radium Institute in Paris, France. In 1946, she named the element *francium* after her homeland, France.

Perey was born in 1909, in Villemomble, France. Her early ambition to become a doctor was thwarted by the death of her father, which left the family with insufficient funds to finance a medical school education. She did study at a technical school for women, where she earned her state diploma in chemistry in 1929. After graduating, she joined the Radium Institute, where she served as a personal assis-

Marguerite Catherine Perey, who was the first to isolate element 87, which was predicted by Mendeleyev as early as 1870. *(AIP Emilio Segrè Visual Archives)*

tant for MARIE SKLODOWSKA CURIE, the institute's director. Curie was so unassuming that Perey mistook her for a secretary when they first met, but she ended up assisting her until Curie's death in 1934. At that point, Perey became a radiochemist at the institute, a position she retained until 1946.

Perey's discovery of element 87 in 1939 fulfilled a long search that began in 1870 with Mendeleyev's proclamation that such an element should exist. He ascertained that it would be the heaviest alkali metal, with an atomic weight of between 210 and 230, and that it would inhabit the first group of the 10th series (group 1 in the new designation) of the periodic table. He also predicted that it should be related to cesium, as an oxide of the type Me_2O. Perey's discovery was prefigured in 1914 by the work of Meyer, Hess, and Paneth, who witnessed the existence of alpha particles in a sample of actinium that they had carefully purified; they attributed these particles to element 87, though they were not able to isolate it. A later identification of element 87 in 1925 proved unfounded, so the search continued.

Previous researchers had investigated sources as various as cigar ash, straw, mushrooms, beet molasses, seawater, and mineral waters in search of element 87. They employed diverse methods as well, utilizing magneto-optical methods, X rays, and cathode-ray analysis. Perey studied the decay of a sample of radioactive actinium by radiometric methods, a strategy that proved successful where other methods failed partly because the half-life of the longest-lived isotope of element 87 lasted a mere 22 minutes. In 1939, Perey managed to isolate the isotope during this brief window of opportunity, establishing that it was indeed of the element 87, which had an atomic mass of 223. She dubbed the isotope *actinium-K* originally. She published her findings that year in an article titled "Sur un element 87, dérive de l'Actinium," in the journal *Comptes Rendus Hebdomadaires des Séances de l'Académie des Sciences.*

In 1946, Perey renamed the element *francium,* in honor of her native country, France. That year, she received her doctor of science degree from the Sorbonne, and she obtained a research position at the National Center for Scientific Research (CNRS). In 1949, the International Union of Chemists assigned the new element, which was the most unstable of the first 101 elements, the symbol Fr for its inclusion on the periodic table. That year, the University of Strasbourg appointed her as a professor of nuclear physics, a position she retained for the rest of her career. She founded a laboratory, and later, in 1958, the CNRS named her the director of the Nuclear Research Center, as it was named.

Perey received numerous honors for her scientific work: In 1960, she was named an officer of the Legion of Honor and was granted the Grand Prize of the City of Paris; in 1962, she was the first woman to be elected as a member of the Academy of Sciences in Paris; in 1964, she won both the Lavoisier Prize of the Academy of Sciences and the Silver Medal of the Chemical Society of France; and in 1974, she received the title of Commander of the National Order of Merit. Perey fought for 15 years against the cancer that eventually claimed her life on May 14, 1975, in Louveciennes, France.

Perlmann, Gertrude E.
(1912–1974)
American
Biochemist

The American Chemical Society awarded Gertrude E. Perlmann the Garvan Medal in 1965 in recognition of her "imaginative and ingenious investigations in protein chemistry," according to the award citation.

Perlmann was born on April 9, 1912, in Liberec, Czechoslovakia. In 1931, she commenced study at the German University of Prague, and five years later, in 1936, she earned her D.Sc. degree. When Adolf Hitler invaded Czechoslovakia in 1937, Perlmann fled to Denmark, where she worked under the protein chemist K. Linderstrøm-Lang as well as F. Lipmann at the Biological Laboratory of the Carlsburg Foundation and Carlsburg Laboratory in Copenhagen. However, the advent of World War II forced her continued migration, and she sought refuge in the United States.

Perlmann conducted research at the Harvard Medical School from 1939 through 1941, when she became a research fellow in medicine at the Massachusetts General Hospital. Her pathological investigations on proteins in bodily fluids utilizing boundary electrophoresis, a new research technique, impressed Lewis G. Longsworth sufficiently to prompt him to arrange a visiting investigator position at the Rockefeller Institute with an Advanced Medical Fellowship of the Commonwealth Fund in 1945. Perlmann remained there for the rest of her career, rising to the ranks of assistant in 1947, associate in 1951, assistant professor in 1957, associate professor the next year, and full professor in 1973, the year before her death.

Perlmann commenced her investigations of phosphate-containing proteins by studying egg albumin with electrophoresis. She succeeded in removing all phosphate from the protein, the first time this had been accomplished without hydrolysis of peptide bonds. She then proceeded to apply phosphoesterases to the proteolytic enzyme pepsin, only to discover that the removal of phosphate had little to no effect on the protein's activities or its structural properties. This discovery fueled Perlmann's long-term focus on pepsin and pepsinogen, its inactive precursor, as she applied every available experimental technique to transform the structure and chemistry of these proteins in order to track any resulting changes in biological functions.

Gertrude Perlmann, a leading biochemist in the investigation of protein chemistry. *(The Rockefeller Archive Center)*

Besides studying pepsin, Perlmann also investigated phosvitin, a phosphoglycoprotein in egg yolks that she found to contain more than 50 percent phosphoserine. Perlmann discovered that the chemical structure of phosvitin differed significantly from other proteins. Among these differences, the most interesting was the fact that phosvitin could form complexes with divalent metal ions.

Perlmann collaborated with Aharon Katchalski and Ephraim Katzir of the Weizmann Institute on her protein research. While Perlmann was working on a paper with Katchalski in Israel, he was killed in a terrorist attack at the Tel Aviv airport. She had other fruitful collaborations, though, as she coedited the text *Proteolytic Enzymes,* volume 19 in the series *Methods in Enzymology,* with Laszlo Lorand in 1970.

Perlmann published extensively in her career, and in presenting her with the 1965 Garvan Medal, the American Chemical Society specifically applauded "her work on the effect of chemical modifications on structural and enzymatic properties of proteins presented in her publications

with lucidity and economy of words." Perlmann remained active in her field up until her death on September 9, 1974, in New York City.

Perrin-Riou, Bernadette
(1955–)
French
Mathematician

Bernadette Perrin-Riou is best known for her ongoing study of p-adic functions, which provide insight into the way mathematical functions operate in other number systems. Perrin-Riou's work has brought her several honors, including the prestigious 1998 Charles-Louis de Saulses de Freycinet Prize of the French Académie des Sciences. A professor at the University of Paris, Perrin-Riou has dedicated herself to abstract mathematical research.

Perrin-Riou was born on August 1, 1955, in Ardeche, France. Both her parents were scientists who encouraged their three daughters to study physics and mathematics. Perrin-Riou's mother was a professor of physics, and her father was employed as a chemist. Perrin-Riou herself would have three sons (in 1981, 1985, and 1989) with her mathematician husband.

After completing her undergraduate work at l'École Normale Supérieur des Jeunes Filles in 1977, Perrin-Riou became a research assistant at the Pierre and Marie Curie University in Paris. There she began to work intensively on advanced mathematics research with Georges Poitou. She received a degree in mathematics from the University of Paris-Sud (South) in 1979 and pursued doctoral studies at the Pierre and Marie Curie University under mathematician John Coates. She was awarded her doctorate in 1983.

Upon receiving her degree, Perrin-Riou accepted an assistant professorship at the Pierre and Marie Curie University. Soon after launching her academic career, though, she left for a yearlong stint as a visiting professor at Harvard University. After her time in the United States, she returned to France and the Pierre and Marie Curie University, where she continued her research. In 1987, Perrin-Riou moved to another campus of the University of Paris, Paris 6, where she remained until 1994, moving up the ranks from assistant professor to professor.

In 1994, Perrin-Riou obtained a position at the University of Paris-Sud in Orsay. Unlike her previous posts, the University of Paris-Sud afforded her the opportunity to focus almost exclusively on her mathematical research rather than having to balance her time between teaching and research responsibilities. Taking advantage of this situation, Perrin-Riou has delved into number theory, specifically the abstract and highly specialized realm of p-adic functions. P-adic functions are constructed with prime numbers (integers that can be divided by no whole numbers other than themselves, such as one, two, three, five, or seven). The P refers to any prime number. Perrin-Riou's work has several meaningful applications. *P*-adic functions have revealed information about the ways functions operate in number systems. The esoteric research also sheds light on the properties of rational numbers and is increasingly being used to address unsolved problems in a new way. For example, *p*-adic functions played a role in unlocking the conundrum of Fermat's last theorem.

Perrin-Riou has remained at the University of Paris-Sud, where she is recognized as a brilliant researcher. Her work has been honored on several occasions. She won the Charles-Louis de Saulses de Freycinet Prize of the French Académie des Sciences in 1998 and the Ruth Lyttle Satter Prize of the American Mathematical Society in 1999. She was also chosen as a speaker at the 1994 International Congress of Mathematicians. Her work on *p*-adic functions is considered to be her most important, though she has also been lauded for her efforts to approach number theory through simple proofs rather than just through big theorems.

Pert, Candace Beebe
(1946–   )
American
Neuroscientist and Biochemist

Candace B. Pert codiscovered, in collaboration with her doctoral adviser, Solomon Snyder, the brain's opiate receptors, or the locations where narcotic drugs such as morphine effect their blockage of pain or their eliciting of pleasure. This discovery led to the identification of endorphins, or the body's own opiates, by the Scottish scientists John Hughes and Hans Kosterlitz. Snyder, Hughes, and Kosterlitz shared the 1978 Lasker Award, while Pert's role in this discovery went unrecognized, leading to accusations of blatant sexism in the determination of Lasker Prize recipients (many of whom go on to receive the Nobel Prize). Pert refused to comment on the controversy but instead continued to allow the excellence of her work to speak for itself, as she moved on to discover peptide T, a potential treatment for AIDS.

Candace Dorinda Beebe was born on June 26, 1946, in Manhattan, New York. Her parents were Mildred and Robert Beebe. After only a few months of study, she transferred from Wheaton College to Hofstra University, which was closer to her Long Island home. In 1966, she dropped out of Hofstra, married Agu Pert, a graduate student, and moved to Bryn Mawr, where he earned his doctorate in psychology and she gave birth to the first of their three children. While cocktail waitressing to support the family, she met an assistant dean from Bryn Mawr College, through whose encouragement she completed her B.A. in biology,

which she received in 1970. That year, she commenced doctoral work in pharmacology at Johns Hopkins University, where her first research project under Snyder's supervision focused on the regulation of the body's production of acetylcholine, the paramount neurotransmitter.

In the summer of 1972, Pert set out in search of opiate receptors by applying technologies from insulin receptor research. She tracked radioactive drugs through animal brain cells, looking for receptor molecules with indentations that accepted opiate drugs. In the March 2, 1973, issue of *Science,* she published her findings with Snyder in the article, "The Opiate Receptor: Demonstration in Nervous Tissue." She proceeded to confirm the existence of opiate receptors in the fetuses of pregnant rats and commenced searching for naturally occurring neurotransmitters that filled the same function as opiates, but her funding dried up before she could identify what Hughes and Kosterlitz later found and dubbed endorphins.

Pert earned her Ph.D. in 1974, and remained at Johns Hopkins as a National Institutes of Health fellow for the next year. Hopkins retained her as a staff fellow from 1975 through 1977 and as a senior staff fellow for the next year before promoting her to the rank of research pharmacologist in 1978. In 1982, the National Institutes of Mental Health appointed her chief of its brain chemistry section, where she investigated receptors of valium and PCP, or angel dust.

In 1986, Pert's team discovered peptide T, an intermediary between proteins and amino acids that regulates emotions, she theorized. She also theorized that peptide T could help cure AIDS, as it could block receptor sites to inhibit the spread of infection and also potentially reverse symptoms. In 1987, she founded her own company, Peptide Design, to research these possibilities. In 1990, she rejoined academia as an adjunct professor in the department of physiology and biophysics at Georgetown University, which later promoted her to the position of research professor.

In 1979, Pert received the Arthur S. Fleming Award. More recently, she published her autobiography in 1997, *Molecules of Emotion: Why You Feel the Way You Feel,* which received considerable attention. Her research has provided the missing link between theories of the emotions and the hard sciences, allowing for the previously mysterious realm of the feelings to be understood at a chemical and biological level.

Péter, Rózsa
(1905–1977)
Hungarian
Mathematician

Rózsa Péter is acknowledged to be the founder of the mathematical field of recursive function theory. She received some recognition for this major contribution during her lifetime, but her role in the development of recursive function and computational theories has since been largely forgotten. The author of three important books—*Playing with Infinity, Recursive Functions,* and *Recursive Functions in Computer Theory*—Péter was the first Hungarian woman mathematician to become an academic doctor of mathematics.

On February 17, 1905, Rózsa Péter was born in Budapest, Hungary. Her name was originally Rosa Politzer, but in 1930, like many other Hungarians, she changed her German-style name to a more authentically Hungarian one. She attended Maria Terezia Girls' School until 1922, when she enrolled at Eötvös Loránd University. Although she initially planned to study chemistry, she discovered that she preferred mathematics. While a student, she worked under some of the great mathematicians of the era, including Lipòt Fejèr and Jòsef Kürschàk. She also met a future collaborator, Làszlò Kalmàr.

After finishing her undergraduate degree in 1927, Péter supported herself by teaching high school and taking tutoring jobs. She also began graduate work in mathematics at Eötvös Loránd University but became discouraged when she found that the results of her first research project—involving number theory—had already been discovered. But her friend Làszlò Kalmàr suggested she explore Gödel's work on the subject of incompleteness. She did so, opting to focus on the recursive functions Gödel employed. (A recursive function is a type of function or expression predicating some concept or property of one or more variables, which is specified by a procedure that yields values or instances of that function by repeatedly applying a given relation or routine operation to known values of the function. The concept was developed by the Norwegian logician Thoralf Albert Skolem as a way to avoid the so-called paradoxes of the infinite that arise in certain contexts when "all" is applied to functions that range over infinite classes. It does so by specifying the range of a function without any reference to infinite classes of entities.)

In 1932, Péter presented a paper on recursive functions at the International Congress of Mathematicians in which she first proposed that recursive functions constituted a separate subfield of mathematics. She further outlined her position in a series of scientific papers. Péter received her Ph.D. summa cum laude in 1935 for her work on recursive functions.

After completing her doctorate, Péter was named a contributing editor of the *Journal of Symbolic Logic* in 1937. However, she was unable to find a teaching position because of the fascist laws passed in 1939. She was confined to the Budapest ghetto, though she continued working during the harsh war years. In 1943, she wrote her book *Playing with Infinity,* which was intended for lay

readers. In it, she discussed such complex topics as number systems, arithmetical progression, and the law of prime numbers in clever and friendly terms. She obtained her first teaching post in 1945 when she was hired at the Budapest Teachers' College. In 1951, she published her second book, *Recursive Function,* which presented her findings on that topic. When the Budapest Teachers' College closed in 1955, Péter became a professor at Eötvös Loránd University, where she remained until she retired in 1975. She completed her third work, *Recursive Functions in Computer Theory,* in 1976.

Like many other eastern European scientists of the communist era, Péter has fallen into obscurity. However, she did receive a number of honors and awards during her lifetime. She won the Kossuth Prize in 1951 for her scientific accomplishments as well as the Mano Beke Prize awarded by the Janos Bolyai Mathematical Society in 1953 and the State Prize Gold Degree in 1973. She became the first female mathematician to be elected to the Hungarian Academy of Sciences in 1973. She died on February 16, 1977.

Petermann, Mary Locke
(1908–1975)
American
Biochemist

Mary Locke Petermann is renowned for her discovery of animal ribosomes. Initially named *Petermann's particles* in her honor, ribosomes are the cellular structures that synthesize proteins. Petermann's scientific contributions also include determining the structure of ribosomes, elaborating on the importance of ions in the cellular structure of ribosomes, and describing ribosomal transformation.

The only daughter of Albert Edward and Anna Grierson Petermann, Mary Locke Petermann was born on February 25, 1908, in Laurium, Michigan. Because Petermann's father was president of the Calumet and Hecla Consolidated Copper Company in Calumet, Michigan, the family enjoyed a high status in the community. Petermann would never marry or have children.

Petermann attended Smith College, graduating with high honors in chemistry in 1929. After a year working at Yale University as a technician, she spent four years studying the acid-base imbalance of psychiatric patients at the Boston Psychopathic Hospital. She returned to school in 1936 and earned her Ph.D. in physiological chemistry from the University of Wisconsin in 1939 for her dissertation on the role of the adrenal cortex in ion regulation.

After completing graduate school, Petermann joined the staff of the Department of Physical Chemistry at the University of Wisconsin, becoming the first woman chemist there. From 1939 to 1945, she remained at Wisconsin as a postdoctoral fellow, working with Alwin Pappenheimer to research the physical chemistry of proteins. During this period, she discovered Petermann's particles, which were renamed ribosomes at the 1958 meeting of the Biophysical Society. She also successfully isolated several types of ribosomes and recorded their properties. While at Wisconsin, she launched a study of antibodies. Her research on this topic would lay the groundwork for Robert Porter's Nobel Prize-winning findings concerning the structure of immunoglobulins.

In 1945, Petermann left the University of Wisconsin for a position as a research chemist at Memorial Hospital in New York City. In her new capacity, she examined the role plasma proteins play in the spread of cancer. In 1946, she was named the Finney-Howell Foundation fellow at the newly created Sloan-Kettering Institute, where she was charged with exploring the role of nucleoproteins in cancer. She was promoted to associate member of the institute in 1960, and she became the first female full member three years later. In addition to fulfilling her obligations at Sloan-Kettering, Petermann concurrently taught biochemistry at the Sloan-Kettering Division of the Graduate School of Medicine at Cornell University. In 1966, she became the first woman to be appointed a full professor at Cornell.

Petermann retired from Cornell in 1973 and founded the Memorial Sloan-Kettering Cancer Center Association for Professional Women the following year. She also served as the association's first president. The author of more than 100 scientific papers, Petermann provided insight into the significance of proteins and nucleoproteins. Her pioneering work on the nature of the cell ribosome was seminal and has been widely recognized. In 1963, Petermann received the Sloan Award, and she also won the 1966 Garvan Medal of the American Chemical Society. In addition, she was honored with the Distinguished Service Award from the American Academy of Achievement. Petermann died on December 13, 1975, of intestinal cancer. In 1976, the Educational Foundation of the Association of Women in Science named a graduate scholarship program in her honor.

Peterson, Edith R.
(1914–1992)
American
Medical Researcher

Edith R. Peterson distinguished herself as the first scientist to grow myelin, a protective sheath encasing nerve cells, in vitro. This discovery proved significant to the research of multiple sclerosis, a disease resulting from the breakdown of the myelin in the nerve cells of the brain and spinal cord,

as well as muscular dystrophy, a disease that degenerates skeletal muscles, and other nervous system diseases. She utilized the organotype culture method, whereby the generated cells mimic the structure and function of the cells they replicate, and Peterson became one of the world's experts on this particular procedure, teaching it to students from throughout the world.

Peterson was born Edith Elizabeth Runne on June 24, 1914, in Brooklyn, New York. Her father was Hermann Runne, the co-owner of a restaurant and catering business. He died unexpectedly in 1920, just a short while before he was to join his family, who were in Germany visiting relatives. Edith, her sister, and her mother, Else Helmke, remained in Germany for the next six years, and upon their return to the United States, her mother supported the family by working as a custom dress designer.

Runne remained in New York City for her undergraduate studies, attending Barnard College, where she earned her B.S. degree in 1937. She continued with graduate studies at Columbia University, where she earned her master's degree in zoology in 1939. Two years later, in September 1941, she married Charles Peterson, a commercial artist. More than a decade after that, they had a son, Wesley, in 1952. Two years later, in 1954, they added a daughter, Rhonda Lea, to the family.

After obtaining her master's degree at Columbia, Peterson remained there to work in Margaret Murray's laboratory. She cultured living nerve cells from samples of chicken embryos through a special method called organotype culture, which distinguished itself from other cell-regeneration methods in that the generated cells retain the function and structure of the original cells from which they were replicated. Besides replicating nerve, brain, and spinal cord cells, she also succeeded in growing myelin for the first time anyone achieved this outside of the organism. This discovery proved significant to multiple sclerosis research, as this disease specifically involves the degeneration of the myelin casing of the brain and spinal cord.

After two and a half decades at Columbia, Peterson moved to the Albert Einstein College of Medicine of Yeshiva University in the Bronx, New York, where she worked with Murray Bornstein starting in 1966. It was here that she focused on muscular dystrophy research while also teaching to disseminate the organotype culture techniques she had developed to students from around the globe.

Peterson continued to work well into her 70s, but a debilitating stroke in 1990 robbed her of the use of her right hand, forcing her to retire. She and her husband moved to upstate New York to enjoy her retirement. Two years later, she suffered another stroke, this time a fatal one: Peterson died on August 15, 1992, in Middletown, New York.

Phelps, Almira Hart Lincoln
(1793–1884)
American
Scientific Writer and Educator

Almira Hart Lincoln Phelps devoted her career to education as well as to scientific writing, authoring textbooks that proved extremely popular due to their accessibility. Phelps's educational philosophy centered around the hands-on approach, reasoning that students learn best when they conduct the experiment themselves. She also prescribed to Pestalozzi's method of teaching, which stated that the teacher should instruct the student first on the rudimentaries before graduating to the intricacies of understanding. Amos Eaton, the founder of Rensselaer Polytechnic Institute, introduced her to both the hands-on as well as the Pestalozzi methods.

Phelps was born in 1793 as Almira Hart. She received her early education in public schools and later in private academies. At the age of 20, she commenced her career as a teacher, working in public schools from 1813 through 1817. She even opened her own boarding school before marrying Simeon Lincoln, a newspaper editor. When he died of yellow fever in 1923, she returned to teaching, this time at the boarding school for girls that her sister, Emma Willard, had established as a model of its kind, the Troy Female Seminary. She remained there until 1831, when she married John Phelps.

Together with her new husband, Phelps founded a series of schools for girls where she ran the academic aspects as principal while he tended to the financial aspects as the business manager. She drew on her teaching experience during this time period as the basis for her "Familiar Lectures" series of successful science textbooks: *Familiar Lectures on Botany,* published in 1829; *Familiar Lectures on Natural Philosophy,* published in 1837; and *Familiar Lectures on Chemistry,* published in 1838. These textbooks sold thousands of copies and went through several reprints. In early editions, she stuck to the Linnaean system of classification, but by the time the fifth editions came out, Phelps had modernized her approach to include the natural classification system that was current. She also published *Chemistry for Beginners* in 1834. Applying Pestalozzi's method to chemistry, Phelps's books moved the students from understanding the basics, such as attractions, to comprehending the complexities of organic chemistry.

Phelps's most successful school was the female institute at Ellicott's Mills, Maryland. There, she challenged the girls with collegiate-level curricula, especially in the sciences. Phelps's laboratory work proved particularly demanding to the girls, as she required each of them not only to perform difficult experiments but also to present a talk on the subject as well. Although this regimen prepared them for the most rigorous of professions, her students typically graduated into teaching or homemaking, as they were effectively barred from all other professions that might better utilize their knowledge and skills.

Phelps's husband, John, died in 1849, but she continued to teach until her retirement in 1856. Three years later, in 1859, the American Association for the Advancement of Science elected her as only its second woman member, behind astronomer Maria Mitchell. Phelps died in 1884. Her biography, *Almira Hart Lincoln Phelps: Her Life and Work,* appeared in 1936, authored by Emma L. Bolzau.

Phillips, Melba Newell
(1907–)
American
Physicist

In 1981, the American Association of Physics Teachers (AAPT) established the Melba Newell Phillips Award in recognition of exceptional contributions to the teaching of physics through work with the association. Appropriately, Phillips herself was the first to receive this honor, as she had devoted her career to the advancement of physics teaching by coauthoring two widely used textbooks, publishing several articles on the history of physics, and actively working within the AAPT to promote the teaching of physics. Needless to say, Phillips was also an inspiring classroom teacher.

Phillips was born on February 1, 1907, in Hazleton, Indiana. She attended Oakland City College in Indiana, where she earned her B.A. in 1926. Two years later, she received her master's degree from Battle Creek College in Michigan. She then pursued her doctorate in physics at the University of California at Berkeley. After receiving her Ph.D. in 1933, she remained at the university until 1935 as an instructor.

In 1935, Phillips and J. Robert Oppenheimer coauthored a paper entitled "Transmutation Function for Deuterons," which the journal *Physics Review* published. In it, the pair advanced the theory of the Oppenheimer-Phillips effect, describing the process of neutron capture in deuteron bombardment that formed a compound nucleus. That year, she was the Helen Huff research fellow at Bryn Mawr College. The next year, from 1936 to 1937, she moved to the prestigious Institute for Advanced Studies at Princeton University, where she was the Margaret Maltby fellow. She then held two instructorships, first at the Connecticut College for Women and then at Brooklyn College.

From 1941 through 1944, Phillips was a lecturer at the University of Minnesota, and in 1944, she contributed to the war effort by working on the staff at the Harvard Radio Research Laboratory. She then returned to Brooklyn College as an assistant professor of physics while concurrently working in the Columbia University Radiation Laboratory.

In 1952, both institutions dismissed her for refusing to testify against her colleagues in Senator Joseph McCarthy's investigations. In 1987, Brooklyn College publicly apologized to Phillips for this action.

Between 1957 and 1962, Phillips acted as the associate director of the Academic Year Institute at Washington University in St. Louis, Missouri. She then moved to the University of Chicago, where she held a professorship for a decade. Phillips graduated to emeritus status in 1972, though she did not fully retire: She held a visiting professorship at the State University of New York at Stony Brook from 1972 through 1975. She also came out of retirement in 1980 as a visiting professor at the Graduate School of the University of Science and Technology of the Chinese Academy of Science in Beijing.

Besides having an award named in her honor, Phillips was recognized extensively for the excellence of her career. In 1964, she received an honorary doctorate from her alma mater, Oakland City College, and a decade later, she received the Oersted Medal from the AAPT. The American Institute of Physics bestowed on her its Karl Taylor Compton Award in recognition of her distinguished ambassadorship in science. In 1988, she received the Guy and Rebecca Forman Award for Outstanding Teaching in undergraduate physics. On May 17, 1997, Brooklyn College ran a symposium in her honor and established a scholarship in her name.

Picotte, Susan La Flesche
(1865–1915)
American
Physician

Susan La Flesche Picotte was the first Native American woman to earn her medical degree from the Women's Medical College of Pennsylvania and the first Native American woman physician. She administered her medical knowledge and skills to her tribe as the head physician on the Omaha reservation for four years. Later, she established a county medical society, headed the county's board of health, and founded a hospital, which bore her name after her death. Her life straddled two dichotomous cultures, and she managed to incorporate the best aspects of each culture into her practices without negating the other culture.

Picotte was born in a tepee on June 17, 1865, on the Omaha reservation in northeastern Nebraska. She was the fourth daughter and the fifth and youngest child born to One Woman, or Mary La Flesche, and Iron Eye, or Joseph La Flesche, the Omaha tribal chief. They raised their daughter in the traditional manner, speaking their native language. However, they also believed in the importance of education, so they sent her at the age of 14 to the Eliza-

beth Institute for Young Ladies in New Jersey, where she learned English. She then matriculated at the Hampton Institute in Virginia, which granted her a gold medal for high academic achievement upon her graduation as the salutatorian in May 1886, after she spent only two years there.

La Flesche then began her medical studies at the Women's Medical College of Pennsylvania in Philadelphia with financial assistance from the Women's National Indian Association, which had been founded in 1880 by Mary Lucinda Bonney and Amelia Stone Quinton to support Native American women training to enter the professions. La Flesche worked on an accelerated track, completing a three-year course of study in two years to graduate with an M.D. in 1889 at the head of her class of 36 students. She remained in Philadelphia to fulfill her internship at the Women's Hospital before returning to the Omaha reservation as the physician for the government school.

In about 1891, the Omaha tribe appointed La Flesche as the head physician for the entire reservation, administering health care and education to some 1,300 Omahas. She committed herself wholeheartedly to this burdensome task for four years, resigning in 1894. That year, she married Henry Picotte, a half-French, half-Sioux farmer, and they set up a homestead in Bancroft, Nebraska. She mothered two sons, Caryl and Pierre, while maintaining an active medical practice. She also cared for her husband during the slow decline of his health; he died in 1905.

In 1906, Picotte moved to the newly founded town of Walthill, Nebraska, where she became an active member of the community by organizing the County Medical Society and chairing the local health board. She also established a hospital there in 1913 as a medical missionary for the Presbyterian Board of Home Missions. Her own health declined due to an infection of her facial bones, which she had suffered for many years. She finally fell prey to the infection and died on September 15, 1915, in Walthill. Posthumously, the hospital she founded was named after her. In recent years, there has been a resurgence of interest in her, as three biographies were published in the 1990s: *Native American Doctor: The Story of Susan Laflesche Picotte* by Jeri Ferris, published in 1991; *A Doctor to Her People: Dr. Susan La Flesche Picotte* by J. L. Wilkerson; and *Susan La Flesche Picotte, M.D.: Omaha Indian and Reformer* by Benson Tong, both published in 1999.

Pierce, Naomi E.
(1954–   )
American
Lepidopterist and Molecular Biologist

As the first Sydney A. and John H. Hessel Professor of Biology at Harvard University, Naomi Ellen Pierce became the first woman to gain tenure in the university's department of organismal and evolutionary biology. She made her name as a scientist studying the symbiotic relationship between mistletoe butterflies (a species of Lycaenid) and farmer ants, whereby the ants protect and transport the cocooned caterpillars in exchange for protein- and carbohydrate-rich secretions. Pierce extended this metaphor of cooperation to the laboratories she heads by replacing the traditional hierarchical model with a collaborative spirit. She supports the entrance of women into the sciences by filling three-quarters of her lab positions with women, and she actively promotes supportive dialogue among women in the sciences as a means of combating the isolation that most women experience as they climb the academic or corporate ladder to ranks that are less populated by fellow women.

Pierce was born on October 19, 1954, in Denver, Colorado. When she entered Yale University, she intended to follow in the footsteps of her grandfather, a novelist in Japan, by studying history and the humanities. However, she changed majors in her third year after a conversation with her evolution professor, Charles Remington, whose infectious enthusiasm for studying butterflies convinced her to pursue a scientific career. She enrolled in summer courses at the Rocky Mountain Biological Laboratory, where Remington conducted field studies, in her home state of Colorado, and in 1976, she graduated from Yale with a bachelor's degree in biology. In this sense, she followed in the footsteps of her father, who also devoted his life to science as a geophysicist.

Pierce then traveled to Southeast Asia as a John Courtney Murray fellow to study the Lycaenids of Australia. This field research formed the basis of her doctoral dissertation, which she wrote under the guidance of the renowned entomologist Edward O. Wilson and under the supervision of her doctoral advisers Bert Hilldobler and R. E. Silberglied. She earned her Ph.D. in biology from Harvard University in 1983.

Pierce returned to Australia to conduct postdoctoral research on carnivorous Lycaenids as a Fulbright fellow at Griffith University in Queensland from 1983 through 1984 and again from 1987 through 1988 as a visiting research fellow at the University of New England in Armidale. She spent the intervening years at Oxford University, first as a NATO postdoctoral fellow in the zoology department from 1984 through 1985 and then as a research lecturer at Christ Church until 1986. That year, Princeton University appointed her as an assistant professor of biology, promoting her to the rank of associate professor in 1989.

In 1988, Pierce became one of only 31 recipients of the MacArthur Fellowship Award, supporting unrestricted research for the next five years. The next year, the Bunting Institute of Radcliffe College appointed her as a science scholar. In 1991, Harvard named her the Hessel Professor and also appointed her curator of lepidoptera at the Harvard Museum of Comparative Zoology, both prestigious

positions. Pierce used her MacArthur monies to take her students to Australia for three months every year to continue her field research there.

In 1997, Pierce gave birth to twin daughters, Kate and Megan. In addition to all her administrative and academic duties, she shares parenting responsibilities with her husband, Andrew Berry. This experience, which she jestingly calls her current "research interest," helps her understand the struggles of women in the sciences to balance professional with familial expectations.

Pinckney, Eliza Lucas
(1722–1793)
American
Horticulturist

Eliza Lucas Pinckney's major contribution to society carried significance to both science and economics: She was the first in the American South to develop a successful means of cultivating the indigo plant for transformation into a marketable blue textile dye. In the wake of her horticultural innovations, the indigo market in the South grew exponentially. She also experimented with various other agricultural crops.

Pinckney was born as Eliza Lucas in Antigua in the West Indies in 1722, the oldest of four children. Her father, George Lucas, was a planter and a lieutenant colonel in the British army. He sent her to England for her education at a finishing school that focused on French, music, and other subjects deemed acceptable for female students at that time, though Lucas preferred the scientific subject of botany most.

In 1738, the family moved to Charles Town (now Charleston), South Carolina. Lucas enjoyed living in England the most, but she preferred South Carolina over the West Indies. When the army dispatched her father to the War of Jenkins Ear with Spain, he left his eldest daughter in charge of his three plantations: the 600-acre Wappoo Creek plantation, where she lived and oversaw 20 slaves; the 1,500-acre Garden Hill plantation, which produced pitch, tar, and pork; and 3,000 acres of rice planted along the Waccamaw River. Lucas rose to the responsibilities placed on her head, conducting the business of the plantations with efficiency.

In 1740, Lucas commenced her experiments on the best way to cultivate indigo for dye production. Her father sent a dye maker to aid in her trials, but she fired him after he sabotaged her early attempts. With the help of a new dye maker, she developed the process of fermenting the leaves in water, draining the fermented product, beating it into a blue sediment, then adding lime and draining it again to dry and cut into cubes. Lucas shared the indigo seeds sent from her father with her neighbors, thereby spreading indigo production, which had been a mere 5,000 pounds in 1745 through 1746 but rose drastically to 138,300 pounds of export in 1747. She also experimented with crops such as fig orchards (for exporting dried figs), flax, hemp, and silk.

In 1744, Lucas married Colonel Charles Pinckney, a widower nearly twice her age. She gave birth to four children in five years, and three of them survived: Charles Cotesworth, who later signed the U.S. Constitution; Harriot, who later married Daniel Horry; and Thomas, who later became a lawyer. Pinckney homeschooled her children according to John Locke's theory that the human mind was a tabula rasa, or "blank slate," to be inscribed with the wisdom of previous generations. In 1753, Charles Pinckney moved the family to England, where he worked as an agent for South Carolina merchants. In 1758, he died of malaria, and she moved the family back to Charles Town.

There, Pinckney again took charge of running the family plantation and reestablished herself in Charles Town society. She was well-known throughout the states, and upon her death from breast cancer on May 26, 1793, President George Washington served as one of her pallbearers.

Pressman, Ada Irene
(1927–   )
American
Engineer

Control systems engineer Ada Irene Pressman has been a leading authority in power plant controls and process instrumentation. She has specialized in creating shutdown systems for nuclear power plants and has also sought to ensure that the turbines, steam engines, and reactors in nuclear power plants work together properly. In addition, she has developed an important technology for emergency systems—a secondary cooling system that operates from a diesel generator in case primary power at a plant is lost. Thanks to efforts such as these, Pressman is credited with vastly improving the safety of both nuclear and fossil fuel power plants for workers and residents living near such plants. Moreover, she has bolstered the status of control systems engineers by successfully lobbying the state engineering board of California to recognize the specialty as a distinct engineering field.

Pressman was born on March 3, 1927, in Sydney, Ohio. Although she planned to become a secretary after high school, her father encouraged her to attend college. Following his advice, she enrolled at Ohio State University. She earned her bachelor's degree in mechanical engineering in 1950, after only two years of course work.

In 1950, Pressman took her first job as a product engineer at Bailey Meter Company in Cleveland, Ohio. At Bai-

ley, Pressman specialized in the areas of process flow, instrumentation, and control engineering for increasingly complex fossil fuel plants. She also designed boiler controls and burner management systems. She left the company in 1955 because of the limited opportunities for advancement that were afforded her.

After severing her ties with Bailey, Pressman accepted a position as a project engineer at Bechtel Corporation in Los Angeles, California. From 1955 to 1964, she played a key role in the industry-wide shift toward more automated controls of equipment and systems at power plants. In particular, Pressman helped to improve the precision and reliability of sensors and controls. In 1964, she was promoted to instrument group leader at Bechtel. With the advancement came new responsibilities, including heading the design and construction of four power generation units. In 1968, Pressman was tapped to head the systems engineering group at Bechtel's Rancho Seco, a 900-megawatt nuclear power plant. (The first nuclear power plant built away from a major body of water, it posed major engineering challenges.) After serving as Bechtel's assistant chief control systems engineer from 1971 to 1974, Pressman was again promoted to chief control systems engineer. In this capacity, she managed 18 design teams, responsible for more than 20 power-generating plants worldwide. While successfully fulfilling her duties at Bechtel, she earned her master's of business administration from Golden Gate University, in San Francisco, California, in 1974.

On top of all of her other responsibilities, Pressman was also active in a number of professional organizations and strove to raise the profile of her branch of engineering. In the 1970s, she campaigned successfully to have control systems engineering classified as a separate discipline by the California engineering board. She then became the first person to register in the new field. These efforts enhanced the professional status of other control systems engineers. In addition, Pressman was dedicated to promoting women's careers in the sciences. From 1979 to 1980, she served as the president of the Society of Women Engineers.

Pressman retired from Bechtel in 1987. Her career was marked by numerous awards, including the E. G. Bailey Award from the Instrument Society of America and the Outstanding Engineer Merit Award. Moreover, the Society of Women Engineers bestowed on her its Annual Achievement Award in 1976.

Prichard, Diana García
(1949–   )
American
Chemical Physicist

Research scientist Diana Garcia Prichard is best known for the groundbreaking research she conducted as a graduate student. While enrolled at the University of Rochester, she constructed the first instrument capable of measuring van der Waals clusters. Subsequently, she has worked at the Eastman Kodak Company for most of her career, studying fundamental photographic materials. Prichard is also an active leader in the Hispanic community and heads the Center for the Advancement of Hispanic Scientists and Engineers.

On October 27, 1949, Prichard was born in San Francisco, California. Her mother, Matilde Dominguez García, immigrated to the United States from Nicaragua, and her father, Juan García, was of Mexican and Native American ancestry. Despite the fact that her parents had received little formal schooling, they encouraged their daughter to pursue an education.

Heeding her parents' advice, Prichard enrolled in nursing school at the College of San Mateo and obtained her LVN (nursing) degree in 1969. After marrying Mark Prichard, she had two children—Erik and Andrea—and devoted herself to caring for her children. However, in 1979, Prichard opted to return to school. Since she had always been interested in science, she chose chemistry/physics as her major and was awarded her bachelor's degree in 1983 from California State University at Hayward. She then enrolled at the University of Rochester, earning her master's degree in physical chemistry in 1985 and her Ph.D. in chemical physics in 1988.

While still a graduate student, Prichard conducted pioneering research. Working with the high-resolution infrared spectrum—which reveals the amount and type of atoms or molecules present in a sample—she created the first instrument able to measure van der Waals clusters. Named in honor of the Dutch Nobel Prize-winning physicist Johannes Diderik van der Waals, the van der Waals equation explains the behavior of nonideal gases. Ideal gas laws originally assumed that gas molecules have zero volume and that no attractive forces exist between them. However, neither assumption is true. In 1881, van der Waals introduced into the ideal gas law two parameters—size and attraction—and thus invented a more precise formula—later dubbed the van der Waals equation—to account for the behavior of gas molecules. The van der Waals clusters that Prichard analyzed are weakly bound complexes that exist in nature but are rare. Prichard's work allowed other scientists to produce these clusters by experimental methods, thereby making it much easier to study them.

After finishing graduate school in 1983, Prichard accepted a position as a research scientist in the photo science research division at the Eastman Kodak Company in Rochester, New York. In this capacity, she has studied silver halide materials for photographic systems and has conducted other research on fundamental photographic materials.

Prichard has also been highly involved in efforts to create opportunities for Hispanics in science. In addition to heading the Center for the Advancement of Hispanic Scientists and Engineers, she founded the Partnership in Education program in Rochester, which provides Hispanic role models for public schools to help teach science and math to students with limited English language skills. Furthermore, Prichard collaborated in launching the Hispanic Organization for Leadership in Advancement at Eastman Kodak. She is an active member in the Society of Hispanic Professional Engineers. Prichard remains at Eastman Kodak. While her role in bringing Hispanic students to scientific fields has brought her much attention, her scientific legacy will stem from her work on van der Waals clusters. More than 100 authors have cited her graduate work in subsequent publications.

Profet, Margie Jean
(1958–)
American
Evolutionary Biologist

Although biomedical researcher Margie Profet has no formal academic credentials, she has promulgated several influential theories. The central tenet of Profet's work is that humans coevolve and adapt to pathogens and toxin-producing substances through defensive actions. For Profet, allergic reactions, as well as nausea and food aversions during pregnancy and menstruation, shield humans from various toxins. She is a prolific writer, publishing not only a popular book, *Protecting Your Baby-to-Be,* but also a number of scholarly articles as well as shorter pieces in mainstream publications such as *Time* and *Omni*. While her ideas have not been readily embraced by the medical community, Profet was awarded a $250,000 fellowship from the MacArthur Foundation in 1993 to pursue her research into evolutionary biology.

The second of four children, Profet was born on August 7, 1958, in Berkeley, California. Both parents were trained physicists who encouraged her to pursue her affinity for science and mathematics. When Profet enrolled at Harvard University, though, she was put off by the regimentation she experienced in her science classes there. She opted to major in classical political philosophy instead and received her bachelor's degree in 1980.

After graduating, Profet migrated to Germany, where she worked as a computer programmer for National Semiconductor Company in Munich from 1980 to 1981. She remained intrigued by scientific questions, so she returned to the United States and enrolled at the University of California at Berkeley to study physics. She was awarded her second bachelor's degree in 1985.

After a miserable six months in graduate school, where she felt stifled by the lack of intellectual freedom, Profet established herself as an independent researcher. Although she had to hold down various part-time jobs to support herself, she was able to devote herself to the topics that interested her. Utilizing Medline—an on-line database dedicated to medical issues—she immersed herself in her work. In 1986, she first promulgated her theory that morning sickness (the nausea and food aversion that most women experience during the first trimester of pregnancies) was not simply an uncomfortable side effect of pregnancy. Instead, Profet postulated, pregnant women's brains and bodies become more sensitive to toxins in food (especially those in plants, such as broccoli and potatoes). For Profet, pregnancy sickness is an evolutionary adaption that actually protects the fetus from potentially damaging toxins during the first trimester, when limbs and organ systems are formed.

Profet gained academic support in 1988 when Bruce Ames, a toxicologist and professor at the University of California at Berkeley, offered her a position in his laboratory. In 1991, Profet outlined her second important theory in an article entitled "The Function of Allergy: Immunological Defense Against Toxins," published in the *Quarterly Review of Biology*. Here again, Profet broke new scientific ground, proposing that humans develop allergic reactions as a means of guarding the body from toxins. Although most medical researchers assumed that allergic reactions were the result of hypersensitivity to specific substances, Profet argued that the reaction was instead an immunological boon—the last line of defense in the body's arsenal of responses to toxic substances. (She noted that allergy sufferers are less likely to develop cancer.) In 1993, Profet examined menstruation, and in another paper published in the *Quarterly Review of Biology*, she again introduced a new theory. Physicians generally hold that menstruation is simply a shedding of the uterine wall with no purpose. Profet, by contrast, posited that menstruation serves to cleanse the uterus and oviducts wall of sperm-borne pathogens that could cause infection or infertility.

Profet's research received a significant boost in 1993 when the MacArthur Foundation appointed her a fellow. She left Berkeley for Seattle and an office in the Department of Molecular Biotechnology at the University of Washington. In 1995, she published her popular book, *Protecting Your Baby-to-Be*, which presented her pregnancy sickness theory in greater detail and offered advice to expecting mothers for preventing birth defects through proper nutrition.

Although some physicians have begun to accept her theories, Profet retains a reputation as a maverick thinker. She has been lauded for her ability to question accepted paradigms and for her contributions to the fledgling field of evolutionary biology. In 1996, Profet began a new line of inquiry—into astronomy and the structure of space-time. She currently serves as a visiting scholar in the astronomy department of the University of Washington.

Q

Quimby, Edith H.
(1891–1982)
American
Biophysicist

Edith H. Quimby developed diagnostic and therapeutic uses for X rays, radium, and radioactive isotopes. The field of radiology was still in its infancy when she began her work, but Quimby's groundbreaking research ascertained the extent of radiation's ability to penetrate an object, thereby allowing physicians to use the smallest possible doses of the potentially dangerous substances. Quimby also cofounded the Radiological Research Laboratory at Columbia University, which studied the medical uses of radioactive isotopes (especially in cancer diagnosis and treatment).

Born on July 10, 1891, in Rockford, Illinois, Quimby was one of Arthur and Harriet Hinckley's three children. The family moved frequently during Quimby's childhood, and she completed high school in Boise, Idaho.

Quimby attended Whitman College in Walla Walla, Washington, majoring in physics and mathematics. After graduating in 1912, she taught high school in Nyssa, Oregon. In 1914, she won a fellowship to study physics at the University of California, where she met and married Shirley L. Quimby (a fellow physics student). Edith Quimby was awarded her master's degree in 1915 and then returned to teaching high school science for four more years.

Quimby's career changed course in 1919 when her husband was offered a faculty position at Columbia University. In New York, she accepted a position at the newly created New York City Memorial Hospital for Cancer and Allied Diseases. Working with chief physicist Dr. Gioacchino Failla, Quimby began to explore the medical uses of X rays and radium, particularly in treating tumors. Quimby and Failla would enjoy a 40-year scientific association.

Although physicians had begun to use radiation by this time to diagnose and treat certain diseases, their efforts were often haphazard at best since no standardized techniques had yet been developed. Quimby's work did much to alleviate this problem. In 1923, she instituted a film badge program so that X-ray film could be employed to gauge radiation exposures accurately. More importantly, she determined the specific radiation doses required to treat tumors. In 1932, she became the first to determine the distribution of radiation doses in tissue from various arrangements of radiation needles. The techniques she devised became the standard in the United States. During this period, she also focused on measuring the penetration of radiation, quantifying the different doses required to produce the same biological effect, such as skin erythema (reddening). In the process, she formulated the concept of biological effectiveness of radiation, which is still employed by radiobiologists today.

After spending more than 20 years at Memorial Hospital, Quimby moved with Failla to the Columbia University College of Physicians and Surgeons in 1943, becoming an associate professor of radiology and later advancing to the rank of full professor. While at Columbia, she and Failla cofounded the Radiological Research Laboratory. At their new laboratory, the pair began working with the newly available artificial radioisotopes being produced by accelerators and reactors. They concentrated on the application of radioactive isotopes, such as radioactive sodium and iodine, to treat thyroid disease and diagnose brain tumors. These early clinical trials established Quimby as a pioneer of nuclear medicine. During her tenure at Columbia, Quimby was also involved with several other projects. In addition to working on the Manhattan Project, which created the atomic bomb, she joined the Atomic Energy Commission and advised the United States Veterans Administration on radiation therapy.

Quimby finished her career at Columbia University by teaching a new generation about radiation physics and the clinical use of radioisotopes. She retired in 1960, though

Edith Hinckley Quimby, who is acknowledged as a founder of radiobiology. *(Center for the American History of Radiology, courtesy of AIP Emilio Segrè Visual Archives)*

she continued to research, write, and lecture. She coauthored a book, *Physical Foundations of Radiology,* and wrote a number of influential scientific papers. Acknowledged as a founder of radiobiology, Quimby received a slew of honors and distinctions. In 1940, she became the first woman to win the Janeway Medal of the American Radium Society. The Radiological Society of North America later awarded her a gold medal, citing her work "which placed every radiologist in her debt." She died at the age of 91 on October 11, 1982.

Quinn, Helena Rhoda Arnold
(1943–   )
Australian/American
Physicist

Theoretical physicist Helena Rhoda Arnold Quinn has made major contributions to the understanding of how particles and forces work at the subatomic level. She was one of the first scientists to elaborate how the three forces—strong, weak, and electromagnetic—that bind subatomic particles together can merge to form a single coupling constant in a grand unified theory. With Robert Peccei, she proposed the Peccei-Quinn symmetry, which explains how strong interactions can maintain the symmetry between subatomic particles even as weak interactions do not. She has also posited the quark–hadron duality, which shows how the physics of quarks can be used to predict the same properties of the physics of hadrons (which are subatomic particles composed of quarks). Quinn was awarded the 2000 Dirac Medal for her work.

Quinn was born on May 19, 1943, in Melbourne, Australia, to Ted Adamson Arnold and Helen (Down) Arnold. When her father moved to Belmont, California, in 1962, Quinn—then in her second year of college—opted to join him in the United States. She transferred to Stanford University and received her bachelor's degree in physics in 1963. She remained at Stanford for graduate studies, earning her Ph.D. in physics in 1967. She married Daniel Quinn in 1966 and had two children, Elizabeth (born in 1971) and James (born in 1974).

From 1967 to 1970, Quinn was a postdoctoral fellow at the Deutsches Elektronen Synchroton in Hamburg, Germany. She then accepted a postdoctoral fellowship at Harvard University in 1971 and became an assistant professor there in 1972. Quinn was promoted to associate professor in 1976. Her work at Harvard focused on determining the implications of theories about the subatomic world. With Joel Primack and Thomas Applequist, she conducted some of the earliest investigations into the predictive power of a theory now called the standard model of fundamental particles and interactions. She then teamed up with Howard Georgi and Steven Weinberg to determine the circumstances under which electromagnetism, the strong force, and the weak force (the three forces that hold atoms intact) become identical at high energies.

Quinn left Harvard in 1977 for a yearlong visiting professorship at Stanford University. She stayed at Stanford thereafter, first as a senior postdoctoral fellow at the Stanford Linear Accelerator Center (SLAC) from 1978 to 1979 and then as a staff scientist at SLAC. (Linear accelerators such as SLAC allow for the study of subatomic particles by breaking apart atoms into their constituent particles.) At SLAC, Quinn tackled the issue of CP symmetry. With Robert Peccei, Quinn arrived at an explanation for the fact that a basic symmetry of the laws of physics called CP holds true for strong interactions but not for weak interactions. CP is the combination of C—also known as particle-antiparticle symmetry—and P or parity (mirror-image) symmetry. In order to explain this phenomenon, Quinn and Peccei proposed another near symmetry of the universe, which they dubbed the Peccei-Quinn symmetry.

This theory predicts the existence of another class of subatomic particles called axions. Neither the Peccei-Quinn symmetry nor the actual existence of axions has yet been verified.

Quinn has remained a staff scientist at SLAC since 1979 and has also served as the education and public outreach manager at SLAC since 1988. Moreover, Quinn is the president of the Contemporary Physics Education Project, a nonprofit group that produces material on contemporary topics in physics for high school and college students. In addition to winning the 2000 Dirac Medal, she was named a fellow of the Alfred Sloan Foundation from 1974 to 1978. She is a fellow of the American Physical Society and was elected to the American Academy of Arts and Sciences in 1998. Her work on CP symmetry is considered to be seminal, as are her pioneering contributions to the ongoing effort to develop a unified theory of the subatomic particles, quarks and leptons.

_R_

Ramart-Lucas, Pauline
(1880–1953)
French
Organic Chemist

The second woman to be appointed a full professor at the University of Paris (MARIE CURIE had been the first), Pauline Ramart-Lucas's research delved into a diverse array of topics in the field of organic chemistry. Her most significant finding involved the discovery of a new kind of isomerism (a type of structural difference) that led to a revision in the electronic structure of carbon in a broad class of organic molecules. After World War II, Ramart-Lucas shifted her focus from research to science policy, and she played a vital role in shaping the teaching of science in France from her seat as vice president of the French Consultative Assembly's educational section. The French government thrice recognized her many contributions, making her a knight in the country's Legion of Honor in 1928, elevating her to an officer a decade later, and conferring the rarified rank of commander upon her in 1953 (one of the few women to that point to be so honored).

Ramart-Lucas was born in Paris on November 22, 1880. Her parents were of very modest means and could not afford to educate her beyond elementary school. At a young age, therefore, Ramart-Lucas worked to supplement her family's income by arranging flowers very near the Sorbonne, France's most prestigious university. She became determined to attend the Sorbonne someday, and to this end, she began taking night classes to earn her secondary school diploma. After taking English lessons from a pharmacist who piqued her interest in chemistry and overcoming significant prejudice due to her gender, Ramart-Lucas obtained a _licence_ in physical sciences (a degree roughly equivalent to an American bachelor of science) at the age of 29.

Upon receiving her degree, Ramart-Lucas took a position in the laboratory of Albin Haller, an organic chemist who himself had followed a nontraditional career path. (He had apprenticed as a woodworker before turning to the sciences.) Under Haller's supervision, Ramart-Lucas obtained her doctorate in 1913. She left Haller's laboratory during World War I but returned after the war, becoming laboratory manager. With the exception of three years during World War II, she remained at the Sorbonne for the duration of her career.

Ramart-Lucas's work initially focused on the manner in which the dehydration of various alcohols (which are organic compounds, including the oxygen-hydrogen radical OH) causes molecular changes. In 1925, she was elevated to the position of lecturer, becoming a professor in 1930. She later shifted her research to examine the structure, ultraviolet absorption spectrum, and chemical reactivity of organic compounds, particularly a wide range of dyes. During the course of this work, she made her important discovery concerning the new type of isomerism that led to a revision in the electronic structure of carbon in a wide class of organic molecules.

Although Ramart-Lucas was removed from her position at the Sorbonne from 1941 to 1944 as France languished under the Vichy regime, she regained her post after the liberation of France by the Allies. After the war, Ramart-Lucas emerged as a leading science administrator in France, becoming vice president of the Consultative Assembly's educational section and sitting on the boards of France's national science museum and premier municipal science school. Along with her positions in France's Legion of Honor, Ramart-Lucas received many awards for her work, including the 1928 Ellen H. Richards Research Prize from the American Association of University Women. The author of more than 200 articles and a lengthy chapter on absorption spectra and molecular structure in Victor Grignard's treatise on organic chemistry, Ramart-Lucas also oversaw 50 doctoral theses and graduate memoirs, taking a particular interest in the work of female students. She never married nor had children. Ramart-Lucas died on March 13, 1953.

Ramey, Estelle
(1917–   )
American
Physiologist

Estelle Ramey is renowned for her research in the field of endocrinology. One of the first scientists to explore gender differences on the physiological level, Ramey studied such diverse topics as the relationship between sex hormones and longevity as well as the effect of hormones on the cardiovascular, pulmonary, and immune systems. Ramey was also committed to advancing the interests of female scientists. In addition to founding the Association of Women Scientists, she served as that organization's president. A prolific writer, Ramey has authored more than 150 articles for both scholarly and popular publications.

On August 23, 1917, Ramey was born in Detroit, Michigan, to Henry and Sarah White. Her father, a businessman, died before Ramey completed college, leaving the family financially strapped. In 1941, Ramey married law student James T. Ramey, with whom she had two children (James and Drucilla).

After graduating from Brooklyn College in 1937 (at the age of 19), the only work Ramey was offered was as a hat model. Desperate to earn more money in order to help her family, Ramey found a position as a teaching fellow in the department of chemistry at Queens College. (A former professor from Brooklyn College had been named the department's head and had remembered Ramey's aptitude in the subject.) Buoyed by this success, Ramey began graduate studies and earned her master's in chemistry from Columbia University in 1940.

After marrying in 1941, Ramey moved with her husband to Knoxville, Tennessee. Although the chairman of the chemistry department at the University of Tennessee refused to hire her (or any other woman), she was ultimately offered a position as a lecturer at the university in 1942. She remained at Tennessee until 1947, when her husband joined the Atomic Energy Commission in Chicago. After the move, Ramey resumed her graduate work at the University of Chicago. She was named a Mergler Scholar in 1949 and was awarded her Ph.D. in physiology in 1950.

Upon completing her studies, Ramey was appointed an instructor of physiology at the University of Chicago. At the same time, she received a U.S. Public Health Service postdoctoral fellowship in endocrinology. The branch of medicine that focuses on the endocrine system, endocrinology is the study of glands—such as the pituitary,

thyroid, and adrenal—that produce internal secretions that are introduced directly into the bloodstream. The secretions are then carried to other parts of the body whose functions they regulate and control.

In 1956, Ramey joined the faculty of Georgetown University Medical School in Washington, D.C., as an assistant professor. She was promoted to associate professor in 1966 and later to full professor. She was the first female tenured professor at Georgetown's Medical School and went on to become the first female chair of the university's department of physiology and biophysics. Her research continued to concentrate on endocrinology, especially on the correlation between sex hormones and longevity. In animal laboratory tests, Ramey examined the outcome of both synthetic and endogenous (internally generated) hormonal exposure and identified distinct differences between the sexes. She determined that because of the effect of estrogen on human systems, women lived longer than men.

Ramey remained at Georgetown until 1987, when she retired and became a professor emerita in the biophysics department. Her numerous scientific accomplishments have been well recognized. A member of the American Physiological Society, the American Chemical Society, and the Endocrine Society, Ramey also served on an executive advisory panel to the chief of naval operations and the General Medical Study Section of the National Institutes of Health. In addition, she was awarded the 1994 Women's Health Research Group Award. Ramey also was devoted to creating opportunities for women scientists. After founding the Association of Women Scientists, she acted as the organization's president from 1972 to 1974. In 1989, *Newsweek* magazine heralded her as one of the "25 Who Made a Difference." Ramey is still active in her field, participating in conferences and publishing articles.

Randoin, Lucie
(1888–1960)
French
Physiologist

Lucie Randoin's greatest scientific contribution arose from research she conducted into the way vitamins affect the body's functioning. Over the course of her career, she helped demonstrate the manner in which vitamins can impact human physiology and how illness and age can affect the way the body uses nutrients. The importance of her work was widely appreciated, and she received many awards and prizes for it, including being named as a commander in the French Legion of Honor (which was particularly noteworthy as few women to that point had been so recognized).

Randoin was born Lucie Fandard, in Boeurs-en-Othe, France, in 1888. She evinced an aptitude for science from an early age, winning particularly high marks in botany, chemistry, and physiology from the Parisian schools she attended. After obtaining a degree in physiology, she broke new ground by becoming the first woman to vie for a Natural Sciences Fellowship. She received one in 1911 and used it to study physiology at the University of Claremont-Ferrand. (Another recipient of a National Fellowship was her future husband, Arthur Randoin.) At the University of Claremont-Ferrand, Randoin came under the tutelage of Dr. A. Dastre, with whom she studied both general and nutritional physiology. Randoin earned her doctorate in 1918.

It was during the course of her work with Dastre that Randoin's interest in the study of vitamins was kindled. The existence of such substances and their importance for proper nutritional health had only recently begun to be elucidated in the work of American biochemists Elmer McCollum and MARGUERITE DAVIS, British biochemist Sir Frederick Gowland Hopkins, and Dutch physician Christian Eijkman. Serving as Dastre's research assistant during World War I, and then independently afterward, Randoin began to expand upon these scientists' work. (Tragically, Dastre disappeared shortly before the conclusion of the war, a loss to Randoin that was both professional and personal, as Dastre had been a mentor and strong advocate for her work.)

After Dastre's disappearance, Randoin continued her study of vitamins at the Oceanographic Institute's laboratory in Paris. In 1920, she moved over to the Ministry of Agriculture's Research Center, working in the physiology laboratory there. She was named the laboratory's director in 1924 and remained there until 1953. It was at the physiology laboratory that she performed some of her most significant work, demonstrating in the 1920s the ways that vitamins B and C impact the body's use of sugars and other chemicals. Randoin also explored vitamins' specific composition, providing insights that aided the general understanding of how these important nutritional elements operate.

In 1942, Randoin obtained a concurrent appointment as the director of the Institute of Nutritional Science. She would hold this post until her death. In both of her capacities, she continued to focus on vitamins, elucidating the particular roles specific vitamins played in human physiology, determining the doses of vitamins necessary for good health, and exploring how vitamins could act as preventive medicine to ward off illness. She also devised quality standards for vitamins to be used in foods and collaborated in founding a national school for dietary studies, which focused on training students to serve as dieticians in cafeterias and hospitals.

Randoin's scientific contributions were widely recognized. In addition to her position in the French Legion of Honor, she was named president of the French Society of Biological Chemistry, general secretary of the Institute of National Hygiene, and was a member of the French Biological Society. Randoin also acted as the official representative of France at international conferences in 1931 and 1934 on the standardization of vitamins. After a protracted illness, she died on September 13, 1960.

Ratner, Sarah
(1903–1999)
American
Biochemist

Sarah Ratner is best known for her pioneering research into amino acids (the subunits of protein molecules), which led her to a number of important findings. In addition to discovering argininosuccinic acid, she played a key role in elucidating the workings of the urea cycle, and she was one of the first scientists to use nitrogen isotopes to analyze amino acids. The recipient of the Carl Neuberg Medal from the American Society of European Chemists, Ratner is also acknowledged as a seminal figure in the development of biochemistry as a distinct discipline.

Born on June 9, 1903, in New York City, Ratner was the youngest of Aaron and Hannah (Selzer) Ratner's five children. Ratner's father had fled anti-Semitic persecution in Russia and immigrated to the United States in the late 19th century. He worked as a manufacturer, while Ratner's mother occupied herself with the family.

Although her parents initially opposed her desire to go to college, Ratner won a scholarship to attend Cornell University, where she received her bachelor's degree in chemistry in 1924. She returned to New York City for her graduate work, earning her master's degree from Columbia University in 1927. To make ends meet while she completed her Ph.D. at Columbia, she worked as an assistant in the university's fledgling biochemistry department. A Geiss Fellowship, which she received from 1934 to 1936, also helped to fund her graduate studies. Ratner's doctoral research laid the foundation for her future groundbreaking work on amino acids. She examined the interaction of formaldehyde with the amino acid cysteine and determined that together the two could create a stable product called thiazolidine carboxylate. In 1937, she was awarded her doctorate in biochemistry.

Ratner remained at Columbia after completing her Ph.D. After serving as a Macy Resident Fellow from 1937 to 1939, she rose to the rank of assistant professor. In 1946, she left Columbia for an assistant professorship at the New York University College of Medicine, where she began her research on the urea cycle. In most fishes, amphibians, and mammals (including humans), the nitrogen by-products emitted by protein metabolism are detoxified in the liver and excreted as urea (a soluble, harmless compound found in urine). The sequence of reactions leading to the formation of urea is called the urea cycle. In 1949, Ratner published the first of 15 articles in the *Journal of Biological Chemistry* on this process. By 1953, she and her colleagues had discovered argininosuccinate (which is a substance that plays a role in the production of urea and is formed when the amino acid citrulline is converted to arginine). Ratner then isolated and purified the enzyme that converts argininosuccinate to orithine and urea.

In 1954, Ratner accepted a post as an associate member of the department of biochemistry at the Public Health Research Institute for the City of New York. At the same time, she preserved her connection to New York University, becoming an adjunct associate professor of biochemistry there. In 1957, she was promoted to full member at the Public Health Research Institute, where she remained for the duration of her career. In this capacity, she continued her research on amino acids until her retirement in 1992. She died in New York City on July 28, 1999.

Ratner's contributions to the field of biochemistry were widely recognized. In 1959, she was awarded the Carl Neuberg Medal from the American Society of European Chemists. The American Chemical Society bestowed its Garvan Medal on Ratner in 1961, and in 1974, she was elected to the National Academy of Sciences. From 1978 to 1979, she received the National Institutes of Health's prestigious Fogarty Scholar-in-Residence fellowship. She also served on the editorial boards of both the *Journal of Biochemistry* and *Analytical Biochemistry*. Ratner is credited with contributing to the development of biochemistry as a distinct discipline, standing apart from both chemistry and physiology.

Ray, Dixy Lee
(1914–1994)
American
Marine Biologist

During her varied career, Dixy Lee Ray integrated her scientific expertise with the formation of public policy. After working as a professor of zoology for 27 years, Ray was appointed to the Atomic Energy Commission (AEC) and later served as the AEC's first female chair. In this capacity, she supported the use of nuclear energy, though she argued for stricter safeguards at nuclear power plants in order to protect the environment. In 1976, Ray was elected governor of her home state of Washington. Upon retiring from both political and academic life, Ray authored two books—*Trashing the Planet* and *Environmental Overkill*.

Dixy Lee Ray was born on September 3, 1914, in Tacoma, Washington, to Alvis and Frances (Adams) Ray. One of five girls in her family, Ray developed an early interest in marine biology during family outings on Fox Island in the Puget Sound. She attended Mills College in Oakland, California, graduating as valedictorian of her class with a degree in zoology in 1937. She went on to earn her master's degree in zoology from Mills in 1938. For the next four years, Ray taught science in Oakland's public schools. In 1942, she won a John Switzer Fellowship and began graduate work at Stanford University, where she received her Ph.D. in biological sciences in 1945.

Upon finishing her studies, Ray began an academic career at the University of Washington. She was promoted from lecturer to assistant professor of zoology in 1947 and to associate professor in 1957. Her research focused on marine invertebrates, especially crustaceans (the class of arthropods—such as shrimps, crabs, barnacles, and lobsters—that live in the water, breathe through gills, and have a hard outer shell and joined appendages). In particular, Ray examined the effects of the isopod Limnoria on submerged wood and the damage caused to marine organisms by boats, dry docks, and wharf pilings. Ray was awarded a prestigious Guggenheim Fellowship in 1952, and she was named an executive committee member of the Friday Harbor Laboratories in 1957. From 1960 to 1962, Ray served as special consultant in biological oceanography to the National Science Foundation. In 1964, her research took her aboard the ship *Te Vega*. As chief scientist and visiting professor at the International Indian Ocean Expedition, Ray studied the environment of the Indian Ocean.

Ray became involved in public policy in 1963 when she was appointed director of the Pacific Science Center in Seattle. In this position, Ray was charged with raising public interest in science. She converted six buildings from the 1962 Seattle World's Fair into an active science center, housing a science museum and a conference center for scientific symposia. Ray's efforts garnered national attention. In 1972, President Richard Nixon named her to the AEC. Within a year, she had been promoted to the organization's chair. Faced with the prospect of long-term fossil fuel shortages, Ray became an outspoken supporter of nuclear power. At the same time, she sought to eliminate safety defects in existing nuclear power plants. She led a public debate on issues such as emergency core cooling systems and nuclear waste, and she held workshops in major American cities to discuss the public's questions and concerns.

When the AEC ceased to exist in 1975—it was folded into the Energy Research and Development Administration—Ray was appointed assistant secretary of state in charge of the Bureau of Oceans and International Environmental and Scientific Affairs. Feeling stifled by her new position, Ray returned to her home state of Washington a few months later. Still interested in the intersection of science and public policy, Ray ran for governor of Washington, winning election in 1976. As governor, she maintained an interest in nuclear issues, publicly supporting some controversial efforts of the Bonneville Power Authority. She was not reelected at the end of her first term.

Ray retired from both politics and research but remained active in environmental issues. She gave lectures on environmental topics and wrote two books—*Trashing the Planet* (1990) and *Environmental Overkill* (1993)—in which she took issue with more strident environmental groups. In these works, Ray continued to voice support for nuclear power and denied the existence of phenomena such as global warming and acid rain. Throughout her career, she received recognition for her work. She was awarded the Clapp Award in Marine Biology in 1958, the Frances K. Hutchinson Medal for Service in Conservation in 1973, and the United Nations Peace Medal in 1973. She died of bronchial complications at the age of 79, on January 3, 1994.

Rees, Mina S.
(1902–1997)
American
Mathematician

Mina S. Rees, past president, the Graduate Center, the City University of New York. *(Courtesy of the Graduate Center CUNY)*

Mina S. Rees holds the distinction of being the first woman elected to the presidency of the American Association for the Advancement of Science. She was also the founding president of the Graduate Center of City University of New York. A highly respected mathematician, Rees brought together other mathematicians to solve complex problems during World War II. She later directed the mathematics branch of the Office of Naval Research (ONR).

The daughter of Moses and Alice Louise (Stackhouse) Rees, Mina Rees was born on August 2, 1902, in Cleveland, Ohio. The family later moved to New York, and Rees received her primary education in that city's public schools.

In 1923, Rees earned her bachelor's degree summa cum laude from Hunter College in New York City. She then taught at Hunter College High School from 1923 to 1926, during which time she also pursued graduate studies. She was awarded her master's degree in mathematics from the Teacher's College of Columbia University in 1925. The following year, she was appointed an instructor in the mathematics department at Hunter College. With the aid of a fellowship, Rees left New York for the University of Chicago, where she received her Ph.D. in mathematics in 1931 for her dissertation on abstract algebra.

Upon completing her studies, Rees returned to Hunter College, where she was promoted to assistant professor in 1932 and to associate professor in 1940. But her academic career was interrupted by the outbreak of World War II. In 1943, Rees joined the Applied Mathematics Panel (AMP) of the Office of Scientific Research and Development. Working as the executive assistant and technical aid to the director, Rees coordinated the office's program to involve leading mathematicians in the war effort. She established contacts with professors at preeminent universities, such as

Brown, Harvard, and Columbia, and recruited them to apply their expertise to military applications, such as shock waves, underwater ballistics, jet engine design, air-to-air gunnery, supply and munitions inspection methods, and computer development. Rees's work in this capacity won her the Certificate of Merit from President Harry Truman in 1948 as well as the Medal for Service in the Cause of Freedom from Britain's King George VI.

After the war, Rees remained a civil servant. From 1946 to 1953, she worked at the ONR—as head of its mathematics branch from 1946 to 1950 and as its deputy science director from 1952 to 1953. Rees guided the ONR in its research into a variety of programs, such as hydrofoils, logistics, computers, and numerical development. She was especially instrumental in shaping the ONR's implementation of projects studying mathematical algorithms for computing. During her time at the ONR, the organization funded university research programs to build computers—such as Project Whirlwind at the Massachusetts Institute of Technology.

In 1953, Rees returned to academia and Hunter College as dean of faculty and professor of mathematics. She married physician Leopold Brahdy in 1955. She was subsequently appointed dean of graduate studies for the City University of New York. In that capacity, she was responsible for establishing graduate programs by pooling faculty from the city colleges. Rees was named provost of the Graduate Division in 1968 and became the first president of the Graduate School and University Center in 1969. She retired in 1972.

Rees died on October 27, 1997, at the Mary Manning Walsh Home in Manhattan. However, her contributions to her field outlive her. By the time she retired, she had overseen the creation of 26 doctoral programs and had enrolled more than 2,000 students. Her efforts at the ONR and the AMP had guided several essential research projects. In addition to her election to the presidency of the American Association for the Advancement of Science in 1971, she received the first ever award for Distinguished Service to Mathematics from the Mathematical Association of America in 1962 and the American Public Welfare Medal of the National Academy of Sciences in 1983.

Richards, Ellen Swallow
(1842–1911)
American
Chemist

Ellen Swallow Richards is considered the founder of home economics, or what she called *oekology* in referring to the application of sanitary chemistry to the domestic sphere. After graduating from the Massachusetts Institute of Technology (MIT) as its first woman recipient of a bachelor of science degree, she remained there for 38 years, teaching chemistry and directing the Women's Science Laboratory. She actively worked to open up avenues for women to enter the sciences, and she succeeded in ushering women from the periphery into the official student body of MIT.

Richards was born on December 3, 1842, in Dunstable, Massachusetts. She was educated at home by her father, Peter Swallow, and her mother, Fanny Gould Taylor, both schoolteachers. In 1859, the family moved to Westford, Massachusetts, where Swallow completed her high-school education at Westford Academy in 1863. In 1868, at the age of 25, Swallow continued her education by enrolling at the newly established Vassar College. There, she studied astronomy under MARIA MITCHELL and chemistry under Charles Ferrand to earn her A.B. degree in 1870.

Richards applied to MIT, which did not admit women students at the time. The university made an exception by admitting her free of tuition, as a special student, and dubbed her enrollment the "Swallow experiment," as if conducting research on a species of bird. She became MIT's first woman graduate when she earned her B.S. degree in chemistry in 1873. That same year, she earned her M.A. degree from Vassar after submitting a thesis analyzing the amount of vanadium in iron ore. Swallow continued studying at MIT for two more years (while simultaneously teaching chemistry there untitled and unsalaried), but the university refused to grant her a doctorate. However, the Swallow experiment did result in a decree on May 11, 1876, allowing the official enrollment of women into the university to study chemistry, commencing Swallow's ca-

Known today as the woman who founded ecology, Ellen Swallow Richards was a 19th-century pioneer in the field of environmental and sanitary engineering. *(Courtesy of the Massachusetts Institute of Technology Museum)*

reer-long contributions to the advancement of women in academia and the professions.

On June 4, 1875, after a two-year engagement, Richards married Robert Hallowell Richards, the head of MIT's Department of Mining Engineering. The marriage yielded no children, though she did bring forth many innovations in the domestic sciences, such as the rerouting of plumbing lines past the furnace for hot water. In 1876, she established a laboratory for women at MIT, funded by the Boston Women's Education Association, a group that employed her as a chemistry teacher at a girls' school in Boston while she was still an undergraduate. With the title of assistant, Richards taught the women students chemical analysis, industrial chemistry, mineralogy, and biology, though MIT did not recognize this instruction as accredited coursework. In 1884, however, MIT allowed work from this laboratory into its official curriculum.

That year, MIT appointed Richards as the first woman on its faculty in the Department of Sanitary Chemistry. She retained this position—teaching the analysis of food, water, air, and sewage—for the next 27 years. She published 10 books in her career, including *The Chemistry of Cooking and Cleaning: A Manual for House-keepers* in 1882 and *Food Ma-*

terials and Their Adulterations in 1885. The latter elicited the passing of the first Pure Food and Drug Act in Massachusetts. In 1887, the Massachusetts Board of Health commissioned Richards to survey its drinking water and sewage, for which she tested 40,000 samples and created the so-called Richards's Normal Chlorine Map, predicting instances of inland water pollution.

Richards participated in many associations, founding many aimed at supporting women in the home and in the professions. In 1879, the American Institute of Mining and Metallurgical Engineers inducted her as its first woman member. In 1881, Richards collaborated with Marion Talbot to form the Association of Collegiate Alumnae, later renamed the American Association of University Women. She established the New England Kitchen in Boston in 1890 to disseminate information about proper food hygiene; she later modeled the Rumford Kitchen at the 1893 Chicago World's Fair, which she supervised with Mary Hinman Abel, on the New England Kitchen example.

In 1899, Richards organized a conference on home economics at Lake Placid, New York; this initiative generated the momentum by 1908 to establish the American Home Economics Association, which she served as president for its first two years and underwrote the publication of its *Journal of Home Economics*. In 1910, she finally received an honorary doctorate from Smith College. She died on March 30, 1911, in Jamaica Plain, Massachusetts, at the age of 69.

Richardson, Jane S.
(1941–   )
American
Biochemist and Crystallographer

Jane S. Richardson and her husband, David C. Richardson, run the Richardson laboratory at Duke University, where they conduct research on the three-dimensional structures of proteins. The Richardsons use X-ray crystallography to determine and describe the folding patterns of proteins, and they create free computer software to manipulate the graphical depiction of protein structures according to their actual physical determinants. Richardson likens this practice of protein folding to the ancient Japanese art of origami, or decorative paper folding. Her work earned her the title of genius from the John D. and Catherine T. MacArthur Foundation in 1985.

Richardson was born in 1941 in Teaneck, New Jersey. She distinguished herself early on in the field of science, placing third in the national Westinghouse Science Talent Search in 1958. She matriculated at Swarthmore College, where she studied philosophy, mathematics, and physics. She graduated cum laude with a bachelor of arts degree in 1962, and she earned a Woodrow Wilson Fellowship that

same year. The next year, she married David C. Richardson, and the couple had two children: one son, Robert, who was born in 1973, and one daughter, Claudia, who was born in 1980.

Richardson continued her education at Harvard University, where she earned her M.A. in philosophy of science and her M.A.T. in natural sciences in 1966. She simultaneously worked as a technical assistant in the Department of Chemistry at the Massachusetts Institute of Technology from 1964 through 1969. That year, she moved to the National Institutes of Health (NIH) in Bethesda, Maryland, where she worked as a general physical scientist at the Laboratory of Molecular Biology. That year also saw her first publication, which she coauthored for the *Proceedings of the National Academy of Sciences*.

Richardson remained at the NIH for only one year before moving to Duke University, where she has remained ever since 1970. She started there as an associate in the Department of Anatomy and worked her way up to become the James B. Duke professor of biochemistry, a medical research associate professor of anatomy at the Duke University Medical Center, and the codirector of the medical center's Comprehensive Cancer Center.

The Richardson lab conceived of *kinemages*, or computer-generated images of protein structures. The Richardsons also invented Mage software, programming in the physical properties of proteins so that computer users could view and manipulate kinemages in ways that reflected the actual reactions of protein structures in reality. Mage software is used extensively in teaching and in other applications for predicting how actual proteins would transform in response to stimuli.

In recognition of her groundbreaking work in protein crystallography, Richardson received the MacArthur "Genius" Award in 1985. The next year, Swarthmore College awarded her an honorary doctor of science degree, and she has received two more honorary doctorates: one from Duke University in 1990 and one from the University of North Carolina in 1994. Richardson was further honored by *Science Digest*, which included her work in the list of its "One Hundred Best Innovations."

Ride, Sally K.
(1951–   )
American
Astronaut and Physicist

Sally Kristen Ride catapulted into the annals of history as the first American woman to travel in space on the space shuttle *Challenger* in June 1983. She went on another *Challenger* flight in 1984 and was training for a third flight when the *Challenger* exploded during liftoff in January 1986. President Ronald Reagan appointed Ride to the Rogers Commission, which was charged with investigating the *Challenger* tragedy, and she concluded that the National Aeronautics and Space Administration (NASA) had knowingly compromised the safety of its astronauts. Ride resigned from NASA in 1987 to become an administrator, physics professor, and children's book author.

Ride was born on May 26, 1951, in Encino, California. Her father, Dale Burdell Ride, was a political science professor at Santa Monica Community College, and her mother, Carol Joyce Anderson Ride, was an English tutor. Ride earned a partial scholarship to the prestigious Westlake School for Girls in Los Angeles on the strength of her tennis prowess (she ranked 18th on the national junior circuit). Dr. Elizabeth Mommaerts, the physiology teacher, inspired Ride's interest in science at a time when the field still discouraged women from becoming scientists. Ride graduated in 1968 and matriculated at Swarthmore College, where she studied physics only briefly while deciding whether to pursue a career as a professional tennis player.

Ride decided to continue her education at Stanford University, splitting her focus between Shakespeare and Einstein to earn dual degrees—a B.A. in English and a B.S. in physics—in 1973. She chose science over literature for her master's work, which she completed in 1975 at Stanford. She remained at Stanford to pursue her doctorate, writing a dissertation entitled "The Interaction of X-Rays with the Interstellar Medium" for her 1978 Ph.D.

That year, Ride whimsically sent in a postcard application to NASA for entrance into its astronaut training program, which was recommencing after laying fallow since the 1960s and was considering women applicants for the first time. NASA admitted Ride as one of 35 trainees chosen from a field of more than 8,000 applicants. On the second space shuttle flight, in November 1981, and the third, in March 1982, Ride worked on the ground-support crew as a capsule communicator (capcom), acting as an intermediary between the ground team and flight crew. On July 26, 1982, Ride married fellow active astronaut Steven Alan Hawley at his family home in Kansas; the marriage yielded no children and ended in divorce in 1987.

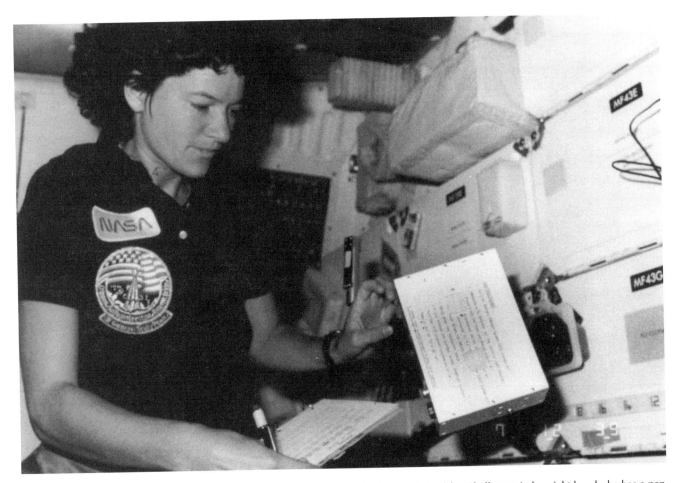

Astronaut Sally K. Ride, a mission specialist with NASA, carrying out a test in the middeck of the *Challenger*. In her right hand, she has a pen and pad for logging test findings. *(Courtesy of NASA)*

Ride became the first American woman in space on flight STS-7 of the space shuttle *Challenger* from June 18 through June 24, 1983. She operated the shuttle's mechanical arm to deploy a satellite, the first such operation ever performed. Ride again worked the robot arm on flight STS-41G of the *Challenger* from October 5 through October 13, 1984—this time using the arm to chip ice off the outside of the shuttle and adjust a radar antenna. Ride's training for her third shuttle flight was interrupted by the midair explosion of the *Challenger* in January 1986, and she was reassigned to the Rogers Commission investigating the circumstances leading up to that explosion. She concluded that NASA had willfully compromised its astronauts' safety in its haste to launch shuttle flights with possible equipment problems.

Before resigning from NASA in 1987, Ride wrote a report entitled *Leadership and America's Future in Space* recommending a redirection of the administration's focus toward international research projects on environmental problems. From NASA, Ride joined Stanford University's Center for International Security and Arms Control for two years before moving to the University of California at San Diego as a physics professor and as director of the California Space Institute. She has written the children's books *To Space and Back* (with childhood friend Susan Okie) and *Voyager: An Adventure to the Edge of the Solar System* (with former tennis partner Tam O'Shaughnessy) to encourage children's interest in the sciences, an area that she finds lacking in the American educational system. Her work has earned her the Jefferson Award for Public Service, and she received the National Spaceflight Medal twice.

Rigas, Harriet B.
(1934–1989)
Canadian/American
Electrical Engineer

Harriet B. Rigas distinguished herself as both an electrical engineer and a computer scientist. She developed improvements in binary coding that increased computer storage capacity, and she innovated automatic patching systems for analog/hybrid computers. She received several important awards in her lifetime, including the Society of Women Engineers Achievement Award, and after she died, several awards were named in her honor.

Rigas was born on April 30, 1934, in Winnipeg, Manitoba. She attended Queen's University in Ontario, where she earned her bachelor's degree in 1956. She spent the next year working at the Mayo Clinic as an engineer before returning to school to pursue graduate study at the University of Kansas. She earned her master of science degree in 1959, and that year she married. In 1963, she became the first woman to receive a doctorate from the

School of Engineering at the University of Kansas, in electrical engineering.

While completing her doctorate, Rigas held a position as an instructor in physics, mathematics, and engineering at Ventura College. When she received her doctorate, she went to work for Lockheed Missile and Space Company as a senior research engineer. However, she remained in the industry for only two years before returning to academia.

Washington State University (WSU) hired her in 1965, commencing a 19-year relationship. By 1968, WSU had named her as manager of its Hybrid Facility. Between 1975 and 1976, she served in a concurrent position as the program director for the National Science Foundation. In 1976, WSU promoted her to full professor, a post she retained until 1984.

Rigas performed the fund-raising and developed the curriculum to establish WSU's electrical and computer engineering department, which she then chaired. She also chaired WSU's Commission on the Status of Women and served as the president of the university's Association for Faculty Women. Throughout her career, she actively supported the advancement of women in academia and in the professions.

In 1980, the Naval Postgraduate School named Rigas as the chair of its electrical engineering department. In 1987, Michigan State University hired her as a professor and chair of its department of electrical engineering, a position she held until her death on July 26, 1989. Rigas received many honors throughout her career and beyond, as numerous awards were named in her honor. In 1982, the Society of Women Engineers granted her its Achievement Award. The next year, the University of Kansas recognized her accomplishments with its Distinguished Engineering Service Award. In 1988, she received the Institute of Electrical and Electronics Engineers (IEEE) Educational Activities Board Accreditation Award, and the next year, the IEEE bestowed its Rare Fellow Award on her posthumously.

Two awards are named after Rigas. WSU's Association for Faculty Women recognizes the academic accomplishments and professional potential of its outstanding woman doctoral student with the Harriet B. Rigas Award. The IEEE, in conjunction with Hewlett-Packard, sponsors the $2,000 Harriet B. Rigas Award to the woman professor who best supports undergraduate education in engineering or computer science.

Roberts, Dorothea Klumpke
(1861–1942)
American
Astronomer

Dorothea Klumpke Roberts achieved many firsts in her career as an astronomer: She earned the first Prix des Dames from the Astronomical Society of France; she was the first woman elected as an officer by the Paris Academy of Sciences; the first woman awarded a D.Sc. degree in mathematics from the Sorbonne; and the first woman to make airborne astronomical observations. She ensured the continuation of her legacy of excellence in astronomy by endowing the Dorothea Klumpke Roberts Prize Fund with the Astronomical Society of the Pacific.

Roberts was born on August 9, 1861, in San Francisco, California. Her father, a German immigrant named John Gerard Klumpke, had moved the family west to cash in on the California Gold Rush by selling his handmade boots for the astronomical price of $100. Her mother, Dorothea Matilda Tolle, believed in equal educational opportunities for her five daughters as for her two sons, so she sent her children to the finest European schools in Germany, Switzerland, and France. Dorothea's four sisters thrived on this intellectual stimulation: Anna became an artist and the biographer of Rosa Bonheur; Julia became a concert violinist and composer; Matilda became a pianist; and Augusta became a neurologist. Dorothea also thrived, first in the study of languages and ultimately in the study of mathematical astronomy at the University of Paris, where she earned her B.S. degree in 1886.

The Paris Observatory hired Roberts in 1887 to measure the positions of stars on photographic plates. This project took on increased importance after the International Astronomical Congress agreed at its Paris meeting to generate a *Carte du Ciel,* or Chart of the Heavens, cataloging all stars to the 11th magnitude after photographing the entire sky. Roberts drew on her previous study of languages to translate the proceedings and papers from the conference into French, and she continued to catalog star measurements from photographs. The excellence of her work prompted the Paris Observatory to choose her over 50 male applicants as director of the Bureau of Measurement, created in 1891 to oversee the photographic star measurements contributed to the *Carte du Ciel* project. She retained this position until 1901.

In 1889, the Astronomical Society of France awarded its first Prix des Dames to Roberts. In 1893, she achieved two more firsts: The Paris Academy of Sciences elected her as its first woman Officier; and on December 14, she became the first woman to earn a doctor of science degree in mathematics from the University of Paris. For her dissertation topic, she took up the unfinished work of SOFIA KOVALEVSKAIA—a mathematical study of the rings of Saturn. Also in 1893, Roberts's paper on astronomical mapping won a $300 prize from the French Academy of Sciences at the World's Columbian Exposition, held in Chicago.

On November 16, 1899, Roberts became the first woman to make airborne astronomical observations when the Société Français de Navigation Aérienne appointed her to the team that launched in a balloon to observe the Leonid meteor showers. The balloon *La Centaure* reached an altitude of 1,640 feet, where the team observed 24 meteors, 11 of them Leonid meteors.

In 1901, Roberts married the 72-year-old astronomical photography pioneer Isaac Roberts and moved into his English estate in Crowborough, Sussex. The two worked collaboratively in the observatory he had built and equipped with a 20-inch reflector and camera until his 1904 death. Roberts continued their work and, in 1929, published *Isaac Roberts' Atlas of 52 Regions, A Guide to William Herschel's Fields of Nebulosity.* In 1932, the French Academy of Sciences awarded Roberts its Helene-Paul Helbronner Prize for this work.

In 1934, the Legion of Honor recognized her 48 years of service to French astronomy by electing her a chevalier; and the president of France honored her with the Cross of the Legion. In 1937, Roberts endowed the Dorothea Klumpke Roberts Prize Fund with the Astronomical Society of the Pacific to award support to outstanding astronomy and mathematics students at the University of California at Berkeley, an extension of the Klumpke-Roberts Lecture Fund that she had already established.

Dorothea Klumpke Roberts died in San Francisco on October 5, 1942. In 1974, the Astronomical Society of the Pacific transformed the Klumpke-Roberts Lecture Fund into the Dorothea Klumpke-Roberts Award "for outstanding contributions to education or popularization in astronomy." Recipients have included HELEN SAWYER HOGG in 1983 and HEIDI HAMMEL in 1995.

Robinson, Julia Bowman
(1919–1985)
American
Mathematician

Julia Bowman Robinson gained renown as the pivotal player in the solution of Hilbert's 10th problem, a challenge that occupied her for 22 years. After conclusive proof demonstrated that integers could not be used to solve a given diophantine equation, the mathematical community showered recognition on Bowman Robinson: the National Academy of Sciences (NAS) elected her as its first woman mathematician in 1975; the University of California at Berkeley appointed her as a full professor in 1976; and the American Mathematical Society (AMS) appointed her as its first woman officer in 1978 and its first woman president in 1982. Robinson preferred not to be remembered as the first woman to achieve her accomplishments but rather as a top-notch mathematician without regard to her gender.

Robinson was born on December 8, 1919, in St. Louis, Missouri. Her mother, Helen Hall Bowman, died when Julia was two years old. Her father, Ralph Bowman, sent Julia and her older sister, Constance, to live with their grandmother near Phoenix, Arizona. He joined them a year later after he had retired from his machine tool and equipment business. In 1925, he married Edinia Kridelbaugh and moved the family to San Diego, California. In 1928, Julia's half sister, Billie, was born.

Robinson contracted scarlet fever at the age of nine, resulting in a monthlong quarantine, and a year later developed rheumatic fever, keeping her bedridden for a year. She first became intrigued with mathematics when her tutor, who taught her curricula covering fifth- through eighth-grade material, pointed out that the square root of two did not repeat decimals, according to available calculations. She entered San Diego High School in the ninth grade and graduated in 1936 with honors in mathematics and science courses as well as the Bausch-Lomb medal for all-around excellence in science.

At the age of 16, Robinson matriculated at San Diego State College (SDSC—now San Diego State University), majoring in mathematics and teaching. Her sophomore

Julia Robinson, a mathematician who played a pivotal role in the solution of Hilbert's 10th problem. *(American Mathematical Society)*

year turned tragic when her father, who had depleted his savings in supporting his family, committed suicide. SDSC charged her a mere $12 tuition per semester to facilitate her continuation there.

For her senior year, Robinson transferred to the University of California at Berkeley, where she blossomed academically and socially. She was elected into the honorary mathematics fraternity, and she met her future husband in the professor for her number theory course, Raphael M. Robinson. After her 1940 graduation, she remained at Berkeley for graduate study, and in December 1941, she married Robinson despite the fact that nepotism rules prevented her from subsequently retaining her teaching assistantship.

During World War II, Robinson worked on secret military projects under Jerzy Neyman at the Berkeley Statistical Laboratory. Heart scarring from her rheumatic fever caused her to lose her only baby, but she found redemption by redirecting her energy back to mathematics. In 1947, she pursued her doctorate under Alfred Tarski, writing her dissertation—"Definability and Decision Problems in Arithmetic"—as a proof that the theory of the rational number field was unsolvable algorithmically. Berkeley conferred a Ph.D. on her in 1948.

Interestingly, it was this proof that played a key role in the solution of David Hilbert's 10th problem, which Tarski introduced to her that very year. Robinson spent the next decade ensconced in the problem until Martin Davis and Hilary Putnam sent her a paper in 1959 that significantly advanced her theorem. In 1961, the three mathematicians collaborated on their own paper, which took Robinson one step closer to the final solution.

It was almost another decade before Robinson arrived at the final solution, when she collaborated in 1970 with Leningrad mathematician Yuri Matijasevic, who took the last step from the foundation set by Robinson to prove the insolubility of Hilbert's 10th problem. It was at this point that Robinson became fully recognized for her mathematical skills, leading to her NAS election in 1975, her full professorship at Berkeley in 1976, and her AMS presidential appointment in 1982. The MacArthur Foundation also granted her a five-year, $60,000 fellowship at this time.

In the summer of 1984, Robinson learned that she had developed leukemia, and she died on July 30, 1985.

Rockwell, Mabel MacFerran
(1902–1979)
American
Electrical Engineer and Aeronautical Engineer

Mabel MacFerran Rockwell was one of the first women aeronautical engineers in the United States, and she contributed to this field by demonstrating the greater effectiveness and efficiency of spot welding as opposed to riveting. She also designed the guidance systems for the Polaris missile and the Atlas guided missile launcher until she resigned due to her antiwar sentiments. She contributed to other fields as well: As an electrical engineer, she devised a means for tracing multiple simultaneous transmission malfunctions, and she helped to design the electrical installations at the Boulder and Hoover Dams. As an aquatic and mechanical engineer, she designed underwater propulsion systems and submarine guidance mechanisms.

Mabel MacFerran was born in 1902 in Philadelphia, Pennsylvania. Her father was Edgar O. MacFerran, and her mother was Mabel Alexander. She attended The Friends' School in Germantown and then matriculated at Bryn Mawr College. She later transferred to the Massachusetts Institute of Technology, where she ranked first in her class when she graduated with a bachelor's degree in science, teaching, and mathematics in 1925. She then traveled west to Stanford University, where she earned her degree in electrical engineering in 1926.

MacFerran remained in California, starting her professional career as a technical assistant at the Southern California Edison Company. There, she proved an excellent problem solver, devising a method for locating multiple simultaneous faults in transmission relays by utilizing symmetrical components in multiple-circuit power lines. The Metropolitan Water District of Southern California then hired MacFerran as an assistant engineer on the Colorado River Aqueduct Project. She served as the only woman to work on the installations of the electrical systems at the Boulder and Hoover Dams. She also worked for the United States Bureau of Reclamation with the Central Valley Project, designing transportation systems to distribute water from dams to Californian irrigation districts.

In 1935, MacFerran married Edward W. Rockwell, a fellow electrical engineer. A daughter, Margaret Alice, was born the next year. At about this time, Lockheed Aircraft Corporation hired Rockwell as a plant electrical engineer, and she was promoted in 1940 to the rank of production research engineer in charge of 25 engineers and technicians. (Apparently, Rockwell received promotions quicker than her husband, and they divorced in 1962.)

Although Rockwell opposed war, her innovations contributed to the war effort in the 1940s. Rockwell conducted research to support the replacement of riveting with spot welding, which proved quicker, cleaner, cheaper, and more effective in airplane production. She published a report on spot welding in the July 1941 issue of *Aviation* magazine, and in the October 1942 issue of the *Welding Journal,* she coauthored a paper entitled "The Effect of Weld Spacing on the Strength of Spotwelded Joints." She also published reports on producing airplane parts from sheet metal: "Mechanics of Deep Drawing Sheet Metal Parts," which she coauthored for the February 1942 issue of *Aero Digest,* and "Stretch-Forming Contoured-Sheet Metal Aircraft Parts," which appeared in the October 1942 issue of *Welding Journal.*

Also during World War II, Rockwell conducted research on underwater propulsion systems and submarine guidance. After the war, Westinghouse hired Rockwell to work on the electrical control system of the Polaris missile launcher. Rockwell resigned from this project due to her antiwar stance, but she then worked on the Atlas guided missile systems at Convair.

In 1958, President Dwight D. Eisenhower recognized these contributions to national defense by naming Rockwell the Woman Engineer of the Year. Also in 1958, the Society of Women Engineers granted Rockwell its Achievement Award. Rockwell spent the end of her career as a consulting technical editor in the electrical engineering department at Stanford University. She continued there into her 70s, earning a reputation from students and faculty alike for her "Mabelizing," or editing papers to her own tough standards. Rockwell died in June 1979.

Roemer, Elizabeth
(1929–)
American
Astronomer

Elizabeth Roemer is best known as a "recoverer" of "lost" comets. Throughout her career, she confirmed the existence of 79 periodic comets, predicting their returns based on calculations from their last visits. In addition, Roemer contributed significantly to the field of astronomy through work on professional commissions and organizations and through extensive publication. In recognition of these diverse contributions, she received several awards distinguishing her career.

Elizabeth Roemer was born on September 4, 1929, in Oakland, California. She attended the University of California at Berkeley, where she was a Bertha Dolbeer Scholar. She graduated in 1950 with a B.A. in astronomy. She continued with graduate study at Berkeley, financing it over the next two years by teaching adult classes in the Oakland public school system. While continuing her doctoral study, she held a fellowship as an assistant astronomer and then as a lab technician at Lick Observatory. The University of California at Berkeley granted her a Ph.D. in astronomy in 1955. Roemer remained at Berkeley as an assistant astronomer for the next year while simultaneously holding a position as research associate at the University of Chicago's Yerkes Observatory.

In 1957, she became an astronomer with the U.S. Naval Observatory's Flagstaff Station, where she remained for almost a decade. During this time, she began to earn her reputation for recovering comets by photographing them through the high-definition 40-inch astronomic reflecting telescope. She used these photographs to demonstrate that short-period comet nuclei could not exceed three to four kilometers and long-period comet nuclei could not exceed seven to eight kilometers.

The Lick Observatory named Roemer as its acting director in 1965. That same year, her colleague, P. Wild, named asteroid 1657 Roemera in her honor. The next year, the University of Arizona hired her as an associate professor of astronomy in its Lunar and Planetary Laboratory, and in 1969, the university promoted her to the rank of full professor. In 1980, Stewart Observatory hired her as a full astronomer, and she remained on the faculty at the University of Arizona.

Roemer published extensively throughout her career. The *Astronomical Journal* published "Astronomic Observations and Orbits of Comets" in its October 1961 edition. *The Publications of the Astronomical Society of the Pacific* published three important papers by Roemer: "An Outburst of Comet Schwassmann-Wachmann I" (1958); "Activity in Comets at Large Heliocentric Distance" (October 1962); and "Comet Notes" (June 1971).

Roemer served on many professional commissions and organizations throughout her career. Between 1973 and 1975, Roemer served on the Space Science Review Panel Associateship Program of the Office of Scientific Personnel for the National Research Council, chairing the panel in 1975. Between 1976 and 1979, Roemer served as the president of Commission 6 Astronomical Telegrams. Then from 1979 through 1982, Roemer served as the vice president of Commission 20 and as its president from 1982 through 1985. In 1988, Roemer returned to the position of president of Commission 6 Astronomical Telegrams.

Roemer's career was distinguished by many honors: She received the Dorothea Klumpke Roberts Prize as an undergraduate at Berkeley; she won the Mademoiselle Merit Award in 1959; in 1971, the National Academy of Sciences granted her the Benjamin Apthorp Gould Prize; and in 1986, she received a NASA Special Award. Roemer retained emerita status at the University of Arizona after retiring from active duty on the astronomy faculty.

Roman, Nancy Grace
(1925–)
American
Astronomer

Nancy Grace Roman led the initiative at the National Aeronautics and Space Administration (NASA) to gather astronomical research from space as a means of avoiding the atmospheric interference inherent in earthbound observation.

Roman was born on May 16, 1925, in Nashville, Tennessee. Her mother was Georgia Frances Smith, and her father, Irwin Roman, was a geophysicist with the U.S. Geological Survey. The family moved to Baltimore, Maryland, where Roman graduated from Western High School in 1943. She then matriculated at Swarthmore College, where she studied astronomy and worked at the Sproul Observatory. She became the Joshua Lippincott Memorial Fellow and graduated with a B.A. in 1946. She conducted her doctoral research at the Yerkes Observatory of the University of Chicago, which granted her a Ph.D. in astronomy in 1949. That year, the *Astrophysical Journal* published her dissertation, "The Ursa Major Group," an investigation of the radial velocities, spectra, and convergent point of the cluster that includes all but the end stars of the Big Dipper.

Roman worked as a summer research associate at the Warner and Swasey Observatory of the Case Institute of Technology in 1949. She then returned to the Yerkes Observatory as a research associate until 1952 and then as an astronomy instructor until 1955, when she was appointed to an assistant professorship. However, she realistically comprehended that women generally did not receive tenured positions at that time, so she left academia for the United States Naval Research Observatory (NRO) in Washington, D.C., in 1955.

At the NRO, she worked in the Radio Astronomy Branch as a physicist, researching radio star spectra on the observatory's 50-foot cast aluminum mirror. In 1956, she was promoted to the rank of astronomer and head of the

Nancy Grace Roman, an astronomer who led the initiative at NASA to gather astronomical research from space in order to avoid the interference inherent in earthbound observations. *(AIP Emilio Segrè Visual Archives, Roman Collection)*

Microwave Spectroscopy Section. The next year, she was invited at the last minute to attend the Soviet Academy of Sciences Symposium in honor of its new Bjuraken Astrophysical Observatory. At this same time, NASA was being formed, and as a result of contacts she made on the Soviet trip, NASA asked Roman to head the office of satellites and sounding rockets.

This job change moved her from research to administration, but she was interested in the challenge of working with people (which she considered more complex than working with stars). Roman organized NASA's program devoted to astronomical research from outside the Earth's atmosphere, a project that had not been attempted before. Roman's programs contributed to the research conducted by the Orbiting Solar Observatories, the *Viking* probe of Mars, the *Copernicus* satellite, the U.S. space station *Skylab*, and the Hubble space telescope.

Over the decade from 1966 through 1976, Roman received honorary degrees from Russell Sage College, Hood College, Bates College, and Swarthmore College. In 1980, Roman retired from NASA to care for her elderly mother, though she maintained a working relationship with the administration as a senior scientist with the Astronomical Data Center. In 1994, NASA promoted her to the head of the Astronomical Data Center. A year earlier, Roman spent a week performing astronomical observations, a task she had not carried out in almost 20 years.

Rothschild, Miriam
(1908–   )
English
Naturalist

Miriam Rothschild became a preeminent entomologist and parasitologist, specializing in the study of fleas, despite the fact that she never earned a degree and received little formal education. Her natural curiosity, unstifled by academia, roamed far and wide, with her interests ranging from marine biology and chemistry to horticulture and zoology.

Miriam Louisa Rothschild was born on August 5, 1908, at Ashton Wold, her family's estate near Peterborough, England. She was the oldest of four children born to Rozsika von Wertheimstein Rothschild, a Hungarian lawn tennis champion, and Nathaniel Charles Rothschild, a renowned banker but an even more avid zoologist and flea expert. He founded the Society for the Promotion of Nature Preserves, which listed 280 sites throughout Britain that were vital to preserve (Miriam later visited 180 of these sites to ascertain the success of the society's efforts). Her uncle, Walter Rothschild, was also an avid naturalist and collector of 2,250,000 butterflies, 300,000 bird skins, and 200,000 bird eggs, among

other things. Miriam gained her love of naturalism from both her uncle and father, though she lost this spark temporarily after her father committed suicide.

Rothschild attended evening classes at a local polytechnic institute before enrolling at the University of London, but her double major in English literature and zoology proved too wide a range to sustain—lectures inevitably overlapped, frustrating her. She eventually opted to study marine biology under G. C. Robson when she was awarded the London University Table, an all-paid position researching marine snails in Naples, Italy. She continued to study snail-borne parasites for the next seven years in England, until her laboratory at the Marine Biological Station in Plymouth was bombed in 1940.

Rothschild joined a team of marine biologists and mathematicians, including the eccentric genius Alan Turing, on Britain's top secret Enigma project. For some odd reason, the marine biologists proved better at cracking Nazi codes than the mathematicians. In 1943, she married George Lane, a Hungarian émigré who became a British soldier, and the couple had four children and adopted two more before divorcing in 1957. Rothschild always prioritized her children over her work, though her chronic insomnia allowed her to work nights after tending to the children and wounded war refugees during the day.

Rothschild collaborated with the Nobel laureate Tadeus Reichstein to demonstrate that the monarch caterpillar and butterfly's diet of milkweed deters predators, who find the milkweed glycosides distasteful, if not poisonous. She later published books on butterflies—*Dear Lord Rothschild: Birds, Butterflies, and History* (dedicated to her Uncle Walter, the famed lepidopterist) and *Butterfly Cooing like a Dove.* She published her first book, *Fleas, Flukes, and Cuckoos,* in 1944, and thereafter she released more than 350 papers on entomology, neurophysiology, chemistry, and zoology.

Throughout her career, Rothschild continued to catalog her father's flea collection at the Natural History Museum in London, spending 20 years to compile six volumes. Rothschild relied not so much on intuition as on keen observation in her research. While studying fleas in Australia, she noted a correlation between the reproductive cycles of the hosts, rabbits, and those of fleas. The fleas migrate from the male rabbit to the female rabbit while these hosts copulate, and then the fleas further migrate to the offspring upon their birth, all the while experiencing syncretic reproductive cycles themselves.

Rothschild's scientific curiosity continued into her 90s. Although temporarily wheelchair-bound due to an accident at 91, she planned experiments to test her hypothesis on the telepathic capabilities of dogs, and she devoted 150 acres of her Ashton Wold estate to the propagation of wildflowers in order to preserve genetic diversity by creating a seed bank.

Rowley, Janet D.
(1925–   )
American
Cytogeneticist

Starting in the early 1960s, Janet Davison Rowley conducted cytogenetic research on the chromosomal abnormalities that cause cancers such as leukemia. In the early 1970s, she discovered several chromosomal translocations, or the transferral of one type of chromosome onto another type of chromosome, that explained the cause of several varieties of leukemia. The importance of her work continued to be recognized and honored more almost three decades later, as she received important awards in her field in the late 1990s.

Janet Davison was born on April 5, 1925, in New York City. Her father, Hurford Davison, was a Harvard M.B.A. who taught college courses on retail store management, and her mother, Ethel Mary Ballantyne, was a high-school English teacher and librarian. In 1940, at the age of 15, Davison entered "The Four Year College," a scholarship program established by Robert Maynard Hutchins at the University of Chicago that treated attendees as college students as they completed their last two years of high school and their first two years of undergraduate study.

Davison earned her Ph.B. degree in 1944, then remained at the University of Chicago to earn her B.S. degree in 1946 and her M.D. in 1948. She married Donald Adams Rowley on December 18, 1948, the day after graduating from medical school. The couple had four sons: Donald Adams Jr., born in 1952; David Ballantyne, born in 1954; Robert Davison, born in 1960; and Roger Henry, born in 1963.

Janet Rowley spent the bulk of her career at the University of Chicago, where she commenced her professional life as a research assistant in 1949. She performed her internship at Marine Hospital in Chicago, her residency at Cook County Hospital, and then a clinical instructorship in neurology at the University of Illinois Medical School. She obtained her medical license in 1951, and she was an attending physician at the Infant and Prenatal Clinics in the Department of Public Health in Montgomery County, Maryland, from 1953 through 1954. From 1955 through 1961, she raised her children and worked part-time with mentally handicapped children as a research fellow at the Dr. Julian Levinson Foundation in Chicago.

During her husband's 1961 sabbatical in Oxford, England, Rowley secured a position through the National Institutes of Health as a special trainee in the radiobiology laboratory at Churchill Hospital. There, she studied chromosome abnormalities in hematological diseases. Upon her 1962 return to the University of Chicago as a part-time research associate in the Department of Hematology, she discovered a great demand for her research into the cytogenetic causes of cancer.

The University of Chicago promoted Rowley to the rank of associate professor in 1969. The next year, she used her own sabbatical to return to Oxford, where she learned the new technique of *banding,* which allowed for viewing chromosomes with greater clarity. Using this technique while researching in Scotland and Sweden, Rowley noticed a translocation of the so-called Philadelphia chromosome, or chromosome 22, into chromosome nine. This observation proved the cause of chronic myeloid leukemia. Rowley went on to identify a translocation of chromosome eight and 21 in patients with acute myeloblastic leukemia.

In 1977, Rowley became a full professor at the University of Chicago medical school and at the Franklin McLean Memorial Research Institute. In 1984, these institutions named her the Blum-Riese Distinguished Service professor. In the 1990s, Rowley shifted her focus from identifying translocation breakpoints to cloning genes that might block abnormal chromosome movement and thus prevent the occurrence or spread of malignancy. Rowley helped disseminate the results of research in her field as the cofounder and coeditor of the journal *Genes, Chromosomes and Cancer.*

Rowley received numerous distinguished awards: the 1983 Esther Langer Award; the 1989 G. H. A. Clowes Memorial Award of the American Association for Cancer Research; and the 1993 Robert de Villiers Award from the Leukemia Society of America. The University of Chicago held a symposium in 1995 to honor her 70th birthday. In 1998, Rowley won two important awards: the Albert Lasker Clinical Medical Research Award, which she shared with two others; and the National Medal of Science & Technology in the biological sciences.

Rubin, Vera Cooper
(1928–   )
American
Astronomer

Vera Cooper Rubin's research, mainly focused on the rotation of galaxies, defined the cutting edge of astronomical theory, though it often took the field several decades to accept its validity. Rubin created controversy before she became a professional, presenting her master's thesis on the rotation of the universe to a hostile audience who accepted the idea only years later, after an experienced male scientist confirmed her results. Her doctoral dissertation posited the clustering of galaxies; this time, it took the astronomical community decades to accept her theory. In her greatest contribution to astronomy, she confirmed the existence of *dark matter,* or matter invisible to the human eye, which accounts for up to 90 percent of the universe.

Vera Cooper was born on July 23, 1928, in Philadelphia, Pennsylvania. Her mother was Rose Applebaum, and

Vera C. Rubin, whose research has focused on the rotation of galaxies, defined the cutting edge of astronomical theory. *(Photo by Jack Parsons)*

her father, Philip Cooper, was an electrical engineer who built his daughter a telescope to fulfill her early passion for stargazing. After graduating with a B.A. from Vassar College in 1948, she married the physical chemist Robert Rubin, and together they had four children—David, Allan, Judith, and Karl—all of whom followed their parents' footsteps into science.

Rubin joined her husband as a graduate student at Cornell University, where she studied galaxy dynamics under Martha Stahr Carpenter. Rubin presented her master's thesis, an analysis of the motion of 108 galaxies ambitiously entitled "Rotation of the Universe," to the 1950 meeting of the American Astronomical Society, which rejected her theory outright. It was not until cosmologist Gérard de Vaucouleurs confirmed her theory of rotational galactic motion in addition to universal expansion several years later that the astronomical community finally accepted her ideas. In the meanwhile, she earned her master of arts degree from Cornell in 1951.

Rubin then conducted doctoral research at Georgetown University, under physicist George Gamow, and theorized in her dissertation that galaxies tend to clump together instead of distributing themselves randomly. Rubin received her Ph.D. in astronomy in 1954, but it took the astronomical community two decades to accept the theory put forth in her dissertation. Georgetown retained Rubin as a research astronomer, then promoted her to the rank of assistant professor in 1960.

In the mid-1960s, Rubin realized the need to support her theoretical research with her own astronomical observations. In 1965, Rubin became the first woman to observe officially at California's Palomar Observatory, which previously forbade female observers due to its lack of women's rest rooms (Martha Burbridge had secretly observed there under her husband's name and presumably managed to relieve herself in the men's rest room.) That year, Rubin also joined the Department of Terrestrial Magnetism (DTM) as a research associate at Carnegie Institution in Washington, D.C., and she remained there the rest of her career.

Rubin collaborated with DTM physicist W. Kent Ford to research the rotation of spiral galaxies. In 1975, they observed that some galaxies, including the Milky Way, travel faster than the expansion rate of the universe. The astronomy community greeted this discovery—the so-called Rubin-Ford effect—with skepticism. By 1982, however, other astronomers confirmed this effect.

In 1978, Rubin and Ford made an even more startling observation—that stars distant from the center of a galaxy, which should travel slower than closer-orbiting stars, in fact traveled the same speed, if not a bit faster. Mathematically, this was impossible, unless the amount of mass far exceeded what was visible. Quite by accident, Rubin and Ford confirmed the existence of what astronomer Fritz Zwicky had previously hypothesized: dark matter. By 1983, Rubin had calculated as much as 90 percent of the universe to be dark matter. This revolutionary discovery became the center of debate in the astronomy community for years to follow.

Rubin has received numerous honors for her work. Between 1978 and 1990, she received honorary degrees from Creighton, Harvard, and Yale Universities. In 1981, she was elected into the National Academy of Sciences. In 1987, she became a president's distinguished visitor at Vassar College. In 1993, Rubin received her highest distinction: the National Medal of Science.

Russell, Elizabeth Shull
(1913–)
American
Geneticist

Elizabeth Shull Russell is world renowned for her skill in breeding laboratory mice. When she discovered a muscular defect similar to human muscular dystrophy in one of her mice, who she named Funnyfoot after her peculiar limp, requests came pouring in from scientists around the globe eager to use genetically related funnyfoot mice in their research. Russell not only bred mice but also performed original research on them as well, specializing in pigmentation.

Elizabeth Shull was born on May 1, 1913, in Ann Arbor, Michigan. She hailed from a scientific family: Her mother, Margaret Jeffrey Buckley, taught zoology at Grinnell College; her father, Aaron Franklin Shull, taught zoology and genetics at the University of Michigan; her paternal uncles included a geneticist, a plant physiologist, and a botanical artist; and a maternal uncle was a physicist. She followed in their footsteps by enrolled at the University of Michigan at the age of 16 to study zoology and genetics. She graduated with an A.B. in genetics in 1933.

At her father's behest, Shull entered a scholarship program at Columbia University, where she earned her M.A. degree in 1934. She went to the University of Chicago to pursue a doctorate under Sewall Wright, who theorized about genetic dominance. Shull wrote her dissertation on the role of genes in the pigmentation of guinea pigs, earning her Ph.D. in 1937. That year, she married William L. Russell, and together they had four children—three sons and a daughter.

Although her husband had been hired by the Roscoe B. Jackson Laboratory in Bar Harbor, Maine, antinepotism rules prevented Russell from working there as well. However, the lab hired her as an independent investigator from 1937 through 1946, during which time she carefully bred almost 100,000 laboratory mice. In 1939, the American Association of University Women named her its Nourse Fellow, and in 1947, she was named the Finney-Howell Fellow. In 1946, Russell finally joined the Jackson Laboratory staff as a research associate.

Fire broke out throughout Bar Harbor in October 1947, engulfing the laboratory as well and destroying almost all of Russell's carefully bred mice. It took several years for Russell and her colleagues to rebuild their mouse population.

In 1951, Russell noticed the limping Funnyfoot mouse, so she bred its brothers and sisters to see if the defect was regenerative. The trait did recur, but the funnyfoot mice died young and the females experienced difficulties reproducing. Russell solved this problem by transplanting the ovaries of funnyfoot mice into healthy female mice, who could reproduce, thereby replicating the genetic defect for scientific study. Interestingly, Russell's mice also contributed to the scientific understanding of organ transplantation.

Jackson Laboratory appointed Russell as its scientific staff director in 1953. In 1958, she received a Guggenheim fellowship, and that same year, she also became a senior staff scientist at the laboratory. She remained at the lab in this capacity until 1978, when she shifted into emerita status. In 1972, she was elected into the National Academy of Sciences. She served as the vice president of the Genetics Society of America in 1974 and as its president from 1975 through 1976. Russell's bibliography is so extensive that the Joan Staats Library at the Jackson Laboratory maintains a complete listing of it.

S

Sabin, Florence Rena
(1871–1953)
American
Anatomist

Florence Rena Sabin broke new ground as an anatomical researcher, a medical school professor, and a supporter of women's rights. She spent the latter part of her career advancing public health in Colorado, and she received abundant recognition of her significant contributions.

Sabin was born on November 9, 1871, in Central City, Colorado. Her father, George Kimball Sabin, was a mining engineer. He moved the family to Denver when Sabin was four. Her mother, Serena Miner, was a teacher who died in childbirth when Sabin was seven. Sabin attended Wolfe Hall boarding school in Colorado before the family moved to Lake Forest, Illinois, where she also attended private school. She then transferred to Vermont Academy in Saxton's River, Vermont, where her grandfather Sabin lived on a farm as the community's doctor.

After graduating from Vermont Academy, she joined her sister, Mary, at Smith College, where she studied mathematics and science. She graduated Phi Beta Kappa with a B.S. degree in 1893. That same year, Johns Hopkins School of Medicine was founded with the generous support of the so-called Women of Baltimore, a group of women's rights activists who stipulated that their contributions support the admission of women with equal consideration as men. To raise tuition, Sabin taught mathematics at Wolfe Hall for two years, then worked as an assistant in the biology department at Smith College, and worked in the summer of 1896 at the Marine Biological Laboratories at Woods Hole, Massachusetts. In October 1896, she was one of 15 women to enter the fourth class at Johns Hopkins School of Medicine.

Sabin conducted research on the central nervous system of newborn infants under Dr. Franklin Paine Mall (the lifelong mentor whose biography she wrote in 1934). She specifically focused on the medulla and midbrain, constructing detailed anatomical models that were later reproduced and used by many medical schools. Sabin published the results of this research in her first book, *An Atlas of the Medulla and Midbrain,* in 1901, a year after earning her M.D. from Johns Hopkins. She conducted her internship at Johns Hopkins, and she remained there as a researcher in the Department of Anatomy under a fellowship created specifically for her by the Baltimore Association for the Advancement of University Education of Women. The medical school then retained her for the next year, 1902, as an assistant instructor in anatomy.

In 1903, Sabin published a series of papers on her original research on the origins of the lymphatic system. By studying pig embryos as small as 23 millimeters long, she determined that the lymphatic system springs from the circulatory system, thus solving an ongoing debate in the field. Her papers won the 1903 Naples Table Association prize, and in 1916, she published her accrued findings in *The Origin and Development of the Lymphatic System.*

Johns Hopkins promoted Sabin to the rank of associate professor of anatomy in 1905, and then in 1917, she became the first woman appointed as a full professor at the medical school. When appointing the head of the anatomy department, Johns Hopkins passed over Sabin in favor of a male, ironically a former student of hers, prompting her to resign in protest. She accepted a position at New York City's Rockefeller Institute for Medical Research (now Rockefeller University), and she remained from 1925 until 1938.

That year, Sabin retired to Denver, where she lived with her older sister, Mary, a former high school mathematics teacher. Sabin spent the end of her career advancing public health in Colorado: She served as vice

Florence Rena Sabin, the first woman to serve as president of the American Association of Anatomists. *(National Library of Medicine, National Institutes of Health)*

president of the board of directors of Denver's Children's Hospital; as head of the Sabin Committee on public health of the state's Post-War Planning Committee; as manager of the Denver Department of Health and Welfare; and as chair of Denver's Board of Health and Hospitals.

Sabin's career received copious recognition, as she received 12 honorary degrees as well as the first Jane Addams Medal for distinguished service by an American woman. She was also the first woman to serve as president of the American Association of Anatomists and the first woman elected into the National Academy of Sciences. Sabin died of a heart attack on October 3, 1953, while watching the Brooklyn Dodgers in the World Series. In 1958, the state of Colorado honored her as one of its two most revered citizens by erecting a statue of her in the National Statuary Hall in the Capitol in Washington, D.C.

Sager, Ruth
(1918–1997)
American
Geneticist

Ruth Sager overturned the law of inheritance posited by Gregor Johann Mendel in the 19th century, which stated that chromosomes alone controlled all heredity. In the early 1950s, Sager discovered an instance of nonchromosomal transmission of genetic traits; however, the scientific community did not fully acknowledge the significance of her research until after the second wave of feminism in the 1970s forced the patriarchal system to take women's scientific work more seriously.

Sager was born on February 7, 1918, in Chicago, Illinois. Her father was an advertising executive, and her mother encouraged the academic growth of Sager and her two sisters. Sager matriculated young at the University of Chicago, at the age of 16, and graduated in 1938 with a B.S. degree. After a brief academic hiatus, she continued with graduate study in plant physiology at Rutgers University to earn her M.S. degree in 1944. Sager won a fellowship in 1946 to study genetics under botanist Marcus Rhoades at Columbia University, where she earned her Ph.D. in 1948.

In 1949, Sager won a three-year Merck Fellowship at the National Research Council. After that, the Rockefeller Institute (now Rockefeller University) hired her in 1951 as an assistant biochemist working with Yoshihiro Tsubo. In 1953, Sager noticed that the pond-scum alga she was studying, *Chlamydomonas* (which she nicknamed Clammy), required a nonchromosomal gene to survive in water laden with streptomycin, an antimicrobial drug. Sager had been trying to prove the possibility of nonchromosomal genetic heredity, so this situation seemed perfect to test her hypothesis.

Sager encouraged sexual reproduction in Clammy, which usually reproduced asexually, by removing nitrogen from its environment. The streptomycin-resistant gene passed from either the male or the female to the offspring, Sager demonstrated experimentally, thus proving the existence of nonchromosomal genetic inheritance. By this time, Sager had moved to Columbia University as a research associate in zoology. However, it was not until 1960, seven years after her discovery, that she announced the results of her research at the first Gilbert Morgan Smith Memorial Lecture at Stanford University and then at the Society of American Bacteriologists in Philadelphia a few months later. *Science* magazine published her results near the end of 1960.

That year, Columbia promoted her to the rank of senior research associate. In 1961, she published her theories in the textbook *Cell Heredity* (coauthored by Francis J. Ryan). She had observed that Clammy's mutations were limited to the nonchromosomal gene, and she further hypothesized that nonchromosomal genes represented a simpler—and thus more primitive—form of life at the cellular level, perhaps even predating the evolution of deoxyribonucleic acid.

Hunter College of the City University of New York hired Sager as a professor of biology in 1966. In 1972, she published her second text, *Cytoplasmic Genes and Organelles*. The Imperial Cancer Research Fund Laboratory in London, England, hosted her as a Guggenheim research fellow for an academic year abroad from 1972 through 1973. When she returned, she married Dr. Arthur B. Pardee. In 1975, the Dana Farber Institute of Harvard University hired Sager as a professor of cellular genetics and to head its Division of Cancer Genetics.

From this point on, Sager finally received the recognition she deserved for her pioneering work in cellular genetics. In 1977, the National Academy of Sciences elected her as a member, securing her reputation as a significant scientist. In 1988, the academy awarded her its Gilbert Morgan Smith Medal, and in 1990, the University of Texas honored her with its prestigious Schneider Memorial Lectureship. On March 29, 1997, Sager died of bladder cancer in her Brookline, Massachusetts, home.

Sanford, Katherine Koontz
(1915–)
American
Biologist and Medical Researcher

Katherine Koontz Sanford was the first scientist to clone in vitro (in a test tube) a single living cell of a mammal—in this instance, a rodent. She performed this feat in search of a means to research how cells transform into malignancy while she was working for the National Cancer Institute (NCI), where she spent the vast majority of her half-century career. She conducted research on cell biology, genetic predisposition to cancer, and neurodegenerative disease, such as Alzheimer's disease.

Sanford was born on July 19, 1915, in Chicago, Illinois. Her parents were Alta Rache and William James Koontz. She had two sisters, with whom she attended Wellesley College. After graduating with a bachelor of arts degree, she pursued graduate study at Brown University. After earning her master of arts degree, she wrote her doctoral dissertation under her advisor, Arthur M. Banta, on "The Effect of Temperature on the Expression of Intersexuality in *Daphnia longispina*" to earn her doctorate in zoology in 1942.

Sanford spent her first two postdoctoral years as a biology instructor at two liberal arts colleges: Western College in Oxford, Ohio, and then Allegheny College in Meadville, Pennsylvania. Johns Hopkins University Nursing School in Baltimore, Maryland, then hired her as the assistant director of its science program. In 1947, Sanford joined the tissue culture section of the NCI's Laboratory of Biology. She spent the remainder of her career there, spanning 49 years.

Almost immediately after joining the NCI, Sanford made her significant discovery of a method for cloning mammalian cells in vitro. Although other scientists found it difficult to reproduce her results, her method did allow them to replicate genetically identical cells from one cell, thus facilitating research on the metabolic and genetic features of that cell. In recognition of the importance of this discovery, she received the 1954 Ross Harrison Fellowship Award.

Sanford married Charles Fleming Richards Mifflin on December 11, 1971. In 1974, the NCI appointed her head of its cell physiology and oncogenesis section of the Laboratory of Biochemistry. Three years later, in 1977, the NCI promoted her to head the in vitro carcinogenesis section of its Laboratory of Cellular and Molecular Biology. In 1985, Sanford developed a test for genetic predisposition to cancer. In the 1990s, she developed another test to distinguish people with a predisposition to cancer from those with Alzheimer's disease.

Sanford published several important papers in her career: "Studies on the Difference in Sarcoma-Producing Capacity of Two Lines of Mouse Cells Derived in Vitro from One Cell," published in the *Journal of the National Cancer Institute* in 1958; "Familial Clustering of Breast Cancer: Possible Interaction Between DNA Repair Proficiency and Radiation Exposure in the Development of Breast Cancer," published in the *International Journal of Cancer* in 1995; and "Fluorescent Light-Induced Chromatid Breaks Distinguish Alzheimer Disease Cells from Normal Cells in Tissue Culture," published in the *Proceedings of the National Academy of Science, USA*, in 1996.

In the 1990s, Sanford succeeded in creating the first laboratory tests to distinguish people with the devastating neurodegenerative disease Alzheimer's and those with a predisposition to cancer. In her test—the cytogenetic assay—she exposed a person's skin fibroblasts in culture to fluorescent light, which damaged the cells' DNA. She then treated the cells with DNA repair inhibitors and compared the cells for chromatid breaks. Alzheimer's and cancer patients' cells displayed far more chromatid breaks under certain conditions—thereby allowing researchers to test for the two diseases. In 1995, Sanford discovered that the phenolic compounds in green tea can inhibit the effects of fluorescent light on the cultured cells' DNA.

Sanford retired from the NCI, but she remains active, working five days a week to complete her research on cytogenetic tests. Although some scientists have criticized her cytogenetic assay because they have had difficulty obtaining the same results, the test has correlated closely with identifying genetic predispositions to certain kinds of cancer. Sanford's significant contributions to her field were recognized in 1954 when she won the Ross Harrison Fellowship Award.

Sarachik, Myriam

(1933–)
American
Solid-State Physicist

Myriam Sarachik distinguished herself very early in her career, publishing research from her doctoral dissertation in two prominent journals—*Physical Review Letters* and *IBM Journal of Research and Development*—the same year she received her doctorate. Since then, she has been conducting groundbreaking research on semiconductors. She divides her research between two areas of solid-state physics: doped semiconductors near the metal-insulator transition and macroscopic quantum tunneling of magnetization.

Sarachik was born in 1933. She attended Barnard College, where she received her B.A. degree in 1954. The same year as her graduation, she married Philip E. Sarachik, and together they have one child. While pursuing graduate studies at Columbia University from 1955 through 1960, Sarachik also worked as a research assistant in experimental solid-state physics at the IBM (International Business Machine) Watson Laboratory, located at Columbia. She earned her master of arts degree in 1957 and her Ph.D. in physics in 1960.

Sarachik focused her doctoral research on the classic semiconductors, tin and lead, measuring the superconducting penetration depth of these elements to determine the superconducting energy gap. She published her results not only in her dissertation but also in two scientific journals: "Observations of the Energy Gap by Low Temperature Penetration Depth Measurement in Lead," coauthored by R. L. Garwin and E. Erlbach, appeared in a 1960 issue of *Physical Review Letters;* and "Measurement of Magnetic Field Attenuation by Thin Superconducting Films," written by the same team of three, appeared in the *IBM Journal of Research and Development* in 1960.

After receiving her doctorate, Sarachik remained at IBM's Watson Lab for one year as a research associate while simultaneously working as an instructor at City College of the City University of New York (CUNY). In 1962, she joined the technical staff at the AT&T Bell Telephone Laboratories in Murray Hill, New Jersey, for two years. While there, she conducted research that established a link between the presence of dilute magnetic moments and the minimum electrical resistivity of an alloy as a function of temperature. She published her results, which helped confirm the theory of the Kondo effect, in an article entitled "Resistivity of Mo-Nb and Mo-Re Alloys Containing 1% Fe" in a 1964 issue of *Physical Review.*

In 1965, Sarachik left the industry to return to academia and City College of CUNY as an assistant professor. She remained there for the rest of her career, steadily rising through the ranks. In 1967, she became an associate professor, and in 1971, she became a full professor. Sarachik also held concurrent positions throughout this time: From 1965 though 1972, she was the principal investigator for a U.S. Air Force research grant, and from 1972 through 1974, she worked on a National Science Foundation grant. In 1975, CUNY appointed her as executive officer of its Ph.D. program in physics.

Sarachik's career came to fruition in the 1990s. In 1994, the National Academy of Sciences elected her as a member, and in 1995, CUNY named her as a distinguished professor. That same year, she received the New York City Mayor's Award for excellence in science and technology. Her research, which continued to break new ground in the 1990s, has been written up in prominent magazines and journals, such as *Physics Today, Science,* and *Nature.* She published more than 75 scientific papers throughout her career and continued to make important contributions to the field of solid-state physics.

Schafer, Alice Turner

(1915–)
American
Mathematician

Alice Turner Schafer is a cofounder of the Association of Women Mathematicians (AWM), an organization devoted to breaking down barriers to women's full participation in the field. She divided her own career between the discipline of abstract algebra and the cause of promoting women in mathematics.

Alice Turner was born on June 18, 1915, in Richmond, Virginia. Her mother died in childbirth, so Turner was raised by Pearl Dickerson, a close family friend. Turner determined to pursue mathematics after a primary-school teacher expressed doubt that she could master long division. Turner studied mathematics at the University of Richmond, which segregated its campus between the sexes. As the only female mathematics major, she had to travel to the men's side of campus to attend classes. In her junior year, Turner won a competition in real analysis, and in 1936, she graduated Phi Beta Kappa with a B.A. degree in mathematics.

In order to earn money for graduate study, Turner taught high-school math in Glen Allen, Virginia, for three years. The University of Chicago granted her a fellowship in 1939, and she conducted her doctoral research on projective differential geometry under Ernest P. Lane. After she received her Ph.D. in 1942, she married fellow mathematician Richard Schafer, and together the couple had two sons: Richard Stone Schafer (born in 1945) and John Dickerson Schafer (born in 1946). She later published her dissertation in two journals: In 1944, the *Duke Mathematical Journal* published "Two Singularities of Space Curves," and in

1948, the *American Journal of Mathematics* published "The Neighborhood of an Undulation Point of a Space Curve."

Schafer first taught linear and abstract algebra and calculus at Connecticut College for two years before working at Johns Hopkins University's Applied Physics Laboratory during the last year of World War II. She then taught at a string of institutions: the University of Michigan, Douglass College, Swarthmore College, Drexel Institute of Technology, and the University of Connecticut. Next, she returned to Connecticut College as a full professor. In 1962, Schafer commenced her long relationship with Wellesley College, where she headed the mathematics department and became the Helen Day Gould Professor of Mathematics before her forced retirement in 1980.

Within a year, Schafer came out of retirement to teach in the Radcliffe College's management program seminars and to consult at Harvard University. She added teaching duties at Simmons College until her husband retired, when they moved to Arlington, Virginia. There she became a professor of mathematics at Marymount University from 1989 through her second retirement in 1996.

In 1971, Schafer collaborated with BHAMA SRINIVASAN, Linda Rothschild, and Linda Gray to found the Association of Women Mathematicians, serving as its president from 1973 through 1975. She has published eight articles on the issue of women in mathematics, including "Guidelines for Equality: A Proposal" with M. W. Gray in the journal *Academe* and "Women and Mathematics" in the book *Mathematics Tomorrow,* both in 1981, and "Mathematics and Women: Perspectives and Progress" in *American Mathematical Notices* in 1991. She also helped to implement a black studies department at Wellesley College.

In 1964, Schafer became the first woman to receive an Honorary Doctor of Science degree from her alma mater, the University of Richmond. The university's Westhampton College honored her with its Distinguished Alumna Award in 1977. In 1989, the AWM established the Alice T. Schafer Mathematics Prize, awarded annually to an outstanding undergraduate mathematics major. In 1998, the Mathematical Association of America honored Schafer by conferring on her its Yueh-Gin Gung and Dr. Charles Y. Hu Award for Distinguished Service to Mathematics.

Scharrer, Berta Vogel
(1906–1995)
German/American
Biologist

Berta Vogel Scharrer, along with her husband, Ernst Scharrer, established the field of neuroendocrinology, based on their discoveries of the biological system that both conducts electrical impulses (as does the nervous system) and secretes hormones (as does the endocrine system.) Late in her career, Scharrer conducted pioneering research in another newly emerging field, neuroimmunology.

Berta Vogel was born on December 1, 1906, in Munich, Germany. Her mother was Johanna Greis, and her father, Karl Phillip Vogel, was a judge and vice president of the Federal Court of Bavaria. She studied biology at the University of Munich under Karl von Frisch, a bee behaviorist who later won the Nobel Prize, in 1973. She wrote her dissertation on the correlation between sweetness and nutrition in sugars to earn her Ph.D. in biology in 1930. She met Ernst Scharrer at the University, and they married in 1934. In the meanwhile, they both worked at the Research Institute of Psychiatry in Munich, Berta as a research associate studying spirochete infections in bird and amphibian brains. They also spent their summers researching at the Zoological Station in Naples, Italy.

In 1928, Ernst Scharrer published a paper reporting his finding of nerve cells that secreted hormones in the brain of a fish, a discovery that rocked the European scientific community's belief in the singular functions of the nervous system and of the endocrine systems. Together, the Scharrers set out to prove the electrical and secretory functions of the neuroendocrine system by experimenting with various species throughout the animal kingdom, which they split between themselves: Ernst handled vertebrates, and Berta covered invertebrates.

The Scharrers moved to the Neurological Institute of the University of Frankfurt, where Ernst directed the Edinger Institute for Brain Research and Berta worked as a research associate. Her invertebrate investigations yielded proof of neuroendocrine cells in mollusks in 1935, in worms in 1936, and in insects in 1937. That year, the Scharrers immigrated to the United States instead of remaining in Germany, where the Nazi regime expected them to support its anti-Semitic policies.

The Scharrers worked at the University of Chicago under Ernst's one-year Rockefeller Fellowship, and then they spent two years at the Rockefeller Institute (now Rockefeller University) in Herbert Gasser's laboratory. At the 1940 meeting of the Association for Research in Nervous and Mental Diseases, the couple presented a paper introducing their theory of neurosecretion, which was received much more readily than in Europe.

Also in 1940, the Scharrers moved to Western Reserve University, where Berta held a fellowship in the histology laboratory. There, she conducted research on the South American woodroach, *Leucophaea maderae,* surgically manipulating its neurosecretory glands to determine their functions. In 1946, the Scharrers moved to the University of Colorado in Denver; during their nine-year stay there, Berta supported her research with a Guggenheim

Fellowship and a Special Fellowship from the National Institutes of Health.

In 1955, the newly established Albert Einstein College of Medicine at Yeshiva University in New York hired Scharrer as a full professor of anatomy and her husband as the department head. This represented Scharrer's first academic appointment, thus representing institutional recognition equal to her husband's (though her salary amounted to only half of his). In 1963, the Scharrers published *Neuroendocrinology,* the seminal textbook in the discipline they established. When Ernst died in a swimming accident in 1965, Einstein granted Berta full salary and appointed her acting chair of the anatomy department (she refused to fill his open position permanently). She again served as acting chair in 1976.

In 1978, Einstein named Scharrer distinguished university professor emerita, but this did not mark the end of her career. In 1982, she initiated a new line of research in collaboration with George Stefano and Georg and Hente Hanson. In 1990, at the age of 84, Scharrer helped establish the newly emerging field of neuroimmunology by conducting new research and acting as associate editor of the journal *Advances in Neuroimmunology.*

Scharrer died at her Bronx, New York, home on July 23, 1995, after a long illness. Her career had been garlanded with praise: President Ronald Reagan granted her the National Medal of Science in 1983, and that same year, the Deutsche Akademic fur Naturforscher Leopoldina bestowed its Schleiden Award on her. She had been inducted into the National Academy of Sciences in 1967, and throughout her career, she received 10 honorary degrees from the likes of Harvard University and the University of Frankfurt.

Scott, Charlotte Angas
(1858–1931)
English
Mathematician

Charlotte Angas Scott sought to "break the iron mold" of the male denial of equality to females, and she accomplished this feat by excelling in her education and in her professional life as a mathematician. She not only solved a complex problem posed by EMMY NOETHER, another prominent female mathematician, but she also advanced the field of mathematics by spearheading the organization of the American Mathematical Society and coediting the *American Journal of Mathematics.* She also spread her influence beyond her field by lobbying to establish the College Entrance Examination Board, serving as one of its first chief examiners.

Scott was born on June 8, 1858, in Lincoln, England. Her mother was Eliza Exley, and her father, Caleb Scott,

was the president of Lancashire College. She commenced her education in mathematics at the age of seven at the hands of tutors. She matriculated at Hitchin College (now Girton College), the women's division of Cambridge University, with an academic scholarship.

In January 1880, she took Cambridge's notoriously difficult, weeklong Tripos oral examinations and placed eighth. When officials overlooked her achievement at the awards ceremony, sympathetic male students shouted "Scott of Girton" in the eighth slot, over the male name. Although Cambridge conferred no degree upon her, she proceeded to earn a B.S. degree from the University of London while lecturing in mathematics at Girton. She remained at the University of London for doctoral work under algebraist Arthur Cayley, writing her dissertation on the analysis of singularities in algebraic curves to earn her doctor of science degree of the highest rank in 1885.

Also in 1885, the Society of Friends founded Bryn Mawr College in Pennsylvania as the first women's college to grant graduate degrees. As one of the few women in the world with a doctorate in mathematics, Scott was invited to join the Bryn Mawr faculty as an associate professor. Bryn Mawr promoted her to the rank of full professor within three years.

Scott specialized in algebraic geometry, and she conducted research on the theory of plane algebraic curves, systems of curves, and the nature of circuits, among other topics. She published her classic textbook, *An Introductory Account of Certain Modern Ideas and Methods of Plane Analytical Geometry,* in 1894, and its popularity saw it through new editions in 1924 and 1961.

In 1895, Scott helped transform the New York Mathematical Society (which she joined as its first woman in 1891) into the American Mathematical Society (AMS) by serving on its council during the transition and again in 1899 through 1901. In 1905, she served as the vice president of the AMS. In 1899, the *American Journal of Mathematics* appointed her as coeditor. That same year, she gained international acclaim by publishing "A Proof to Emmy Noether's Fundamental Theorem" in *Mathematische Annalen.*

Scott's reforms to Bryn Mawr's admissions policies influenced the establishment of the College Entrance Examination Board, which standardized college requirements nationally. She acted as the board's chief examiner from 1902 through 1903. In 1909, Bryn Mawr honored Scott with an endowed chair and a formal citation. At the age of 67, she retired from Bryn Mawr to return to England, where she took up gardening to relieve her rheumatoid arthritis and horse betting to continue practicing mathematics. She died in November 1931, in Cambridge, England.

Seibert, Florence Barbara
(1897–1991)
American
Biochemist

Florence B. Seibert contributed to the worldwide fight against tuberculosis by developing the skin test from tuberculin protein that was first adapted by the United States and almost immediately thereafter adapted by the World Health Organization (WHO). She also contributed to the improvement of clinical medicine by discovering the cause of protein fever, believed to result from the proteins in intravenous injections. Seibert's research revealed, however, that the culprit resided in the distilled water used to prepare intravenous injections, and so she invented an improved distilling apparatus to prevent the introduction of contaminants into the water. This apparatus subsequently saw standard use in blood transfusions during surgical procedures.

Florence Barbara Seibert was born on October 6, 1897, in Easton, Pennsylvania, one of three children. Her mother was Barbara Memmert, and her father, George Peter Seibert, was a rug manufacturer and merchant. Seibert contracted polio at the age of three, leaving her handicapped. However, she did not allow this difficulty to slow her down. She studied zoology and chemistry under Jessie E. Minor at Goucher College in Baltimore, where she earned her A.B. degree in 1918.

After graduating, Seibert joined Minor at the chemistry laboratory of the Hammersley Paper Mill in Garfield, New Jersey, where they studied the chemistry of wood pulp and cellulose and even coauthored papers on these topics. Seibert longed to attend medical school, but pure research suited her physical condition better, so she pursued graduate studies in biochemistry at Yale University under Lafayette B. Mendel, who had discovered vitamin A.

Mendel was confounded by the mysterious fevers that patients contracted when injected with protein solutions intravenously, so he encouraged Seibert to focus her doctoral research on an inquiry into which protein caused this reaction. If she had stuck slavishly to this hypothesis, Seibert would never have discovered the cause, which resided not in the proteins but in bacteria contaminating the distilled water used to prepare the solution. Seibert not only identified the problem, but she also solved it by inventing an apparatus to prepare distilled water free of contaminants. Hospitals continued to use this tool long after Seibert earned her Ph.D. in physiological chemistry for this research in 1923.

Seibert worked on her postdoctoral fellowship at the University of Chicago, where she conducted research, taught pathology, and also attended seminars under H. Gideon Wells. She met her scientific collaborator of the next 31 years, Esmond R. Long, in a Wells seminar. When her fellowship ended, Wells employed her part-time as an assistant at the Otho S. A. Sprague Memorial Institute, which he directed.

Seibert and Long focused their research on the tuberculosis disease, specifically searching for an effective and standardized diagnostic test. First, they isolated the protein that caused tuberculosis by means of crystallization, but this method proved impractical in the limited amounts it generated. Seibert and Long eventually generated a dry powder called tuberculin protein trichloracetic acid precipitated through ultrafiltration, and they further purified this substance into purified protein derivative (PPD). In 1941, they arrived at a standard dose of 107 grams (PPD-S, or standard), and when this test proved effective, both the United States and the WHO adapted it as their standard.

In 1932, Long moved to the Henry Phipps Institute of the University of Pennsylvania in Philadelphia, and Seibert accompanied him as an assistant professor. She continued her research to standardize the tuberculosis skin test on a Guggenheim Fellowship in the laboratory of Nobel laureate Theodor Svedberg at the University of Uppsala in Sweden in 1937. When she returned in 1938, the University of Pennsylvania promoted her to the rank of associate professor and, in 1955, again promoted her to the rank of full professor. She retired to emerita status in 1959.

Seibert received many awards recognizing her achievements: the Trudeau Medal from the National Tuberculosis Association in 1938; the Garvan Gold Medal of American Chemical Society in 1942; the Gimbal Award in 1945; the Scott Award in 1947; and the Elliot Memorial Award of the American Association of Blood Banks in 1962. She also received five honorary degrees. She continued to conduct research in her retirement, directing the cancer research laboratory at the Mound Park Hospital Foundation in St. Petersburg, Florida. The National Women's Hall of Fame in Seneca Falls, New York, inducted her as a member in 1990. On August 23, 1991, she died in St. Petersburg of complications from her childhood polio.

Sessions, Kate Olivia
(1857–1940)
American
Horticulturist

Kate Olivia Sessions founded Balboa Park in San Diego, California, not by official decree but by planting the trees that turned the land into an urban oasis. She ran a nursery in the San Diego area for almost 60 years, introducing to the region new and exotic plant varieties, such as queen palm, flame eucalyptus, and camphor trees. She helped to found the San Diego Floral Association and contributed about 250 articles to the association's magazine, *California Garden*.

Sessions was born on November 8, 1857, in San Francisco, California. Her father, the fine-horse breeder Josiah Sessions, moved the family to a farm in Oakland the next year. Her mother, Harriet Parker, gave birth there to Sessions's only sibling, a younger brother. Sessions grew up riding ponies and gardening. After graduating from the Oakland public school system, she traveled to Hawaii, where she further developed her love of plants. In 1877, she enrolled in the scientific course at the University of California at Berkeley. She submitted her thesis, "The Natural Sciences as a Field for Women's Labor," and graduated in 1881 with a Ph.B. in chemistry.

Sessions taught in the Oakland school system for the remainder of 1881 before moving to San Diego, where she taught at the Russ School. She then served as the principal of what is now San Diego High School. However, she could not ignore the call of her own thesis and devoted the rest of her career to the natural sciences, specifically to botany and horticulture. In 1885, she established a nursery business in Coronado, an island in San Diego harbor, and opened a floral shop in the city.

In 1892, Sessions struck a deal with the city of San Diego: In exchange for the use of 30 acres to grow plants for her nursery, she donated 300 trees to the city yearly and planted 100 more on the property each year. In this way, Sessions endowed San Diego with a natural haven near downtown—Balboa Park—and contributed to the naturalization of the rest of the city. She closed her floral shop in 1909, but she retained her nursery business until her death.

Sessions joined together with Alfred D. Robinson and other horticulturists to found the San Diego Floral Association, and she served on its board or as an officer for the next 20 years. She also contributed articles regularly to the association's monthly magazine, *California Garden*. In 1915, the San Diego grammar schools hired Sessions as supervisor of agriculture, a post she held for three years. In 1939, the University of California Extension Division hired her to teach an adult course entitled "Gardening Practice and Landscape Design."

Sessions's contributions to floriculture and horticulture received ample recognition. The 1935 California-Pacific Exposition held K. O. Sessions Day in her honor. In 1939, she became the first woman awarded the Meyer Medal by the American Genetic Association. Honors continued even after her March 24, 1940 death, at age 82 from bronchial pneumonia following a broken hip. In 1956, the Pacific Beach section of San Diego named an elementary school in honor of Sessions, and in 1957, Pacific Beach dedicated a memorial park to her memory.

Shattuck, Lydia White
(1822–1889)
American
Botanist and Chemist

Lydia White Shattuck was considered one of the outstanding scientists of her day, distinguishing herself not by making major discoveries or authoring important papers but rather by teaching generations of women scientists, classifying and cataloging plants, and corresponding with other influential scientists. In fact, her correspondences serve as key historical documents of the state of scientific inquiry in the 19th century. Shattuck spent her entire career at Mount Holyoke Seminary (now Mount Holyoke College), helping to advance the education of women and establish her institution as a preeminent scientific center.

Shattuck was born on June 10, 1822. Her parents, Betsey Fletcher and Timothy S. Shattuck, had moved to East Landaff (now Easton) in the mountains of New Hampshire to avoid the scandal of being first cousins. Lydia was their fifth child but the first to survive into maturity. Shattuck completed her schooling at age 15 and spent the next 11 years teaching in the district school.

Shattuck supplemented her education with short stints at academies in Newbury, Vermont, and Haverhill, New Hampshire, but she recommenced her schooling in earnest at the age of 26 when she matriculated at Mount Holyoke Seminary. Mary Lyon, the founder of this, one of the first women's academies, introduced Shattuck to the laboratory method of teaching, which she later employed as a teacher. Shattuck supported herself by working four hours a day for the school baking bread, an activity involving complex chemical reactions that she studied with fascination and later employed as examples in her chemistry lectures.

Shattuck graduated from Mount Holyoke in 1851, and she remained there throughout the rest of her career. She also taught at the Anderson School of Natural History on Penikese Island in Buzzards Bay, Massachusetts. Her affiliation with this institution dated back to 1873, when founders Elizabeth and Louis Agassiz handpicked 50 scientists (Shattuck was one of 15 women tapped) to attend the inaugural summer session. Shattuck later recommended her own Mount Holyoke students to attend the Anderson School, which was one of the few scientific institutions at the time that promoted the education of women.

Shattuck mentored many future eminent women scientists, including CORNELIA CLAPP, who attended Anderson and later earned dual doctorates in zoology from Syracuse University in 1889 and from the University of Chicago in 1896, and Henrietta Hooker, who earned one of the first doctorates in botany from Syracuse University in 1888. She also corresponded with leading scientists of the day, including botanist Asa Gray of Harvard University, who shared her support of evolution as a valid scientific theory. In the 1880s, Shattuck participated as a corresponding member of the Torrey Botanical Club in New York City.

Shattuck also contributed to the greater field of science by helping found or preside over societies and associations. She attended the 1874 Priestley Centennial, the gathering that gave birth to the American Chemical Society, but she and a handful of other women chemists were asked to step aside when a picture of the "founders" was taken. Thankfully, not all of Shattuck's contributions were similarly obscured; she unquestionably served as president of the Connecticut Valley Botanical Association, and she definitely gathered and cataloged the thousands of plants that started the botanical collection housed in the Mount Holyoke Seminary herbarium.

Mount Holyoke conferred emerita status on Shattuck upon her retirement in 1889, one year after the seminary became a college. Also in 1889, the Woods Hole Marine Biological Laboratory Corporation elected her as a member. Shattuck remained active up until her death on November 2, 1889, canvassing for donations to fund the construction of a new building to house the departments of chemistry and physics at Mount Holyoke. The college christened this building Shattuck Hall upon its 1893 opening; when this building needed to be demolished in 1954, Mount Holyoke retained the name Shattuck Hall for a new physics building.

Shaw, Mary
(1943–)
American
Computer Scientist

Mary Shaw helped to advance the field of computer software engineering into a true academic discipline. She speeded up digital computations by developing the Shaw-Traub algorithm. She also created the programming language Alphard, which abstracted computer commands further from the binary coding of 1s and 0s and closer to human language.

Shaw was born on September 30, 1943, in Washington, D.C. Her mother was Mary Holman, and her father, Eldon Shaw, was a civil engineer and a Department of Agriculture economist who sparked her interest in science early on with electronics kits and science texts. She attended high school in Bethesda, Maryland, where International Business Machine (IBM) employee George Heller ran an extracurricular activity that introduced students to the IBM 709 computer. During her high school summers, Shaw worked at the Research Analysis Corporation of the Johns Hopkins University Operation Research Office through a program established by Jean Taylor to expose students to fields outside their standard curricula.

Shaw matriculated at Rice University to study topology, but after her freshman year, she happened upon the Rice Computer Project, where she worked thereafter with its computer, the Rice I, under the head programmer Jane Jodeit. Shaw attended the University of Michigan summer school, where she met Carnegie Mellon University computer science professor Alan Perlis. After graduating cum laude from Rice with a B.A. degree in mathematics in 1965, Shaw conducted graduate work in computer science under Perlis at Carnegie Mellon. She wrote her doctoral dissertation on compilers, or the interface between human and computer languages, and received her Ph.D. in computer science in 1971. Two years later, she married civil and software engineer Roy R. Weil.

Carnegie Mellon retained Shaw as an assistant professor of computer science. She collaborated with Joseph Traub to develop the Shaw-Traub algorithm, increasing the speed with which a computer could compute. She developed abstract data types, or object-oriented computer programs organized by the type of information to improve the efficiency with which they would run. Working in collaboration with William A. Wulf and Ralph L. London between 1974 and 1978, she developed the Aphard programming language, which utilized abstract data types to extrapolate computer commands further from computer language and closer to human language.

One of Shaw's major goals was to advance the field of software engineering (the term was coined in 1968) from its infancy into its maturity as an academic discipline. She researched other engineering fields, such as civil or chemical engineering, and noticed that each went through three stages: the craft stage, when it is practiced by amateurs; the commercial stage, when it is practiced by skilled craftspeople as a business; and the scientific stage, when it is studied by professionals.

In Shaw's November 1990 *IEEE Software* article entitled "Prospects for an Engineering Discipline in Software," she placed the present state of the software engineering field somewhere between the craft and commercial stages. However, she had been working toward advancing the field already, pushing for the establishment of the Software Engineering Institute at Carnegie Mellon, which she served as chief scientist from 1984 through 1988. She simultaneously worked with the Information Systems Division of IBM to implement a software-engineering curriculum in-house.

Sherrill, Mary Lura
(1888–1968)
American
Chemist

Mary Lura Sherrill gained renown for her organic synthesis of antimalarial compounds. Equal to her achievements as a researcher, though, were her accomplishments as a teacher at

Mount Holyoke College, where she worked for three decades. She inherited the chair of the chemistry department from her close colleague, EMMA PERRY CARR, and, like Carr, received the American Chemical Society's Garvan Medal.

Sherrill was born on July 14, 1888, in Salisbury, North Carolina. She was the youngest of seven children born to Sarah Bost, daughter of a Confederate captain killed in the Civil War, and Miles Sherrill, also a Confederate soldier, who later served as the state librarian, as a trustee of Davenport College for Women, and on the state senate. One of her brothers, Joseph Garland Sherrill, helped to found the American College of Surgeons, and another, Clarence, engineered the reconstruction of the Lincoln Memorial and served as an aide to Presidents Theodore Roosevelt, Warren Harding, and Calvin Coolidge.

After attending local public schools, Sherrill matriculated in 1906 at Randolph-Macon Women's College in Lynchburg, Virginia. There, she studied chemistry under Fernando Wood Martin to earn her bachelor of arts degree in 1909. She remained there, pursuing graduate study in physics. She supported herself as an assistant in chemistry, and after receiving her master of arts degree in 1911, she became an instructor in chemistry, a position she retained until 1916.

That year, Sherrill enrolled in the doctoral program at the University of Chicago, where she studied chemistry under the supportive mentorship of Julius Stieglitz. After attending for one academic year, she then spent summers studying at the university while teaching during the school years. She served as an adjunct professor at Randolph-Macon during the 1917–18 year, then as an associate professor at the North Carolina College for Women from 1918 through 1920.

Sherrill's mentor, Stieglitz, enlisted her help as a research associate at the Chemical Warfare Service (CWS) during World War I. After the war, Sherrill joined the CWS full-time as an associate chemist from 1920 through 1921 to synthesize a sneezing gas (to prevent asphyxiation by inducing breathing). In 1921, she commenced her longtime relationship with Mount Holyoke College, which hired her as an assistant professor in chemistry. In her first two years there, she finished up her doctoral research on barbiturate and methylenedisalicylic acid synthesis to earn her Ph.D. from the University of Chicago in 1923.

The next year, Mount Holyoke promoted Sherrill to the rank of associate professor. She contributed to the chemistry department's team research by synthesizing and purifying the organic compounds used in ultraviolet spectroscopy experiments. From 1928 through 1929, she used a fellowship to visit Jacques Errera's laboratory in Brussels and Johannes van der Waals's lab in Amsterdam, where she learned novel purification methods and techniques for determining molecular structure through the dipole moments of organic compounds. In 1931, Mount Holyoke promoted her to a full professorship.

During World War II, Sherrill conducted her most famous research, synthesizing new antimalarial drugs to make up for the scarcity of quinine. After the war, Sherrill replaced Emma Carr, her colleague, housemate, and partner, as the head of the chemistry department at Mount Holyoke. Mount Holyoke students conceived of Sherrill and Carr as inseparable, as they were such close professional collaborators. Sherrill followed in the footsteps of Carr, who had won the first Garvan Medal from the American Chemical Society in 1937, by winning the Garvan Medal a decade later, in 1947, for her antimalarial research as well as for her teaching excellence.

Sherrill retired from the Mount Holyoke faculty in 1954 and returned to her home state of North Carolina in 1961 when her health began to fail her. She died on October 27, 1968, in High Point, North Carolina.

Shiva, Vandana
(1952–)
Indian
Physicist and Ecofeminist

Vandana Shiva has been called "one of the world's most prominent radical scientists." She abandoned her career path as a quantum physicist to devote herself to championing diversity in nature as well as in society. She is an ardent critic of environmental colonialism, or the corporate control of agriculture and other environmental arenas, such as logging. She won the Right Livelihood Award, or the so-called alternative Nobel Prize, in recognition of her work in support of traditional agricultural and cultural practices.

Shiva was born in 1952 in Dehra Dun, a north Indian city in the foothills of the Himalaya Mountains. She had a familial heritage of activism, education, and environmentalism: Her grandfather went on a hunger strike in support of the building of a women's college; her mother was an inspector of schools; and her father was a forester. Shiva studied physics at Punjab University, obtaining her bachelor of science degree in 1972 and her master of science degree in particle physics in 1974. However, she found the male domination at the university's Atomic Research Institute stifling, so she transferred to Western Ontario University in Canada for her doctoral work. She found the atmosphere at the Foundation of Physics Program there much more conducive to learning. She earned her Ph.D. in the philosophy of science in 1979.

In 1980, Shiva returned to India to join the Indian Institute of Management in Bangalore. Two years later, she became a consultant for the United Nations University. However, as she became more successful, she became increasingly disillusioned with titles and the hegemony of expertise. Her husband, on the other hand, became increasingly jealous of her titles and expertise, leading to their separation. Shiva retained custody of their son, but her husband refused to grant her a divorce; she had to fight a long battle to get their marriage dissolved.

Shiva sacrificed a promising career in quantum physics to devote herself to ecofeminism, or the intersection between environmentalism and women's issues. Shiva sees the inextricable link between the two, as women in her culture traditionally practiced sustainable agriculture and conserved seed stocks, sharing the best seeds with their communities. Shiva helped start the Chipko movement, spearheaded by Himalayan women to halt corporate deforestation and chemical-reliant agriculture. This movement succeeded in banning logging above 1,000 meters in the Himalayas.

In 1990, Shiva formalized her political activism by founding and directing the Research Foundation for Science, Technology, and Natural Resource Policy in her home of Dehra Dun. The foundation promotes its goals through practical action by establishing seed banks (to counter corporate seed monopolies) and training farmers in traditional, chemical-free, sustainable farming methods, among other initiatives.

Shiva disseminates her political message through intensive lecturing as well as through book writing. She has published 11 books, including *The Violence of the Green Revolution* in 1992, *Monocultures of the Mind* in 1993, and *Biopiracy: The Plunder of Nature and Knowledge* in 1997. In honor of her work, she received the Global 500 Award from the United Nations Environmental Program in 1992, and in 1993, she won the Earth Day International Award. Most significantly, she won the 1994 Right Livelihood Award, which split a $2 million prize between five women (including Shiva) from other countries.

Shockley, Dolores Cooper
(1930–   )
American
Pharmacologist

Dolores Cooper Shockley was the first African-American woman to earn a Ph.D. from Purdue University and the first African-American woman to earn a doctorate in pharmacology in the United States. When she entered the professional world, she had to confront skepticism about her status and capabilities, as she was the first black woman pharmacologist that her colleagues had encountered. The excellence of her research and her administrative and teaching abilities quickly dispelled any doubts, though. She later became the head of the pharmacology department at Meharry Medical College in Nashville, Tennessee.

Shockley was born on April 21, 1930, in Clarksdale, Mississippi. She attended Louisiana State University to study pharmacy with the intention of starting her own drugstore. However, her academic experience swayed her

away from retail and toward research, so after she earned her bachelor of science degree in 1951, she pursued graduate study in pharmacology at Purdue University.

Shockley worked as an assistant in Purdue's Department of Pharmacology from 1951 until 1953, when she earned her master of science degree. In 1955, Purdue awarded a Ph.D. to an African-American woman for the first time when it hooded Shockley as a doctor of pharmacology. Shockley conducted postdoctoral research under a Fulbright Fellowship at the University of Copenhagen from 1955 through 1956. In 1957, she married, and her family grew over the years to include four children.

In 1955, Meharry Medical College hired Shockley as an assistant professor in its department of pharmacology. She retained this title until 1967, when the medical college promoted her to the rank of associate professor. She held this position for a decade, at which point she was appointed as the head of the Department of Microbiology. Shockley has continued to serve as the department chair since 1977. Shockley has spent her entire professional career with Meharry save one interruption, when she served as a visiting assistant professor at the Albert Einstein Medical College of Yeshiva University in the Bronx, New York, from 1959 through 1962.

Shockley focuses her research on the effects of certain drugs on stress, specifically tracking the drug's action in response to stress and the reaction of stress levels to the drug. She also studies how the use of certain drugs affects the nutrition of the drug taker. In other areas of her research, she takes measurements of non-narcotic analgesics, or painkillers. Shockley conducts research not only on human-made pharmaceuticals but also on substances produced by the body; for example, she studies the effects of hormones on connective tissue.

In addition to Shockley's research, teaching, and departmental duties, she also maintains responsibilities within the college and in connection with the larger academic community: She acts as Meharry's foreign student adviser, and she also represents the college with the Association of American Medical Colleges coordinating international activities. Shockley is also a member of the American Pharmaceutical Association and the American Association for the Advancement of Science. For her indefatigable work, Shockley received the Lederle Faculty Awards from 1963 through 1966.

Shreeve, Jean'ne Marie
(1933–   )
American
Inorganic Chemist

Jean'ne Marie Shreeve gained prominence for her discovery of the compound perfluorourea, an oxidizer. She also developed methods for synthesizing other oxidizers, namely chlorodifluoroamine and difluorodiazine. These compounds and methods facilitate the synthesis of rocket oxidizers, which was previously a much more arduous task. Shreeve received the prestigious Garvan Medal from the American Chemical Society in 1972 in recognition of these important contributions.

Shreeve was born on July 2, 1933, in Deer Lodge, Montana. She attended the University of Montana, where she earned her bachelor of arts degree in 1953. She then pursued graduate study at the University of Minnesota under a teaching assistantship, earning her master of science degree in analytical chemistry in 1956. She subsequently conducted her doctoral study as an assistant at the University of Washington, which granted her a Ph.D. in inorganic chemistry in 1961.

Shreeve commenced her professional career in 1961 at the University of Idaho as an assistant professor of chemistry, and she has remained there ever since. The university hired her on the strength of her research skills, as it had recently received doctoral-degree-granting status from the state, and thus it had to enhance its research facilities, faculty, and curriculum. Shreeve obliged by better equipping the laboratories and by conducting research that garnered national and even international attention.

Since she brought distinction to the school, Shreeve quickly rose through the ranks, from assistant professor to associate professor in 1965 and two years later to full professor. That year, in 1967, she received several significant fellowships: a Cambridge University Fellowship, a National Science Foundation Fellowship, and an honorary U.S. Ramsey Fellowship. She also received an Alfred P. Sloan Fellowship from 1970 through 1972.

Shreeve served as acting chair of the chemistry department from 1969 to 1970 and again in 1973, after which she was appointed as the head of the department. She served as a visiting professor at Britain's University of Bristol in 1977, and the next year, she held a guest professorship from Germany's University of Göttingen. She retained her position as department head at Idaho until 1987, when the university appointed her vice provost of research and dean of the college of graduate studies in recognition of her role in improving both of these areas of the university.

Shreeve's research focuses on fluorine compounds containing nitrogen, sulfur, and phosphorus. She investigates modes of fluorine synthesis and tests fluorine compounds for reactions. Despite the fact that she is an inorganic chemist, she also researches organic compounds that contain fluorine.

Besides the 1972 Garvan Medal, Shreeve has also received the University of Montana's 1970 Distinguished Alumni Award as well as an honorary doctorate from her alma mater in 1982. The University of Minnesota granted her its Outstanding Achievement Award in 1975. In 1978,

she won both the Senior U.S. Scientist Award from the Alexander Von Humboldt Foundation and the American Chemical Society's Fluorine Award. In 1980, the Chemical Manufacturers Association honored her with its Excellence in Teaching Award.

Silbergeld, Ellen Kovner
(1945–   )
American
Toxicologist

Ellen Kovner Silbergeld has combined a career in research science, focusing specifically on environmental toxicology, and public policy, advocating the implementation of laws based on hard scientific research. She commenced her scientific career studying lead neurotoxicity as well as the toxicology of food dyes.

Ellen Kovner was born on July 29, 1945, in Washington, D.C. Her mother, Mary Gion, was a journalist who had already given birth to Ellen's older brother. Her father, Joseph Kovner, was a liberal lawyer who was ousted from his governmental job by the House Un-American Activities Committee. He established a private practice in Concord, New Hampshire, from when Kovner was seven until she reached eighth grade, when the political climate had changed enough for the family to return to Washington, D.C.

Kovner studied modern history at Vassar College, where she earned her A.B. degree in 1967. She then studied quantitative history at the London School of Economics on a Fulbright Fellowship. In 1968, she returned to Washington, D.C., where she spurned history and economics in favor of working as a secretary and program director for the Committee on Geography at the National Academy of Sciences of the National Research Council until 1970. In 1969, she married Mark Silbergeld, and together they had two children—a daughter, Sophia, who was born in 1981, and a son, Reuben Goodman, who was born in 1985.

Silbergeld's exposure to reports at the National Academy of Sciences peaked her interest in science, so she decided to pursue a doctorate in environmental engineering starting in 1968 at Johns Hopkins University, where she was the only woman in the program. She earned her Ph.D. in 1972, and she remained at Johns Hopkins under a postdoctoral fellowship conducting biochemical research on lead neurotoxicity.

In 1975, the National Institutes of Health (NIH) hired Silbergeld as a staff scientist, and by 1979, the NIH appointed her as the laboratory chief of her section, the Unit on Behavioral Neuropharmacology of the National Institute of Neurological Disorders and Stroke. She continued her research on lead while also investigating the toxicity of

Ellen Silbergeld, an environmental toxicologist who has combined public policy work with scientific research. *(Courtesy of Ellen Silbergeld)*

food dyes and conducting research on neurological disorders such as Huntington's and Parkinson's diseases.

At the risk of compromising her scientific credibility, Silbergeld assumed a public policy position in 1982 as the Chief Toxic Scientist and Director of the Toxic Chemicals Program for the Environmental Defense Fund, a position she retained until 1991. Although some oppose the mingling of science and politics, Silbergeld believes the two can be symbiotic, with politics cueing science on real-life problems to investigate and science providing hard facts for political decision making.

During the 1980s, Silbergeld held several concurrent positions at the National Institute of Child Health Development, at the University of Maryland School of Medicine, and at Johns Hopkins Medical Institutions. In the 1990s, she decided to return to a more academic focus, accepting more intensive appointments at the University of Maryland (as an affiliate professor of environ-

mental law and a professor of pathology and epidemiology) and Johns Hopkins (as an adjunct professor of health policy and management). She retained her affiliation with the Environmental Defense Fund as a senior consultant toxicologist.

Silbergeld received a prestigious MacArthur Foundation Fellowship in 1993, validating the significance of her career. She has also received the 1987 Warner-Lambert Award for Distinguished Women in Science, the 1991 Abel Wolman Award, and the 1992 Barsky Award. In 1994, her early work was rewarded with a patent for her lead detection procedure.

Simmonds, Sofia
(1917–)
American
Biochemist

Sofia Simmonds spent the majority of her career at the Yale University School of Medicine, where she helped to organize the Department of Molecular Biophysics and Biochemistry. In the 1950s, she published a text that helped to define the field of biochemistry. In the late 1960s, she received the Garvan Medal from the American Chemical Society, distinguishing her as one of the top woman chemists in the United States.

Simmonds was born on July 31, 1917, in New York City. She remained in the city for her undergraduate study at Barnard College of Columbia University, where she earned her bachelor of arts degree in chemistry in 1938. While there, she married fellow biochemist Joseph S. Fruton in 1936. After graduating from Barnard, she stayed in New York City to pursue doctoral study as an assistant biochemist at the Medical College of Cornell University. She earned her Ph.D. in biochemistry in 1942 and remained there as a research associate until 1945.

That year, Simmonds commenced her long relationship with Yale University School of Medicine, starting out as an instructor in physiological chemistry. In 1946, she became a microbiologist there. Yale promoted her to the rank of assistant professor in 1950 and to associate professor in 1954.

That year, Simmonds coauthored *General Biochemistry* with her husband, Dr. Joseph S. Fruton. The text became a standard in the field, prompting demand for a revised edition in 1958. In 1962, Yale named her a biochemist, and she added the title of molecular biophysicist in 1969. That year, she won the Garvan Medal from the American Chemical Society, the top prize awarded specifically to women chemists.

In 1973, Simmonds added administrative duties to her other academic responsibilities as the director of undergraduate studies for the Department of Molecular

Biophysics and Biochemistry at Yale. Two years later, she obtained full professorship in the department.

Simmonds conducted her research on the amino acid metabolism of bacteria. She studied the efficiency of certain peptides by mutant strains that grow only in the presence of a particular component of amino acid. She also studied amino acid and protein metabolism at the level of microorganisms. She published her findings in more than 40 articles throughout her career in the pages of scientific journals such as the *Journal of Biological Chemistry, Science,* and *Biochemistry.* She also served on the editorial board of the *Journal of Biological Chemistry* for five years.

In 1988, Simmonds retired to emerita status as a professor, but she picked up new administrative responsibilities as the associate dean of the Yale University School of Medicine and the dean of undergraduate studies. Two years later, she returned to the classroom as a lecturer from 1990 through 1991 and retained her post as dean of undergraduate studies. She thus continued to contribute the workings of the School of Medicine at Yale even in her retirement.

Simon, Dorothy Martin
(1919–   )
American
Physical Chemist

Dorothy Martin Simon contributed significantly to the scientific understanding of heat, helping to establish the theory of flame propagation and quenching as well as developing the ablative coatings that shielded intercontinental ballistic missiles (ICBMs) from the excessive heat of atmospheric reentry. Later in her career, she distinguished herself as one of the first women to climb to the top of the corporate ladder.

Dorothy Martin was born on September 18, 1919, in Harwood, Missouri. Her mother was Laudell Flynn, and her father, Robert William Martin, headed the chemistry department at Southwest Missouri State College. She graduated as the valedictorian of Greenwood Laboratory School in Springfield, Missouri, and of Southwest Missouri State College, where she received her A.B degree with honors in 1940. She continued with graduate study at the University of Illinois, where she conducted her doctoral research on active deposits from radon and thoron gases, thus pioneering the investigation of radioactive fallout. She earned her Ph.D. in chemistry in 1945, and on December 6, 1946, she married fellow scientist Sidney L. Simon, who later became vice president of Sperry Rand.

Simon divided her career between corporate and governmental assignments. She commenced her career in 1945 as a research chemist at the DuPont Company, investigating chemical reactions resulting from the manufacture of the synthetic fiber Orlon. She then worked for the government from 1946 on the Atomic Energy Commission at Oak Ridge Laboratory in Tennessee, where she isolated a new isotope of calcium, and at the Argonne National Laboratory in Illinois as an associate chemist.

Simon remained with the government from 1949 through 1955, working as an aeronautical research scientist for the National Advisory Committee for Aeronautics (the prototype for the National Aeronautics and Space Administration). She won a Rockefeller Public Service Award for her findings on the properties of flame, and she utilized the $10,000 fellowship to conduct research at Cambridge University in England as well as in France and the Netherlands from 1953 through 1954.

In 1955, Simon returned to the private sector as a group leader in combustion at the Magnolia Petroleum Company in Dallas, Texas, for one year. The Avco Corporation then hired her as a principal scientist and technical assistant to the president of the research and advanced development division. There, she worked on ablation cooling for Avco's ICBMs from 1956 through 1962. She then rose through Avco's corporate structure at a steady pace: She served as the director of corporate research from 1962 through 1964, as the vice president of the defense and industrial products group from 1964 through 1968, and as the corporate vice president and director of research from 1968 through 1981.

Simon received recognition for her achievements in both the public and private sectors. Pennsylvania State University asked her to deliver the Marie Curie Lecture in 1962. The Society of Women Engineers conferred its Achievement Award on Simon in 1966. Worcester Polytechnic Institute recognized Simon as "perhaps the most important woman executive in American industry today" in bestowing an honorary doctorate on her in 1971.

Simpson, Joanne Malkus
(1923–   )
American
Meteorologist

As the first woman to receive a doctorate in meteorology, Joanne Gerould Malkus Simpson opened the door for women meteorologists in academia and in the government as well as in the private sector. She encountered opposition to her advancement at every step, entrenching her resolve to achieve to the best of her abilities. Despite this opposition, she proved the capability of women by rising to the top of each sector: She was named to an endowed chair in academia, became a senior fellow at the National Aeronautics and Space Administration (NASA), and she founded her own company, working there as its chief scientist.

Simpson was born on March 23, 1923, in Boston, Massachusetts. When she entered the University of Chicago, most of its male students were shipping off to fight in World War II, so the university welcomed female students. After she had studied in the meteorology training program for nine months, the military services called upon her to train its weather forecasters at New York University. She received her B.S. degree in 1943 and her M.S. degree in 1945.

However, once the war ended, women were expected to return to their domestic roles to make way for the men returning from the war to their professional roles. Simpson refused to leave the graduate program, despite losing all financial support through fellowships.

In order to support her doctoral study, she taught physics and meteorology at the Illinois Institute of Technology (IIT), where she could then take courses for free. She completed most of the coursework required for her doctorate there and transferred the credits to the University of Chicago. She wrote her dissertation on tropical meteo-

Joanne Simpson, the first woman to receive a doctorate in meteorology, shown here flying a C-130 in a research experiment in the tropics off West Africa. *(Courtesy of NASA)*

rology under the only professor that would advise her. She earned her Ph.D. in meteorology in 1949.

Simpson remained at IIT as an assistant professor for two more years before becoming a meteorologist at the Woods Hole Oceanographic Institute for the next decade. In 1960, she moved to the University of California at Los Angeles, where she remained for five years as a professor of meteorology. During her last year there, she received a Guggenheim Fellowship, allowing her to fill an honorary lectureship at Imperial College of the University of London.

In 1965, the National Oceanographic and Atmospheric Administration hired Simpson to head its atmospheric physics and chemistry laboratory and then its experimental meteorological laboratory in Coral Gables, Florida. During this time, she innovated new methods for experimenting with cloud seeding and manipulating the dynamics of cumulus clouds. She also held an adjunct professorship at the University of Miami.

In 1974, Simpson returned full-time to academia as an environmental science professor at the University of Virginia's Center for Advanced Studies. At this time, she and her second husband, Robert Simpson, started their own private firm, Simpson Weather Associates, where she served as chief scientist. She had married Simpson in 1965 after divorcing her previous husband, William Malkus, whom she had married in 1948. She had three children from her marriages.

In 1976, the University of Virginia named Simpson the W. W. Corcoran Professor, though she resigned from this position in protest against the accusations that her appointment to an endowed chair merely filled affirmative action quotas. She left academia for good to return to governmental work in 1979 as head of the Severe Storms Branch of NASA's Goddard Space Flight Center. In 1988, Goddard appointed her chief scientist of the meteorological and earth sciences directorate and senior fellow.

Simpson's proliferate honors serve as testament to her success in meteorology despite severe obstacles. The American Meteorological Society granted her its Meisinger Award in 1962 and, in 1983, its highest award, the Rossby Research Medal. The Department of Commerce awarded her its Silver Medal in 1967 and its Gold Medal in 1972. The Weather Modification Association bestowed its V. J. Schaefer Award on her, and NASA honored her with its Exceptional Science Achievement Medal in 1982.

Sinclair, Mary Emily
(1878–1955)
American
Mathematician

Mary Emily Sinclair was the first woman to earn a doctorate in mathematics from the University of Chicago. She spent 37 years teaching at Oberlin College, serving as the head of the mathematics department and occupying an endowed chair. She maintained interest in her scholarship through numerous leaves of absence to study at some of the most prestigious institutions in the world, including the Sorbonne and Princeton's Institute for Advanced Studies.

Sinclair was born on September 27, 1878, in Worcester, Massachusetts. Her mother was Marietta Survetta Fletcher, and her father, John Elbridge Sinclair, was a professor of mathematics at Worcester Polytechnic Institute. She attended Oberlin College, graduating Phi Beta Kappa in 1900 with a bachelor's degree. She then pursued graduate study at the University of Chicago, where she earned her master's degree in mathematics in 1903.

Sinclair commenced her professional career as an instructor at the University of Nebraska from 1904 through 1907. All the while, however, she continued with doctoral work under the supervision of Oscar Bolza. She wrote her dissertation on the calculus of variations. In 1908, she became the first woman to earn a Ph.D. in mathematics from the University of Chicago. In January of the next year, the *Annals of Mathematics* published her dissertation, entitled "Concerning a Compound Discontinuous Solution in the Problem of the Surface Revolution of Minimum Area."

In the meanwhile, Sinclair had returned to her alma mater, Oberlin College, in 1907 to become an instructor in mathematics. The next year, when Sinclair had received her doctorate, the college promoted her to an associate professorship. In 1914, the single Sinclair adopted a daughter, naming her Margaret Emily. She took a sabbatical that year to care for the infant, and in 1915, she adopted a son. She also took advantage of her leave to study at Columbia University and Johns Hopkins University. In 1922, the American Association of University Women granted her a Julia C. G. Piatt Fellowship, which she utilized to study at the University of Chicago and Cornell University.

In 1925, Oberlin promoted her to the rank of full professor. That year, she took another sabbatical to pursue studies at the University of Rome and at the Sorbonne. In the spring of 1935, she again took a leave to study at the Institute for Advanced Studies at Princeton University. In 1939, she assumed the chair of the Department of Mathematics at Oberlin, and two years later, in 1941, the college named her the Clark Professor of Mathematics, a title she retained until her 1944 retirement.

Sinclair did not sit idly in retirement but rather taught part time at Berea College in Kentucky. In 1947, she moved back to Oberlin, and in 1953, she moved to Maine. She died there two years later, in 1955. Before she died, she endowed the Fellowship Fund of the American Association of University Women with a generous gift in symbolic repayment for the Piatt Fellowship that she had received, which fueled "the best period of creative scholarship in my life," Sinclair commented.

Singer, Maxine
(1931–)
American
Biochemist and Geneticist

Maxine Singer voiced the most rational views for the scientific community during the public debate over experimentation with recombinant DNA, or the alteration of deoxyribonucleic acid to transform genetic characteristics. She developed a logical compromise between unfettered experimentation on the one hand and complete banning on the other hand by charting a cautious approach. Singer exhibited her levelheaded thinking in other arenas of her professional life, writing clearly about complex issues and managing groups effectively.

Maxine Frank was born on February 15, 1931, in New York City. Her mother, Henrietta Perlowitz, was a hospital admissions officer, children's camp counselor, and model, and her father, Hyman Frank, was an attorney. Frank earned her B.A. degree from Swarthmore College in 1952. That year, she married Daniel Singer, and together the couple had four children: Amy Elizabeth, Ellen Ruth, David Byrd, and Stephanie Frank.

Maxine Singer went on to receive her Ph.D. in biochemistry from Yale University in 1957. She then commenced her long career with the National Institutes of Health (NIH) working on a U.S. Health Service postdoctoral fellowship from 1956 through 1958 at the National Institute for Arthritis, Metabolism, and Digestive Diseases. The NIH retained Singer as a biochemical research chemist from 1958 through 1974. During that time, she conducted research on both DNA and its sister, RNA (ribonucleic acid), trying to identify virus-causing tumors. The Department of Genetics at the Weizmann Institute of Science in Rehovot, Israel, also invited her to its campus as a visiting scientist in the early 1970s.

Paul Berg, coauthor of two books with Singer, first recombined DNA in 1972, and then he voluntarily halted this controversial line of research. Singer cochaired the 1973 Gordon Conference, a gathering of significant scientists that she used as a forum to discuss the potential wonders, as well as the incumbent moral dilemmas, of recombinant DNA. By ballot, the scientists agreed to address the ethical concerns in a letter, written by Singer and conference cochair Dieter Söll of Yale University, to the National Academy of Sciences and to *Science* magazine, which published it.

In 1976, Singer drafted a list of four principles concerning recombinant DNA experimentation: Overly risky experiments should be banned; less risky experiments that require the use of recombinant DNA should be allowed with safeguards; the more risky an experiment, the stricter the safeguards; and these guidelines should be reviewed annually. The NIH committee charged with deciding the protocol ratified Singer's guidelines.

Meanwhile, in 1974, the NIH promoted Singer to head the Section of Nucleic Acid Enzymology of the Division of Cancer Biology and Diagnosis (DCBD) at the National Cancer Institute. In 1979, the NIH again promoted her to become the chief of the DCBD's Laboratory of Biochemistry. She retained this position until 1988, when the NIH allowed her emerita status (the first person to receive this honor) so that she could become president of Carnegie Institution.

Singer wrote more than 100 publications in her career. She served on the editorial board of *Science* magazine and also published numerous articles in its pages, including "The Recombinant DNA Debate" in 1977 and "Recombinant DNA Revisited" in 1980. With Paul Berg, she coauthored two books: *Genes and Genomes: A Changing Perspective,* a graduate-level textbook hailed at its 1990 publication as "superbly written"; and *Dealing with Genes: The Language of Heredity,* published in 1992 and aimed at a more general readership with clear explanations of complex concepts.

Singer's career has been highly honored with more than 40 awards, including 15 honorary degrees. The National Academy of Sciences elected her as a member in 1979. Her highest honor came in 1992 when the National Science Foundation granted her the National Medal of Science. More important than what she has received, Singer has given back to her community: She established a program called First Light to improve science education for inner-city youths.

Sinkford, Jeanne C.

(1933–   )
American
Dentist and Physiologist

Jeanne Frances Craig Sinkford became the first African-American woman to head a university dentistry department in the United States when she assumed the chair of Howard University's Department of Prosthodontics in 1964. She later became dean of Howard's College of Dentistry. After retiring from Howard, she continued to contribute to the field of dentistry as the assistant executive director of the Division of Women and Minority Affairs for the American Association of Dental Schools.

Jeanne Frances Craig was born on January 30, 1933, in Washington, D.C. She married Stanley M. Sinkford in 1951 in the midst of her undergraduate study at Howard University, and she eventually had three children. She received her B.S. degree from Howard in 1953, and that year she worked as a research assistant in psychology for the United States Department of Health, Education, and Welfare. She continued at Howard toward her D.D.S. degree, which she received in 1958.

Sinkford remained at Howard's College of Dentistry as an instructor in dentistry for two years, from 1958 through 1960. Then she decided to pursue graduate study in physiology at Northwestern University in Chicago. She earned her M.S. degree in 1962, and she worked as a clinical instructor in the university's Dentistry School from 1963 through 1964 while conducting doctoral research. She received her Ph.D. in physiology in 1963.

In 1964, Sinkford returned to Howard to head its Department of Prosthodontics as an associate professor. Howard promoted her to the rank of full professor in 1968, and the year before had appointed her as associate dean of its College of Dentistry. In 1975, Howard named her dean of the College of Dentistry and the next year appointed her as a professor in the Department of Physiology of its Graduate School of Arts and Sciences.

The first African-American woman to head a university dentistry department in the United States, Jeanne Sinkford became the chair of Howard University's Department of Prosthodontics in 1964 and later the dean of its College of Dentistry. *(Photo by Twin Lens Photo, Courtesy of Jeanne Sinkford)*

Throughout this time, Sinkford practiced dentistry at several Washington, D.C., hospitals: Howard University Hospital, Children's Hospital of the National Medical Center, and the District of Columbia General Hospital. She also conducted research on substances that reduced or prevented inflammation internally, chemicals that facilitated healing, agents that retracted the gums, hereditary dental and oral secretion defects, and neuromuscular problems.

In 1991, Sinkford took up a special assistantship at the division of women and minority affairs with the American Association of Dental Schools. When she retired from Howard the next year, she became the assistant executive director of the division. There, she drew from her experience as a woman and a minority to assist others in their entrance into the field of dentistry.

Sinkford has received numerous honors for her work. Howard University granted her two awards: its 1969 College of Dentistry Alumni Award for Dental Education and Research and its 1971 Alumni Federation Outstanding Achievement Award. She also received Northwestern University's 1970 Alumni Achievement Award. In 1971, the American Prosthodontic Society granted her its Certificate of Merit. She also received two honorary doctorates, from Georgetown University in 1978 and from the University of Medicine and Dentistry of New Jersey in 1992.

Sitterly, Charlotte Emma Moore
(1898–1990)
American
Astrophysicist

Charlotte Emma Moore Sitterly worked for the majority of her career as an astrophysicist at the National Bureau of Standards, where she compiled three volumes of *Atomic Energy Levels as Derived from the Analysis of Optical Spectra*. Astrophysicists regarded these works as essential references in their work and praised Sitterly for her exhaustive research.

Charlotte Moore was born on September 24, 1898, in Ercildoun, Pennsylvania. Her mother, Elizabeth Palmer Walton, was a schoolteacher, and her father, George Winfield Moore, was a superintendent of the Chester County schools; both were Quakers. Moore graduated from high school in 1916. In 1920, she graduated Phi Beta Kappa with a B.A. in mathematics from Swarthmore College.

Moore became a mathematical computer at the Princeton Observatory from 1920 through 1925. There, her collaboration with astrophysicist Henry Norris Russell resulted in the 1928 publication of a monograph, *Presence of Predicted Iron Lines in the Solar Spectrum of Iron*. In the meanwhile, Moore had moved to the Mount Wilson Observatory in Pasadena, California, where she collaborated with Dr. Charles E. St. John and others on the *Revision of*

Charlotte Moore Sitterly, whose exhaustive research produced three volumes of astrophysical data that are regarded as essential references in the field. *(AIP Emilio Segrè Visual Archives, John Irwin Collection)*

Rowland's Preliminary Table of the Solar Spectrum Wavelengths, With an Extension to the Present Limit of Infrared, also published in 1928.

That year, the University of California at Berkeley recognized the excellence of Moore's work with its Lick Fellowship. She received her Ph.D. in astronomy in 1931, and the next year, the Carnegie Institution published excerpts from her doctoral dissertation as a monograph, *Some Results from the Study of the Atomic Lines in the Sun-Spot Spectrum*. Moore returned to the Princeton Observatory in 1931 as an assistant spectroscopist and was promoted to the rank of research associate in 1936. The next year, she married the astronomer and physicist Bancroft W. Sitterly. Although she took his name, she also continued to publish under her own name.

Sitterly remained at the Princeton Observatory until 1945, when she became a physicist in the spectroscopy division at the National Bureau of Standards. There, William F. Meggers assigned her to compile listings on atomic energy levels. Sitterly valued reliable accuracy, so instead of simply gathering previously published information, she also used her influence as a respected astrophysicist to cajole spectroscopists to confirm or even take new readings of some 485 atomic species. Her efforts resulted in the publication of three volumes of *Atomic Energy Levels as Derived from the Analysis of Optical Spectra,* released in 1949, 1952, and 1958.

Sitterly retired from the Bureau of Standards in 1968, but she continued to work, first at the Office of Standard Reference Data for three years, then at the U.S. Naval Research Laboratory from 1971 through 1978. There, she collaborated with Richard Tousey, analyzing ultraviolet solar spectra using data collected from V-2 rockets.

Sitterly received recognition for her excellent work starting early in her career. In 1937, the American Astronomical Society awarded her its ANNIE JUMP CANNON Prize. The U.S. Department of Commerce gave her its Silver Medal in 1951 and its Gold Medal in 1960. The next year, she was one of six women to receive the first Federal Woman's Award from the U.S. government. In 1972, the Optical Society of America granted her an honor bearing the name of her former colleague, the William F. Meggers Award. Sitterly received three honorary degrees, from Swarthmore College in 1962, from the University of Kiel in Germany in 1968, and from the University of Michigan in 1971. On March 3, 1990, Sitterly died of heart failure at her home in Washington, D.C.

Slye, Maud Caroline
(1879–1954)
American
Pathologist

Known in the popular press as the American Curie (after the renowned French physicist MARIE SKLODOWSKA CURIE), Maud Caroline Slye advanced the understanding of the inheritance of cancer through her breeding of mice. Throughout her career, Slye bred, raised, and then performed autopsies on approximately 140,000 mice, meticulously tracking the spontaneous recurrence of cancerous tumors in subsequent generations. She lobbied (unsuccessfully) for tracking of the recurrence of cancer in humans, though physicians began to request family histories of cancer in response to her research. More controversially, Slye proposed the eradication of human cancer by matchmaking based on genotype (again, she was unsuccessful).

Slye was born in Minneapolis, Minnesota, on February 8, 1879, the middle child of three. She inherited a love of poetry and an academic leaning from her mother, Florence Alden Wheeler, who was the granddaughter of the Reverend Timothy Alden, the founder and first president of Allegheny College. Her father, James Alvin Slye, was an attorney and an author who died upon her high-school graduation. Slye supported her family for the next decade, from 1886 through 1895, as a stenographer in Saint Paul.

Slye then pursued a full course load of undergraduate study at the University of Chicago, financing it by working almost full time as a clerk for university president William Harper. This overload led to a nervous breakdown, from which she convalesced with relatives in Woods Hole, Massachusetts. She took courses at the Woods Hole Marine Biological Laboratory, peaking her interest in biology, and completed her A.B. degree at Brown University in 1899.

Slye taught as a professor of psychology and pedagogy at the Rhode Island State Normal School from 1899 until 1905, when she returned to the University of Chicago as a graduate student. She received a postgraduate grant in 1908 allowing her to conduct research under Charles Whitman on six Japanese waltzing mice to discover the hereditary neurological disorder that caused them to "waltz." She then applied the same mice-breeding techniques to track the hereditary recurrence of spontaneous cancerous tumors in mice.

In 1911, the University of Chicago hired Slye at its newly established Sprague Memorial Institute, providing her with a salary and a laboratory. As a woman scientist investigating a controversial theory, Slye's requests for funding to support her cancer research were often rejected, so she paid for an assistant, Harriet Holmes, and for the mouse feed out of her own pocket.

At the May 1913 meeting of the American Society of Cancer Research, Slye presented her paper, "The Incidence and Inheritance of Spontaneous Cancer in Mice," chronicling the results of her investigation: She had raised 5,000 mice, 298 of which developed cancer. She suggested seven conclusions, most importantly that if both parents had cancer, the majority of their offspring would develop cancer, and that cancer is not contagious. Slye made a similar presentation to the June 1914 meeting of the American Medical Association, which awarded her its Gold Medal for her exhibit.

In 1919, the University of Chicago appointed Slye as the director of its Cancer Laboratory, promoting her to the rank of assistant professor of pathology in 1922 and associate professor of pathology in 1926. In 1936, Slye distilled her years of research into a theory on the inheritance of cancer, noting that trauma or chronic irritation of susceptible tissue must accompany hereditary predisposition in order to develop cancer. A male cancer researcher, Dr. Clarence Cook Little, attacked both her theory and her status as a woman scientist. Slye's theory, which acknowledged the complexity of cancer inheritance, proved credible, though later research showed cancer inheritance to be even more complex than she had proposed.

In 1944, Slye reached the retirement age of 65, as mandated by university rules. That same year, funding for her research dried up, so she activated her own insurance policy to pay for food to see the last generation of mice through to natural deaths.

Though Slye received little funding and much criticism, she also received numerous honors recognizing the significance of her work. She received the Ricketts Prize in 1915 and the Gold Medal of the American Radiological Society in 1922. She also received an honorary doctorate from Brown University in 1937. Slye died of a heart attack on September 17, 1954, and was buried in Chicago's Oak Woods Cemetery.

Solomon, Susan
(1956–)
American
Atmospheric Chemist

In the mid-1980s, when the threat of atmospheric ozone depletion first gained national attention, Susan Solomon led scientific expeditions to Antarctica, where the ozone hole had been observed. The ozone layer is crucial to life on Earth, as it blocks much of the Sun's lethal ultraviolet radiation. The data she gathered supported Solomon's theory that chlorofluorocarbons (CFCs), which charge refrigerators and aerosol cans, also deplete the ozone. Subsequent research has further validated her theory, which has gained wide acceptance.

Solomon was born on January 19, 1956, in Chicago, Illinois. Her mother, Alice Rutman, was a fourth-grade teacher, and her father, Leonard Solomon, was an insurance agent. Jacques Cousteau's television programs sparked her interest in science. She studied chemistry at the Illinois Institute of Technology, where she earned her B.S. degree in 1977. She decided to specialize in atmospheric chemistry after a senior project called on her to measure a process that occurs in Jupiter's atmosphere when ethylene reacts with hydroxyl radical.

In the summer of 1977, Solomon worked at the National Center for Atmospheric Research (NCAR) in Boulder, Colorado, where research scientist Paul Crutzen introduced her to the study of stratospheric ozone. In the fall of 1977, she commenced graduate study in chemistry at the University of California at Berkeley, where chemistry professor Harold Johnston peaked her interest in the negative atmospheric effects of supersonic transport. Solomon earned her M.S. degree in 1979, and then she returned to NCAR to conduct research with Crutzen for her doctoral dissertation. She earned her Ph.D. in 1981.

That year, Solomon returned to Boulder again but this time to work as a research chemist for the Aeronomy Laboratory of the National Oceanic and Atmospheric Admin-

Atmospheric chemist Susan Solomon, who led scientific expeditions to Antarctica in the mid-1980s to study the ozone hole. *(American Geophysical Union, courtesy of AIP Emilio Segrè Visual Archives)*

istration (NOAA). She developed computer models of the ozone layer located between 32,000 and 74,000 feet in altitude. This theoretical work prepared her for action in 1985 when scientists first reported a hole in the ozone layer rapidly expanding over Antarctica in the spring there. Solomon devised a hypothesis to explain the phenomenon at a lecture she attended at about that time on polar stratospheric clouds.

Solomon volunteered to lead an otherwise all-male expedition to McMurdo Sound, Antarctica, from August through October, 1986. As Solomon expected, they measured unusually high levels of stratospheric chlorine dioxide, a result of reactions between CFCs and ozone. Solomon's return trip to Antarctica in August 1987 confirmed her theory. She explained her hypothesis in an article, "On Depletion of Antarctic Ozone," in the June 19, 1986, edition of *Nature* and followed this up with a 1990 *Nature* article, "Progress Towards a Quantitative Understanding of Antarctic Ozone Depletion."

On September 20, 1988, Solomon married Barry Lane Sidwell, who brought a son with him into the marriage. In

1989, the U.S. Department of Commerce (the parent of the NOAA) granted her a Gold Medal for her explanation of ozone depletion. In 1985, the American Geophysical Union had bestowed on her the J. B. MacElwane Award, and *R&D Magazine* named her its 1992 Scientist of the Year. That year also, the National Academy of Sciences elected her as a member, validating her significance as an important scientist.

Somerville, Mary Fairfax
(1780–1872)
Scottish
Mathematician and Astronomer

Mary Fairfax Somerville distinguished herself as a writer of scientific texts. Her first book translated an incomprehensible French mathematics text into understandable terms; her second book made an offhand suggestion that led to the discovery of the planet Neptune; and her third book became a standard university textbook for half a century. She was one of the first women elected into the Royal Astronomical Society, and Oxford University later named its women's college after her.

Mary Fairfax was born on December 26, 1780, in the church manse in Jedburgh, Scotland. Her mother was Margaret Charters, and her father, Sir William George Fairfax, was a naval officer who later became vice admiral. Her parents supported education for their two sons but not for their two daughters, and thus she received only one year of schooling at Miss Primrose's boarding school for girls in Musselburgh. However, she educated herself, first reading Euclid's *Elements* after overhearing her art teacher, Alexander Nasmyth, recommend it to another student. She later took up algebra after noticing its peculiar symbols in a women's fashion magazine. Her parents removed the candles from her bedroom to discourage her from reading math texts, but she simply memorized problems and worked them out in her head at night.

In 1804, Fairfax married Samuel Greig, a distant cousin who was a captain in the Russian Navy. He died in 1807, soon after Mary gave birth to their second son. After only three years of marriage, widowhood liberated Mary from the social and financial constraints of pursuing her academic interests. Through the encouragement of Edinburgh University natural philosophy professor John Playfair, she initiated correspondence with William Wallace, a professor of mathematics at the Royal Military College at Great Marlow. He directed her attention to mathematical problems in the Mathematical Repository; in 1811, she received a silver medal for her solution to one of these problems.

In 1812, Mary married another cousin, William Somerville, a surgeon with the British Navy who grew up in the church manse where Mary was born. Together, the couple had four children. Unlike her former husband, Somerville approved of Mary's intellectual pursuits. He urged his wife to continue her correspondence with Wallace, who recommended she read J. Ferguson's *Astronomy,* Isaac Newton's *Principia,* and Laplace's *Mécanique Céleste.*

In 1826, Somerville presented a paper entitled "The Magnetic Properties of the Violet Rays of the Solar Spectrum" to the Royal Society. Her findings, though disproved much later, exhibited "ingenuity of original speculation" according to the society, which published her paper in its *Proceedings.* When Lord Henry Brougham discovered in 1827 that Somerville had read *Mécanique Céleste* (it was said that only 10 men in England could understand Laplace), he asked her to translate it for the Society for the Diffusion of Useful Knowledge. She translated not only its language but also its concepts into comprehensible explanations, making *The Mechanism of the Heavens* a popular text upon its 1831 publication.

During an 11-month trip in Europe from 1832 through 1833, Somerville wrote her second book, *The Connection of the Physical Sciences,* which proved a popular success upon its 1834 publication. That same year, both the Société de Physique et d'Histoire Naturelle de Genève and the Royal Irish Academy elected her as an honorary member. The next year, the Royal Astronomical Society elected Somerville and CAROLINE HERSCHEL as its first women members. These accomplishments helped convince men that women could excel in science. Somerville further influenced the future of women in science by tutoring ADA LOVELACE, daughter of the poet Lord Byron and mother of computer science.

In 1834, Somerville became one of the first woman scientists to receive a salary when the British prime minister Sir Robert Peel awarded her an annual civil pension of 200 pounds; in 1837, his successor, William Lamb, increased her pension to 300 pounds. In 1842, Somerville reiterated her speculation that an undiscovered planet perturbed the orbit of Uranus in her sixth edition of *The Connection of the Physical Sciences.* An undergraduate named Adams discovered the planet Neptune soon thereafter. In 1848, Somerville published her most popular book yet, *Physical Geography.*

In 1869, Somerville finished both an autobiography (published in excerpts posthumously) and her fourth scientific book, *Molecular and Microscopic Science.* In 1870, the Royal Geographic Society awarded her its Victoria Gold Medal. Two years later, Somerville died peacefully on November 29, 1872, in Naples, Italy. In 1879, Oxford University named Somerville College in honor of her support for women's education.

Spaeth, Mary
(1938–   )
American
Physicist

Mary Dietrich Spaeth invented the tunable dye laser, which allowed for midstream changes to the color of light emitted. Lasers were invented six years earlier, but the technology limited them to one color at a time, thus also limiting their usefulness in scientific experimentation. Spaeth's invention opened up the spectrum to use with lasers, thus allowing for important scientific and commercial uses, such as isotope separation.

Spaeth was born as Mary Dietrich on December 17, 1938, in Houston, Texas, the eldest of two children born to Louise Dittman. Her father, Fred Dietrich, was an insurance salesman who encouraged the tomboy in his daughter with gifts of footballs and tools. She aspired to become a doctor and became interested in science in seventh grade.

Dietrich attended Valparaiso University, where she earned her B.S. degree in physics and mathematics in 1960. She then pursued graduate study at Wayne State University, where she earned her M.S. degree in nuclear physics in 1962. That year, she joined the technical staff at Hughes Aircraft Company.

In about 1960, Theodore Maiman of Hughes Aircraft introduced the first working laser. *Laser* is actually an acronym for "light amplification by stimulated emission of radiation," meaning that the excitation of atoms increases their energy level, causing them to radiate energy as a beam of light. Maiman's laser emitted monochromatic light, or light of one color, thus excluding scientific applications that relied on different color readings.

Spaeth (who retained the name from her marriage that produced three children and ended in divorce in 1988) had conducted research on saturable dyes. She conceived of the idea of applying this knowledge to lasers, but she did not work in the same division of Hughes Aircraft as Maiman. In about 1966, during a two-week research hiatus while awaiting the delivery of equipment, Spaeth built a dye laser prototype out of glass tubes and Duco cement. She borrowed a larger laser to charge her own laser, which worked on her first attempt. She was not certain if it were a true laser because of the shifting colors, but it fit all the other definitions of a laser.

At this same time, Peter Sorokin had published an article in a scientific journal proposing the use of dyes to affect color in laser projections. A Hughes Aircraft colleague showed Spaeth this article, which helped to validate her invention as a laser. She coauthored an article with D. P. Bortfeld, entitled "Simulated Emission from Polymethine Dyes" and published in a 1966 edition of the journal *Applied Physics Letters*. Other scientists agreed that Spaeth's invention was indeed a laser. Spaeth did not receive a patent for her invention because she was working at the time on a contract for the U.S. Army, which retained the patent.

Spaeth later disobeyed direct orders from Hughes and continued work on a key component in the manufacture of ruby range finders. She did receive the patent for this invention, and Hughes manufactured huge numbers of them. However, Spaeth was much less interested in obtaining patents than in inventing useful tools. Her problem-solving skills led to her promotion at Hughes to the rank of senior scientist and later project manager.

In the 1970s, Spaeth innovated a means for separating elemental isotopes, namely plutonium and uranium, using her dye laser. In preparing these radioactive elements as fuel for light-water nuclear power reactors, the isotopes, or similar varieties of the same element, must be separated. Spaeth applied tunable dye lasers to the problem, using different colors to alter the electrical charges of one isotope distinct from its sibling isotopes. This process greatly reduced the cost of preparing fuel for nuclear reactors.

In 1974, Spaeth left Hughes for the Lawrence Livermore National Laboratory, which was more actively involved in isotope separation. She coauthored, with J. I. Davis and J. Z. Holtz, an article on her process entitled "Status and Prospects for Lasers in Isotope Separation," published in the September 1982 edition of the journal *Laser Focus*. The laboratory promoted her from the rank of group leader to the rank of associate program leader, and in 1986, she was appointed deputy associate director of the isotope separation project. Spaeth continues to work on applications for her tunable dye laser, such as on the guide star project, which aspires to achieve the resolution of the Hubble Space Telescope in earthbound telescopes with the help of Spaeth's tunable dye laser.

Sparling, Rebecca Hall
(1910–1996)
American
Mechanical Engineer

Rebecca Hall Sparling developed new methods for examining metal components by visible penetration inspection without destroying them and for testing structural elements for short periods at high temperatures. Much of the research she conducted was classified, as she worked in the aerospace industry selecting and testing materials for use in missiles, airplanes, and other aerospace structures. She also became one of the first women to gain registration as a professional engineer in the state of California.

Rebecca Hall was born on June 7, 1910, in Memphis, Tennessee. She was the 10th child born to Kate Sampson and Robert Meredith Hall. Both her parents graduated from college, and her father continued on to earn a law degree, though he practiced business instead. Hall attended Hollins

College in Virginia before transferring to Vanderbilt University. There, she earned her B.A. degree in chemistry in 1930, and she remained there for graduate study, earning her M.S. degree in physical chemistry in 1931.

Hall commenced her professional career in the metallurgical departments of American Cast Iron Pipe Company for one year and then at Lakeside Malleable Castings Company for two years. She then worked as a technical writer for William H. Baldwin in New York. In 1935, she married Edwin K. Smith. For the next nine years, she worked out of her home, consulting and writing for the American Foundrymen's Association, the International Nickel Company, and the Naval Gunn Factory, while caring for her son, Douglas, who was born in 1938. When he turned four, she started writing *American Malleable Iron,* the first such source to appear in 20 years and thus a standard reference in the iron castings industry from its 1944 publication. She divorced Smith in 1947 and married Joseph Sparling in 1948.

In 1944, Northrop Aircraft hired Sparling as chief materials and process engineer in its turbodyne division, where she stayed for the remainder of the 1940s. In 1951, General Dynamics hired her as a materials design specialist. In both of these positions, Sparling's main objective was to procure and then test materials used in the construction of missiles and other aerospace applications. One challenge was to recreate the extreme conditions that these materials would experience, such as highly elevated temperatures. Another challenge was to test the materials without destroying them. One of her greatest accomplishments was the design and casting of a highly tolerant magnesium wing with the dimensions of 16 feet by five feet by a mere quarter-to-half inch. Sparling retired from General Dynamics in 1968 and continued to freelance as a consultant, writer, and speaker.

Sparling received several honors throughout her career. The Society of Women Engineers granted her its Achievement Award in 1957. In 1978, she received two distinctions: the Outstanding Engineering Merit Award from the Institute for the Advancement of Engineering and the Engineering Merit Award from the Orange County Engineering Council. Sparling died in 1996, and in 1998, the Society of Women Engineers received a bequest of $68,593 from the Rebecca Sparling estate.

Sperry, Pauline
(1885–1967)
American
Mathematician

Pauline Sperry was the first woman promoted to the rank of assistant professor in the Department of Mathematics at the University of California at Berkeley. She distinguished

herself in her teaching of geometry at Berkeley, but she distinguished herself even more by refusing to sign a loyalty oath mandated by the University of California Board of Regents in the midst of the McCarthy era of anti-Communist hysteria. Sperry defended her intellectual freedom, even in the face of dismissal, and she was ultimately vindicated with reinstatement and back pay.

Sperry was born on March 5, 1885, in Peabody, Massachusetts. Her mother was Henrietta Leoroyd, and her father, Willard Sperry, was a Congregational minister. She attended Smith College from 1904 through 1906, when she graduated Phi Beta Kappa with a bachelor of arts degree. She taught mathematics at Hamilton Institute in New York City for one year before returning to Smith to pursue graduate study in music and mathematics. After earning her master's degree in music in 1908, she remained at Smith as an instructor in mathematics.

In 1912, Smith granted Sperry a traveling fellowship for further graduate study in mathematics at the University of Chicago. She studied projective differential geometry under the geometer Ernest Julius Wilcznski, who advised her in the writing of her master's thesis—"On the Theory of a One-to-Two Correspondence with Geometric Illustrations." She earned her master's degree in mathematics in 1914, and then she conducted doctoral research to write her dissertation, "Properties of a Certain Projectively Defined Two-Parameter Family of Curves on a General Surface." She received her Ph.D. in 1916, and two years later, the *American Journal of Mathematics* published her dissertation.

Smith promoted Sperry to an assistant professorship in mathematics upon her return, but in 1917, she traversed the country to become an instructor at the University of California at Berkeley. Berkeley promoted her to the rank of assistant professor in 1923, the first time a woman held this position in the department of mathematics there. In the 1920s, Sperry published two textbooks: *Short Course in Spherical Trigonometry* and *Plane Trigonometry and Tables* (with H. E. Buchanan). Berkeley promoted her to the rank of associate professor in 1931. That year, she published *A Bibliography of Projective Differential Geometry.*

In 1949, as Sperry neared retirement from Berkeley, the Board of Regents of the University of California system adopted a mandatory loyalty oath for all faculty and staff denying membership in any Communist or antigovernment organizations. Thirty-one faculty members at Berkeley and Los Angeles, including Sperry, refused to sign the oath, and the Board of Regents dismissed all of them in 1950. Interestingly, none of the dissenters had any proven connection to Communist or antigovernment organizations, and they all had professional records unstained by charges of incompetence.

The California Supreme Court ruled against the University of California Board of Regents in 1952 and ordered the reinstatement of all dismissed faculty. However, Sperry had already reached retirement age, so Berkeley appointed her associate professor of mathematics emerita as of July 1, 1952. In 1956, the Regents were further required by law to back-pay the salary lost by the nonsigners between 1950 and 1952. Sperry worked with the American Civil Liberties Union in her retirement. She died in 1967 in a retirement home in Pacific Grove, California.

Sponer, Hertha
(1895–1968)
German/American
Physicist

Hertha Sponer contributed most to physics with her two-volume text, *Molekul Spektren I and II (Molecular Spectra and Their Application to Chemical Problems)*. She also was indirectly responsible for the development of the Franck-Condon principle: James Franck sent Sponer, his former University of Göttingen physics student, the paper containing proofs that her University of California at Berkeley colleague E. U. Condon immediately extended to more general applications. Late in her life, she married Franck, a Nobel laureate in physics.

Sponer was born in 1895. She studied for her doctorate under Franck at the University of Göttingen, which granted her a Ph.D. in physics in 1920. She remained at Göttingen's Physics Institute as an assistant from 1921 until 1925, when Göttingen appointed her as a private docent.

Also in 1925, Sponer received an International Education Board Fellowship to conduct research at the University of California at Berkeley, where her colleague Condon expressed interest in a paper her former professor, Franck, had presented at the Faraday Society meeting in London. Franck sent the paper, which Sponer generously shared with Condon, who within a week generalized it into the Franck-Condon principle in his 1926 paper.

In 1932, the University of Göttingen named Sponer an "unofficial, extraordinary professor," an unpaid though significant position created by Albert Einstein for mathematician EMMY NOETHER (who had been teaching under David Hilbert's title until then since women could not hold professorships.) The Nazi regime opposed women professors, resulting in Sponer's dismissal. She fled the Nazification of Germany in 1934, emigrating first to Norway as a visiting professor at Oslo University and then to the United States.

In 1935 and 1936, Sponer published the two volumes of *Molekul Spektren,* the texts that earned her fame in the physics world by providing a complete tabulation of the important aspects of all molecular spectra known then. President Few of Duke University hired Sponer as a "star" physicist, despite the urging by Robert A. Millikan of the California Institute of Technology to hire a man to get more mileage out of the institutional investment.

Besides publishing her two-volume book, she published extensively in physics journals. Some of her important publications include the following: "Heat of Dissociation of Non-Polar Molecules," cowritten by R. Birge, in a 1926 *Physics Review;* "Lattice Energy of Solid CO_2," cowritten by M. Bruch-Willstatter; "Analysis of Near U.V. Electronic Transition Benzene," with L. Nordheim, A. L. Sklar, and E. Teller; and "On the Application of the Franck-Condon Principle to the Absorption Spectrum of $HgCl_2$," all the latter three in the *Journal of Chemistry and Physics.*

In 1946, at the age of 51, Sponer married her former professor, James Franck, who had won the 1925 Nobel Prize for physics for his experimental verification of quantum physics. She herself worked on experimental applications of quantum mechanics to atomic and molecular physics. In her atomic physics investigations, she focused on electron impact, mean free path, and spectroscopic studies.

Sponer remained at Duke University until 1966, when she retired to emerita status. Her husband, Franck, had died in 1965, and she returned to Germany to live with relatives. She died there in 1968.

Spurlock, Jeanne
(1921–   )
American
Psychiatrist

Jeanne Spurlock's psychiatric research focuses on psychological stress suffered by individuals and groups as a result of discrimination: single mothers, gays, racial and ethnic minorities, and children, among others. She investigates how poverty, racism, sexism, and other forms of prejudice can conspire to oppress these groups not only physically but also psychologically. Spurlock also devises means of overcoming the stresses resulting from discrimination and prejudice.

Spurlock was born on July 21, 1921, in Sandusky, Ohio. She was the eldest of seven children born to Godene Anthony and Frank Spurlock. They moved to Detroit when Spurlock was merely six months old. When she was nine, she broke her leg and received such incompetent treatment at the hospital that she vowed to improve how doctors treated patients, especially those from minority groups.

In 1940, Spurlock entered Spelman College on scholarship and worked almost full-time to support herself.

After two years of this intense schedule, she transferred to Roosevelt University in Chicago. Instead of completing her bachelor's degree there, she transferred to Howard University in the spring of 1943 to enroll in its accelerated medical program, earning her M.D. in 1947. She remained in Chicago for her one-year internship in psychiatry at Provident Hospital, and then she fulfilled her residency in general psychiatry at Cook County Hospital from 1948 through 1950.

Spurlock next held a one-year fellowship in child psychiatry at the Institute for Juvenile Research. She remained on the staff there for the next two years while simultaneously working as a staff psychiatrist at a mental hygiene clinic at Women's and Children's Hospital. From 1953 through 1962, Spurlock studied psychoanalysis part time at the Chicago Institute for Psychoanalysis.

Spurlock simultaneously directed the Children's Psychosomatic Unit at Chicago's Neuropsychiatric Institute and taught as an assistant professor of psychiatry at the University of Illinois College of Medicine. From 1960 through 1968, Spurlock concurrently held a clinical assistant professorship in psychiatry at the Illinois College of Medicine and also directed the child psychiatry clinic at the Michael Reese Hospital. She also maintained a private practice in psychiatry from 1951 through 1968.

That year, Meharry Medical College appointed her as head of its Department of Psychiatry, a position she retained until 1973. She then spent one year as a visiting scientist at the National Institute of Mental Health, after which the American Psychiatric Association named her its deputy medical director. During her tenure there, she held two concurrent academic appointments as clinical professor of psychiatry at the colleges of medicine at George Washington University and Howard University. She resigned from the American Psychiatric Association in 1991.

Spurlock championed psychiatric evaluation of subcultures that often escaped critical evaluation due to their peripheral status in the larger culture. Interestingly, these groups suffered psychological stresses very particular to their situations. Spurlock considered the specific problems of women who have never married, of single mothers and children with absent fathers, and of gays.

For example, in 1985, she published "Survival Guilt and the Afro-American" in the *Journal of the National Medical Association,* in which she discussed the particular guilt experienced by African Americans who elevate themselves into higher economic classes. Instead of simply identifying the problem, Spurlock also suggested ways to mediate the stress, such as evaluating whether the individual prioritizes financial success or inclusion in their cultural group.

The American Academy of Child and Adolescent Psychiatry and the National Institute on Drug Abuse honored Spurlock by naming the Jeanne Spurlock Research Fellowship in Drug Abuse and Addiction for Minority Medical Students after her.

Srinivasan, Bhama
(1935–   )
Indian/American
Mathematician

Bhama Srinivasan presented the prestigious EMMY NOETHER Lectures in 1990 to the Association for Women in Mathematicians (AWM) on her research specialty, the representation theory of finite groups. She had previously served as president of the AWM in the early 1980s. Her research produced interesting crossovers into quantum physics, placing her in demand; she served as a visiting professor in France, Germany, Australia, and Japan as well as teaching and conducting research in England, Canada, and India.

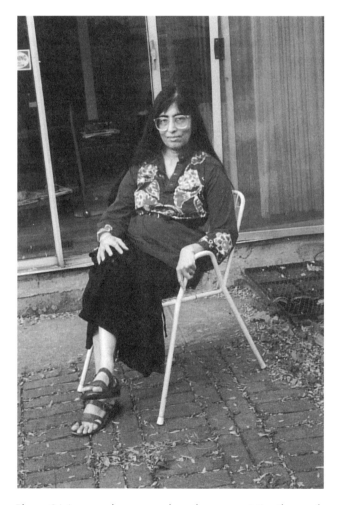

Bhama Srinivasan, whose research on the representation theory of finite groups has intersected in significant ways with quantum physics. *(Courtesy of Bhama Srinivasan)*

Srinivasan was born on April 22, 1935, in Madras, India. She attended the University of Madras, where she earned her bachelor of arts degree in 1954 and her master of science degree in 1955. She traveled to England for her doctoral study, writing her dissertation under the direction of J. A. Green at the University of Manchester. She received her Ph.D. in physics in 1959.

Srinivasan remained in England to commence her professional academic career as a lecturer in mathematics at the University of Keele from 1960 through 1964. She then pursued a postdoctoral fellowship at the University of British Columbia through the National Research Council of Canada from 1965 through 1966. She returned home to India to teach at the Ramanujan Institute of Mathematics of her alma mater, the University of Madras, from 1967 though 1970.

Srinivasan then immigrated to the United States, where she taught for the next decade at Clark University in Worcester, Massachusetts, as an associate professor. In 1977, she became a naturalized citizen of the United States. That year, she was a member of the Institute for Advanced Studies at Princeton. In 1980, she commenced her long-standing tenure at the University of Illinois as a professor of mathematics at the Chicago Circle campus.

Srinivasan has distinguished herself in her field throughout her career. From 1981 through 1983, she served as president of the AWM. In January, 1979, she delivered the Invited Address to the American Mathematical Society (AMS) at the Joint Mathematics Meetings in Biloxi, Mississippi. She has also been invited to fill visiting professorships internationally at the École Normale Supérieure in Paris, the University of Essen in Germany, Sydney University in Australia, and the Science University of Tokyo in Japan. She has served as an editor for several journals in her field: *Proceedings of the AMS* (from 1983 through 1987); *Communications in Algebra* (from 1978 through 1984); and *Mathematical Surveys and Monographs* (from 1991 through 1993). From 1991 through 1994, she served on the Editorial Boards Committee of the AMS.

In 1990, Srinivasan delivered the prestigious Noether lecture to the AWM. She used this opportunity to give an overview of her subspecialty, the representation theory of finite groups, specifically informing her audience how certain geometric methods were adopted by group theory and what future direction this research might head. Srinivasan collaborated with Paul Fong on finite groups of the Lie type, and this work has been linked to Lusztig's research on quantum groups, thus crossing over between mathematics and physics. Although Srinivasan generally advocates pure mathematical research, resisting the temptation to find a practical application for all mathematics, she nevertheless got excited by the application of her research to physics.

Stanley, Louise
(1883–1954)
American
Chemist and Home Economist

Louise Stanley became the first woman to direct a bureau of the U.S. Department of Agriculture when the USDA established the Bureau of Home Economics in 1923. She promoted the scientific study of nutrition, housekeeping, rural housing, and consumer purchasing. She exerted her influence internationally through home economics education in Latin America and as a consultant for the Office of Foreign Agricultural Relations at the end of her career.

Stanley was born in 1883. She attended Peabody College, where she earned her A.B. degree in 1903. She then matriculated at the University of Chicago, where she earned a bachelor of education degree in 1906. The next year, she pursued graduate studies at Columbia University, earning her master's degree in 1907. She continued with doctoral work at Yale University, which granted her a Ph.D. in 1911.

At that time, home economics was emerging as an academic field of study, applying scientific methods to research on the domestic sphere. In 1917, the University of Missouri hired Stanley as an instructor in home economics. Within six years, she had climbed the academic ladder to the rank of full professor, and she was also appointed chair of the Department of Home Economics. Besides her teaching and administrative duties, she conducted research on food chemistry, specifically focusing on the proper temperatures for baking. She also investigated organic and inorganic phosphorus and purin enzymes.

In 1923, the USDA established a new bureau in this emerging field, the Bureau of Home Economics. The government appointed Stanley as the first chief of this bureau, making her the first woman to head a bureau of the USDA. The USDA later chose her as its official representative to the American Standards, so she became the first woman to hold this appointment. In this capacity, she promoted the standardization of clothing sizes.

Stanley remained the chief of the home economics bureau for more than a quarter of a century. Under her direction, the bureau conducted research on nutrition, resulting in the adaptation of dietary guidelines for families at four different economic levels. In an effort to streamline efficiency in homemakers' work, the bureau conducted time and motion studies of housekeeping routines. The bureau also instituted the first national surveys of housing in rural areas and of spending by consumers.

In 1935, Stanley disseminated her wealth of knowledge in the book *Foods, Their Selection and Preparation*. Five years later, she received an honorary degree from the institution where she formerly sat on the faculty, the University of Missouri. During World War II, she promoted home economics education in Latin America, a part of the world neglected by the focus on the European and Asian arenas.

In 1950, Stanley retired from her post as director of the Bureau of Home Economics. For the next three years, she consulted in the Office of Foreign Agricultural Relations on home economics issues. In 1954, Stanley passed away. The University of Missouri honored her by naming its home economics building after her in 1961. In 1966, she was the topic of a Helen T. Finneran's American University master's thesis, entitled "Louise Stanley: A Study of the Career of a Home Economist, Scientist and Administrator."

Steitz, Joan Argetsinger
(1941–)
American
Biochemist and Geneticist

Joan Argetsinger Steitz is world renowned for her discovery of small nuclear ribonucleoproteins, or snRNPs (pronounced "snurps"), which aid in mammalian reproduction. The transfer of genetic coding takes place in the splicing of double-stranded DNA (deoxyribonucleic acid) into single-stranded RNA (ribonucleic acid). snRNPs come into play by helping to transcribe the necessary genetic information (and blocking unnecessary information) when the RNA recomposes into a new double strand of DNA.

Joan Argetsinger was born on January 26, 1941, in Minneapolis, Minnesota. Her mother, Elaine Magnusson, was a speech pathologist, and her father, Glenn Davis Argetsinger, was a high school guidance counselor. Argetsinger studied chemistry at Antioch College in Yellow Springs, Ohio, but she found her professional calling in molecular genetics courses—it fascinated her that tiny molecules controlled genetics. She earned her B.S. degree in chemistry in 1963 and then pursued a doctorate at Harvard University.

The field of molecular genetics exploded with the 1953 discovery by James Watson, Francis Crick, Maurice Wilkins, and Rosalind Franklin of the double helical structure of DNA. Nobel laureate Watson guided Argetsinger's doctoral research, which involved the in vitro assembly of R17, an RNA bacteriophage (or virus that attacks certain bacteria). Harvard granted her a Ph.D. in biochemistry and molecular biology in 1967. The year before, she married fellow scientist Thomas A. Steitz, and together the couple had one son.

Between 1967 and 1970, Steitz conducted postdoctoral research under Nobel laureate Crick at the Medical Research Council Laboratory of Molecular Biology in Cambridge, England. She investigated how bacterial ribosomes, or intracellular organelles, recognize the correct location to trigger protein synthesis on messenger RNA molecules. Upon her return, Yale University hired Steitz as an assistant professor of molecular biophysics and biochemistry. The university promoted her to associate professor in 1974 and to full professor in 1978.

Steitz considers her discovery of snRNPs as her most significant scientific contribution. In genetic transcription, double-stranded DNA splices into single-stranded RNA, which carries the DNA's genetic code. This code contains both essential and nonessential information. As Steitz discovered, snRNPs supervise the process of slicing up the RNA into sections, or introns, winnowing out the nonsense while retaining the important information. They then reassemble the remaining sections in the correct order to create exons, or molecules ready for protein synthesis.

Steitz discovered this process while working with rheumatic diseases, which generate autoantibodies, or antibodies against their own snRNPs. Her discovery helps clinicians to determine which antibodies patients have, thus aiding in diagnosis and treatment. Steitz has also applied her snRNP research to the understanding of how certain herpes viruses infect their host cells.

Steitz continued her research away from Yale on two sabbaticals. First, in 1976 through 1977, she researched at the Max Planck Institute for Biophysical Chemistry in Göttingen, Germany, as a Josiah Macy Scholar. Then, in 1984 through 1985, she conducted research at the California Institute of Technology in Pasadena, California, as a Fairchild Distinguished Fellow. In 1986, Yale assigned Steitz to a concurrent position as an investigator at its Howard Hughes Medical Institute. Then in 1992, Yale named her its Henry Ford II Professor of Molecular Biophysics and Biochemistry.

Steitz's honors read like a grocery list of prestigious awards. This list commences in 1975 with the Young Scientist Award from the Passano Foundation and the Eli Lilly Award in Biological Chemistry in 1976. In 1983, the National Academy of Sciences elected her as a member, and in 1986, President Ronald Reagan conferred on her the National Medal of Science. In 1989, she shared the Warren Triennial Prize with Thomas R. Cech, and in 1992, she received the Christopher Columbus Discovery Award in Biomedical Research. She has received countless other honors, but she is most proud of her 1994 Weizmann Women and Science Award, because it specifically recognizes the accomplishments of women in a field that was slow to accept their capabilities as equal to men's.

Stephenson, Marjory
(1885–1948)
English
Biochemist and Microbiologist

Marjory Stephenson helped to establish the study of bacterial chemistry as a branch of biochemistry. She conducted her research on bacterial metabolism and was considered the leading authority on the enzymes that control this process. She was one of the first two women elected as fellows of the Royal Society. The Society of General Microbiology named its principal prize after Stephenson, a monetary award that also included an invitation to present a paper at the society's meeting that would later be published in one of the society's journals.

Stephenson was born in 1885 in Cambridgeshire, England. Her father, a farmer whose reading of Darwin and Mendel inspired him to grow fruit in a region previously devoid of orchards, peaked his daughter's interest in science at an early age by explaining nitrogen fixation to her during strolls through clover fields. Her mother supervised her formal education and urged her to pursue university study at a time when women rarely attended college. Stephenson abided by her mother's advice and studied at Newnham College from 1903 through 1906.

Stephenson taught domestic science briefly after her collegiate career. In 1911, she commenced work in biochemical research in London. During World War I, she volunteered with the Red Cross, which stationed her in France and in Salonika. After the war, she returned to Cambridge to conduct research under F. G. Hopkins in the Biochemical Laboratory, where she remained for the rest of her career. Her first investigations involved vitamins. In 1922, she started her lifelong study of bacterial metabolism, focusing specifically on the enzymatic control of this process. She demonstrated that the activity and nature of these enzymes was essentially the same as the activity and nature of higher organisms.

In 1930, Stephenson helped establish the study of bacterial physiology with her publication of *Bacterial Metabolism*, considered the first authoritative monograph on the subject. However, she made no mention of photosynthesis in the text. In the meanwhile, van Niel demonstrated that bacteria utilize light to transform carbon dioxide into sugar, or in other words, that bacteria perform photosynthesis just as plants do. Stephenson called this discovery "an event in bacterial chemistry of first-class biological importance" that amounted to "the revelation of a new mode of life." Stephenson devoted an entire new chapter to bacterial photosynthesis in the second edition of *Bacterial Metabolism*.

In 1945, the Royal Society opened its membership to women for the first time, inducting two women into its ranks. The society elected Stephenson as its first woman member in the biological sciences and the crystallographer KATHLEEN YARDLEY LONSDALE as its first woman physical scientist. Three years later, in 1948, Stephenson died.

The Society for General Microbiology honored Stephenson by naming its principal prize after her. Granted biennially, the prize recognized outstanding contributions of current importance in microbiology and awarded the winner 1,000 British pounds. In addition, the winner delivered a lecture at the society's meeting, and the Society generally published this paper in one of its journals.

Stern, Frances
(1873–1947)
American
Nutritionist

Frances Stern pioneered the study of applied nutrition, as she was the first to study scientifically the effects of nutrition on human beings. On the practical level, she taught many underprivileged people how to feed themselves in nutritious but affordable ways. On the theoretical level, she studied the correlation between nutrition and sociology, often noticing problems with the former related to problems with the latter.

Stern was born in 1873. She attended the Garland Kindergarten Training School in Boston, graduating in 1897. While teaching in the North End of Boston, she began to notice a correlation between the food her students ate and their demeanor. At this time, she landed a position at the Massachusetts Institute of Technology (MIT) as the secretary and research assistant for ELLEN SWALLOW RICHARDS, who is credited with founding the field of home economics. Stern studied food chemistry and sanitation at MIT in 1909. She also created a visiting housekeeping program for the Boston Association for the Relief and Control of Tuberculosis (similar to the program she later started for the Boston Provident Association.)

In 1912, the Massachusetts State Board of Labor and Industries appointed Stern as the industrial health inspector, a position she retained until 1915. With the outbreak of World War I, she contributed to the war effort as a member of the U.S. Food Administration's Division of Home Conservation. She also worked for the U.S. Department of Agriculture, as an investigator of the adequacy of the food eaten by industrial workers, and for the Federal Food Conservation in Washington, D.C. Near the end of the war, she served the American Red Cross in France. After the war ended, she remained in Europe, traveling to England to study economics and politics at the London School of Economics.

Upon returning to the United States, Stern established the Boston Dispensary Food Clinic, the first such organization in the world. There she worked with immigrants adapting their native recipes to affordable foods available in this country. She also utilized color-coded pictures to convey her advice to non-English speakers. In 1925, she founded a Nutrition Education Department to disseminate her knowledge of nutrition to doctors, nurses, dentists, and social workers.

Throughout this time, Stern taught nutrition and dietetics at Simmons College, Tufts College Medical School, MIT, and the State Teachers College at Framingham. She also published numerous books on nutrition: *Food for the Worker* in 1917; *Food and Your Body* in 1932; *Applied Dietetics* in 1936; *How to Teach Nutrition to Children* in 1942; and *Diabetic Care in Pictures*. She was one of the first people to take a holistic approach to the issue of nutrition, taking into consideration the whole person (including their emotions and food inhibitions) instead of treating symptoms.

In 1943, the name of the Boston Dispensary was changed to the Frances Stern Food Clinic. In 1947, Stern died, and in 1957, the name was again changed to the Frances Stern Nutrition Center. The center eventually merged with the New England Medical Center and Tufts University.

Stevens, Nettie Maria
(1861–1912)
American
Geneticist

Nettie Maria Stevens first theorized that specific chromosomes in sperm determine the sex of embryos. The renowned male biologist Edmund Wilson reached similar conclusions at about the same time through research independent of Stevens. Wilson dismissed Stevens's research, and, interestingly, the annals of scientific history generally agree, attributing the discovery to Wilson. However, current scientific opinions consider Stevens's evidence and argument stronger than Wilson's, and Stevens is finally receiving deserved recognition nearly a century after her discovery.

Stevens was born on July 7, 1861, in Cavendish, Vermont. Her mother, Julia Adams, gave birth to three other children, but only Nettie and her sister Emma survived. Their mother died when Nettie was two years old, and her father, a carpenter named Ephraim Stevens, married Ellen Thompson. He moved the family to Westford, Massachusetts, where Stevens excelled in both the public schools and then at Westford Academy. She graduated in 1880 and proceeded to teach Latin, English, mathematics, physiology, and zoology at the Lebanon, New Hampshire, high school.

Stevens continued her education from 1881 to 1883 at the Westfield Normal School, where she earned the highest scores in her class. She returned to teaching at the Minot's Corner School in Westford and then at her alma mater, Westford Academy, from 1885 through 1892. She then worked as a public librarian until 1896, when she matriculated at Stanford University, where she studied physiology under Oliver Peebles Jenkins. She continued her

research even in the summers, performing experiments on the life cycles of Boveria, or protozoan parasites of sea cucumbers, at the Hopkins Seaside Laboratory in Pacific Grove, California. She published her findings in the 1901 edition of the *Proceedings of the California Academy of Sciences*.

Stevens graduated from Stanford in 1899 with a B.A. degree in physiology. She remained at Stanford for graduate study and the next year earned her M.A. degree. In 1901, she traveled overseas to conduct research at the Naples Zoological Station in Italy and at the Zoological Institute of the University of Würzburg in Germany. She then pursued a doctorate at Bryn Mawr College as a research fellow in biology. She earned her Ph.D. in biology in 1903, and that year, she received a grant from the Carnegie Institution to continue investigating chromosomes and sex determination through 1905. In 1904, Bryn Mawr also appointed her as a reader in experimental morphology, and the next year she was promoted to an associate in experimental morphology, a position she retained until her death.

Stevens conducted her chromosomal research on the common mealworm, *Tenebrio molitor*. She discovered that mealworm sperm contained one of two chromosomes, a large X chromosome or a small Y chromosome, whereas unfertilized mealworm eggs contained only X chromosomes. Stevens hypothesized that the type of chromosome in the male sperm determined the sex of the offspring: X-chromosome sperm produced females, while Y-chromosome sperm produced males. Further research on other species confirmed this theory. In 1905, Carnegie Institution published Stevens's findings on chromosomal inheritance in her paper entitled "Studies in Spermatogenesis with Especial Reference to the 'Accessory Chromosome.'"

Stevens wrote extensively, publishing 38 papers in 11 years. In 1905, one paper, entitled "A Study of the Germ Cells of *Aphis rosae* and *Aphis oenotherae*," earned Stevens the Ellen Richards Research Prize, awarded by the Naples Zoological Station in recognition of outstanding scientific research by a woman. In 1908 and 1909, Stevens returned to the Zoological Institute of the University of Würzburg. In 1912, the trustees of Bryn Mawr created a special research professorship for Stevens, but before she could occupy it, she died. Breast cancer took her life on May 4, 1912, in Baltimore, Maryland.

Steward, Susan Smith McKinney
(1847–1918)
American
Physician

Susan Smith McKinney Steward helped to found the Women's Hospital and Dispensary (now Memorial Hospital for Women and Children) in Brooklyn, New York. She was that state's first African-American woman physician and the third black woman doctor in the United States.

Steward was born in 1847 in Brooklyn, New York. Her mother was Anna Eliza Springsteel, and her father, Sylvanus Smith, was a hog farmer whose success elevated the family into the upper echelon of African-American society in Brooklyn, which had not yet become urban. Growing up, Smith studied organ under John Zundel and Henry Eyre Brown, and she played in churches. At some point, her interest in medicine eclipsed her interest in the organ (perhaps in reaction to the deaths by cholera of her two brothers in 1866), though she continued to play music throughout her life.

Steward studied medicine (a rarity for African Americans and for women, particularly African-American women) at the New York Medical College and Hospital for Women. She supported her own education, though her family surely could have afforded it. She graduated as the valedictorian of her class, earning her M.D. in 1870. She then married the Reverend William G. McKinney, a traveling preacher, and together the couple had two children: a daughter named Anna, and a son named after his father.

Steward spent the first decade after becoming a doctor establishing a private practice, which proved challenging in an age of rampant racism and sexism. However, many of her patients suffered from the same discrimination and prejudice, so she became an important resource for treating African Americans and other ethnicities, women, children, the poor. In 1881, Steward helped to found the Women's Hospital and Dispensary, serving on its staff for the next 15 years. She also served on the staff at the New York Medical College and Hospital for Women, where she had received her M.D., from 1882. In 1892, she added duties as a manager and medical staff member (one of two women) at the Brooklyn Home for Aged Colored People.

Steward presented at least two papers to the Kings County Homeopathic Society. The first, which she read in 1883, described the case history of one of her pregnant patients who died (along with her baby) from exposure to carbolic acid. The second, presented in 1886, concerned childhood diseases resulting from malnutrition, her specialty.

In 1887, Steward returned to school for a year of postgraduate study as the only woman at Long Island Medical College. Her husband died on November 25, 1895, from complications due to a cerebral hemorrhage suffered five years earlier. In 1896, Steward married another minister, the Reverend Theophilus Gould Steward. The U.S. Army, which Steward's husband served as a chaplain, relocated him regularly, so she followed. They spent stints in numerous locations: Forts Missoula, Niobrara, and McIntosh (in Montana, Nebraska, and Texas, respectively); Ohio; Cuba; and the Philippines.

In Ohio, Steward worked at Wilberforce University as an instructor and as the college doctor, first from 1898 through 1902 and again from 1906 through 1911. That year, she and her husband traveled to Europe. In London, Steward delivered a paper, "Colored Women in America," at the first Interracial Conference. Upon their return, she delivered another paper, "Women in Medicine," to the Colored Women's Club in Wilberforce. This paper proved influential, as it traced a comprehensive history of African-American women's contributions to medicine, and it was widely distributed in pamphlet form.

After Steward died suddenly on March 7, 1918, in Wilberforce, W. E. B. DuBois delivered the eulogy at her Brooklyn burial. In honor of her pioneering work in medicine, the African-American female medical societies in New York, Connecticut, and New Jersey all adopted her name. Later, in 1975, a junior high school in Brooklyn renamed itself the Susan Smith McKinney Junior High on the urging of her grandson.

Stoll, Alice Mary
(1917–)
American
Biophysicist

Alice Mary Stoll's research focused on the effects of high-speed flight and the dangers of exposure to intense heat and cold. She developed her own instrumentation—much of which she patented—to fill the demands of her innovative experiments, and her research on heat- and flame-resistance led to the development of new, fire-retardant fibers.

Stoll was born on August 25, 1917, in New York City. She remained in the city for her undergraduate study, earning her B.A. degree from Hunter College in 1938. She then worked for five years as an assistant in allergy, metabolism, and infrared spectroscopy at the New York Hospital and Medical College of Cornell University. Stoll joined the U.S. Naval Reserve in 1943 and remained active until 1946. She returned to Cornell as a graduate student to earn dual M.S. degrees in physiology and biophysics in 1948.

Stoll remained at Cornell for the next five years, working as a physiological research associate in environmental thermal radiation at the medical college and as an instructor in the school of nursing. From 1952 through 1953, she held a concurrent position as a consultant to the Arctic Aero-Medical Laboratory at Ladd Air Force Base in Alaska.

In 1953, Stoll left Cornell for the U.S. Naval Air Development Center (NADC) in Warminster, Pennsylvania, to take up a position as a physiologist in the medical research laboratory. The NADC promoted her several times: to special technical assistant in 1956; to head of the thermal laboratory in 1960; to head of the biophysical and bioastronautical division in 1964; and finally, to head of the biophysical laboratory in the crew systems department from 1970 until her retirement.

While with the NADC, Stoll researched how best to manage the physiological effects of supersonic flight, such as the physiological stress of breaking the sound barrier, and the effects of this degree of acceleration on the cardiovascular system. She also tested for human endurance to extremes of cold at high altitudes and extremes of heat in the case of a fire, an ever-present threat. Her studies on human tissue damage and pain threshold demonstrated that the source of the heat did not matter and that the amount of heat exposure determined the degree of damage. Her research into heat-resistant fabrics resulted in the invention of the Nomex fire-resistant fabric by DuPont.

Stoll published many of her findings in professional journals. In 1959, she published the article "Relationship Between Pain and Tissue Damage due to Thermal Radiation" (coauthored by Leon C. Greene) in the *Journal of Applied Physiology*. She published three important articles with coauthor Maria A. Chianta: "Flame Contact Studies" (also coauthored by L. R. Munroe) in a 1964 edition of the *Journal of Heat Transfer;* "Method and Rating System for Evaluation of Thermal Protection" in a 1969 *Aerospace Medicine;* and "Thermal Analysis of Combustion of Fabric in Oxygen-Enriched Atmospheres" in a 1973 *Journal of Fire & Flammability.*

Stoll received significant recognition for her achievements, including the 1965 Federal Civil Service Award, the 1969 Society of Women Engineers' Achievement Award, and the 1972 Paul Bert Award from the Aerospace Medical Association. Stoll retired from the NADC in 1982.

Stone, Isabelle
(1868–1944?)
American
Physicist

History remembers Isabelle Stone as the first woman to receive a doctorate in physics from the University of Chicago and as a founding member of the American Physical Society. However, she contributed much more significantly to the field as a teacher of science to generations of young women. In the early 20th century, Stone helped to expand educational horizons by introducing science into the curricula for young women, a radical step at the time. Stone eventually established her own girls' school, where she remained until her death.

Stone was born in 1868. She attended Wellesley College, where she received her A.B. degree in 1890. She pursued graduate studies at the University of Chicago, earning her M.S. degree in 1896. She then conducted her doctoral research under the direction of A. A. Michelson. Her dissertation, "On the Electrical Resistance of Thin Films," was published by the scientific journal *Physical Review* in 1898, a year after she became the first woman to receive a Ph.D. in physics from the University of Chicago. The next year, in 1899, Stone and Marcia Keith were the only two women physicists to become founding members of the American Physical Society by attending its first meeting.

After earning her doctorate, Stone taught for one year at the Bryn Mawr Preparatory School in Baltimore, Maryland. She also spent a postdoctoral year in Berlin, Germany, though it is unclear when. Interestingly, when Vassar College hired Stone in 1898, it was not as a physics professor but rather as an instructor, even though she held a doctorate from a prestigious university. Furthermore, she remained in this instructorship for eight years without a promotion.

It is not surprising that Stone left Vassar for an endeavor more commensurate with her abilities. In 1907, Stone and her sister established the School for American Girls in Rome, Italy. Together, they ran the school for the next seven years. In 1916, Stone left Rome to accept a position as a physics professor at Sweet Briar College in Virginia. She was later promoted to head the college's physics department. In this position, she not only fulfilled her administrative duties but also conducted research in her area of concentration—the electrical and optical properties of thin metal films. More specifically, she investigated the color in platinum films and the properties of thin films in a vacuum.

In 1923, Stone and her sister moved their School for American Girls from Rome to Washington, D.C., where they renamed it the Misses' Stone's School for Girls. Here, Stone continued her mission of introducing the study of science and physics in particular. At that time, young women were still being denied scientific education under the pretense that the information was too difficult or that the experimentation was too intense for their "fragile" bodies and psyches. Stone refused to accept such rationalizations for sexism.

Stone continued to head the Misses' Stone's School for Girls until her death, the date of which remains uncertain.

Stubbe, JoAnne
(1946–   )
American
Chemist

JoAnne Stubbe focuses her research on enzyme catalysis and, in particular, nucleotide reductases—specific enzymes that catalyze, or cause, a key step in the biosynthesis of nucleotides into deoxyribonucleic acid (DNA), the genetic foundation of biological life. She has synthesized nucleotide analogs, or derivatives, that could function as antitumor, antiviral, and antiparasitical agents, pending further experimentation. Stubbe distinguished herself early in her career, winning the Pfizer Award in 1986, given by the American Chemical Society to the enzyme chemist under 40 years of age whose work stands out most.

Stubbe was born on June 11, 1946. She studied chemistry at the University of Pennsylvania, where she earned her B.S. in 1968 with high honors. She then pursued doctoral studies in chemistry under George Kenyon at the University of California at Berkeley, earning her Ph.D. in 1971. Stubbe spent the next year as a postdoctoral fellow conducting chemical research under Julius Rebek at the University of California at Los Angeles.

Stubbe entered her profession as an assistant professor of chemistry at Williams College in 1972. She spent her last two years with Williams, from 1975 through 1977, on yet another postdoctoral fellowship, this time at Brandeis University. Stubbe then spent three years at Yale University as an assistant professor in the Department of Pharmacology. In 1980, she left Yale for the University of Wisconsin in Madison, which promoted her from assistant professor to full professor in 1985. In 1987, the Massachusetts Institute of Technology (MIT) hired Stubbe as a professor of

JoAnne Stubbe, who received the Pfizer Award from the American Chemical Society in 1986, which is given to the enzyme chemist under 40 years of age whose work stands out most in the field. (*Courtesy of JoAnne Stubbe*)

chemistry, and in 1992, it named her the John C. Sheehan Professor of Chemistry and Biology, a title she has retained ever since.

Ribonucleotide reductases are enzymes that play an important role in DNA biosynthesis, determining the rate of cell growth. Blocking their functions could inhibit the spread of tumors and viruses, according to Stubbe's research. Stubbe has also conducted research on the antitumor antibiotic called bleomycin. In collaboration with John Kozarich, Stubbe investigated how bleomycin's anticancer functions relate to its ability to degrade DNA by binding to it.

Stubbe is a prolific writer, who has published more than 80 papers in her career. The first two papers she published after earning her doctorate concerned the enzymes enolase (a carbohydrate metabolizer) and pyruvate kinase. Later papers, published in important scientific journals such as *Advances in Enzymology, Science,* and *Biochemistry,* focused on ribonucleotide reductases in general and specifically addressed issues such as the mechanism involving intermediates that contain at least one paired electron, or radical intermediates.

Stubbe has also received much recognition of the significance of her research in the advancement of scientific understanding. Before winning the Pfizer Award, the National Institutes of Health granted her a career development award. In 1989, Stubbe won the ICI-Stuart Pharmaceutical Award for excellence in chemistry, and the next year, MIT bestowed an award on her for her outstanding teaching. The National Academy of Sciences elected her as a member in 1992, and in 1993, Stubbe received the Arthur C. Cope Scholar Award.

Sudarkasa, Niara
(1938–   )
American
Anthropologist

In 1987, Niara Sudarkasa became the first female president of Lincoln University, and she summarily transformed one of the oldest African-American colleges in the United States from an all-male into a coeducational institution. She was also the first African-American woman to gain tenure as a full professor at the University of Michigan. Her anthropological research focuses on comparative studies, specifically seeking to establish the ties that bind African culture with African-American culture.

Sudarkasa was born Gloria Albertha Marshall on August 14, 1938, in Fort Lauderdale, Florida. Her grandparents, who came from the Bahamas, raised her in the West Indian tradition. At the age of 14, she won a Ford Foundation Early Entrant Scholarship to Fisk University, a predominantly African-American college, where she first encountered anthropology. She spent her junior year as an exchange student at Oberlin College, which introduced her to comparative studies between the Yoruba people of Africa and her Caribbean relatives. She remained at Oberlin to earn her B.A. in sociology in 1957 at the age of 19.

Sudarkasa then pursued graduate study at Columbia University, where she earned her M.A. degree in 1959. Between 1960 and 1963, she conducted her doctoral research on Yoruba language, sociology, and economy with a Ford Foundation Foreign Area Training Fellowship, first at the University of London School of Oriental and African Studies and then in Nigeria. She returned to a Carnegie Foundation Study of New Nations Fellowship at the University of Chicago from 1963 through 1964. That year, she received her Ph.D. in anthropology from Columbia University.

Sudarkasa commenced her professional academic career as an assistant professor of anthropology at New York University from 1964 through 1967. That year, she began two decades of work for the University of Michigan as an assistant professor of anthropology and research associate with the Center for Research in Economic Development. In 1970, the university promoted her to the rank of associate professor, and in 1976, she became the first African-American woman to attain the rank of full, tenured professor at the University of Michigan.

On an anthropological field trip to Africa in the early 1970s, Sudarkasa decided to adopt an African name in recognition of her sense of reconnection to her ancestry. She chose Niara, a Swahili word meaning, approximately, "woman of high purpose." The name Sudarkasa belonged to her first husband; she later married John Clark, an inventor, sculptor, and contractor, with whom she has one son, Michael.

Sudarkasa's academic work seeks to reclaim African-American ties to African culture in opposition to the Western culture that has been imposed upon African Americans. For example, women in many African cultures participate in the economic and social spheres outside the home, whereas many African-American women have been relegated to the domestic sphere by the sexism of Western culture and by its racism that perpetuates poverty on many African Americans.

In 1981, the University of Michigan appointed Sudarkasa director of the Center for Afro-American and African Studies. Three years later, she became associate vice president for academic affairs. These more administrative roles prepared her for her next step when she became the president of Lincoln University in 1987. Her experience as an undergraduate at Fisk prepared her well for leading an African-American university, as she stresses the important role African-American institutions play in academia: African-American graduates of African-American colleges are seven times as likely to pursue doctorates as African-American graduates of "white" colleges. Sudarkasa herself holds 13 doctorates, though 12 of them are honorary degrees from American and African universities and colleges.

Sullivan, Betty J.
(1902–)
American
Protein Chemist

Betty J. Sullivan stands out as one of the few women who entered American industry as a scientist before the exodus of men to fight in World War II opened up the professions to women. She conducted important research on the chemistry of wheat protein; she investigated the effect of almost every aspect of processing on the nutritional value of the wheat that was transformed into flour and then baked into foods. She was the first woman to win the Thomas Burr Osborne Medal from the American Association of Cereal Chemists, and she also won the prestigious Garvan Medal from the American Chemical Society.

Sullivan was born on May 31, 1902, in Minneapolis, Minnesota. She attended the University of Minnesota, where she earned her bachelor of science degree in 1922. She entered the industry immediately, working as a laboratory assistant for Russell Miller Milling Company until 1924. She spent the next academic year, from 1924 through 1925, as a scholar in biochemistry at the University of Paris.

Sullivan furthered her education while continuing to work. She pursued doctoral studies and research in biochemistry at the University of Minnesota, which granted her a Ph.D. in 1935. She returned from Paris to her assistantship at Russell Miller in 1926. Within a year, she had worked herself into the chief chemist position, which she retained for the next two decades. Russell Miller then promoted her to director of research in 1947, and she filled that position for the next decade.

Sullivan focused her research on all the aspects of transforming wheat into edible food. She analyzed the grain itself; she measured the moisture levels in the wheat and flour; she determined how specific amino acid chains reacted in the processing of flour; she also researched the chemistry of wheat gluten and the biochemistry of baking. She published approximately 40 papers on these topics in chemistry and biochemistry journals.

As her career progressed, Sullivan devoted more of her time to administrative responsibilities as she climbed the corporate ladder to the very top of her field, an unusual feat for a woman in her time. In 1958, the Peavey Company hired her as its vice president and director of research, concurrent positions that she filled for the next decade. She continued to rise when, in 1967, Experience, Inc., appointed her as its vice president. Within two years, she ran the corporation as its president. In 1973, she became chairperson of the board, and in 1975, she became its director.

Sullivan retired from the industry in 1992. Her career had been well recognized: The American Association of Cereal Chemists elected her as its president in 1943 and 1944, and then in 1948, the association granted her its Osborne Medal, the first woman so honored. In 1954, the American Chemical Society awarded her its Garvan Medal. She was a longtime member of the American Chemical Society as well as the American Association for the Advancement of Science.

Swope, Henrietta Hill
(1902–1980)
American
Astronomer

Henrietta Hill Swope discovered more than 2,000 variable stars while working at the Harvard College Observatory, making her the next most successful variable-star finder there after HENRIETTA LEAVITT. Later in her career, she calculated the distance to the Andromeda galaxy. Throughout her career, she fought against the second-class treatment of women by academic institutions. Near the end of her career, she won the prestigious ANNIE JUMP CANNON Prize from the American Astronomical Society.

Swope was born on October 26, 1902, in St. Louis, Missouri. She was one of five children born to Mary Dayton Hill and Gerard Swope, an electrical engineer and president of General Electric. She attended Barnard College of Columbia University, where she earned her bachelor's degree in 1925. She then pursued graduate study at the School of Commerce and Administration at the University of Chicago, but she found the atmosphere too fierce, so she matriculated at Radcliffe College of Harvard University. She received her master's degree there in 1928.

On the advice of MARGARET HARWOOD, Swope applied for and received a position on the staff at the Harvard College Observatory, where she remained for the next 14 years. Swope discovered within herself an innate capacity for finding variable stars, as she located more than 2,000 variable stars during her time at the Harvard College Observatory.

In 1943, Swope joined the United States Navy Hydrographic Office as a mathematician during World War II. She also contributed to the war effort on the scientific staff at the Radiation Laboratory at the Massachusetts Institute of Technology. In 1947, after the war had ended, she reapplied to the Harvard College Observatory, which offered her only a pittance of a salary. Other prominent women astronomers, such as DORRIT HOFFLEIT and Annie Cannon, had essentially volunteered their work at Harvard (Cannon

donated her salary back to the astronomy department). Although Swope had independent means of supporting herself comfortably, she refused to work for a token salary on principle, as she believed women should receive pay commensurate with their capabilities (and she had already proven her extraordinary expertise).

Swope thus returned to her alma mater, Barnard College, where she taught as an associate in astronomy from 1947 through 1952. That year, Walter Baade invited her to assist in his research at the Mount Wilson and Palomar Observatory in Pasadena, California. She accepted the job on condition that the antiquated title of her position be changed from computer to assistant. Eventually, the observatory named her a research fellow.

Working with Baade, Swope conducted research on the Andromeda nebula as well as several dwarf galaxies. In 1962, she calculated the distance to the Andromeda galaxy as 2.2 million light-years, based on plates photographed through the 200-inch Hale Reflector. In 1968, she deduced the existence of an intermediate class of stars, falling between so-called Population I and II sequenced variable stars, in the dwarf galaxies she observed.

That year, the American Astronomical Society awarded Swope its Annie Jump Cannon Prize. In 1975, the University of Basel bestowed on her an honorary doctorate. That same year, Barnard College granted her its Distinguished Alumna Award, and five years later, in 1980, Barnard awarded her its Medal of Distinction. Swope died on November 24, 1980, in Pasadena, California. The Los Campanas Observatory in Chile named its 40-inch telescope (purchased with funds gifted by Swope upon her retirement) after Henrietta Hill Swope.

Szkody, Paula
(1948–)
American
Astronomer and Astrophysicist

Paula Szkody's astronomical research focuses on cataclysmic variables, or twin-star systems that include a white dwarf star and another star (oftentimes dwarf novae) that feeds the white dwarf with energy from the close distance of approximately one diameter of the Sun. Her work has contributed to the understanding of the transfer and accretion of mass between these binary stars and of magnetic and nonmagnetic white dwarfs. For her work, she won the ANNIE JUMP CANNON Award in 1978.

Szkody was born on July 17, 1948, in Detroit, Michigan. She studied astrophysics at Michigan State University, where she earned her B.A. degree in 1970. She then pursued doctoral study at the University of Washington, where she received her Ph.D. in astronomy in 1975. She has remained affiliated with the university ever since, though she

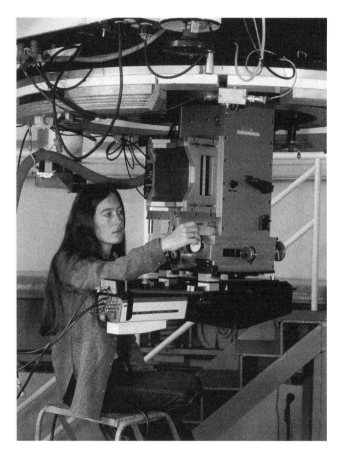

Astronomer and astrophysicist Paula Szkody, who won the Annie Jump Cannon Award in 1978, at the age of 30. *(Courtesy of Paula Szkody)*

has worked at other institutions as well throughout her career.

In 1976, Szkody worked at Kitt Peak National Observatory as a visiting scientist. The next year, she moved to the University of California at Los Angeles, where she served as a visiting instructor in 1977 and as an adjunct assistant professor in 1980 through 1981. In the intervening years, she visited the University of Hawaii in 1978 as an assistant professor and the California Institute of Technology from 1978 through 1980 as a visiting associate.

Throughout this entire time period, Szkody retained her connection to the University of Washington as a research associate and lecturer from 1975 through 1983. That year, the university appointed her as a research associate professor, a position she retained for the next eight years. In 1991, she became a research professor, and in 1993, the university named her a full professor in addition to her research status.

Szkody focused her research on cataclysmic variables, or binary systems with the two stars separated by less than the distance across our Sun. About half of the stars in the universe exist in pairs, so she further specialized in binary systems containing white dwarfs, which creates the constant possibility of novae. Szkody prioritizes the study of dwarf novae over the study of classical novae (both of which can recur within the same star).

Besides her work using multiple wavelengths to make observational studies on close binary stars, Szkody has conducted research on stellar atmospheres, the evolution of star systems, stars that degenerate, stellar jets, globular clusters of stars, stellar populations, and the Milky Way. She studies the instrumentation of her work as well as the techniques employed in observation. She also addresses the issue of women in astronomy and supports the teaching of astronomy. She has published steadily since 1981 in these areas of expertise.

In 1978, the American Association of University Women in conjunction with the American Astronomical Society honored Szkody with its Annie Jump Cannon Award, given biannually then (it is now an annual award) to a woman for "distinguished contributions to astronomy" or a like field. In 1994, she became a fellow of the American Association for the Advancement of Science.

T

Taussig, Helen Brooke
(1898–1986)
American
Pediatric Cardiologist

Helen Brooke Taussig helped to pioneer a new surgical procedure diverting a major heart artery to the lungs to oxygenate the blood of cyanotic children, or blue babies, whose color resulted from a hereditary cardiac defect. In 1944, Taussig collaborated with surgeon Dr. Alfred Blalock and surgical technician Vivien T. Thomas to perform the first Blalock-Taussig shunt, a surgical procedure that subsequently saved tens of thousands of children's lives around the world. Taussig is considered the founder of pediatric cardiology, or the study of children's heart diseases.

Taussig was born on May 24, 1898, in Cambridge, Massachusetts. She hailed from an academic family: Her father, a Harvard professor, is considered the father of modern economics, and her grandfather was a Civil War doctor. Her mother died of tuberculosis when Taussig was 11 years old. In addition to suffering this grief, Taussig contended with dyslexia, a condition that was not yet recognized by doctors. She studied at Radcliffe College for two years before transferring to the University of California, graduating in 1921.

After Harvard Medical School reiterated to Taussig its policy excluding women, she enrolled in anatomy courses at Boston University. Her professor, Dr. Alexander Begg, strongly recommended her at Johns Hopkins University School of Medicine, which had recently decided to admit women. She matriculated there in 1923, specializing in cardiology, and received her M.D. in 1927.

A mere three years later, in 1930, Dr. Edwards A. Park appointed Taussig as a pediatrics professor and head of the Johns Hopkins Pediatric Cardiac Clinic. She focused her research on how cardiac diseases, especially congenital (or inherited) conditions, manifested themselves in the heart. The regularity of incidence of heart defects in children at her clinic distressed Taussig, and she determined to find solutions to these problems that claimed such young lives on a regular basis. She came to realize the cause of blue-baby syndrome, or tetralogy of Fallot—insufficient blood flow through the lungs, where oxygenation occurs.

In 1943, a surgical solution to this problem occurred to Taussig during a conversation with Park and Blalock: join the subclavian artery into the pulmonary artery in order to feed blood from the heart into the oxygen-rich lungs. On November 29, 1944, Drs. Taussig and Blalock performed the first surgical procedure that bore their names, the Blalock-Taussig shunt, at the Johns Hopkins Hospital on a frail cyanotic child who survived several months, long enough to confirm the success of the operation.

Taussig and Blalock coauthored a paper published in the *Journal of the American Medical Association* in 1945 chronicling the first three operations incorporating their shunt. In 1947, Taussig published a comprehensive textbook, *Congenital Malformations of the Heart*. Pediatric surgeons from around the globe read this paper and the textbook and consulted Taussig and Blalock personally. Their anastomosis (or arterial joining) procedure was subsequently performed worldwide, saving tens of thousands of cyanotic children. One European family learned of the procedure from an American newspaper wrapped around a mailed package.

Taussig made other significant contributions to medicine. In the 1960s, she traveled throughout Europe to observe babies suffering from phocomelia, or malformation of the limbs. She traced the cause to thalidomide and warned pregnant women against taking this sleeping pill. Later in her career, she continued to practice medicine even after losing her hearing unexpectedly—she used her hands to "listen" to her patients.

Helen Brooke Taussig, who helped to pioneer a new surgical procedure to divert a major heart artery to the lungs to help oxygenate the blood of so-called blue babies, whose color results from a hereditary heart defect. *(National Library of Medicine, National Institutes of Health)*

For her lifesaving work, Taussig received international recognition: the French Chevalier Légion d'Honneur, the Italian Feltrinelli Prize, the Peruvian Presidential Medal of Honor, and the Albert Lasker Medical Research Award. In September 1964, President Lyndon B. Johnson awarded her the U.S. Medal of Freedom. A year later, the American Heart Association elected her as its first woman president. She received honorary doctorates from 17 international institutions, most notably Harvard University, which had denied her access to a doctoral degree as a student. On May 20, 1986, Taussig was involved in a fatal car accident. Johns Hopkins honored her by naming her former workplace the Helen B. Taussig Children's Heart Center.

Taussky-Todd, Olga
(1906–1995)
Austro-Hungarian
Mathematician

Olga Taussky-Todd's first love was number theory, but she exerted an even larger influence on the field of her second love, matrix theory. One of her 300-plus papers, a 1949 survey of the history of matrix theory, gained very wide readership through its erudite discussion, helping to establish the field as its own discipline. She was the first woman to earn tenure and to attain full professorship at the California Institute of Technology (Caltech), where she spent the majority of her career. Her homeland of Austria bestowed on her its highest scientific award, the Cross of Honor, in recognition of her status at the forefront of mathematics in the 20th century.

Olga Taussky was born on August 30, 1906, in Olmütz in the Austro-Hungarian Empire (now Olomouc in the Czech Republic). Her mother was Ida Pollach, and her father, Julius David Taussky, was an industrial chemist and director of a vinegar factory who opposed the scientific education of his daughters. Ironically, all three became scientists: Olga's older sister Ilona became a consulting chemist in the glyceride industry, and her younger sister Bertha became a pharmacist and a clinical chemist at Cornell University Medical College. Her father died the year before Taussky matriculated at the University of Vienna, forcing her to do tutoring in order to earn funds for her tuition. Taussky first studied chemistry before switching to mathematics. She wrote her dissertation on the emerging discipline of class field theory to earn her doctorate in 1930. In 1932, her dissertation was published in the *Crelle Journal*.

In 1931, Taussky moved to the University of Göttingen, working as an assistant editing the first volume of David Hilbert's complete works on number theory. The next year, she edited Emil Artin's lecture notes into a volume on class field theory while also assisting EMMY NOETHER in her class field theory course. In 1933, Girton College, the women's college of Cambridge University in England, offered Taussky a three-year fellowship. She deferred for one year in order to pursue a fellowship at Bryn Mawr College in the United States, where the head of the mathematics department, ANNA JOHNSON PELL WHEELER, had invited her to join Noether. Taussky returned to Girton in 1935, after Noether had died.

In 1937, Taussky obtained a junior-level teaching position at the University of London. There she met John (Jack) Todd, a fellow mathematician specializing in classical analysis. The couple married on September 29, 1938. The next year, the University of Cambridge granted her an M.A. *ad eundem* after the British Parliament enacted a law allowing women to receive degrees. World War II interrupted their teaching, and from 1943 until 1946, Taussky-

Todd worked at the National Physical Laboratory in Teddington with the Flutter Group, so called because it researched the mathematics of flutter at supersonic speed.

In September 1947, the Todds immigrated to the United States, where they worked for the National Bureau of Standards at the National Applied Mathematics Laboratory and the Institute of Numerical Analysis. A decade later, Caltech invited the couple to join its faculty—Todd as a full professor, Taussky-Todd as a research associate "of equal academic rank." She gained tenure in 1963, and in 1971, Caltech promoted her to the rank of full professor. She remained there until her 1977 retirement, when she continued as a professor emerita.

The *American Mathematical Monthly* published Taussky-Todd's most influential paper, "A Recurring Theorem on Determinants," in 1949. In 1970, the same journal published "Sums of Squares," a paper that earned her the 1971 Ford Prize from the Mathematical Association of America. Her 1981 Noether Lecture to the Association of Women in Mathematics was published in the journal *Linear Algebra and Applications* the next year as "The Many Aspects of the Pythagorean Triangles."

Taussky-Todd received many other honors and distinctions. *The Los Angeles Times* voted her one of its nine Women of the Year in 1963. In 1978, the Austrian government granted her its highest distinction, the Cross of Honor in Science and Arts, First Class. The University of Vienna renewed her doctorate in 1980 as a Golden Doctorate. In 1985, Taussky-Todd served as the vice president of the American Mathematical Society, and in 1988, the University of Southern California awarded her an honorary doctorate. On October 7, 1995, Taussky-Todd passed away peacefully in her sleep in her Pasadena, California, home.

Taylor, Lucy Hobbs
(1833–1910)
American
Dentist

Lucy Hobbs Taylor was the world's first woman to graduate from an accredited dental college as a doctor of dental surgery. She was also the first woman to practice dentistry in Kansas. She achieved these distinctions despite repeated obstacles; at the time she entered the field, there was scant precedent for women doctors or even for women dentists. Her persistence and excellence thus set an example for future generations of women dentists, and an award was named after her by an association of women dentists.

Lucy Hobbs was born on March 14, 1833, in a log cabin in Ellenburg, New York. She was the seventh of 10 children; her mother, Lucy Beaman, died when Hobbs was 10 years old, and her father, Benjamin Hobbs, died in 1865. She graduated in 1849 from Franklin Academy in Malone,

New York, after which she caught the railroad west to Michigan and supported herself as a seamstress and a schoolteacher.

Hobbs began studying medicine while teaching in Brooklyn, Michigan, and she moved to Cincinnati, Ohio, which hosted the only medical school that admitted women at the time. Once she arrived, however, the Eclectic College of Medicine announced its new policy prohibiting women from study there. On the advice of one of the school's professors under whom she had apprenticed, she switched her ambition to dentistry (considered less demanding than medicine) and applied to the Ohio College of Dental Surgery. This school, too, denied her admission based on her gender, so she set up a practice in Cincinnati in 1861 despite her lack of full credentials, a common practice at the time.

Hobbs then moved her practice to Iowa—first to Bellevue in 1862 and then to McGregor until 1865. The Iowa State Dental Society, which had elected her a member in July 1865 and had sent her as its delegate to the American Dental Association conference in Chicago, supported her pursuit of a dental degree. She joined the senior class at the Ohio College of Dental Surgery in November 1865, and on February 21, 1866, she became the first woman in the world to graduate as a doctor of dental surgery from an accredited dental college.

Hobbs practiced dentistry briefly in Chicago, during which time she met James Myrtle Taylor, a Civil War veteran to whom she taught dentistry after they married. In December 1867, they moved to Lawrence, Kansas, where they established a joint practice. They continued to practice together until her husband's death in 1886, after which Taylor retired.

Her retirement lasted only until 1895, when she reopened and practiced dentistry until she died of a stroke on October 3, 1910, in her home in Lawrence. In 1983, the American Association of Women Dentists named the Lucy Hobbs Taylor Award after her in recognition of her having established the practice of dentistry by women. The association presents the award annually to the association member who most exemplifies Taylor's contribution to the profession of dentistry and to society in general.

Telkes, Maria
(1900–1995)
Hungarian/American
Physical Chemist

Maria Telkes was known as the Sun Queen in reference to her innovations in harnessing solar energy, such as a solar oven and a solar water-distilling system for converting seawater into freshwater. She is best remembered for her design of a new type of solar-heating process, which stored the collected energy chemically rather than as heated water or rocks. Both the Dover House (built in Dover, Massachusetts, in 1948) and the Carlisle House (built in Carlisle, Massachusetts, in 1980) employed her solar-heating system designs, drawing attention to this alternative, renewable energy source.

Telkes was born on December 12, 1900, in Budapest, Hungary. Her parents were Aladar and Maria Laban de Telkes. The concept of capturing energy from the Sun peaked her interest in high school. She studied physical chemistry at the University of Budapest and earned her B.A. in 1920. She continued there with doctoral study, earning her Ph.D. in 1924. In 1925, Telkes visited her uncle in the United States, and she remained there for the rest of her life. She was sworn in as an American citizen in 1937.

In the United States, Telkes worked in association with a number of business and academic institutions. She commenced her professional career in 1925 at the Cleveland Clinic Foundation, researching more broadly on life-transformative energy, such as when a cell dies or develops into a cancer cell. She remained with the clinic for a dozen years until the year of her citizenship, when Westinghouse Electric hired her as a research engineer investigating the conversion of heat energy into electric energy.

Telkes diverted her thermoelectric research toward solar-energy applications when she joined the Massachusetts Institute of Technology (MIT) Solar Energy Conversion Project in 1939. While there, she developed a new solar-heating design for an entire house, the so-called Dover House, built in 1948 by architect Eleanor Raymond with funds provided by the sculptor Amelia Peabody. Telkes's system differed from others, whereby the Sun heated rocks or water, in that it stored the solar-generated energy chemically via crystallization of a sodium sulfate solution.

The United States government called upon Telkes's inventiveness to devise a solar system for desalinating seawater, thus converting it into drinkable freshwater. Telkes responded by designing a portable water evaporator, which removed salt from the seawater by vaporizing it into steam and then recooling it into freshwater. This system could condense small enough to carry onboard a life raft or expand large enough to supply freshwater for the Virgin Islands, whose natural freshwater sources sometimes dry up.

In 1953, Telkes moved from MIT to New York University College of Engineering, where she organized a laboratory devoted to the study of solar energy. Five years later, she left academia for the industry. The Curtiss-Wright Company recruited her to direct its solar energy laboratory, where she applied her solar know-how to practical matters, from powering dryers and water heaters to gathering energy in space with airborne thermoelectric generators. She also designed the architecture for heating and storing the energy to fuel the laboratory building at Curtiss-Wright's Princeton, New Jersey, site.

Telkes contributed to the Apollo and Polaris projects by developing space- and sea-proof materials for protecting temperature-sensitive devices while working at Cryo-Therm from 1961 through 1963. She returned to the question of water purification upon joining the MELPAR company as director of its solar energy laboratory. She then returned to academia in 1969, researching the most efficient means and materials for capturing, storing, and converting solar heat energy into fuel at the Institute of Energy Conversion at the University of Delaware.

Early in her career, Telkes received a $45,000 Ford Foundation grant that she used to design a solar oven that could be adapted to almost any situation, allowing for its use in any country to cook any food. She also won the first Society of Women Engineers Achievement Award winner when the prize was born in 1952, and in 1977, the American Section of the International Solar Energy Society granted her its Charles Greely Abbot Award. The next year, she retired to emeritus status at the University of Delaware, but she continued to consult until three years before her death. She died in Budapest, her Hungarian homeland, on December 2, 1995.

Tereshkova, Valentina Vladimirovna Nikolayeva
(1937–)
Russian
Cosmonaut

Valentina Vladimirovna Nikolayeva Tereshkova became the first woman to travel in space when she orbited Earth 48 times in the 1963 *Vostok VI* flight. She married fellow cosmonaut Adrian Nikolayev, who had orbited Earth 64 times in the 1962 *Vostok III* flight, and together they had a daughter, the first child born of parents who had both experienced the physical duress of space travel. Doctors studied the girl's physiology for adverse effects but found no damage passed down from her parents. Tereshkova later entered politics, supporting Soviet communism and feminism.

Tereshkova was born on March 6, 1937, in Maslennikovo, on the Volga River in western Russia, near Yaroslavl. Her mother, Elena Fyodorovna, raised Tereshkova and her brother and sister after their father, Vladimir Tereshkov, a tractor driver who fought in the Red Army, was killed at the outset of World War II when Tereshkova was a mere two years old. The family's poverty prevented her from starting school until she was 10, and by the age of 18, she was living with her grandmother in Yaroslavl, working as an apprentice in a tire factory.

By 1955, while working in the Krasny Perekop cotton mill as a loom operator with her mother and sister, Tereshkova had earned a diploma through a correspondence course from the Light Industry Technical School. She enthusiastically joined the Youth Communist League and soon became a full-fledged member of the Communist Party. In 1959, she joined the Yaroslavl Air Sports Club to learn how to parachute. By the time she applied for the Soviet spaceflight training program, she had parachuted a record 126 times.

In 1961, the Soviet Air Force chose Tereshkova as one of four candidates for the first-time woman-in-space flight and commissioned her as a junior lieutenant. After 18 months of training, the air force selected her to pilot the *Vostok VI* (despite her lack of experience as an airplane pilot). She launched from the Tyuratam Space Station at 12:30 in the afternoon on June 16, 1963, thus becoming the first woman to exit the Earth's atmosphere into space. She navigated the spacecraft manually to orbit the Earth 48 times, once every 88 minutes. She traveled a total of 1.2 million miles in 70 hours 50 minutes before parachuting from the craft upon reentering the atmosphere, landing 380 miles northeast of Qaraghandy in the central Asian country of Kazakhstan on June 19.

Soviets hailed Tereshkova's success; at a Kremlin ceremony on June 22, 1963, Presidium Chairman Leonid Brezhnev decorated her with the Order of Lenin and the Gold Star Medal while naming her a hero of the Soviet Union. She had proven that a woman could withstand the psychological and physical stress of spaceflight as well as a man, and she thus came to symbolize the Soviet tenet of equality of the sexes. She actively promoted feminism on a world tour of the United Nations (where she received a standing ovation), Cuba, and Mexico. Tereshkova married her fellow Soviet cosmonaut, Colonel Adrian Nikolayev, on November 3, 1963. Doctors closely monitored the birth of their daughter, Yelena Adrianovna Nikolayeva, on June 8, 1964, watchful for abnormalities that might have resulted from the union of two persons who had experienced the transformation of spaceflight. They found no defects.

Tereshkova remained with the space program as an aerospace engineer, but she devoted increasing time to politics. From 1966 through 1989, she served as a Deputy to the Supreme Soviet, and from 1989 until the collapse of Soviet Communism in 1991, she served as a People's Deputy. She concurrently served as a member of the Supreme Soviet Presidium from 1974 through 1989, and she chaired the U.S.S.R.'s International Cultural and Friendship Union from 1987 through 1991. She continued to promote feminism as a member of the Soviet Women's Committee, which she served from 1968 until 1987, heading the committee in 1977. After the collapse of communism, she chaired the Russian Association of International Cooperation.

Tesoro, Giuliana Cavaglieri
(1921–)
American
Polymer Chemist

Giuliana Cavaglieri Tesoro conducted research on polymers and served as an administrator in the textile industry as well as worked in academia as a research professor near the end of her career. Her many innovations include development of chemicals to prevent static in synthetic fibers; improvement of the permanent press properties of textiles; development of flame-resistant fibers; and the translation of the results from pilot plants, or experimental, small-scale versions of factories, into full-scale commercial operations. In addition to her polymer research, she also conducted research on the synthesis of pharmaceuticals. She holds more than 100 patents in the fields of organic chemistry and textile processing.

Tesoro was born in 1921. She attended Yale University, where she studied organic chemistry and earned her Ph.D. in 1943. That same year, she married, and she subsequently mothered two children.

After receiving her doctorate, Tesoro worked for the next two summers as a research chemist at Calco Chemical Company. In 1944, the Onyx Oil and Chemical Company hired her as a research chemist. Within two years, Onyx promoted her to head its organic synthesis department. She retained this position for the next decade. In 1955, Onyx again promoted her, this time to the position of assistant director of research. Two years later, she again rose in rank to associate director of research.

In 1958, Tesoro moved from Onyx to J. P. Stevens & Company, which hired her as the assistant director of organic research in its central research laboratory. She remained in this position for the next decade. In 1968, the Textile Research Institute hired her as a senior chemist, and she stayed there for one year. In 1969, she moved to Burlington Industries, Inc., again as a senior chemist. Two years later, she became the director of chemical research at Burlington, a position she retained until 1972.

That year, Tesoro left the textiles industry to enter academia as a visiting professor at the Massachusetts Institute of Technology (MIT) for four years. MIT retained her as an adjunct professor and senior research scientist until 1982. That year, the Polytechnic Institute appointed her as a research professor, a position she retained for the remainder of her career.

Tesoro served in other capacities in her field throughout her career. In 1974, the Fiber Society appointed her its president. From 1979 through 1982, she served as a member of the National Research Council's committee on military personnel supplies. She also served on the National Research Council's committee on toxic combustion products from 1984 through 1989. At other points in her career, she served on the National Academy of Sciences' committee on fire safety of polymeric materials and on the editorial board of the *Textile Research Journal.*

Tesoro received numerous awards for her contributions to polymer research. In 1978, the Society of Women Engineers granted her its Achievement Award. She also received the American Dyestuff Reporter Award and the Olney Medal from the Association of Textile Chemists and Colorists.

Tharp, Marie
(1920–   )
American
Geologist and Oceanographic Cartographer

Marie Tharp prepared maps of the ocean floor that provided proof for the theory of plate tectonics, or continental drift, a much-maligned hypothesis that the Earth was actually expanding, instead of cooling and contracting, as the conventional wisdom of the time held. In the early 1950s, Tharp helped confirm the existence of a rift valley splitting the Mid-Atlantic Ridge, a mountain range rising from the seafloor. Seismographic data showed that the epicenters of the region's earthquakes all fell in the rift valley, confirming the theory of seafloor spreading. In the 1950s, Tharp published the first comprehensive map of the North Atlantic seafloor, and later in her career, she published a map of the entire world's ocean floor.

Tharp was born on July 30, 1920, in Ypsilanti, Michigan. Her mother, Bertha Louise Newton, taught German and Latin, and her father, William Edgar Tharp, worked as a soil surveyor for the Bureau of Chemistry and Soils of the United States Department of Agriculture. His job moved the family wherever soil needed surveying: Iowa,

Geologist and oceanographic cartographer Marie Tharp, c. 1950s, taken in her Lamont Hall office at the Lamont-Doherty Earth Observatory, Columbia University. *(Courtesy of Lamont-Doherty Earth Observatory, Columbia University)*

Michigan, Indiana, Alabama, Washington, D.C., New York, and Ohio. Tharp attended 24 public schools (almost flunking out of fifth grade in Selma, Alabama) before matriculating at Ohio University, where she received her first bachelor's degree in 1943.

The exodus of young men to service during World War II opened university and professional doors to women like Tharp. She took advantage of the men's absence to study in the master's program in geology at the University of Michigan, graduating in 1944 with a M.S. degree. She entered the professional field as a junior geologist for Tulsa, Oklahoma-based Stanolind Oil & Gas, which barred women from fieldwork, so Tharp prepared maps from the geological data collected by men. During this time, Tharp earned a second bachelor's degree, this one in mathematics from the University of Tulsa in 1948.

That year, Tharp relocated to Columbia University as a research assistant to geologist Maurice Ewing, who assigned her to help his graduate students. Her assistance to graduate student Bruce Heezen transformed into collaboration over time, and the pair worked together for more than two decades. Their work took on increasing significance when the laboratory moved from Columbia to the Lamont Geological Observatory in Palisades, New York, in 1950.

Heezen and Tharp worked with data on the North Atlantic seafloor gathered by the U.S. Navy with echo-sounding equipment (Tharp could not gather data herself, as the Navy barred women from its ships). Starting in 1952, they translated these data into a physiographic map, which gave a more realistic, three-dimensional sense of the seafloor than two-dimensional contour maps (which were classified by the Navy, anyway). The map's revelation of the valley rift (and the coincidence of earthquakes centered in it) confirmed continental drift in the minds of Tharp and Heezen, but they realized the difficulty they would encounter convincing others of the implications of this physical fact.

The Lamont group sat on their discovery until 1956, when Heezen presented the information to the incredulous scientific community. Acclaimed French oceanographer and filmmaker Jacques-Yves Cousteau erased all remaining doubt by releasing underwater footage of the rift valley in 1959. That year, Heezen and Tharp reprinted the first edition of their map of the North Atlantic.

Tharp and Heezen spent the next decade updating their North Atlantic map in response to new data. Thereafter, they gathered data globally to prepare the World Ocean Floor Panorama, a map of the ocean floors for the entire Earth published in 1977, just three weeks prior to Heezen's death. The next year, the National Geographic Society awarded Tharp and Heezen with its Hubbard Medal. Tharp remained with Lamont until her 1983 retirement, after which she consulted for oceanographers and founded a map-distributing business based in South Nyack, New York. In 1998, the Library of Congress honored Tharp for her major contributions to the field of cartography and to the Library of Congress, and in 1999, she received the Women Pioneers in Oceanography Award from the Women's Committee of the Woods Hole Oceanographic Institution.

Thomas, Martha Jane Bergin
(1926–   )
American
Chemist and Engineer

Martha Jane Bergin Thomas specialized in phosphors, physical substances that illuminate when stimulated by external energy. She conducted this research at the General Telephone and Electronics (GTE) Corporation, where she became the first woman director of a division (technical quality control). Her work also earned her the distinction of being the first woman to receive the New England Award for engineering excellence from the Engineering Societies of New England.

Martha Jane Bergin was born on March 13, 1926, in Boston, Massachusetts. Both of her parents—Augusta Harris and John A. Bergin—were teachers. She went to high school at Boston Girl's Latin School and remained in Boston for her undergraduate study at Radcliffe College. She majored in chemistry as a premed to earn her bachelor's degree with honors in 1945, at age 19.

However, before she could pursue her medical degree, Sylvania Electric Products (now GTE) hired Bergin as a junior technician, and she never turned back to medicine. In fact, Bergin spent her entire 45-year career at the very same company. While working there, she pursued graduate studies at Boston University (BU), where she earned first a master's degree in 1950 and then a Ph.D. in chemistry in 1952. At BU, Bergin met fellow chemist George R. Thomas, and in 1955, the couple married. Together, they had four daughters: Augusta, Abigail, Anne, and Susan.

By 1959, Thomas had risen to the rank of senior engineer in charge of GTE's chemical laboratory. For the next half-dozen years, she worked as a group leader in the lamp material engineering laboratories of GTE's lighting products division. In 1966, she became the section head of the chemical phosphor laboratory, a position she retained for another half-dozen years. GTE promoted her to manager of the technical assistance laboratories for the next decade, from 1972 through 1981.

Thomas's managerial position inspired her to return to school, earning her M.B.A. from Northeastern University in 1981. GTE then promoted her to technical director of the technical services laboratories, and finally she became the first woman director of a GTE division, technical quality control, and she held the position for the next decade. Throughout her career, she maintained her academic connection by teaching evening courses at BU from 1952 until

1970 and from 1974 to 1993 as an adjunct professor at the University of Rhode Island.

Thomas received 23 patents for her innovations, most involving phosphorescent illumination. Her first patent protected her innovation of a method of etching fine tungsten coils that improved the lighting of telephone switchboards. She developed a natural white phosphor that generated daylike light, and she discovered a phosphor that increases the brightness of mercury lamps by 10 percent. She organized the setup of two pilot plants (or prototype factories that test the feasibility of full-scale production) for producing phosphorus.

Thomas won several prestigious awards: The Society of Women Engineers granted her its National Achievement Award in 1965, and the next year, she won the Gold Plate of the American Academy of Achievement. In 1991, she was named the New England Inventor of the Year jointly by the Boston Museum of Science, the Inventors Association of New England, and the Boston Patent Law Association. In addition, each educational institution she attended—Northeastern, BU, Radcliffe, and even Boston Girl's Latin—granted her distinguished graduate awards.

Tinsley, Beatrice Muriel Hill
(1941–1981)
English
Astronomer

Beatrice Muriel Hill Tinsley led a brief but bright career in astronomy, as she died prematurely. She contributed to her field as the first person to generate accurate computer models of galactic evolution. Her work synthesized diverse aspects of astronomy into a more coherent and comprehensive cosmology than was previously held. In recognition of her immense contributions to astronomy, she received the ANNIE JUMP CANNON Prize.

Beatrice Muriel Hill was born in 1941 in England. She was premature but lived through the war-torn winter through the generosity of neighbors, who donated heating coal to Hill's family. After the war, the family moved to New Zealand, where Hill attended New Plymouth Girls High School. She then matriculated at Canterbury University, where she earned her bachelor of science degree in 1961. That year, she married fellow physicist Brian Tinsley, and together, the couple adopted two children. Tinsley continued at Canterbury University, earning her master of science honors degree in physics in 1963.

That year, Tinsley moved to the United States, where her husband had secured a position in atmospheric physics at the University of Texas at Dallas. There she pursued doctoral study, distinguishing herself from her peers by earning marks of 99 and 100 percent on all her papers, whereas no previous student had achieved above 80 percent scores in the six-year

program. Her doctoral thesis, "Evolution of Galaxies and its Significance for Cosmology," was called "one of the profound doctoral dissertations of the 20th century" and commenced the modern study of the evolution of massive island universes. In 1967, she earned her Ph.D. in astronomy.

Unable to secure a position in astronomy in Dallas, Tinsley commuted 400 miles back and forth to the University of Texas at Austin, where she held a yearlong fellowship. In 1969, she returned to the University of Texas at Dallas as a visiting scientist in physics under partial funding from the National Science Foundation. In 1973, she joined the faculty as an assistant professor. Although she had helped to design the astronomy department there, when she applied for the position of chair of the department, the university did not even so much as reply to her application.

In 1974, the astronomy community began to recognize Tinsley's immense contributions by granting her the Annie Jump Cannon Prize. That year, Tinsley turned down offers from Britain's Cambridge University and other prestigious universities to accept an associate professorship in astronomy at Yale University. In 1978, Yale promoted her to full professor status, and she retained this position for the remainder of her foreshortened career.

Tinsley's research encompassed all aspects of astronomical knowledge. Starting with her doctoral dissertation, she compiled the colors and luminosity of evolving stars in the entire galaxy. She eventually extrapolated that the evolution of galaxies is so dynamic that they cannot be considered static, thus complicating comparative measurements. However, she also simplified astronomical study by demonstrating how diverse aspects of astronomy could be synthesized into more coherent models of galactic evolution. She reported her findings in more than 100 scientific papers, an astonishing output considering the brevity of her career.

In about 1978, Tinsley learned of a melanoma on her leg that turned out to be malignant. This skin cancer spread throughout her body, and she died in 1981. After her death, the University of Texas at Austin named a visiting professorship in astronomy after Tinsley. The American Astronomical Society created a biennial prize named after Tinsley to recognize individuals who contribute exceptionally creative or innovative research to the fields of astronomy or astrophysics.

Tolbert, Margaret E. M.
(1943–)
American
Chemist

Margaret E. M. Tolbert has contributed to the advancement of science not only as a researcher, investigating the chemical reactions taking place in the liver, but also as an administrator, heading the Carver Research Foundation, the

Research Improvement in Minority Institutions Program of the National Science Foundation (NSF), and the New Brunswick Laboratory of Argonne National Laboratory.

Tolbert was born Margaret E. Mayo on November 24, 1943, in Suffolk, Virginia, the third of six children. Her mother, Martha Artis, was a domestic worker, and her father, J. Clifton Mayo, was a World War II veteran who worked as a landscape gardener. Her parents separated when she was young, and soon thereafter her mother died. Her paternal grandmother, Fannie Mae Johnson Mayo, housed and helped raise the family. Margaret worked as a baby-sitter and maid; one family that employed her, the Simon Cooks, helped her apply to colleges and offered some financial assistance as well.

Tolbert received a small scholarship from Tuskegee Institute (now University), where she studied chemistry as a research assistant under C. J. Smith and L. F. Koons. She conducted experiments on the electrical conductivity of different chemicals dissolved in water solutions. She interned summers at Central State College in Durham, North Carolina, and at the Argonne National Laboratory in Illinois. She earned her bachelor of science degree in 1967 and went on to earn a master's degree in analytical chemistry in one year from Wayne State University in Detroit, Michigan. She then returned to Tuskegee to continue her electrochemical experiments as a supervisor for three years.

In 1970, Brown University recruited Tolbert into its doctoral program in chemistry. She studied biochemistry under John N. Fain, focusing on the chemical reactions in the liver. The Southern Fellowship Fund partially supported her research, and she further supported herself teaching basic science to nurses and mathematics to welders at the Opportunities Industrialization Center in Providence. In 1973, Tuskegee hired her as an assistant professor in chemistry, and the next year, she received her Ph.D. from Brown.

Tolbert continued her liver cell research at the Carver Research Foundation laboratories, and she remained at Tuskegee until 1977. That year, the College of Pharmacy and Pharmaceutical Sciences at Florida A&M University hired her as an associate professor. The next year, she divorced her husband and continued to raise their one son, Lawson Kwia Tolbert.

In 1979, Florida A&M offered Tolbert the position of associate dean of its School of Pharmacy, but she chose instead to spend five months researching drug metabolism in liver cells at the International Institute of Cellular and Molecular Pathology in Brussels, Belgium, funded by faculty scholarships provided by the National Institute of General Medical Sciences. On her return, she served as a visiting associate professor at Brown for seven months.

Later in 1979, Tuskegee hired her to direct the Carver Research Foundation. She established collaborative programs with the Battelle, Oak Ridge, and Lawrence Liver-more National Laboratories as well as with academic institutions in West Africa. In 1987, the British Petroleum Corporation hired Tolbert as a senior planner and senior budget and control analyst. In 1990, the NSF hired her as director of the Research Improvement in Minority Institutions Program.

Tolbert consulted for the Howard Hughes Medical Institute for three months in 1994, and then she worked as the director of Argonne National Laboratory's Division of Educational Programs. Two years later, she became director of Argonne's New Brunswick Laboratory, overseeing the country's stockpile of nuclear material. In 1998, the American Association for the Advancement of Science elected Tolbert as a fellow.

Trotter, Mildred
(1899–1991)
American
Anatomist

Mildred Trotter commenced her career studying hypertrichosis, or excessive facial-hair growth, and she made important discoveries that disproved many myths surrounding this phenomenon. However, she truly distinguished herself with her bone studies, especially conducting research in physical anthropology, investigating the bones of the dead. She made groundbreaking discoveries correlating bone length in younger ages with the eventual height of the subject.

Trotter was born on February 3, 1899, in Monaca, Pennsylvania, to Jennie Zimmerly and James R. Trotter, farmers of German and Irish descent, and had two sisters, Sarah Isabella and Jeannette Rebecca, and a brother, Robert James. Trotter graduated from her one-room grammar school in 1913 and attended high school in Beaver, Pennsylvania, where she insisted on studying geometry over home economics.

Trotter studied zoology at Mount Holyoke College, where she earned her bachelor's degree in 1920. She spurned a more lucrative job teaching high-school biology to work as a research assistant in the anatomy department at Washington University in St. Louis, Missouri, under Dr. C. H. Danforth. He had procured funding from a donor whose wife and daughter suffered from hypertrichosis, or excessive facial-hair growth, and he assigned the research to Trotter. She earned her master's degree in 1921 based on this research, and she continued to conduct this research with renewed funding to earn her Ph.D. in 1924.

Trotter remained at Washington University as an instructor. Her collected papers on hypertrichosis, in which she ascertained that hair follicles grow, rest, and shed in fixed patterns, were published serially first by the American Medical Association, then as a book under Danforth's name in 1925. The next year, she received a National Research Council Fellowship to study at Oxford University in England, where she turned her attention to studying bones, namely museum specimens from ancient Egypt and Roman-era Britain. Although she received another fellowship, Washington University enticed her back with a promotion to assistant professor.

Trotter remained at this rank for 16 years until she simply confronted the department chair to answer why she had been passed over for promotions. He responded in 1946 by promoting her to the rank of full professor, the first woman to reach this level at Washington University. She spent the year of 1948 on unpaid sabbatical volunteering as director of the Central Identification Laboratory at Schofield Barracks in Oahu, Hawaii. There, she identified dead soldiers from the war based on their skeletal remains.

In 1949, she returned to Washington University, where the new chair of the department removed the word *gross* from her title, truncating it to professor of anatomy. In 1955, she served as president of the American Association of Physical Anthropologists, which she had helped to found in 1930. She spent the latter part of her career contributing to her field by writing articles on the skin and exoskeleton for the 1953 and 1956 editions of the *Encyclopædia Britannica*, for example, or by writing other reference books, such as a lab guide, an anatomical atlas, and a dictionary of scientific terms in Latin. She also served as visiting professor at Makarere University College in Uganda.

Trotter retired from her long and distinguished career in 1967 to become a professor emerita and a lecturer in anatomy and microbiology. She was the first woman to win the Viking Fund of the Wenner-Gren Foundation, in 1931. Washington University Medical School named a lectureship after Trotter, the first woman to be so honored. She suffered a stroke in early 1991 and died on August 23 of that year. She donated her body to research.

Trotula of Salerno
(c. 11th century)
Italian
Physician

It is believed that Trotula occupied the chair of medicine at the School of Salerno, which was famous as a spa and health resort. It was also home to the first known hospital and was known around the world for its excellent medical services. Many women were trained in Salerno as physicians and professors of medicine.

Trotula, who lived in the 11th century, hailed from nobility as a member of the di Ruggiero family. She purportedly married the physician Johannes Platearius, a member of the faculty of the University of Salerno. Their two sons, Mattias and Johannes, followed in their mother's footsteps as medical writers.

The University of Salerno was the first university to admit women, resulting in the cultivation of a large group of women scholars known collectively as the Mulieres Salenitanae, or the women of Salerno. Trotula stands out from her contemporaries as the only surviving voice, recorded through her writings, which crystallized her medical advice so succinctly that her words passed through much reprinting and anthologizing.

Four texts attributed to Trotula's authorship establish her significance as a medical writer, disseminating her general medical knowledge as well as her obstetrical and gynecological expertise. Two texts seem to have been composed by her whole family: the medical encyclopedia *Practica Brevis*, or "Brief Handbook," and *De Compositione Medicanmentorum*, or "On the Preparation of Medicines."

Trotula is best known for her pair of single-authored texts—*Passionibus Mulierium Curandorum* (or "Cures for Women's Diseases") and *Ornatu Mulierum* (or "Women's Cosmetics"), which became known as *Trotula Maior* and *Trotula Minor*, respectively. In the former, Trotula addresses fundamental aspects of female health and sexuality. The latter, with its less medical approach, was eventually fused into *Trotula Maior*, consisting of 63 chapters detailing menstruation, conception, pregnancy, childbirth, and postpartum care.

Trotula favored a holistic approach to medicine, pointing out the extraphysical causes of disease as well as identifying in descriptive detail the physical realities of injury and its care. She derived her remedies from plants, and she made certain diagnoses based on the interaction between humans and plants: She tested for both male and female impotence by mixing urine samples with bran and observing the relative reactions, and she cured impotence with a powder of pig's liver and testicles. Many of her remedies, far from being this radical, have been adapted into modern medicine.

Trotula practiced obstetrics and gynecology as a faculty member at the University of Salerno's medical school. She also practiced surgery, suturing postpartum perineal tears and other causes of bleeding. She also dabbled in pharmacology, administering opiates to still the pain of childbirth, a practice opposed by the church since the Bible decreed women's suffering in reparation for the original sin. She translated her commonsense approach to medical practice into her writing, resulting in no-nonsense descriptions of symptoms and procedures. Her frank approach to the female body and its sexuality maintains objectivity instead of clouding the issue with sentimental moralism, such as the assumption that women could not know and describe their bodies intimately. It is believed that Trotula died in 1097.

_U_

Uhlenbeck, Karen Keskulla
(1942–   )
American
Mathematician

In 1990, Karen Keskulla Uhlenbeck was the second woman, after EMMY NOETHER in 1932, to deliver the Plenary Lecture to the International Congress of Mathematicians, which met in Kyoto, Japan, that year. Two years earlier, Uhlenbeck delivered the Noether Lecture to the Association of Women Mathematicians. She served as the vice president of the American Mathematical Society, and she received a prestigious MacArthur Fellowship in 1983.

Karen Uhlenbeck was the second woman, after Emmy Noether in 1932, to deliver the Plenary Lecture to the International Congress of Mathematicians, which she did in 1990 when the Congress met in Kyoto, Japan. *(Photo by Marsha Miller, Courtesy of Karen Uhlenbeck)*

Uhlenbeck was born on August 24, 1942, in Cleveland, Ohio. She was the eldest of four children. Her mother was an artist, and her father was an engineer. Both her parents were first-generation college graduates, so they supported her education but could not afford to send her to expensive institutions such as the Massachusetts Institute of Technology (MIT) or Cornell University. Instead, Uhlenbeck attended the honors program at the University of Michigan, graduating in 1964.

Uhlenbeck then spent a year at the Courant Institute of New York University before marrying a biochemist from Harvard, so she transferred to Brandeis University under her four-year National Science Foundation Fellowship. She performed her doctoral research under Richard Palais to earn her Ph.D. in 1968. She remained in Boston, teaching at MIT for a year while her husband finished his doctorate at Harvard. She then spent two years at the University of California at Berkeley. Throughout this time, Uhlenbeck was told that the prestigious universities such as MIT, Stanford, and Princeton wooing her husband would not hire a woman.

The University of Illinois at Champaign-Urbana hired Uhlenbeck, and her husband generously followed her instead of forcing her to subvert her career while he fulfilled his at a more high-powered university. While at Champaign-Urbana, Uhlenbeck received a Sloan Fellowship that relieved her from teaching duties temporarily. She then moved to Chicago, where she taught briefly at Northwestern University and Chicago Circle before landing a job at the University of Chicago in 1982. The next year, she received a MacArthur Fellowship, validating the success of her career thus far. Following on the heels of this distinction, the American Academy of Arts and Sciences elected her as a member in 1985, and the next year, the National Academy of Sciences inducted her as a member.

Uhlenbeck moved from Chicago to the University of Texas at Austin, where she has held the Sid W. Richardson Regents Chair in Mathematics since 1987. The next year, Knox College granted her an honorary doctorate, and the Association of Women Mathematicians chose her to deliver its Noether Lecture in Atlanta, Georgia. She titled her address, "Moment Maps in Stable Bundles: Where Analysis Algebra and Topology Meet." At this point, Uhlenbeck began to realize the mathematical fields were not discrete entities but, rather, that they intersected one another in interesting ways.

Uhlenbeck's success reached the international sphere in 1990 when the International Congress of Mathematicians chose her to deliver its Plenary Lecture. Uhlenbeck has taken these distinctions in stride by reminding herself that success does not remove one from the realm of human imperfection.

-V-

Van Dover, Cindy Lee
(1954–   )
American
Oceanographer and Biologist

As pilot of the *Alvin* submersible laboratory for two years, Cindy Lee Van Dover familiarized herself intimately with seafloor hydrothermal vents, or deep-sea hot-water springs that had been predicted by tectonic theories but had not been confirmed to exist until their 1977 discovery. Van Dover, who believes that these cracks into the Earth's interior may hold the key to the origins of life on the planet, became the foremost authority of this phenomenon, publishing a comprehensive book on the topic in the year 2000.

Van Dover studied environmental science at Rutgers University, focusing on basic zoology to earn her B.S. degree in 1979. She pursued graduate study at the University of California at Los Angeles, which emphasized mathematics in the ecology program that granted her its M.Sc. degree in 1985. She then conducted doctoral research at the Woods Hole Oceanographic Institution (WHOI) and Massachusetts Institute of Technology joint program, earning her Ph.D. in biological oceanography in 1989.

That year, Van Dover became a pilot of the *Alvin*, navigating journeys deep to the ocean's floor, where she used the craft's clawlike manipulators to conduct oceanographic experiments by taking soil and biological samples. In order to fill this position, she had learned how to break down *Alvin* and reassemble her to learn all her components. She then had to pass a series of oral exams and ultimately the navy boards, administered in San Diego, where she had to describe how *Alvin* functions and tell how she would respond to hypothetical situations. During her two years navigating *Alvin*, she dove more than 100 times, reaching depths greater than 2,000 meters, where she made constant observations of deep-sea hydrothermal vents.

Van Dover returned to WHOI as a visiting investigator from 1992 through 1994. The Duke University School of the Environment named her Mary Derrickson McCurdy Visiting Scholar at its marine laboratory in Beaufort, North Carolina, for the next academic year. From 1995 through 1998, she held concurrent appointments as the science di-

Oceanographer and biologist Cindy Lee Van Dover, an authority on deep-sea hot-water springs. *(Courtesy of Cindy Lee Van Dover)*

rector of the West Coast National Undersea Research Center and research associate professor at the Institute of Marine Science at the University of Alaska in Fairbanks. In 1998, she joined the College of William and Mary faculty as an assistant professor in its biology department.

Van Dover focuses her laboratory's research on the chemical makeup and ecology of underwater ecosystems. She particularly pays attention to the geographical distribution of a biological system and its reproductive cycles as well its diversity. She also selects certain invertebrate species to study their individual ecology. She has supported this research with grants from the National Aeronautics and Space Administration, the National Science Foundation, and the College of William and Mary.

In 1996, Van Dover published *The Octopus's Garden* (reissued in paperback as *Deep Ocean Journey*), a personal account of her experiences as a pilot and scientist, geared toward a general audience. She addresses her 2000 text, *The Ecology of Deep-Sea Hydrothermal Vents,* to a more academic audience, conveying an authoritative accounting of the phenomenon of hydrothermal vents. Reviewers applauded Van Dover's lucidity and accessibility, and they labeled this text as the standard against which future hydrothermal vent research would be measured. Most importantly, she provides yet more information to help answer the question of the evolution of life.

Vassy, Arlette
(1913–)
French
Atmospheric Physicist

Arlette Vassy is an atmospheric physicist world renowned for her understanding of ozone, the gas related to the oxygen humans breathe and as vital to human survival. Ozone contains one more O atom than oxygen, and it resides predominantly in the upper atmosphere, blocking the Sun's deadly ultraviolet light from reaching the lower atmosphere. Vassy's studies on ozone became increasingly significant as scientists realized that the protective ozone layer of the atmosphere was being depleted by human pollution. Vassy also participated in France's rocket-launching projects to advance atmospheric research.

Vassy was born in 1913 in St.-Nexans, in the Dordogne region of France. Her mother was Jeanne Vitrac, and her father, Pierre Tournaire, was a physics teacher licensed by the state after passing the difficult *agrégation* pedagogical procedure. Vassy followed in her father's footsteps by studying physics at the University of Paris. In 1934, the Sorbonne granted her a license in physics, equivalent in rank (though more advanced in reality) to a bachelor's degree. The next year, Vassy earned her *diplôme d'études supérieures* in physics, equivalent in rank (though again more advanced in reality) to a master's degree. In 1936, she married Etienne Vassy, a fellow physicist.

The couple conducted many of their studies in collaboration throughout their careers, starting in 1937, when they conducted a five-month study of atmospheric light absorption from the geophysical station in Ifrane, recently established in the Moroccan mountains. Both used this work as the basis for their doctoral dissertations; Vassy earned a Ph.D. in 1941, while her husband had earned his in 1937.

In 1939, the government of France founded the Centre National de la Recherche Scientifique (CNRS), roughly equivalent to the National Science Foundation in the United States. The CNRS funded research, outfitting laboratories and employing scientists without saddling them with teaching responsibilities. Vassy maintained a career-long relationship with the CNRS, which funded her doctoral research and then hired her as a scientist once she had completed her doctorate. In 1954, she was named *maître de recherche,* or senior scientist. In 1968, she took over the directorship of the CNRS atmospheric ozone laboratory.

Vassy specialized not only in laboratory research but also in field observations. In the field, Vassy worked on almost 40 rocket launches conducted by the French space program starting in 1954 in order to learn more about the atmosphere. She directed the country's high-altitude ballistic rocket research program from 1963 until 1967. In 1975, she presented a paper at the ninth History of Astronautics Symposium in Lisbon, Spain, outlining "Early French Upper Atmosphere Research Using Rockets." A decade earlier, she had published an important article in *Advances in Geophysics* surveying the study of "Atmospheric Ozone."

In 1959, the French government appointed Vassy as an officer in its Order of Academic Palms. In 1988, the Society for the Encouragement of Progress granted her its gold medal. Vassy's work formed the foundation of ozone study that ballooned in the 1980s after the discovery of a hole in the ozone layer above Antarctica. Without Vassy's researches, this dire environmental problem would not have been understood as well as it has.

Vennesland, Birgit
(1913–)
American
Enzymologist

Birgit Vennesland helped to pioneer the use of radioactive carbon 11 to study carbohydrate metabolism. Her researches also focused on the enzymology and mechanism of photosynthesis as well as the carboxylation reactions in animals and plants. She participated in research teams with the National Science Foundation and the Public Health Service. In the mid-1960s, the American Chemical Society awarded her its principal prize for women chemists, the Garvan Medal. The Max Planck Society of Germany honored her when it named its Vennesland Research Institute, which she directed for more than a decade in the 1970s.

Vennesland was born in 1913. She had a twin sister, Kirsten Vennesland, who became a physician and an authority on tuberculosis. Both sisters attended the University of Chicago, where they earned bachelor of science degrees in 1934. Both remained at the university for graduate study: Kirsten attended the medical school, and Birgit pursued doctoral studies in biochemistry. She earned her Ph.D. in 1938.

The University of Chicago retained Vennesland for one year as an assistant biochemist before she moved to the Harvard University School of Medicine as a research fellow from 1939 through 1941. She then returned to the University of Chicago, where she remained for the next 27 years. She worked as an instructor for three years before her appointment as an assistant professor in 1944. That year, at the height of World War II, she contributed to the war effort as a civilian with the Office of Scientific Research and Development.

In 1948, the University of Chicago promoted Vennesland to an associate professorship. Although she rose to this rank rather quickly, the university delayed her promotion to a full professorship for a decade, an experience shared by many women academicians. Finally, in 1957, the university appointed her to a full professorship, and she retained this chair for the next decade.

In 1968, the Max Planck Institute of Cell Biology in Germany hired Vennesland as its director. Two years later, the Max Planck Society named her the director of the Vennesland Research Institute. She directed the institute named after her throughout the 1970s. She retired in 1981 and joined her twin sister in Hawaii, where Kirsten led the tuberculosis branch of the state health department. In 1987, Birgit came out of retirement to join the faculty of the John A. Burns School of Medicine in the Department of Biochemistry and Biophysics at the University of Hawaii at Manoa as an adjunct professor. She later retired to emerita status.

Besides having a research institute named after her, Vennesland received other honors distinguishing her career. In 1950, she received the Hales Award. In 1964, the American Chemical Society granted Vennesland its Garvan Medal, given annually to a woman chemist who had contributed distinguished service to the field. The award consisted not only of the gold medal but also a $5,000 cash gift. She was also elected a fellow of the New York Academy of Sciences as well as the American Association for the Advancement of Science.

Vivian, Roxana Hayward
(1871–19??)
American
Mathematician

Roxana Hayward Vivian was the first woman to receive a doctorate in mathematics from the University of Pennsylvania. She enjoyed a 26-year career at Wellesley College, though she took several leaves of absence. On one leave, she traveled to Constantinople, Turkey, where she taught at a women's college and served as its president for two years.

Vivian was born on December 9, 1871, in Hyde Park, Massachusetts. Her parents were Roxana Nott and Robert Hayward Vivian. She attended Hyde Park High School, and then, in 1890, she matriculated at Wellesley College. There, she majored in Greek and mathematics to earn her bachelor of arts degree in 1894. Vivian then taught high school in suburban Boston for four years, first at the Stoughton Public High School for one year and then at Walnut Hill preparatory school in Natick for three years.

In 1898, Vivian commenced graduate study in mathematics as the alumnae fellow for women at the University of Pennsylvania. She wrote her doctoral thesis on the "Poles of a Right Line with Respect to a Curve of the Order n," using analytic methods to discuss poles in relation to higher-plane curves. She submitted this dissertation to earn her Ph.D. in mathematics, the first woman to do so at the University of Pennsylvania, in 1901.

That year, Vivian commenced her professional academic career at her alma mater, Wellesley College, as an instructor in mathematics. In 1906, she took a leave of absence from Wellesley to pursue her dual interests of philanthropy and women's education in the Near East by teaching at the American College for Girls of Constantinople College in Turkey. From 1907 though 1909, she served as the acting president of the college. Wellesley lured her back to the United States by promoting her to an associate professorship, and when she returned, she gave lectures on her experiences in Constantinople and Turkey.

Wellesley again promoted Vivian in 1918, this time to a full professorship, a title she retained for the remainder of her time there. However, she also took concurrent positions elsewhere. From 1913 through 1914, she lectured in statistics at the University Extension in Boston, and from 1913 through 1915, she served the Women's Educational and Industrial Union of Boston as its financial secretary. At Wellesley, she added to her teaching responsibilities by directing the Graduate Department of Hygiene and Physical Education there from 1918 through 1921.

Vivian described her one-year visiting professorship (1925–26) at Cornell University as "delightful." When she returned to Wellesley, though, academic jealousy and politics forced her to resign from her professorship after teaching at Wellesley for 26 years. She taught mathematics in a temporary position in a Vassalboro, Maine, private school. In 1929, Hartwick College of Oneonta, New York, hired her as a professor of mathematics and dean of women, dual positions she held until 1931. That year, she moved to the Rye Public High School in New York as an instructor in mathematics and dean of girls. She retired from this position in 1935. The date of her death is unknown.

Vold, Marjorie Jean Young
(1913–)
American
Colloid Chemist

Marjorie Jean Young Vold conducted research throughout her long career at the University of Southern California on kinetic and thermodynamic studies of soaps and liquid crystals. She specialized in the study of colloid chemistry, focusing on association colloids and colloidal suspension. Later in her career, she generated computer simulations of colloidal processes. In 1967, the American Chemical Society awarded her its prestigious Garvan Medal.

Marjorie Jean Young was born on October 25, 1913. She attended the University of California at Berkeley, where she earned her bachelor of science degree in 1934. She remained at Berkeley for her doctoral studies, receiving her Ph.D. in chemistry in 1936. That year, she married Robert D. Vold, and together the couple had three children: Mary, Robert, and Wylda.

In 1937, at the height of the depression, when men experienced difficulty finding jobs and women found it nearly impossible, Vold landed a position as a junior research associate in the chemistry department at Stanford University, where she remained for the next four years. In 1941, she moved to the University of Southern California (USC) as a research associate. In 1942, she took on a concurrent position as a research chemist with the Union Oil Company, a position she retained until 1946. During World War II, she also worked as a civilian with the Office of Naval Research.

After the war, Vold returned to her research associate position (as well as lecturing in the chemistry department) at USC, where she remained for the rest of her career. However, she traveled abroad twice to work at international institutions. In 1953, she received a one-year Guggenheim Fellowship, which she used at the University of Utrecht in the Netherlands. The next year, she traveled to Bangalore, India, where she served as a professor of physical chemistry at the Indian Institute of Science from 1955 through 1957.

In 1958, soon after returning to USC from abroad, Vold was promoted to the rank of adjunct professor. She retained this title for 15 years, at a time when few women held academic appointments at universities. While working at USC in the early 1960s, she produced two of her most important works. She coauthored *An Introduction to the Physical Sciences*, published in 1961, and *Colloid Chemistry*, published in 1964.

Vold received some recognition of the significance of her research career. In 1934, in her senior year of undergraduate study, the University of California at Berkeley granted her its University Medal. More than 30 years later, in 1967, the American Chemical Society gave her its Garvan Medal, awarded yearly to the woman chemist who contributed distinguished service to her field. The gold medal was accompanied by a $5,000 cash prize. In 1997, Sharon Sue Kleinman published a paper in the *Journal of Women and Minorities in Science and Engineering* using Vold's career as an example of the experience of women in the sciences. The piece was entitled "Unpacking the Gendering of Chemistry: A Biographical Case Study of Marjorie Vold."

Von Mises, Hilda Geiringer
(1893–1973)
Austrian/American
Mathematician

Despite having to overcome a number of serious obstacles, Hilda Geiringer von Mises was a highly respected mathematics professor and researcher. During Adolph Hitler's rise to power in Nazi Germany, von Mises was forced out of German academics because she was Jewish. She eventually settled in the United States but was unable to find an appointment at a top-rated university because she was a woman. Nevertheless, she persisted in her research, which focused primarily on probability theory and mathematical theories of plasticity.

Hilda Geiringer was born on September 28, 1893, in Vienna, Austria, to Ludwig Geiringer, a Hungarian-born textile manufacturer, and Martha Wertheimer. Because Hilda showed an acute mathematical ability as a young student, her parents supported her financially so that she could study mathematics at the University of Vienna.

After completing her undergraduate education at the University of Vienna, Geiringer remained at that institution to pursue doctoral studies. She was awarded her Ph.D. in 1917 for her thesis on Fourier series in two variables. From 1917 to 1919, Geiringer worked as Leon Lichtenstein's assistant in editing the mathematics review journal *Jarbuch über die Fortschritte der Mathematik*.

Geiringer left Vienna in 1921 for an appointment as an assistant to Richard von Mises at the Institute of Applied Mathematics in Berlin. That same year, she married Felix Pollaczek, a fellow Jewish mathematician who earned his doctorate in 1922. The couple had a daughter, Magda Pollaczek, in 1922, but their marriage ended soon thereafter. Educated as a pure mathematician, Geiringer sought to develop her expertise in applied mathematics while working at the Institute of Applied Mathematics. She began to explore statistics—especially probability theory and mathematical theories of plasticity. She competed a second thesis, which she submitted to the University of Berlin. It was not immediately accepted, however, a fact which engendered a great deal of controversy between 1925 and 1927. During this period, Geiringer remained at the Institute of Applied Mathematics.

By 1933, Geiringer had been nominated for a professorship at the University of Berlin. However, Hitler implemented the civil service law the same year, which did not allow Jews to teach at universities. Geiringer was fired, and she left Berlin with her daughter for a position at the Institute of Mechanics in Brussels, Belgium, where she worked on the theory of vibrations. Her mentor, Richard von Mises, had fled Germany as well, and he was named chair of the mathematics department at the University of Istanbul in Turkey. Geiringer joined him there in 1934, becoming a professor of mathematics.

Both Geiringer and von Mises came to the United States in 1939. She was appointed a lecturer at Bryn Mawr College and was also recruited to conduct classified research for the National Research Council as part of the American war effort. In addition, she was selected to give a series of lectures on mechanics at Brown University as part of an effort to raise the level of American advanced education to German standards. Her lectures were used widely in the United States for many years after. Geiringer married von Mises in 1943 and left Bryn Mawr in 1944. She took a post as chairperson and professor at Wheaton College in Norton, Massachusetts, so that she could be closer to her new husband (who taught at Harvard University). She felt isolated at her small new school and longed to work in a collaborative research environment. However, she was unable to obtain appointments at more prestigious universities and continued to teach and conduct independent research at Wheaton.

After von Mises died in 1953, Geiringer began editing his works. In 1956, the University of Berlin at last elected her a professor emerita with a full salary. She retired from Wheaton in 1959. Although Geiringer was forced to overcome a slew of obstacles—persecution, prejudice, and having to learn two new languages—she always remained committed to her work. She was also sure that her efforts would help advance the cause of future women scientists. She died on March 22, 1973, in Santa Barbara, California.

Vrba, Elisabeth
(1942–   )
South African/American
Paleontologist

Elisabeth Vrba developed the *turnover-pulse hypothesis* which posits how major climatic changes trigger biologi-

cal evolution explosions, whereby species adapt radically to the altered environment or else they disappear. This theory overturned the prevailing theory that human evolution had been proactive, in order to improve life, rather than reactive. Based on her comparison of fossil records with current evidence, she has proposed the habitat theory of evolution to replace the competition paradigm, thus attributing evolution not to survival of the fittest but rather to the ability to respond to major environmental shifts physically through rapid biological evolution.

Vrba studied zoology and mathematical statistics at the University of Cape Town to earn her undergraduate degree. She remained there for doctoral study in zoology and paleontology to earn her Ph.D. After receiving her doctorate, Vrba conducted her early research on African fossil records over the last several million years, tracking the sequence of fossils from analyzing the geological strata and analyzing the morphology of the fossils.

She focused her research particularly on African bovids and noticed that this family exploded in diversity about two and a half million years ago, adding a host of new species: the hartebeest, wildebeest, gazelle, impala, springbok, and others. Coincidental with this burst in biological evolution was a great climatic change, suggesting a cause and effect link between biological evolution and global climate change. Her hypothesis of effect macroevolution attracted attention from Stephen Jay Gould and Niles Eldredge, who helped her to expand her theory to refute the idea of species selection. By 1992, she confidently extended her turnover-pulse hypothesis to contest the biological paradigm holding competition as the driving force behind evolution. Vrba proposed that biological responses to Earth changes account for evolution much better than the idea of interspecies combat for dominance.

Not only did Vrba's theory send shock waves through the biological community, as it overturned the assumption of the competitive paradigm for species evolution, but also it struck home in the present, offering startling implications for current and future climate changes. Moreover, Vrba's dating of this evolutionary and climatic shift coincides with the emergence of the species *Homo erectus*.

Vrba became a professor in the Department of Geology and Geophysics at Yale University while simultaneously directing the Yale Institute for Biospheric Studies as well as the Ecosave Center. She curates the vertebrate paleontology and zoology collections at the Peabody Museum, and she is a research associate with the American Museum of Natural History in New York. Her research continued to try to shed light on the reality of evolution, as she investigated brain evolution and applied evolutionary paradigms to anthropological, neuroscientific, and physiological research.

Vrba writes prolifically, having authored more than 100 scientific papers and four books. She has also received

Paleontologist Elisabeth Vrba, who developed the turnover-pulse hypothesis, which posits that major climatic changes trigger biological evolution explosions. *(Photo by Barrle-Kent Photographers, Courtesy of Elisabeth Vrba)*

numerous honors in her career. She won the British Association Medal for research conducted by a scientist under age 40. South Africa honored her as its professional woman of the year in 1982. Three years later, in 1985, the Royal Society of South Africa elected her a fellow.

Vyssotsky, Emma T. R. Williams
(1894–1975)
American
Astronomer

Emma Williams Vyssotsky collaborated with her husband, Alexander Vyssotsky, on observations of stellar spectra and stellar parallaxes. She worked for more than a dozen years at the University of Virginia before being promoted to a professorship, but by then, a debilitating illness prevented her from conducting research. Soon thereafter, though, she won the ANNIE JUMP CANNON Medal of the American Astronomical Society.

Emma Williams was born on October 23, 1894, in Media, Pennsylvania. She attended Swarthmore College, where she received a bachelor of arts degree in mathematics in 1916. She remained at Swarthmore as a demonstrator in mathematics for one year before joining the Fidelity Mutual Life Insurance Company as an actuary, using her mathematical skills to compute insurance and annuity premiums and dividends. After two years, she traveled to Germany to volunteer in the Child Feeding Program from 1919 through 1921, and she then returned to actuarial work in the United States for Provident Mutual Life Insurance Company.

In 1925, Williams returned to her alma mater as a mathematics instructor for two years, and then she pursued doctoral study in astronomy at Radcliffe College, where she took course work at Harvard University. There she met the Russian astronomer Alexander Vyssotsky; the couple married in 1929 and eventually had one son, Victor, who grew up to become a mathematician at Bell Labs. Although Vyssotsky finished her doctoral dissertation,

"A Spectrophotometric Study of Stars," in 1929, she did not receive her Ph.D. until February 1930. She was the second woman to receive a doctorate in astronomy from Radcliffe and only the third person to receive an astronomy doctorate from Harvard.

Vyssotsky followed her husband to the University of Virginia, which had hired him as an assistant professor and her as an instructor. The couple worked together at the Leander McCormick Observatory, where they studied stellar parallaxes by applying trigonometric functions to observations made on multiple photographic exposures. They discovered many of these parallaxes by attaching a special objective prism to the observatory's astrograph. Their research led to accurate calculations of stellar motions and the determination of the structure of galaxies.

Illness forced Vyssotsky to take a leave of absence from her research at the University of Virginia in 1944. The next year, the university promoted her to the rank of professor, but she was never able to fill this position actively, as her illness kept her from the observatory throughout the remainder of her career. However, she continued to exert influence on her field. In 1948, she and her husband jointly published their monograph, *An Investigation of Stellar Motions,* which became a standard in the field. She also continued to coauthor journal articles with her husband, publishing them with her name listed first sometimes, his name first at other times.

In 1946, the American Astronomical Society awarded Vyssotsky the Annie Jump Cannon Medal in recognition of her contributions to the understanding of stellar spectra. However, she never recovered enough from her illness to contribute individually to her field. Her illness was eventually diagnosed as chronic brucellosis, also known as undulant or Malta fever, which recurs regularly as malaria. She recovered with the help of antibiotics but did not have enough strength to resume her career. She died in a retirement community in Winter Park, Florida, on May 12, 1975, two years after her husband had passed away.

W

Waelsch, Salome Gluecksohn Schoenheimer
(1907–   )
German/American
Geneticist

Salome Gluecksohn Schoenheimer Waelsch was the first geneticist to teach courses in the discipline in the United States. She specialized in studying the genetic difference between normal and abnormal cells, specifically focusing on inherited abnormalities to investigate how these traits passed from generation to generation.

Salome Gluecksohn was born on October 6, 1907, in Danzig, Germany. Her father, Ilya Gluecksohn, died in the 1918 flu epidemic, when she was 11, and her mother, Nadia, lost the family's savings in the devaluation of the deutsche mark following World War I. However, Gluecksohn persevered, attending university at Königsberg, Berlin, and Freiburg. She commenced studying classical languages, but she found her calling in biology, studying under the renowned geneticist Hans Spemann at Freiburg to earn her Ph.D. in zoology in 1932.

The University of Berlin proved to be the only institution that would hire Gluecksohn, a Jewish woman, in the midst of Nazi Germany. She worked there as a research assistant in cell biology for one year, during which time she met and married Rudolf Schoenheimer, a biochemist. In 1933, the couple emigrated from the Nazi regime to the United States, where she became a naturalized citizen in 1938. Rudolf landed a job with Columbia University's College of Physicians and Surgeons upon arrival to the United States. Throughout the academic community at that time, institutional rules prevented nepotism but allowed for exploitation—in this instance, Salome worked unsalaried in Leslie Dunn's genetics laboratory.

Schoenheimer focused her genetic research on the embryonic differentiation of body parts, specifically searching for clues to explain how congenital defects and diseases pass along. She commenced these studies in Dunn's lab, where she worked for 17 years before receiving an official appointment at the College of Physicians and Surgeons, as a research associate in the department of obstetrics, in 1953. She maintained this focus on the genetics of cellular differentiation throughout the rest of her research career.

In 1941, Schoenheimer's husband died, and on January 8, 1943, she married another biochemist, Heinrich B. Waelsch. Together, the couple had two children, a daughter named Naomi Barbara and a son named Peter Benedict, before Waelsch died in 1966.

The Albert Einstein College of Medicine, founded in the early 1950s, finally recognized Waelsch commensurately with her contributions by appointing her as an associate professor of anatomy in 1955. Within three years, the college named her full professor of anatomy. At this time, she designed the first courses in genetics to be taught in the United States, and Einstein responded by establishing a genetics department for her to head in 1963.

In 1978, Waelsch retired to emerita status, but she continued to conduct genetic research. The next year, the National Academy of Sciences elected her a member. In 1982, Waelsch declined to receive the gold doctorate offered by her alma mater, the University of Freiburg, because she did not want to revisit the painful past of the Holocaust. She did accept the 1993 National Medal of Science from President Bill Clinton and the 1999 Genetics Society of America Medal.

Walker, Mary Edwards
(1832–1919)
American
Physician

Mary Edwards Walker was one of the first women doctors in the United States and, as such, spent her life not only caring for the sick but also fighting for acknowledgment of her actions equal to that accorded to male doctors. Throughout her life, titles and medals were denied her based solely on her gender; she wanted neither the titles nor the medals themselves but, rather, the respect that they carried.

Walker was born on November 26, 1832, on a farm in Oswego, New York, the fourth of five children born to Vesta Walker. Walker's father, Alvah, was a carpenter who read vociferously and exerted significant influence on his daughter in his liberal thinking. He championed the education of his four daughters, which led Walker in 1853 to Syracuse Medical College, the first American medical school to admit women (Walker was the only woman in her class). In 1855, she graduated with a M.D.

That same year, Walker married fellow physician Albert Miller in an unorthodox ceremony where she wore pantaloons and omitted the vow to obey her husband. She also retained her maiden name, and within five years, his infidelity broke up their marriage. Her actions outside the medical arena shed much light on her behavior within her field, as well. She fought for women's dress reform as a matter of freedom but also because her active lifestyle prevented her from wearing women's fashions.

When the Civil War erupted in 1861, Walker sought a commission as a surgeon in the Union army. However, the army did not allow women officers. Walker petitioned this decision repeatedly, appealing directly to President Abraham Lincoln in January 1864, to no avail. All the while, she contributed to the war relief as an unpaid and untitled assistant physician and surgeon.

In April 1864, the Confederate army captured Walker and held her as a prisoner of war in Castle Thunder in Richmond, Virginia. The Union freed her in August by exchanging a Confederate officer for her. In September 1864, the army contracted her as an acting assistant surgeon, the first time the army appointed a woman as an assistant surgeon. She also received a salary of $100 a month in addition to back pay.

After the war, Walker continued to petition, now for the rank of major in the army. The secretary of war refused on the grounds that she had not risen up the ranks. However, on January 24, 1866, President Andrew Jackson conferred on her the Congressional Medal of Honor in recognition of her meritorious service.

In 1917, Congress emended the guidelines governing the Medal of Honor to include only combatants as candidates, and thus the army revoked Walker's medal. She continued to wear the medal in active defiance of the law. On a trip to Washington, D.C., to demand her medal reinstatement, she fell on the steps of the Capitol Building and died from complications of this fall on February 21, 1919, at the age of 86. In 1977, one of her descendants secured the medal for her, and Walker then officially received the Congressional Medal of Honor.

Washburn, Margaret Floy
(1871–1939)
American
Psychologist

Margaret Floy Washburn was the first woman to earn a doctorate in the emerging field of psychology. She helped to define the field with her motor theory of consciousness, which held that people's motor reactions express their thoughts and perceptions. She was the second woman to serve as president of the American Psychological Association and the second woman elected into the National Academy of Sciences.

Washburn was born on July 25, 1871, in New York City. She was the only child of the Reverend Francis and Elizabeth Floy Washburn. She graduated from high school at the age of 15 in 1896 and then matriculated at Vassar College, where she majored in chemistry and French. She graduated with a bachelor of arts degree in 1891.

Margaret F. Washburn, who in 1894 became the first woman to earn a doctorate in the emerging field of psychology. *(National Library of Medicine, National Institutes of Health)*

In the late 19th century, the new field of experimental psychology was developing, and Columbia University had just opened a psychology laboratory. Washburn determined to study and research there under James Cattell, but she had to petition Columbia's board of trustees for three months straight before it allowed her special dispensation to attend Cattell's classes as a hearer.

After a year at Columbia, Washburn was advised by Cattell to apply for a scholarship to Cornell University's Sage School of Philosophy, where she could, as a woman, earn a graduate degree in psychology. She studied there under E. B. Tichener, and in the meanwhile, earned her master of arts degree in 1893 from Vassar College for work done in absentia. The next year, Washburn became the first woman doctor of psychology when Cornell awarded her a Ph.D. in 1894.

Washburn spent the next half-dozen years as a professor of psychology, philosophy, and ethics at Wells College. Washburn lectured in philosophy while serving as a woman's dormitory warden at the Sage School after the turn of the century, but within two years, she realized that she was overqualified. She accepted an assistant professorship at the University of Cincinnati for the next year.

In 1903, Washburn returned to Vassar College as an associate professor of philosophy. That year she also made Cattell's list of the thousand most important "men of science," and the *American Journal of Psychology* appointed her as a cooperating editor, a title she retained the rest of her career. Over the next five years, Washburn helped architect a department of psychology at Vassar, which it inaugurated in 1908 by appointing Washburn to a professorship. She remained there for the rest of her career, over the next 30 years.

In 1908, she published the work for which she is most famous, *The Animal Mind*. In it, she points out that animals judge each other by their actions, whereas human judgments are based on the perceived mental states of the other. However, the only mind humans can know are their own, so their perceptions of others amount to projections of their own self-perceptions.

In 1916, Washburn presented her motor theory of consciousness in her second text, *Movement and Mental Imagery*. In addition to her books, she published some 200 scientific papers. She introduced the controversial practice of crediting the students in the Vassar Psychological Laboratory, who conducted much of the research reported in her papers, as coauthors. The common practice allowed professors to claim authorship of all work performed in their laboratories.

The American Psychological Society elected Washburn as its president in 1921. That same year, the Edison Phonograph Company awarded her a $500 prize for the best research on the effects of music with her paper, "The Emotional Effects of Instrumental Music," coauthored by a member of Vassar's music faculty. In 1932, she acted as the U.S. delegate to the International Congress of Psychology in Copenhagen.

In 1937, Washburn retired to emerita status at Vassar. On March 17 of that year, she suffered a cerebral hemorrhage, and two years later, on October 29, 1939, she died in her Poughkeepsie, New York, home.

Wattleton, Alyce Faye
(1943–   )
American
Nurse-Midwife and Public Health Executive

Alyce Faye Wattleton was the first African-American woman president of the Planned Parenthood Federation of America (PPFA). During her 14-year tenure in this post, she navigated the pro-choice organization through bitter opposition, and in the process, politicized Planned Parenthood as a means of defending its position in favor of educating people on their family planning options.

Wattleton was born on July 8, 1943, in St. Louis, Missouri. Her father was George Wattleton, and her mother, Ozie Garret Wattleton, was an itinerant preacher in the fundamentalist Church of God. She attended Ohio State University, where she earned her bachelor of science degree in nursing in 1964.

Wattleton commenced her professional career as an instructor at the Miami Valley School of Nursing in Dayton, Ohio, from 1964 through 1966. She then pursued graduate study at Columbia University in New York City, where she received her master of science degree in maternal and infant health care as well as her nurse-midwife certification in 1967. She returned to Dayton as the assistant director of nursing at the Dayton Public Health Nursing Association from 1967 through 1970.

In 1970, the Planned Parenthood Association of Miami Valley in Dayton appointed Wattleton as its executive director, a position she held for the majority of the decade. During her tenure there, the United States Supreme Court legalized abortion in its *Roe* v. *Wade* ruling, thus allowing Planned Parenthood to promote freedom of choice in family planning decisions. In the 1970s, Wattleton married Franklin Gordon, and together they had one daughter, Felicia. The couple divorced in 1981.

In 1978, the PPFA named Wattleton as its youngest president, in charge of 191 affiliates in 43 states staffed by 3,000 workers and 20,000 volunteers serving more than one million patients annually. Her greatest challenges there—besides fighting racism, sexism, clinic bombings, and death threats—were Title X in 1981 and the Supreme Court's *Webster* v. *Reproductive Health Services* case in 1989.

In the case of Title X, Wattleton's testimony before the Senate Committee on Labor and Human Resources in

March 1981 helped to retain federal control over health care funding, instead of handing this control to the states (which would have eroded family-planning funding). In the Webster case, the Supreme Court allowed states some latitude in restricting abortion, a decision that Wattleton vowed to fight.

In 1992, Wattleton resigned as president of Planned Parenthood. She next hosted a syndicated television show and then founded a think tank on women's issues, the Center for Gender Equality, in 1995. In 1996, she published her autobiography, *Life on the Line.*

Wattleton received ample recognition of the significance of her career. In 1990, she received the Claude Pepper Humanitarian Award as well as the Boy Scouts of America Award. In 1991, the Yeshiva University's Albert Einstein College of Medicine granted her its Spirit of Achievement Award. In 1992, she received three important honors: the Jefferson Public Service Award, the Margaret Sanger Award, and the Dean's Distinguished Service Award of the Columbia School of Public Health. She has also received six honorary degrees at last count.

Weertman, Julia
(1926–   )
American
Physicist

Julia Randall Weertman conducted research on metals under stress, such as dislocation (a condition enhancing the malleability of metals) and very high temperatures. She conducted much of her important research in the 1950s, before taking a break from her professional career to raise her children. She continued to produce important research after returning to her career in the 1970s and gained increasing recognition through the next two decades.

Julia Randall was born on February 10, 1926, in Muskegon, Michigan. She and her family—her mother, Louise Neumeister, her father, Winslow Randall, and her sister, Louise—moved to Pittsburgh, where she graduated from Mount Lebanon High School in 1943. She remained in Pittsburgh to study aeronautical engineering at the Carnegie Institute of Technology (now Carnegie-Mellon University), a prerequisite imposed by her parents before they would allow her to fly as a pilot. However, her interests veered to physics, in which she earned her bachelor of science degree in 1946.

Randall remained at Carnegie for graduate study, earning her master of science degree in 1947 and her doctorate in 1951. The year before, on her birthday (February 10, 1950), she married Johannes Weertman. She pursued postdoctoral work on a Rotary International fellowship at the École Normale Supérieure in Paris. In 1952, Weertman returned to the United States to work for the Naval Research Laboratory in Washington, D.C., where she investigated magnetism and specifically focused on the resonance created by ferromagnetic spin.

In 1958, Weertman accompanied her husband to London, where he worked for the Office of Naval Research while she raised their daughter, Julia. In 1960, the family returned to Evanston, Illinois, where Johannes worked for Northwestern University while Julia raised both their daughter and a new son, Bruce. Although she devoted most of her time to the duties of child rearing, such as leading the Girl Scout troop, she carved out time to collaborate with her husband in writing *Elementary Dislocation Theory,* which was published in 1964.

Weertman returned to her professional career in 1972 as a visiting assistant professor at Northwestern University, which retained her after three years of teaching by granting her tenure. In 1982, she became a full professor in the Department of Material Science and Engineering, and upon returning from a stint teaching in Switzerland in 1986, she took the helm of the department until 1992. Meanwhile, in 1988, the university promoted her to a distinguished professorship, naming her the Walter P. Murphy Professor of Material Sciences.

Weertman focused her research on metals under stress in order to discover ways to manipulate them more easily. She investigated the fatigue and failure thresholds by heating metals and alloys to very high temperatures, and under these conditions, she also studied the tensile strength and relative brittleness of metals. Her research on small-angle neutron scattering helped other scientists understand mechanical properties and identify boundary interactions.

In 1979, the city of Evanston recognized Weertman with its Environmental Award. The National Science Foundation twice granted its Creativity Award to Weertman, first in 1981 and then in 1986. In 1988, she won the National Academy of Engineering Award, and in 1991, the Society of Women Engineers granted her its Achievement Award "for pioneering research on the failure of materials at elevated temperatures."

Weisburger, Elizabeth Amy Kreiser
(1924–   )
American
Chemist

Elizabeth Amy Kreiser Weisburger led a long career researching carcinogens, or cancer-inducing substances, for the National Cancer Institute (NCI) of the National Institutes of Health (NIH). The American Chemical Society recognized the importance of her cancer research by awarding her the Garvan Medal, one of the society's highest honors.

Elizabeth Amy Kreiser was born on April 9, 1924, in Finland, Pennsylvania. Her parents were schoolteachers of German descent. Her father, Raymond Samuel Kreiser, also sold Prudential life insurance in Bucks County before the family moved back to Lebanon County soon after Elizabeth's birth. Her mother, Amy Elizabeth Snavely, homeschooled her until the age of eight. Her aunt, Lottie Snavely, taught her English and Latin at the high school in Jonestown. She had nine sisters and brothers.

Kreiser received scholarships from the state and from Lebanon Valley College, which she entered in 1940. She studied chemistry, mathematics, and physics, and graduated cum laude in 1944 with a bachelor of science degree in chemistry. She then pursued doctoral study at the University of Cincinnati as a graduate assistant in Dr. Francis Earl Ray's cancer research laboratory. In 1947, she earned her Ph.D. in organic chemistry. On April 7, 1947, she married John Hans Weisburger, a native German and fellow graduate student. The couple eventually had three children: William Raymond, born in 1948; Diane Susan, born in 1955; and Andrew John, born in 1959.

After receiving her doctorate, Weisburger continued to work in Ray's laboratory as a research associate synthesizing carcinogen analogs for the NCI. Starting in 1949, she worked directly for the NCI in Bethesda, Maryland, under a two-year postdoctoral research fellowship (her husband was also a postdoctoral fellow there). She remained with the NCI for the rest of her career as a commissioned officer in the U.S. Public Health Services.

Weisburger focused her research on carcinogens, studying them by means of aromatic amines and aminoazo dyes. She conducted her research on rodents, drawing a correlation between their development of tumors in the liver and other tissues in reaction to chemicals and human tumorigenicity.

From 1972 through 1974, Weisburger acted as a consultant for the National Academy of Sciences of the National Research Council. In the year between, the NCI appointed her as the chief of the Laboratory of Carcinogen Metabolism, a post she retained until 1981. During that time, from 1977 through 1979, she chaired the NCI's Carcinogenesis Working Group. In 1985, she became the assistant director of the Division of Cancer Etiology. That year, she also became the first woman appointed president of her alma mater's board of trustees.

In 1989, Weisburger retired from the NCI, drawing to a close her distinguished career, in which she published more than 255 papers and reviews in journals and books. She continued to write, publishing nine more papers after 1990. The significance of her work began to receive recognition early in her career. In 1973, the U.S. Public Health Service of the NIH awarded Weisburger its Meritorious Service Medal. In 1981, she received an honorary doctorate from the University of Cincinnati as well as the Hillebrand

Prize of the Chemical Society of Washington, D.C. That year she also received the most distinguished of her honors, the Garvan Medal from the American Chemical Society. She was also the 1996 recipient of the Herbert E. Stokinger Award, given each year to an individual who has made a significant contribution in the broad field of industrial and environmental toxicology.

Wethers, Doris L.
(1927–   )
American
Pediatrician

As the founder and director of the Comprehensive Sickle Cell Program at St. Luke's/Roosevelt Hospital Center in New York City, Doris L. Wethers has greatly advanced the medical community's understanding of sickle-cell anemia, a potentially fatal disease that almost exclusively affects the African-American population. Earlier in her career, she was the second African-American woman to graduate from the Yale University School of Medicine.

Wethers was born in 1927 in Passaic, New Jersey. Her mother, Lilian Wilkenson, was a schoolteacher, and her father, William Wethers, was a family-practice physician. After World War II, the family moved to Harlem, where her father established his practice in the family home. Wethers and her sister, Agnes, emulated their father's work by setting up their dolls in sickbeds. She continued to follow in her father's footsteps by studying biology as a premed at Queens College, graduating in 1948.

Wethers then matriculated at the Yale University School of Medicine, and in 1952, she became its second African-American woman graduate. In 1953, she earned her license to practice medicine from the state of New York. That year, she married Garval H. Booker, a dentist, and together they had three sons. She performed her residency at King's County Hospital in Brooklyn, New York, and in 1955, she became the chief resident in pediatrics there.

For the first decade of her career, Wethers joined her father's practice, where she mostly encountered poverty-related conditions: lead poisoning, tuberculosis, babies born afflicted by the poor prenatal nutrition and drug addiction of their mothers, and sickle-cell disease. In 1965, she joined a health maintenance organization, which allowed her to confront these afflictions not only directly as a doctor but also systematically as an administrator and educator. Also in 1965, she began work as the director of pediatrics at three New York hospitals—Knickerbocker Hospital, Sydenham Hospital, and St. Luke's/Roosevelt Hospital Center—positions she retained until 1979, when she founded and became the director of the Comprehensive Sickle Cell Program at St. Luke's/Roosevelt Hospital Center.

This program conducted research on sickle-cell conditions, in which the body's red blood cells transform shape, become trapped in the blood vessels (thus inhibiting the delivery of oxygen and nutrients to the body), and eventually burst (resulting in a plummeting blood count and anemia). Sickle cell trait, a minor form of the disease, afflicts 10 percent of the African-American population. Sickle-cell anemia, the advanced form of the disease, can lead to childhood strokes, heart attacks, and gallbladder disease.

Also in 1979, Wethers commenced her teaching career, and by 1987, she rose to the rank of professor of clinical pediatrics at the College of Physicians and Surgeons at Columbia University. In addition to her teaching and administrative responsibilities, Wethers contributed to her field in numerous capacities, for example, as the chair of the GENES Sickle-Cell Advisory Committee from 1978 through 1992 and as the director of pediatrics for the Manhattan Medical Group from 1983 through 1992.

When Wethers first started researching sickle-cell disease, most children with the condition did not live past childhood. Thanks in large part to 25 years of research initiated by Wethers, 85 percent of children with sickle-cell disease now live past the age of 20, and the average life expectancy for sickle-cell carriers is 48 years for women and 42 for men. Wethers continues to serve as chair of the National Institutes of Health Consensus Conference on Newborn Screening for Sickle Cell Disease and Other Hemoglobinopathies, and she cochairs the New York State Sickle Cell Implementation Committee.

In recognition of these advances, the Columbia University College of Physicians and Surgeons granted her its Charles Drew Memorial Award in 1984, and in 1993, St. Luke's/Roosevelt Hospital Center bestowed on her its Community Service Award. In 1991, the New York Health Research Training Program named her its Preceptor of the Year.

Wexler, Nancy Sabin
(1945–   )
American
Neuropsychologist

Nancy Sabin Wexler is the world's leading authority on the genetics of Huntington's disease, a degenerative condition identified in 1872 by George Huntington but only understood more than a century later due to Wexler's work. She traveled to the highest concentration of Huntington's, in Venezuela, to collect a pool of blood samples for genetic studies. From these samples, a tag gene was identified, and later, the Huntington's gene itself was identified. Wexler promoted support for the study of Huntington's as the longtime president of the Hereditary Disease Foundation.

Wexler was born on July 19, 1945, in Los Angeles, California. Her mother, Leonore Sabin, was diagnosed with Huntington's disease in 1968. Her father, psychoanalyst Milton Wexler, responded that very year by founding the Hereditary Disease Foundation, dedicated to studying Huntington's in search for diagnostic procedures and potential cures.

The previous year, 1967, Wexler graduated from Radcliffe College with bachelor's degrees in English and social relations. The next year, she traveled to Jamaica and studied at London's Hampstead Clinic Child Psychoanalytic Training Center on a Fullbright scholarship. When she returned, she pursued doctoral studies in clinical psychology as an intern and teaching fellow at the University of Michigan. She used her Ph.D. dissertation, entitled "Perceptual-motor, Cognitive, and Emotional Characteristics of Persons-at-Risk for Huntington's Disease," as a means of specializing in the study of Huntington's to earn her Ph.D. in 1974.

Wexler held a personal stake in her research, as she had a 50 percent chance of being genetically predisposed to develop Huntington's, which had claimed not only her mother but also her maternal grandfather and three uncles. She thus continued her research on Huntington's through the National Institutes of Health (NIH) while also teaching psychology as an assistant professor in the graduate faculty at the New School for Social Research in New York City. In 1976, she expanded her influence to policy decision making in her congressional appointment as chair of the NIH's Commission for the Control of Huntington's Disease and its Consequences.

Starting in 1979, Wexler traveled annually to the community of Lake Maracaibo, Venezuela, where she studied the highest concentration of Huntington's disease population. She took blood and skin samples and recorded the results of genetic testing performed by James Gusella of the Massachusetts General Hospital. Gusella identified a tag DNA (deoxyribonucleic acid) close to the Huntington's gene and thus developed a test that predicted genetic propensity for Huntington's within 96 percent accuracy.

In 1983, Wexler assumed the presidency of the Hereditary Disease Foundation, thus further facilitating her implementation of policies to support her findings in her research on the causes and possible cures for Huntington's. In 1985, she resumed her academic career as an associate professor of clinical neuropsychology in the College of Physicians and Surgeons of Columbia University. In 1992, the university promoted her to full professorship.

In 1993, the gene that predisposes individuals toward Huntington's disease was discovered in the gene pool from Wexler's Lake Maracaibo samples. In October of that same year, Wexler received the Lasker Public Service Award from the Albert & Mary Lasker Foundation.

Wheeler, Anna Johnson Pell
(1883–1966)
American
Mathematician

Anna Johnson Pell Wheeler wielded influence in her field as the head of the mathematics department at Bryn Mawr College. She was only the second woman to receive a doctorate in mathematics from the University of Chicago. She became the first woman to deliver the prestigious Colloquium Lectures to the American Mathematical Society in 1927, a distinction not conferred on another woman until 1980, when JULIA ROBINSON was so honored.

Anna Johnson was born on May 5, 1883, in Calliope (now Hawarden), Iowa. Her father, Andrew Gustav, moved the family (mother Amelia Friberg, sister Esther, and brother Elmer) to Akron, Iowa, where he established himself as a furniture dealer and undertaker. There, Johnson attended public school until 1899, when she enrolled as a subfreshman at the University of South Dakota. She studied mathematics under Alexander Pell and graduated in 1903 with an A.B. degree.

Johnson then pursued graduate study on scholarship, obtaining dual master's degrees. First, at the University of Iowa, she taught freshman mathematics and wrote her thesis, "The Extension of the Galois Theory to Linear Differential Equations," to earn her M.A. degree in 1904. Next, she studied under Maxime Bôcher, Charles Bouton, and William Osgood at Radcliffe College, where she received her A.M. degree in 1905.

Johnson then studied mathematics under David Hilbert at the University of Göttingen on an Alice Freeman Palmer Fellowship, which stipulated that she remain single throughout the yearlong fellowship. She had maintained contact with Pell, who remained her mentor since her undergraduate days. Despite familial opposition based on the age disparity between the two mathematicians (he was 25 years her senior), Pell traveled to Göttingen to marry Johnson in July 1907, after her fellowship officially ended.

The Pells started their marriage in South Dakota, where the university had appointed Alexander as the first dean of its College of Engineering. Pell taught courses at South Dakota for one term, then returned unaccompanied to Göttingen to earn her doctorate. She apparently had a falling out with Hilbert and did not receive her Ph.D. However, she continued her doctoral study at the University of Chicago under Eliakim Moore and submitted the dissertation she had written in Göttingen independent of Hilbert, "Biorthogonal Systems of Functions with Applications to the Theory of Integral Equations." The University of Chicago conferred upon her a Ph.D. magna cum laude in 1909, making her only the second woman to receive a doctorate in mathematics from the institution.

Anna Pell Wheeler, who in 1927 delivered the prestigious Colloquium Lectures of the American Mathematical Society, the only woman to be invited to do so until 1980. *(Courtesy of Bryn Mawr College Archives)*

Unable to secure a full-time position as a woman, Pell taught part-time at the University of Chicago until her husband suffered a paralytic stroke in the spring of 1911, and she assumed his teaching duties at the Armour Institute of Technology. In order to support her disabled husband, though, she had to find full-time work, so she moved to Mount Holyoke College in the fall of 1911 as an instructor, and she was promoted to associate professor in 1914. In 1918, she moved to Bryn Mawr College as an associate professor, where she remained for the rest of her career (with a few breaks).

In 1921, Alexander Pell died. In 1924, Anna succeeded CHARLOTTE ANGAS SCOTT as chair of the mathematics department, and in 1925, Bryn Mawr promoted her to a full professorship. That year, she married Arthur Leslie Wheeler, a classics scholar and professor of Latin at Princeton University, where the couple moved. Anna Wheeler maintained her affiliation with Bryn Mawr as a lecturer until her second husband passed away in 1932, when she returned to Bryn Mawr full time.

Throughout the remainder of her career, Wheeler worked to transform Bryn Mawr's mathematics department into an internationally recognized program. She offered mathematics luminary EMMY NOETHER refuge from Nazi Germany in her department, a move that drew OLGA TAUSSKY-TODD there as well until Noether died in 1935. She continued to head the mathematics department until her 1948 retirement to emerita status, though she still conducted research and contributed to mathematical associations and societies.

In 1927, Wheeler became the first woman to deliver the Colloquium Lectures to the American Mathematical Society; unfortunately, the talks were never published. From 1927 through 1945, she served as an editor of the *Annals of Mathematics,* and in 1939, she joined other mathematicians to establish the *Mathematical Reviews* to continue the work done by the German journal *Zentralblatt für Mathematik und ihre Grenzgabiete,* which had fallen to Nazi opposition. In early 1966, at the age of 82, Wheeler suffered a stroke; she died in Bryn Mawr on March 26.

Wheeler, Emma Rochelle
(1882–1957)
American
Physician

Emma Rochelle Wheeler founded and ran the first African-American hospital in Chattanooga, Tennessee. She not only practiced medicine and surgery at the facility, but she also administered its school of nursing. She later established an innovative system of prepaying for hospitalization and in-home nursing services, an initiative that reduced the financial strain on families at the time of unexpected sickness.

Wheeler was born on February 8, 1882, in Gainesville, Florida. Her father was a farmer and veterinarian. Her interest in medicine peaked when, suffering from an eye problem at the age of six, she visited a white woman diagnostician, who befriended Wheeler and encouraged her to pursue the study of medicine. She thus attended Cookman Institute in Jacksonville, graduating at the age of 17 in 1899.

In 1900, Wheeler married Joseph R. Howard, a teacher who died a year later of typhoid fever while she was pregnant with a son, whom she named after his dead father. She moved them to Nashville, Tennessee, to attend Meharry Medical, Dental, and Pharmaceutical College of Walden University, graduating with a medical degree in 1905. During commencement week, she married a fellow doctor, John N. Wheeler, and eventually the couple had two daughters, Thelma and Bette, and they adopted Wheeler's nephew, George.

The Wheelers moved to Chattanooga, Tennessee, where they established a joint practice. After a decade of saving money and planning, Wheeler founded the Walden Hospital, a facility dedicated to serving the African-American community's medical needs (which went largely unmet by the existing medical establishment). On July 30, 1915, she opened the three-story building she had commissioned, complete with 30 beds, nine private rooms, and a 12-bed medical dispensary ward. Two staff doctors and three nurses worked the hospital's surgical, maternity, and nursery departments, and 17 physicians and surgeons with the Mountain City Medical Society (to which Wheeler herself belonged) admitted patients there. Although Wheeler's husband practiced at the hospital, Wheeler herself managed, operated, and funded the hospital.

After the hospital's first decade, Wheeler established the Nurse Service Club of Chattanooga in 1925. This radically new concept, an initiative independent from the hospital, provided its members with two weeks of prepaid hospitalization and subsequent in-home nurse care afterward. Also in 1925, she collaborated with Emma Henry, Zenobia House, and Marjorie Parker to establish the Pi Omega chapter of the Alpha Kappa Alpha sorority. By this time, Wheeler had given up on performing surgery at the hospital in order to devote herself to her administrative responsibilities, including the hospital's school of nursing, which she maintained for over two decades.

In 1949, the Chattanooga branch of the National Association for the Advancement of Colored People granted Wheeler its Negro Mother of the Year award. Two years later, failing health forced her to retire from administering the hospital, which foundered without her leadership and closed on June 30, 1953. Wheeler continued her private medical practice, but at the age of 75, she entered Hubbard Hospital in Nashville. She died there on September 12, 1957.

Five years after Wheeler's death, the Chattanooga Housing Authority named its new housing project the Emma Wheeler homes in her memory. On February 16, 1990, the Tennessee Historical Commission officially recognized the historical significance of Walden Hospital by placing a historical marker at its former site.

Whiting, Sarah Frances
(1847–1927)
American
Physicist

Sarah Frances Whiting established the second undergraduate physics department in the United States, at Wellesley College, in 1876. Through this program, she introduced laboratory experimentation to the first generation of women research scientists. She realized the importance of X-ray technology immediately after its discovery in 1895; she made the first X-ray pictures in the United States as well as incorporating X-ray theory and practice into her curriculum.

Whiting was born in 1847. She attended Ingham University, where she earned her bachelor of arts degree in 1865. She spent the next dozen years teaching classics and mathematics at Brooklyn Heights Seminary as well as at her alma mater.

In 1876, Wellesley College appointed Whiting as its first professor of physics and charged her with establishing a physics department by supporting her study for two years in the only existing undergraduate physics department in the United States at the time, at the Massachusetts Institute of Technology (MIT). There she studied not only the subject of physics but also the pedagogy for teaching physics and the process of outfitting a physics laboratory. Many of the necessary instruments were not available by catalog and had to be ordered from Germany by way of existing clients.

In 1877, one of Whiting's physics professors at MIT, Edward Pickering, became director of the Harvard College Observatory. Through his auspices, Whiting observed Harvard's pedagogical and observational techniques in astronomy, imparting this knowledge back to Wellesley, where she introduced astronomy courses into the science curriculum in 1880, using only a celestial globe and a four-inch portable telescope. She also established a weather station at the college, where her students gathered meteorological information and forwarded it to the U.S. Weather Bureau. Whiting continued her education throughout her career: In 1888, she traveled to Europe to study at the University of Berlin for one year, and in 1896, she returned to Europe for a year of study at Edinburgh University.

Up until 1893, Wellesley required its students to pass physics classes for eligibility to receive an undergraduate degree, thus heavily taxing Whiting in terms of her teaching course load. However, this demand also inspired Whiting to innovate new experimental techniques for demonstrating physical principles, especially since she did not have a laboratory assistant until 1885.

In 1895, Wilhelm Conrad Röntgen discovered X rays, and Whiting was the first scientist in the United States to take X-ray pictures. Her incorporation of X rays, both theoretically and experimentally, into the Wellesley science curriculum exhibits Whiting's grasp of the cutting edge of physics at the time. She ended her tenure as professor of physics in 1912, and she later wrote a monograph, "History of the Physics Department of Wellesley College from 1878 to 1912," which is now housed in the Wellesley College Archives.

In 1900, a Wellesley trustee, Mrs. John C. Whitin, endowed the school with the funds necessary to build an observatory. Whiting oversaw the building of the Whitin Observatory that year, outfitting the new facility with spectrophotometric equipment. She then acted as director of the observatory, overseeing its expansion in 1906. In 1912, she published the popular text *Daytime and Evening Exercises in Astronomy for Schools and Colleges*. Whiting retired from Wellesley in 1916, maintaining emerita status until her 1927 death.

Widnall, Sheila E.
(1938–)
American
Aeronautical Engineer

Sheila Evans Widnall became the first woman to head a branch of the United States military when President Bill Clinton named her secretary of the U.S. Air Force. Widnall had established herself as an expert in fluid dynamics, specifically conducting research on noise, instability, and vibrations generated by rotating helicopter blades. She also investigated noise and related issues on V/STOL aircraft, or aircraft that make vertical, short takeoffs and landings. In order to carry out these investigations, she oversaw the construction of an anechoic wind tunnel at the Massachusetts Institute of Technology (MIT).

Sheila Evans was born on July 13, 1938, in Tacoma, Washington. Her mother, Genevieve Alice, was a juvenile probation officer, and her father, Rolland John Evans, was a rodeo cowboy before becoming a production planner for Boeing Aircraft Company and then a teacher. Evans won first prize in her high school's science fair, exhibiting her early interest and proficiency in science. She matriculated at MIT in 1956 as one of 21 women in a class of 900 incoming students. She earned her bachelor of science degree in aeronautics and astronautics in 1960, and in June of that year, she married fellow aeronautical engineer William Soule Widnall. Together, the couple had two children, William and Ann Marie.

Widnall continued with graduate study at MIT. She earned her master of science degree in 1961 and her doctorate in 1964, both in aeronautics and astronautics. She remained at MIT as an assistant professor in mathematics and aeronautics. MIT promoted her to associate professor in 1970, and in 1974, she rose to the rank of full professor. She was the first alumna to teach in MIT's school of engineering.

Also in 1974, Widnall's career branched out into administration, as she became the first director of university research for the U.S. Department of Transportation. She increased her administrative responsibilities the next year when she assumed the chair of MIT's division of fluid mechanics, a position she retained until 1979. That year, MIT appointed her as director of the fluid dynamics laboratory (until 1990) and also voted her as the first woman to chair its faculty. In 1978, she continued her administrative service in the public sector when President Jimmy Carter appointed her to the U.S. Air Force Academy's board of visitors, which she chaired from 1980 through 1982.

Aeronautical Engineer Sheila Widnall, former secretary of the U.S. Air Force, was the first woman to head a branch of the U.S. military. *(Courtesy Sheila Widnall)*

Widnall continued to advance her career, both within MIT and in her field. In 1986, MIT named her the Abby Rockefeller Mauze Professor, an endowed chair reserved for a faculty member who promoted the status of women in industry, the arts, or the professions. In 1988, the American Association for the Advancement of Science elected her its fifth woman president. In 1991, she chaired MIT's committee on academic responsibility, and the next year, MIT appointed her as associate provost of the university.

In August 1993, President Bill Clinton appointed Widnall as secretary of the U.S. Air Force, thus making her the first woman to head a branch of the United States military. She was responsible for the 380,000 active members of the air force and its 251,000 reservists. She devoted her term to raising the quality of life for service members as well as modernizing the air force based on the latest technological advances. Widnall retired as secretary of the air force in 1997 to return to MIT.

Widnall's career has been honored extensively. In 1972, the American Institute of Aeronautics and Astronautics granted her its Lawrence Sperry Award. She won the 1975 Outstanding Achievement Award from the Society of

Women Engineers. In 1985, the National Academy of Engineering inducted her as a member, and in 1987, the Boston Museum of Science bestowed on her its Washburn Award. In 1998, the Women's International Center presented its Living Legacy Award to Widnall.

Williams, Anna Wessels
(1863–1954)
American
Bacteriologist

Anna Wessels Williams helped to discover several vaccines against serious diseases, including diphtheria and rabies. She also developed a method for diagnosing rabies that reduced the pronouncement waiting period from 10 days to a matter of minutes. She searched throughout her career for an antitoxin against streptococcus, or the condition of strep throat. Her ability to detect the influenza bacillus led to her designation as the "influenza picker."

Williams was born on March 17, 1863, in Hackensack, New Jersey, the second daughter of six children. Her mother, Jane Van Saun, supported the missionary activities of the First Reformed Dutch Church, and her father, William Williams, was a private school teacher who educated his children at home. At age 12, she entered the State Street Public School, where she first encountered a wondrous microscope. She then attended the New Jersey State Normal School, graduating in 1883.

Williams taught school for the next two years. In 1887, her sister almost died from the complications of delivering a stillborn child; Williams vowed to become a doctor to prevent such incidents from happening. She attended the Women's Medical College of the New York Infirmary for Women and Children, where she studied obstetrics and gynecology under Dr. ELIZABETH BLACKWELL to earn her M.D. in 1891. She remained at the New York Infirmary as an instructor in pathology and hygiene until 1893 and as assistant to the chair of this department until 1895. From 1892 through 1893, she continued her medical education in Europe, studying at the universities of Vienna, Heidelberg, and Leipzig and interning at the Royal Fräuen Klinik of Leopold in Dresden.

In 1894, New York City established its Department of Health, where Williams volunteered to work under the directorship of William Hallock Park. Together, Park and Williams conducted research on diphtheria. While Park was on vacation, Williams discovered *Corynebacterium diphtheriae* from a mild case of tonsillar diphtheria, and this bacillus became the stock strain from which the first effective diphtheria antitoxin (called Park-Williams #8) was prepared and made available free to patients who could not afford it. In 1895, the health department promoted Williams to assistant bacteriologist.

Williams turned her attention to streptococcal and pneumococcal infections, as well as scarlet fever, in search of antitoxins. In 1896, she brought her diphtheria bacillus to the Pasteur Institute in Paris, and in return, the Pasteur scientists helped her antitoxin research. She collaborated with Alexander Marmorek in streptococcal investigations (which preoccupied her the rest of her career) and with Emile Duclaux as well on rabies diagnosis and vaccinations. She returned to the United States in 1898 with a rabies vaccine in hand, and it was mass-produced immediately.

Collaborating with Alice G. Mann from 1902 on, Williams searched for ways to improve the means of diagnosing rabies and strep, among other diseases. She discovered a brain cell aberration specific to rabid animals, but an Italian physician, Adelchi Negri, working independently, published the same discovery in 1905, before she could announce her finding (hence the name Negri bodies for these aberrant brain cells). That year, the New York City Department of Health named her assistant director of its research laboratory.

Later in 1905, Williams announced her discovery of a method of staining the brain cells she and Negri had discovered, thus allowing for near-instant diagnosis of rabies. Also in that year, she and Parks published the second edition of their classic text, *Pathogenic Microorganisms Including Bacteria and Protozoa: A Practical Manual for Students, Physicians and Health Officers,* which saw 11 editions. In 1907, the American Public Health Association named Williams chair of the Committee on the Standard Methods for the Diagnosis of Rabies, her area of expertise.

Williams increased her administrative influence in her field, becoming president of the Women's Medical Association in 1915. In 1931, the American Public Health Association named her vice chair of its laboratory section, and the next year she chaired the section, the first woman to hold this position. In 1932, she published her life's work, the monograph *Streptococci in Relation to Man in Health and Disease.*

In 1934, New York City mayor Fiorella La Guardia enforced the mandatory retirement age of 70, notwithstanding pleas from the scientific community to make an exception to allow Williams continue with her work. She retired to Woodcliff Lake, New Jersey, then joined her sister, Amelia Wilson, in Westwood, New Jersey. There, she died of heart failure on November 20, 1954. She was 91 years old.

Williams, Cicely Delphin
(1893–1992)
Jamaican/English
Physician

Cicely Delphin Williams identified the disease of kwashiorkor as a form of malnutrition distinct from other forms.

She also discovered the cause of Jamaican vomiting disease. By the end of her career, she had worked in 70 countries on five continents fighting disease. She became the first woman appointed by the British Colonial Office as head of one of its sections.

Williams was born on December 2, 1893, in Jamaica. Her father, James Towland Williams, hailed from a Welsh plantation family and directed the department of education in Jamaica. Williams wished to follow her father's footsteps through Oxford University, a desire denied until the advent of World War I opened up Oxford's doors to women. She studied tropical medicine and hygiene under Sir William Osler and passed her finals in 1923, but she did not complete her doctoral thesis until much later.

Williams held a two-year surgical gynecology residency at South London Hospital for Women and Children while petitioning the British Colonial Office for an overseas appointment. In 1929, the office assigned her to the Gold Coast (now Ghana) in Africa. There, she observed toddlers with swollen legs and bellies, as well as scaly rashes and red-tinged hair, and concluded that the condition must relate to malnutrition. Despite other doctors' opinion that she was misdiagnosing pellagra, she identified what the Ghanaians called weaning disease, or kwashiorkor, to be a distinct disease caused by protein deficiency in the period between nursing and eating solids.

Williams published her findings in the *Gold Coast Colony Annual Medical Report* of 1931–32. In 1933, she published a clinical description of kwashiorkor in *Archives of Disease in Children,* and in 1935, she distinguished between kwashiorkor and pellagra in a *Lancet* article. The next year, she submitted her delayed doctoral thesis, "Child Health in the Gold Coast," and Oxford granted her an M.D. Throughout this time, she was recovering from blood poisoning she had contracted while performing kwashiorkor autopsies, which almost proved fatal.

In 1936, the British Colonial Office transferred Williams to Trengganu, Malaya (now Malaysia), which the Japanese invaded in 1941. As a prisoner of war, Williams suffered from dysentery, undernourishment, and psychological torture. She survived, and in 1946, the British Colonial Office appointed her as head of the Maternity and Child Welfare Services, the first woman to hold a superscale post in the Colonial Service. In 1948, the newly formed World Health Organization enlisted her services as chief of its Maternal and Child Health Section.

In 1951, Williams heeded the beckoning of the Jamaican government to return to her homeland to help find the cause of Jamaican vomiting sickness. She isolated the root as spoiled, green akee fruit, and she designed a glucose therapy as treatment. In 1955, she returned to England as a senior lecturer in nutrition at London University, and in 1959, the American University in Beirut appointed her professor of maternal and child health. In 1964, she returned to England to advise the Family Planning Association on its training programs. In 1968, she accepted her appointment as professor of international family health at the Tulane University School of Public Health.

Williams collected numerous honors in recognition of her lifesaving efforts throughout her career. In 1965, the British Pediatric Association awarded her its James Spence Memorial Gold Medal. In 1967, the American Medical Association granted her both its Joseph Goldberger Award and its Council on Foods and Nutrition Award. In 1969, she received an honorary doctorate from the University of the West Indies, and in 1971, she received the Martha May Eliot Award from the American Public Health Association. The British Medical Association awarded her its Dawson Williams Prize in 1972, and the next year, she became an honorary member of the American Academy of Pediatrics. On July 13, 1992, she died at the age of 98.

Williams, Heather
(1955–   )
American
Ornithologist

Heather Williams specializes in ornithological neuroethology, or the study of the relationship between behavior and the nervous system of birds. Her research has pinpointed links between birds' song perception and their brain, nervous system, and even sexual behaviors. She received research support from numerous grants, including a five-year National Institutes of Health (NIH) grant and a prestigious MacArthur Foundation Fellowship. In addition to her scientific research, Williams also finds time to compete on the international level in orienteering, the sport combining cross-country running with compass and map reading.

Williams was born on July 27, 1955, in Spokane, Washington. Her mother was Maria Greig, and her father, James Edwards Williams, worked for the U.S. foreign service, necessarily moving the family of six (Williams had three siblings) to such countries as Laos, Turkey, and Bolivia. Williams studied biology at Bowdoin College, where she graduated summa cum laude in 1977.

Williams spent the next academic year studying marine biology on a Thomas J. Watson Fellowship at Hebrew University in Eilat, Israel. She returned to the United States to pursue doctoral study at Rockefeller University in New York City. Working under her adviser Fernando Nottebohm, a specialist in canary ethology, she wrote her doctoral dissertation, "A Motor Theory for Bird Song Perception," to earn her Ph.D. in behavioral neuroscience in 1985. That same year, she was a finalist for the Society of Neuroscience's Lindsey Award.

Williams remained at Rockefeller University to conduct postdoctoral research at the Field Research Center for Ethology and Ecology in collaboration with Nottebohm, Jeffrey Cynx, and David Vicario. In 1986, the university appointed her as an assistant professor, and she received a two-year award from the Air Force Office for Scientific Research. That year, she also married Patrick D. Dunlavey, and together the couple had two children, Maria Greig and Alan Peter Dunlavey.

In 1988, Williams joined the faculty of Williams College in Williamstown, Massachusetts, which was close enough to New York City to allow her to continue her research at Rockefeller. That year, she also received a five-year grant from the NIH to support her research, extending the studies she performed for her doctorate. She studied the zebra finch, *Taenioppygia guttata,* specifically focusing on how the brain and nervous system collaborate to recognize and respond to songs specific to a species. Williams discovered a clustering effect in the finch's song production, organized around three-syllable "chunks" that the birds rearranged corresponding to the situation, including expressions of sexual preference (or sexual dimorphism).

At the termination of her NIH grant in 1993, Williams received a MacArthur Foundation Fellowship. The next year, Williams College promoted her to associate professor (she retained her association with Rockefeller University as an adjunct professor). She honed her research skills through orienteering (she ranked third in the United States in this sport from 1980 through 1989). She saw a correlation between the brain-body connection of orienteering, which requires participants to navigate a racecourse, and bird songs. In 1993, Williams contributed to the World Orienteering Championships as a member of the course-setting team.

Willson, Lee Anne Mordy
(1947–   )
American
Astrophysicist

Based on her research of stellar evolution and mass loss from stars, Lee Anne Mordy Willson experimentally extrapolated that the Sun will become a red giant, engulfing much of its solar system in flames. Willson constructs computer models to recreate the growth and decline of stars based on the known variables. According to her calculations, as the Sun transforms from a red giant into a white dwarf, it will scorch Earth, unless variables not taken into consideration intervene. Earlier in her career, in 1980, Willson won the ANNIE JUMP CANNON Award in Astronomy from the American Astronomical Society and the Association of American University Women.

Lee Anne Mordy was born on March 14, 1947, in Honolulu, Hawaii, where she lived until 1956, when she moved to Saltsjöbaden, Sweden, for four years. When she returned to the United States in 1960, she settled in Reno, Nevada, until she matriculated at Harvard University. There she earned her A.B. in physics in 1968, after which she spent the next year in Sweden studying at the University of Stockholm on Fulbright and American Scandinavian Foundation Fellowships. On July 19, 1969, she married mathematician Stephen J. Willson, and together the couple had two children: a daughter, Kendra, born in 1972, and a son, Jeffrey, born in 1976.

After returning from abroad, Willson pursued doctoral study in astronomy at the University of Michigan, where she served as a planetarium demonstrator. She earned her M.S. there in 1970, and in 1973, the University of Michigan conferred on her a Ph.D. in astronomy. That same year, she started working as an instructor at Iowa State University (ISU), commencing a relationship that has lasted the rest of her career. She rose steadily through the ranks there: to assistant professor in 1975; to associate professor in 1979; to full professor in 1988; and in 1993, to university professor, a title created that year to honor significant contributions by senior faculty.

While at ISU, Willson served as the coordinator of the astronomy and astrophysics program from 1983 through 1985 and again from 1987 through 1990. She also held summer, sabbatical, and visiting positions throughout her time at ISU. She spent several summers at the University of Washington: as a research associate in 1974; as a visiting scientist in 1980; and as a visiting professor in 1982. She traveled extensively during her two sabbaticals: The first, in 1980, she spent at the University of Colorado, the University of Texas, and Kitts Peak National Observatory; the second, in 1985, she spent as a visiting fellow at the Canadian Institute for Theoretical Astrophysics and at the University of Toronto Department of Astronomy. In 1991, she returned to Sweden as a visiting professor at Uppsala University.

Willson has served her field throughout her career, especially focusing on the status of women in astronomy. She served on the working group on the status of women in astronomy of the American Astronomical Society in 1972 through 1973, and from 1987 through 1989, she chaired the group (which had committee status by then.) She participated in the American Association of Variable Star Observers for 20 years until the organization named her senior vice president in 1997 and then promoted her to president in 1999.

Witkin, Evelyn Maisel
(1921–   )
American
Geneticist

Evelyn Maisel Witkin specialized in the study of bacterial mutation, especially focusing on the common bacteria *Es-*

cherichia coli. She conducted research on both spontaneous and induced bacterial mutation, and she also investigated the effects of radiation on genes. In addition, she researched how enzymes could repair DNA damage.

Evelyn Maisel was born on March 9, 1921, in New York City. Her mother was Mary Levin, and her father was Joseph Maisel. She conducted her undergraduate study at New York University, graduating magna cum laude in 1941 with a bachelor of arts degree. She remained in New York City for graduate work in zoology at Columbia University, where she earned her master of arts degree in 1943. On July 9, 1943, she married Herman A. Witkin, and together the couple had two sons: Joseph, born in 1949, and Andrew, born in 1952.

In 1947, Columbia conferred on Witkin a Ph.D. in zoology. In the meanwhile, she had started working as a research associate in bacterial genetics at the Carnegie Institution in 1946 while completing her doctorate. She conducted two years of postdoctoral research at the American Cancer Society, from 1947 through 1949. In 1950, Witkin officially joined the staff in the Department of Genetics at Carnegie, where she remained until 1955. That year, she commenced her long-standing relationship with the Downtown Medical Center of the State University of New York, which appointed her as an associate professor of medicine. She retained this position until 1969, when the university promoted her to full professorship.

In 1971, Douglass College of Rutgers University hired Witkin as a professor of biological sciences. While there, she published one of her more important papers, "UV Mutagenesis and Inducible DNA Repair in *E. coli,*" which appeared in the journal *Bacteriological Review* in 1976. In 1979, the Waksman Institute of Microbiology named her the BARBARA MCCLINTOCK professor of genetics. In July of that same year, she lost her husband.

In 1991, Witkin retired to the title of Barbara McClintock Professor Emerita, though she did not abandon her scientific career altogether. For example, in 1992, the *Journal of Bacteriology* published her article entitled "Overproduction of DnaE Protein (Alpha Subunit of DNA Polymerase III) Restores Viability in a Conditionally Inviable *Escherichia coli* Strain Deficient in DNA Polymerase I." This capped her career of publishing more than 45 papers in prestigious scientific journals, such as the *Proceedings of the National Academy of Sciences* and *The Cold Spring Harbor Symposia of Quantitative Biology.* She also served as the editor of the journal *Microbial Genetics* from 1950 through 1964 and as a member of the editorial board of the journal *Mutation Research* from 1960 on.

Witkin received numerous honors in her career. In 1977, the French Academy of Sciences awarded her its Prix Charles Leopold Mayer, and that same year, the National Academy of Sciences elected her a member. The next year, she received an honorary degree from the New York Medical College, and in 1979, she won the Lindback Award.

Wong-Staal, Flossie
(1947–   )
Chinese/American
Retrovirologist

Flossie Wong-Staal is one of the world's leading retrovirologists and experts on human immunodeficiency virus (HIV) and acquired immune deficiency syndrome (AIDS). In 1983, she and her colleague Robert Gallo discovered HIV simultaneously with a French researcher, and in 1985, she was the first to clone HIV. In 1990, the Institute for Scientific Information listed Wong-Staal as the top woman scientist of the past decade and the fourth-ranking scientist under the age of 45.

Wong-Staal was born Yee Ching Wong on August 27, 1947, in Canton, China. Her mother was a homemaker, and her father was a businessman. In 1952, the family fled communist China for Hong Kong, where her father anglicized her name after Typhoon Flossie, which had just struck Hong Kong. She graduated from an all-girls Catholic high school in 1965 and immigrated to the United States to study molecular biology at the University of California at Los Angeles (UCLA). She graduated magna cum laude with a bachelor's degree in bacteriology in 1968.

Wong-Staal continued at UCLA, pursuing graduate study as a teaching assistant from 1969 through 1970 and as a research assistant from 1970 through 1972. In the latter year, UCLA granted her a Ph.D. in molecular biology as well as recognizing her with one of its Women Graduates of the Year Awards. She then filled a postdoctoral fellowship at the University of California at San Diego (UCSD) Medical Center from 1972 through 1973. During this time, she married and added her husband's last name to hers. The couple had two children before they divorced.

In 1973, Wong-Staal joined the National Cancer Institute as a Fogerty Fellow, and she subsequently worked her way up through the ranks from visiting associate in 1975 to cancer expert in 1976 to senior investigator in 1978. In 1982, she became the section chief of Molecular Genetics in Hematopoietic Cells in the Laboratory of Tumor Cell Biology. In 1985, she held a visiting professorship at the Institute of General Pathology of the First University of Rome in Italy.

In 1990, UCSD named Wong-Staal to the Florence Seeley Riford Chair in AIDS Research. In 1994, UCSD appointed Wong-Staal director of the Center for AIDS Research (CFAR), newly established through a National Institutes of Health grant. In September 1996, the University of California Board of Regents approved the formation of the AIDS Research Institute, the equivalent of an academic department, which incorporated the CFAR.

Wong-Staal has pursued a two-pronged approach in her AIDS research: She tries to gain as much knowledge of the virus on the one hand, and on the other hand, she investigates ways to destroy the virus. She searches actively for a vaccine, and she simultaneously explores gene therapy for HIV infection. Gene therapy can either strengthen the immune system of the host or weaken the virus; Wong-Staal focuses on the latter tactic.

Besides being named the top woman scientist of the 1980s, Wong-Staal has received many other honors. In 1987, the Chinese Medical and Health Association granted her its Outstanding Scientist Award. In 1991, she received the Excellence 2000 Award from the United States Pan Asian American Chamber of Commerce and the Organization of Chinese American Women. Significantly, Wong-Staal has received the most recognition in the form of requests for her to impart her knowledge actively: She delivered seven honorary lectures between 1985 and 1996.

Wood, Elizabeth Armstrong
(1912–   )
American
Crystallographer

Elizabeth Armstrong Wood gained esteem for the clarity of her writing in reporting the findings of her crystallographic research. She spent practically her entire professional career working for Bell Telephone Laboratories, where she conducted research in X-ray crystallography, contributing significantly to the developing field of solid-state physics and chemistry. Her research also included work on optical mineralogy as well as the geology and petrology of igneous and metamorphic rocks.

Wood was born in 1912 in New York City. Growing up, she attended the Horace Mann School of Columbia University. She then matriculated at Barnard College, where she received her bachelor of arts degree in 1933. She next pursued graduate studies in geology at Bryn Mawr College, earning her master of arts degree in 1934. She remained at Bryn Mawr as a demonstrator in geology while working on her doctoral dissertation to earn her Ph.D. in geology in 1939.

Throughout her doctoral studies, she bounced back and forth between Barnard and Bryn Mawr: In 1936, she returned to Barnard as an assistant; the next year, she went back to Bryn Mawr as a demonstrator; in 1938, she retraced her steps to Barnard, this time as a lecturer. After she received her doctorate, Barnard promoted her to a research assistantship in 1941.

In 1941, Wood received a National Research Council fellowship as a research assistant. The next year, Bell Telephone Laboratories hired her onto its technical staff in crystal research. She remained with Bell for the rest of her 24-year career.

In 1947, Wood married Ira E. Wood. That same year, as secretary of the American Society for X-Ray and Electron Diffraction (ASXRED), she collaborated with William Parrish, the secretary of the Crystallographic Society of America (CSA), to invite the International Union of Crystallography to hold its first meeting in the United States. When the union agreed, Harvard University agreed to host the 350 crystallographers from 11 countries at the First General Assembly and Congress of the union in late July through early August in 1948. The next year, the CSA and ASXRED merged to form the American Crystallographic Association (ACA), and Wood served as its president in 1957.

Wood gained her reputation for clear, concise writing with her book publications. In 1963, she published the *Crystal Orientation Manual* based on her experimentation with the orientation of quartz crystal oscillator plates. The next year, she had two major publications: *Crystals and Light: An Introduction to Optical Crystallography,* which was reprinted in 1977; and *Experiments with Crystals and Light,* a booklet and kit that was translated into six different languages. In 1969, she published her last book, *Science for the Airplane Passenger,* which was reprinted in 1975 as *Science from Your Airplane Window* (a Japanese edition was published in 1972).

Before Wood retired in 1967, she received three honorary degrees, from Wheaton College and Worcester Polytechnic Institute in Massachusetts and from Western College in Ohio. In honor of her work, the ACA established the Elizabeth Armstrong Wood Science Writing Award to recognize writing that imparts an understanding of crystallography to a wider audience. She attended the St. Louis meeting of the association to present the first award to Roald Hoffman of Cornell University, the 1981 Nobel laureate in chemistry.

Woods, Geraldine Pittman
(1921–1999)
American
Embryologist and Health Services Administrator

Geraldine Pittman Woods conducted embryological research before putting her career on hold to raise her family. She later recommenced her career as an administrator with the National Institutes of Health (NIH), spearheading two important initiatives: the Minority Biomedical Research Support (MBRS) program, which educated minority institutions on how to compete for governmental and philanthropic grants; and the Minority Access to Research Careers (MARC) program, which promoted the recruitment and retention of minority students and scholars in the sciences.

Geraldine Pittman was born on January 29, 1921, in West Palm Beach, Florida. Her parents, Susie King and Oscar Pittman, conducted business in the farming and lumber industries, as well as owning restaurants and rental properties. In the fourth grade, Pittman transferred from private Episcopal school to the only public school that allowed African-American students, Industrial High School, where she graduated without particular distinction in 1938. She then attended Talladega College, an African-American institution in Talladega, Alabama.

In 1940, Pittman's mother came down with a serious illness, and doctors admitted her at the Johns Hopkins Hospital in Baltimore, Maryland. Pittman transferred to Howard University in Washington, D.C., a mere 35 miles from her mother. The death of her father when she was a teenager combined with her mother's illness inspired her to excel in her studies. With support from her biology and zoology professor, Dr. Louis Hansborough, Radcliffe College and Harvard University accepted her for graduate study in its biology program. She matriculated there after her 1942 graduation from Howard with a bachelor of science degree.

Within one year, Pittman earned her master of science degree from Radcliffe College, and she then began her doctoral research. She focused on the embryonic development of spinal cord nerves, questioning what factors determined the differentiation of these nerve cells from the rest of the cells—internal factors within the nerve cells or external factors from the surrounding cells. She discovered that both these factors collaborated and also that the number of muscle cells determined to some degree the number of nerve cells developed. A mere two years after receiving her master's degree, she earned her Ph.D. from Harvard in 1945 and was elected into Phi Beta Kappa.

Pittman returned to Howard as an instructor, and after one semester there, she married Robert Woods, who was studying dentistry at Meharry Medical School. The couple commuted between Washington, D.C., and Nashville, Tennessee, until Robert graduated from dentistry school and moved his wife to California to set up a dental practice. Woods put her career on hold to raise three children into their teenage years.

Woods gradually recommenced her career, first by volunteering in equal opportunity and civil rights initiatives. She earned a reputation for championing minority rights on the personnel board of the California Department of Employment, drawing the attention of Lady Bird Johnson, wife of United States President Lyndon B. Johnson. In 1965, the First Lady invited Woods to the White House to participate in the launching of Project Head Start, a preschool initiative for low-income families.

Thenceforth, Woods's administrative career snowballed: In 1968, she chaired the Defense Advisory Commission on Women in the Services; from 1968 through

1972, she served as vice chair of the Community Relations Conference of Southern California; and she served on the board of directors of the National Council of Negro Women from 1969 through 1972. In 1969, the National Institute of General Medicine of the NIH appointed Woods as a special consultant. For the next two decades, she initiated the MBRS and MARC programs to promote opportunities for minorities in the sciences. Woods retired from the NIH in 1991 and moved to Aliso Viejo, California. She died on December 27, 1999, at her home.

Wright, Jane Cooke
(1919–   )
American
Physician

Jane Cooke Wright distinguished herself as a cancer researcher, specifically investigating chemotherapies. She also contributed to the study of cancer as director of the Cancer Research Foundation, an organization started by her father.

Wright was born on November 20, 1919, in New York City. Her mother was Corinne Cooke Wright. On her father's side, she hailed from a family of eminent physicians: Her father, Dr. Louis Tompkins Wright, was one of the first African-American graduates of Harvard Medical School; her father's father was one of the first graduates of Meharry Medical College, founded in Tennessee to teach the practice of medicine to former slaves; and other relatives also distinguished themselves in medicine.

Wright attended private school before going to Smith College on a four-year scholarship. After she graduated from Smith in 1942, New York Medical School granted her a four-year scholarship. There, she served as vice president of her class and president of the honor society. She graduated with honors (third in a class of 95) in 1945. Two years later, she married Harvard Law School graduate David D. Jones Jr. on July 27, 1947. She retained her maiden name professionally but preferred her married name in her personal life. Together, the couple had two daughters, Jane and Alison.

Wright interned at New York City's Bellevue Hospital, where she remained for an assistant residency. She then performed a two-year residency in internal medicine at Harlem Hospital. She commenced her professional career in 1949 as the New York City school physician, as well as holding the post of visiting physician at Harlem Hospital. She also studied the efficacy of certain drugs in staying the growth of tumors as a clinician at the hospital's Cancer Research Foundation, headed by her father at the time. When he died in 1952, she took over as director of the foundation.

In 1955, the New York University Medical Center hired her to direct its cancer chemotherapy research department and teach surgery. In 1961, the university promoted her to the rank of adjunct professor in the department of surgery. That year, the African Research Foundation appointed her vice president, entailing her traveling to East Africa to survey the organization's medical mission. In 1964, the president appointed her to his Commission on Heart, Disease, Cancer, and Stroke, which established a national network of centers for treating these conditions.

In 1967, Wright returned to the New York Medical College as associate dean and professor of surgery, practicing at the three affiliated hospitals. She remained at the college until 1987, when she retired to emerita status.

Wright's influence as a cancer researcher received much recognition. In 1952, *Mademoiselle* magazine granted her its Merit Award for outstanding contribution to medical science. In 1965, she won the Spirit of Achievement Award from the women's division of the Albert Einstein College of Medicine. In 1967, she received the Hadassah Myrtle Wreath, and the next year, she was awarded the Smith College Medal. In observation of International Women's Year in 1975, the journal *Cancer Research* named her one of eight outstanding women scientists. In 1980, Ciba Geigy included her in a poster of exceptional black scientists.

Wrinch, Dorothy Maud
(1894–1976)
English
Mathematician and Biochemist

Dorothy Maud Wrinch fused mathematics with biochemistry by explaining a structure of protein mathematically. Her areas of expertise were incredibly broad: She taught and conducted research not only in mathematics but also in all three major branches of the sciences—physics, chemistry, and biology. The popular press dubbed her the Woman Einstein for her far-reaching brilliance. She was the first woman to earn a doctorate from Oxford University.

Wrinch was born on September 13, 1894, in Rosario, Argentina. Her parents, Hugh Edward Hart and Ada Minnie Souter Wrinch, returned the family to Surbiton, near London, England. After graduating from Surbiton High, she attended Girton College of Cambridge University on scholarship. In 1916, she graduated as a Wrangler in mathematics (the highest rank attainable on the final examination). She remained at Cambridge to earn her master of arts degree in mathematics in 1918, after which she lectured in mathematics at University College of the University of London for two years.

Wrinch continued to collect academic degrees. She earned a master of science degree in 1920 and a doctorate in 1921 from the University of London. In 1923, she moved to Balliol College of Oxford University. That same year, she married John William Nicholson. The next year, she earned yet another master of arts degree, this time from Oxford. In 1927, she gave birth to her only child, Pamela, and in 1929, she earned a second doctorate, but it was the first doctorate awarded by Oxford to a woman.

While studying at Oxford, she also worked there as a lecturer in mathematics and as the director of studies for women, as well as sitting on the physical sciences faculty. As well, she tutored mathematics at the five women's colleges in the area. In the early 1930s, she studied abroad twice: first at the University of Vienna from 1931 through 1932 and then at the University of Paris from 1933 through 1934.

In 1935, Wrinch secured a six-year Rockefeller Research Fellowship to support her application of mathematics to biochemical structures. During this same period, her husband's chronic alcoholism became insupportable, and in 1938, their marriage was dissolved. The next year, Wrinch immigrated with her daughter to the United States, where she landed a position lecturing in chemistry at Johns Hopkins University for the next two years.

In 1941, Wrinch's Rockefeller Fellowship expired, and she served as a visiting professor in natural sciences concurrently at three of the five colleges in the Amherst region of Massachusetts: Smith, Amherst, and Mount Holyoke Colleges. That year, she also wed Otto C. Glaser, and their marriage lasted nine years, until his 1950 death. Smith College retained Wrinch as a lecturer in physics from 1941 through 1954, when the college promoted her to a visiting professorship, a title she retained until her 1971 retirement.

In 1935, Wrinch proposed a theory, which became known as the cyclol structure, whereby she applied the principle of mathematical symmetry to the physical bonds between two adjacent, covalent amino acids. This theory generated both renown (for its interdisciplinary genius) and controversy (for countering the existing genetic theories.) Later experimentation demonstrated that her theory did not apply across the board, but it did confirm her proposed structure in some alkaloids.

Wrinch published prolifically in her career. Her early output was almost evenly divided between math and science; between the years of 1918 and 1932, she published 20 papers on pure and applied mathematics and 16 on scientific methodology and on the philosophy of science. She later wrote three books in her various fields: *Fourier Transforms and Structure Factor,* published in 1946; *Chemical Aspects of the Structure of Small Peptides: An Introduction,* published in 1960; and *Chemical Aspects of Polypeptide Chain Structure and the Cyclol Theory,* published in 1965. Throughout her career, she also authored or coauthored 192 professional papers on topics as diverse as X-ray crystallography, mineralogy, and the structure of globular protein molecules. Wrinch spent the end of her life in Woods Hole, Massachusetts, and she died in 1976.

Wu, Chien-Shiung
(1912–1997)
Chinese
Experimental Physicist

Chien-Shiung Wu performed the experiment that shattered one of the fundamental laws of nature, the principle of the conservation of parity. Physicists believed that nature does not differentiate between right and left, but Wu demonstrated experimentally that it does. Wu received numerous distinctions honoring her work, including the National Medal of Science. Interestingly, though, Wu did not share in the Nobel Prize in physics awarded in 1957 to her male collaborators, Dr. Tsung-Dao Lee and Dr. Chen Ning Yang, despite support from the physics community in favor of including her in the prize.

Chien-Shiung Wu, whose hyphenated name means "strong hero" in Chinese, was born on May 29, 1912, near Shanghai. Her father, Wu Zhong-Yi, supported his daughter's scientific career regardless of any obstacles. He demonstrated his belief in opportunity for women by opening the first school for girls in China. Wu then attended the National Central University in Nanjing, China, where she earned her bachelor's degree in 1936.

That year, Wu traveled to the United States to pursue graduate study at the University of California at Berkeley. She received her doctorate in 1940 and remained there for two years as a resident fellow and lecturer. In 1942, she married a fellow experimental physicist, Luke C. L. Yuan, and the couple had one son, Vincent Yuan, who became a research scientist at the Los Alamos National Laboratory in New Mexico.

In 1942, Wu taught at Smith College as an assistant professor and the next year at Princeton University as the first female instructor in its physics department. In 1944, she joined the Division of War Research at Columbia University, otherwise known as the Manhattan Project. There she developed a process of gaseous diffusion for enriching uranium ore as fuel for the atomic bomb. At the request of Enrico Fermi, she solved the mystery behind an aborted chain reaction at an atomic stockpile in Hanford, Washington.

After World War II, Columbia University retained Wu as a research associate, and she remained there for the rest of her career. In 1952, the university promoted her to the rank of associate professor, and in 1958, she became a full professor of physics. In 1972, Columbia named her its first Michael I. Pupin Professor of Physics, a title she retained until her 1981 retirement to emerita status.

The 1975 recipient of the National Medal of Science, Chien-Shiung Wu disproved the long-accepted law of parity, which had held that in nature there is no real difference between right and left. *(Courtesy of AIP Niels Bohr Library)*

In the spring of 1956, Lee approached Wu with questions concerning her specialty, beta decay, or the atomic emission of electrons. He proposed several experimental methods of testing the principle of the conservation of parity under weak-force circumstances, or in radioactive decay. She made a countersuggestion that became the method for her famous 1957 experiment. Using equipment at the National Bureau of Standards, Wu supercooled cobalt 60 to 0.01 degree above absolute zero, then placed this radioactive isotope in a strong electromagnetic field to set its nuclei spinning along the same axis. According to the law of conservation of parity, the nuclei should release equal numbers of particles in each direction, but Wu measured a release of larger numbers of particles in the opposite direction of the spin, thus overturning parity. In 1966, Wu published *Beta Decay,* which became the standard reference for physicists. They acknowledged her as an expert in the field of beta decay after she established Fermi's theory of weak interactions in the nucleus with experimental proof by measuring the emission of low-energy electrons.

Although the Nobel Prize was denied her, Wu received many other awards. In 1958, she became the first woman to be awarded an honorary doctorate in science from Princeton University (she eventually accrued honorary degrees from 11 institutions internationally). That same year, she won the Research Corporation Award and the National Academy of Sciences elected her as a member. She received achievement awards from the American Association of University Women in 1960 and from the Taiwanese Chi-Tsin Culture Foundation in 1965. The year before, she became the first woman to receive the Cyrus B. Comstock Award from the National Academy of Sciences, awarded only once every five years. In 1975, the American Physical Society elected her as its first woman president and also awarded her its Tom Bonner Prize. That same year, she received the National Medal of Science. In 1978, she became the first woman to receive the Wolf Prize in Physics from the state of Israel and, in 1990, the first living scientist to have an asteroid named after her. On February 16, 1997, Wu died of a stroke at St. Luke's-Roosevelt Hospital in Manhattan.

Wu, Sau Lan
(1940–)
Chinese/American
Physicist

Sau Lan Wu participated in the Massachusetts Institute of Technology (MIT) team that discovered a new particle, the charm quark (or J/psi), in 1974. Five years later, at the University of Wisconsin, she was part of another team that discovered gluon, a kind of glue that holds together quarks to form particles of protons and neutrons. For this work, she received the 1995 European Physical Society High Energy and Particle Physics Prize.

Wu was born in Hong Kong on May 11, 1940. Growing up, she aspired to be a painter until she read the biography of MARIE CURIE, which turned her aspirations toward physics. She traveled to the United States where she received a scholarship from Vassar College for undergraduate study. She graduated summa cum laude and Phi Beta Kappa with a B.A. in physics in 1963.

Wu then received a fellowship for her first year of graduate study at Harvard University. The next year, she received her M.A. from Harvard as well as a Leopold Schepp Foundation Fellowship to continue with doctoral study. While pursuing her doctorate, she held a research assistantship in the physics department at MIT. In the middle of her doctoral studies, she married T. T. Wu in 1967. She earned her Ph.D. in physics from Harvard in 1970. Wu remained at MIT as a research associate until 1972 and then as a research physicist until 1977.

While at MIT, Wu's team discovered the J/psi particle, or a bound state of charm and anticharm quarks, in 1974. This discovery paved the way to comprehending the fundamental structure of matter in terms of leptons and quarks. In 1977, the University of Wisconsin at Madison hired Wu as an assistant professor. While in this position, she conducted research with a team using an electron-positron colliding beam to detect three jet events that confirmed the existence of gluon. The university promoted her to the rank of associate professor in 1980, and in 1983, she became a full professor.

Wu has held several visiting scientist positions in her career: at Deutsche Elektronen Synchrotron from 1970 through 1972 and from 1977 through 1986; and at CERN in Geneva, Switzerland, from 1975 through 1977 and from 1986 on. She also held concurrent positions with Brookhaven National Laboratory from 1972 through 1975, with the European Organization for Nuclear Research from 1975 through 1977 and from 1981 on, and with the U.S. Department of Energy from 1977 on.

Wu has received much recognition in her career. In 1980, the U.S. Department of Energy granted her its Outstanding Junior Investigator Award. The next year, the University of Wisconsin at Madison gave her its Romans Faculty Award. The university named her the Enrico Fermi Professor of Physics in 1990, and the next year, it awarded her the Hilldale Professorship. In 1995, she shared the High Energy and Particle Physics Prize from the European Physical Society with her teammates for the first direct observation of gluon. The American Physical Society elected her as a member in 1992, and the American Academy of Arts and Sciences made her a fellow in 1996. In 1998, the University of Wisconsin conferred on her the Vitas Professorship. Wu holds three named chairs at the university.

Wyse, Rosemary
(1957–)
Scottish
Astrophysicist

Rosemary Wyse has weighed in with her expert opinion on some of the most problematic issues in modern astrophysics—galaxy bulges and dark matter. With both issues, she has not taken sides with competing theories; rather, she has continued to conduct research that reveals the complexity of galactic structure, which individual theories cannot fully contain. In 1986, the American Association of University Women granted her the ANNIE JUMP CANNON Award, the most prestigious prize for women astronomers.

Wyse was born on January 26, 1957, in Scotland. She studied physics and astrophysics at Queen Mary College of the University of London, where she earned her bachelor of science degree with first-class honors in 1977. She then pursued graduate study in theoretical physics in the department of applied mathematics at the University of Cambridge, passing part three of the mathematics Tripos examination (the equivalent of a master's degree) with distinction in 1978.

Wyse supported her doctoral study at the Institute of Astronomy at the University of Cambridge through a series of scholarships and fellowships: She held the Bachelor Scholarship at Emmanuel College from 1978 through 1981; the Amelia Earhart Fellowship of ZONTA International from 1981 through 1982; and the Lindemann Fellowship of the English Speaking Union of the Commonwealth from 1982 through 1983. That year, she received her Ph.D. from Cambridge in astrophysics.

Wyse spent the next five years in postdoctoral fellowships at the University of California at Berkeley. She first held the Parisot Postdoctoral Fellowship from 1983 through 1985; then from 1985 through 1987, she held two subsequent University of California President's Fellowships. In 1986, she held a concurrent postdoctoral research fellowship in academic affairs with the Space Telescope Science Institute.

In 1988, Johns Hopkins University hired Wyse as an assistant professor in its Department of Physics and

Astronomy. In 1990, Hopkins promoted her to associate professor. That same year, the Alfred P. Sloan Foundation granted her a two-year fellowship. In 1993, Hopkins appointed her to a full professorship, a title she has retained since then.

Wyse has collaborated fruitfully throughout her career, starting at Cambridge, when she worked with mentor Bernard T. Jones on the structure and formation of elliptical galaxies. Together, they published "The Formation of Disc Galaxies" in the April 1983 edition of *Astronomy and Astrophysics,* among other journal articles. Even as she studied at Berkeley, Wyse maintained her Cambridge connections, collaborating with Gerard Gilmore to determine the distribution of metals heavier than hydrogen and helium in the Milky Way. They published "The Chemical Evolution of the Galaxy" under his name first in the August 28, 1986 edition of *Nature* and "The Structure of the Galaxy" under her name first in the June 1990 edition of the *Annals of the New York Academy of Science.*

At Berkeley (and thereafter), Wyse collaborated with Joseph Silk on star formation rates, which build from her work on the chemical composition and structure of galaxies. At Johns Hopkins, she has contributed to the academic dialogue on galaxy bulges and dark matter. In 1994, she published her own theories on the latter issue in an article entitled, "Dark Matter in the Galaxy," which appeared in the *International Journal of Modern Physics.* Wyse continues to search for answers to the perplexities of the universe.

Xie, Xide
(1921–2000)
Chinese
Physicist

Xide Xie helped establish the study of solid-state physics in China. She wrote two important texts on the subject and taught for years at Fudan University. After the Cultural Revolution, she helped reestablish the educational system as the vice president and then president of Fudan University. She also founded the Institute of Modern Physics at the university.

Xie was born in 1921 in South China. Her family moved to Peking, where her father was a professor of physics at Yenching University (now part of Peking University). The Japanese occupation of Peking in July 1937 forced the family to flee. Xie contracted tuberculosis in her hip joint in flight to Guiyang, in the Guizhou Province. During her four-year recuperation, she taught herself English, calculus, and physics.

Xie studied physics at Amoy University, where her father taught. She earned her bachelor of science degree in physics in 1946. She taught for the next year at the University of Shanghai and then traveled to the United States in 1947 to pursue graduate study. She earned her master's degree from Smith College in 1949 and then moved to the Massachusetts Institute of Technology for doctoral study. She wrote her dissertation on the wave function of electrons in very compressed gases to earn her Ph.D. in 1951.

On her way back to China, Xie spent time in England, where she met and married Cao Tianquin, a biochemist. Together they returned to China, where she lectured in physics at Fudan University and he taught at the Shanghai

branch of the Chinese Academy of Science. In 1956, Fudan University promoted her to the rank of associate professor of physics.

In 1958, the Chinese Academy of Sciences named Xie adjunct director of the Shanghai Institute of Technical Physics. That same year, she coauthored with Kun Huang the textbook entitled *Semi-conductor Physics,* which introduced this field in China. She published her next book, *Solid Physics,* in 1962, the same year that she became a professor of physics at Fudan University.

In 1966, the Cultural Revolution erupted, and Mao Tse-tung waged an anti-intellectual war. Both Xie and her husband were removed from their posts and incarcerated; Xie was locked into the Low Temperature Laboratory for nine months; and their 10-year-old son was left alone in their apartment to care for himself. In 1972, she returned to teaching, but her students were ill-educated peasants and soldiers.

As soon as the Cultural Revolution ended in 1976, Xie worked to advance China's standing in physics. Within one year, she founded the Institute of Modern Physics at Fudan University, and a year later, the university appointed her vice president. In 1983, she assumed the presidency of Fudan University. In 1985, she became director of the Center for American Studies at the university, and after ending her presidency in 1988, she remained an adviser to Fudan University.

Xie received honorary doctorates from nine universities in the United States, Canada, Britain, and Japan. She was a fellow of the American Physical Society and the Chinese Academy of Sciences and a foreign honorary member of the American Academy of Arts and Sciences. She died on March 4, 2000, in Shanghai after a long struggle with cancer.

Y

Yalow, Rosalyn Sussman
(1921–   )
American
Medical Physicist

Rosalyn Sussman Yalow was the first woman to receive the Albert Lasker Basic Medical Research Award and the first American woman to receive the Nobel Prize for medicine. These awards recognized her 1959 discovery, in conjunction with her longtime collaborator Dr. Solomon Berson, of radioimmunoassay (RIA), a method of radioactive tagging of minuscule amounts of certain hormones or proteins in blood, allowing for detailed measurements of physiological changes.

Rosalyn Sussman was born on July 19, 1921, in Bronx, New York. Her mother was Clara Zipper and her father was Simon Sussman. She became interested in chemistry at Walton High School and in physics at Hunter College. She graduated from Hunter with honors in both chemistry and physics in 1941. Purdue University rejected her application for graduate study because she was a Jewish woman (unless her Hunter sponsors could guarantee her a job afterward, which they could not).

Sussman worked briefly as a secretary until the University of Illinois offered her a teaching assistantship in physics. On her first day at Champaign-Urbana, she met fellow physics student A. Aaron Yalow, and their 1943 marriage later yielded two children—Benjamin and Elanna. Yalow conducted her doctoral research under Maurice Goldhaber, husband of GERTRUDE SCHARFF GOLDHABER, and earned her Ph.D. in physics in 1945.

Yalow returned to New York City to work as an assistant engineer at the Federal Telecommunications Laboratory for a year. Her alma mater, Hunter College, hired her in 1946 as a lecturer and then an assistant professor in physics. She also worked part time as a consultant for the Radiotherapy Service at Bronx Veterans Administration (VA) Hospital as a consultant. In 1950, she quit Hunter to become a full-time physicist at the VA and the assistant chief of the Radioisotope Services in charge of its development. That year, Berson joined the VA, and they commenced their collaborative research that led to the development of RIA and lasted for 22 years. They alternated whose name appeared first in their joint publications so that they shared equal credit for their work.

Yalow and Berson focused their research on the tracking of insulin in the human body. They realized in 1959 that injected insulin disappeared from the circulation because the human body develops antibodies to the animal insulin. This discovery opened the door for other scientists to utilize RIA methods for other purposes: to screen blood-bank samples for hepatitis; to track levels of drugs or foreign substances in the blood; and to correct hormone levels in infertile couples, to name just a few of the applications. Yalow and Berson chose not to patent RIA in order to make this method available to any who could benefit from it.

In 1968, Mt. Sinai School of Medicine appointed Yalow as a research professor. That same year, she became the acting chief of the radioisotope unit, and two years later, she became chief of the nuclear medicine service at the VA. In 1972, Berson died prematurely, so Yalow became the senior medical investigator. She arranged for the renaming of her lab to the Solomon A. Berson Research Laboratory so that her collaborator's name would continue to grace her papers.

In 1974, Mt. Sinai named her distinguished service professor. In 1979, Albert Einstein College of Medicine of Yeshiva University appointed Yalow as distinguished professor at large, and the next year, she became the chair of the department of clinical science at the Montefiore Hospital and Medical Center. Yalow retired to emerita status at Albert Einstein in 1986.

Yalow received some of the most prestigious awards a scientist can win: the 1961 Eli Lilly Award from the American Diabetes Association; the 1976 Albert Lasker Basic Medical Research Award; the 1977 Nobel Prize for

Rosalyn S. Yalow, who in 1977 became the first American woman to receive the Nobel Prize for medicine. *(AIP Emilio Segrè Visual Archives, Physics Today Collection)*

medicine; and the National Medal of Science in 1988. In 1978, she received the Rosalyn S. Yalow Research and Development Award, named after her. She also received honorary doctorates from 11 universities internationally. However, she refused the 1961 Federal Woman's Award, as she did not wish to be considered for an award as a woman but rather as a scientist. Similarly, she refused a special woman's award from the *Ladies Home Journal,* for she considered such awards reverse discrimination for not judging women's work in direct competition with men's work.

Yener, Kutlu Aslihan
(1946–   )
Turkish/American
Archaeologist

Kutlu Aslihan Yener exploded the Assyrian accounts of the Bronze Age metal trade by locating tin ore, the key to metallurgy, in the Taurus Mountains (or the legendary Silver Mountains of the Akkadians). She thus established Anatolia, not Hindu Kush, as the source of the metals that fueled Bronze Age smelting and artistry. She utilized chromatic dating techniques to pinpoint the actual age of the artifacts she discovered, thus helping her to map the path of metal trade.

Yener was born in Turkey. When she was six months old, her family moved to New York. She entered Adelphi University in 1964 to study chemistry. As the sixties progressed, she grew restless and traveled to her homeland of Turkey. She then transferred to Robert College in Istanbul to study art history.

While visiting Roman ruins on a class trip, the context from which the art was produced fascinated Yener more than the artwork itself, so she decided to fuse her earlier interest in science with her interest in art to become an archaeologist. She determined "to apply atomic bomb techniques to archaeology" by tracing the age of archaeological artifacts through radioactive isotope dating. She obtained her doctorate from Columbia University in 1980.

Yener's goal was to discover who first had the idea to smelt tin-and-copper mixtures, thus identifying the origins of the global economy of the metal trade in the Bronze Age, about 250,000 years before the birth of Christ. She traveled to the Taurus Mountains of Anatolia, which were reputed to be the source of the precious metals that fueled Bronze Age metallurgy, according to Sumerian myth. She led her archaeological team above 8,500 feet, the highest any archaeologist had ever gone. There, she discovered a complex series of mines that had been tapped out. She did, however, find traces of tin, the key to ancient metallurgy.

She spent the next five years searching for virgin tin ore, eventually finding it in a neighboring valley. (Turkish tin ore is burgundy instead of black, complicating her search.) She then applied chromatic dating techniques to archaeological artifacts to discover their age as a means of mapping the routes of metal trade in the ancient world.

In 1994, the University of Chicago hired Yener as an assistant professor in its Department of Near-Eastern Languages and Civilizations and in the Oriental Institute. She was later promoted to associate professor of Anatolian archaeology. Through the university, she negotiated an affiliation with the Argonne National Laboratory whereby she and her Oriental Institute colleagues could use Advanced Photon Source (APS) technology to examine archaeological artifacts with X-ray fluorescence to discover their chemical composition through light spectra.

Yener also uses APS technology to test soil samples for evidence of pollutants resulting from copper and iron smelting, thus allowing her to pinpoint locations where metallurgy took place. Besides performing this research, Yener also teaches, writes, and lectures extensively on her findings. Yener is invigorated by the fact that her research essentially rewrites history, as she discovers concrete evidence to inform our understanding of our past.

Young, Grace Chisholm
(1868–1944)
English
Mathematician

Grace Chisholm Young was the first woman to earn a doctoral degree from the prestigious University of Göttingen. However, she did not enter mathematics directly but rather influenced the field in collaboration with her husband, William Young. Often, they submitted joint work under his name alone in order to get it published. He openly acknowledged her mathematical contributions to their partnership, thus establishing the importance of Young in her own right as an important mathematician.

Grace Emily Chisholm was born on March 15, 1868, in Haslemere, Surrey, England, the youngest of three surviving children. Her mother was Anna Louisa Bell, and her father, Henry William Chisholm, worked his entire career for the British civil service, reaching the rank of warden of the standards of the Department of Weights and Measurements. Her brother, Hugh, served as editor of the 11th edition of the *Encyclopædia Britannica*. She was homeschooled by her mother and a governess.

In 1889, at the age of 21, Chisholm entered Cambridge University's Girton College (Britain's first undergraduate institution devoted to educating women) as the Sir Francis Goldschmid Scholar. Although women could not earn university degrees at the time, she would have graduated first-class according to her score on Cambridge's Tripos examination. On a dare, she took Oxford's final examination unofficially, and she scored higher than any other Oxford student in 1892.

In 1893, Chisholm traveled to Germany to study at the University of Göttingen. She wrote her doctoral dissertation, "The Algebraic Groups of Spherical Trigonometry," under her adviser, the renowned mathematician Felix Klein. Both the decision to admit a woman and the decision to grant her a degree had to be approved by the Berlin Ministry of Culture, and in 1895, Chisholm became the first woman to earn a doctorate from the University of Göttingen, which granted her the degree magna cum laude.

In 1896, Chisholm married William Henry Young, who had tutored her at Girton. Young returned to Göttingen with her new husband, who had secured a position in the mathematics department there. Young raised a family that grew to six children, and she studied anatomy at the university (she eventually completed all requirements for a medical degree except the internship). She also served as her husband's professional collaborator in a unique arrangement where he provided the inspiration and she provided the discipline to follow each idea to its conclusion.

Together with her husband, Young published two influential books, *The First Book of Geometry* in 1905 and *The Theory of Sets of Points* the next year. In 1908, the Youngs moved to Geneva, Switzerland, where Grace would remain with the children while her husband traveled internationally to whatever academic positions could support the family. Before World War I, he taught in Calcutta, India, and after the war, he taught at the University of Wales in Aberystwyth.

Between 1914 and 1916, Young wrote a series of important papers on derivatives of real functions that contributed to the formulation of the Denjoy-Saks-Young theorem. One paper, "A Note on Derivatives and Differential Coefficients," was published under her own name in *Acta Mathematica* in 1914. In 1915, Cambridge University awarded her the Gamble Prize for a paper she wrote on the foundations of calculus. She also tried her hand at writing historical fiction and children's books—in the latter, she introduced principles of science to her young readers.

In the spring of 1940, Young traveled to England just before the Nazi occupation of France, which severed her from her husband in Switzerland. Thus separated, he went senile and died in 1942. Young died of a heart attack in 1944. Ironically, the fellows of Girton College had just named her to receive an honorary degree, which she was unable to accept.

Young, Judith Sharn
(1952–)
American
Astronomer

Judith Sharn Young has been honored as the most promising young woman astronomer with the ANNIE JUMP CANNON Prize and as the best young physicist in the world with the MARIA GOEPPERT-MAYER Award. Over the course of some 14 years, she gathered the most extensive body of data on galaxies. Besides leading a distinguished career in astronomy, she has devoted her later career to biochemistry, looking for a cure for cancer.

Judith Sharn Rubin was born on September 15, 1952, in Washington, D.C. She sprang from scientific genes: Her father, Robert J. Rubin, was a physical chemist, and her mother, VERA COOPER RUBIN, was a prominent astronomer who proved the existence of dark matter. She enjoyed chemistry and biochemistry but was turned on to astronomy when her mother taught her high-school science class about black holes. She attended Radcliffe College and Harvard University, where a professor advised her in her junior year to quit school and get married. Instead, she earned her bachelor of arts degree with honors in 1974.

That year, Rubin enrolled in the graduate program in astronomy at the University of Minnesota. When she informed her professors of her plan to marry, again she had to ignore their advice to quit school. Instead, she switched to the physics department, where she wrote her dissertation, "The Isotopic Composition of the Cosmic Rays Neon Through Nickel," under Phyllis Freier to earn her Ph.D. in 1979. In 1975, she married Michael Young, a graduate geology student, and the couple had one daughter, Laura Rose, born in 1983. When the couple divorced in 1990, Judith retained the name Young, which she had adopted to deflect attention away from her famous lineage.

Young served as a postdoctoral research associate at the Five College Radio Astronomy Observatory of the University of Massachusetts at Amherst from 1979 through 1982. There, she collaborated with Nick Z. Scoville measuring carbon monoxide and cold gas content of galaxies. They discovered that the distribution of gas and light in galaxies is proportional. In 1982, the American Astronomical Society recognized the significance of this work by awarding her the Annie Jump Cannon Prize for promising research by a young woman astronomer.

The University of Massachusetts retained Young as a visiting assistant professor from 1982 through 1984, when the university dropped the *visiting* term from her title. In 1986, the American Physical Society honored Young as the best young physicist in the world by naming her the first recipient of the Maria Goeppert-Mayer Award. That year, she also received a three-year Sloan Research Fellowship. The next year, in 1987, she was promoted to an associate professorship. In 1991, she served as the visiting JCMT fellow at the University of Hawaii, and in 1993, the University of Massachusetts appointed her to a full professorship.

As a full professor, Young had the freedom to choose her own research topics. Consequently, she diverted her research into biochemistry, searching for a cure for cancer. Of course, she maintained her teaching and administrative responsibilities as well as her research duties in the astronomy department. She also promoted a hands-on approach to astronomy by designing and installing a Sunwheel, for observing sunrises and sets.

Young, Lai-Sang
(1952–   )
Chinese
Mathematician

Lai-Sang Young won the 1993 Ruth Lyttle Satter Prize from the American Mathematical Society (AMS) for her work on statistical (or so-called ergodic) properties of dynamical systems. Her research identified a model generated in random that nevertheless followed statistical norms. She thus pinpointed some degree of order amid chaos, an unexpected breakthrough in mathematics. She also won a National Science Foundation Faculty Award for Women Scientists and Engineers. In 1997, she received a prestigious Guggenheim Fellowship, given for "unusually distinguished achievement and exceptional promise for future accomplishment."

Young was born in 1952 in Hong Kong. She received a Cantonese education growing up that including classes taught in English. She immigrated to the United States, where she attended the University of Wisconsin at Madison. She earned her Bachelor of Science degree there in 1973 and then pursued graduate studies at the University of California at Berkeley. She earned her master's degree there in 1976 and her doctorate in 1978.

That year, Young commenced her professional academic career at Northwestern University. From there, she went to Michigan State University in 1980. She then held two concurrent positions at the University of Arizona and at the University of California at Los Angeles (UCLA). She returned to Berkeley in a visiting position at the Mathematical Sciences Research Institute in 1983 through 1984. In 1985, she received a Sloan Fellowship that she used to travel to the Universitat Bielefeld in Germany, and she later taught at the University of Warwick in England. In 1989, she held a visiting position at the Institute for Advanced Studies at Princeton University.

Young's breakthrough involved her identification of a model that is both random and yet statistically normal. In this sense, she discovered some degree of order in chaos. She came about this by studying the probability of repetition in a model as well as the possibility of replication at different scales. Young also focused her attention on "strange attractors," specifically Henon attractors that graph out to look like a boomerang. The simplicity of this model startled her mathematical colleagues. She later generated work with quadratic maps that reproduced similarly startling and simple results that walked the line between random and predictable.

The AMS invited Young to give the keynote address at its 1985 meeting. In 1994, she returned to the University of Warwick in England to present a workshop on lattice dynamics and ergodic theory at the Mathematics Research Centre. She continued the tradition of aiding in the selection of the next Ruth Lyttle Satter Prize winner in 1995, SUN-YOUNG ALICE CHANG, just as the first recipient, MARGARET DUSA MCDUFF, had helped pick Young.

In 1997, Young received a prestigious Guggenheim Fellowship that helped finance her research while she worked as a professor of mathematics at UCLA. That year, she also organized the Summer Research Institute on differential geometry and control at the University of Colorado as a member of the AMS Committee on Summer Institutes and Special Symposia.

Young, Roger Arliner
(1889–1964)
American
Zoologist and Biologist

Roger Arliner Young was hailed as a "genius in zoology" early in her career by her mentor, Ernest Everett Just. She won international acclaim for her first published paper, and she was the first African-American woman to earn a doctorate in zoology. However, her career derailed under professional and personal pressures.

Young was born in 1889 in Clifton Forge, Virginia. She grew up in Burgettstown, Pennsylvania, and in 1916, she matriculated at Howard University to major in music. Just's general biology course, which she attended in her junior year as her first science course, peaked her interest in the subject. Although she earned only a C, Just recognized her potential and took her under his wing. When she graduated with a baccalaureate degree in 1923, Just hired her as an assistant professor of zoology.

Young conducted pioneering research on paramecium, publishing her results in an article entitled "On the Excretory Apparatus in Paramecium" in the September 12, 1924, edition of *Science*. The renowned Russian cell physiologist Dimitry Nasconov published similar results two months later, casting a favorable light on Young for anticipating such important research as a young scientist who had yet to earn a doctorate. She commenced graduate study part-time at the University of Chicago in the summer of 1924 and received her master's degree in zoology in 1926.

In the summers of 1927 and 1928, Young served as Just's research assistant at the Marine Biological Laboratory in Woods Hole, Massachusetts, where she was the first African-American woman to work in the scientific community there. She collaborated with Just in research on the effects of ultraviolet light on the development of marine eggs of the genera *Nereis, Plat nereis, Arbacia,* and *Chaetopterus.* When Just traveled to Europe to conduct research in 1929, Young assumed his teaching duties as well as his responsibilities as head of the zoology department at Howard.

During Just's absence, Young collaborated with his former mentor, University of Chicago embryologist Frank Lil-lie, at Woods Hole in the summer of 1929. That fall, she received a General Education Board Fellowship to pursue doctoral study at the University of Chicago. However, Young's assumption of Just's responsibilities left scant time for her to prepare for her qualifying examinations, which she failed, leading to her dismissal from the doctoral program.

Just had been priming Young for a prominent role in Howard's zoology department, which had received a five-year, $80,000 Julius Rosenwald Fund Grant to create a new professorship, among other things. Although rumors of a romantic relationship between Just and Young cannot be substantiated, this would help to explain her academic collapse, as he left for Europe yet again just prior to her examination. Whatever the cause of Young's failure in Chicago, she fell from grace with Just, her main supporter. Just fired her in 1936, ostensibly for missing classes and mishandling laboratory equipment.

In the meanwhile, Young had continued to conduct significant research, collaborating with V. L. Heilbrunn and Donald P. Costello in Just's absence from Woods Hole in the summer of 1930. This threesome took up the marine egg studies Young had commenced there with Just, and they published the results in a series of articles. After Young's dismissal from Howard, Heilbrunn invited her into the doctoral program at the University of Pennsylvania, securing a two-year General Education Board Grant for her. She wrote her dissertation on "The Indirect Effects of Roentgen Rays on Certain Marine Eggs" to earn her Ph.D. in 1940.

Instead of launching into a successful career, Young began to experience many personal and professional difficulties. She could not hold a job: She worked first as an assistant professor at the North Carolina College for Negroes, then headed the biology department at Shaw University, then returned to North Carolina College in 1947 as a professor of biology. She then bounced from Paul Quinn College in Texas to a temporary position at Jackson State College in Mississippi to a temporary visiting lectureship at Southern University in Louisiana.

While at Jackson State, Young admitted herself into the Mississippi Mental Asylum. She was released in 1962 and lived the last two years of her life in poverty. She died on November 9, 1964.

Z

Zakrzewska, Marie Elizabeth
(1829–1902)
German
Physician

Marie Elizabeth Zakrzewska benefited women and children not only through her own ministrations as an obstetrician, gynecologist, and general practitioner but also by training several generations of woman doctors at the hospitals she founded: the New York Infirmary for Women and Children (now Beth Israel Medical Center) and the New England Hospital for Women and Children (now the Dimock Community Health Center). She is credited with opening the door for women to practice medicine in the United States through her dogged determination and high standards.

Zakrzewska was born on September 6, 1829, in Berlin, Germany. She was the oldest child of five daughters and one son born to Frederika C. W. Urban, a gypsy of the Lombardi tribe, and Martin Ludwig Zakrzewska, a Prussian army officer discharged for his liberal views. Zakrzewska accompanied her mother to study midwifery at Berlin's Charité Hospital, where she came to the attention of Dr. Joseph Hermann Schmidt.

Eighteen-year-old Zakrzewska's application for admission to the midwifery program was at first rejected due to her young age, but Schmidt secured a position for her when she was 20. In her second year, she became his teaching assistant, and a year after she graduated in 1851, Schmidt appointed Zakrzewska chief midwife and professor of midwifery just hours before he died. Suspicious of Schmidt's motives for appointing his young female protégé, Zakrzewska's colleagues ostracized her despite her excellent teaching and midwifery. She resigned after only half a year, and in 1853, she immigrated to the United States with two of her sisters.

Zakrzewska and her sisters survived for a year on meager proceeds from knitting. In 1854, Zakrzewska met Dr.

ELIZABETH BLACKWELL, a pioneer in women's medicine who recommended the young German for admission to Cleveland Medical College (or Western Reserve College). Zakrzewska earned her M.D. in 1856 and returned to New York. There, she set up an office in Blackwell's back parlor, and the two woman doctors pursued plans to found a hospital devoted to employing and training women doctors, who could find no such opportunities elsewhere.

On May 1, 1857, the New York Infirmary for Women and Children opened. Zakrzewska had handled fund-raising from Boston to Philadelphia, and she acted as both resident physician and general manager for the hospital's first two years. In 1859, she moved to Boston where the New England Female Medical College had appointed her professor of obstetrics and diseases of women and children, and as resident physician, when it opened its new hospital. Samuel Gregory had founded the college the year before, but unfortunately, his mismanagement sunk the school's standards and financial status.

When Zakrzewska resigned in 1862, Gregory abolished the college's clinical department as well as its board of lady managers. Unbeknownst to Gregory, Zakrzewska and the board had prepared the clinical department for conversion into a teaching hospital, the New England Hospital for Women and Children, which opened on July 1, 1862. What started as a 10-bed facility grew into a larger building two years later, and in 1872, the hospital moved to a nine-acre site in Roxbury.

Soon thereafter, Zakrzewska handed over her title of resident physician to Dr. Lucy Sewall to become the attending physician until 1887, when she became advisory physician. She also built up an extensive private gynecological practice, and she continued as a consulting physician at her hospital into her seventies despite suffering from arteriosclerosis and heart disease. In 1899, Zakrzewska retired, and on May 12, 1902, she died of apoplexy in her Jamaica Plains home.

Zoback, Mary Lou
(1952–   )
American
Geophysicist

Mary Lou Zoback specializes in the study of active tectonics, and she has influenced her field not only through her research but also through her administrative work. She chaired the World Stress Map Project of the International Lithosphere Program. In 1995, the National Academy of Sciences voted her a member, and President Bill Clinton later appointed her to the committee that chose the National Medal of Science winners.

Mary Lou Chetlain was born on July 5, 1952, in Sanford, Florida. Her father was a county commissioner and a sports editor for the local newspaper. She commenced her undergraduate study in oceanography, but a mechanical engineering professor teaching a course on elasticity and plate tectonics inspired her to transfer to Stanford University in her junior year to study geology and geophysics. In 1973, during her senior year in college, she married Mark D. Zoback, and together the couple eventually had two children, Eli and Megan.

A year after getting married, Zoback graduated from Stanford with a bachelor of science degree, and she remained there for graduate study. She earned her master of science degree in 1975 and her Ph.D. in geophysics in 1978. She conducted postdoctoral research on a National Research Council Fellowship with the United States Geological Survey (USGS) Heat Flow Studies Group from 1978 through 1979. The USGS retained her as a geophysicist thereafter at its Menlo Park, California, office of the Western Earthquake Hazards Team. She spent one year on sabbatical from 1990 through 1991 in Karlsruhe, Germany, under a USGS Gilbert Fellowship Award.

Zoback has focused her research on active tectonics, specifically considering the relationship between the stresses in the Earth's crust and the incidence of earthquakes. Living in California, she conducts much of her research on the San Andreas fault system. Her expertise has prompted her appointment to many influential positions in the field of geophysics. From 1986 through 1992, she chaired the World Stress Map Project of the International Lithosphere Program, which required her to coordinate the efforts of some 40 scientists from 30 different countries to gather and interpret data in order to generate a comprehensive picture of the lithosphere according to the tectonic stresses of various magnitudes.

Zoback has served her field in other capacities as well. She served on the United States Geodynamics Committee of the National Research Council from 1985 through 1989. She worked as a member of the National Research Council's Panel on Coupled Hydrologic/Tectonic/Hydrothermal Systems at Yucca Mountain in Nevada from 1990 through 1992. She served as the president of the Tectonophysics Section of the American Geophysical Union as well as presiding over the Cordilleran Section and the Geophysics Division of the Geological Society of America.

The importance of Zoback's work has received ample recognition. In 1987, the American Geophysical Union granted her its Macelwane Award. In 1995, the National Academy of Sciences inducted her as a member, one of the most prestigious memberships in the scientific community in the United States. President Clinton appointed her to serve on his committee to choose the recipients of the National Medal of Science.

Zuber, Maria T.
(1958–   )
American
Geophysicist

Maria T. Zuber was named to an endowed chair at Johns Hopkins University even before she reached the rank of full professor, and then three years after joining the Massachusetts Institute of Technology (MIT) as a professor, MIT named her to an endowed chair there. She has received numerous awards for her undergraduate teaching and research mentoring as well as for her contributions to the Mars Global Surveyor Science Team. In 1987, she discovered an asteroid.

Born on June 27, 1958, in Norristown, Pennsylvania, Zuber grew up in Pennsylvania and attended the University of Pennsylvania for her undergraduate study. She wrote her honors thesis, entitled "Velocity-Inclination Correlations in

Geophysicist Maria Zuber, who discovered an asteroid in 1987, one year before her 30th birthday. *(Courtesy of Maria Zuber)*

Galactic Clusters," to earn her bachelor of arts degree in astrophysics and geology. She then pursued graduate study at Brown University as a research assistant in the Department of Geological Sciences, earning her master of science degree in 1983. She conducted doctoral research under her advisor, E. M. Parmentier, and wrote her doctoral thesis on "Unstable Deformation in Layered Media: Application to Planetary Lithospheres" to earn her Ph.D. in geophysics from Brown in 1986.

Zuber commenced her career-long working relationship with the National Aeronautical and Space Administration (NASA) at its Goddard Space Flight Center as a National Research Council research associate in the geodynamics branch during her final year of doctoral study. Upon becoming a doctor, Zuber remained at Goddard as a geophysicist, a position she retained until 1992. In 1990, she served as a visiting assistant professor of geophysics in Johns Hopkins's Department of Earth and Planetary Sciences, and the next year, she joined the university's faculty full-time as an associate research professor of geophysics.

In 1993, Johns Hopkins honored Zuber by naming her the first Second Decade Society term associate professor, a title endowed by Johns Hopkins alumni from the years 1970 through 1986 to recognize tenure-track assistant or associate professors committed to undergraduate teaching and scholarly research. In 1995, the university promoted Zuber to the rank of professor before her title, which lasted from three to five years, even expired.

That same year, though, MIT hired Zuber as a professor of geophysics and planetary science. In the meanwhile,

Zuber had rejoined NASA in 1994 as a senior research scientist in Goddard's laboratory for terrestrial physics, a concurrent position she retained thereafter. During the summers from 1996 through 1999, she served as a guest investigator at the Woods Hole Oceanographic Institution. In 1998, MIT named her to an endowed chair as the E. A. Griswold professor.

In her research, Zuber generates theoretical models of geophysical processes. She also analyzes the geophysical conditions of altimetry, gravity, and tectonics to uncover the evolution and structure of Earth as well as other similar solid planets. In her teaching, she offers courses such as Introduction to Planetary Science, Techniques in Remote Sensing, Numerical Geodynamical Modeling, Current Topics in Planetary Sciences, and Marine Geology and Geophysics.

Zuber's career was amply recognized through honors and awards. NASA granted her numerous awards: the Outstanding Performance Award every year from 1988 through 1992; and the Group Achievement Award in 1991, 1993, 1994, 1998, and 2000. She won the Johns Hopkins University Oraculum Award for Excellence in Undergraduate Teaching in 1994 and the university's 1995 David S. Olton Award for Outstanding Contributions to Undergraduate Student Research. In 1999, she served as the distinguished leaders in science lecturer of the National Academy of Sciences. In January of that same year, she brought her family—her husband, Jack, and her two sons, Jack Jr. and Jordan—to the launching of the Mars Polar Lander, a project that she had worked on extensively.

BIBLIOGRAPHY

Abir-Am, P. G., and Dorinda Outram, eds. *Uneasy Careers and Intimate Lives: Women in Science 1789–1979.* New Brunswick, N.J.: Rutgers University Press, 1987.

Agnes Scott College. "Biographies of Women Mathematicians." Available online. URL: http://www.agnesscott.edulriddle/women/women.htp.

Ainley, Marianne Gosztonyi, ed. *Despite the Odds: Essays on Canadian Women and Science.* Montreal: Véhicule Press, 1990.

Alic, Margaret. *Hypatia's Heritage: A History of Women in Science from Antiquity Through the Nineteenth Century.* Boston: Beacon Press, 1986.

Ambrosser, Susan, ed. *No Universal Constants: Journeys of Women in Science and Engineering.* Philadelphia: Temple University Press, 1990.

American Association for the Advancement of Science. *Science in Africa: Women Leading from Strength.* Washington, D.C.: American Association for the Advancement of Science, 1993.

American Men and Women of Science, 1995–96: A Biographical Directory of Today's Leaders in Physical, Biological, and Related Sciences. New York: R. R. Bowker, 1994.

Arnold, Lois Barber. *Four Lives in Science: Women's Education in the Nineteenth Century.* New York: Schocken Books, 1984.

Asimov, Isaac. *Asimov's Biographical Encyclopedia of Science and Technology: The Lives and Achievements of 1510 Great Scientists from Ancient Times to the Present Chronologically Arranged.* New York: Doubleday, 1982.

Association of Women in Mathematics. "Women in Mathematics." Available online. URL: http://www.awm-math.org.

Bailey, B. *The Remarkable Lives of 100 Women Healers and Scientists.* Holbrook, Mass.: Bob Adams, 1994.

Bailey, M. J. *American Women in Science.* Santa Barbara, Calif.: ABC-CLIO, Inc., 1994.

Barnhart, John H. *Biographical Notes upon Botanists.* Boston: G. K. Hall, 1965.

The Biographical Dictionary of Scientists. New York: P. Bedrick Books, 1983–85.

Bleier, Ruth, ed. *Feminist Approaches to Science.* New York: Pergamon Press, 1986.

———. *Science and Gender.* New York: Pergamon Press, 1984.

Brooke, Elisabeth. *Women Healers: Portraits of Herbalists, Physicians, and Midwives.* Rochester, Vt.: Healing Arts Press, 1995.

Camp, Carole A. *American Astronomers: Searchers and Wanderers.* Springfield, N.J.: Enslow, 1996.

Chaff, Sandra L., et al., eds. *Women in Medicine: A Bibliography of the Literature on Women Physicians.* Metuchen, N.J.: Scarecrow Press, 1977.

Chang, Hsiao-lin. *Vital Force on the Development of the Chinese Nation: Women Scientists and Technicians in China.* China: China Intercontinental Press, 1995.

Chipman, Susan F., et al., eds. *Women and Mathematics: Balancing the Equation.* Mahwah, N.J.: Lawrence Erlbaum Associated, 1985.

Concise Dictionary of Scientific Biography. New York: Scribner, 1981.

Cooney, Miriam P. *Celebrating Women in Mathematics and Science.* Reston, Va.: National Council of Teachers of Mathematics, 1996.

Current Biography. New York: H. W. Wilson, 1940– .

CWP (Contributions of Women to Physics). Available online. URL: http://www.physics.ucla.edu/~cwp.html.

Daintith, John, Sarah Mitchell, Elizabeth Tootill, and Derek Gjertsen, eds. *Biographical Encyclopedia of Scientists.* Philadelphia: Institute of Physics Publishing, 1994.

Dash, Joan. *The Triumph of Discovery: Women Scientists Who Won the Nobel Prize.* Englewood Cliffs, N.J.: Julian Messner, 1991.

Dictionary of Scientific Biography. New York: Scribner's, 1970–1980.

Discover. "Special Issue: A Celebration of Women in Science." December 1991.

Eric's Treasure Trove of Scientific Biographies. Available online. URL: http:// www.treasure-troves.com.

Epstein, Vivian Sheldon. *History of Women in Science for Young People.* Denver, Colo.: VSE Publisher, 1994.

Etzkowitz, Henry, et al. "The Paradox of Critical Mass for Women in Science." *Science,* October 7, 1994, 51–54.

Fausto-Sterling, Anne, and Lydia L. English. *Women and Minorities in Science: Course Materials Guide.* Providence, R.I.: Brown University, 1990.

Fort, Deborah C., ed. *A Hand Up: Women Mentoring Women in Science.* Washington, D.C.: Association for Women in Science, 1993.

Gacs, Uta, et al., eds. *Women Anthropologists: A Biographical Dictionary.* New York: Greenwood Press, 1988.

Gaillard, Jacques. *Scientists in the Third World.* Lexington: University Press of Kentucky, 1991.

Gornick, Vivian. *Women in Science: One Hundred Journeys into the Territory.* New York: Simon & Schuster, 1990.

Griffiths, Sian, ed. *Beyond the Glass Ceiling: Forty Women Whose Ideas Shape the Modern World.* Manchester, England: Manchester University Press, 1996.

Herzenberg, Caroline L. *Women Scientists from Antiquity to the Present: An Index.* West Cornwall, Conn.: Locust Hill, 1986.

Holloway, Marguerite. "A Lab of Her Own." *Scientific American,* November 1993, 94–103.

Hsiao, T. C., ed. *Who's Who in Computer Education and Research: U.S. Edition.* Latham, N.Y.: Science and Technology Press, 1975.

Kass-Simon, G. and Patricia Farnes. *Women of Science: Righting the Record.* Bloomington: Indiana University Press, 1990.

Keller, Evelyn Fox. *Reflections on Gender and Science.* New Haven, Conn.: Yale University Press, 1985.

Kelly, Farley, ed. *On the Edge of Discovery.* Melbourne, Australia: Text Publishing Company, 1993.

Kessler, James H., Jerry S. Kidd, Renee A. Kidd, Katherine Morin, and Tracy More, eds. *Distinguished African American Scientists of the 20th Century.* Phoenix, Ariz.: Oryx Press, 1996.

Krapp, Kristine M. *Notable Black American Scientists.* Detroit, Mich.: Gale Research, 1998.

Krishnaraj, Maithreyi. *Women and Science: Selected Essays.* Bombay, India: Himalaya Publishing House, 1991.

Lindop, Laurie. *Dynamic Modern Women: Scientists and Doctors.* New York: Holt, 1997.

McGrayne, Sharon Bertsch. *Nobel Prize Women in Science: Their Lives, Struggles, and Momentous Discoveries.* Seacaucus, N.J.: Carol Publishing Group, 1993.

McMurray, Emily J., and Donna Olendorf, eds. *Notable Twentieth Century Scientists.* Detroit, Mich.: Gale Research, 1995.

Millar, David, Ian Millar, John Millar, and Margaret Millar, eds. *The Cambridge Dictionary of Scientists.* New York: Cambridge University Press. 1996.

Morse, Mary. *Women Changing Science: Voices from a Field in Transition.* New York: Plenum Press, 1995.

The Nobel Foundation. "Nobel Laureates." Available online. URL: http://www.nobel.com.

Ogilvie, Marilyn Bailey. *Women in Science: Antiquity Through the Nineteenth Century.* Cambridge, Mass.: MIT Press, 1986.

Parker, Lesley H., et al., eds. *Gender, Science, and Mathematics: Shortening the Shadow.* Norwell, Mass.: Kluwer Academic Publishers, 1995.

Perl, Teri H. *Women and Numbers: Lives of Women Mathematicians.* San Carlos, Calif.: Wide World Publishing/Tetra, 1993.

Porter, Roy, ed. *The Biographical Dictionary of Scientists.* New York: Oxford University Press, 1994.

Read, Phyllis J., and Bernard L. Whitleib. *The Book of Women's Firsts.* New York, Random House, 1992.

Reed, Elizabeth Wagner. *American Women in Science Before the Civil War.* Minneapolis: University of Minnesota Press, 1992.

Rosser, Sue V. *Biology and Feminism: A Dynamic Interaction.* New York: Twayne, 1992.

———. *Female-Friendly Science.* New York: Pergamon Press, 1995.

Rubin, Vera. "Women's Work." *Science 86,* July–August 1986, 58–65.

Shearer, B. F. and B. S. Shearer, eds. *Notable Women in the Life Sciences.* Westport, Conn.: Greenwood Press, 1996.

———. *Notable Women in the Physical Sciences.* Westport, Conn.: Greenwood Press, 1997.

Shepard, Linda. *Lifting the Veil: The Feminine Face of Science.* Boston: Shambhala, 1993.

Sirch, Willow Ann. *Eco-Women: Protectors of the Earth.* Golden, Colo.: Fulcrum Publishing, 1996.

St. Andrews. "History of Science." Available online. URL: http://www-history.mcs.st-andrews.ac.uk/history/.

Uglow, Jennifer S. *International Dictionary of Women's Biography.* New York: Continuum, 1985.

University of California-Los Angeles. "Women in Science and Engineering." Available online. URL: http://www.physics.ucla.edu/~cwp/Phase2/.

Veglahn, Nancy J. *Women Scientists.* New York: Facts On File, 1991.

Weisberger, Robert A. *The Challenged Scientists: Disabilities and the Triumph of Excellence.* New York: Praeger, 1991.

Who's Who in Science and Engineering, 1994–1995. New Providence, N.J.: Marquis Who's Who, 1994.

Who's Who in Science in Europe. Essex, England: Longman, 1994.

Yount, Lisa. *A to Z of Women in Science and Math.* New York: Facts On File, 1998.

Zuckerman, Harriet, Jonathan R. Cole, and John T. Bruer, eds. *The Outer Circle: Women in the Scientific Community.* New York: W. W. Norton, 1991.

ENTRIES BY FIELD

ENTRIES BY COUNTRY OF BIRTH

ENTRIES BY COUNTRY OF MAJOR SCIENTIFIC ACTIVITY

ENTRIES BY YEAR OF BIRTH

CHRONOLOGY

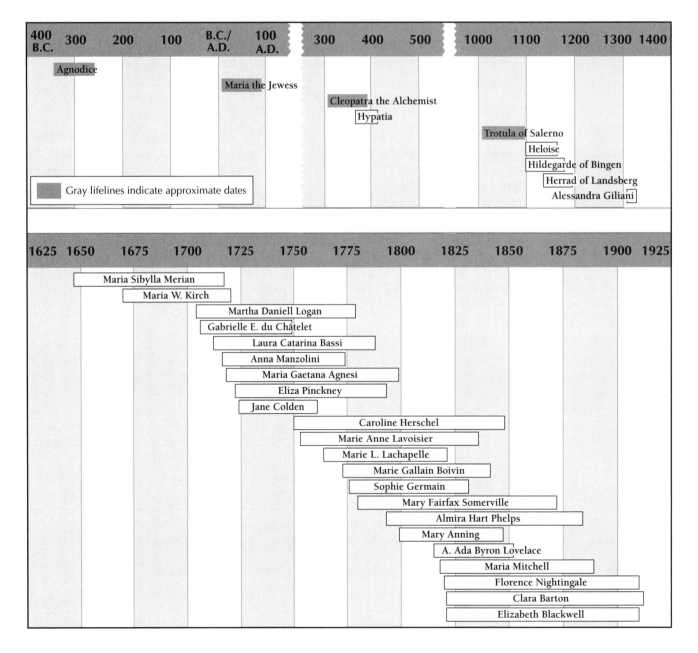

| 400 B.C. | 300 | 200 | 100 | B.C./ A.D. | 100 A.D. | | 300 | 400 | 500 | | 1000 | 1100 | 1200 | 1300 | 1400 |

Agnodice

Maria the Jewess

Cleopatra the Alchemist

Hypatia

Trotula of Salerno

Heloise

Hildegarde of Bingen

Herrad of Landsberg

Alessandra Giliani

Gray lifelines indicate approximate dates

| 1625 | 1650 | 1675 | 1700 | 1725 | 1750 | 1775 | 1800 | 1825 | 1850 | 1875 | 1900 | 1925 |

Maria Sibylla Merian

Maria W. Kirch

Martha Daniell Logan

Gabrielle E. du Châtelet

Laura Catarina Bassi

Anna Manzolini

Maria Gaetana Agnesi

Eliza Pinckney

Jane Colden

Caroline Herschel

Marie Anne Lavoisier

Marie L. Lachapelle

Marie Gallain Boivin

Sophie Germain

Mary Fairfax Somerville

Almira Hart Phelps

Mary Anning

A. Ada Byron Lovelace

Maria Mitchell

Florence Nightingale

Clara Barton

Elizabeth Blackwell

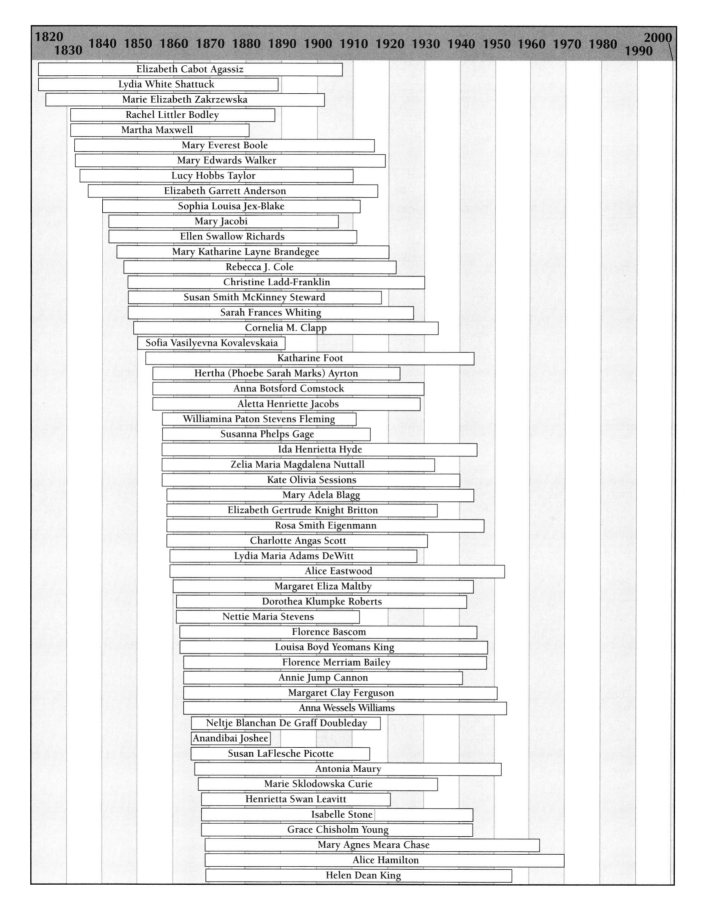

| 1820 | 1840 | 1850 | 1860 | 1870 | 1880 | 1890 | 1900 | 1910 | 1920 | 1930 | 1940 | 1950 | 1960 | 1970 | 1980 | 2000 |
| 1830 | | | | | | | | | | | | | | | 1990 | |

Elizabeth Cabot Agassiz
Lydia White Shattuck
Marie Elizabeth Zakrzewska
Rachel Littler Bodley
Martha Maxwell
Mary Everest Boole
Mary Edwards Walker
Lucy Hobbs Taylor
Elizabeth Garrett Anderson
Sophia Louisa Jex-Blake
Mary Jacobi
Ellen Swallow Richards
Mary Katharine Layne Brandegee
Rebecca J. Cole
Christine Ladd-Franklin
Susan Smith McKinney Steward
Sarah Frances Whiting
Cornelia M. Clapp
Sofia Vasilyevna Kovalevskaia
Katharine Foot
Hertha (Phoebe Sarah Marks) Ayrton
Anna Botsford Comstock
Aletta Henriette Jacobs
Williamina Paton Stevens Fleming
Susanna Phelps Gage
Ida Henrietta Hyde
Zelia Maria Magdalena Nuttall
Kate Olivia Sessions
Mary Adela Blagg
Elizabeth Gertrude Knight Britton
Rosa Smith Eigenmann
Charlotte Angas Scott
Lydia Maria Adams DeWitt
Alice Eastwood
Margaret Eliza Maltby
Dorothea Klumpke Roberts
Nettie Maria Stevens
Florence Bascom
Louisa Boyd Yeomans King
Florence Merriam Bailey
Annie Jump Cannon
Margaret Clay Ferguson
Anna Wessels Williams
Neltje Blanchan De Graff Doubleday
Anandibai Joshee
Susan LaFlesche Picotte
Antonia Maury
Marie Sklodowska Curie
Henrietta Swan Leavitt
Isabelle Stone
Grace Chisholm Young
Mary Agnes Meara Chase
Alice Hamilton
Helen Dean King

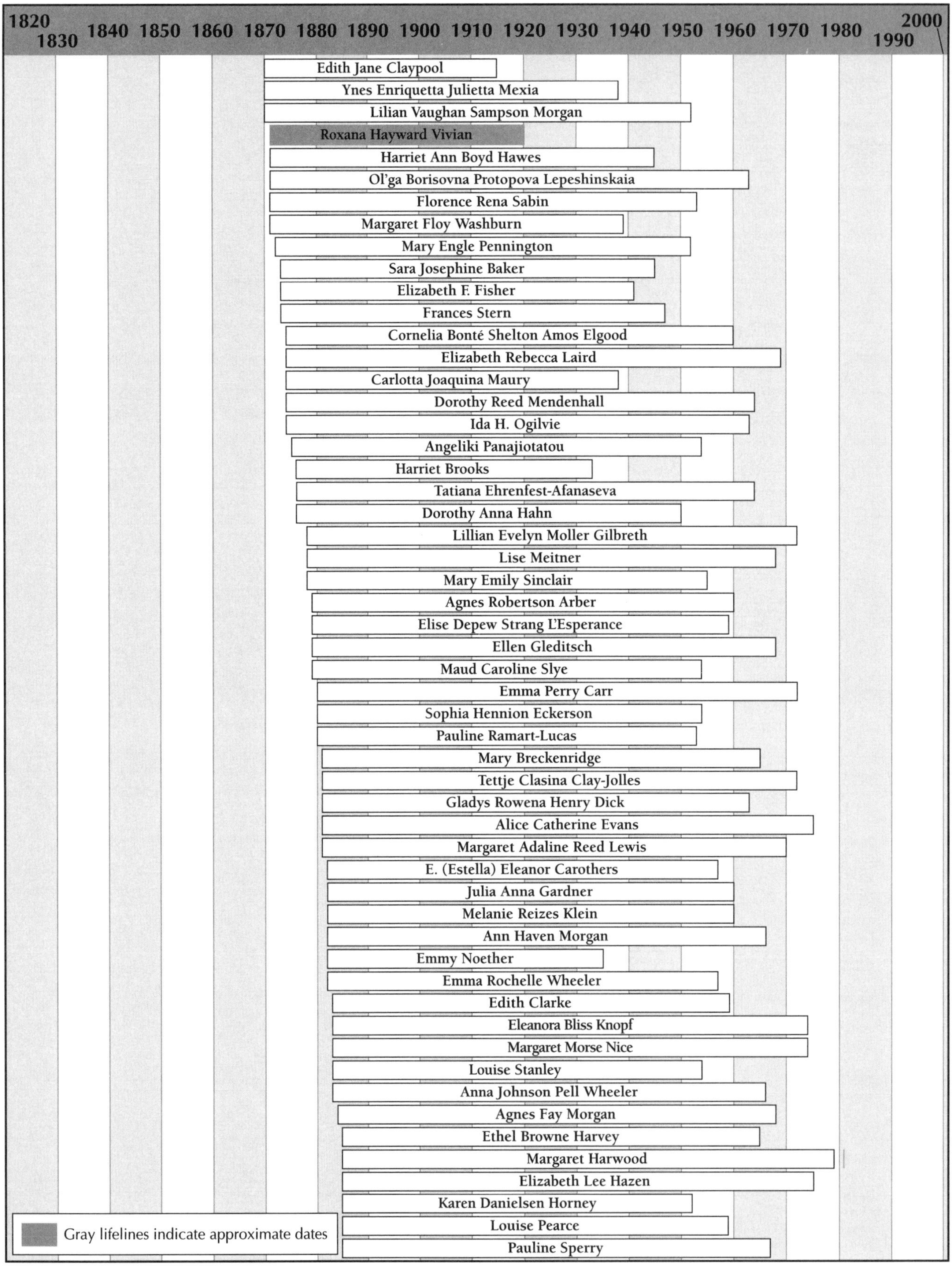

1820	1840	1850	1860	1870	1880	1890	1900	1910	1920	1930	1940	1950	1960	1970	1980	2000
1830															1990	

Edith Jane Claypool

Ynes Enriquetta Julietta Mexia

Lilian Vaughan Sampson Morgan

Roxana Hayward Vivian

Harriet Ann Boyd Hawes

Ol'ga Borisovna Protopova Lepeshinskaia

Florence Rena Sabin

Margaret Floy Washburn

Mary Engle Pennington

Sara Josephine Baker

Elizabeth F. Fisher

Frances Stern

Cornelia Bonté Shelton Amos Elgood

Elizabeth Rebecca Laird

Carlotta Joaquina Maury

Dorothy Reed Mendenhall

Ida H. Ogilvie

Angeliki Panajiotatou

Harriet Brooks

Tatiana Ehrenfest-Afanaseva

Dorothy Anna Hahn

Lillian Evelyn Moller Gilbreth

Lise Meitner

Mary Emily Sinclair

Agnes Robertson Arber

Elise Depew Strang L'Esperance

Ellen Gleditsch

Maud Caroline Slye

Emma Perry Carr

Sophia Hennion Eckerson

Pauline Ramart-Lucas

Mary Breckenridge

Tettje Clasina Clay-Jolles

Gladys Rowena Henry Dick

Alice Catherine Evans

Margaret Adaline Reed Lewis

E. (Estella) Eleanor Carothers

Julia Anna Gardner

Melanie Reizes Klein

Ann Haven Morgan

Emmy Noether

Emma Rochelle Wheeler

Edith Clarke

Eleanora Bliss Knopf

Margaret Morse Nice

Louise Stanley

Anna Johnson Pell Wheeler

Agnes Fay Morgan

Ethel Browne Harvey

Margaret Harwood

Elizabeth Lee Hazen

Karen Danielsen Horney

Louise Pearce

Pauline Sperry

Gray lifelines indicate approximate dates

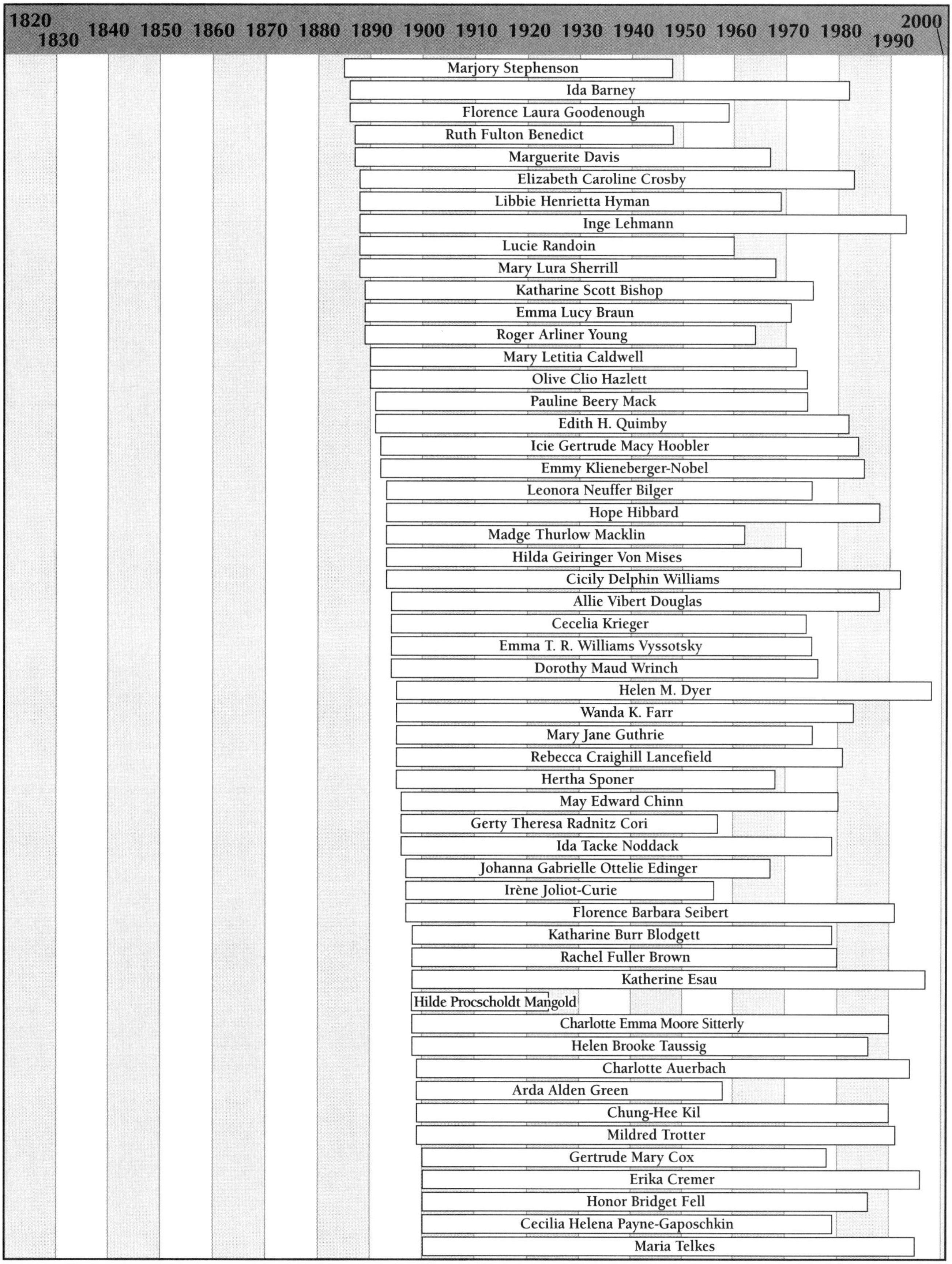

1820	1840	1850	1860	1870	1880	1890	1900	1910	1920	1930	1940	1950	1960	1970	1980	2000
1830															1990	

Marjory Stephenson

Ida Barney

Florence Laura Goodenough

Ruth Fulton Benedict

Marguerite Davis

Elizabeth Caroline Crosby

Libbie Henrietta Hyman

Inge Lehmann

Lucie Randoin

Mary Lura Sherrill

Katharine Scott Bishop

Emma Lucy Braun

Roger Arliner Young

Mary Letitia Caldwell

Olive Clio Hazlett

Pauline Beery Mack

Edith H. Quimby

Icie Gertrude Macy Hoobler

Emmy Klieneberger-Nobel

Leonora Neuffer Bilger

Hope Hibbard

Madge Thurlow Macklin

Hilda Geiringer Von Mises

Cicily Delphin Williams

Allie Vibert Douglas

Cecelia Krieger

Emma T. R. Williams Vyssotsky

Dorothy Maud Wrinch

Helen M. Dyer

Wanda K. Farr

Mary Jane Guthrie

Rebecca Craighill Lancefield

Hertha Sponer

May Edward Chinn

Gerty Theresa Radnitz Cori

Ida Tacke Noddack

Johanna Gabrielle Ottelie Edinger

Irène Joliot-Curie

Florence Barbara Seibert

Katharine Burr Blodgett

Rachel Fuller Brown

Katherine Esau

Hilde Procscholdt Mangold

Charlotte Emma Moore Sitterly

Helen Brooke Taussig

Charlotte Auerbach

Arda Alden Green

Chung-Hee Kil

Mildred Trotter

Gertrude Mary Cox

Erika Cremer

Honor Bridget Fell

Cecilia Helena Payne-Gaposchkin

Maria Telkes

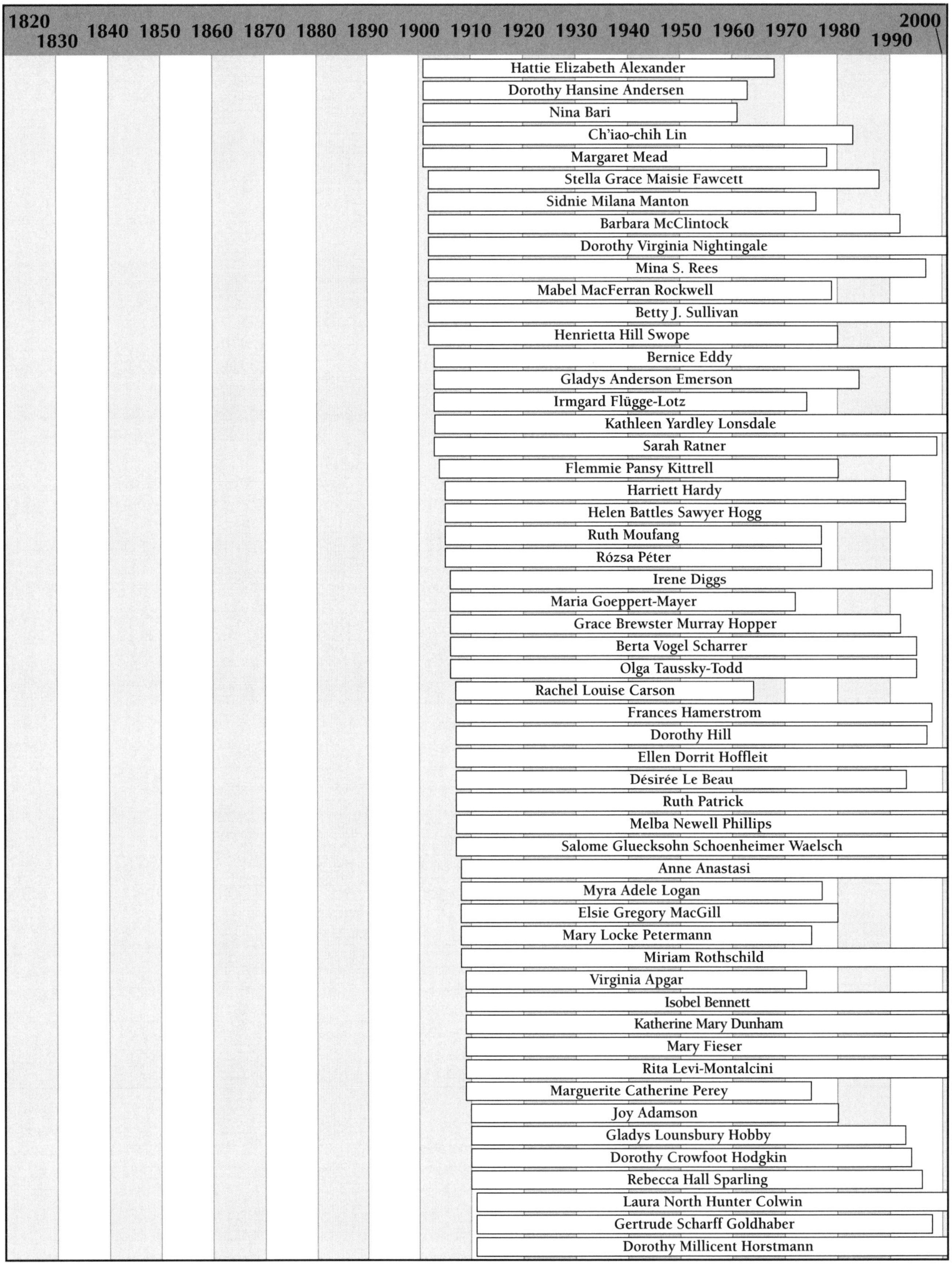

	1820	1840	1850	1860	1870	1880	1890	1900	1910	1920	1930	1940	1950	1960	1970	1980	2000
		1830														1990	

Hattie Elizabeth Alexander

Dorothy Hansine Andersen

Nina Bari

Ch'iao-chih Lin

Margaret Mead

Stella Grace Maisie Fawcett

Sidnie Milana Manton

Barbara McClintock

Dorothy Virginia Nightingale

Mina S. Rees

Mabel MacFerran Rockwell

Betty J. Sullivan

Henrietta Hill Swope

Bernice Eddy

Gladys Anderson Emerson

Irmgard Flügge-Lotz

Kathleen Yardley Lonsdale

Sarah Ratner

Flemmie Pansy Kittrell

Harriett Hardy

Helen Battles Sawyer Hogg

Ruth Moufang

Rózsa Péter

Irene Diggs

Maria Goeppert-Mayer

Grace Brewster Murray Hopper

Berta Vogel Scharrer

Olga Taussky-Todd

Rachel Louise Carson

Frances Hamerstrom

Dorothy Hill

Ellen Dorrit Hoffleit

Désirée Le Beau

Ruth Patrick

Melba Newell Phillips

Salome Gluecksohn Schoenheimer Waelsch

Anne Anastasi

Myra Adele Logan

Elsie Gregory MacGill

Mary Locke Petermann

Miriam Rothschild

Virginia Apgar

Isobel Bennett

Katherine Mary Dunham

Mary Fieser

Rita Levi-Montalcini

Marguerite Catherine Perey

Joy Adamson

Gladys Lounsbury Hobby

Dorothy Crowfoot Hodgkin

Rebecca Hall Sparling

Laura North Hunter Colwin

Gertrude Scharff Goldhaber

Dorothy Millicent Horstmann

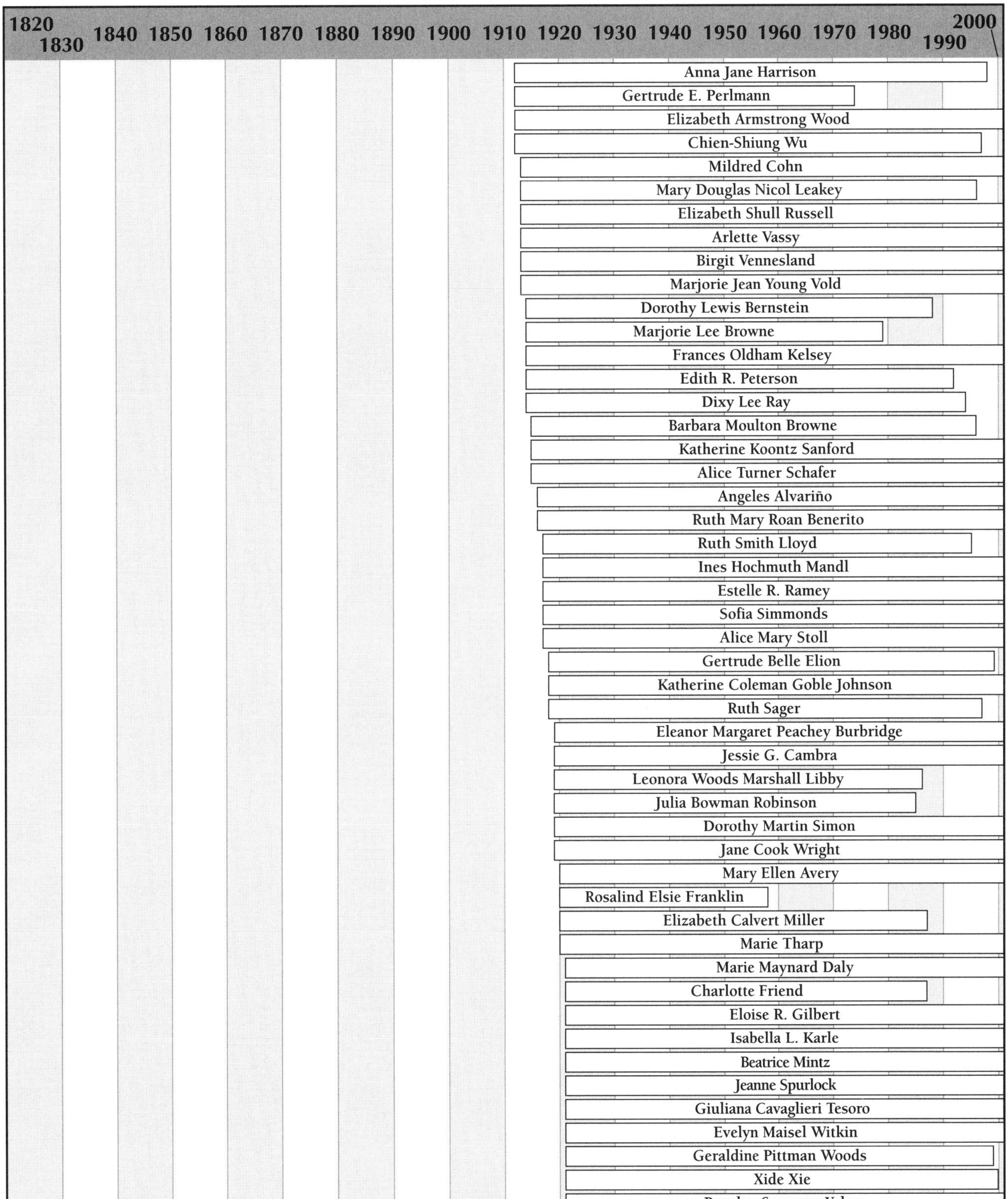

1820 1830	1840	1850	1860	1870	1880	1890	1900	1910	1920	1930	1940	1950	1960	1970	1980	2000 1990

Anna Jane Harrison

Gertrude E. Perlmann

Elizabeth Armstrong Wood

Chien-Shiung Wu

Mildred Cohn

Mary Douglas Nicol Leakey

Elizabeth Shull Russell

Arlette Vassy

Birgit Vennesland

Marjorie Jean Young Vold

Dorothy Lewis Bernstein

Marjorie Lee Browne

Frances Oldham Kelsey

Edith R. Peterson

Dixy Lee Ray

Barbara Moulton Browne

Katherine Koontz Sanford

Alice Turner Schafer

Angeles Alvariño

Ruth Mary Roan Benerito

Ruth Smith Lloyd

Ines Hochmuth Mandl

Estelle R. Ramey

Sofia Simmonds

Alice Mary Stoll

Gertrude Belle Elion

Katherine Coleman Goble Johnson

Ruth Sager

Eleanor Margaret Peachey Burbridge

Jessie G. Cambra

Leonora Woods Marshall Libby

Julia Bowman Robinson

Dorothy Martin Simon

Jane Cook Wright

Mary Ellen Avery

Rosalind Elsie Franklin

Elizabeth Calvert Miller

Marie Tharp

Marie Maynard Daly

Charlotte Friend

Eloise R. Gilbert

Isabella L. Karle

Beatrice Mintz

Jeanne Spurlock

Giuliana Cavaglieri Tesoro

Evelyn Maisel Witkin

Geraldine Pittman Woods

Xide Xie

Rosalyn Sussman Yalow

Eugenie Clark

Esther Marly Conwell

Cécile Andrée Paule Dewitt-Morette

Mary Ellen Jones

Doris Kuhlman-Wilsdorf

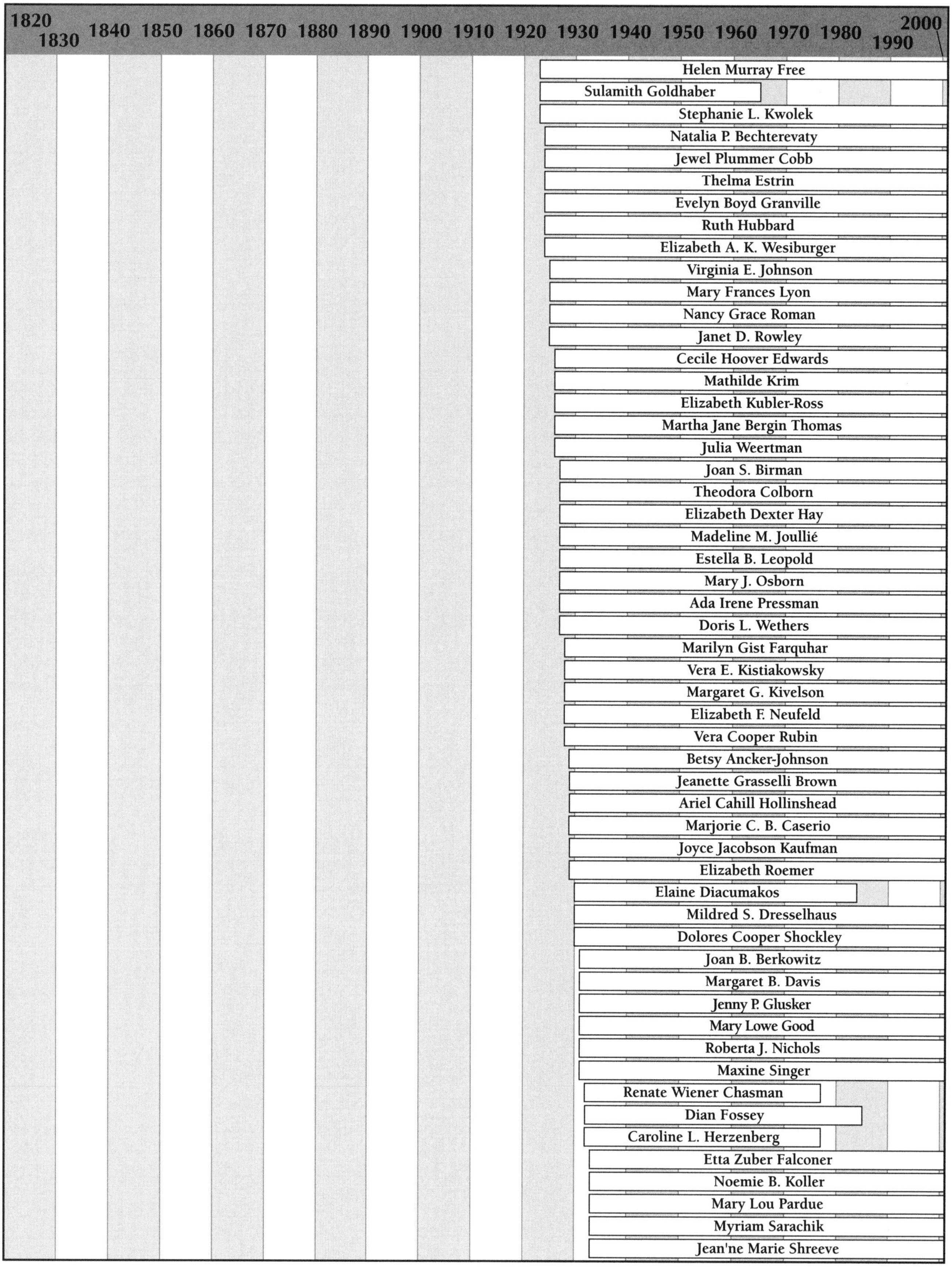

1820	1830	1840	1850	1860	1870	1880	1890	1900	1910	1920	1930	1940	1950	1960	1970	1980	1990	2000

Helen Murray Free

Sulamith Goldhaber

Stephanie L. Kwolek

Natalia P. Bechterevaty

Jewel Plummer Cobb

Thelma Estrin

Evelyn Boyd Granville

Ruth Hubbard

Elizabeth A. K. Wesiburger

Virginia E. Johnson

Mary Frances Lyon

Nancy Grace Roman

Janet D. Rowley

Cecile Hoover Edwards

Mathilde Krim

Elizabeth Kubler-Ross

Martha Jane Bergin Thomas

Julia Weertman

Joan S. Birman

Theodora Colborn

Elizabeth Dexter Hay

Madeline M. Joullié

Estella B. Leopold

Mary J. Osborn

Ada Irene Pressman

Doris L. Wethers

Marilyn Gist Farquhar

Vera E. Kistiakowsky

Margaret G. Kivelson

Elizabeth F. Neufeld

Vera Cooper Rubin

Betsy Ancker-Johnson

Jeanette Grasselli Brown

Ariel Cahill Hollinshead

Marjorie C. B. Caserio

Joyce Jacobson Kaufman

Elizabeth Roemer

Elaine Diacumakos

Mildred S. Dresselhaus

Dolores Cooper Shockley

Joan B. Berkowitz

Margaret B. Davis

Jenny P. Glusker

Mary Lowe Good

Roberta J. Nichols

Maxine Singer

Renate Wiener Chasman

Dian Fossey

Caroline L. Herzenberg

Etta Zuber Falconer

Noemie B. Koller

Mary Lou Pardue

Myriam Sarachik

Jean'ne Marie Shreeve

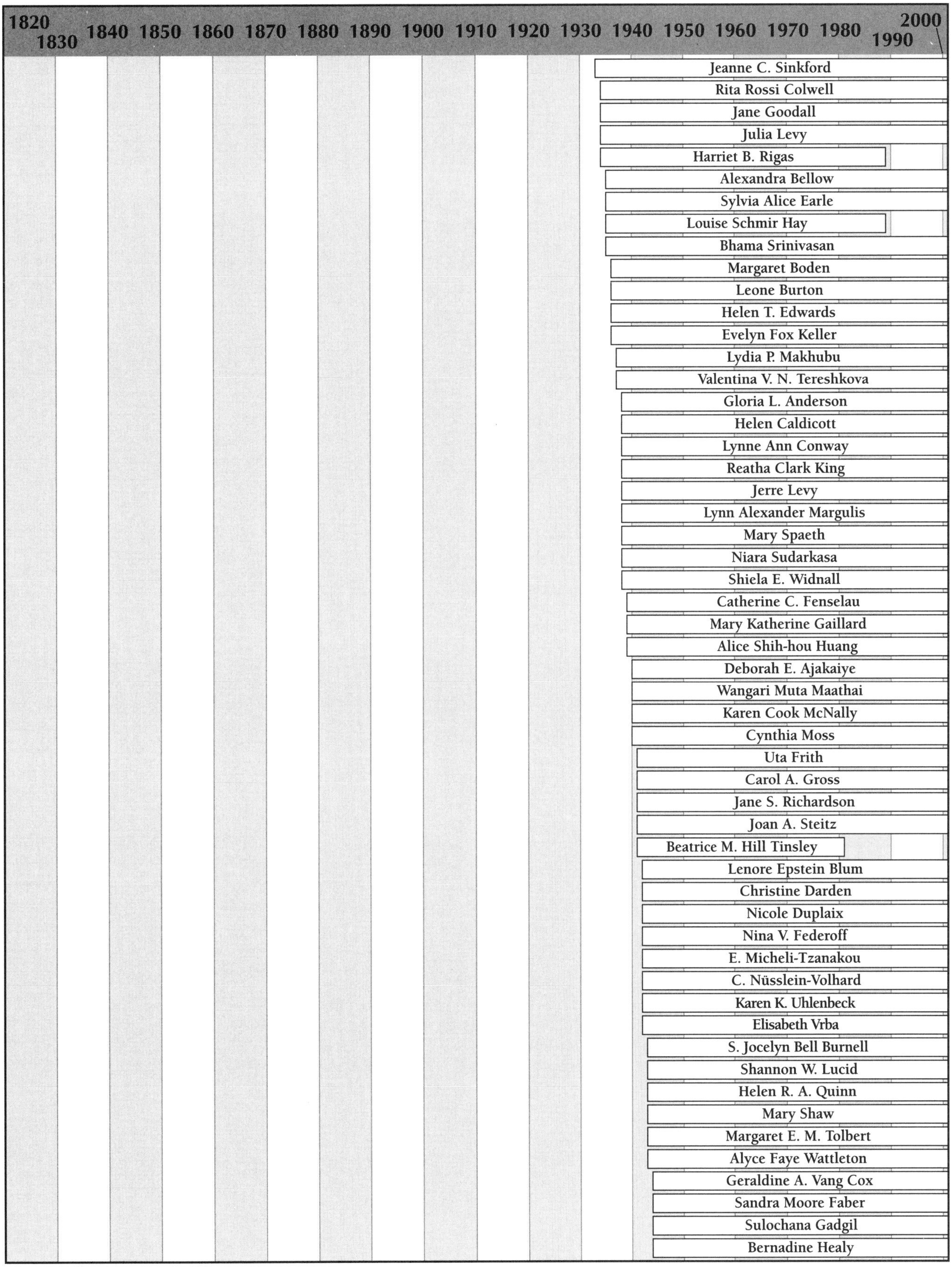

1820 1830	1840	1850	1860	1870	1880	1890	1900	1910	1920	1930	1940	1950	1960	1970	1980	1990 2000

Jeanne C. Sinkford

Rita Rossi Colwell

Jane Goodall

Julia Levy

Harriet B. Rigas

Alexandra Bellow

Sylvia Alice Earle

Louise Schmir Hay

Bhama Srinivasan

Margaret Boden

Leone Burton

Helen T. Edwards

Evelyn Fox Keller

Lydia P. Makhubu

Valentina V. N. Tereshkova

Gloria L. Anderson

Helen Caldicott

Lynne Ann Conway

Reatha Clark King

Jerre Levy

Lynn Alexander Margulis

Mary Spaeth

Niara Sudarkasa

Shiela E. Widnall

Catherine C. Fenselau

Mary Katherine Gaillard

Alice Shih-hou Huang

Deborah E. Ajakaiye

Wangari Muta Maathai

Karen Cook McNally

Cynthia Moss

Uta Frith

Carol A. Gross

Jane S. Richardson

Joan A. Steitz

Beatrice M. Hill Tinsley

Lenore Epstein Blum

Christine Darden

Nicole Duplaix

Nina V. Federoff

E. Micheli-Tzanakou

C. Nüsslein-Volhard

Karen K. Uhlenbeck

Elisabeth Vrba

S. Jocelyn Bell Burnell

Shannon W. Lucid

Helen R. A. Quinn

Mary Shaw

Margaret E. M. Tolbert

Alyce Faye Wattleton

Geraldine A. Vang Cox

Sandra Moore Faber

Sulochana Gadgil

Bernadine Healy

1820	1840	1850	1860	1870	1880	1890	1900	1910	1920	1930	1940	1950	1960	1970	1980	2000
1830															1990	

	Krystyna M. Kuperberg
	Antonia Coello Novello
	Roberta Lynn Bondar
	Adele Goldberg
	Philippa Marrack
	Margaret Dusa McDuff
	Ellen K. Silbergeld
	Nancy Sabin Wexler
	Biruté Galdikas
	Catharine D. Garmany
	Sarah Blaffer Hrdy
	Shirley Ann Jackson
	Mary-Claire King
	Candace Beebe Pert
	JoAnne Stubb
	Kutlu Aslihan Yener
	Sylvia Bozeman
	Nance K. Dicciani
	Margaret Joan Geller
	Jane Lubchenco
	Polly Matzinger
	Francine Patterson
	Lee Anne M. Willson
	Flossie Wong-Staal
	Elizabeth Helen Blackburn
	Patricia Suzanne Cowings
	Mimi A. R. Koehl
	Susan Love
	Constance Tom Noguchi
	Alice Chang Sun-Young
	Paula Szkody
	Mary Styles Harris
	Mary K. Hudson
	Diana García Prichard
	Alexa I. Canady
	Louise Ann Dolan
	Irene Duhart Long
	Sally K. Ride
	Susan Ildestad
	Marcia Kemper McNutt
	Vandana Shiva
	Judith Sharn Young
	Lai-Sang Young
	Mary Lou Zoback
	Ingrid Daubechies
	Cindy Lee Van Dover
	Dale Brown Emeagwali
	Naomi E. Pierce
	Bernadette Perrin-Riou
	Heather Williams
	Mae Carol Jemison
	Mona Kessel
	Susan Solomon
	Margarita H. Colmenares

1820 1830	1840	1850	1860	1870	1880	1890	1900	1910	1920	1930	1940	1950	1960	1970	1980 1990	2000
															Wendy L. Freedman	
															Yvonne Pendleton	
															Deborah L. Penry	
															Rosemary Wyse	
															Jacqueline N. Hewitt	
															Ellen Ochoa	
															Margie Jean Profet	
															Maria T. Zuber	
															Andrea Bertozzi	

INDEX

Page numbers in **boldface** indicate main entries; page numbers in *italic* indicate illustrations.